MOLECULAR MECHANISMS OF BACTERIAL VIRULENCE

Developments in Plant Pathology

VOLUME 3

The titles published in this series are listed at the end of this volume.

Molecular Mechanisms of Bacterial Virulence

Edited by

C.I. KADO

University of California
Davis, CA 95616
U.S.A.

and

J.H. CROSA

Oregon Health Sciences University
Portland, OR 97201
U.S.A

SPRINGER-SCIENCE+BUSINESS MEDIA, B.V.

Library of Congress Cataloging-in-Publication Data

Molecular mechanisms of bacterial virulence / edited by C.I. Kado and
  J.H. Crosa.
      p.   cm. -- (Developments in plant pathology ; v. 3)
   Outgrowth of a conference held at Fallen Leaf Lake, South Lake
Tahoe, California, Sept. 10-13, 1992.
   Includes index.
   ISBN 978-94-010-4322-9
   1. Bacterial diseases of plants--Molecular aspects--Congresses.
2. Phytopathogenic bacteria--Molecular aspects--Congresses.
3. Phytopathogenic bacteria--Host plants--Congresses. 4. Plant
-pathogen relationships--Molecular aspects--Congresses.   I. Kado,
Clarence I.   II. Crosa, Jorge H.   III. Series.
SB734.M64 1994
632'.32--dc20   93-11842

ISBN 978-94-010-4322-9   ISBN 978-94-011-0746-4 (eBook)
DOI 10.1007/978-94-011-0746-4

Printed on acid-free paper

Cover illustration by Eeva-Liisa Nurmiaho-Lassila,
University of Helsinki, Helsinki, Finland

Contents

II. PATHOGEN INGRESSION AND INVASIVE MECHANISMS

III. ELABORATION OF PATHOGENIC FACTORS

IV. REGULATION OF VIRULENCE GENES AND SIGNAL TRANSDUCTION

V. MECHANISMS AGAINST HOST DEFENSES

Contributors

Luis A. Actis, Department of Microbiology and Immunology L220, Oregon Health Sciences University, 3181 SW Sam Jackson Park Road, Portland, Oregon 97201–3098, USA

Matthieu Arlat, Laboratoire de Biologie Moléculaire des Interactions Plantes-Microorganismes, CNRS-INRA, Chemin de Borde Rouge, BP.27, 31326 Castanet Tolosan cedex, France

Naveen Arora, Laboratory of Microbial Ecology, National Institute of Dental Research, National Institutes of Health, Bethesda, Maryland 20892, USA

Michael Bagdasarian, Department of Microbiology, Michigan State University, S-110 Plant Biology Bldg., East Lansing, Michigan 48824–1312, USA

Patrick A. Barberis, Laboratoire de Biologie Moléculaire des Interactions Plantes-Microorganismes, CNRS-INRA, Chemin de Borde Rouge, BP.27, 31326 Castanet Tolosan cedex, France

Blaine L. Beaman, Department of Medical Microbiology and Immunology, University of California, School of Medicine, Davis, California 95616, USA

Steven V. Beer, Department of Plant Pathology, Cornell University, Ithaca, New York 14853, USA

Thomas Bergman, Department of Cell and Molecular Biology, University of Umea, S-901 87 Umea, Sweden

Frank Bernhard, Max-Planck-Institut für medizinische Forschung, Jahnstr. 29, D-6900 Heidelberg, Germany

Nicholas H. Carbonetti, Department of Microbiology and Immunology, U.M.A.B., 655 West Baltimore Street, Baltimore, Maryland 21201–1559, USA

Christian Boucher, Laboratoire de Biologie Moléculaire des Interactions Plantes-Microorganismes, CNRS-INRA, Chemin de Borde Rouge, BP.27, 31326 Castanet Tolosan cedex, France

N. Bunce, Department of Biological Sciences, University of Warwick, Coventry, CV4 7AL, United Kingdom

Trevor C. Charles, Department of Microbiology, SC-42, University of Washington, Seattle, Washington 98195, USA

J. Chen, Plant Pathology Department, University of Florida, Gainesville, Florida 32611, USA

Seong Ho Choi, Department of Plant Pathology, Throckmorton Hall, Kansas State University, Manhattan, Kansas 66506–5502, USA

D. E. Clements, Hawaii Biotechnology Group, Inc., Aiea, Hawaii 96701, USA

David Coplin, Department of Plant Pathology, The Ohio State University, Columbus, Ohio 43210–1087, USA

T. Cox, Department of Biological Sciences, University of Warwick, Coventry, CV4 7AL, United Kingdom

Jorge H. Crosa, Department of Microbiology and Immunology L220, Oregon Health Sciences University, 3181 SW Sam Jackson Park Road, Portland, Oregon 97201–3098, USA

Anath Das, Department of Biochemistry and Plant Molecular Genetics Institute, University of Minnesota, 1479 Gortner Avenue, St. Paul, Minnesota 55108, USA

Fernando de la Cruz, Departamento de Biologia Molecular, Universidad de Cantabria and Instituto de Fisica Moderna y Biologia Aplicada, C.S.I.C. Santander 39011, Spain

Heike Deppisch, Lehrstuhl für Mikrobiologie, Theodor-Boveri-Institut für Biowissenschaften (Biozentrum), Am Hubland, 8700 Wurzburg, Germany

Timothy P. Denny, Department of Plant Pathology, University of Georgia, Athens, Georgia 30602, USA

Elena Dubinina, Department of Cell and Molecular Biology, University of Umea, S-901 87 Umea, Sweden

Firous Ebrahim-Nesbat,Institut für Pflanzenpathologie und Pflanzenschutz, Universität, 3400 Göttingen, Germany

Wagih El-Shouny, Institut für Pflanzenpathologie und Pflanzenschutz, Universität, 3400 Göttingen, Germany

Dominique Expert, Laboratoire de Pathologie Végétale, INRA, Institut National Agronomique, 16 rue Claude Bernard, 75231 Paris Cedex 08, France

Stanley Falkow, Department of Microbiology and Immunology, Stanford University School of Medicine, Stanford, California 94305–5402, USA

Gottfried J. Feistner, Beckman Research Institute of the City of Hope, Duarte, California 91010, USA

Sergei A. Filichkin, Department of Biological Sciences, Purdue University, West Lafayette, Indiana 47907, USA

Ake Forsberg, Department of Cell and Microbiology, FOA 4, S-901 82, Umea, S-901 82 Umea, Sweden

Joachim Frey, Institute for Veterinary Bacteriology, University of Berne, Langgassstrasse 122, CH-3012 Berne, Switzerland

Barbara J. Froehlich, Department of Microbiology and Immunology, Emory Health Sciences Center, Atlanta, Georgia 30322, USA

Laura S. Frost, Department of Microbiology, University of Alberta, Edmonton, Alberta, Canada T6G 2E9

Thilo M. Fuchs, Lehrstuhl für Mikrobiologie, Theodor-Boveri-Institut für Biowissenschaften (Biozentrum), Am Hubland, 8700 Wurzburg, Germany

Ichiro Fukuda, Department of Bacteriology, The Institute of Medical Science, The University of Tokyo, Minato-ku, Shiroganedai-Machi, Tokyo 108, Japan

Dean W. Gabriel, Plant Pathology Department, University of Florida, Gainesville, Florida 32611, USA

Ann H. Gabrik, Beckman Research Institute of the City of Hope, Duarte, California 91010, USA

Edouard Galyov, Department of Cell and Molecular Biology, University of Umea, S-901 87 Umea, Sweden

Klaus Geider, Max-Planck-Institut für medizinische Forschung, Jahnstr. 29, D-6900 Heidelberg, Germany

Stanton B. Gelvin, Department of Biological Sciences, Purdue University, West Lafayette, Indiana 47907, USA

Stephane Genin, Laboratoire de Biologie Moléculaire des Interactions Plantes-Microorganismes, CNRS-INRA, Chemin de Borde Rouge, BP.27, 31326 Castanet Tolosan cedex, France

M. Gibson, Department of Biological Sciences, University of Warwick, Coventry, CV4 7AL, United Kingdom

P. Golby, Department of Biological Sciences, University of Warwick, Coventry, CV4 7AL, United Kingdom

Clare L. Gough, Laboratoire de Biologie Moleculaire des Interactions Plantes-Microorganismes, CNRS-INRA, Chemin de Borde Rouge, BP.27, 31326 Castanet Tolosan cedex, France

Stephen J. Gracheck, Experimental Therapy Department, Parke Davis Pharmaceutical Research, Warner Lambert Company, 2800 Plymouth Road, Ann Arbor, Michigan 48105, USA

Dennis C. Gross, Department of Plant Pathology, Washington State University, Pullman, Washington 99164–6430, USA

Michael Gross, Institut für Pflanzenpathologie und Pflanzenschutz, Universität, 3400 Göttingen, Germany

Roy Gross, Lehrstuhl für Mikrobiologie, Theodor-Boveri-Institut für Biowissenschaften (Biozentrum), Am Hubland, 8700 Wurzburg, Germany

Ailan Guo, Department of Plant Pathology, Throckmorton Hall, Kansas State University, Manhattan, Kansas 66506–5502, USA

Kielo Haahtela, Department of General Microbiology, P. O. Box 41, SF-00014 (Mannerheimintie 172), University of Helsinki, Finland

Sebastian Hakansson, Department of Cell and Molecular Biology, University of Umea, S-901 87 Umea, Sweden

Volker Haring, Department of Microbiology, Monash University, Clayton 3168, Australia

Douglas P. Henderson, Department of Microbiology, University of Texas, Austin, Texas 78712–1095, USA

Ursula Hettwer, Institut für Pflanzenpathologie und Pflanzenschutz, Universität, 3400 Göttingen, Germany

Sunggi Heu, Department of Botany, University of Maryland, College Park, Maryland 20742, USA

Sarah K. Highlander, Department of Microbiology and Immunology, Baylor College of Medicine, One Baylor Plaza, Houston, Texas 77030, USA

Barbara Hohn, Friedrich Miescher-Institut, P.O. Box 2543, CH-4002 Basel, Switzerland

Wim G. J. Hol, Department of Biological Structure, School of Medicine, University of Washington, Seattle, Washington 98195, USA

Yasuko Homma, Department of Bacteriology, The Institute of Medical Science, The University of Tokyo, Minato-ku, Shiroganedai-machi, Tokyo 108, Japan

Christopher M. Hopkins, Department of Plant Pathology, Throckmorton Hall, Kansas State University, Manhattan, Kansas 66506–5502, USA

Hsiou-Chen Huang, Department of Botany, University of Maryland, College Park, Maryland 20742, USA

Jianzhong Huang, Departments of Microbiology and Plant Pathology, University of Georgia, Athens, Georgia 30602, USA

Scott J. Hultgren, Department of Molecular Microbiology, Washington University School of Medicine, Box 8230, 660 S. Euclid Avenue, St. Louis, Missouri 63110, USA

Tom Humphreys, Pacific Biomedical Research Center, University of Hawaii, Honolulu, Hawaii 96822, USA

Steven W. Hutcheson, Department of Botany, University of Maryland, College Park, Maryland 20742, USA

Bradley D. Jones, Department of Microbiology and Immunology, Stanford University School of Medicine, Stanford, California 94305–5402, USA

Susan E. Jones, Department of Biological Sciences, University of Warwick, Coventry, CV4 7AL, United Kingdom

Antonio Juarez, Departamento de Microbiologia, Universidad Central de Barcelona, Barcelona, Spain

Yolanda Jubete, Laboratory of Molecular Biology, National Cancer Institute, National Institutes of Health, Bethesda, Maryland 20992, USA

Clarence I. Kado, Davis Crown Gall Group, University of California, Davis, California 95616, USA

Chris C. Kao, Department of Plant Pathology, University of Wisconsin, Madison, Wisconsin 53706, USA

Margaret E. Katz, Department of Animal Science, Monash University, Clayton 3168, Australia

Noel T. Keen, Department of Plant Pathology, University of California, Riverside, California 92521, USA

Mark T. Kingsley, Plant Pathology Department, University of Florida, Gainesville, Florida 32611, USA

Thomas G. Kinscherf, Department of Plant Pathology, University of Wisconsin, Madison, Wisconsin 53706, USA

Zoltan Klement, Institut für Pflanzenpathologie und Pflanzenschutz, Universität, 3400 Göttingen, Germany

Kurt R. Klimpel, Laboratory of Microbial Ecology, National Institute of Dental Research, National Institutes of Health, Bethesda, Maryland 20892, USA

D. Y. Kobayashi, Department of Plant Pathology, Rutgers University, New Brunswick, New Jersey 08903, USA

Syoko Kojima, Department of Biology, School of Science, Nagoya University, Chikusa-ku, Nagoya 464–01, Japan

Meta J. Kuehn, Department of Molecular Microbiology, Washington University School of Medicine, Box 8230, 660 S. Euclid Avenue, St. Louis, Missouri 63110, USA

Keiko Komatsu, Department of Bacteriology, The Institute of Medical Science, The University of Tokyo, Minato-ku, Shiroganedai-Machi, Tokyo 108, Japan

J. Pierre Laulhère, Laboratoire de Biologie Moléculaire Végétale, Université Joseph Fournier, BP 53X, F-38041 Grenoble Cedex, France

Jan E. Leach, Department of Plant Pathology, Throckmorton Hall, Kansas State University, Manhattan, Kansas 66506–5502, USA

Stephen H. Leppla, Laboratory of Microbial Ecology, National Institute of Dental Research, National Institutes of Health, Bethesda, Maryland 20892, USA

Michael C. Lidell, Department of Botany, University of Maryland, College Park, Maryland 20742, USA

J. Lorang, Department of Plant Pathology, University of California, Riverside, California 92521, USA

Jari Louhelainen, Department of General Microbiology, P. O. Box 41, SF-00014 (Mannerheimintie 172), University of Helsinki, Finland

Yasunori Machida, Department of Biology, School of Science, Nagoya University, Chikusa-ku, Nagoya 464–01, Japan

Francis L. Macrina, Department of Microbiology and Immunology, Virginia Commonwealth University, Richmond, Virginia 23298–0678, USA

Bruno Mahé, Laboratoire de Pathologie Végétale, INRA, Institut National Agronomique, 16 rue Claude Bernard, 75231 Paris Cedex 08, France

Morton Mandel, Department of Biochemistry and Biophysics, University of Hawaii, Honolulu, Hawaii 96822, USA

Céline Masclaux, Laboratoire de Pathologie Végétale, INRA, Institut National Agronomique, 16 rue Claude Bernard, 75231 Paris Cedex 08, France

Syogo Matsumoto, Institute for Cell Biology and Genetics, Faculty of Agriculture, Iwate University, Ueda 3–18–8, Morioka 020, Japan

Ken Matsuoka, Department of Biology, School of Agriculture, Nagoya University, Chikusa-ku, Nagoya 464–01, Japan

Ann G. Matthysse, Department of Biology, CB 3280, University of North Carolina, Chapel Hill, North Carolina 27599, USA

Doris Majerczak, Department of Plant Pathology, The Ohio State University, Columbus, Ohio 43210–1087, USA

Mark Mazzola, Department of Plant Pathology, Throckmorton Hall, Kansas State University, Manhattan, Kansas 66506–5502, USA

V. Mulholland, Department of Biological Sciences, University of Warwick, Coventry, CV4 7AL, United Kingdom

Cindy Munro, Department of Medical-Surgical Nursing, Virginia Commonwealth University, Richmond, Virginia 23298–0678, USA

Kenzo Nakamura, Department of Biology, School of Agriculture, Nagoya University, Chikusa-ku, Nagoya 464–01, Japan

Noboru Nakata, Department of Bacteriology, The Institute of Medical Science, The University of Tokyo, Minato-ku, Shiroganedai-Machi, Tokyo 108, Japan

Claire Neema, Laboratoire de Pathologie Végétale, INRA, Institut National Agronomique, 16 rue Claude Bernard, 75231 Paris Cedex 08, France

Eugene W. Nester, Department of Microbiology, SC-42, University of Washington, Seattle, Washington 98195, USA

Michael Neugebauer, Institut für Pflanzenpathologie und Pflanzenschutz, Universität, 3400 Göttingen, Germany

Matthias Nöllenburg, Institut für Pflanzenpathologie und Pflanzenschutz, Universität, 3400 Göttingen, Germany

Roland Nordfelth, Department of Cell and Molecular Biology, University of Umea, S-901 87 Umea, Sweden

Eeva-Liisa Numiaho-Lassila, Department of Electron Microscopy, P.O. Box 41, SF-00014 (Mannerheimintie 172), University of Helsinki, Finland

Tuula Ojanen, Department of General Microbiology, P.O. Box 41, SF-00014 (Mannerheimintie 172), University of Helsinki, Finland

Nobuhiko Okada, Department of Bacteriology, The Institute of Medical Science, The University of Tokyo, Minato-ku, Shiroganedai-machi, Tokyo 108, Japan

Norihiro Okada, Faculty of Bioscience and Biotechnology, Tokyo Institute of Technology, 4259 Nagatsuta-cho, Midori-ku, Yokohama 227, Japan

Kazuhiko Ohshima, Faculty of Bioscience and Biotechnology, Tokyo Institute of Technology, 4259 Nagatsuta-cho, Midori-ku, Yolohama 227, Japan

Linda J. Overbye, Department of Microbiology, Michigan State University, East Lansing, Michigan 48824–1312, USA

Suresh S. Patil, The Biotechnology Program, University of Hawaii, Honolulu, Hawaii 96822, USA

Shelly M. Payne, Department of Microbiology, University of Texas, Austin, Texas 78712–1095, USA

Cathrine Persson, Department of Cell and Molecular Biology, University of Umea, S-901 87 Umea, Sweden

Neil B. Quigley, Department of Plant Pathology, Washington State University, Pullman, Washington 99164–6430, USA

Peter J. Reeves, Department of Biological Sciences, University of Warwick, Coventry, CV4 7AL, United Kingdom

Peter Reimers, Department of Plant Pathology, Throckmorton Hall, Kansas State University, Manhattan, Kansas 66506–5502, USA

Jessica J. Rich, Department of Molecular Biology, University of Wisconsin, Madison, Wisconsin 53706, USA

Marja Rimpilainen, Department of Cell and Molecular Biology, University of Umea, S-901 87 Umea, Sweden

P. Roberts, Plant Pathology Department, University of Florida, Gainesville, Florida 32611, USA

Elina Roine, Department of General Microbiology, P.O. Box 41, SF-00014 (Mannerheimintie 172), University of Helsinki, Finland

Martin Romantschuk, Department of General Microbiology, P. O. Box 41, SF-00014 (Mannerheimintie 172), University of Helsinki, Finland

Julian I. Rood, Department of Microbiology, Monash University, Clayton 3168, Australia

Luca Rossi, Friedrich Miescher-Institut, P.O. Box 2543, CH-4002 Basel, Switzerland

Roland Rosqvist, Department of Cell and Microbiology, FOA 4, S-901 82 Umea, Sweden

K. B. Rowley, The Biotechnology Program, University of Hawaii, Honolulu, Hawaii 96822, USA

Klaus W. E. Rudolph, Institut für Pflanzenpathologie und Pflanzenschutz, Universität, 3400 Göttingen, Germany

Patricia Salinas, Department of Microbiology and Immunology L220, Oregon Health Sciences University, 3181 SW Sam Jackson Park Road, Portland, Oregon 97201–3098, USA

George P. C. Salmond, Department of Biological Sciences, University of Warwick, Coventry, CV4 7AL, United Kingdom

Maria Sandkvist, Unit for Laboratory Animal Medicine, University of Michigan, 105 ARF, Ann Arbor, Michigan 48109, USA

Chihiro Sasakawa, Department of Bacteriology, The Institute of Medical Science, The University of Tokyo, Minato-ku, Shiroganedai-machi, Tokyo 108, Japan

Mark A. Schell, Department of Microbiology and Department of Plant Pathology, University of Georgia, Athens, Georgia 30602, USA

June R. Scott, Department of Microbiology and Immunology, Emory Health Sciences Center, Atlanta, Georgia 30322, USA

Luis Segueira, Department of Plant Pathology, University of Wisconsin, Madison, Wisconsin, 53706, USA

H. Shen, Department of Plant Pathology, University of California, Riverside, California 92521, USA

Titia K. Sixma, BIOSON Research Institute and Department of Chemistry, University of Groningen, 9747 AG Groningen, The Netherlands

Lynn N. Slonim, Department of Molecular Microbiology, Washington University School of Medicine, Box 8230, 660 S. Euclid Avenue, St. Louis, Missouri 63110, USA

John Simon, Department of Microbiology, University of Alberta, Edmonton, Alberta, Canada T6G 2E9

Bernd Sonnenberg, Institut für Pflanzenpathologie und Pflanzenschutz, Universität, 3400 Göttingen, Germany

S. K. Stephens, Department of Biological Sciences, University of Warwick, Coventry, CV4 7AL, United Kingdom

John Swanson, National Institutes of Health, National Institute of Allergy and Infectious Diseases, Laboratory of Microbial Structure and Function, Rocky Mountain Laboratories, Hamilton, Montana 59840, USA

Yoshito Takahashi, Department of Biology, School of Science, Nagoya University, Chikusa-ku, Nagoya 464–01, Japan

Bruno Tinland, Friedrich Miescher-Institut, P.O. Box 2543, CH-4002 Basel, Switzerland

Toru Tobe, Department of Bacteriology, The Institute of Medical Science, The University of Tokyo, Minato-ku, Shiroganedai-machi, Tokyo 108, Japan

Marcelo E. Tolmasky, Department of Microbiology and Immunology L220, Oregon Health Sciences University, 3181 SW Sam Jackson Park Road, Portland, Oregon 97201–3098, USA

Timothy S. Uphoff, Department of Medical Microbiology and Immunology, University of Wisconsin School of Medicine, Madison, Wisconsin 53706, USA

Joop Van Doorn, Department of Plant Quality, Bulb Research Centre, Vennestraat 22, Postbus 85, 2160 AB Lisse, The Netherlands

Frederique Van Gijsegem, Laboratoire de Biologie Moléculaire des Interactions Plantes-Microorganismes, CNRS-INRA, Chemin de Borde Rouge, BP.27, 31326 Castanet Tolosan cedex, France

Vincent T. Wagner, Department of Biology, CB 3280, University of North Carolina, Chapel Hill, North Carolina 27599, USA

Lillian S. Waldbeser, Department of Microbiology and Immunology L220, Oregon Health Sciences University, 3181 SW Sam Jackson Park Road, Portland, Oregon 97201–3098, USA

Alison Weiss, Department of Microbiology and Immunology, Virgina Commonwealth University, Richmond, VA 23298, USA

Rodney A. Welch, Department of Medical Microbiology and Immunology, University of Wisconsin School of Medicine, Madison, Wisconsin 53706, USA

Frank F. White, Department of Plant Pathology, Throckmorton Hall, Kansas State University, Manhattan, Kansas 66506–5502, USA

D. Kyle Willis, Agriculture Research Service, U.S.D.A., Department of Plant Pathology, University of Wisconsin, Madison, Wisconsin 53706, USA

Sara A. Wold, Experimental Therapy Department, Parke Davis Pharmaceutical Research, Warner Lambert Company, 2800 Plymouth Road, Ann Arbor, Michigan 48105, USA

Catherine L. Wright, Department of Microbiology, Monash University, Clayton 3168, Australia

Kerstin Wydra, Institut für Pflanzenpathologie und Pflanzenschutz, Universität, 3400 Göttingen, Germany

Yingxian Xiao, Department of Botany, University of Maryland, College Park, Maryland 20742, USA

Y. Yang, Plant Pathology Department, University of Florida, Gainesville, Florida 32611, USA

Masanosuke Yoshikawa, Department of Bacteriology, The Institute of Medical Science, The University of Tokyo, Minato-ku, Shiroganedai-machi, Tokyo 108, Japan

Yasushi Yoshioka, Department of Biology, School of Science, Nagoya University, Chikusa-ku, Nagoya 464–01, Japan

Juan Carlos Zabala, Departamento de Biologia Molecular, Universidad de Cantabria and Instituto de Fisica Moderna y Biologia Aplicada, C.S.I.C. Santander 39011, Spain

Y. X. Zhang, The Biotechnology Program, University of Hawaii, Honolulu, Hawaii 96822, USA

Claudine Zischek, Laboratoire de Biologie Moléculaire des Interactions Plantes-Microorganismes, CNRS-INRA, Chemin de Borde Rouge, BP.27, 31326 Castanet Tolosan cedex, France

Alim Zomorodian, Institut für Pflanzenpathologie und Pflanzenschutz, Universität, 3400 Göttingen, Germany

Preface

The growing body of information on bacteria pathogenic for humans, mammals and plants generated within the past ten years has shown the interesting conservation of newly identified genes that play a direct role in the pathogenic mechanism. In addition to these genes, there are also genes that confer host specificities and other traits important in pathogenesis on these pathogens.

In this volume, we have organized the subject areas to best fit the concept on the way bacterial pathogens recognize, interact and invade the host, on the regulation of genes involved in virulence, on the genes involved in the elaboration of toxins and other pathogenic components such as iron sequestering proteins, and on the mechanisms of circumventing the host defense systems. These areas are divided into Sections. Section I covers the first step when the pathogen seeks its host, and Sections II through VI cover subsequent steps leading to pathogenesis while avoiding host defenses. We conclude this work with a chapter summarizing information on examples of virulence mechanisms that are highly conserved.

This volume was made possible by the gracious sponsorship of the Cooperative State Research Service, USDA, the University of California, Davis, College of Agriculture and Environmental Sciences, the Agricultural Research Service, USDA, and the National Science Foundation (Grant MCB-9205467) to whom we are deeply indebted for their continuous support. We also thank the following sustaining corporations: E. I. Dupont de Nemours & Company (Biomedical Products Division), Campbell Institute for Research & Technology, Monsanto Company (Agricultural Group), Ciba-Geigy Corporation (Biotechnology Division), Blackwell Scientific. In addition, we gratefully thank all the contributors of this volume, who presented the results of their research at the Fallen Leaf Lake Conference held at Fallen Leaf Lake, South Lake Tahoe, California, September 10–13, 1992. The Fallen Leaf Lake Conferences provide unique opportunities for experts in the field as well as junior scientists to interact and exchange ideas in a serene setting combined with elegant dining, social and recreational events. This volume is not a proceedings of the Fallen Leaf Lake Conference, but represents the logical extension of the very recent observations and findings presented at the conference.

The contributions in this volume, therefore, were individually peer reviewed by experts in the field and we gratefully acknowledge the following reviewers for their expert assistance: Luis Actis, Christian Boucher, Trevor Charles, David Coplin, Lidia Crosa, Anath Das, Fernando de la Cruz, Gottfried Feistner, Laura Frost, Dean Gabriel, Klaus Geider, Stanton Gelvin, Dennis Gross, Sarah Highlander, Barbara Hohn, Steven Hutcheson, Haresh Kamdar, Noel Keen, Jan Leach, Yasunori Machida, Frank Macrina, Ann Matthysee, Eugene Nester, Nick Panopoulos, Suresh Patil, Martin Romanschuk, Pam Ronald, George Salmond, Mark Schell, Luis Sequeira, Marcelo Tolmasky, Alison Weiss, Frank White, and others who wished to remain anonymous. The papers published in this volume have also been reviewed by the editors.

We thank Fernando de la Cruz, Francis L. Macrina and Luis Sequeira for their excellent cooperation as members of the organizing committee. We are indebted to members of the Davis Crown Gall Group, and to Thanh Maxwell, Valinda Stagner, Louisa Ruesda, Guyla Yoak, Elizabeth Jeffery-Noring and Judy Martin of the Department of Plant Pathology for their invaluable assistance. We also thank Diana Arington Evans for providing an exceptional atmosphere that fostered a great deal of scientific interactions among all of the contributors.

Clarence I. Kado
Jorge H. Crosa

Foreword

CLIFFORD J. GABRIEL, CSRS, Washington D.C.

Introduction to the Fallen Leaf Lake Conference

The Fallen Leaf Lake Conference series was initiated in 1985 to address molecular and genetic aspects of plant pathogenic bacteria and fastidious prokaryotes as well as related areas associated with these organisms. Although the Conference has been sponsored by numerous organizations, the primary sponsor has been the U.S. Department of Agriculture's Cooperative State Research Service (CSRS). CSRS is one if the primary research agencies of the Department and can be viewed as a much smaller agricultural version of the National Institutes of Health's extramural research program. CSRS administers competitive grants, training grants, special research grants (congressional line-items), and formula funds (block grants to specific institutions, the amount of which is predetermined by formula) to Land-Grant Colleges and other private and public institutions.

Additional sponsors for this Conference have included the Department's Agricultural Research Service (the Department's intramural research program) and the Animal and Plant Health Inspection Service, the National Science Foundation, the University of California, U.S. Environmental Protection Agency, numerous industrial sponsors, and others. Through the generous support of these organizations, the Fallen Leaf Lake Conferences have been able to continue and thrive.

The initial discussions that led to the creation of the Fallen Leaf Lake Conference were between the late CSRS Chief Scientist John F. Fulkerson, to whom this volume is dedicated, and University of California-Davis Plant Pathologist Clarence I. Kado. These discussions focused on the need for an informal international forum to focus on new developments in plant pathology research. Their objective was to encourage creative thinking and collaboration. By design, the primary participants would be young, promising scientists and outstanding senior scientists. This was in keeping with the philosophy of John Fulkerson, long viewed as a champion for young scientists and innovative research.

The first three Conferences dealt with specific genera of plant pathogenic bacteria. These included conferences on *Xanthomonas, Erwinia,* and *Agrobacterium* in 1985, 1986, and 1987, respectively. Subsequent Conferences addressed more cross-cutting topics such as the molecular biology of bacterial plant pathogens in 1989 and promiscuous plasmids of gram-positive and -negative bacteria in 1990. The later conferences began to attract researchers with a wider diversity of expertise.

The Fallen Leaf Lake Conference on which this volume is based was held in September 1992. It followed in the tradition of past Conferences, but took the pioneering step to bring together a critical mass of scientists with expertise in plant, animal, and human bacterial pathology. The objective was to explore the potential for bacterial virulence mechanisms that might be common to these different groups of pathogenic bacteria. The results of this Conference highlighted similarities and differences as well as the fact that we know so little about many of the fundamental mechanisms underlying bacterial pathogenesis.

It is hoped that through CSRS support and through the support of others, the Fallen Leaf Lake Conference will continue to provide a unique forum for scientists to discuss their most recent research findings that will lead to new insights into bacterial pathogens and control of the disease they incite.

[This volume is dedicated to Dr. John F. Fulkerson (1922–1991). Dr. Fulkerson's continuous dedication to the application of biological principles to agricultural helped shape the current Federal-State partnership for agricultural research. Through Dr. Fulkerson's vision, the Fallen Leaf Lake Conference series on molecular and genetic aspects of plant pathogenic bacteria was established. His vision is missed greatly.]

Sponsors

We gratefully acknowledge the generous grant support provided by the following agencies and corporate sponsors:

Cooperative State Research Service, United States Department of Agriculture

Agricultural Research Service, United States Department of Agriculture

National Science Foundation

College of Agricultural and Environmental Sciences, University of California, Davis, California

E. I. Dupont de Nemours & Company (Biomedical Products Division)

Campbell Institute for Research & Technology

Monsanto Company (Agricultural Group)

Ciba-Geigy Corporation (Biotechnology Division)

Blackwell Scientific

SECTION I

Host recognition and attachment mechanisms

1. Phenotypic and genetic aspects of host cell invasion by *Salmonella* species

BRADLEY D. JONES and STANLEY FALKOW

Abstract. *Salmonella* species are enteric pathogens that successfully infect a host by crossing the epithelial cell barrier of the small bowel and subsequently invade the deeper tissue of the host. The events that occur at the epithelial surface of host cells during *Salmonella* entry have been the subject of considerable study in animal models and in cell culture. The development of *in vitro* tissue culture models, which successfully reproduce the interactions that occur between bacteria and host cells during invasion, has been invaluable in evaluating the contribution of bacterial and host cell factors in *Salmonella* entry. Tissue culture experiments have shown that only actively growing bacteria are capable of invading mammalian cells, while the host cell response is characterized by cytoskeletal changes that are accompanied by alterations in calcium ion flux and host cell protein phosphorylation. Bacterial growth in both low-oxygen and high-osmolarity conditions seem to induce the expression of proteins, which influence the ability of *Salmonella* to enter cells. Loci necessary for bacterial entry into mammalian cells have been identified using genetic techniques. It is not yet clear whether these genes encode structural or regulatory elements. Currently, work is underway to identify the host cell signal pathway affected by microbial invasion.

Abbreviations: EGF, Epidermal Growth Factor; LPS, Lipopolysaccharide; MDCK, Madin-Darby Canine Kidney

Introduction

Pathogenic bacteria rely on their ability to circumvent host defenses and establish a niche for growth. One successful survival tactic employed by bacterial pathogens is to enter and grow within the intracellular environment of a host cell. *Salmonella* species are facultative intracellular pathogens that use this strategy to penetrate the intestinal epithelium of a host during infection (Rubin and Weinstein, 1977). Once the epithelial surface of the bowel has been crossed, the organisms move into the organs of the lymphatic system where they disseminate causing a systemic infection. Since the cells lining the mucosal surface of the bowel are nonphagocytic enterocytes, bacterial entry of the epithelium must be an active process directed, at least in part, by *Salmonella*.

The current understanding of the ability of *Salmonella* to enter a host cell is the result of research carried out by several laboratories, including our own. Previously, the main focus of research was on the characterization of various aspects of the bacterial-host cell interactions that occur during invasion, while very little was done to understand the molecular mechanism of entry. Within the

C.I. Kado and J.H. Crosa (eds.), Molecular Mechanisms of Bacterial Virulence, 3–16.
© 1994 *Kluwer Academic Publishers.*

last decade research efforts have begun to unravel the genetic details of *Samonella* invasion and have shown that this invasion process is multi-faceted requiring many structural and regulatory factors. Many pieces of this complex interaction remain to be identified and understood before we can achieve an understanding of the *Salmonella* invasion process.

Bacterial-host cell interactions during *Salmonella* entry

Description of Salmonella *invasion at the cellular level*

Takeuchi (1967) first reported a detailed electronmicroscopic study that described the interactions of *Salmonella typhimurium* with guinea pig cells during the development of acute enteritis. Several hours following oral challenge, organisms were observed in the gut lumen, as well as in contact with the apical surface of the ileal mucosa. When a bacterium was found to be in close proximity to the brush border, it was observed that the host cell microvilli had begun to deteriorate. In many instances, the cytoplasm of the host cell bulged or projected into the region of microvilli degeneration. Following deterioration of the apical surface of the cell, cavities in the host cell membrane were observed forming around invading bacteria. At later time points, membrane-bound intracellular organisms were observed detached from the luminal membrane and seemed to be in the process of migrating from the apical surface toward the basolateral surface. While *Salmonella* clearly disrupted the brush border of the host cell prior to entry, such damage was not irreversible. Many cells that contained membrane-bound organisms were observed with normal brush borders. These fundamental observations have been confirmed and expanded by other research groups, albeit using different systems. Turnbull and Richmond (1978) microscopically examined tissue from infected loops of chickens and found that the cecal epithelium degenerated after interacting with invasive *S. enteriditis* in a fashion similar to that observed by Takeuchi. In another study (Kohbata *et al.*, 1986), the effect of virulent *S. typhi* on the ileal epithelium of murine ligated loops was examined. As seen by others, bacterial adherence to, and entry into, epithelial cells was accompanied by the dissolution of the enterocyte microvilli. In addition, *S. typhi* seemed to preferentially attach to and enter the M cells of the Peyer's patches.

M cell interactions

M cells are specialized epitheloid cells that overlay lymphoid follicles in the gut known as Peyer's patches. The function of M cells seems to be the transportation of antigens, including bacteria, to phagocytes and lymphocytes, which lay immediately beneath the epithelium of the Peyer's patches (Owen and Jones, 1974; Owen and Ermak, 1990). As described above, *S. typhi* seems to preferentially enter the M cells of mice. In addition, an autopsy of a chimpanzee

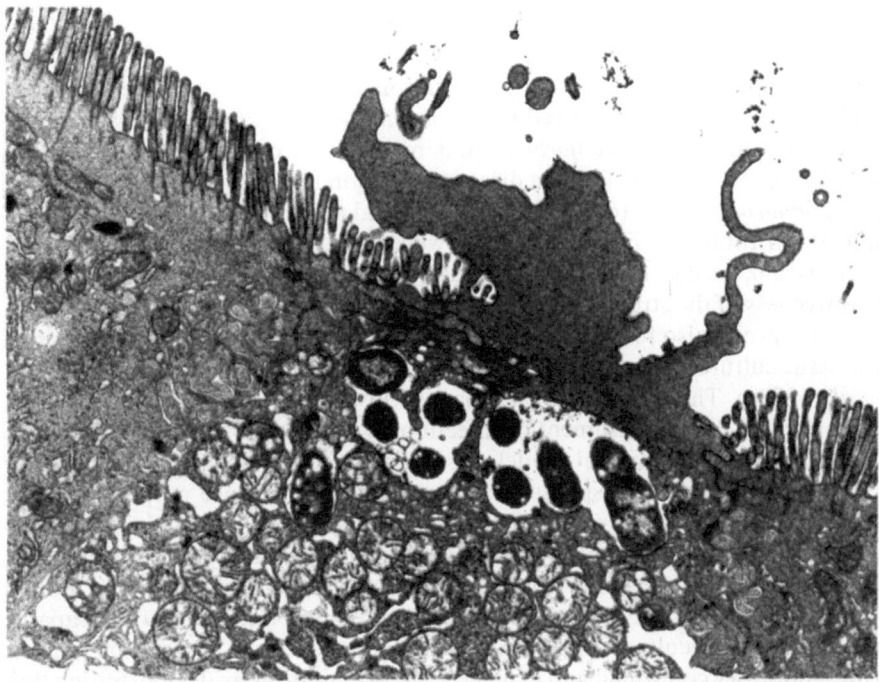

Figure 1. Electron micrograph of *S. typhimurium* interacting with Peyer's patch tissue from the ileum of a Balb/c mouse. Individual organisms have entered an M cell causing membrane ruffling and cytotoxicity.

that had been experimentally infected with *S. typhi* indicated that many more organisms were associated with the ileal Peyer's patches and the mesenteric lymph nodes than the cell wall of the ileum (Sprinz *et al.*, 1956). Studies with *S. typhimurium* in rodent models confirm that following oral inoculation the majority of organisms surviving passage through the stomach are associated with the Peyer's patch tissue, rather than the ileum intestinal wall or the cecum (Carter and Collins, 1974; Carter, 1975; Hohman *et al.*, 1978). Our laboratory has recently studied the interactions of *S. typhimurium* with murine tissue. Intestinal ligated loops were inoculated with an invasive strain of *S. typhimurium* and Peyer's patch tissue from the mouse was microscopically examined. We found that invasive *S. typhimurium* selectively invaded and disrupted the apical surface of the M cells within 30 minutes of infection (Fig. 1, B. D. Jones, unpublished). These data confirm that *Salmonella* species are able to actively enter the M cells of the Peyer's patch, presumably to gain access to the underlying lymphatic system. Following entry and passage through the M cells, host-adapted serotypes of *Salmonella*, which possess the ability to persist in the lymphatic system, disseminate causing systemic disease. Growth of non-adapted *Salmonella* serotypes, which seem to lack a specific survival strategy, is limited to the localized intestinal infection known as gastroenteritis.

Use of in vitro *assays to study invasion factors*

The development of a quantitative invasion assay has greatly increased our ability to study factors involved in *Salmonella* invasion. In 1973, it was clearly demonstrated that *S. typhimurium* could invade tissue culture cells (Giannella *et al.*, 1973b). The authors found a direct correlation between the ability of several *S. typhimurium* strains to invade HeLa cells and the ability to penetrate the ileal mucosa of rabbits in ligated loops. However, the murine LD_{50} determined for each strain, using the intraperitoneal route, did not correlate with the invasiveness of the strains for either HeLa cells or the rabbit mucosa. However, the same group also found that strains of *S. typhimurium* that were noninvasive for tissue culture cells were avirulent for mice when inoculated orally (Giannella *et al.*, 1973a). This result indicates that the invasiveness of *Salmonella* is most important in the establishment of infection by the oral route.

Since no suitable animal model exists for the study of typhoid fever, the *in vitro* assay has been indispensable in studying the virulence properties of the invasive human pathogen *S. typhi*. Yokoyama *et al.* (1987) confirmed that *S. typhi* induces shortening of microvilli and membrane disruption in HeLa cells concomitant with bacterial entry, as observed for *S. typhimurium* in animal models. A second study (Yabuuchi *et al.*, 1986) found that *S. typhi* was capable of entering tissue culture cells within 10 minutes of addition to a cell monolayer. Our laboratory confirmed and extended this observation by demonstrating that *S. typhimurium* strains could invade tissue culture cells within fifteen minutes of infection if grown under conditions that induce the expression of invasion factors (Francis *et al.*, 1992).

The tissue culture assay has been used to examine different aspects of *Salmonella* invasion in detail. Kihlstrom and Nilsson (1977) quantitatively examined the effect of various drugs on *S. typhimurium* entry into HeLa cells. The presence of the glycolytic inhibitors, iodoacetic acid, and N-ethylmaleimide greatly reduced the numbers of bacteria recovered from invasion assays, while sodium azide had no effect on the number of organisms recovered. Drugs that interfere with host cell function also affect the ability of *Salmonella* to invade. *S. typhi*, *S. cholerasuis*, and *S. typhimurium* were unable to enter eukaryotic cells in the presence of the microfilament inhibitors cytochalasin B and D (Finlay and Falkow, 1988; Kihlstrom and Nillson, 1977; Mroczenski-Wildey *et al.*, 1989). In contrast, drugs such as ammonium chloride and methylamine, which prevent the acidification of the endosome and normal microtubule function, had no effect on *Salmonella* entry as shown in Table 1 (Finlay and Falkow, 1988). Ginocchio and colleagues (1992) used cultured epithelial cells to demonstrate that an intracellular Ca^{2+} flux occurred after exposure of the monolayer to invasive *S. typhimurium*. Presumably, a calcium flux is part of the signal that causes depolarization of the microvilli on the apical membrane during bacterial entry.

Polarized epithelial monolayers display many of the same properties possessed by a tissue epithelium (Simons and Fuller, 1985). A polarized cell line,

Table 1. Effect of potential inhibitors of *Salmonella* invasion in MDCK cells.

Agent	Conc.	Yersinia enerocolitica	Salmonella cholerae–suisi	Shigella flexneri
None		100	100	100
Chloroquine	200 μg/ml	1.4	0.5	97
NH$_4$Cl	50 mM	115	104	134
Methylamine	50 mM	105	78	162
Monensin	20 μM	110	104	85
Cytochalasin B	5 μg/ml	56	8.5	3.7
Cytochalasin D	5 μg/ml	0.81	0.83	0.25
Colchicine	5 μg/ml	109	104	74

Data from Finlay and Falkow, 1988.

Madin-Darby canine kidney (MDCK), was first used by our laboratory to reproduce many of the events that occur during *S. cholerasuis* invasion of host tissue in an *in vitro* tissue culture assay (Finlay *et al.*,1988). We found that close association of *S. cholerasuis* organisms to the MDCK cells caused dissolution of the microvilli and bulging of the cytoplasm. The bacteria were observed entering the cell in the region of the membrane that was disrupted in a manner similar to that described for *Salmonella* invasion *in vivo* (Kohbata *et al.*, 1986; Takeuchi, 1967; Turnbull and Richmond, 1978; Worton *et al.*, 1989). Furthermore, microscopic examination of intracellular bacteria revealed that they usually entered the cells singly and then replicated within vacuoles that coalesced over time (Finlay and Falkow, 1988; Leung and Finlay, 1991). Given enough time, bacterial replication within the vacuoles would often fill the host cell cytoplasm entirely with organisms. In addition, we have used this model to demonstrate that intracellular bacteria can transcytose from the apical surface of the MDCK cells to the basolateral membrane. In fact, it was possible to recover viable *S. cholerasuis* from the basolateral media of the MDCK cells. As was suggested by previous work (Kihlstrom and Edebo, 1976), neither formalin- or UV-treated *Salmonella* organisms were able to enter the polarized epithelial monolayer. Collectively, the data indicate that *Salmonella* invasion is a complex, energy-requiring process for both the bacterial pathogen and the host cell.

Role of motility in invasion

Motility and chemotaxis are known to have an indirect effect on the ability of *Salmonella* to invade. Nonmotile *S. typhimurium* strains have a reduced ability to efficiently attach to and enter HeLa cells (Jones *et al.*, 1981) and have a reduced capacity to be taken up by cultured macrophages (Tomita and Kanegasaki, 1982). In both instances, the invasion defects of the nonmotile strains can be overcome by centrifugation of the bacteria onto the cultured cells. In contrast, *S. typhi* Fla$^-$, Mot$^-$, and Che$^-$ mutants were found to be noninvasive in tissue culture assays, even when the bacteria were impacted directly onto the tissue culture monolayer by centrifugation (Liu *et al.*, 1988). Recently, we showed, as did Khoramian *et al.*(1990), that *S. typhimurium* smooth swimming Che⁻ mutants were more invasive than the wild-type parent *in vitro* (Table 2) (Jones *et al.*, 1992). Moreover, we found that these same strains had an increased ability to enter the Peyer's patches of mice. Conversely, Fla⁻, Mot⁻, and tumbly swimming Che⁻ strains were less invasive than the parent strain for tissue culture cells and murine Peyer's patches. We found that centrifugation of the Mot⁻ and tumbly swimming Che⁻ strains onto the cellular monolayer did not increase their ability to enter eukaryotic cells. While there is little doubt that motility increases the probability that organisms will come into contact with host cells, we believe that the data indicate that the presence of flagella on nonmotile and some non-chemotactic strains has other effects on *Salmonella* pathogenicity. It is likely that these appendages interfere with the interactions that allow efficient invasion. However, there may be still unknown effects of *Salmonella* motility and flagellar structure that play a role in the invasive ability of these organisms.

Table 2. Invasiveness of Che⁻, Fla⁻ and Mot⁻ derivatives of *S. typhimurium* SL1344 for HEp-2 cells and murine Peyer's patches.

SL1344 or derivative	Swimming phenotype	*In vitro* bacterial invasiveness, %[a]		Relative *in vivo* invasiveness for Peyer's patch tissue[b]
		Aerobic growth	Low-oxygen growth	(mutant strain/SL 1344)
SL1344	wild type	0.001	0.538	1.0 ± 0.0
BJ11 (*cheA52*)	smooth	0.028	0.638	3.0 ± 2.6
BJ17 (*cheB111*)	tumbly	0.00001	0.088	0.4 ± 0.2
BJ32 (*motB275*)	nonmotile	0.00004	0.008	0.1 ± 0.1
χ3420 (*fli-8007*)	nonflagellated	0.01	0.306	0.8 ± 0.2

[a] Values are the percentage of 1×10^7 organisms surviving exposure to 100 μg/ml of gentamicin for 90 minutes after addition to HEp-2 monolayers for 15 minutes.
[b] Values are the ratio of bacteria recovered from the Peyer's patches of mice after injecting equal numbers (5×10^6) of a mutant strain and SL1344 into intestinal ligated loops.

Genetic aspects of *Salmonella* invasion

Role of the virulence plasmid

Within the last ten years, efforts have been made to understand the molecular genetics of *Salmonella* invasion. It was noted that virulent strains of *S. typhimurium* harbored large plasmids (Jones *et al.*, 1982) that were similar in size to the "cryptic" plasmid originally described in strain LT2 (Dowman and Meynell, 1970). The role of this 90 kb plasmid in adhesion, invasion, and virulence has been examined (Jones *et al.*, 1982). Plasmid "cured" derivatives of two *S. typhimurium* strains were obtained and examined in an HeLa cell tissue culture assay and a mouse virulence model. For each *S. typhimurium* strain tested, the elimination of the "cryptic" plasmid significantly reduced the ability of the organisms to attach to and invade HeLa cells. In addition, both strains became much less virulent for mice. Others have confirmed the role of this plasmid in virulence (Gulig and Curtiss, 1987; Hacket *et al.*, 1986; Heffernan *et al.*, 1987), but none were able to reproduce the finding that plasmid genes are involved in adherence or invasion of tissue culture cells. It is now generally agreed that the "virulence" plasmid is essential for penetration into, and survival within, the deeper tissues of the lymphatic system, but does not encode genes for adherence or invasion factors (Gulig, 1990).

Regulation of invasion

Until recently, little was known about environmental conditions that regulate the invasive phenotype of *Salmonella*. We observed that there was a significant lag between the time stationary phase *S. cholerasuis* and *S. typhimurium* were added to MDCK monolayers and the time that the bacteria were able to productively interact with the cells (Finlay *et al.*, 1989). The ability to adhere and to invade was found to correlate with the expression of several new bacterial proteins. This led us to propose a model that epithelial cell surfaces induced the synthesis of proteins necessary for adherence and invasion of MDCK cells. However, this model did not stand up under experimental scrutiny. Subsequently, we showed that bacterial growth conditions, rather than the presence of eukaryotic cells, modulated the expression of the *Salmonella* invasive phenotype (Lee and Falkow, 1990). Aerobically growing bacteria, which are relatively non-adherent and non-invasive, could be induced to become adherent and invasive if grown under oxygen-limiting conditions (Table 3). Organisms that had reached the stationary phase of growth, whether grown in high- or low-oxygen conditions, were non-invasive for mammalian cells. Other groups have similarly reported that the levels of oxygen in the growth environment affect the ability of *Salmonella* to invade tissue culture cells (Ernst *et al.*, 1990; Schiemann and Shope, 1991). Environmental conditions have also

Table 3. Effect of oxygen availability on *S. typhimurium* invasiveness.

Bacterial species	Bacterial culture	Bacterial invasiveness (%)
S. typhimurium	Nonagitated, late log	100 ± 0
	Agitated, midlog	0.1 ± 0
	0% O_2	41 ± 1
	1% O_2	218 ± 18
	20% O_2	3 ± 0.3

Data from Lee CA & Falkow S, 1990.

been found to regulate expression of *invA*, a *Salmonella* invasion gene discussed below (Galan and Curtiss, 1990). High osmolarity growth conditions were found to increase transcription of this gene, as well as to increase the *in vitro* invasiveness of *S. typhimurium*. Thus, it appears that the invasion pathway of *Salmonella* species is regulated by osmolarity, the growth state of the bacterial culture, and oxygen levels in the growth medium.

Genetic loci involved in invasion

The initial approach used to identify chromosomal genes involved in *Salmonella* invasion was based on the likelihood that the microbial factors mediating entry into a eukaryotic cell are located in the bacterial membrane; hence, the probability of identifying the genes for such surface macromolecules can be increased by examining Pho⁺ Tn*phoA* insertion mutants. Such mutants have insertions in genes encoding cell envelope proteins (Manoil and Beckwith, 1985). An analysis of *S. cholerasuis* Tn*phoA* mutants revealed that 6.9% (42 of 626) of the Pho⁺ isolates were less invasive (Finlay *et al.*, 1988). Preliminary analysis of the Tn*phoA* strains revealed that many of the transposon insertions had disrupted lipopolysaccharide (LPS) biosynthesis, while the effects of other insertions are unknown. The absence of intact LPS seemed to nonspecifically disrupt the ability of *S. cholerasuis* to attach to and invade tissue culture cells. Interestingly, some of the *S. cholerasuis* non-invasive Tn*phoA* mutants retained virulence for mice, while others lost virulence as expected. More detailed characterization of these Tn*phoA* mutants is needed before it will be possible to determine which mutations have a direct effect on invasion and what that effect might be.

Efforts have been made to directly clone the invasion genes of *Salmonella* into *E. coli* and confer the ability to enter tissue culture cells. Elsinghorst *et al.* (1989) successfully cloned a 33 kb region of the *S. typhi* Ty2 chromosome, which conferred low levels of invasiveness on *E. coli* HB101. Analysis of the cloned DNA by transposon mutagenesis identified four separate loci that were necessary for invasion. During characterization of the invasion genes, it was determined that they were closely linked to *recA* and *srlC*, which localized the cluster of invasion genes to 58 minutes on the *Salmonella* chromosome. Cosmids containing homologous regions of the *S. typhimurium* chromosome were obtained; however, they did not confer an invasive phenotype upon *E. coli* HB101.

Another approach to identify invasion genes involved restoration of the *in vitro* invasiveness of a noninvasive *S. typhimurium* LT2 laboratory strain (Galan and Curtiss, 1989). A cosmid that complemented the inability of the LT2 strain to enter Henle-407 cells was identified from a chromosomal gene bank of the invasive *S. typhimurium* strain SR-11. The cosmid carried four open reading frames, designated *invA*, *inv-B*, *inv-C*, and *inv-D*, which were necessary for complementation of the invasion defect. Polar *invA* mutations, transduced into two wild-type strains of *S. typhimurium*, significantly increased the LD$_{50}$ values for mice following oral challenge. Hybridization studies demonstrated that virtually all *Salmonella* strains carry genes homologous to *invA-D*; strains of *E. coli*, *Yersinia* species, and *Shigella* species do not carry homologous genes (Galan and Curtiss, 1991). A new isolated gene, *invE*, has been identified upstream of *invA* (Ginocchio *et al.*, 1992), which also belongs to this cluster of invasion genes. The nucleotide sequences of *invA* and *invE* were used to search the GenBank data bank for similar sequences (Galan *et al.*, 1992a; Ginocchio *et al.*, 1992). The predicted amino acid sequence of InvA was found to be similar to *Caulobacter crescentus* FlbF, *Yersinia* LcrD, *S. flexneri* VirH, and *E. coli* FlhA, while InvE was similar to LcrE of *Yersinia*. The exact functions of these proteins are unknown, although it is believed that they are involved in regulating gene expression or in the translocation of proteins through the bacterial membrane.

In an effort to identify invasion genes regulated by oxygen levels, our laboratory generated *S. typhimurium* transposon mutants and selected for the ability of these mutants to enter HEp-2 cells when grown under repressing, aerobic growth conditions (Lee *et al.*, 1992). The transposon used in this study, Tn*5*B50, carries the *neo* promoter cloned into the left insertion sequence of the transposon. Upon chromosomal insertion, the *neo* promoter within the transposon constitutively expresses downstream genes. Using Tn*5*B50, a class of mutants that possessed the ability to enter HEp-2 cells during aerobic growth was obtained. Analysis of these mutants identified a locus, *hil* (hyperinvasive locus), which either regulates or is directly involved in *S. typhimurium* epithelial cell entry. Deletion of the DNA region containing the *hil* locus conferred a noninvasive phenotype on *S. typhimurium*. This locus has been mapped to chromosome minute 59.5 between *srl* and *mutS* in a region adjacent to other genes implicated in the invasive phenotype of *Salmonella* strains (Elsinghorst *et al.*, 1989; Galan and Curtiss, 1989).

As an alternative approach to identifying oxygen-regulated invasion genes, we have recently created *lac* operon fusions in *S. typhimurium* (B. D. Jones, unpublished). Initially, pools of Tn*5 lac* transposon mutants were screened for isolates that only expressed β-galactosidase in oxygen-limiting conditions. This was accomplished by plating bacteria containing Tn*5 lac* insertions on MacConkey lactose agar and looking for colonies that only turned red in the center, or oxygen-limiting portion, of the colony. The invasiveness of mutants with such oxygen-regulated *lac* fusions was measured in an *in vitro* tissue culture assay. We identified several independent *S. typhimurium* mutants that were reduced approximately 1000-fold in their ability to enter HEp-2 cells. A cosmid that fully restores the invasive phenotype of one of the noninvasive oxygen-regulated *lac* fusion strains has recently been identified. The oxygen-regulated genes have been localized to a 4.8 kb segment of this cosmid by subcloning and deletion analysis. A search of the GenBank data bank with a partial sequence from this complementing DNA revealed no similar sequences. Identification of the proteins and complete sequence analysis of the complementing DNA fragment are underway.

Models for Salmonella entry

Figure 2. Electron micrograph of *S. typhimurium* inside of a polarized Caco-2 cell. Initial bacterial interactions with the microvilli induced ruffling of the cell membrane which is still present after internalization of the organism.

As first suggested by Takeuchi's work, entry of *Salmonella* into non-phagocytic cells appears to be dependent upon their ability to cause changes in the structural integrity of the cell membrane (Fig. 2). Microscopic observations of tissue culture cells infected with the adherent and noninvasive *hil*, *invA*, and *invE* mutants have shown that these strains are unable to initiate a signal that causes the characteristic disruption of the host cell membrane during *Salmonella* invasion (Galan *et al.*, 1992a; Ginocchio *et al.*, 1992; Francis *et al.*, 1993). Interestingly, each of these noninvasive mutants has been found to enter tissue culture cells in the presence of wild-type invasive organisms. However, this phenomenon may not be specific for *Salmonella* since we have found that *E. coli* HB101 enters HEp-2 cells in the presence of invasive *Salmonella* (Francis *et al.*, 1993). Conversely, Ginocchio *et al.* (1992) saw no internalization of an *E. coli* RDEC strain under the conditions that allowed a noninvasive *Salmonella* strain to enter Henle-407 cells in the presence of invasive *Salmonella*.

Galan *et al.* (1992b) recently examined protein phosphorylation of host cells during *S. typhimurium* invasion. Phosphorylation of several proteins occurred as *Salmonella* entered host cells; this phenomenon was not observed following infection with a noninvasive *invA* mutant. One of the proteins, with a molecular mass of 170,000, was shown to be the epidermal growth factor (EGF) receptor by immunoprecipitation with a monoclonal antibody. The role of the EGF receptor in *Salmonella* entry was tested by examining the effect of EGF on the ability of an *invA S. typhimurium* mutant to invade tissue culture cells. The presence of EGF in the invasion assay did, in fact, "rescue" the invasive phenotype of *invA* mutants, while the invasiveness of the *E. coli* RDEC strain did not change in the presence of EGF. In a similar set of experiments, our laboratory also found that a noninvasive *Salmonella* mutant could enter HEp-2 cells in the presence of EGF (Francis *et al.*, 1993). However, we found that EGF promoted the internalization of noninvasive *E. coli* strains as well. We now have examined the ability of *S. typhimurium* to enter the fibroblast cell line Swiss 3T3 and its derivative, NR-6 (Pruss and Herschman, 1977). The NR-6 cell line is an epidermal growth factor-nonresponsive variant of the Swiss 3T3 cell, which does not synthesize EGF receptor protein or produce mRNA from the EGF receptor gene (Schneider *et al.*, 1986). We measured the ability of invasive *S. typhimurium* strain SL1344 to enter each of these cell lines and found no difference in the levels of invasion (Francis *et al.*, 1993). The results published by Galan *et al.* (1992) suggest that *Salmonella* invasion may occur via an interaction with the EGF receptor. However, the results obtained by our laboratory suggest that the ability of invasive *Salmonella* and EGF to promote entry of noninvasive organisms is not a *Salmonella*-specific event. In addition, we have data that indicate that the EGF receptor is not directly involved in *Salmonella* entry. Perhaps, the ability of invasive *S. typhimurium* or EGF to "rescue" the invasive phenotype of the noninvasive *S. typhimurium* mutants is an indirect result of membrane ruffling activity in the host cell caused by invasive *Salmonella* and EGF. It is well known that EGF elicits a variety of effects on host cells including increased membrane ruffling, pinocytosis, and calcium fluxes (Rozengurt, 1986). Thus, it is possible that increased host cell membrane activity caused by invasive *Salmonella* and EGF allow otherwise noninvasive organisms to enter the host cell. Whether the fate of these internalized noninvasive bacteria is the same as invasive *Salmonella* remains to be seen.

Conclusions

Early work on *Salmonella* invasion largely described nonquantitative events that could be visualized as the bacteria approached and entered host cells. The development of *in vitro* models, which mimic events in natural infections, has proved to be invaluable in identifying some of the specific bacterial and host cell factors necessary for *Salmonella* entry. Several genetic loci essential for the invasive phenotype have now been identified and are presently being characterized. Efforts to understand the host cell responses that occur following interaction with *Salmonella* organisms are underway. As data accumulate we will begin to get our first detailed view of *Salmonella* entry into eukaryotic cells. Eventually, an understanding of the mechanism and regulation of *Salmonella* invasion will be reached. Undoubtedly, the process will prove to be both complex and interesting and will reveal much about the host and the bacterium.

Acknowledgements

We would like to thank Carol Francis and Sara Fisher for proofreading the manuscript and providing thoughtful insights. We thank C. L. Francis for sharing unpublished results. This work was supported by the Digestive Disease Program Project grant DK38707 and NIH grant no. AI26195 to Stanley Falkow, and by NIH postdoctoral fellowship no. AI08404 to Bradley D. Jones.

References

Carter P and Collins F (1974) The route of enteric infection in normal mice. J Exp Med 139: 1189–1203.

Carter PB (1975) Spread of enteric fever bacilli from the intestinal lumen. *In*: Schlessinger D (ed.) Microbiology-1975. (pp. 182–187) American Society for Microbiology, Washington, D.C.

Dowman JE and Meynell GG (1970) Pleiotropic effects of derepressed bacterial sex factors on colicinogeny and cell wall structure. Mol Gen Genet 109: 57–68.

Elsinghorst EA, Baron LS and Kopecko DJ (1989) Penetration of human intestinal epithelial cells by *Salmonella*: molecular cloning and expression of *Salmonella typhi* invasion determinants in *Escherichia coli*. Proc Natl Acad Sci USA 86: 5173–5177.

Ernst RK, Dombroski DM and Merrick JM (1990) Anaerobiosis, type 1 fimbriae, and growth phase are factors that affect invasion of HEp-2 cells by *Salmonella typhimurium*. Infect Immun 58: 2014–2016.

Finlay BB and Falkow S (1988) Comparison of the invasion strategies used by *Salmonella choleraesuis*, *Shigella flexneri*, and *Yersinia enterocolitica* to enter cultured animal cells: endosome acidification is not required for bacterial invasion or intracellular replication. Biochimie 70: 1089–1099.

Finlay BB, Gumbiner B and Falkow S (1988) Penetration of *Salmonella* through a polarized Madin-Darby canine kidney epithelial cell monolayer. J Cell Biol 107: 221–230.

Finlay BB, Heffron F and Falkow S (1989) Epithelial cell surfaces induce *Salmonella* proteins required for bacterial adherence and invasion. Science 243: 940–943.

Finlay BB, Starnbach MN, Francis CL, Stocker BAD, Chatfield S, Dougan G and Falkow S (1988) Identification and characterization of Tn*phoA* mutants of *Salmonella* that are unable to pass through a polarized MDCK epithelial cell monolayer. Mol Microbiol 2: 757–766.

Francis CL, Ryan TA, Jones BD, Smith SJ and Falhow S (1993) Ruffles induced by *Salmonella* and other stimuli direct macropinocytosis of bacteria. Nature (London) 364: 639–642.

Francis CL, Starnbach MN and Falkow S (1992) Morphological and cytoskeletal changes in epithelial cells occur immediately upon interaction with *Salmonella typhimurium* grown under low-oxygen conditions. Mol Microbiol 6: 3077–3087.

Galan JE and Curtiss R, III (1989) Cloning and molecular characterization of genes whose products allow *Salmonella typhimurium* to penetrate tissue culture cells. Proc Natl Acad Sci USA 86: 6383–6387.

Galan JE and Curtiss R, III (1990) Expression of *Salmonella typhimurium* genes required for invasion is regulated by changes in DNA supercoiling. Infect Immun 58: 1879–1885.

Galan JE and Curtiss R, III (1991) Distribution of the *invA*, *–B*, *–C*, and *–D* genes of *Salmonella typhimurium* among other *Salmonella* serovars: *invA* mutants of *Salmonella typhi* are deficient for entry into mammalian cells. Infect Immun 59: 2901–2908.

Galan JE, Ginocchio C and Costeas P (1992a) Molecular and functional characterization of the *Salmonella* invasion gene *invA*: homology of InvA to members of a new protein family. J Bacteriol 174: 4338–4349.

Galan JE, Pace J and Hayman MJ (1992b) Involvement of the epidermal growth factor receptor in the invasion of cultured mammalian cells by *Salmonella typhimurium*. Nature (London) 357: 588–589.

Giannella RA, Formal SB, Dammin GJ and Collins H (1973a) Pathogenesis of salmonellosis. Studies of fluid secretion, mucosal invasion, and morphologic reaction in the rabbit ileum. J Clin Invest 52: 441–453.

Giannella RA, Washington O, Gemski P and Formal SB (1973b) Invasion of HeLa cells by *Salmonella typhimurium*: A model for study of invasiveness of *Salmonella*. J Infect Dis 128: 69–75.

Ginocchio C, Pace J and Galan JE (1992) Identification and molecular characterization of a *Salmonella typhimurium* gene involved in triggering the internalization of *Salmonellae* into cultured epithelial cells. Proc Natl Acad Sci USA 89: 5976–5980.

Gulig PA (1990) Virulence plasmids of *Salmonella typhimurium* and other salmonellae. Microb Pathog 8: 3–11.

Gulig PA and Curtiss R, III (1987) Plasmid-associated virulence of *Salmonella typhimurium*. Infect Immun 55: 2891–2901.

Hackett J, Kotlarski I, Mathan V, Francki K and Rowley D (1986) The colonization of Peyer's patches by a strain of *Salmonella typhimurium* cured of the cryptic plasmid. J Infect Dis 153: 1119–1125.

Heffernan EJ, Fierer J, Chikami G and Guiney D (1987) Natural history of oral *Salmonella dublin* infection in BALB/c mice: effect of an 80-kilobase-pair plasmid on virulence. J Infect Dis 155: 1254–1259.

Hohmann AW, Schmidt G and Rowley D (1978) Intestinal colonization and virulence of *Salmonella* in mice. Infect Immun 22: 763–770.

Jones BD, Lee CA and Falkow S (1992) Invasion by *Salmonella typhimurium* is affected by the direction of flagellar rotation. Infect Immun 60: 2475–2480.

Jones GW, Rabert DK, Svinarich DM and Whitfield HJ (1982) Association of adhesive, invasive, and virulent phenotypes of *Salmonella typhimurium* with autonomous 60-megadalton plasmids. Infect Immun 38: 476–486.

Jones, GW, Richardson LA and Uhlman D (1981) The invasion of HeLa cells by *Salmonella typhimurium*: Reversible and irreversible bacterial attachment and the role of bacterial motility. J Gen Microbiol 127: 351–360.

Khoramian FT, Harayama S, Kutsukake K and Pechere JC (1990) Effect of motility and chemotaxis on the invasion of *Salmonella typhimurium* into HeLa cells. Microb Pathog 9: 47–53.

Kihlstrom E and Edebo L (1976) Association of viable and inactivated *Salmonella typhimurium* 395 MS and MR10 with HeLa cells. Infect. Immun. 14: 851–857.

Kihlstrom E and Nilsson L (1977) Endocytosis of *Salmonella typhimurium* 395 MS and MR10 by HeLa cells. Acta Path Microbiol Scand Sect B 85: 322–328.

Kohbata S, Yokoyama H and Yabuuchi E (1986) Cytopathogenic effect of *Salmonella typhi* GIFU 10007 on M cells of murine ileal Peyer's patches in ligated ileal loops: An ultrastructural study. Microbiol Immunol 30: 1225–1237.

Lee CA and Falkow S (1990) The ability of *Salmonella* to enter mammalian cells is affected by bacterial growth state. Proc Natl Acad Sci USA 87: 4304–4308.

Lee CA, Jones BD and Falkow S (1992) Identification of a *Salmonella typhimurium* invasion locus by selection for hyperinvasive mutants. Proc Natl Acad Sci USA 89: 1847–1851.

16

Leung KY and Finlay BB (1991) Intracellular replication is essential for the virulence of *Salmonella typhimurium*. Proc Natl Acad Sci USA 88: 11470–11474.

Liu S, Ezaki T, Miura H, Matsui K and Yabuuchi (1988) Intact motility as a *Salmonella typhi* Invasion-related factor. Infect Immun 56: 1967–1973.

Manoil C and Beckwith J (1985) Tn*phoA*: a transposon probe for protein export signals. Proc Natl Acad Sci USA 82: 8129–8133.

Mroczenski-Wildey MJ, Di Fabio JL and Cabello FC (1989) Invasion and lysis of HeLa cell monolayers by *Salmonella typhi*: the role of lipopolysaccharide. Microbiol Path 6: 143–152.

Owen RL and Ermak TH (1990) Structural specializations for antigen uptake and processing in the digestive tract. Springer Semin Immunopathol 12: 139–152.

Owen RL and Jones AL (1974) Epithelial cell specialization within human Peyer's patches: An ultrastructural study of intestinal lymphoid follicles. Gastroenterology 66: 189–203.

Pruss RM and Herschman HR (1977) Variants of 3T3 cells lacking mitogenic response to epidermal growth factor. Proc Natl Acad Sci USA 74: 3918–3921.

Rozengurt E (1986) Early signals in the mitogenic response. Science 234: 161–166.

Rubin RH and Weinstein L (1977) Salmonellosis: Microbiologic, Pathologic, and Clinical Features. Stratton Intercontinental Medical Book Corporation, New York.

Schiemann DA and Shope SR (1991) Anaerobic growth of *Salmonella typhimurium* results in increased uptake by Henle 407 epithelial and mouse peritoneal cells *in vitro* and repression of a major outer membrane protein. Infect Immun 59: 437–440.

Schneider CA, Lim RW, Terwilliger E and Herschman HR (1986) Epidermal growth factor-nonresponsive 3T3 variants do not contain epidermal growth factor receptor-related antigens or mRNA. Proc Natl Acad Sci USA 83: 333–336.

Simons K and Fuller SD (1985) Cell surface polarity in epithelia. Annu Rev Cell Biol 1: 243–288.

Sprinz H, Landy M, Gaines S and Edsall G (1956) Experimental typhoid fever in chimpanzees. III. Pathogenesis. Fed Proc 15: 614–615.

Takeuchi A (1967) Electron microscope studies of experimental salmonella infection. I. Penetration into the intestinal epithelium by *Salmonella typhimurium*. Am J Path 50: 109–136.

Tomita T and Kanegasaki (1982) Enhanced phagocytic response of macrophages to bacteria by physical impact caused by bacterial motility or centrifugation. Infect Immun 38: 865–870.

Turnbull PCB and Richmond JE (1978) A model of Salmonella enteritis: The behaviour of *Salmonella enteriditis* in chick intestine studied by light and electon microscopy. Br J Exp Path 59: 64–75.

Worton KJ, Candy DCA, Wallis TS, Clarke GJ, Osborne MP, Haddon SJ and Stephen J (1989) Studies on early association of *Salmonella typhimurium* with intestinal mucosa *in vivo* and *in vitro*: relationship to virulence. J Med Microbiol 29: 283–294.

Yabuuchi E, Ikedo M and Ezaki T (1986) Invasiveness of *Salmonella typhi* strains in HeLa S3 monolayer cells. Microbiol Immunol 30: 322–328.

Yokoyama H, Ikedo M, Kohbata S, Ezaki T and Yabuuchi E (1987) An ultrastructural study of HeLa cell invasion with *Salmonella typhi* GIFU 10007. Microbiol Immunol 31: 1–11.

2. CS1 pili of enterotoxigenic *E. coli*

JUNE R. SCOTT and BARBARA J. FROEHLICH

Abstract. In the establishment of bacterial infections, the first step is attachment to host tissue. For most Gram-negative bacteria, this is accomplished by surface appendages called pili. Different strains of enterotoxigenic *E. coli* (ETEC), which are a major cause of intestinal disease in humans, utilize a limited number of serologically different pili for attachment to the human gut. We are investigating one of these, CS1, whose morphogenesis appears to be totally unrelated to that of the better-studied pili of *E. coli* associated with urinary tract infections. The CS1 pilin operon, which is located on a large plasmid, requires *rns*, a *trans*-acting positive regulator encoded on a different plasmid, for its expression. Although the G + C content of *E. coli* DNA is about 50%, *rns* has a G + C content of only 28%. The major CS1 pilin subunit, encoded by *cooA*, has no cysteines and therefore can have no disulfide loop. Upstream of *cooA* we have identified *cooB*, which is not required for stability or transport of the major pilin subunit, but is required for its assembly into pili. CooB is not structurally similar to chaperonins or other proteins in the GENBANK database except its counterpart in another ETEC pilus, CFA/I. Although there is no detectable DNA hybridization, *rns*, *cooA* and *cooB* are closely related at the amino acid sequence level to their counterparts in the operon of the serologically different ETEC pilus CFA/I. We suspect that the CS1 locus will serve as the prototype for a new type of pilus which is common among ETEC strains.

Abbreviations: ETEC, enterotoxigenic *Escherichia cdi*

Introduction

Colonization of a host by a pathogen requires attachment as a first step. This process is usually specific on the part of both the host and the pathogen, that is, there is usually a ligand on the pathogen that interacts specifically with a receptor on the host. In the case of many Gram negative bacteria, the ligand is borne on long thin hairlike structures called pili or fimbriae. These structures consist predominantly of multiple copies of a single protein subunit, usually referred to as a pilin. In some cases, the major pilin protein is not the specific bacterial ligand; instead, a minor protein acts as the adhesin. For the Pap pili of *E. coli* associated with urinary tract infections, which are among the best-studied of all pili, the adhesin is located at the tip (Lindberg *et al.*, 1987). For *E. coli* type 1 pili, on the other hand, the adhesin seems to be interspersed along the shaft of the structure, but it is thought that only the copy of the adhesin protein that is exposed on the tip is functional for attachment (Ponniah *et al.*, 1991).

C.I. Kado and J.H. Crosa (eds.), Molecular Mechanisms of Bacterial Virulence, 17–30.
© 1994 *Kluwer Academic Publishers.*

We have been studying pili of human enterotoxigenic *E. coli* (ETEC) strains, which are presumed to be required for adherence of this pathogen to the human gut. ETEC are an important cause of diarrhea in humans, especially in travelers and in infants (Levine, 1987). The disease caused by these bacteria is a serious world health problem because of the large excess mortality in which it results (Schoolnik, this volume). The severity of the human health problem is related to the hygiene standards of the community. Thus, in rural parts of developing countries where the sanitary standards are not as high as in our country, diarrheal diseases account for a very high proportion of child deaths. Among bacteria, ETEC strains are one of the predominant causes of diarrheal diseases in very young children. Thus, it is a high priority of the World Health Organization to develop effective vaccines against these organisms. Among the major candidates for such vaccines are the proteins that comprise the pili of these bacteria.

ETEC strains specific for animals are also serious pathogens for their hosts. In veterinary medicine, pili have been used successfully as vaccines against these diseases (Acres *et al.*, 1979; Morgan *et al.*, 1978; Nagy *et al.*, 1978; Rutter & Jones, 1973). It is therefore hoped that pili of human ETEC strains will be useful for disease prevention as well. Because there are a small number of different serological types of pili associated with human ETEC strains, it seems possible that a vaccine could be developed with multiple antigen-specific immunological determinants or that a single immunogenic determinant might be found to be common to the pili of different serotypes. For this reason, a knowledge of the morphogenesis of ETEC pili, including an understanding of the role of the different possible minor proteins, would be valuable.

Transport and assembly of proteins to form the pilus structure is an interesting problem in basic biology. Proteins are usually transported through the inner (cytoplasmic) membrane of bacterial cells by the signal peptidase systems (Inouye & Halegoua, 1980; Perlman & Halvorson, 1983). These enzymes recognize certain motifs at the N-terminus of the transported protein and cleave a signal peptide from it during transport. Pilin proteins use this system for transport into the periplasm. However, transport of proteins through the outer membrane of *E. coli* is less common and less studied and several alternative mechanisms seem to be used by different proteins. Thus, an understanding of this transport process for pilin proteins would make a basic contribution to our knowledge of cell biology.

In addition, study of the assembly of pilin proteins into a functional structure should provide important contributions to our knowledge of biological morphogenesis. For all the well-studied *E. coli* pili, the process of morphogenesis appears to be similar. The prototype and best understood is the pyelopnephritis-associated Pap pilus (see Slonim *et al.*, this volume). The assembly process seems to require a chaperonin protein for each of the pilin components (Hultgren *et al.*, 1991). This protein prevents incorrect folding of the structural subunits of the pili and is involved in transport of these subunits through the outer membrane of the *E. coli* cell.

The pili of human ETEC strains that we are studying are called CS1 and CS2 (coli surface antigens 1 and 2). Originally, they were included with the CS3 short flexible fimbriae in the designation CFA/II (colonization factor antigen II) because all three may be produced by ETEC strains and the original antiserum used to identify them recognized them all (Evans & Evans, 1978). CS1 and CS2 are antigenically different and their genes do not cross-hybridize (Cravioto *et al.*, 1982; Smyth, 1982; Levine, *et al.* 1984; our unpublished results). Other pili of human ETEC strains include the single antigenic type designated CFA/I (Evans *et al.*, 1975), which appears to be very closely related to CS1 (see below) and CFA/IV which, like CFA/II, consists of three antigenically distinct fimbriae, called CS4, CS5, and CS6 (Thomas *et al.*, 1985; McConnell *et al.*, 1988; Wolf *et al.*, 1989). Unlike CS5 and CS3, the N-terminal amino acid sequences of CS1, CS2, CS4, and CFA/I all have a substantial degree of homology (Smyth *et al.*, 1990) and, as we learn more about their genetic organization, we find analogies and homologies among them (see below). These types of pili, however, appear quite different from the better-studied ones exemplified by Pap. The differences include their mode of regulation of synthesis and their morphogenesis. It seems, therefore that CS1, the best understood of these human ETEC strain pili, may represent a second prototype for pilus synthesis and assembly.

Regulation of expression of CS1 and CS2

The CS1 and CS2 pili, which are antigenically distinguishable, are found on human ETEC strains of specific serotypes and biotypes, but both types of pili have never been found together on the same strain (Smyth, 1986). Like most other *E. coli* pili, CS1 and CS2 are only found on bacteria grown at temperatures above about 28°C (Smyth, 1982). Expression of these pili requires the presence of a large plasmid (Smith *et al.*, 1983; Mullany *et al.*, 1983). This was demonstrated when curing a CS1- or CS2-producing strain of the plasmid led to loss of expression of CS1 pili (Boylan & Smyth, 1985). When the plasmid was re-introduced, pili were again found on the surface of the cells. However, introduction of the plasmid from the CS1 strain into the CS2 strain led to production of CS2 pili, not CS1 pili (Boylan & Smyth, 1985). This indicated that genetic information from two unlinked regions of DNA appeared to be required to produce either CS1 or CS2. Thus, there are clearly interesting lessons about regulation of virulence determinants to be learned from a study of these organelles.

It was originally believed that the large plasmid encoded the major pilin protein for CS1 and CS2. This idea was based on hybridization between this plasmid and a mixed oligonucleotide probe derived from the N-terminal amino acid sequence of CS1 and CS2 (Boylan *et al.*, 1987; Klemm *et al.*, 1985). However, the smallest DNA fragment of the plasmid required for CS1 or CS2 production was found to be too small to encode these pilins. Its DNA sequence showed that it was not the structural gene for either major pilin antigen (Caron *et al.*, 1989). This led to the idea that the plasmid fragment positively regulated synthesis of the pilin genes, either by binding a negative regulator to the plasmid DNA or by synthesis from this fragment of a *trans*-acting protein. The open reading frame on the plasmid DNA was shown to produce a protein of the size expected (30 kda) from its sequence *in vitro* (Caron *et al.*, 1989). To demonstrate that this protein was responsible for pilin synthesis, a mutation was constructed within this open reading frame and found to be unable to synthesize the protein *in vitro*. When the mutated plasmid was used in the wild type CS1 strain, no CS1 was produced and in the CS2 strain, no CS2 was detectable. The gene defined by this mutation is called *rns* (regulation of CS antigens; Caron *et al.*, 1989).

A search of the GENBANK database showed that Rns has homology with the AraC family of positive regulatory proteins (Caron *et al.*, 1989). There are two apparent helix-turn-helix domains in Rns and both of these are in regions of homology with regions of AraC. At least one of these regions in AraC which are homologous to Rns seems to be important for DNA binding since an amino acid substitution in it alters the AraC regulatory activity *in vitro* (Francklyn & Lee, 1988). By analogy then, we believe that Rns is a positive regulator acting at the level of transcription. We do not know whether it acts directly on the CS1 and CS2 operons or is part of a regulatory cascade.

Since the identification of Rns, similar proteins have been found in many other Gram-negative pathogens. These positive regulators are all required for expression of virulence genes. The list of related proteins includes: *cfaR* (Savelkoul *et al.*, 1990) and *fapR* (Klaasen *et al.*, 1990) of *E. coli*, *virF* (Cornelis *et al.*, 1989) of *Yersinia*, *virF* (Sakai *et al.*, 1986) of *Shigella*, and *tcpN* (Ogierman & Manning, 1992) of *Vibrio cholerae*.

What advantage accrues to the pathogen as a result of having virulence determinants under positive regulation? Presumably energy could be saved if these determinants were not expressed unless needed. For example, when an ETEC strain is in the open environment instead of being in its human host, it might be disadvantageous to synthesize pili for two reasons: first, if the pili served to attach the strain to some molecule outside the human gut the strain might become attached in an inhospitable niche. Second, synthesis of pili probably requires a significant input of energy because there are many of them per cell. Thus, if the pili are not needed, it probably would be a selective advantage not to produce them. In some of the better studied systems of this family, including *araC*, there is negative as well as positive regulation involved. This gives more sensitive control over expression of the gene than regulation of either type alone. At the present time, we do not know the mechanism of regulation by *rns* and cannot tell whether a negative component of the control system is also present.

A very unusual feature of the *rns* gene is its extremely low G + C content: 28% The *E. coli* genome has an average G + C content of 50% (McClelland *et al.*, 1987), so this suggests that *rns* evolved in a foreign organism. Furthermore, the DNA region upstream of *rns* has a G + C content more typical of *E. coli* (see Fig.

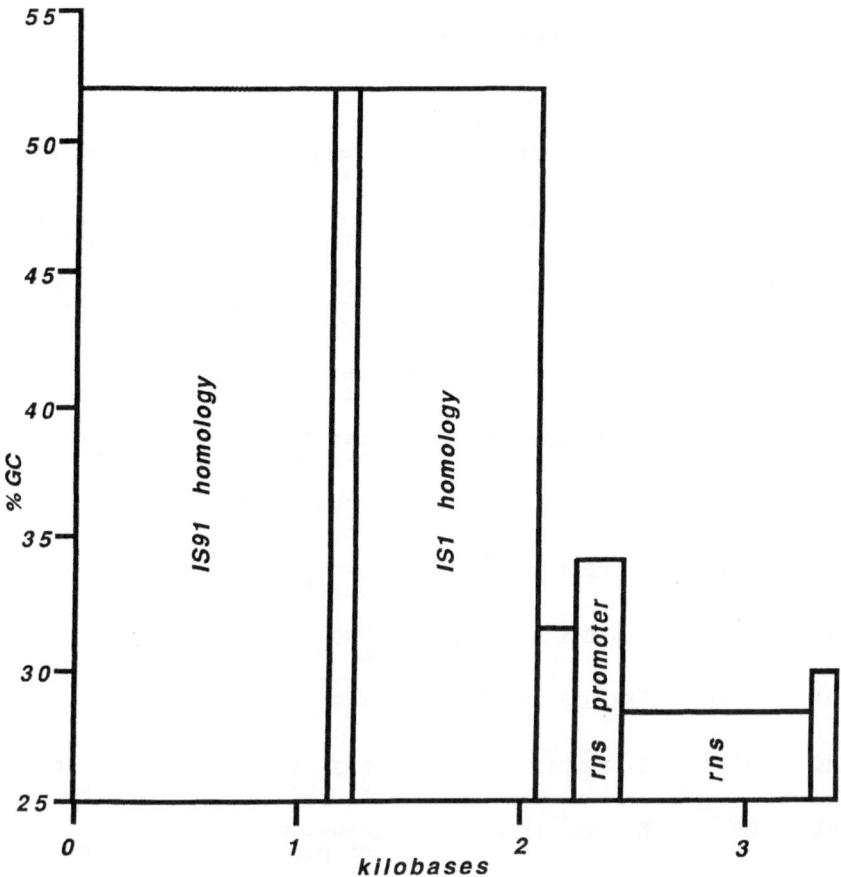

Figure 1. Percent G + C of *rns* and neighboring DNA. The IS91 homology extends from bp 162 (of IS91) to the end of IS91 but the DNA region adjacent to *rns* is missing bp 1121–1365 of IS91.

1). Of the group of *araC*-related genes, in addition to *rns*, *cfaR*, *fapR* and *csvR* of ETEC, *virF* of *Shigella*, *tcpN* of *V. cholera* and *appY* of *E. coli* K-12 have very low G + C content (28%–32%, see Table 1). This low G + C content is not a feature of all members of the *araC* family (Table 1), which suggests that although some of these regulatory genes were recently acquired by their hosts, others have co-evolved with their hosts for a much longer time.

The lack of genetic linkage between *rns* and the CS operons it controls is consistent with the possible recent introduction of *rns* into ETEC strains. Furthermore, a search of the GENBANK database identified a nearby region upstream of *rns* with homology to two insertion elements (Fig. 1 and 2), suggesting the possibility that the *rns* gene was introduced on a transposon. The region immediately upstream of *rns* is 98% homologous with the DNA sequence of IS1. However, translation of this DNA sequence suggests that it is not functional since it has no open reading frame that might correspond to a

Table 1. Percent G + C of genes from the *araC* family which are related to *rns*.

Gene	%G+C	Species	Function	Reference
rns	28	E. coli (ETEC)	+ regulation of CS1&CS2	Caron et al., 1989
cfaR	27.5	E. coli (ETEC)	+ regulation of CFA/I	Savelkoul et al., 1990
csvR	30.7	E. coli (ETEC)	? (partially replaces cfaR & rns)	de Haan et al., 1991
fapR	31.6	E. coli (ETEC)	+ regulation of 987P	Klaasen et al., 1990
virF	30	S. flexneri	+ regulation vir genes	Sakai et al., 1986
tcpN	28	V. cholerae	+ regulation of pilus	Ogierman & Manning, 1992
appy	31.6	E. coli (K12)	global regulation	Kemp et al., 1987
virF	41.4	Y. enterocolitica	+ regulation of vir genes	Cornelis et al., 1989
araC	49.6	Erwinia carotovora	regulation of ara	Lei et al., 1985
araC	53.9	S. typhimurium	regulation of ara	Horwitz et al., 1981
envY	50	E. coli (K12)	regulation of porin	Lundrigan et al., 1989
rhaS	51.9	E. coli (K12)	regulation of rha	Tobin & Schleif, 1987
rhaR	49.7	E. coli (K12)	regulation of rha	Tobin & Schleif, 1987

transposase gene. Upstream of the IS1 homology is a 1251 bp region that is 91% homologous to IS91 (Fig. 2; Mendiola *et al.* 1992). Compared with IS91, our sequence shows several deletions, including one of 245 bp in the region which is thought to code for the transposase. Our DNA sequence does not extend far enough to determine whether the beginning of IS91 is present. Thus, if the *rns* gene was introduced by transposition, it must have happened long enough ago to allow these transposons to acquire several mutations.

The question of the host in which the *rns* gene originated has not been satisfactorily answered. Other enteric pathogens with low G + C content (mycoplasma, Clostridia, Campylobacter; Caron *et al.*, 1990) were tested for hybridization to *rns*, but all results were negative. There have been no good suggestions as to the host with low G + C genomic DNA in which these genes originated.

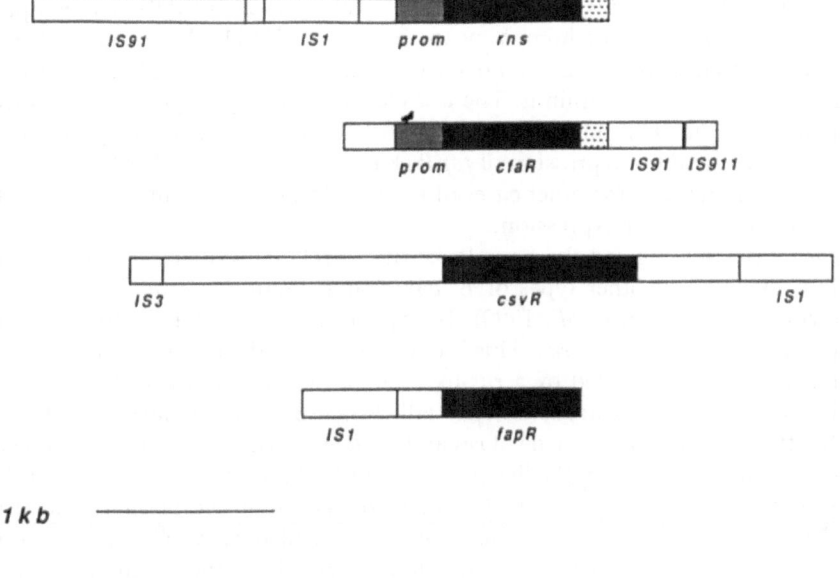

HOMOLOGY WITH RNS

	DNA	AMINO ACIDS	
		IDENTICAL	CONSERVED
cfaR	99%	94%	98%
csvR	80%	77%	89%
fapR	50%	32%	55%

Figure 2. Comparison of *rns*-related positive regulators and sequences surrounding them. The three shaded sections are the parts of *rns* and *cfaR* which are homologous. The black boxes delineate the extent of the coding regions. The grey boxes in *rns* and *cfaR* labeled "prom" contain the promoter (Caron *et al.*, 1989; Savelkoul *et al.*, 1990; unpublished results). The stippled region probably contains the transcription terminator (Caron *et al.* 1989; Savelkoul *et al.*, 1990). Not all of the IS-elements are complete (see text). The DNA and amino acid homologies were calculated for the coding region. In the case of *csvR*, only the region which is homologous to *rns* (first 801 bp of the open reading frame) was used in the calculation.

The Rns protein is highly homologous to the regulatory protein for CFA/I (CfaR: Savelkoul *et al.*, 1990) and to CsvR (isolated from a CS5, CS6-producing ETEC strain: de Haan *et al.*, 1991). This homology is seen both at the protein and the DNA level and all three have low G + C content (see Fig. 2). This suggests that these genes are very closely related and may have been introduced recently into all three types of ETEC strains. It is also consistent with the close similarity between CS1 and CFA/I (see below).

Like *rns*, *cfaR* and *csvR* are adjacent to DNA homologous to IS-elements (Fig. 2) A search of the GENBANK database with the published DNA sequence (Savelkoul *et al.*, 1990) showed that the DNA region downstream from *cfaR* has homology with IS91 (90%: Mendiola *et al.*, 1992) and with IS911 (98%: Prere *et al.*, 1990). The 500 bp region of homology to the right end of IS91 is interrupted by the 157 bp region of homology to the end of IS911. Since the published sequence extends no further, it is not clear whether there is a complete IS91 which has a complete IS911 within it. The less closely related gene *fapR*, the positive regulator of the 987P pilin operon, has an IS1 element immediately upstream which is required for expression of *fapR* (Klaasen & de Graaf, 1990; Klaasen *et al.*, 1990). In none of the other cases of *rns*-related positive regulators are the IS elements required for expression.

We have investigated the prevalence of *rns*-related sequences among other ETEC strains and other types of *E. coli* that are enteric pathogens by DNA hybridization (Caron *et al.*, 1990). It appears that most ETEC strains carry sequences homologous to *rns*. This is consistent with the possibility that ETEC pili are usually controlled by a *rns*-like regulator. In support of this, we have found that *cfaD* and *rns* are at least partially functionally interchangeable (Caron & Scott, 1990). The same is true of *rns* and *csvR* as well as a *rns*-homologous gene in a CS4 producing strain (Willshaw *et al.*, 1991). Thus, it seems possible at this time that serologically distinct ETEC pili have a common evolutionary ancestor.

Most or all of the virulence determinants controlled by AraC-like regulatory proteins are environmentally regulated. Like the others, environmental regulation of *rns*-regulated genes occurs. It is manifested by lack of production of CS1 or CS2 pili at temperatures below about 28°C. The *araC*-type system itself has not been shown to be responsive to environmental change, but at least one other member of the *araC* family, *envY*, is involved in temperature regulation. Thus, it is possible either that AraC-like proteins will be found to respond to such changes or that other environmental-sensitive components of more complex regulatory controls are involved in expression of these virulence determinants.

Most of the other positive regulators of the *araC* family regulate more than one virulence determinant. Often these comprise more than one operon. At this time we do not know whether Rns regulates operons other than those needed for pilus production.

The structural gene for CS1 pilin, cooA

Using an *E. coli* K12 strain containing the cloned *rns* gene, we constructed and screened libraries made from CS1 or CS2 ETEC strains and isolated DNA that encoded production of CS1 and CS2 pili (Perez-Casal *et al.*, 1990). Hybridization experiments indicate that CS1 is encoded on a large plasmid separate from the one that carries *rns* and that CS2 is encoded in the chromosome of their respective strains (Perez-Casal *et al.*, 1990). The CS1 locus has been subcloned and its analysis begun.

The structural gene for the major CS1 pilin protein, called *cooA*, encodes a leader sequence presumably used for transport through the cytoplasmic membrane (Perez-Casal *et al.*, 1990). The translated CooA protein (mature size of 15.2 kda) is not homologous to other proteins in the database, except the major CFA/I pilin subunit. Furthermore, unlike many other pilin proteins, there are no disulfide loops in CooA because it has no cysteine residues. It is, however, very similar to the major subunit of CFA/I, and, like the gene for the CFA/I protein, *cooA* is the second gene in the operon.

The first gene in the operon: cooB

The sequence upstream of *cooA* revealed an open reading frame for a protein with a signal sequence (Jordi *et al.*, 1991). This DNA produced a protein of the expected size (26 kda) in an *in vitro* transcription-translation system that was not present in a mutant constructed by inserting an omega fragment into the open reading frame (Scott *et al.*, 1992). This defines the *cooB* gene. Sequence analysis and database searching indicated that the only known similar protein in the database is the first protein in the CFA/I operon, whose function has not yet been defined.

To learn about the function of *cooB*, the *cooB*::omega mutant was introduced by allele replacement into the CS1 plasmid of the wild type ETEC strain. To assay function of CS1 pili, hemagglutination of bovine erythrocytes in the presence of mannose is used. The *cooB* mutant differed from the wild type strain by its inability to cause hemagglutination. Since the omega element terminates transcription, the downstream CooA protein was not expressed unless *cooA* was provided in *trans*. This was demonstrated by the absence of CooA antigen in cell extracts, and demonstrates that CooA and B are encoded on the same transcript.

To determine the function of CooB, we provided *cooA* and the downstream cloned CS1 DNA in *trans* in the *cooB*::omega mutant strain. This mutant still did not cause hemagglutination. However, in the *cooB* mutant, CooA pilin was present both intracellularly and in cell surface extracts. It was also present in the medium. Therefore, the *cooB* gene product is not needed for *cooA* synthesis, stability, or transport, so it does not function like the Pap chaperonin proteins (Hultgren *et al.*, 1991; Scott *et al.*, 1992).

In collaboration with Paul Orndorff and Walker Russell, the *cooB*::omega mutant strain was examined by electron microscopy (Scott *et al.*, 1992). The pili present on the wild type cell surface are completely absent on the mutant. We conclude, therefore, that *cooB* is required for assembly of the CooA protein into CS1 pili. No such function has previously been described for a pilin gene. It suggests the possibility that the morphogenesis of CS1 represents a new type of assembly mechanism and that the further study of this system will be very rewarding.

Evolutionary considerations

Because *rns* has such a low G + C content, we determined whether the genes it regulates are similar in G + C content. Although they are lower in G + C than the average *E. coli* gene (50%: Fig. 3), they are similar to the G + C content of other virulence factors and much higher (around 45% for the pilin gene, *cooA*, and around 35% for *cooB*, C, D) than *rns* (Fig. 3). Thus, it appears that the genes regulated by *rns* have been present in *E. coli* much longer than the *rns* gene. It

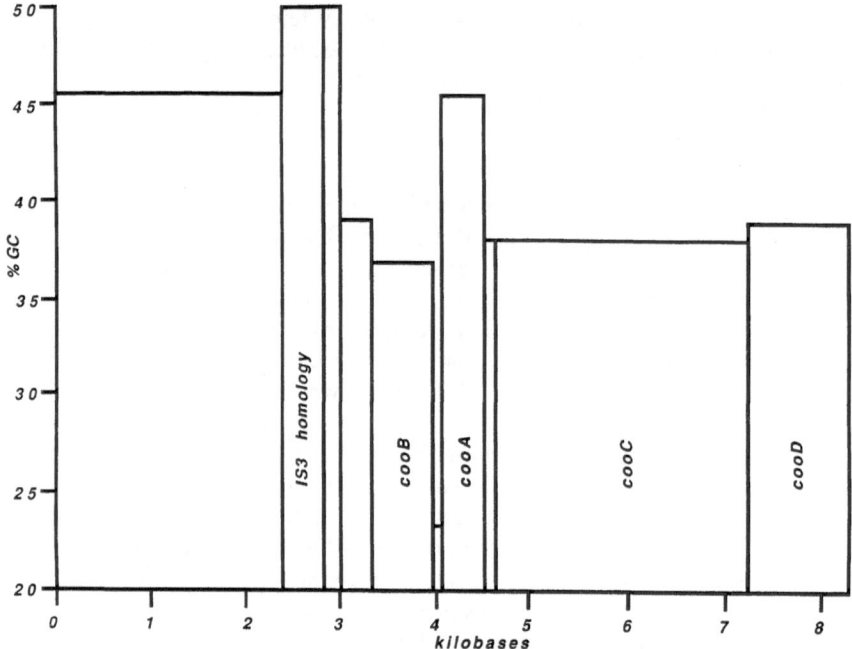

Figure 3. Percent G + C of the *coo* operon. The IS150 homology region is an incomplete element consisting of a 438 bp region within *orfB* of IS150 (Schwartz *et al.*, 1988). The regions labeled *cooA*, *cooB*, *cooC* and *cooD* are the coding regions (Perez-Casal *et al.*, 1990; Froehlich & Scott, in preparation). These genes are transcribed from left to right as drawn.

should be kept in mind, however, that there is no evidence indicating that *rns* regulates the *coo* operon directly.

Promoter of coo *operon*
Upstream of the *coo* operon is a region showing homology with IS150, a member of the IS3 family (Schwartz *et al.*, 1988; Scott *et al.*, 1992), consistent with the possibility that this whole coding region was introduced on a transposon. The promoter for the *coo* operon does not reside within this IS3-related DNA. Primer extension analysis indicates that the 5'-end of the *cooB-A* transcript is located 14 bp upstream from the start of CooB translation (our unpublished data, see Fig 3). There is a sequence located at −10 and −35 from the start of transcription which resembles a consensus promoter sequence.

Conclusion

Pili of ETEC strains, including CS1 and CFA/I, appear to have significant differences from pili of other types of organisms. These differences include the mode of regulation of their synthesis and their mechanism of morphogenesis.

In many bacterial pathogens, several virulence factors are co-regulated (Miller *et al.*, 1989). In addition, environmental regulation of such factors is often found (Maurelli, 1989). The CS1 and related pili of ETEC strains fall into both of these categories. The *trans*-acting positive regulator of the CS1 operon, called Rns, is related to AraC of *E. coli* and its relatives. They both seem to be members of a growing family of related proteins that serve as global regulators of virulence genes in enteric organisms. The mode of action of these proteins is not yet understood and the possibility of indirect regulation by a cascade of factors still exists for the CS1 pili.

The well-studied morphogenesis mechanisms characteristic of type 1 and Pap pili of *E. coli* may represent only one approach to assembly of pili. CS1 and its relatives seem to differ significantly from this and may represent a new class of assembly mechanism. For CS1, there is a protein, CooB, required for assembly of pili, but not needed for stability or transport of the major pilin protein, CooA (Scott *et al.*, 1992). No such protein is found for the other pilus class represented by Pap. Further analysis of ETEC pilin genes should lead to a better understanding of this alternative method for protein assembly. In addition, we hope to learn whether there is a separate protein that serves as the adhesin which interacts specifically with the eukaryotic host cell. Although studies of ETEC pili have just begun, they promise to reveal important new approaches to common biological problems.

Acknowledgements

The work on this subject in our lab was supported by grant AI24870 from the National Institutes of Health. We appreciate the assistance of Judy Caron, Kathy Veilleux, Jeffrey Wakefield, Robert Lloyd, and Chris Haynes sequencing. We are especially indebted to Jeffrey Wakefield for his tireless assistance in analysis of the CS1 operon.

References

Acres, S.D., R.E. Isaacson, L.A. Babiuk, and R.A. Kapitany. 1979. Immunization of calves against enterotoxigenic colibacillosis by vaccinating dams with purified K99 antigen and whole cell bacterins. Infect. Immun. 25: 510–550.

Boylan, M., D.C. Coleman, and C.J. Smyth. 1987. Molecular cloning and characterization of the genetic determinant encoding CS3 fimbriae of enterotoxigenic *Escherichia coli*. Microbiol. Pathog. 2: 195–209.

Boylan, M., and C.J. Smyth. 1985. Mobilization of CS fimbriae-associated plasmids of enterotoxigenic *E. coli* of serotype O6:K15:H16 or H− into various wild-type hosts. FEMS Microbiol. Lett. 29: 83–89.

Caron, J., L.M. Coffield, and J.R. Scott. 1989. A plasmid-encoded regulatory gene, *rns*, required for expression of the CS1 and CS2 adhesins of enterotoxigenic *Escherichia coli*. Proc. Nat. Acad. Sci. USA 86: 963–967.

Caron, J., D.R. Maneval, J.B. Kaper, and J.R. Scott. 1990. Association of *rns* homologs with colonization factor antigens in clinical *Escherichia coli* isolates. Infect. Immun. 58: 3442–3444.

Caron, J., and J.R. Scott. 1990. A *rns*-like regulatory gene in CFA/I that controls expression of CFA/I pilin. Infect. Immun. 58: 874–878.

Cornelis, G., C. Sluiters, C. Lambert De Rouvroit, and T. Michiels. 1989. Homology between *VirF*, the transcriptional activator of the Yersinia virulence regulon, and *AraC*, the *Escherichia coli* arabinose operon regulator. J. Bacteriol. 171: 254–262.

Cravioto, A., S.M. Scotland, and B. Rowe. 1982. Hemagglutination activity and colonization factor antigens I and II in enterotoxigenic and non-enterotoxigenic strains of *E. coli* isolated from humans. Infect. Immun. 36: 189–197.

De Haan, L.A.M., G.A. Willshaw, B.A.M. Van Der Zeijst, and W. Gaastra. 1991. The nucleotide sequence of a regulatory gene present on a plasmid in an enterotoxigenic *Escherichia coli* strain of serotype O167:H5. FEMS Microbiol. Lett. 83: 341–346.

Evans, D.G., and D.J. Evans. 1978. New surface-associated heat-labile colonization factor antigen (CFA/II) produced by enterotoxigenic *Escherichia coli* of serogroups 06 and 08. Infect. Immun. 21: 638–647.

Evans, D.G., R.P. Silver, D.J. Evans, D.G. Chase, and S.L. Gorbach. 1975. Plasmid-controlled colonization factor associated with virulence in *Escherichia coli* enterotoxigenic for humans. Infect. Immun. 12: 656–667.

Francklyn, C.S., and N. Lee. 1988. AraC proteins with altered DNA sequence specificity which activate a mutant promoter in *Escherichia coli*. J. Biol. Chem. 263: 4400–4407.

Horwitz, A.H., L. Heffernan, C. Morandi, J.-H. Lee, J. Timko, and G. Wilcox. 1981. DNA sequence of the *araBAD-araC* controlling region in *Salmonella typhimurium* LT2. Gene 14: 309–319.

Hultgren, S.J., S. Normark, and S.N. Abraham. 1991. Chaperone-assisted assembly and molecular architecture of adhesive pili. Annu. Rev. Microbiol. 45: 383–415.

Inouye, M., and S. Halegoua. 1980. Secretion and membrane localization of proteins in *Escherichia coli*. Crit. Rev. Biochem. 7: 339–371.

Jordi, B.J.A.M., A.H.M. Van Vliet, G.A. Willshaw, B.A.M. Van Der Zeijst, and W. Gaastra. 1991. Analysis of the first two genes of the CS1 fimbrial operon in human enterotoxigenic *Eschericia coli* of serotype 0139:H28. FEMS Microbiol. Lett. 80: 265–270.

Kemp, E.H., N.P. Minton, and N.H. Mann. 1987. Complete nucleotide sequence and deduced aminoacid sequence of the M5 polypeptide gene of *escherichia coli*. Nucleic Acids Res. 15: 3924.

Klaasen, P., and F.K. De Graaf. 1990. Characterization of FapR, a positive regulator of expression of the 987P operon in enterotoxigenic *Escherichia coli*. Mol. Microbiol. 4: 1779–1783.

Klaasen, P., M.J. Woodward, F.G. Van Zijderveld, and F.K. De Graaf. 1990. The 987P gene cluster in enterotoxigenic *Escherichia coli* contains an ST_{pa} transposon that activates 987P expression. Infect. Immun. 58: 801–807.

Klemm, P., W. Gaastra, M.M. McConnell, and H.R. Smith. 1985. The CS2 fimbrial antigen from *E. coli*, purification, characterization and partial covalent structure. FEMS Microbiol. Lett. 26: 207–210.

Lei, S.-P., H.-C. Lin, L. Heffernan, and G. Wilcox. 1985. *AraB* gene and nucleotide sequence of the *araC* gene of *Erwinia carotovora*. J. Bacteriol. 164: 717–722.

Levine, M.M. 1987. *Escherichia coli* that cause diarrhea: enterotoxigenic, enteropathogenic, enteroinvasive, enterohemorrhagic, and enteroadherent. J. Infect. Dis. 155: 377–389.

Levine, M.M., P. Ristaino, G. Marley, C. Smyth, S. Knutton, E. Boedeker, R. Black, C. Young, M.L. Clements, C. Cheney, and R. Patnaik. 1984. Coli surface antigens 1 and 3 of colonization factor antigen II-positive enterotoxigenic *E. coli*: morphology, purification and immune responses in humans. Infect. Immun. 44: 409–420.

Lindberg, F., B. Lund, L. Johansson, and S. Normark. 1987. Localization of the receptor-binding protein adhesin at the tip of the bacterial pilus. Nature (London) 328: 84–87.

Lundrigan, M.D., M.J. Friedrich, and R.J. Kadner. 1989. Nucleotide sequence of the *Escherichia coli* porin thermoregulatory gene *env*Y. Nucleic Acids Res. 17: 800.

Maurelli, A.T. 1989. Temperature regulation of virulence genes in pathogenic bacteria: a general strategy for human pathogens? Microb. Pathog. 7: 1–10.

McClelland, M., R. Jones, Y. Patel, and M. Nelson. 1987. Restriction endonucleases for pulse field mapping of Bacterial Genomes. Nucleic Acids Res. 15: 5985–5991.

McConnell, M.M., L.V. Thomas, D.G. Willshaw, H.R. Smith, and B. Rowe. 1988. Genetic control and properties of coli surface antigens of colonization factor antigen IV (PCF8775) of enterotoxigenic *Escherichia coli*. Infect. Immun. 56: 1974–1980.

Mendiola, M.V., Y. Jubete, and F. De La Cruz. 1992. DNA sequence of IS91 and identification of the transposase gene. J. Bacteriol. 174: 1345–1351.

Miller, J.F., J.J. Mekalanos, and S. Falkow. 1989. Coordinate regulation and sensory transduction in the control of bacterial virulence. Science 243: 916–922.

Morgan, R.L., R.E. Isaacson, H.W. Moon, C.C. Brinton, and C.C. To. 1978. Immunization of suckling pigs against enterotoxigenic *Escherichia coli*-induced diarrheal disease by vaccination dams with purified 987 or K99 pili: protection correlates with pilus homology of vaccine and challenge. Infect. Immun. 22: 771–777.

Mullany, P., A.M. Field, M.M. McConnell, S.M. Scotland, H.R. Smith, and B. Rowe. 1983. Expression of plasmids coding for colonization factor antigen II (CFA/II) and enterotoxin production in *E.. coli*. J. Gen. Microbiol. 129: 3591–3601.

Nagy, B.H., H.W. Moon, R.E. Isaacson, C.C. To, and C.C. Brinton. 1978. Immunization of suckling pigs against enterotoxigenic *Escherichia coli* infection by vaccinating dams with purified pili. Infect. Immun. 21: 269–274.

Ogierman, M.A., and P.A. Manning. 1992. Homology of TcpN, a putative regulatory protein of *Vibrio cholerae*, to the AraC family of transcriptional activators. Gene 116: 93–97.

Perez-Casal, J., J. Swartley, and J.R. Scott. 1990. Gene encoding the major subunit of CS1 pili of human enterotoxigenic *Escherichia coli*. Infect. Immun. 58: 3594–3600.

Perlman, D., and H.O. Halvorson. 1983. A putative signal peptidase recognition site and sequence in eukaryotic and prokaryotic signal peptides. J. Mol. Biol. 167: 391–409.

Ponniah, S., R.O. Endres, D.L. Hasty, and S.N. Abraham. 1991. Fragmentation of *Escherichia coli* type 1 fimbriae exposes cryptic D-mannose-binding sites. J. Bacteriol. 173: 4195–4202.

Prere, M.F., M. Chandler, and O. Fayet. 1990. Transposition in *Shigella dysenteriae*: Isolation and analysis of IS911, a new member of the IS3 group of insertion sequences. J. Bacteriol. 172: 4090–4099.

Rutter, J.M., and G.W. Jones. 1973. Protection against entric disease casued by *Escherichia coli* – a model for vaccination with a virulence determinant. Nature (London) 242: 531–532.

Sakai, T., C. Sasakawa, S. Makino, and M. Yoshikawa. 1986. DNA sequence and product analysis of the *vir*F locus responisble for congo red binding and cell invasion in *Shigella flexneri* 2a. Infect. Immun. 54: 395–402.

Savelkoul, P.H.M., G.A. Willshaw, M.M. McConnell, H.R. Smith, A.M. Hamers, B.A.M. Van Der Zeijst, and W. Gaastra. 1990. Expression of CFA/I fimbriae is positively regulated. Microb. Pathog. 8: 91–99.

Schwartz, E., M. Kroger, and B. Rak. 1988. IS150: distribution, nucleotide sequence and phylogenetic relationships of a new *E. coli* insertion element. Nucleic Acids Res. 16: 6789–6800.

Scott, J.R., J.C. Wakefield, P.W. Russell, P.E. Orndorff, and B.J. Froehlich. 1992. CooB is required for assembly but not transport of CS1 pilin. Mol. Microbiol. 6: 293–300.

Smith, H.R., S.M. Scotland, and B. Rowe. 1983. Plasmids that code for production of colonization factor antigen II and enterotoxin production in strains of *E. coli*. Infect. Immun. 40: 1236–1239.

Smyth, C.J. 1982. Two mannose-resistant haemagglutinins on enterotoxigenic *E. coli* of serotype O6:K15:H16 or H − isolated from travellers' and infantile diarrhoea. J. Gen. Microbiol. 128: 2081–2096.

Smyth, C.J. 1986. Fimbrial variation in *Escherichia coli*. In: *Antigenic Variation in the Course of Infectious Diseases – A Survival Strategy for Pathogenic Micro-Organisms*, Soc. Gen. Micro. Spec. Publ., IRL Press.

Smyth, C.J., M. Boylan, H.M. Matthews, and D.C. Coleman. 1990. Fimbriae of human enterotoxigenic *Escherichia coli* and control of their expression. In: E.Z. Ron and S. Rottem, eds., *Microbial Surface Components and Toxins in Relation to Pathogenesis*, Plinon Press, London.

Thomas, L.V., M.M. McConnell, B. Rowe, and A.M. Field. 1985. The possession of three novel coli surface antigens by enterotoxigenic *E. coli* strains positive for the putative colonization factor PCF8775. J. Gen. Microbiol. 131: 2319–2326.

Tobin, J.F., and R. Schleif. 1987. Positive regulation of the *Escherichia coli* L-Rhamnose operon is mediated by the products of tandemly repeated regulatory genes. J. Mol. Biol. 196: 789–799.

Willshaw, G.A., H.R. Smith, M.M. McConnell, and B. Rowe. 1991. Cloning of regulator genes controlling fimbrial production by enterotoxigenic *Escherichia coli*. FEMS Microbiol. Lett. 82: 125–130.

Wolf, M.K., G.P. Andrews, B.D. Tall, M.M. Mcconnell, M.M. Levine, and E.C. Boedeker. 1989. Characterization of CS4 and CS6 antigenic components of PCF8775, a putative colonization factor complex from enterotoxigenic *Escherichia coli* E8775. Infect. Immun. 57: 164–173.

3. Molecular escorts required to present bacterial adhesins to eukaryotic receptors

LYNN N. SLONIM, META J. KUEHN and SCOTT J. HULTGREN

Abstract. Presentation of bacterial adhesins to eukaryotic receptors facilitates bacterial attachment and colonization of host mucosal surfaces and in many cases the subsequent invasion of these tissues (Schoolnik et al., 1987; Williams et al., 1988; Hultgren et al., 1989). Microbial attachment and colonization allow bacteria to survive the mechanical cleansing of mucosal surfaces. Heteropolymeric surface fibers called pili on bacteria often mediate microbial attachment (Normark et al., 1986; Williams et al., 1988; Hultgren et al., 1989; Hultgren et al., 1991a). The adhesins which mediate binding are typically minor components of pili, often located at their distal tips (Hultgren et al., 1985; Abraham et al., 1988). We describe here some of the molecular details on how Gram negative bacteria assemble adhesive fibers and how adhesins are correctly presented in these fibers so that they can recognize receptors on eukaryotic epithelial cell surfaces to promote microbial attachment. P pilus biogenesis is used as a prototype system to investigate how protein subunits fold into domains that serve as assembly modules for building adhesive pili on the surface of bacteria.

Abbreviations: CDR, Complementary-Determining Region

The pap proteins

P pili are complex structures composed of six different Pap proteins encoded in the *pap* operon (Fig. 1). PapA subunits compose the bulk of the pilus. The PapK, PapE, PapF, and PapG subunits are minor components of the pilus structure, but play essential roles in the structure, assembly, and function of the adhesive pilus. PapH is involved in anchoring pili to the bacterial cell surface. Two proteins, PapC and PapD, encoded in the *pap* operon are required for the assembly of adhesive pili but are not a part of the final structure. PapD is a periplasmic chaperone (Hultgren et al., 1991b; Kuehn et al., 1991) which binds pilus subunits and guides their correct folding into pili (Norgren et al. 1987; Lindberg et al., 1989). In the absence of PapD the pilus subunit proteins misfold and are consequently proteolytically degraded. PapC is an usher. In other words it functions as a molecular doorkeeper which receives the various pilus subunits delivered by the chaperone and ushers them into pili in a defined order to form the architecturally distinct composite fibers (Dodson et al., 1993).

The P pilus system is ideal for studying the general principles in which interactive monomeric subunits are secreted, folded, and assembled into

C.I. Kado and J.H. Crosa (eds.), Molecular Mechanisms of Bacterial Virulence, 31–45.
© 1994 *Kluwer Academic Publishers.*

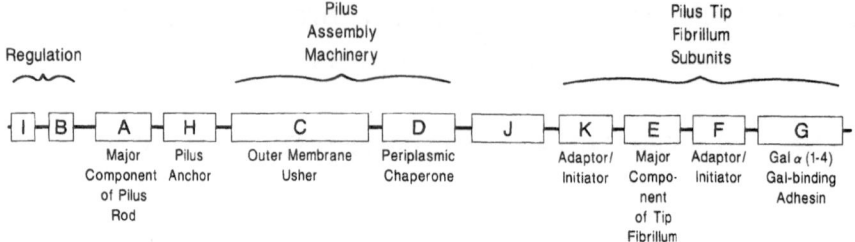

Figure 1. The *pap* operon. Summary of the structure and function of the 11 genes in the *pap* operon. Comparisons with other Gram negative pilus operons have revealed a similar genetic organization.

adhesive fibers on the cell surface and relating these processes to pathogenesis. The studies that are reviewed in this chapter have blended the well-understood genetics of the *pap* system with X-ray crystallography and biochemistry to investigate the mechanisms which regulate the correct protein-protein interactions required for ordered assembly of pili.

PapG adhesin

Most uropathogenic *E. coli* strains express P pili which contain an adhesin that specifically binds to the α-D-galactopyranosyl-(1–4)-ß-D galactopyranoside (Galα(1–4)Gal) moiety present in the globoseries of glycolipids in the urinary tract (Uhlin *et al.*, 1985). The binding event is the result of a stereochemical fit between the microbial adhesin and specific surface glycolipid and glycoprotein receptors located on epithelial cells . Lund *et al.* (1987) demonstrated that the *papG* gene of the *pap* operon encoded the Galα(1–4)Gal binding adhesin. Deletion of the *papG* gene had no effect on pilus formation but abolished the adhesive properties of bacteria and purified pili. The pilus binding specificity could only be restored by *trans*-complementation with *papG*. Complementation with a gene encoding a related adhesin, PrsG, changed the binding specificity of the pilus from human erythrocytes to sheep erythrocytes (Lund *et al.*, 1987). Furthermore, purified PapG adhesin (in a preassembly complex with the PapD chaperone) was able to bind to Galα(1–4)Gal (Hultgren *et al.*, 1989). These results demonstrated that PapG was the adhesive moiety of the pilus.

Uropathogenic *E. coli* bind with different affinities to 2 different digalactoside containing isoreceptors. The restriction of specific isoreceptors to cell types, tissues and species results in tissue and host tropism for pyelonephritic strains of *E. coli* (Stromberg *et al.*, 1990; Stromberg *et al.*, 1991). In addition, variation in PapG binding specificity among wild-type strains correlates with digalactoside containing isoreceptors and the observed host and tissue tropisms (Stromberg *et al.*, 1991).

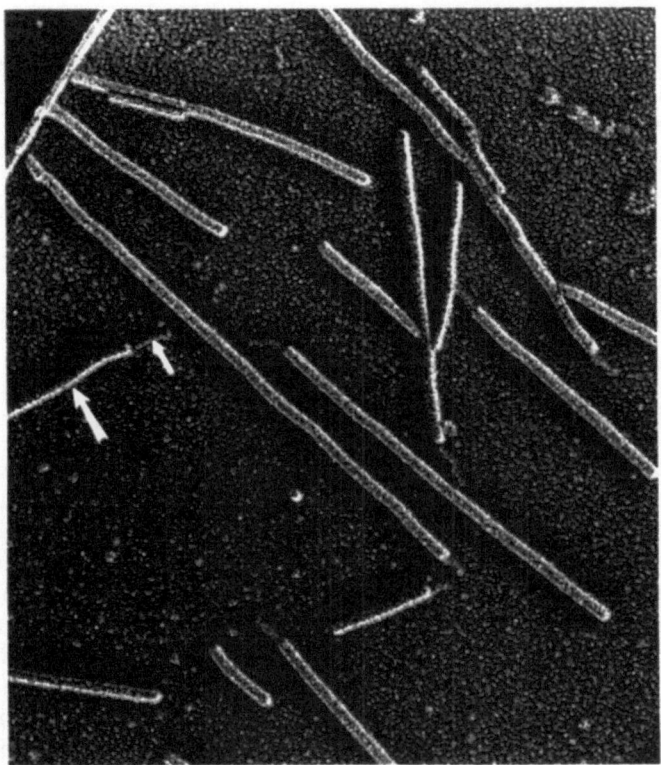

Figure 2. P pili are composite heteropolymeric structures. Freeze-etch electron micrograph of purified P pili. The P pilus consists of a rigid pilus rod (large arrow) connected to a flexible tip fibrillum (small arrow) containing the PapG adhesin (Photo courtesy of John Heuser).

Adhesin presentation to eukaryotic receptors

In order to mediate a binding event, a bacterial adhesin molecule must be exposed or "presented" in an active binding conformation apart from the interfering molecular structures and negatively charged molecules present on the microbial surface. Examination of P pili by a high-resolution electron microscopy technique has revealed the effective strategy uropathogenic *E. coli* use to present an adhesin (Fig. 2). The PapG adhesin is a component of the tip fibrillum which is located at the distal end of the P pilus rod (Kuehn *et al.*, 1992). The pilus rod was shown to be 10 nm in diameter and composed of repeating monomers of the major pilin subunit, PapA, arranged in a right-handed helical conformation. The tip fibrillum, joined end to end to the pilus rod, is approximately one third the diameter of the pilus shaft and is composed mainly of PapE subunits arranged in an open helical configuration (Kuehn *et al.*, 1992). The Galα(1–4)Gal binding PapG adhesin was localized to the distal end of the fibrillum. PapF and PapK are minor components of the tip fibrillum that have specialized adaptor and initiator functions. PapF joins PapG to the tip fibrillum and PapK joins the tip fibrillum to the pilus rod (Jacob-Dubuisson *et al.*, 1993). Both PapF and PapK are also required to initiate the formation of tip fibrillae and pilus rods (Jacob-Dubuisson *et al.*, 1993). The composite architecture of the P pilus fiber revealed a general strategy used by pathogenic bacteria to present adhesins to eukaryotic receptors. The rigid PapA rod extends the adhesin beyond steric interference by LPS and other components at the bacterial cell surface. In addition, the flexible tip fibrillum allows PapG steric freedom to recognize and bind to the digalactoside moiety on the surface of epithelial cells.

Chaperone-assisted pilus assembly

P pili are heteropolymeric structures composed of approximately 1000 proteins of six different types (Hultgren *et al.*, 1991a). With 200–300 pili per bacterium, a system must be in place to prevent premature non-productive interactions of the $\sim 3 \times 10^5$ different subunit types prior to their delivery to outer membrane assembly sites. Moreover, to ensure that functional, adhesive pili are formed, PapA monomers must only be added into the growing pilus after the adhesive fibrillar tip has been formed. PapD has been shown to form periplasmic complexes with each subunit type of protein, many of which have now been purified (Hultgren *et al.*, 1989; Striker *et al.*, Jacob-Dubuisson *et al.*, unpublished data). We used purified PapD and the complex that PapD forms with the PapG adhesin to study the mechanisms which regulate the correct protein-protein interactions required for the formation of distinct composite pilus fibers. The PapD-PapG (DG) complex has been purified from the periplasm by Galα(1–4)Gal affinity chromatography and extensively characterized *in vitro* (Hultgren *et al.*, 1989; Kuehn *et al.*, 1991). PapD and PapG exist in a 1 : 1 molar ratio in the purified periplasmic complex. The ability to purify the complex utilizing the binding specificity of PapG for its receptor suggested that the adhesin exists in a native-like conformation even when bound to PapD. By contrast, cytoplasmic chaperones have been found to bind to their targets and maintain them in an unfolded state (Bochkareva *et al.*, 1988; Crooke *et al.*, 1988; Lecker *et al.*, 1989; Viitanen *et al.*, 1992). Since PapG must be released into the growing pilus, it was not suprising to find that PapD bound to PapG in a reversible manner. Interestingly, chaperone uncapping (the separation of the chaperone from its bound subunit) is probably ATP independent. PapD does not contain a typical ATP binding site, and chaperone uncapping is probably driven by events at the outer-membrane pilus assembly site. The DG complex was dissociated *in vitro* under reducing conditions in the presence of urea and dithiothreitol. Dilution of the denaturant however failed to allow reformation of the complex but instead lead to protein aggregation. However, if the denaturant was diluted in the presence of native PapD the native PapD present in the diluent bound to and capped the interactive surfaces on

PapG to prevent the aggregation of the proteins by reforming the soluble complex (Kuehn *et al.*, 1991). The ability of PapD to bind and cap interactive surfaces on the subunit presumably allowed proper folding by blocking nonproductive interactions. This *in vitro* activity may reflect the *in vivo* role of PapD and other members of the chaperone superfamily to bind to newly translocated unfolded proteins and maintain them in assembly competent conformations.

In vivo, we envision two competing pathways for each interactive subunit as it crosses the cytoplasmic membrane into the periplasm (Fig. 3). In one pathway, interactive surfaces of a pilus subunit protein drive the formation of insoluble aggregates by premature and inappropriate interactions in the periplasmic space. This pathway leads to kinetically dead end pathways and proteolytic degradation. In a competing pathway, PapD physically caps or covers the interactive surfaces on the subunits by directly binding to them as they emerge from the cytoplasmic membrane (Kuehn *et al.*, 1991). As a consequence of chaperone recognition and binding, the subunits are imported into the periplasmic space and stabilized in an unpolymerized, folded form. Subsequently, the preassembly complex is targeted to the outer membrane where uncapping of the chaperone and pilus assembly take place.

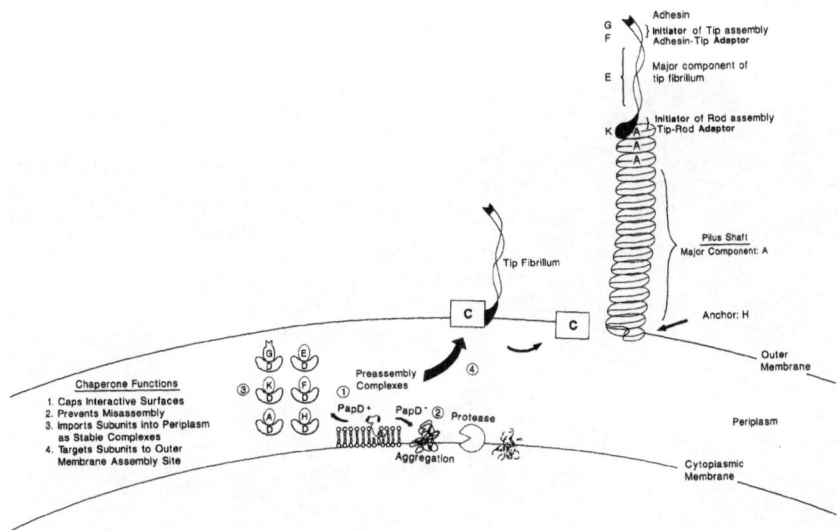

Figure 3. Model of chaperone-assisted P pilus biogenesis. The PapD chaperone performs at least 4 functions to modulate pilus assembly: PapD binds to nascently translocated pilus subunits, capping their potentially interactive surfaces and imports subunits into the periplasm as stable complexes (Step 1, Step 3). PapD then targets subunits to the outer membrane assembly usher, PapC (step 4), where the subunits are incorporated into the growing pilus. In the absence of PapD (step 2), nascent subunits aggregate and are subjected to proteolytic degradation.

Immunoglobulin-like pilus chaperones

PapD is the prototype member of a large family of periplasmic pilus chaperones found in several species of Gram negative bacteria. In most pilus systems analyzed so far, a periplasmic protein with sequence homology to PapD has been identified (Holmgren *et al.*, 1992). In all cases this protein is essential for pilus expression but is not a part of the final structure (Holmgren *et al.*, 1992). The three dimensional structure of PapD has been solved to 2.5 angstroms resolution by Holmgren and Brändén (1989) and recently to 2.0 angstroms by Ogg and Holmgren (personal communication). PapD consists of two globular domains oriented towards one another with the overall shape similar to a boomerang and joined by a hinge region such that a cleft is formed by the two domains (Fig. 4A). Each domain is a ß-barrel structure formed by two antiparallel ß-pleated sheets and has a topology similar to an immunoglobulin fold.

The sequences of twelve periplasmic pilus chaperones required for pilus assembly in *E. coli*, *Klebsiella pneumoniae*, *Haemophilus influenzae*, *Bordetella pertussis*, *Salmonella enteriditis*, and *Yersinia pestis* have been aligned and found to be 30–40% identical and 60% similar when considering conservative substitutions (Galyov *et al.*, 1991; Holmgren *et al.*, 1992; Locht *et al.*, 1992; Clouthier *et al.*, 1993). A consensus sequence derived from the alignment was superimposed onto the crystal structure of PapD to examine the

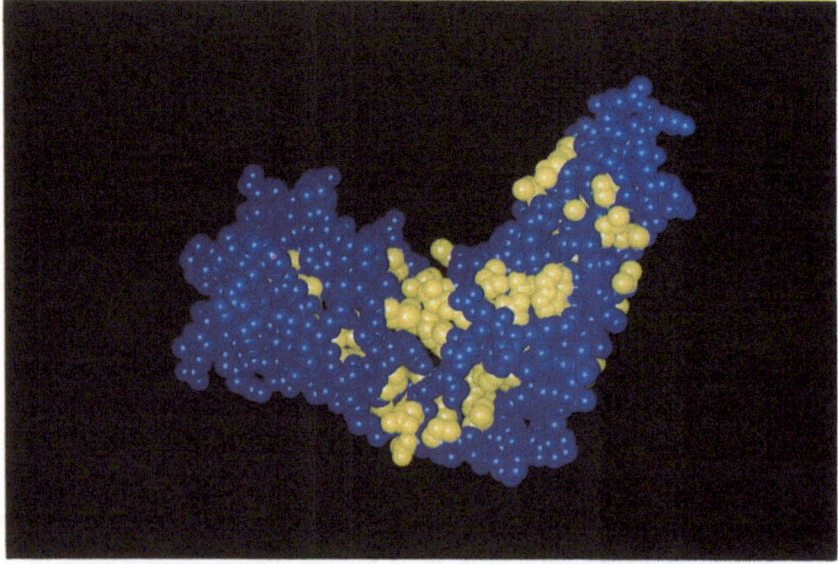

Figure 4A. Three-dimensional structure of the PapD chaperone. Space filling model of the PapD chaperone where invariant and conserved residues are shown in yellow while all other residues are shown in blue. Note that many of the conserved and invariant residues are clustered in the cleft region of PapD.

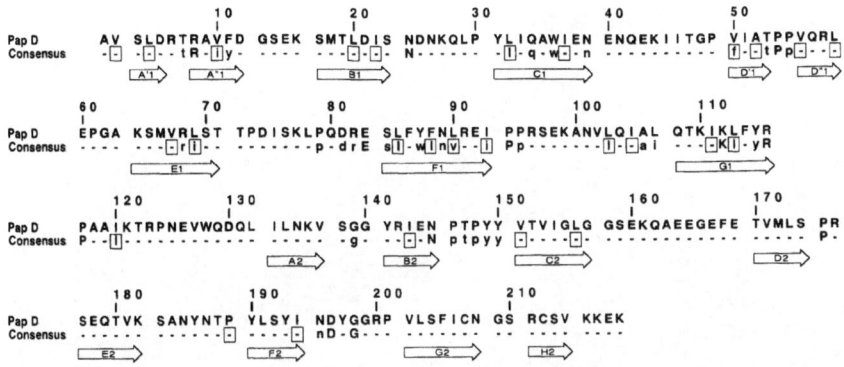

Figure 4B. Alignment of PapD amino acid sequence and periplasmic chaperone superfamily consensus sequence. The consensus sequence was generated by the alignment of 12 chaperone sequences based on the crystal structure of PapD (Holmgren *et al.,* 1992; Jacob-Dubuisson *et al.,* 1993b). Upper case letters in consensus sequence are invariant in 12 out of 12 chaperone sequences; lower case letters in consensus sequence are highly conserved (found in at least 8 out of 12 chaperone sequences); boxes indicate positions of hydrophobic residues in 12 out of 12 chaperone sequences. The arrows below the sequences indicate the name and position of ß-strands of PapD.

possible structural and functional significances of the conserved amino acids (Fig. 4A and 4B). This analysis revealed that most of the strongly conserved residues in the chaperone family participated in maintaining the structural integrity of the protein and were clustered in ß-strands. All of the residues which make up the hydrophobic core of the chaperones were found to be conservatively substituted in the superfamily. In contrast, seventy percent of all residues found in loop regions were variable in this protein family, suggesting that some of the loop regions may be important in the specificity of chaperone-subunit interactions within each individual pilus family. Three classes of invariant residues occur in pilus chaperones. One class of residues occupies critical points in loops or forms intramolecular interactions which serve to orient the loops. Another class of invariant residues forms a salt bridge which connects the two domains. The third class consists of residues that are conserved for no apparent structural reason. These residues are surface exposed with their side chains oriented towards the solvent and do not make any specific interactions with other side chains. The side chains of four of these conserved residues (T7, R8, K112, and M172) point into the empty cleft between the domains (Fig. 5). Four other residues (I93, P94, W36, and 106A) comprise part of a conserved hydrophobic surface on the first domain (Holmgren *et al.,* 1989). These surface exposed invariant residues have been proposed to be important in chaperone function.

Three surface exposed cleft residues that have no apparent structural function, R8, M172, and E167 were targeted for site directed mutagenesis in order to investigate their role in subunit binding. Site directed mutations in the invariant arginine-8 (R8) cleft residue (Fig. 5) abolished PapD function,

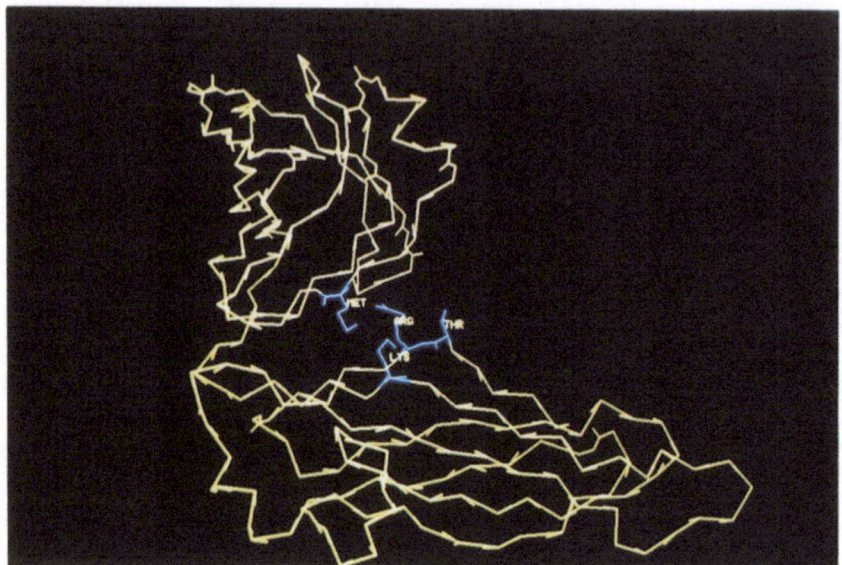

Figure 5. PapD point mutations in invariant and conserved cleft residues. Alpha-carbon backbone photograph of PapD (yellow) highlighting residues which were subjected to site-directed mutagenesis (blue). These residues are all solvent exposed and have side chains that extend into the chaperone cleft.

suggesting that it may form part of an interactive surface in the PapD cleft required for subunit recognition and binding (Fig. 6). Although the glycine, alanine, or methionine mutations constructed at this residue may have altered the conformation of PapD, this was much less likely for several reasons. The mutated R8 residue is situated in a bulge in the first ß strand (A1) where that strand switches from one ß sheet to another. The side chain of this residue points straight into the empty cleft and does not make any specific interactions with other side chains of PapD. The R8 mutants localized correctly to the periplasmic space and maintained their ability to bind PapG, albeit with a weaker affinity (Slonim *et al.*, 1992). These results indicated that site directed mutations in the invariant R8 residue probably only changed the surface properties of this restricted region and did not cause any other structural alterations.

PapD binding and chaperone activity was strongly dependent on the invariant R8 cleft residue. Mutant PapD's bound in a similar manner to each pilus subunit type (Slonim *et al.*, 1992). The binding efficiency of PapD to PapG, PapA, and PapK increased as the steric volume of the mutated residue 8 approached that of the wild type arginine (from glycine to alanine to methionine). All Gram negative pilus proteins that require a PapD-like chaperone for assembly share extensive sequence homology in their carboxyl termini. The C-terminus of PapG previously has been shown to be important for PapD binding to PapG (Hultgren *et al.*, 1989). The binding of PapD to a subunit

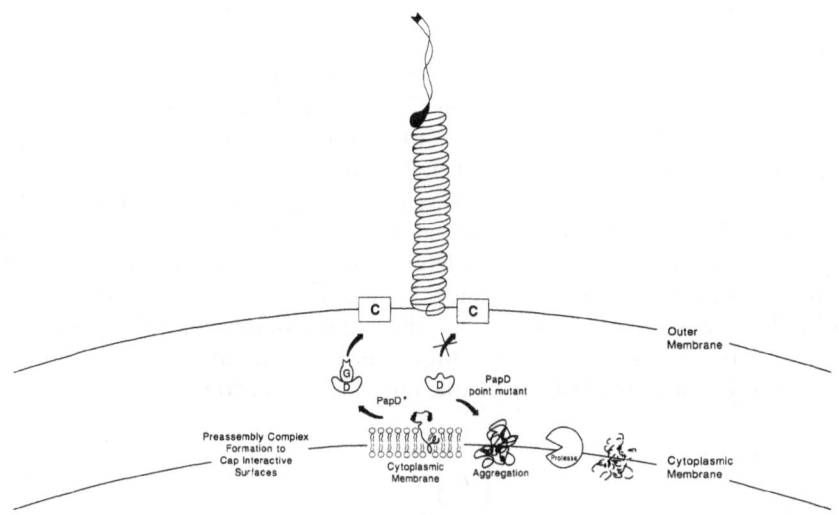

Figure 6. PapD point mutations affect pilus assembly. This model of pilus biogenesis shows that in the presence of a specific PapD point mutant protein, pilus subunits are not targeted to the PapC outer membrane usher (crossed-out arrow). Instead, the subunits undergo nonproductive interactions which lead to their aggregation and subsequent proteolytic degradation.

is most likely dependent upon several of the subunit side chains fitting into a subset of pockets present in the PapD cleft and forming a critical interaction with the invariant residue R8 (Kuehn *et al.*, in press).

Site directed mutations in other invariant and conserved cleft residues (K112 and M172) of PapD also reduced or abolished its ability to bind subunits and modulate pilus assembly (Kuehn *et al.*, in press). In contrast, mutations in a residue that is variable amongst members of the chaperone superfamily, E167, did not abolish PapD activity, suggesting that this residue was not critical for the ability of PapD to modulate pilus assembly. We propose that the highly conserved cleft of the pilus chaperone superfamily may universally function in subunit binding.

Binding paradigms of immunoglobulin domains

The use of the immunoglobulin fold for recognition represents a binding paradigm in many systems. The two domain PapD chaperone is built up of immunoglobulin-like ß-barrel motifs. The use of immunoglobulin-like domains for specific protein-protein interactions occurs in bacteria as well as in higher organisms. In eukaryotes, both the immunoglobulin superfamily (Williams *et al.*, 1988; Bazan, 1990) whose members include antibodies, cell surface adhesion molecules and T cell receptors, and the cytokine receptor superfamily (Bazan, 1990; DeVos *et al.*, 1992) use ß barrel motifs for molecular recognition processes. The basic structure of this motif is best described as two antiparallel ß-sheets packed against each other to form a hydrophobic core (Williams *et al.*, 1988). An immunoglobulin constant domain contains seven ß-strands arranged in two ß-sheets bound together by a disulfide bond. All antibody heavy and light chain constant domains have the same structure. An Ig variable domain contains two additional strands as well as a different strand order of the sheets as compared to a constant domain (for review see Branden and Tooze, 1991). The two additional strands are joined by a loop region which makes up a hypervariable or complementarity-determining region (CDR). Each variable domain contains three CDR's which form a large flat surface which is the antibody combining site (Fig. 7). The immunoglobulin fold provides a structural framework to support unique recognition surfaces formed by the CDR loops which are clustered at one end of the ß-barrel.

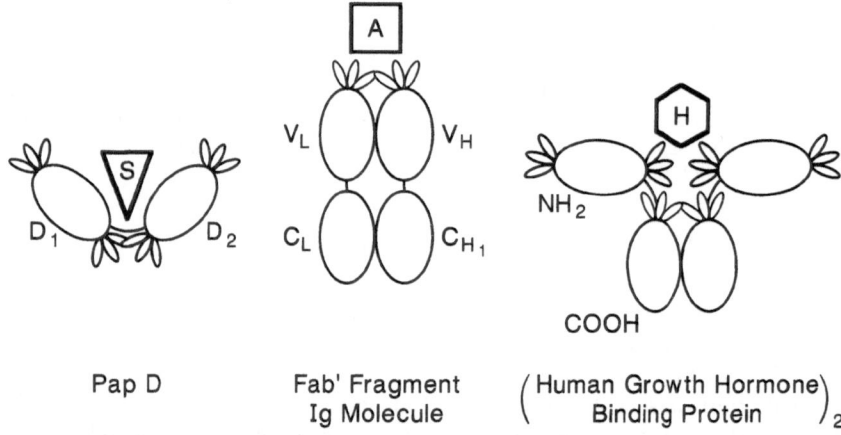

Figure 7. Immunoglobulin domain binding paradigms. A comparison of the binding surfaces used by three immunoglobulin-fold containing proteins: PapD, an antibody molecule, and the human growth hormone binding protein. The binding sites for antigen (A) and the human growth hormone (H) are established. The binding site for pilus subunits (S) has been deduced by site-directed mutagenesis (Adapted from Jones *et al.*, 1992).

The recently reported structure of the human growth hormone receptor, (hGHbp)$_2$, bound to its hormone has shown that it is a cytokine receptor superfamily member, a family which also makes use of the immunoglobulin fold for molecular recognition (DeVos *et al.*, 1992). The two domains of hGHbp are variations of the Ig fold and are identical to the second domain of CD4 and the second domain of PapD. Residues essential for the interaction between the human growth hormone and its receptor are mainly localized in loop residues from both domains that form a binding surface near the hinge region between the two domains (Fig. 7). The relative orientation of the two ß-barrel domains in hGHbp is different from that between the domains of immunoglobulin molecules. Unlike immunoglobulins, hGHbp possesses a salt bridge between the amino and carboxy terminal domains which may be important in stabilizing the structure.

We hypothesize that PapD and other members of the chaperone superfamily rely on a different binding paradigm from that used by antibodies or the growth hormone receptor (Fig. 7). The second domain of PapD has a structural topology similar to the second domain of the HIV receptor, CD4 (Ryu *et al.*, 1990). CD4-like domains differ from the classical constant domain organization by strand switching of the first ß-strand from the upper sheet to the lower sheet. The first domain of PapD has a ß-strand order more similar to that of an immunoglobulin variable region (Holmgren *et al.*, 1989). Unlike other immunoglobulin-like domain containing proteins, PapD binds to proteins by utilizing the cleft between the two linked immunoglobulin domains. In this model PapD binds subunits via their side chain interactions with conserved residues in the PapD cleft. Our current model suggests that the ß-barrel structure also stabilizes variable loop regions that surround the cleft and that residues in these loops may impart specificity to the chaperone.

The outer membrane usher

In the final step in biogenesis of adhesive pili, the periplasmic chaperone-subunit complexes are dissociated and the subunits are converted into the architecturally distinct surface fibers. This process requires the targeting of the chaperone-subunit complexes to an outer membrane assembly site and the uncapping of polymerization surfaces which permit pilus assembly. The targeting and polymerization functions probably reside in PapC which is an 88 kDa outer membrane protein. Genetic lesions in PapC abolish pilus assembly and lead to an accumulation of chaperone-subunit complexes in the periplasm (Norgren *et al.*, 1987).

The role of PapC in correctly targeting the chaperone-subunit complexes to outer membrane uncapping and assembly sites was recently investigated. An *in vitro* assay was developed which tested the ability of the various chaperone-subunit complexes to bind partially purified PapC (Dodson *et al.*, 1993). PapD alone did not bind PapC *in vitro*. In contrast the chaperone-tip fibrillar complexes, PapD-PapG, PapD-PapE, and PapD-PapF, all specifically bound PapC. The ability of PapD to bind to pilus subunits and target them to PapC suggests that PapD has an effector function (Dodson, *et al.*, 1993). Remarkably, PapD-PapA complexes did not bind to PapC *in vitro*. Based on these results Dodson *et al.* proposed that *in vivo*, PapD-PapA complexes are not targeted to empty PapC sites, but instead only recognize PapC in the context of a growing tip fibrillum. Supporting this hypothesis, a deletion of the tip fibrillar genes abolished the ability of bacteria to produce pili, probably due to the inability of PapD-PapA complexes to be targeted to empty PapC sites (Jacob-Dubuisson *et al.*, 1993). These results suggest that the different abilities of different subunits to bind to PapC insures that every pilus rod is joined end to end to an adhesive tip fibrillum.

PapD and PapC are molecular escorts which regulate the correct protein-protein interactions required for pilus assembly. The PapD chaperone binds pilus subunits and guides them along productive folding and assembly pathways. The PapC usher acts as a molecular doorkeeper, modulating the ordered targeting of chaperone-subunit complexes to outer membrane assembly sites and thus ensuring the formation of an adhesive composite structure. The ordered incorporation of each pilus subunit type into the pilus also seems to be controlled by stereochemical fits between complementary surfaces on each subunit type (Jacob-Dubuisson *et al.*, 1993).

The uncapping of PapD at the outer membrane probably exposes polymerization sites which drives assembly. The mechanism of chaperone uncapping is poorly understood but is seemingly ATP-independent and may involve PapC. PapC is a representative member of a family of outer membrane ushers, including FanD, FaeD, FimD, and MrkC (Mooi, *et al.*, 1986; Klemm and Christiansen, 1990; Allen, *et al.*, 1991) which are required for the assembly of a variety of different kinds of pili in Gram negative bacteria. It is likely that all of these proteins have similar ushering functions acting in concert with their respective chaperone partners to regulate the correct interactions necessary for the production of adhesive pili which are important in colonization and infection of susceptible hosts.

Perspectives and discussion

Bacteria rely on high affinity binding to surface receptors located on epithelial and mucosal surfaces to survive the mechanical cleansing and flow associated with mucosal surfaces (Hultgren *et al.*, 1989; Hultgren *et al.*, 1991b). Epithelial cells lining the mucosal surfaces provide a source of surface glycolipids and glycoproteins which are recognized by some pathogens via specific adhesin molecules (Lindberg *et al.*, 1986; Normark *et al.*, 1986; Stromberg *et al.*, 1990; Stromberg *et al.*, 1991). Uropathogens expressing P pili preferentially bind glycolipids containing the digalactoside Galα(1–4)Gal (Uhlin *et al.*, 1985).

Interaction of pyelonephritic strains of *E. coli* with their cell surface isoreceptors is mediated through the P pilus composite fiber which terminates in a flexible tip structure (Kuehn *et al.*, 1992). The location of the pilus adhesin in the tip fibrillum places it in an environment free of obstructing bacterial cell surface components and allows maximum flexibility for interaction with the receptor. The composite pilus structure is probably a general structural feature of most adhesive pili and represents a strategy to present an adhesin to a eukaryotic receptor where it can promote attachment.

Investigating the role of chaperones and molecular ushers in pilus assembly has revealed several general biological principles which describe the pathway monomeric subunits follow from synthesis to incorporation into extracellular organelles. Molecular chaperones on both sides of the cytolasmic membrane are essential in guiding subunits down biologically productive pathways (Kuehn *et al.*, 1991). Periplasmic chaperones are probably required in the assembly of a wide array of heteropolymeric surface structures in Gram negative bacteria. The fact that the periplasmic chaperone family is required for the assembly of at least twelve different structures in six different organisms indicates that periplasmic chaperones are very general phenomena (Holmgren *et al.*, 1992).

Another basic principle of pilus biogenesis is that the chaperone must be displaced from the pilin subunit, a process we refer to as "uncapping", to allow interactive surfaces of the subunit to be exposed at the site of pilus polymerization (Ellis *et al.*, 1989; Kuehn *et al.*, 1991). The role of uncapping and directing subunit incorporation into the pilus is probably carried out by a membrane-associated molecular usher. In contrast to the chaperones' role in preventing inappropriate interactions in the periplasm, the molecular usher directs the appropriate "meeting" of subunits both in time and location (Dodson *et al.*, 1993).

In this review we have described the crucial roles that the molecular escorts play in the presentation of microbial adhesins to cell surface receptors. Adhesin presentation is essential for the critical first step in association with epithelial mucosa which leads to colonization and further disease pathology.

References

Abraham SN, Sun D, Dale JB, and Beachy EH (1988) Conservation of the D-mannose-adhesin proteins among type 1 fimbriated members of the family Enterobacteriaceae. Nature (London) 336: 682–684

Allen BL, Gerlach GF, and Clegg S (1991) Nucleotide sequence and functions of mrk determinants necessary for expression of type 3 fimbriae in *Klebsiella pneumoniae*. J. Bacteriol. 173: 916–920

Bazan FJ (1990) Structural design and molecular evolution of a cytokine receptor superfamily. Proc. Natl. Acad. Sci. USA 87: 6934–6938

Bochkareva ES, Lissin NM, and Girshovich AS (1988) Transient association of newly synthesized unfolded proteins with the heat shock GroEL protein. Nature (London) 336: 254–257

Branden CI, and Tooze J (1991) Introduction to Protein Structure Garland Publishing, New York

Clouthier SC, Müller K-H, Doran IL, Collinson SK, and Kay WW (1993) Characterization of three fimbrial genes, *sef* ABC, of *Salmonella enteritidis*. J. Bacteriol. 175: 2523–2533.

Crooke E, Brundage L, Rice M, and Wickner W (1988) Pro OmpA spontaneously folds into a membrane assembly competent state which trigger factor stabilizes. EMBO J., 7: 1831–1835

DeVos A, Ultsch M, Kossiakoff A (1992) Human growth hormone and extracellular domain of its receptor: crystal structure of the complex. Science 255: 306–312

Dodson K, Jacob-Dubuisson F, Striker, R, and Hultgren SJ (1993) Outer membrane PapC usher discriminately recognizes periplasmic chaperone-pilus subunit complexes Proc. Natl. Acad. Sci. USA 90: 3670–3674.

Ellis RJ, and Hemmingsen SM (1989) Molecular chaperones: proteins essential for the biogenesis of some macromolecular structures. Trends Biochem. Sci. 14: 339–342

Galyov EE, Karlishev AV, Chernovskaya TV, Dolgikh DA, Smirnov OY, Volkovoy KI, Abramov VM, and Zav'yolov VP (1991) Expression of the envelope antigen F1 of *Yersinia pestis* is mediated by the product of caf1M gene having homology with the chaperone protein PapD of *Escherichia coli*. FEBS 286: 79–82

Holmgren A, Kuehn MJ, Branden CI, and Hultgren SJ (1992) Conserved immunoglobulin-like features in a family of periplasmic pilus chaperones in bacteria. EMBO J. 11: 1617–1622

Holmgren A, and Branden CI (1989) Crystal structure of chaperone protein PapD reveals an immunoglobulin fold. Nature (London) 342: 248–251

Hultgren SJ, Abraham SN, and Normark S (1991a) Chaperone-assisted assembly and molecular architecture of adhesive pili. Annu. Rev. Microbiol. 45: 383–415

Hultgren SJ, and Normark S (1991b) Biogenesis of the bacterial pilus. Curr. Opin. Gen. Dev. 1: 313–318

Hultgren SJ, Lindberg F, Magnusson G, Kilberg J, Tennent JM, and Normark S (1989) The PapG adhesin of uropathogenic *Escherichia coli* contains separate regions for receptor binding and for the incorporation into the pilus. Proc. Natl. Acad. Sci. USA 86: 4357–4361

Hultgren SJ, Porter TN, Schaeffer AJ, and Duncan JL (1985) Role of Type 1 pili and effects of phase variation on lower urinary tract infections produced by *Escherichia coli* . Infect. Immun. 50: 370–377

Jacob-Dubuisson F, Heuser J, Dodson K, Normark S, and Hultgren SJ (1993) Initiation of assembly and association of the structural elements of a bacterial pilus depend on two specialized tip proteins. EMBO J. 3: 837–847

Jacob-Dubuisson F, Kuehn M, Hultgren SJ (1993b) A novel secretion apparatus for the assembly of adhesive bacterial pili. Trends in Microbiology 1: 50–55

Jones CH, Jacob-Dubuisson FJ, Dodson K, Kuehn MJ, Slonim L, Striker R, and Hultgren SJ (1992) Adhesin presentation in bacteria requires molecular chaperones and ushers. Infect. Immun. 60: 4445–4451

Klemm P, and Christiansen G (1990) The fimD gene is required for cell surface localization of *Escherichia coli* type 1 fimbriae. Mol. Gen. Genet. 220: 334–338

Kuehn MJ, Heuser J, Normark S, and Hultgren SJ (1992) P pili in uropathogenic *E. coli* are composite fibres with distinct fibrillar adhesive tips. Nature (London) 356: 252–255

Kuehn MJ, Normark S, and Hultgren SJ (1991) Immunoglobulin-like PapD chaperone caps and uncaps interactive surfaces of nacently translocated pilus subunits. Proc. Natl. Acad. Sci. USA 88: 10586–10590

Lecker, SH, Lill R, Ziegelhoffer T, Georgopoulos C, Bassford PJ, Kumamoto CA, and Wickner W (1989) Three pure chaperone proteins of *Escherichia coli* – SecB, trigger factor, and GroEL-form soluble complexes with precursor proteins *in vitro*. EMBO J. 8: 2703–2709

Lindberg FP, Lund B, and Normark S (1986) Gene products specifying adhesion of uropathogenic *Escherichia coli* are minor components of pili. Proc. Natl. Acad. Sci. USA 83: 1891–1895

Lindberg F, Tennent JM, Hultgren SJ, Lund B, and Normark S (1989) PapD, a periplasmic transport protein in P-pilus biogenesis. J. Bacteriol. 171: 6052–6058

Locht C, Geoffroy MC, and Renauld G (1992) Common accessory genes for the *Bordetella pertussis* filamentoud hemagglutinin and fimbriae share sequence similarities with the *papC* and *papD* gene families. EMBO J. 11: 3175–3183

Lund B, Lindberg FP, Marklund BI, and Normark S (1987) The PapG protein is the a-D-galactopyranosyl-(1–4)-b-D-galactopyranose-binding adhesin of uropathogenic *Escherichia coli*. Proc. Natl. Acad. Sci. USA 84: 5898–5902

Mooi FR, Classen I, Baaker D, Kuipers H, de Graaf FK (1986) Regulation and structure of an *Escherichia coli* gene coding for an outer membrane protein involved in export of K88ab fimbrial subunits. Nucleic Acids Res. 14: 2443–2457

Norgren M, Baga M, Tennent JM, and Normark S (1987) Nucleotide sequence, regulation and functional analysis of the papC gene required for cell surface localization of Pap pili of uropathogenic *Escherichia coli*. Mol. Microbiol. 1: 169–178

Normark S, Baga M, Goransson M, Lindberg FP, Lund B, Norgren M, and Uhlin BE (1986) Genetics and biogenesis of *Escherichia coli* adhesins. *In* D. Mirelman (ed.) Microbial lectins and agglutinins: properties and biological activity (pp. 113–143) Wiley Interscience, New York

Ryu SE, Kwong PD, Truneh A, Porter TG, Arthos J, Rosenberg M, Dai X, Xuong N, Axel R, Sweet RW, and Hendrickson WA (1990) Crystal structure of an HIV-binding recombinant fragment of human CD4. Nature (London) 348: 419–426

Schoolnik GK, O'Hanley P, Lark D, Normark S, Vosti K, and Falkow S (1987) Uropathogenic *Escherichia coli*: molecular mechanism of adherence. Adv. Exp. Med. Biol. 224: 53–62

Slonim LN, Pinkner JS, Branden CI, and Hultgren SJ (1992) Interactive surface in the PapD chaperone cleft is conserved in pilus chaperone superfamily and essential in subunit recognition and assembly. EMBO J. 13: 4747–4756

Stromberg N, Nyholm PG, Pascher I, and Normark S (1991) Saccharide orientation at the cell surface affects glycolipid receptor function. Proc. Natl. Acad. Sci. USA 88: 9340–9344

Stromberg N, Marklund BI, Lund B, Ilver D, Hamers A, Gaastra W, Karlsson KA, and Normark S (1990) Host-specificity of uropathogenic Escherichia coli depends on differences in binding specificity to Gala(1–4)Gal-containing isoreceptors. EMBO J. 9: 2001–2010

Uhlin BE, Norgren M, Baga M, and Normark S (1985) Adhesion to human cells by *Escherichia coli* lacking the major subunit of a digalactoside-specific pilus adhesin. Proc. Natl. Acad. Sci. USA. 82: 1800–1804

Viitanen PA, Gatenby AA, and Lorimer GH (1992) Purified chaperonin 60 (groEL) interacts with the nonnative states of a multitude of *Escherichia coli* proteins. Protein Sci. 1: 363–369

Williams AF, and Barclay AN (1988) The immunoglobulin superfamily. Domains for cell surface recognition. Annu. Rev. Immunol. 6: 381–405

Williams PH, Roberts M, and Hinson G (1988) Stages in bacterial invasion. J. Appl. Bacteriol. (Symp. Suppl.) 131S–147S

Note added in proof

The paper referred to as Kuehn *et al.*, in press has the following reference.

Kuehn MJ, Ogg DJ, Kihlberg J, Slonim LN, Flemmer K, Bergfors T and Hultgren SJ (1993) Structural basis of pilus subunit recognition by the PapD chaperone. Science, in press.

4. Studies on the pili of the promiscuous plasmid RP4

LAURA S. FROST and JOHN SIMON

Abstract. The pili of the promiscuous plasmid RP4 (or RP1, a close relative) are composed of a subunit of 8 kDa in size and have been purified in small quantities to allow an amino acid composition to be done. In addition, a variety of conditions known to affect F plasmid transfer, a narrow host range plasmid, were assayed for their effect on mating by RP4. RP4 mating was found to be insensitive to mutations in integration host factor (*himA* and *himD/hip*), *crp* (cyclic AMP receptor protein)*sfrA* or *sfrB*, *ompA* (the major outer membrane protein of *Escherichia coli*), or mutations in the inner core of the lipopolysaccharide. The RP4 mating system was most efficient at 37°C and increased in efficiency 10-fold upon increased aeration of the mating mixture by vigorous shaking. This is in contrast to the previous observation that conjugation was more efficient on a solid surface. It was also relatively insensitive to the addition of 0.1% SDS and mating pair formation was shown to require about 20 min for stabilization as measured by the addition of SDS with time. A second mating system, supplied by derepressed F-like plasmids, demonstrated that RP4 mating efficiency could be enhanced by the presence of another transfer system in the donor cell.

Abbreviations: LPS, Lypopolysaccharide; SDS, Sodium Dodecyl Sulphate

Introduction

Conjugative plasmids have been divided into two main subgroups, the promiscuous or broad host range (BHR) and the narrow host range (NHR) plasmids. While considerable information is known about the mating system of the F plasmid (reviewed in Ippen-Ihler and Maneewannakul, 1991), little is known about the promiscuous plasmids which have been used extensively in the development of a number of shuttle vectors.

In this study, the most prominent of the promiscuous plasmids, RP4 (RP1, RK2 and RP4 appear to be identical), was compared to the F plasmid in order to clarify the similarities and differences between the two classes of mating systems. The RP4 transfer functions are divided between two regions Tra1 and Tra2 (formerly Tra2 and Tra3) which are separated by an IS*21* element. The *oriT* (origin of transfer) is found in Tra1 as are the proteins required for DNA metabolism during transfer (reviewed in Guiney and Lanka, 1989). Tra2 appears to have the majority of genes required for the expression of pili, an extracellular fiber required for mating pair formation, as well as the genes for surface exclusion, a process whereby redundant mating between donor cells is reduced (Barth *et al.*, 1978; Palombo *et al.*, 1989). The entire transfer region of

C.I. Kado and J.H. Crosa (eds.), Molecular Mechanisms of Bacterial Virulence, 47–65.
© 1994 *Kluwer Academic Publishers.*

F has been sequenced and no homology with any of the sequence available for RP4 has been demonstrated (unpublished data).

The mating system of RP1 has been characterized as surface obligatory since mating on solid surfaces was over 2000-fold more efficient than in stationary liquid media (Bradley *et al.*, 1980). The pili of RP1 are rigid filaments of 8 nm in diameter with a knob-like structure at the base of the pilus which is also found on purified F pili. They have pointed tips typically found on pili from many incompatibility groups other than IncF (reviewed in Paranchych and Frost, 1988). RP1 pili are the site of attachment for the bacteriophages PRR1 and Pf3 which resemble the F-specific bacteriophages R17 and f1 respectively, and are also, presumably, the site of attachment of the bacteriophages IKe, X and PR4 (electron microscopy demonstrating attachment has not been done) which also infect cells carrying plasmids from IncI,N,M,U,W, and X groups (reviewed in Frost, 1993). Like F pili, RP1 pili are constitutively expressed, however, no fertility inhibition system has been found for the RP1-like plasmids.

RP1 and its relatives are known to transfer to almost all members of the Gram-negative bacteria (Thomas and Helinski, 1989). In contrast, F transfers within the enteric bacteria alone. If a suitable replicative origin is supplied, F can transfer to *Pseudomonas* (Guiney, 1982), suggesting that it is the inability of the plasmid to replicate within foreign recipient cells that limits its host range and that the mating system is at least as promiscuous as that of RP1. This was further exemplified by the demonstration that both the IncP plasmid R751 (a distant relative of RP1) and F-derived plasmids could transfer to yeast if suitable replication origins and selectable markers were supplied (Heinemann and Sprague, 1989); reinforcing the notion that the transfer systems of F and RP1 may not be completely responsible for the difference in promiscuity of the two different classes of plasmids and that the RP1 mating system may be more like that of F than originally expected.

This study is a preliminary characterization of the pili and mating system of RP4 (or RP1) by comparing it to what is known for F. The picture that is emerging is one of a plasmid that is less dependent than F on host functions; it does form mating pairs which are stabilized with time, but whose mating pair stabilization mechanism may be less robust than that of F.

I. Do other mating systems affect the mating efficiency of the RP4 plasmid?

The apparently poor ability of RP4-like plasmids to mate in liquid culture was a primary concern of this study. The F plasmid mates equally well in liquid or on solid media and mutations that affect the ability of a cell to act as a recipient (*ompA* mutations, for example) can be overridden by performing the mating on solid media rather than in liquid culture (Manoil and Rosenbusch, 1982). This suggested that perhaps two separate systems are at work in establishing a mating pair involving F-like plasmids – one could act at a distance (perhaps through the pilus tip) in "lasso-ing" a recipient cell as demonstrated by video microscopy by M. Durrenberger and presented at the EMBO Workshop on

Gene Transfer in 1991, while the other is the result of close contacts between cells growing side-by-side on a solid support (Durrenberger *et al.*, 1991). No visual evidence for similar conjugation junctions has been reported for mating pairs involving the RP4-like plasmids or other promiscuous plasmids. Alternatively, RP4-like plasmids may not have a mating pair stabilization system similar to the one expressed by F-like plasmids, the function of the F*traNG* genes (Ippen-Ihler and Skurray, 1993). In this case, the solid surface would support a more fragile conjugation junction, such as an extended pilus, for example. Thus, it was of interest to examine whether the RP4 mating system would benefit from the presence of an F-like mating system and a mating pair stabilization apparatus.

The mating efficiency assays were performed as described by Finnegan and Willetts (1971). Unless otherwise stated, the recipient cell was *E. coli* ED24 (Table 1). Donor and recipient cells were grown up overnight as standing

Table 1. Strains used in this study.

Strains	Relevant genotype	Source or reference [a]
E. coli		
JC3272	F⁻ *his trp lys tsx gal malA lac* Δ*X74 strA*	1
ED2601	JC3272, Pil⁻, Fla⁻	W. Paranchych
ED24	F⁻ Lac⁻ Spc⁻	2
AB257	Hfr *met ton strA*	3
MC253	Δ(*lac pro*)xIII *argEam gyrA rpoB supF* Δ82*himA*::Tn*10* Δ3*hip*::Cm^R	4
LS854	F⁻ *trpA9605 his85 crp3 metE70 trpR55 rpsL136*	W. Paranchcych
M1174	JC3272, *his⁺*	5, K. Ippen-Ihler
M1163	M1174, *sfrA4*	5, K. Ippen-Ihler
M1164	M1174, *sfrA5*	5, K. Ippen-Ihler
CC102	JC3272, *galE*	6, C. Manoil
CC209	CC102 *ompA893* (OmpA⁻)	C. Manoil
CC253	CC102, *ompA889* (point mutant)	C. Manoil
CC263	CC102, *ompA888* ([OmpA] decreased)	6, C. Manoil
MC4100	F-*araD139* Δ(*argF--lac*)*U169 rpsL150 relA1 flbB3501 deoC1 ptsF25 rbsR*	C. Manoil
CC650	MC4100, *ompA886* (point mutant)	C. Manoil
CC651	MC4100, *ompA889* (point mutant)	C. Manoil
CS180	F⁻ *thr leuB6 proA argE his thi galK lacY trpE mtl xyl ara-14* Su⁺	7, C. Schnaitman
CS1562	CS180, *tolC*::Tn*10*	C. Schnaitman
CS1858	CS180, *sfrB11*	7, C. Schnaitman
CS1959	CS180, Δ*rfa*(*G--I*)1::ΔCm^R	C. Schnaitman
CS2057	CS1959, *cps-5*::Tn*10*	C. Schnaitman
CS2058	CS180, Δ*rfa*(*Q--J*)2::ΔCm^R*cps-5*::Tn*10*	C. Schnaitman

[a] 1. Achtman *et al.*, 1971; 2. Finnegan and Willetts, 1971; 3. Krahn *et al.*, 1972; 4. Gamas *et al.*, 1987; 5. Silverman *et al.*, 1991; 6. Manoil and Rosenbusch, 1982; 7. Austin *et al.*, 1990.

cultures in the presence of antibiotics as needed. The cells were diluted 50-fold into fresh L-broth media and grown to an OD_{600} of 1.5 (approximately 5×10^8 cells/ml). 100 μl of donor and recipient cultures were added to 1.0 ml of fresh L-broth and the mating mixture was incubated at 37°C without shaking for 30 min. The mating period was terminated by vigorous vortexing and immediate dilution by 100-fold into 1 $\times$ SSC, pH 7.0. The samples were further diluted appropriately and 10 μl spots were dropped onto plates containing the appropriate antibiotics/nutrients to select for transconjugants. Duplicate samples were also plated out on plates that selected for donor or recipient cells and colony counts were performed to ensure that the strains were being stably maintained and had not suffered plasmid loss. The results are reported either as the number of transconjugants/donor cell or as a percentage of mating compared to the control strain after a correction for differences in cell density had been made.

Cells containing RP4 alone or with JCFL0 (Achtman et al., 1971), ColB2-K77 (Frost et al., 1985), F_olac, $F_olacdrd$ or pED208 (Di Laurenzio et al., 1991), which expressed the Lac$^+$ phenotype, as well as the Hfr strain E. coli AB257 containing RP4 were mated with the recipient strain ED24 (Table 1). In addition, cells containing RP4 and mutants of JCFL0 in the traJ,M,A genes (traJ90, traM102, traA1; Achtman et al., 1971; 1972) were also used. Kan^R Lac$^+$ colonies were mated with E. coli ED24 (Lac$^-$ Spc^R) and the transconjugants were scored for Kan^R Spc^R colonies (assaying for RP4 transfer) and Kan^R Spc^R Lac$^+$ colonies (assaying for RP4 and F-like plasmid transfer) to indicate the level of transfer of RP4 alone or together with the F-like plasmids (Table 2). The level of F-like plasmid transfer was normal in all cases (data not shown).

The rate of transfer of RP4 in liquid medium was approximately 40–50-fold less than for JCFL0 alone. The rate of transfer of RP4 in the presence of JCFL0, either wild type or carrying mutations in the traA,J or M genes, was further reduced by about 200-fold (Table 2). This was most likely due to the fertility inhibition of F exerted on RP4 through the action of the pifC gene (Miller et al., 1985). Interestingly, the F mutations that affected expression of the transfer operon (traJ90) further reduced transfer nearly 10 fold. The traM102 and traA1 mutations had little to no effect (a 5-fold difference in mating efficiency was considered the threshold).

Other F-like plasmids, (many plasmids could not be screened because of the presence of genes for kanamycin resistance), did not inhibit RP4 mating. These included ColB2 (IncFII) and the derivatives of F_olac (IncFV). F_olac is a fully repressed plasmid which expresses 1 pilus for every 100–1000 cells, while $F_olacdrd$ is a derepressed plasmid expressing 5–6 pili/cell (Di Laurenzio et al., 1991). pED208 is a mutant of F_olac which is multi-piliated by virtue of an IS2 element inserted at the beginning of its transfer operon (Finlay et al., 1986). The presence of F_olac had no effect on RP4 transfer while both $F_olacdrd$ and pED208 increased RP4 transfer by 100-fold (Table 2). This increase did not appear to be due to cointegrate formation between RP4 and the F_olac-like plasmids since the number of recipient cells receiving both plasmids was 100-

Table 2. The effect of F-like plasmids on the mating efficiency of RP4.

Donor	Mating efficiency T/Donor cell[a]	Compared to RP4[b]
E. coli JC3272 containing		
RP4	0.013	1
pRK2013	0.018	1.4
pOX38::Km	0.60	46.2
JCFL0	0.53	40.8
$F_{o}lac$	4.0×10^{-4}	0.03
$F_{o}lacdrd$	0.20	15.4
pED208	0.78	60
RP4 + JCFL0	9.4×10^{-5}	0.0072
RP4 + JCFL*traA1*	3.4×10^{-5}	0.0026
RP4 + JCFL*traJ90*	1.0×10^{-5}	0.0008
RP4 + JCFL*M102*	5.0×10^{-5}	0.0038
E. coli AB257/RP4	5.7×10^{-5}	0.0044

	T/Donor cell[c]		Compared to RP4[b]
	Km Spc	Km Spc Lac[+]	
E. coli JC3272 containing			
RP4 + $F_{o}lac$	0.03	5.8×10^{-5}	2.3
RP4 + $F_{o}lacdrd$	0.29	4.1×10^{-5}	22.3
RP4 + pED208	0.24	2.2×10^{-5}	18.5
RP4 + ColB2	0.12		9.2

[a] Transconjugants per donor cell. *E. coli* ED24 (Lac⁻ Spc[R]) was the recipient cell in all cases. The matings were performed at 37°C in 1.0 ml of stationary L-broth. RP4 and pRK2013 were detected by kanamycin resistance, F- and $F_{o}lac$-like plasmids were detected by the ability to grow on lactose. In matings where RP4 and F-like plasmids are both present in the donor cell, the mating efficiency for the RP4 plasmid is reported.

[b] Compared to RP4 in *E. coli* JC3272 as the donor cell and ED24 as the recipient, arbitrarily given a value of 1.

[c] Km[R] Spc[R] transconjugants are a measure of mating ability of RP4 into ED24. Km Spc[R] Lac[+] colonies are a measure of the number of recipient cells that received both plasmids.

fold less than the number receiving RP4 alone. This suggests that RP4 is aided by the presence of another mating system (although the number of pili per cell was immaterial) and that the conjugation junction, which covers a wide surface area between mating cells involving the F plasmid (Durrenberger *et al.*, 1991), can be utilized by other transfer systems. This would have implications for mobilizable plasmids which could utilize a conjugation bridge independent of the transfer machinery provided by the co-residing, self-transmissible plasmid.

II. Is RP4 mating influenced by chromosomal genes that affect F transfer?

The chromosomally encoded protein, Integration Host Factor (IHF), has been implicated in a wide range of processes including lambda integration, plasmid replication and F plasmid transfer (Friedman, 1988). Homologues of IHF have also been found in other Gram-negative bacteria (Haluzi *et al.*, 1991), suggesting that it might be available throughout the host range of RP4 and would not be a limiting factor in its promiscuity. Because RP4-like plasmids express resistance for kanamycin, tetracycline and ampicillin in *E. coli*, the

Table 3. The effect of host mutations in the donor or recipient cell on plasmid transfer.

Donor	Recipient	Mating efficiency T/Donor cell[a]	%WT[b]
E. coli JC3272 (WT)/			
pOX38::Km	ED24	0.50	100
pRK2013	ED24	0.013	100
RP4	ED24	0.009	100
E. coli MC253 (IHF⁻)/			
pOX38::Km	ED24	5.5×10^{-5}	0.001
pRK2013	ED24	0.012	92
RP4	ED24	0.011	122
E. coli CS180 (WT)/			
pOX38::Km	ED24	0.86	100
pRK2013	ED24	0.02	100
E. coli LS854(CRP⁻)/			
pOX38::Km	ED24	0.008	0.93
pRK2013	ED24	0.0027	6
E. coli CS1562(TolC⁻)/			
pOX38::Km	ED24	0.61	71
pRK2013	ED24	0.02	100
E. coli CS1858(SfrB11)/			
pOX38::Km	ED24	2.2×10^{-5}	0.0026
pRK2013	ED24	0.016	80
E. coli M1174 (SfrA⁺)/			
pOX38::Km	ED24	0.23	100
pRK2013	ED24	0.012	100
E. coli M1163 (SfrA4)/			
pOX38::Km	ED24	0.0015	0.65
pRK2013	ED24	0.004	33.3
E. coli M1164 (SfrA5)/			
pOX38::Km	ED24	0.29×10^{-4}	0.013
pRK2013	ED24	0.002	16.7

[a] Transconjugants per donor cell.

[b] %WT is the percentage of the mating efficiency for the wild type strain isogenic for that set of mutants.

kanamycin-resistant derivative of RK2, pRK2013 (Figurski and Helinski, 1979), containing the RK2 transfer region cloned into ColE1, was used in these studies where needed. Therefore, both RP4 and pRK2013 were mated into the IHF⁻ strain *E. coli* MC253 which is mutated in both genes for the heterodimeric protein, IHF (Gamas *et al.*, 1987). Because IHF may have an overriding effect on RP4 replication rather than conjugation, both RP4 and pRK2013 were tested for their ability to mate out of an IHF⁻ strain; this was compared to the effect of the IHF⁻ mutation on the mating efficiency of the F derivative pOX38::Km (Gamas *et al.*, 1987). While the mating efficiency of pOX38::Km was drastically reduced in MC253, as previously described, there was little to no effect on RP4 or pRK2013, suggesting that IHF does not play a role in RP4 conjugation (Table 3). This experiment also suggests that the vegetative origin of replication of RP4 is not dependent on IHF activity.

Another mutation known to affect F mating efficiency is the *crp* mutation in the cyclic AMP receptor protein (Harwood and Meynell, 1975; Kumar and Srivistava, 1983). This mutation had a strong effect on pOX38::Km (100-fold decrease), it also had the strongest effect of any of the host mutations on pRK2013 transfer (10–15-fold) (Table 3). Similar experiments were performed using *E. coli* strains mutated in the *sfrA* and *sfrB* loci which were initially described as chromosomal mutations that affect F mating efficiency (Beutin and Achtman, 1981). SfrA is a "two component regulator" which acts through a signal transduction pathway to affect the control of aerobic respiration as well as F plasmid transfer (Silverman *et al.*, 1991). SfrB has been less well studied but has been shown to be the same as the *rfaH* locus which controls expression of lipopolysaccharide and may act as an anti-terminator of transcription of the long operons found for F pilus expression and LPS biosynthesis (Rehemtulla *et al.*, 1986; Pradel and Schnaitman, 1991). Both of these mutations did not effect pRK2013 mating ability to any significant extent, confirming the earlier report by Beutin and Achtman (1979). *E. coli* CS1562, carrying a Tn*10* insertion in the *tolC* gene (Webster, 1991), which is part of the transport system for macro-molecules in *E. coli* , did not affect the mating efficiency of either plasmid.

III. Do mutations in the ompA or rfa Loci affect RP4 mating?

An attempt to define the receptor on the recipient cell for the F transfer system was carried out in a number of laboratories during the 70's (Skurray *et al.*, 1974; Schweizer and Henning, 1977; Reiner, 1974; Havekes *et al.*, 1976; 1977; Achtman *et al.*, 1978). Unlike resistance to bacteriophages or colicins, no mutation in the recipient cell was identified which totally blocked transfer. The Con⁻ mutants initially identified, which were suppressed when the mating was performed on solid media, involved the *ompA* gene product as well as the inner core of the lipopolysaccharide (LPS). These mutants were still able to bind F pili, suggesting that the mutations did not affect the receptor for the pilus. However, it was never demonstrated that this pilus binding was identical to the process of recipient cell identification. The Con⁻ mutants appeared to affect mating pair stabilization as measured by electron microscopy and Coulter counter techniques as well as the colony sectoring assay devised by Skurray *et al.* (1974). Similar results were also found for other F-like plasmids and for IncI plasmids, however, no information on IncP plasmids was available.

Since then, carefully characterized mutants in the *ompA* gene have become available and have been shown to affect F conjugation (Manoil and Rosenbusch, 1982; Manoil, 1983; Morona *et al.*, 1984). Similarly, the genetics of lipopolysaccharide expression in *E. coli* has been described and has been shown to be quite similar to the LPS synthetic pathway in *Salmonella typhimurium* (Schnaitman *et al.*, 1991). Mutations in many of the genes in the *rfa* locus of *E. coli* are now available and a study of the effect of these mutations on mating efficiency was carried out for both F (pOX38::Km) and pRK2013. The description of the mutant strains used in this study are given in Table 1.

As previously reported by Manoil (1983), mutations resulting in the deletion of OmpA as well as point mutations in the *ompA* gene greatly affected F mating (Table 4). However, these mutations had little to no effect on pRK2013-mediated plasmid transfer. Cells deficient in the OmpA protein were slightly less proficient as recipient cells than the wild type, however, the small decrease was not considered to be due to a specific effect on the mating apparatus.

Mutants containing a deletion of the LPS *rfaG–I* genes (CS1959, CS2057) or a deletion of the *rfaQ–J* genes (CS2058) were tested as recipients in matings with cells containing pOX38::Km or pRK2013. Both CS2057 and CS2058 had an additional mutation (*cps*) which prevented the development of the mucoid phenotype, induced by the loss of the *rfaP* gene (Parker *et al.*, 1992), which is known to interfere with mating.

The results were surprising on two accounts. Firstly, none of the mutants or constructs mentioned above affected pRK2013 mating, suggesting that LPS (the core region) is not involved in mating pair formation with cells containing RP4-like plasmids (Table 4). All of the *rfa* mutants and constructs were checked for the presence of the OmpA protein, which appeared to be present in normal amounts (data not shown). Strains carrying mutations in the *rfaG–I* loci did affect mating efficiency with F⁺ cells (the larger deletion of *rfaQ–J* did not further reduce the mating efficiency). The mating efficiency was partially restored by complementation *in trans* by the *rfaP* gene carried on a plasmid vector (data not shown), suggesting that the *rfaP* gene product may be important in F mating pair formation. RfaP is thought to be involved in modification of the heptose residues I and II in the inner core either directly or indirectly through another gene product (Parker *et al.*, 1992). This would corroborate earlier reports that F plasmid transfer was decreased with heptose-deficient recipient cells (Havekes *et al.*, 1976). The role of *rfaP* in F conjugation is being characterized in greater detail.

We conclude from these studies that RP4-like plasmids are not affected by the same alterations in the surface of the recipient cell that affect the F plasmid. Thus, it is possible that RP4-like plasmids do not have a mating pair stabilization system similar to that for F and therefore, mutations in OmpA and the LPS do not affect their mating efficiency.

Table 4. The effect of mutations in the *ompA* and *rfa* loci in the recipient cell on the mating efficiency of F (pOX38::Km) and RP4 (pRK2013).

Donor	Recipient	(T/Donor cell)[a]	% WT[b]
E. coli ED24/pOX38::Km			
	CC102	0.68	100
	CC209	0.0075	1.1
	CC253	0.051	7.5
	CC263	0.0085	1.25
	MC4100	0.90	100
	CC650	0.013	1.4
	CC651	0.017	1.9
	CS180	0.50	100
	CS1861	0.26	52
	CS1959	0.0046	0.9
	CS2057	0.0076	1.5
	CS2058	0.0078	1.6
E. coli ED24/pRK2013			
	CC102	0.021	100
	CC209	0.009	42.9
	CC253	0.032	152
	CC263	0.009	42.9
	MC4100	0.036	100
	CC650	0.022	61.1
	CC651	0.018	50.0
	CS180	0.028	100
	CS1861	0.036	129
	CS1959	0.016	57
	CS2057	0.040	143
	CS2058	0.019	67.8

[a] Transconjugants per donor cell.

[b] %WT is the percentage of the mating efficiency for the wild type strain isogenic for that set of mutants.

IV. Does the solubilization of RP4 pili with SDS affect mating pair formation?

The effect of various concentrations of SDS (sodium dodecylsulfate) on the mating efficiency of pOX38::Km and RP4 was determined before it was used to study mating pair formation. While cells containing pOX38::Km were greatly affected by the addition of SDS before donor and recipient cells were mixed (adding the SDS to the donor cells before the addition of recipient cells was more efficient at reducing mating than the other way around), RP4-containing

56

cells were fairly insensitive to the addition of SDS (7-fold as compared to 2000-fold for pOX38::Km (Fig. 1). In both cases the addition of SDS, up to 0.1% in concentration, did not affect the viability of either pOX38::Km or RP4-containing cells. Whether this difference reflects the resistance of RP4 pili to disaggregation by SDS treatment is not known.

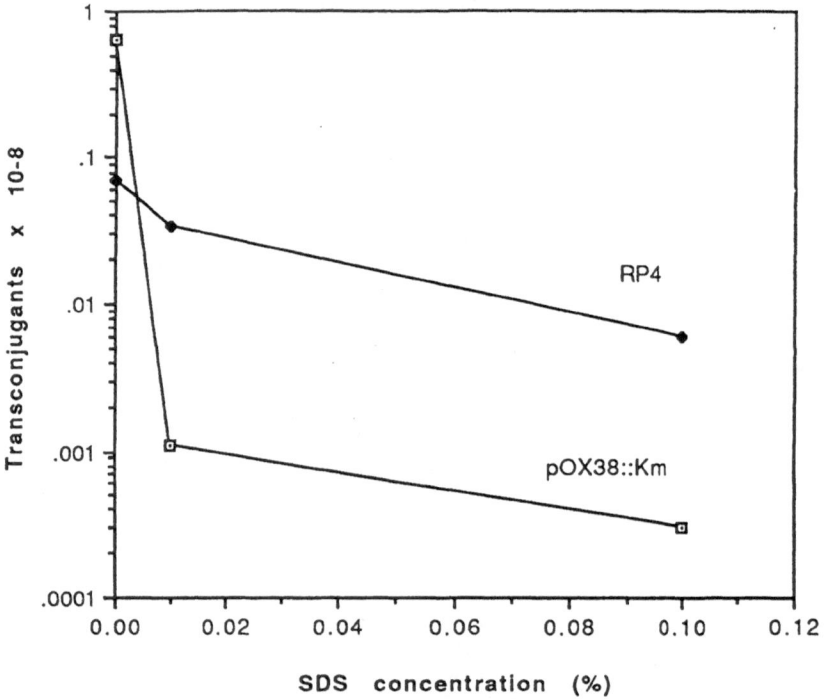

Figure 1. Strains used in this study.

In order to determine the time required for mating pair formation, 0.1% SDS was added at various times after the addition of donor cells to recipients. If mating pair formation was complete, the number of transconjugants should not be affected by the addition of SDS. Previously, Achtman *et al.* (1978) had demonstrated that SDS addition reduced mating efficiency by 10,000-fold when added at zero minutes. They also showed that SDS addition after 19–20 min (but not at 15 min) did not affect the number of aggregated cells or the number of transconjugants. This suggested that extended pili were not required for transfer and that it required 15–20 min for large numbers of stable mating pairs to form.

If RP4 does not have a mating pair stabilization system, the number of transconjugants should be affected by the addition of SDS thoughout the time course of the experiment. The results shown in Fig. 2 demonstrate that the addition of SDS to mating cells has no effect after 20 min, suggesting that the kinetics of mating pair formation is approximately the same as for the F plasmid and that RP4-mediated conjugation does undergo a transition from an SDS-sensitive to an SDS-insensitive step.

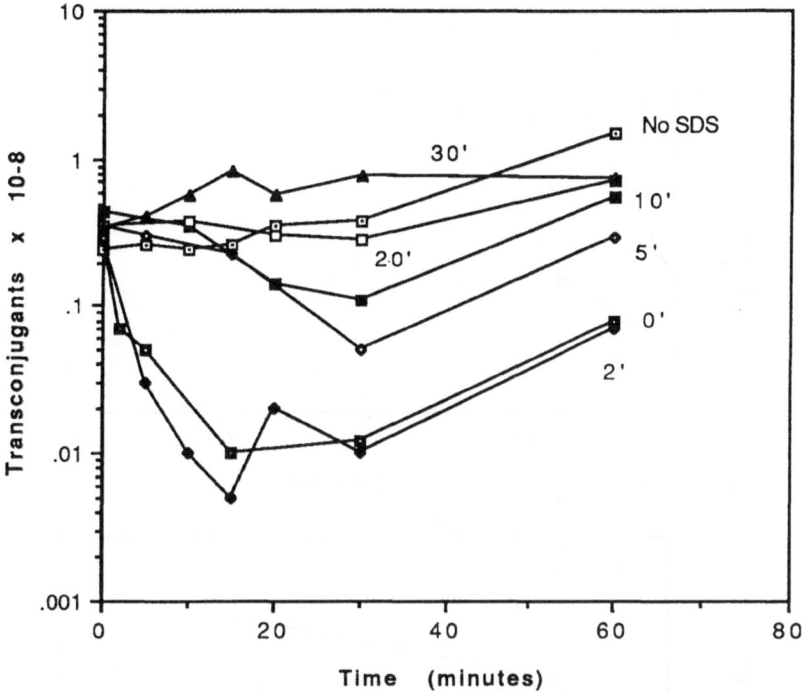

Figure 2. The effect of the addition of 0.1% SDS at various times after the initiation of mating. 0.1% SDS was added to mating bacteria at 0, 2, 5, 10, 20 and 30 min after mixing the donor (RP4/JC3272) and recipient cells (ED24) together. Details of the protocol are given in Fig. 1. The zero time sample was taken 30 seconds after the addition of recipient cells and SDS, in that order.

V. Does temperature or agitation have an effect on RP4-mediated transfer?

The rate of pOX38::Km transfer was compared to that for RP4 at room temperature, 30, 37 and 42°C. All cells were grown at 37°C until mid-log phase and then pre-incubated separately at the mating temperature for one hour before the donor and recipient cells were mixed together. The highest level of transfer was at 37°C and the lowest level was at room temperature for both plasmids (Fig. 3). RP4 was more affected by changes in temperature than pOX38::Km which eventually attained approximately the same number of transconjugants at all temperatures. Both plasmids had a greatly reduced mating efficiency at time zero for the 22°C and 30°C matings as compared to the matings at higher temperatures. This reduction could be due to the loss of pili at lower temperatures (Novotny and Lavin, 1971). However, the large increase in mating efficiency with time in these samples (especially pOX38::Km) suggests

58

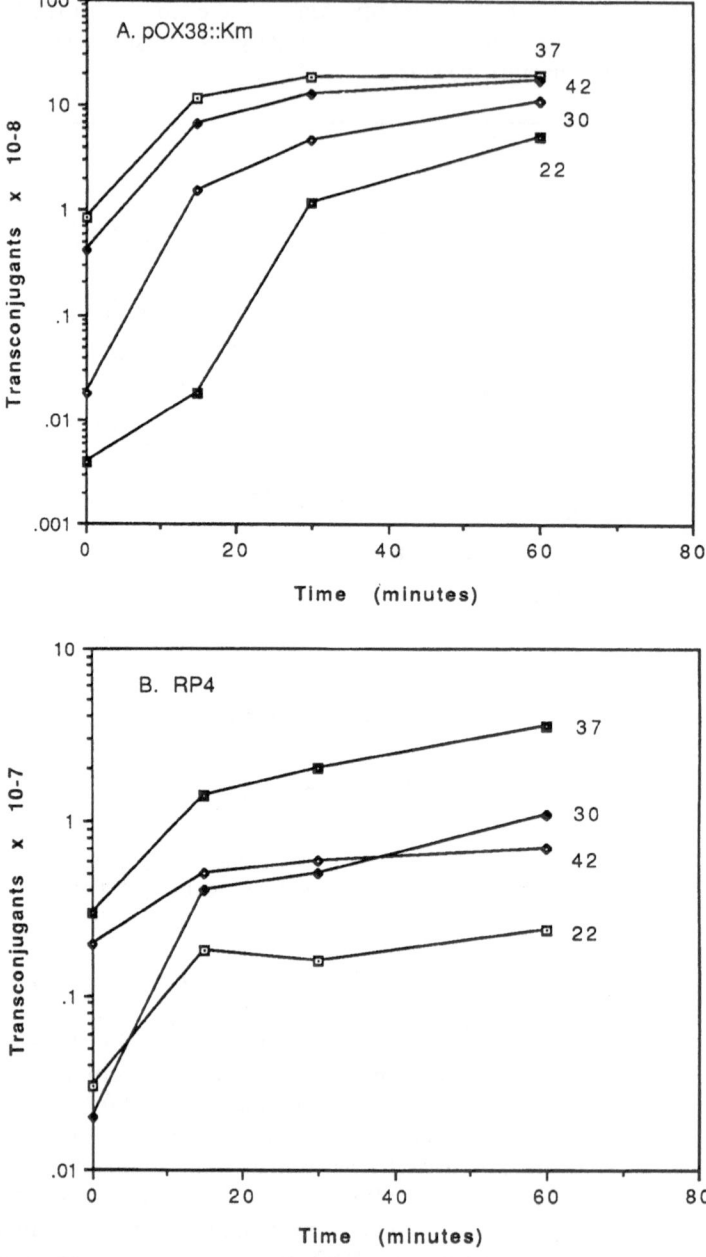

Figure 3. The effect of temperature on mating efficiency of RP4 and pOX38::Km. The donor cells (A. pOX38::Km or B. RP4) were grown at 37°C and diluted in 1 ml L-broth to 1×10^8 cells/ml. The cells were then incubated at room temperature, 30°C, 37°C and 42°C for one hour before the addition of recipient cells (ED24). Samples were taken at 0 (30 seconds), 15, 30 and 60 min and the number of transconjugants and donor cells were assayed as described in Fig. 1.

that mating pairs were capable of forming at a slower rate. Currently, we have no information on the level of piliation of RP4 cells at these temperatures since RP4 pili are extremely difficult to detect by electron microscopy even under optimal conditions.

While assaying the mating efficiency of RP4 plasmids using the protocol described by Manoil (1983), we noticed that agitation of the culture seemed to increase the mating efficiency. This was more carefully determined by comparing the numbers of transconjugants of pOX38::Km and RP4 mated with ED24 in 1.0 ml stationary, shaking (200 rpm) and rolling cultures. Agitation of pOX38::Km matings did not greatly affect the level of transfer as compared to stationary liquid cultures while vigorously shaking or gently rolling the mating mixture of the RP4 matings resulted in a 10-fold increase in the number of transconjugants (Fig. 4). This suggests that the level of aeration may be important in RP4-mediated plasmid transfer and that the mating contacts which are supposedly rather delicate and require support on a solid medium are more

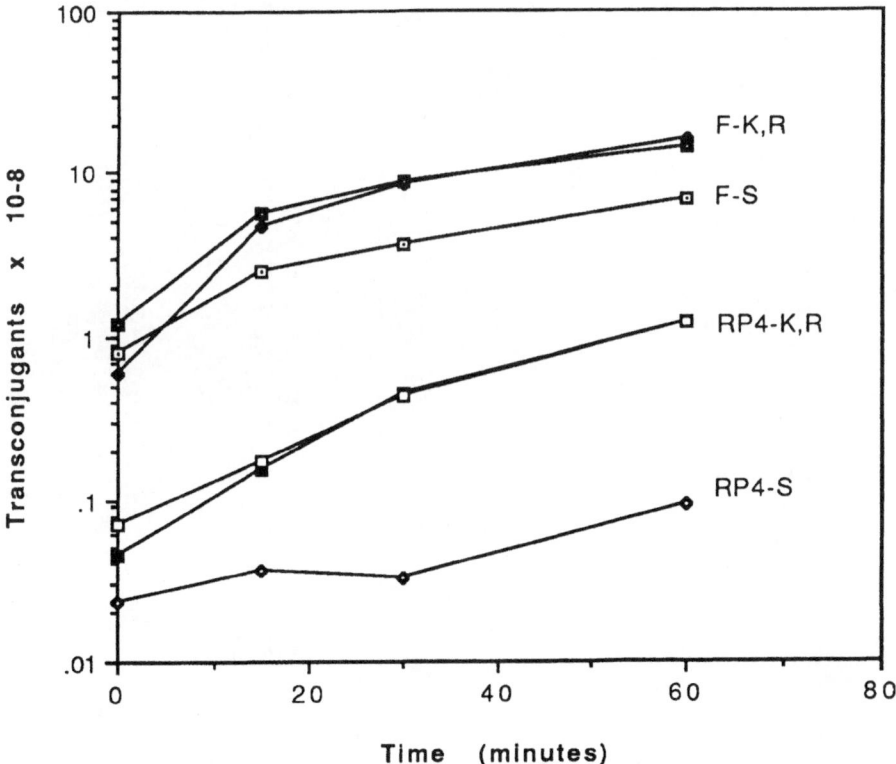

Figure 4. The effect of agitation on the mating efficiency of RP4 and pOX38::Km. Mating mixtures were incubated at 37°C as stationary (S), shaking (K) (200 rpm) or rolling (R) cultures. Samples were taken at 0, 15, 30 and 60 min and the number of transconjugants and donor cells were assayed as described in Fig. 1.

robust than originally thought. The preference for a solid medium may, therefore, be due in part to the higher oxygen concentration on a solid surface.

VI. Are RP4-like pili made up of pilin subunits of similar size to F pilin?

For reasons which are not understood, RP1 pili were more prolific and easier to purify than RP4 pili and were used in the characterization of the pilin subunit. RP1 was introduced into the bald *E. coli* strain ED2601 and was purified essentially as described by Armstrong *et al.* (1980) from 36 aluminum cookie sheets containing L-broth agar grown at 37°C overnight. The cells were scraped from the surface of the agar and gently resuspended and stirred with a

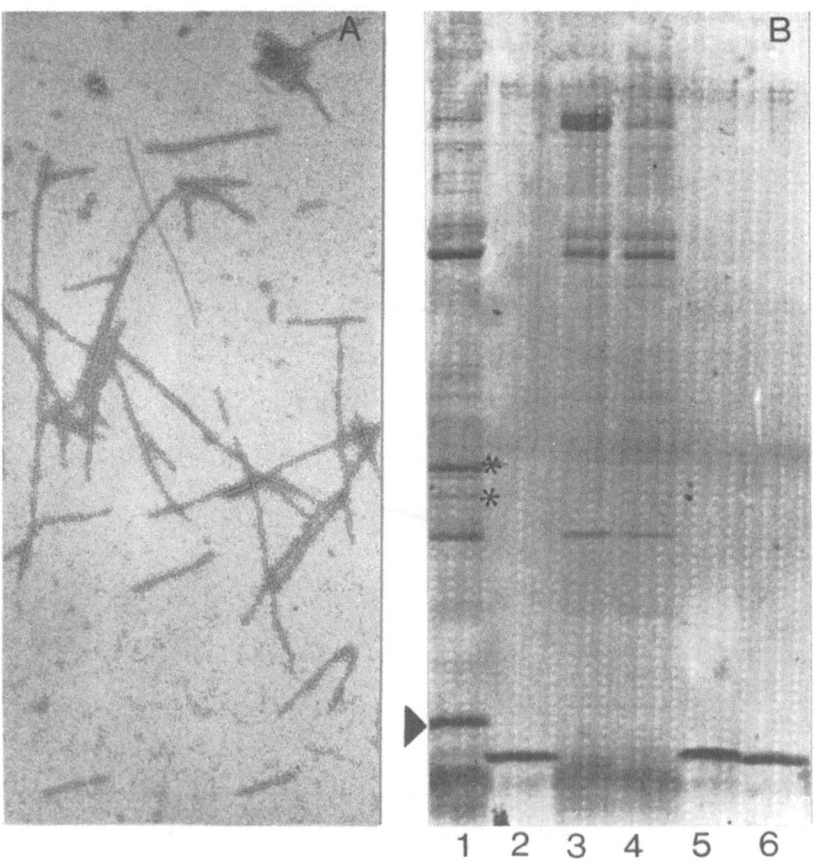

Figure 5. A. Electron micrograph of semi-purified RP1 pili stained with 1% sodium phosphotungstic acid, pH 7.0. B. Silver-stained SDS polyacrylamide gel of RP1 pili with various standards. 1, RP1 pili (marked with arrowhead); 2, purified pED208 pili; 3 and 4, material from a blank pili preparation of ED2601; 5, ColB2 pili; 6, F pili. The asterisks mark the two proteins near 20 kDa discussed in the text.

magnetic stirrer in cold 1 × SSC, pH 7.0 for one hour. The cells were pelleted by centrifugation at 8,000 rpm until the supernatant was clear. Polyethylene glycol 8000 and NaCl were added to a final concentration of 2% PEG 8000, 0.5 N NaCl and the pili were left to precipitate in the cold overnight. The precipitate was collected by centrifugation and resuspended in 8.0 ml of cesium chloride in 1 × SSC, pH 7.0 of refractive index n=1.3610 which corresponded to a density of 1.3 g/cc^3. The pili were banded in an SW 60.1 Ti rotor at 40, 000 rpm at 5°C for 16 hours and the pili appeared as fine flocculent band at a density of 1.3 g/cc^3 above the middle of the gradient. The pili were dialysed extensively against water and their purity was judged by SDS gel electrophoresis on a 15% gel, using silver stain to detect the pilin.

The pili were found to band in a cesium chloride density gradient at a density of 1.3 g/cm^3. An electron micrograph of the purified pili preparation is shown in Fig. 5A. Less than 1 mg of pilin was obtained from approximately 100 g wet weight of cells. A prominent band at 8000 daltons was visible upon silver staining a 15% SDS polyacrylamide gel which was barely visible when stained with Coomassie blue; this is also a characteristic of F-like pilin (Fig. 5B). The pili preparation was contaminated with other proteins, probably derived from the cell envelope as well as lipopolysaccharide which stained as a blurry band beneath the pilin band. Most of these extra bands were present in a blank preparation of material, purified from ED2601, that banded at 1.3 g/cm^3 and were judged not to be associated with the RP1 pili. Two bands near 20kDa in size were not reproducibly seen in various pili preparations and were not considered to be pilin.

Because of the small amount of material, efforts to further purify intact pili were unsuccessful. Instead, several SDS polyacrylamide gels were run and a thin band corresponding to 8kDa in size was excised from the gels. The material was electroeluted using the apparatus and procedure described by Hunkapillar et al. (1983). The material was checked for purity by running a small portion on another SDS gel. An amino acid composition was done on the material recovered (5 nmol (40 µg) of protein) after hydrolysis in 6 N constant boiling HCl at 110°C (Table 5). A second, equivalent amount of protein was transferred to a PVDF membrane (Immobilon polyvinylidene difluoride membrane, Millipore) and an N-terminus determination was performed using automated sequence analysis techniques.

No N-terminal sequence information was obtained. F pili also have a blocked N-terminus which may be characteristic of conjugative pilin (Frost et al., 1984). While the sequence of the complete RP4 plasmid has been finished, it is not yet published and identification of the pilin gene awaits the release of the sequence.

Table 5. The amino acid composition of RP1 pilin.

Amino acid	nmol/nmol pilin[a]		
	RP1	F	pED208
Asx	7.8(8)	4	5
Glx	10.1(10)	2	1
Ser	7.3(7)	6	1
Gly[b]	8.6(9)	7	6
His	1.0(1)	0	0
Arg	2.0(2)	0	0
Thr	4.7(5)	4	6
Ala	6.6(7)	9	5
Pro	2.9(3)	0	0
Tyr	2.0(2)	1	1
Val	4.7(5)	14	9
Met	1.0(1)	5	3
Ile	3.2(3)	3	8
Leu	4.9(5)	5	9
Phe	2.6(3)	4	4
Lys	3.9(4)	5	3
MW	8200	7200	6800

[a] The F and pED208 amino acid compositions are derived from their DNA sequences (Frost et al., 1984; Finlay et al., 1986). An analysis for Trp and Cys was not performed.

[b] The value for Gly is an overestimation because of contamination of the sample with the glycine in the SDS gel running buffer.

Discussion

The mating system of RP4-like plasmids appears to be similar in many respects to that of the more familiar F-like plasmids. The pili of both systems are composed of small subunits of 7–8 kDa which have amino acid compositions rich in hydrophobic residues. Both mating systems appear to operate maximally at 37°C and increased aeration as supplied by rolling or shaking the mating culture, increases the mating efficiency of RP4. While the expression of the F transfer region appears to be affected by a number of chromosomally encoded gene products (IHF, CRP, SfrA, SfrB), RP4 does not share these requirements; this may contribute to the promiscuity of this plasmid transfer system.

RP4 (or pRK2013) does not appear to require intact OmpA or LPS on the recipient cell for successful transfer since mutations which result in the loss of OmpA or the deep rough phenotype do not affect RP4-mediated mating but do affect pOX38::Km (F) mating. This suggests that RP4-like pili recognize another moiety on the recipient cell to establish a mating pair.

OmpA and LPS have been implicated in the formation of stable mating pairs (or aggregates) at a step beyond initial contact between the F pilus and recipient cell. While RP4 does not appear to have a stabilization system based on OmpA and/or LPS, it does appear to undergo a conversion from SDS-sensitive to SDS-insensitive mating pairs at a rate similar to that previously determined for F (Achtman *et al.*, 1978). Whether RP4 mating mixtures form the mating junctions visualized in the electron microscope for F mating pairs (Durrenberger *et al.*, 1991) remains to be determined. The notion that RP4-mediated conjugation is a more fragile event than F plasmid transfer because of the preference for a solid surface may be due to a less effective mating pair stabilization system in RP4. This is confirmed by the increased mating efficiency of RP4 in the presence of derepressed plasmids of the IncF complex (ColB2, pED208, $F_olacdrd$) while a repressed plasmid (F_olac) has no effect. The preference for a solid surface in RP4-mediated mating may also be due to a requirement for more oxygen than is supplied in a stationary culture of 1.0 ml volume in a 13 mm culture tube, the conditions usually employed in mating assays.

The mating systems of the promiscuous plasmids are reaching a stage whereby a more sophisticated series of shuttle vectors, based on the properties of their transfer systems, can be designed. It is hoped that these preliminary studies lead to a greater appreciation of the role of the pilus in establishing a mating pair and will eventually identify the nature of the receptor(s) that the pilus recognizes.

Acknowledgements

We are very grateful for the excellent technical assistance of Stasia Paczkowski. We wish to thank Colin Manoil and Carl Schnaitman who shared their strain collections with us. I would also like to thank Chris Thomas for alerting me to the presence of *pifC*, Loren Day and David Bradley for initial discussions about RP1 pili and phage attachment and Clarence Kado and William Paranchych for helpful discussions. This work was supported by the Medical Research Council of Canada.

References

Achtman, M., S. Schuwchow, R. Helmuth, G. Morelli and P.A. Manning. 1978. Cell-cell interactions in conjugating *Escherichia coli*: Con⁻ mutants and stabilization of mating aggregates. Mol.Gen.Genet. **164**: 171–183.

Achtman, M., N.S. Willetts and A.J. Clark. 1971. Beginning a genetic analysis of conjugational transfer determined by the F factor in *E. coli* by isolation and characterization of transfer-deficient mutants. J.Bacteriol. **106**: 529–538l.

Achtman, M., N. Willetts and A.J. Clark. 1972. Conjugational complementation analysis of transfer-deficient mutants of *Flac* in *Escherichia coli*. J.Bacteriol. **110**: 831–842.

Armstrong, G.D., L.S. Frost, P.A. Sastry and W. Paranchych. 1980. Comparative biochemical studies on F and EDP208 conjugative pili. J.Bacteriol. **141**: 333–341.

Austin, E.A., J.F. Graves, L.A. Hite, C.T. Parker and C.A. Schnaitman. 1990. Genetic analysis of lipopolysaccharide core biosynthesis by *Escherichia coli* K-12: insertion mutagenesis of the *rfa* locus. J Bacteriol. **172**: 5312–5325.

Barth, P.T., N.J. Grinter and D.E. Bradley. 1978. Conjugal transfer system of plasmid RP4: Analysis by transposon insertion. J.Bacteriol. **133**: 43–52.

Beutin, L., P.A. Manning, M. Achtman and N. Willetts. 1981. *sfrA* and *sfrB* products of *Escherichia coli* K-12 are transcriptional control factors. J.Bacteriol. **145**: 840–844.

Bradley, D.E., D.E. Taylor and D.R. Cohen. 1980. Specification of surface mating systems among conjugative drug resistance plasmids in *Escherichia coli* K-12. J.Bacteriol. **143**: 1466–1460.

64

Di Laurenzio, L., B.B. Finlay, L.S. Frost and W. Paranchych. 1991. Characterization of the *oriT* region of the IncFV plasmid pED208. Mol.Microbiol. **5**: 1779–1790.

Durrenberger, M.B., W. Villiger and T. Bachi. 1991. Conjugational junctions: Morphology of specific contacts in conjugating *Escherichia coli* bacteria. J.Struct.Biol. **107**: 146–156.

Figurski, D.H. and DR. Helinski. 1979. Replication of an origin-containing derivative of plasmid RK2 dependent on a plasmid function provided *in trans*. Proc.Natl.Acad.Sci. U.S.A. **76**: 1648–1652.

Finlay, B.B., L.S. Frost and W. Paranchych. 1986. Nucleotide sequence of the *traYALE* region from IncFV plasmid pED208. J.Bacteriol. **168**: 990–998.

Finnegan, D.J. and N.S. Willetts. 1971. Two Classes of F*lac* Mutants insensitive to transfer inhibition by an F-like R factor. Mol.Gen.Genet. **111**: 256–264.

Friedman, D.I. 1988. Integration Host Factor: A protein for all reasons. Cell **55**: 545–554.

Frost, L.S. 1993. Conjugative pili and pilus-specific phages. *In*: D.B. Clewell, (ed.), Bacterial Conjugation. Plenum Press, New York, p. 189-221.

Frost, L.S., B.B. Finlay, A. Opgenorth, W. Paranchych and J.S. Lee. 1985. Characterization and sequence analysis of pilin from F-like plasmids. J.Bacteriol. **164**: 1238–1247.

Frost, L.S., W. Paranchych and N.S. Willetts. 1984. DNA sequence of the F *traALE* region that includes the gene for F pilin. J.Bacteriol. **160**: 395–401.

Gamas, P., L. Caro, D. Galas and M. Chandler. 1987. Expression of F transfer functions depends on the *E.coli* Integration Host Factor. Mol.Gen.Genet. **207**: 302–305.

Guiney, D.G. 1982. Host range of conjugation and replication functions of the *Escherichia coli* sex plasmid F′ *lac*: Comparison with the broad-host-range plasmid RK2. J.Mol.Biol. **162**: 699–703.

Guiney, D.G. and E. Lanka. 1989. Conjugative transfer of IncP plasmids, p. 2756. In C.M. Thomas, (ed.) Promiscuous Plasmids of Gram-Negative Bacteria. Academic Press, London.

Haluzi, H., D. Goitein, S. Koby, I. Mendelson, D. Teff, G. Mengeritsky, H. Giladi and A.B. Oppenheim. 1991. Genes coding for Integration Host Factor are conserved in Gram-negative bacteria. J.Bacteriol. **173**: 6297–6299.

Harwood, C.R. and E. Meynell. 1975. Cyclic AMP and the production of sex pili by *E. coli* K-12 carrying derepressed sex factors. Nature **254**: 628–630.

Havekes, L.M., W. Hoekstra and H. Kempen. 1977. Relation between F, R1, R100 and R144 *Escherichia coli* K-12 donor strains in mating. Mol.Gen.Genet. **155**: 185–189.

Havekes, L.M., B.J.J. Lugtenberg and W.P.M. Hoekstra. 1976. Conjugation deficient *E. coli* K12 F- mutants with heptose-less lipopolysaccharide. Mol.Gen.Genet. **146**: 43–50.

Heinemann, J.A. and G.F. Sprague,Jr.. 1989. Bacterial conjugative plasmids mobilize DNA transfer between bacteria and yeast. Nature **340**: 205–209.

Hunkapiller, M., E. Lujan, F. Ostrander and L. Hood. 1983. Isolation of microgram quantities of proteins from polyacrylamide gels for amino acid sequence analysis. Methods Enzymol. **91**: 227–236.

Ippen-Ihler, K. and S. Maneewannakul. 1991. Conjugation among enteric bacteria: Mating systems dependent on expression of pili, p. 35–69. In M. Dworkin, (ed.) Microbial cell-cell interactions. American Society for Microbiology, Washington, D.C..

Ippen-Ihler, K. and R.A. Skurray. 1993. Genetic organization of transfer-related determinants on the sex factor F and related plasmids, *In:* D. Clewell (ed.) Bacterial Conjugation. Plenum Press, New York, p. 23–52.

Krahn, P.M., R.J. O'Callaghan and W. Paranchych. 1972. Stages in phage R17 infection. VI. Injection of A protein and RNA into the host cell. Virology. **47**: 628–637.

Kumar, S. and S. Srivastava. 1983. Cyclic AMP and its receptor protein are required for expression of transfer genes of conjugative plasmid F in *Escherichia coli*. Mol.Gen.Genet. **190**: 27–34.

Manoil, C. 1983. A genetic approach to defining the sites of interaction of a membrane protein with different external agents. J.Mol.Biol. **169**: 507–519.

Manoil, C. and J.P. Rosenbusch. 1982. Conjugation-deficient mutants of *Escherichia coli* distinguish classes of functions of the outer membrane OmpA protein. Mol.Gen.Genet. **187**: 148–156.

Miller, J., E. Lanka and M. Malamy. 1985. F-factor inhibition of conjugal transfer of broad-host-range plasmid RP4: Requirement for the protein product of *pif* operon regulatory gene *pifC.*. J.Bacteriol. **163**: 1067–1073.

Morona, R., M. Klose and U. Henning. 1984. *Escherichia coli* K-12 outer membrane protein (OmpA) as a bacteriophage receptor: Analysis of mutant genes expressing altered proteins. J.Bacteriol. **159**: 570–578.

Novotny, C.P. and K. Lavin. 1971. Some effects of temperature on the growth of F pili. J.Bacteriol. **107**: 671–682.

Palombo, E.A., K. Yusoff, V.A. Stanisch, V. Krishnapillai and N.S. Willetts. 1989. Cloning and genetic analysis of *tra* cistrons of the Tra2/Tra3 region of plasmid RP1. Plasmid **22**: 59–69.

Paranchych, W. and L.S. Frost. 1988. The physiology and biochemistry of pili. Adv.Microb.Physiol. **29**: 53–114.

Parker, C.T., A.W. Kloser, C.A. Schnaitman, M.A. Stein, S. Gottesman and B.W. Gibson. 1992. Role of the *rfaG* and *rfaP* genes in determining the lipopolysaccharide core structure and cell surface properties of *Escherichia coli* K-12. J Bacteriol. **174**: 2525–2538.

Pradel, E. and C.A. Schnaitman. 1991. Effect of *rfaH* (*sfrB*) and temperature on expression of *rfa* genes of *Escherichia coli* K-12. J Bacteriol. **173**: 6428–6431.

Rehemtulla, A., S.K. Kadam and K.E. Sanderson. 1986. Cloning and analysis of the *sfrB* (Sex Factor Repression) gene of *Escherichia coli.*. J.Bacteriol. **166**: 651–657.

Reiner, A. 1974. *Escherichia coli* females defective in conjugation and in adsorption of a single-stranded deoxyribonucleic acid phage. J.Bacteriol. **119**: 183–191.

Schnaitman, C.A., C.T. Parker, J.D. Klena, E.L. Pradel, N.B. Pearson, K.E. Sanderson and P.R. MacClachlan. 1991. Physical maps of the *rfa* loci of *Escherichia coli* K-12 and *Salmonella typhimurium*. J Bacteriol. **173**: 7410–7411.

Schweizer, M. and U. Henning. 1977. Action of a major outer cell envelope membrane protein in conjugation of *Escherichia coli* K-12. J.Bacteriol. **129**: 1651–1652.

Silverman, P.M., S. Rother and H. Gaudin. 1991. Arc and Sfr functions of the *Escherichia coli* K-12 *arcA* gene product are genetically and physiologically separable. J Bacteriol. **173**: 5648–5652.

Skurray, R.A., R.E.W. Hancock and P. Reeves. 1974. Con⁻ mutants: Class of mutants in *Escherichia coli* K-12 lacking a major cell wall protein and defective in conjugation and adsorption of a bacteriophage. J.Bacteriol. **119**: 726–735.

Thomas, C.M. and D.R. Helinski. 1989. Vegetative replication and stable inheritance of IncP plasmids, p. 1–25. In C.M. Thomas, (ed.) Promiscuous plasmids of Gram-negative bacteria. Academic Press, London.

Webster, R.E. 1991. The *tol* gene products and the import macromolecules into *Escherichia coli*. Mol. Microbiol. **5**: 1005–1011.

Note added in proof

The sequence of Tra2 of RP4 has been reported in Lessl *et al.*, J. Biol. Chem. **267**: 20471–20480 (1992).

5. Fimbria (pilus) mediated attachment of *Pseudomonas syringae, Erwinia rhapontici* and *Xanthomonas campestris* to plant surfaces

MARTIN ROMANTSCHUK, ELINA ROINE, TUULA OJANEN, JOOP VAN DOORN, JARI LOUHELAINEN, EEVA-LIISA NURMIAHO-LASSILA and KIELO HAAHTELA

Abstract. The foliar surfaces of plants are colonized by both more or less opportunistically pathogenic, as well as non-pathogenic epiphytes. Since plant specific bacteria have received more attention only in association with disease or forms of symbiosis where the bacteria have a clearly distinguishable effect on the plant, relatively little is known about the plant-bacterial surface interactions both as a factor in colonization and as an early step in disease induction. Bacterial structures involved in such surface interactions have, however, been described in a number of cases.

Strains of the bacterial plant pathogens *Pseudomonas syringae, Xanthomonas campestris* and *Erwinia rhapontici* express fimbriae or pili that function as adhesins mediating attachment of bacterial cells to plant surfaces. *P. syringae* pathovar phaseolicola and *X. campestris* pathovar hyacinthi appear to attach preferentially to stomata whereas *P. syringae* pathovar syringae is found evenly distributed over the leaf surface. *Erwinia rhapontici* attach to wheat grains and rhubarb leaves without apparent site specificity on the plant surface. Adhesion of *E. rhapontici* is mediated by N-acetyllactosamine specific fimbriae. Adhesion to rhubarb and wheat surfaces as well as production of pink wheat grains is inhibited by lactose functioning as a receptor analog. Non-piliated mutants of *Pseudomonas syringae* lose their ability to attach which may affect the capacity to cause disease in certain conditions. Binding is restored upon complementation of the mutation with isolated genes. The *Xanthomonas campestris* pathovar hyacinthi expresses type-4 fimbriae constitutively. In pathovar vesicatoria expression is variable, apparently induced in certain growth phases as well as *in planta*.

Abbreviations: EPS, Extracellular Polysaccharides

Epiphytic fitness of plant pathogenic bacteria

Many bacterial plant pathogens grow on surfaces of plants without inducing disease symptoms. Gram-negative bacteria in this ecological niche include pathogenic *Pseudomonas, Xanthomonas*, and *Erwinia* species, and may thus be considered facultative pathogens.

Such facultatively phytopathogenic bacteria have received more attention as pathogens than as epiphytes, but the growth of virulent and nonvirulent variants of these bacteria on the surfaces of healthy host and nonhost plants has also been studied because of the interest in the development and spread of disease inoculum (reviewed by Hirano & Upper, 1983). Epiphytic colonization ability of a virulent strain – traits affecting population size, persistence, and ability to survive in various environmental conditions – is likely to correlate positively with successful invasion of the plant when conditions favor disease induction.

C.I. Kado and J.H. Crosa (eds.), Molecular Mechanisms of Bacterial Virulence, 67–77.
© 1994 *Kluwer Academic Publishers.*

Several plant related bacteria, including strains of the plant pathogens *Pseudomonas syringae* and *Xanthomonas campestris* and the nonpathogenic species *Erwinia herbicola* possess an ice nucleation activity (reviewed by Lindow 1983), causing frost damage to sensitive plants at relatively high temperatures. This phenomenon underlines the potential importance of epiphytic symptomless bacterial growth.

Bacteria in a wide variety of environments tend to attach to solid surfaces. In most environments this is a selective advantage permitting the bacterial cells to colonize nutritionally and environmentally favourable locations. This is the case for e.g. animal pathogens (Jann & Jann, 1990) and aquatic bacteria (Fletcher, 1986). Attachment of plant-specific bacteria has most extensively been studied in the case of pathogenicity of *Agrobacterium* and the symbiotic interaction of *Rhizobium* with the host plant. In both cases attachment to a host plant has been shown to be an essential initial step in invasion. A wounded plant surface is required for tumor formation by *Agrobacterium tumefaciens*. Attachment of bacteria to exposed plant cells is then an early and necessary step, preceeding tranfer of the T-DNA of the tumour inducing Ti-plasmid to the plant cell. Non-attaching mutants are avirulent, but the function and relative importance of bacterial genes and factors involved in the attachment process has not been clearly established (see Smit and Stacey, 1990; Romantschuk, 1992 for discussion). In *Rhizobium leguminosarum* a bacterial protein called rhicadhesin was shown to be required for attachment to root hairs and nodule formation (Smit *et al.* 1989). Among the nodulating bacteria *Rhizobium* and *Bradyrhizobium* also other structures, such as pili/fimbriae and cellulose fibrills have been shown to take part in the attachment process (Smit *et al.* 1987; Vesper & Bauer, 1986), but they appear not to be required for nodulation.

In the case of foliar pathogens that apparently also grow as epiphytes the role of attachment is not as clear. After invasion of the plant tissue pathogenic bacteria that can be seen with electron microscopy entrapped in close contact with the plant cell wall are no longer actively growing (Hildebrand *et al.* 1980), but in the epiphytic growth phase, or during initial plant-bacterium interactions, the situation is likely to be different. In the case of epiphytically growing bacteria plant surface adhesion (resistance to displacement by rain and wind) has received relatively little attention.

Traits that have been considered as epiphytic fitness factors include drought and UV-tolerance, motility, and production of various exported compounds such as extracellular polysaccharides (EPS), siderophores, and antibiotics, etc. (Lindow, 1991), but also attachment deserves in our view a place in this list.

In most cases the putative plant receptors or structural components that take part in the attachment process are unknown. If the interaction is based on hydrophobicity, target plant-surface specificity may not play a significant role. However, specific inhibition of binding of bacterial cells or a bacterial adhesin to target plant surfaces, has been described, indicating binding-specificity in at least some plant-microbe interactions (reviewed by Romantschuk, 1992; Smit & Stacey, 1990; and see below).

Bacterial fimbriae and pili

A bacterial structure that is able to bridge the distance between a bacterial cell and a solid surface such as a plant leaf is exemplified by the bacterial fimbriae (pili). Indeed these structures have been shown to mediate adhesion of bacteria to various biological and other surfaces. The filaments are apparently in many cases hydrophobic, suggesting a rather nonspecific hydrophobic interaction with the plant surface wax. On the other hand, specific interactions between the fimbrial adhesin of a plant-specific bacterium and a carbohydrate group have been reported in several cases (Korhonen *et al.* 1987; 1988; Haahtela *et al.* 1989).

Fimbriae or pili are synthesized by a large proportion of bacteria in apparently all environments, including plant-specific bacteria (Romantschuk, 1992). Thus, fimbriae have been observed by electron microscopy or by other means in plant specific strains of at least *Pseudomonas solanacearum, P. syringae, P. savastanoi, Agrobacterium tumefaciens, Rhizobium trifolii, Bradyrhizobium japonicum, Xanthomonas campestris* and *Erwinia* (Stemmer & Sequeira, 1987; Romantschuk & Bamford, 1985; Haahtela *et al.* 1989; Korhonen *et al.* 1987; 1988; van Doorn *et al.*, 1991).

The fimbriae or pili and their suggested role for the plant pathogens *Pseudomonas syringae, Xanthomonas campestris* and *Erwinia rhapontici* is discussed below.

Pseudomonas syringae

About 60% of all *P. syringae* pathovars and strains that we tested are sensitive to bacteriophage $\phi6$ (Romantschuk & Bamford, 1985; Romantschuk, unpublished). The pili produced by these strains function as the primary $\phi6$ phage-receptor. Phage-resistant mutant strains were isolated and found in most cases to be either nonpiliated and thus unable to adsorb phages, but some resistant strains expressed a higher number of phage-adsorbing pili. The latter form of mutant was suggested to be unable to retract the pilus resulting in the phage-resistant phenotype (Romantschuk & Bamford, 1985). Piliated strains, both phage sensitive and resistant ("super-piliated"), adsorbed more efficiently to the surface of plant leaves in an assay monitoring radioactivity associated with plant leaf disks after incubation with radioactively labelled bacteria. In particular the strains producing more pili than the wild type, as evidenced by their phage-adsorption ability, attached to plant leaf surfaces with an efficiency that was several times higher than the respective nonpiliated mutant (Romantschuk & Bamford, 1986; Rantala *et al.* 1992). This was also the case for nonpiliated transposon-mutants complemented with the wild type gene carried in a cosmid. In this case the higher expression of pili may result from a higher gene dosage of the complementing genes (Roine, unpublished).

Attachment of *P. syringae* to the plant surface was not host specific. Different $\phi6$ sensitive pathovars of *P. syringae* adsorb about equally well to

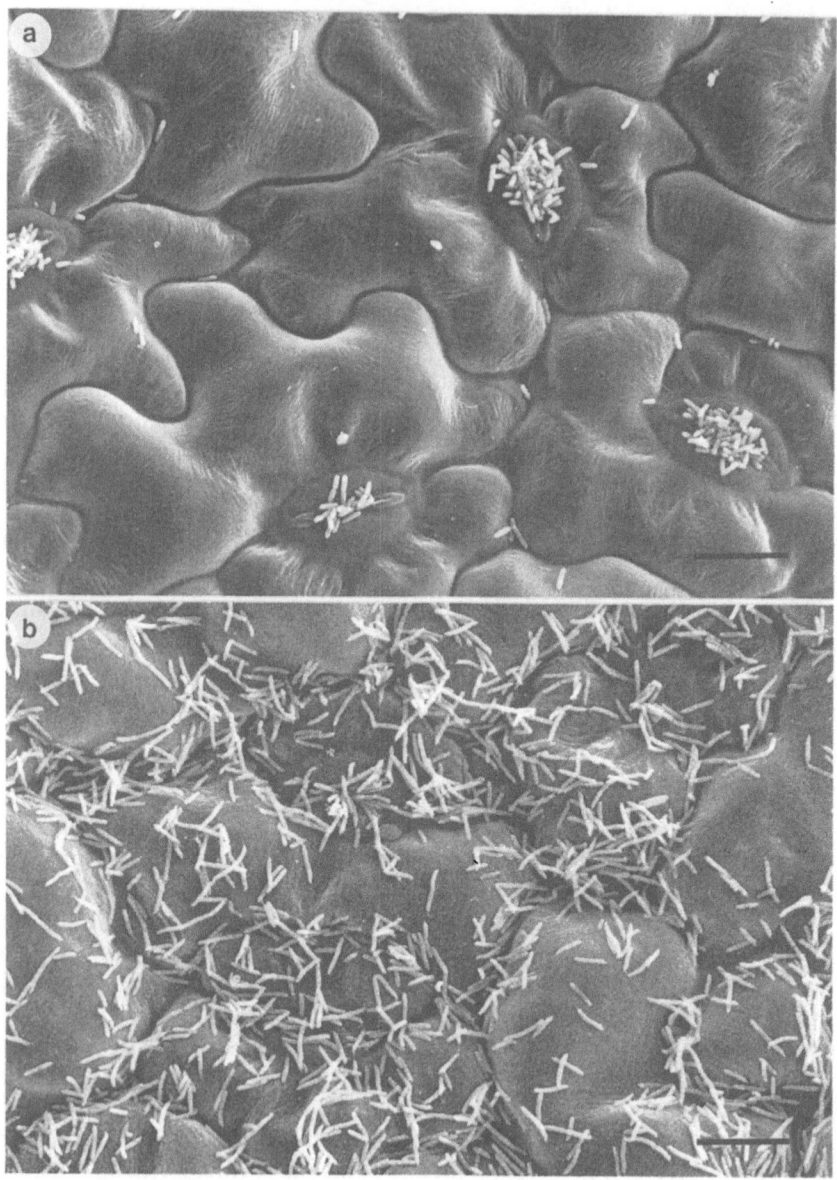

Figure 1. Scanning electron micrographs of *Pseudomonas syringae* on the lower surface of bush bean prepared as in Nurmiaho-Lassila *et al.* 1991. a) *P. s.* pv. phaseolicola HB10Y adhered to stomata; b) *P. s.* pv. syringae R32 evenly distributed on the leaf surface. Bar 10 μm.

both the susceptible host and to nonhost plant leaves. The different wild type strains differed considerably, however, in their leaf adsorption efficiency. For all except one of the tested pathovars leaf adsorption was significantly lower for nonpiliated mutants unable to adsorb the phage than for the wildtype or for superpiliated strains (Korhonen *et al.*, 1986; Rantala *et al.* 1992).

When inoculation of bean plants with *P. s.* pv. phaseolicola was done by spraying bacterial suspension followed by spraying with buffer the piliation also correlated positively with frequency of infection (Romantschuk & Bamford, 1986). No difference in virulence was observed, however, when the different piliation mutants of *P. s.* pv. phaseolicola were injected into bean leaves (Romantschuk & Bamford, 1986). The pilus can thus be considered as a conditional virulence factor, but it is probably more directly a factor in epiphytic colonization ability.

In a plant-bacterium adsorption assay the attachment efficiency was not affected by changing the ionic strength of the adsorption medium, nor were divalent cations required for adhesion (Romantschuk, *et al.*, 1993). Although binding receptors have not been identified, the site of attachment on the plant surface appears to be, in certain cases, specifically localized. The *P. s.* pv. phaseolicola strain used adheres preferentially to the stomata of bean (Romantschuk & Bamford, 1986) whereas some bean- and corn-pathogenic pv. syringae strains are seen evenly distributed over the leaf surface of both the susceptible host plant and non-host plants (Fig. 1, Korhonen *et al.*, 1986; Nurmiaho-Lassila, *et al.*, 1991).

Variants expressing a different degree of piliation were isolated for the *P. syringae* pv. phaseolicola strain HB10Y and pathovar syringae strain R32. In static broth cultures only the piliated strains of the motile strain R32 grow as a pellicle at the air-liquid interphase. Nonpiliated mutants of R32 form no pellicle and are not able to grow in the static liquid culture. None of the differently piliated nonmotile *P. s.* pv. phaseolicola strains were able to grow as air-liquid interphase pellicles. *P. syringae*, being an obligate aerobe is able to grow only close to the surface of an unaerated broth culture. The results suggest that bacterial flagella are required for accumulation of bacterial cells at the air-water interphase, and that the pili are required for maintaining the cells at the surface – neither factor alone is sufficient.

Xanthomonas campestris

Like *P. syringae, Xanthomonas campestris* can be found growing epiphytically without causing visible symptoms. In suitable conditions bacteria invade the plant tissue through natural leaf openings such as stomata and hydathodes, as well as through wounds. Over one hundred pathovars infecting a wide variety of plants have been isolated. Pathovar hyacinthi causes yellow disease in *Hyacinthus*, whereas pv. vesicatoria causes bacterial spot disease in tomato and pepper.

Fimbriae have been observed in strains of pathovars hyacinthi (van Doorn *et al.* 1991) and vesicatoria. In static bacterial liquid cultures of both pathovars a firm pellicle is formed at the air-liquid interphase. Both in the pellicle and on solid media pathovar hyacinthi cells are heavily fimbriated as seen in

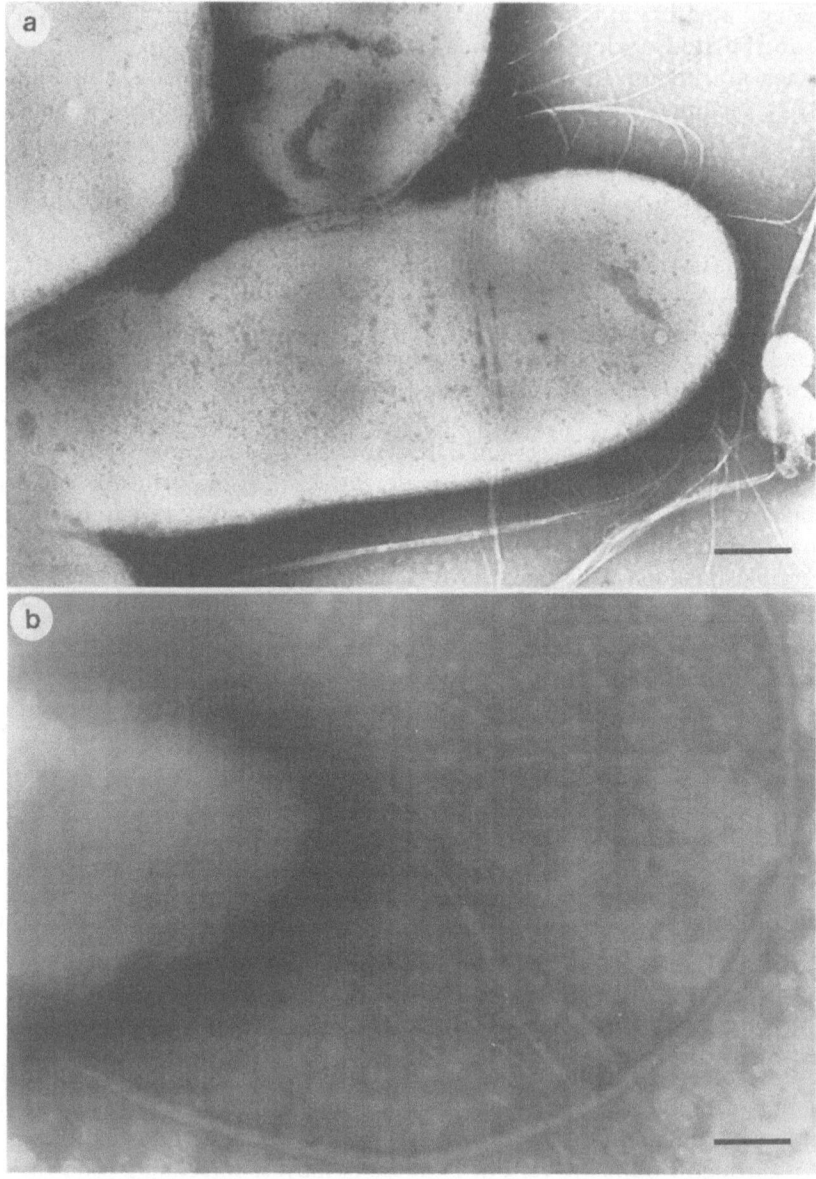

Figure 2. Negative staining electron micrographs of fimbriated *Xanthomonas campestris* cells . a) pv. vesicatoria (2% phosphotungsitc acid; Bar, 200 nm); b) pv. hyacinthi (1% ammonium molybdate; Bar, 100 nm). The flexible fimbriae form bundels, whereas the thick single filament in b) is the bacterial flagellum.

negative staining electron microscopy. Pathovar hyacinthi fimbriae are readily isolated and purified by acid precipitation from bacteria grown on agar plates, whereas vesicatoria fimbriae, showing a lower expression level, are difficult to purify. Expression of fimbriae in pv. hyacinthi appears constitutive, whereas in pv. vesicatoria fimbria production is dependent on growth phase so that fimbria are expressed at late exponential growth phase in shaken cultures or by adding plant material to the growth media. In both pathovars the long flexible fimbriae form bundles (Fig. 2).

The amino-terminus of the isolated hyacinthi fimbrillin shows homology with so called type-4 fimbriae (van Doorn *et al.* 1991) expressed by *Pseudomonas aeruginosa, Moraxella bovis,* and *Neisseria gonorrhoeae* (Dalrymple *et al.* 1987). In preliminary immunological experiments hyacinthi antiserum crossreacted with a vesicatoria fimbrillin preparation, suggesting that the vesicatoria fimbriae are also of type-4.

Pathovar hyacinthi fimbriae as well as the whole cells attach to the stomata of *Hyacinthus* (van Doorn *et al.* 1991), suggesting a role of the fimbriae in plant infection. The equivalent studies have not been done with pathovar vesicatoria. Although bacterial cells and pure *X. c.* pv. hyacinthi fimbriae bind efficiently to the host plant leaf surface a definite role for the fimbriae in plant-bacterial interaction has not been confirmed. Nonfimbriated mutants of the *Xanthomonas* strains are not available, and receptor analogs inhibiting binding have not been identified.

Erwinia rhapontici

Erwinia rhapontici causes crown rot in rhubarb *(Rheum rhaponticum)*, pink grains in *Triticum aestivum,* and browning in *Hyacinthus* sp. Sixteen *E. rhapontici* strains isolated from various plant materials were assayed for haemagglutination. All strains possessed an N-acetyllactosamine-specific haemagglutination activity (Table 1; Korhonen *et al.* 1988).

Table 1. Inhibition of hemagglutination of human O-erythrocytes by *Erwinia rhapontici* strain 139.

Inhibitor	MIC [1](mM)	
	Bacteria	Purified fimbriae
N-acetyllactosamine	0,07	0,1
Lactose	0,15	1,6
Galactose	1	3,1
Glucose	>75	>100
Sucrose	>75	>100

[1] MIC = Minimal inhibitory concentration to prevent hemagglutination

Purified fimbriae from *E. rhapontici* caused a similar lactose-inhibitable haemagglutination, and all strains reacted with an antiserum raised against purified fimbriae. The N-terminal sequence of the 18 kD *E. rhapontici*

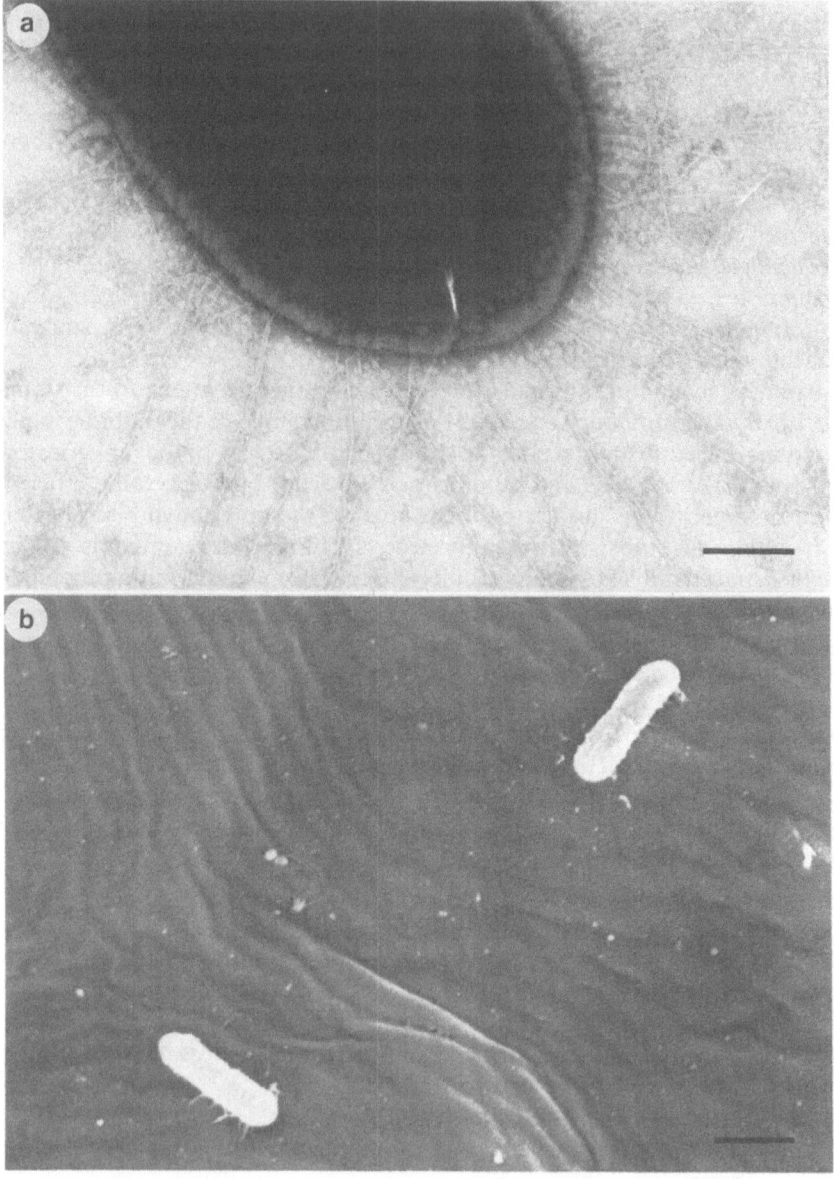

Figure 3. Electron micrographs of *Erwinia rhapontici*. a) Negative staining. Fimbriae visible as single filaments extending from the cell (Bar, 200nm); b) scanning electron micrograph of cells on the surface of rhubarb leaf. Fimbriae seen in contact with the plant surface (Bar, 1 μm).

fimbrillin protein shows some homology to *E. carotovora* type-1 fimbrillin but the tryptic peptide map of the two fimbrillins differed completely, and no immunological cross-reaction was observed between antifimbria sera

(Korhonen *et al.* 1987). The *E. rhapontici* type of fimbriae was not detected in *Erwinia carotovora* subsp. carotovora, *E.c.* atroseptica, *E.c.* betavasculorum, or *E. chrysanthemi* (Louhelainen *et al.* 1992). As with *E. coli* type-1 fimbriae the actual adhesin that determines the attachment specificity of *E. rhapontici* may be a minor protein at the tip of the filament. No observation of such a protein has, however, been made. The *E. rhapontici* fimbriae extend from the cell surface as relatively rigid single filaments (Fig. 3a). In scanning electron microscopy *E. rhapontici* cells can be seen with the fimbriae apparently in contact with the rhubarb leaf surface (Fig. 3b).

Erythrocyte haemagglutination was used to identify the receptor structure for bacterial binding. Inhibition of hemagglutination was achieved with N-acetyl-lactosamine and other β-galactosides (Korhonen *et al.*, 1988). This haemagglutination mediates bacterial attachment to plant surfaces as measured by binding of radioactively labelled cells to wheat grains or to rhubarb leaves. Lactose inhibited binding by 60–80% compared to sucrose (Louhelainen *et al.*, 1992).

Wheat grains were also infected with *E. rhapontici* in the presence of adhesion-inhibiting and non-inhibiting sugars followed by rinsing with buffer. Appearance of pink grains after a 10 day incubation period was reduced when the infection was done in the presence of lactose (25% infected grains) compared to the control where the infection was done in the presence of sucrose that does not inhibit adhesion of bacterial cells (75% infected grains).

In conclusion the results suggest that a galactoside-containing structure of the plant cell surface functions as the receptor for fimbrial adhesion of *E. rhapontici*, and that this adhesion may play a role in disease induction and/or surface colonization.

Conclusions

Among pathogenic or symbiotic interactions of bacteria with their plant hosts the importance of attachment has been established only in certain cases where attachment is absolutely required for invasion. Thus, *Agrobacterium* and *Rhizobium* mutants unable to attach are noninvasive. In the case of the bacteria described here, confirmation of the role of fimbriae or pili, as well as attachment in general awaits further studies. The attachment described takes place on the outer surface of the target plants, suggesting that the phase where the fimbrial adhesins are important is during the epiphytic growth phase, but possibly also in initial stages of plant disease.

If the capacity to bind to plant surfaces only represents an ecological fitness factor its importance may remain undetected in experiments performed under laboratory conditions. No direct evidence exists showing that adhesion provides a selective advantage for bacteria during epiphytic growth. As yet field or microcosm growth experiments comparing the epiphytic colonization and survival properties of adhesion negative mutants and the wild type parental strains have not been performed.

76

When designing bacteria to compete with the natural microbial population of plant surfaces for various biocontrol purposes, such as protection against frost injuries caused by ice nucleation active bacteria, or disease caused by epiphytic opportunistic pathogens, success will depend on comprehensive knowledge of the factors relevant to bacterial colonization of this ecosystem. Thus, modification of the epiphytic bacterial flora may be facilitated if the biocontrol agent is optimal in as many epiphytic fitness traits as possible, including adhesion properties. Furthermore, in order to replace the deleterious (pathogenic) bacterial population rather than coexisting with it, the biocontrolling organism should compete for the epiphytical resources, including the sites being occupied on the plant surface. In this context the knowledge of the bacterial adhesins and their binding specificity is likely to be important.

Acknowledgements

This work was supported in part by the Foundation for Biotechnology and Fermentation Research (MR, ER), the Finnish Ministry of Agriculture and Forestry (MR, KH), the Academy of Finland (JL), and Suomen Kulttuurirahasto (TO).

References

Dalrymple B & Mattick JS (1987) An analysis of the organization and evolution of type 4 fimbrial (MePhe) subunit proteins. J Mol Evol 25: 261–269.

Fletcher M (1986) Measurement of glucose Utilization by *Pseudomonas fluorescens* that are free-living and that are attached to surfaces. Appl. Environ. Microbiol. 52: 672–676.

Haahtela K, Kukkonen M, Rönkkö R, Nurmiaho-Lassila EL, Karjalainen R & Korhonen TK (1989) Fimbriae in adhesion of *Erwinia carotovora* subsp. *carotovora* to plant surfaces. *In*: Klement Z (ed.) Proc. 7th Int. Conf. Plant Path. Bact. Part B (pp. 767–772), Budapest, Hungary.

Hildebrand DC, Alosi MC & Schrot MN (1980) Physical entrapment of pseudomonads in bean leaves by films of air-water interphases. Phytopathology 70: 98–109.

Hirano SS & Upper CD (1983) Ecology and epidemiology of foliar bacterial plant pathogens. Annu. Rev. Phytopathol. 21: 243–269.

Jann K & Jann B (1990) Bacterial adhesins. Current Topics in Microbiology and Immunology. Vol. 151. Springer Verlag. Berlin Heidelberg.

Korhonen TK, Haahtela K, Pirkola A & Parkkinen J (1988) A N-acetyllactosamine-specific cell-binding activity in a plant pathogen, *Erwinia rhapontici*. FEBS Letters 236: 163–166.

Korhonen TK, Haahtela K, Romantschuk M & Bamford DH (1986) Role of fimbriae and pili in the attachment of *Klebsiella, Enterobacter*, and *Pseudomonas* to plant surfaces, p. 229–241. *In*: Lugtenberg B (ed.) Recognition in microbe-plant symbiotic and pathogenic interactions. Springer-Verlag, Heidelberg.

Korhonen TK, Kalkkinen N, Haahtela K, & Old DC (1987) Characterization of type 1 and mannose-resistant fimbriae of *Erwinia* ssp. J.Bacteriol. 169: 2281–2283.

Lindow SE (1983) The role of bacterial ice nucleation in frost injury to plants. Annu. Rev. Phytopatol. 21: 363–384.

Lindow SE (1991) Determinants of epiphytic fitness in bacteria. p. 295–315. *In*: Andrews JH & Hirano SS (ed.) Microbial ecology of leaves. Springer Verlag, New York.

Louhelainen J, Haahtela K, Lindroos O, Nurmiaho-Lassila EL & Korhonen TK (1992). Adhesion of *Erwinia rhapontici* to plant surfaces. 6th International Symposium on the Molecular Plant-Microbe Interactions (Abstract). July 11–16. 1992, Seattle, Washington.

Nurmiaho-Lassila EL, Rantala E & Romantschuk M (1991) Pilus-mediated adsorption of *Pseudomonas syringae* to the surface of bean leaves. Micron Microscopica Acta. 22: 71–72.

Rantala E Nurmiaho-Lassila EL & Romantschuk M (1992) Cloning and characterization of genes involved in pilusproduction of *Pseudomonas syringae* pv. *phaseolicola*. 6th International Symposium on Molecular Plant Microbe Interactions. July 11–16. 1992, Seattle, Washington.

Romantschuk M (1992) Attachment of plant pathogenic bacteria to plant surfaces. Annu. Rev. Phytopathol. 30: 225–243.

Romantschuk, M & Bamford DH (1985) Function of pili in bacteriophage φ6 penetration. J. Gen. Virol. 66: 2461–2468.

Romantschuk M & Bamford DH (1986) The causal agent of halo blight in bean, *Pseudomonas syringae* pv. *phaseolicola*, attaches to stomata via its pili. Microb. Pathogen. 1: 139–148.

Romantschuk M, Nurmiaho-Lassïla El, Roine E & Suoniemi A (1993) Pilus-mediated adsorption of *Pseudomonas syringae* to the surface of host and non-host plant leaves. J. Gen. Microbiol. 139 (in press).

Smit G, Kijne JW & Lugtenberg BJJ (1987) Both cellulose fibrils and Ca^{2+}-dependent adhesin are involved in the attachment of *Rhizobium leguminosarum* to pea root hair tips. J. Bacteriol. 169: 4292–4301.

Smit G, Logman TJJ, Boerrigter MTEI, Kijne JW & Lugtenberg BJJ (1989) Purification and partial characterization of the Ca^{++}-dependent adhesin from *Rhizobium legominosarum* biovar *viciae*, which mediates the first step in attachment of *Rhizobiaceae* to plant root hair tips. J. Bacteriol. 171: 4054–4062.

Smit G & Stacey G (1990) Adhesion of bacteria to plant cells: role of specific interaction versus hydrophobicity. p. 179–210. *In:* Doyle RJ & Rosenberg M (ed.) Microbial Cell Surface Hydrophobicity. American Society for Microbiology, Washington.

Stemmer WPC & Sequeira L (1987) Fimbriae of phytopathogenic and symbiotic bacteria. Phytopathology 77: 1633–1639.

Vesper SJ & Bauer WD (1986) Role of pili (fimbriae) in attachment of *Bradyrhizobium japonicum* to soybean roots. Appl. Environ. Microbiol. 52: 134–141.

Van Doorn J, Boonekamp PM & Oudega B (1991) Characterization of fimbriae, expressed by the plant pathogenic bacterium *Xanthomonas campestris* pathovar *hyacinthi*. In Abstracts from the FEMS Symposium: Molecular recognition in host-parasite interactions. August 5–7 1991. Porvoo, Finland.

6. Attachment of *Agrobacterium tumefaciens* to host cells

ANN G. MATTHYSSE and VINCENT T. WAGNER

Abstract. Attachment of *A. tumefaciens* to host cells is a two step process. In the first step the bacteria bind loosely to a receptor on the surface of the plant cell. In the second step, substances from the plant induce the elaboration of cellulose fibrils by the bacteria. These fibrils bind the bacteria tightly to the plant cell surface. They also cause the formation of bacterial aggregates.

    A. tumefaciens binds to cells of many species of plants. The receptor on the plant cell surface to which the bacteria bind is sensitive to proteases. Bacterial binding to carrot cells can be inhibited by the addition of vitronectin or antivitronectin antibodies. Wild type bacteria bind human vitronectin. Non-attaching mutants show reduced binding of vitronectin. These results suggest that the receptor on the plant cell surface is a vitronectin-like protein.

Abbreviations: T-DNA, Transferred DNA

I. Introduction

Agrobacterium tumefaciens is a Gram negative bacterium found in soil with a world-wide distribution. Infections of wound sites on dicotyledonous plants with this bacterium result in the formation of tumors. Since the wound sites most often infected are those near the surface of the soil (the crown of the plant) the disease caused by this bacterium is called crown gall. This disease is particularly a problem on plants which become injured through cultivation practices such as grafting or pruning or by frost damage since these wound sites easily become contaminated with soil and infected with the bacteria.

    The mechanism by which *A. tumefaciens* and other members of the genus *Agrobacterium* cause tumors is, so far as is known, unique. It is a naturally occurring case of the transfer of DNA between organisms which belong to different kingdoms. The bacteria contain a large plasmid (the Ti plasmid). A piece of this plasmid (the T-DNA) is transferred from the bacterium to the host cell where it becomes integrated into the host chromosome(s) (Chilton *et al.*, 1977). Genes in this DNA have constitutive eukaryotic promoters. They include genes for enzymes involved in the synthesis of the plant growth hormones auxin and cytokinin and genes for enzymes involved in the synthesis and secretion of opines (compounds made only in crown gall tumor cells which are metabolized by the bacteria and serve as carbon and/or nitrogen sources). It is the

C.I. Kado and J.H. Crosa (eds.), Molecular Mechanisms of Bacterial Virulence, 79–92.
© 1994 *Kluwer Academic Publishers.*

unregulated synthesis of growth hormones which results in the formation of a tumor. The advantage to the bacterium of this complex infection process appears to be the provision to the bacteria by the plant cell of opines which are used by the bacteria (Kado, 1991; Ream, 1989).

It is not surprising that this pathogenic mechanism requires that the bacteria attach to the host cells before the transfer of the T-DNA occurs. Once the bacteria have bound to plant cells, genes located in the *vir* region of the Ti plasmid are required for the actual DNA transfer. These *vir* genes are induced by phenolic compounds such as acetosyringone released by the plant at wound sites. They include *vir*D genes which encode an endonuclease which cuts the T-DNA at 24 base border repeats, *vir*E2 which encodes a single stranded DNA binding protein, and other *vir* genes that encode various other proteins of unknown function. The mechanism of DNA transfer between the bacterium and the plant cell is thought to be similar to the mechanism of DNA transfer between bacteria during bacterial conjugation (Ream, 1989).

II. Description and measurement of bacterial attachment

Although in nature the bacteria attach to cells exposed in wound sites, this is an inconveniently complex system to use to study bacterial attachment. Most studies of attachment of *A. tumefaciens* to host cells have used suspension culture cells which provide a relatively uniform population of cells to which additions can be made at known concentrations and times (Douglas *et al.*, 1982; Matthysse *et al.*, 1978). Some studies have used callus cultures or cells released from root caps (Deasey & Matthysse, 1984; Hawes & Pueppke, 1987). Bacteria binding to these cells can be observed in the light or scanning electron microscopes. Numbers of bacteria bound to suspension culture cells can be determined by separating the free bacteria from the plant cells and bound bacteria by filtration using Miracloth or Whatman No. 1 filter paper which have pore sizes which permit the passage of free bacteria but not of plant cells. The number of free and attached bacteria can then be measured by viable cell counts or, if the bacteria were radioactively labelled, by measuring the radioactivity in a scintillation counter.

When the bacteria are added to plant cell cultures they are observed to begin binding to the surface of the plant cells within 10 to 30 min. The number of bacteria attached after various times of incubation was examined and it was observed that the bacteria attach rapidly for the first 20 to 40 min. After this time the numbers of attached bacteria increase at a slower rate over the next few hours, but the attachment never saturates and longer incubation times up to several hours continue to result in an increase in the number of attached bacteria (Matthysse *et al.*, 1978).

In the scanning electron microscope it can be seen that initially bacteria attach individually directly to the surface of the plant cell. After longer incubation times the plant cell surface is observed to be covered with a network of fibrillar material. These fibrils wind around the bacteria on the plant cell surface. In addition large aggregates of bacteria are seen which appear to be trapped in the fibrils and are only indirectly attached to the host cell surface (Fig. 1). It is probably these indirectly attached bacteria which account for the continued increase in the total number of attached bacteria with longer times of incubation. If the bacteria are left with the suspension culture plant cells overnight, the entire plant cell culture becomes one large aggregate held together by bacteria and bacterial fibrils. These aggregates are easily visible to the unaided eye (Matthysse *et al.*, 1981).

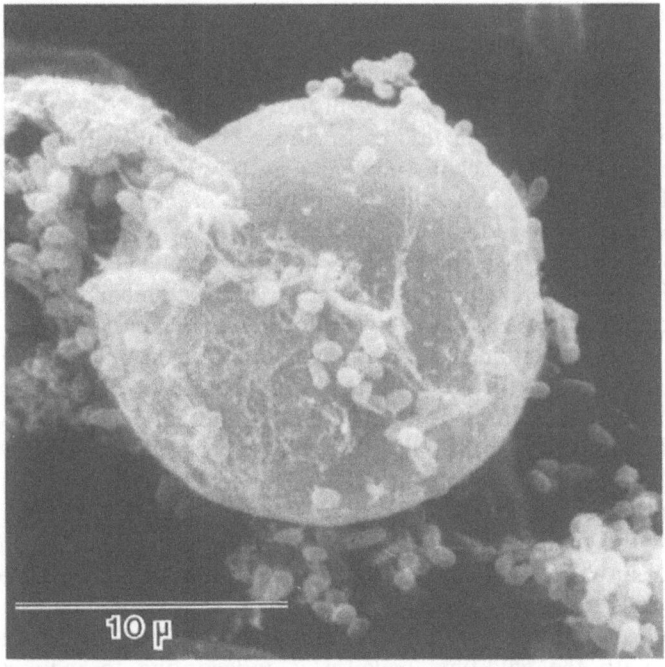

Figure 1. Scanning electron micrograph of *A. tumefaciens* bound to the surface of carrot cells. Note that both directly attached bacteria and indirectly attached bacteria held in place by cellulose fibrils are visible.

These fibrils are made by the bacteria since they were observed in scanning electron micrographs of plant cells which were killed by heat or glutaraldehyde, washed and then incubated with the live bacteria. *A. tumefaciens* also produces these fibrils when the bacteria are grown in the presence of plant extracts. This fact allowed the purification of the fibrils and the analysis of their chemical composition. The fibrils are made of cellulose synthesized by the bacteria in response to the presence of plant cells (Matthysse *et al.*, 1981).

The use of cellulose fibrils by *A. tumefaciens* to aid in adherence to plant cells is a pathogenic mechanism which the plant would have difficulty in overcoming, since removal of the bacteria would require digestion of the fibrils by cellulase which would also damage the plant cell wall.

III. The role of bacterial cellulose fibrils in attachment

A. Properties of cellulose-minus mutants

To examine the role of the cellulose fibrils made by the bacteria in attachment and virulence, cellulose-minus mutants of the bacteria were obtained as follows: *A. tumefaciens* was subjected to transposon mutagenesis using Tn5 carried on a conjugative plasmid which was incapable of replication in *A. tumefaciens*. The resulting random mutants were plated on minimal medium with neomycin (to select for the presence of Tn5) and cellufluor (Matthysse, 1983). Cellufluor is a commercial whitener which fluoresces under UV light in the presence of β-linked polysaccharides such as cellulose. Several mutants were isolated which produced dark colonies on cellufluor plates examined under UV light. The mutant bacteria were tested for the presence of UV fluorescent material with cellulfluor stain using bacteria grown in a variety of media including both solid and liquid complex and minimal media. Mutants which did not appear to produce cellulose under any of these conditions were incubated with plant cells and the mixture examined in the scanning electron microscope. Bacteria attached individually to the plant cells were observed. No fibrils were seen; bacterial aggregates were also absent. When the mutant bacteria were incubated overnight with plant suspension culture cells they failed to cause the formation of visible aggregates of the plant cells. The kinetics of the binding of cellulose-minus mutant bacteria to carrot cells were examined. For the first 20 min of incubation the binding of the wild-type parent bacteria and of the mutant were indistinguishable. However, after this time the parent strain continued to bind to carrot cells while the cellulose-minus mutant showed no additional binding (Matthysse, 1983). This result suggests that the non-saturatable binding observed with the wild-type bacteria is indeed due to indirect bacterial attachment by cellulose fibrils.

The cellulose-minus mutants are all virulent when inoculated as a paste into wound sites on leaves of *Bryophyllum daigremontiana* or in tobacco stems. When the relative virulence of the mutants was compared to that of the parent strain using a semi-quantitative virulence assay the mutants were found to range in virulence from a level indistinguishable from that of the parent strain to less than 1/100 that of the parent strain (Minnemeyer & Matthysse, 1991).

All wild-type strains of *A. tumefaciens* examined produce cellulose. This suggests that although cellulose production is not required for virulence under laboratory conditions it may play a role in the real world. When cellulose-minus mutant bacteria bound to the surface of carrot suspension culture cells, we observed that, unlike the wild type bacteria, the mutant bacteria could easily be removed from the plant cells by shear forces such as swirling or vortexing the flask. In order to examine whether this lack of tight binding of cellulose-minus mutants might be a factor in virulence, leaves of *B. daigremontiana* were wounded and inoculated with the parent strain on one side and a cellulose-minus mutant on the other side. The inoculated leaves were washed gently with water 90 min later and the plants observed for tumor formation. All of the sites inoculated with the wild-type strain formed tumors, but very few of the sites inoculated with the cellulose-minus mutants showed any tumor formation (Matthysse, 1983). Analysis of the plant tissue from the inoculated sites immediately after the water wash showed that most of the mutant bacteria had been removed from the leaf while the numbers of wild type bacteria were unaffected (Sykes & Matthysse, 1986). These results suggest that bacterial production of cellulose fibrils may play a role in adherence of the bacteria to wound sites in the presence of such factors as rain or heavy dew.

B. Genes required for cellulose synthesis

All of the cellulose-minus mutants which we obtained were chromosomal mutations. Indeed strains of *A. tumefaciens* which lack the Ti plasmid such as NT1 and ACH5C3 still produce cellulose when incubated with plant extracts suggesting that genes on the Ti plasmid are not required for cellulose synthesis. The location on the chromosome of the Tn5 insertions which resulted in a failure to make cellulose was determined using a mapping technique suitable for a bacterium in which an Hfr chromosomal mapping system is not available. The plasmid pJB3JI was introduced into the *A. tumefaciens* mutants by conjugation. This plasmid has a tendency to form R' plasmids in which a piece of host cell DNA is included in the plasmid. After the *A. tumefaciens* carrying the plasmid were allowed to grow so that some of the cells would contain R' plasmids, the plasmids were moved to *Escherichia coli* by conjugation. The bacteria were selected by at least 4 passages in liquid for rapid growth at 37° and the antibiotic resistance markers carried on the plasmid. This population of bacteria carrying various R' plasmids was then conjugated to various strains of *A. tumefaciens* carrying chromosomal auxotrophic markers of known map position. The resulting population of transconjugants was selected for growth in the absence of one of the required substances and the resulting colonies replica plated on the same medium containing neomycin. The coinheritance frequency of Tn5 and the auxotrophic marker wild type gene was determined and used to locate the Tn5 on the chromosomal map of *A. tumefaciens* (Robertson et al., 1988).

Using this mapping technique all of the cellulose-minus mutants were found to map near to each other and to be closely linked to the *met-6* gene (Fig. 2) (Robertson *et al.*, 1988). When an attempt was made to complement the cellulose-minus mutants by introducing a cosmid from a library of *A. tumefaciens* DNA, one cosmid was found to be capable of complementing all of the known mutants (Matthysse & Lightfoot, unpublished observations). Thus the genes required for cellulose synthesis appear to be clustered in one location on the *A. tumefaciens* chromosome.

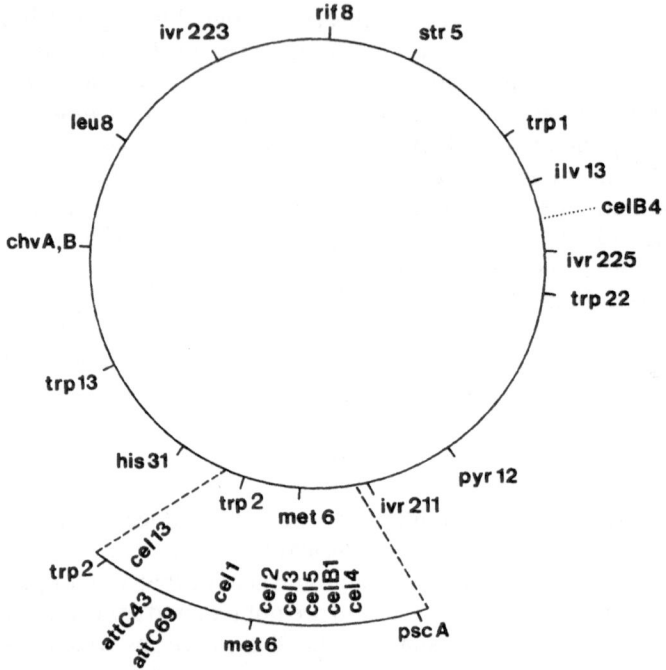

Figure 2. Chromosomal map of *A. tumefaciens*. The position of genes required for cellulose synthesis (*cel*) and of genes required for attachment to plant cells (*chvA, chvB, pscA* and *att*) are shown. This map is taken from Robertson *et al.*, 1988.

C. Regulation of synthesis of cellulose fibrils

Although small amounts of cellulose appear to be made by the bacteria under most growth conditions, the incubation of the bacteria with plant cells or plant extracts results in an increase in the amount of cellulose produced. In order to study the regulation of the expression of genes required for cellulose synthesis insertions of the promoter-probe transposon Tn3HoHo1 (Stachel *et al.*, 1985) in the cloned genes were examined. Those insertions in which the promoterless β-galactosidase gene was oriented in the same direction as the operon into which it had inserted appeared to fall into two groups. The first group showed increased expression in the presence of plant extracts and no change in expression on the addition of metal ions. The second group showed no change in expression on the addition of plant extracts and a decrease in expression in the presence of metal ions. These two groups of insertions are located in different regions of the cloned genes (Rosche & Matthysse, unpublished observations). The results suggest that there are at least two operons involved in cellulose synthesis in *A. tumefaciens* and that the expression of these operons is regulated separately. The significance of this separate regulation is unknown.

IV. The initial step in bacterial attachment

A. Attachment-minus mutants

Three classes of nonattaching mutants of *A. tumefaciens* have been isolated (Table 1), ChvA and ChvB (Douglas *et al.*, 1982), PscA which is also known as ExoC (Thomashow *et al.*, 1987; Cangelosi *et al.*, 1987) and Att (Matthysse, 1987a). ChvA and ChvB mutants are altered in the synthesis of β-1-2-D-glucan. *chvB* encodes the enzyme for the synthesis of this polysaccharide and *chvA* encodes a protein responsible for the transport of the polysaccharide across the plasma membrane. These mutants are pleiotropic; they overproduce acidic exopolysaccharides; they have reduced motility and reduced ability to transfer certain plasmids conjugatively (Douglas *et al.*, 1985; Puvanesarajah *et al.*, 1985). Thus it is uncertain whether the inability of these mutants to attach to plant cells is a direct affect of their inability to synthesize β-1-2-D-glucan or whether the defect in attachment is an indirect affect of the mutation. β-1-2-D-Glucan seems unlikely to play a direct role in the attachment of *A. tumefaciens* to host cells since it is made by many other bacteria such as rhizobia and azospirilla which do not bind to the same range of plant cells as do agrobacteria.

PscA mutants fail to synthesize any exopolysaccharides; they may also have difficulty in exporting extracellular proteins (Thomashow *et al.*, 1987). In addition they show reduced motility and grow slowly on agar surfaces. The *pscA* gene has a direct analogue in *R. meliloti* which has a much more restricted ability to bind to plant cells (Cangelosi *et al.*, 1987). This suggests that while *pscA* may be required for attachment to plant cells, it is not sufficient to cause attachment to the wide variety of plant species to which agrobacteria bind.

Att mutants are missing one or more of three minor outer membrane proteins. They have normal amounts of lipopolysaccharide, cellulose, and other exopolysaccharides. The sizes of all of these molecules also appear to be unaltered; their composition has not been examined. Att mutants are motile (Matthysse, 1987a). It is unknown whether the outer membrane proteins missing in Att mutants participate directly in bacterial binding or whether they act as enzymes making the binding molecules or as scaffolding for the binding site.

All three classes of nonattaching mutants are avirulent suggesting that the initial binding of the bacteria to plant cells is required for the transfer of T-DNA from the bacteria to plant cells.

B. The nature of the bacterial binding site

For convenience in this article we will refer to the site on the surface of the bacterium which binds to the plant cell surface as the bacterial binding site and the site on the plant cell surface to which the bacteria bind as the plant receptor.

Examination of the three types of nonattaching mutants described above suggests that the bacterial binding site may be composed of exopolysaccharides

Table 1. Binding of radioactive vitronectin and attachment of carrot cells to various bacterial strains.

Bacterial strain	Vitronectin Binding (per cent control)[a]	Binding to Carrot Cells (per cent control)[b]
R. meliloti 1021	2 ± 2	10 ± 10[c]
A. tumefaciens		
A6 parent strain	100 (control)	100 (control)
mutants		
Cel-12	91 ± 26	60 ± 15
Ivr-225	98 ± 17	100 ± 25
Att-339	15 ± 15	10 ± 10
PscA	24 ± 15	20 ± 20
C58 parent strain	100 (control)	100 (control)
mutants		
Att-C43	30 ± 15	7 ± 7
Att-C69	35 ± 30	7 ± 7
A1045	4 ± 4	28 ± 28

[a] The control bound between 400 and 800 picograms vitronectin per 10^9 bacteria.

[b] Binding data for carrot are for 20 min. incubation in M&S medium. The per cent of the bacterial inocula bound for the controls was 35% of the A6 strain and 20% of the C58 strain.

[c] The control used in calculating the relative binding of R. meliloti was the A6 strain of A. tumefaciens.

This table is from Wagner & Matthysse, 1992.

and/or outer membrane proteins. None of the exopolysaccharides have been found to inhibit bacterial attachment when they were added to carrot suspension culture cells prior to the addition of the bacteria. Lipopolysaccharides do inhibit bacterial attachment under these conditions, but only when used at relatively high concentrations (0.04 micrograms ketodeoxyoctonate ml^{-1} or more, Matthysse, 1987b). Thus at one time the bacterial binding site was thought to be composed of lipopolysaccharide (Whatley *et al.*, 1976). However, lipopolysaccharide mutants show normal binding to carrot cells (Metts *et al.*, 1991). In addition, inhibitors which prevent the synthesis of lipopolysaccharide have no affect on the ability of the bacteria to attach to carrot cells (Goldman *et al.*, 1992). Thus it is unlikely that lipopolysaccharide plays a direct role in the binding of *A. tumefaciens* to plant cells.

At the present time the nature of the bacterial binding site remains uncertain. It is probable that it is composed of exopolysaccharides and/or outer membrane proteins.

C. Regulation of bacterial attachment

Live bacteria are required for binding to suspension culture cells. Bacteria which have been killed by heat or UV light or azide fail to attach to plant cells. However, bacteria grown in Luria broth can be added to plant cells in the presence of protein synthesis inhibitors without affecting their ability to bind to the plant cells (Matthysse, 1987a & b). Thus it does not appear that the requirement for live bacteria is due to a requirement for gene induction before the bacteria can bind. It is possible that the treatments used to kill the bacteria also alter the cell surface and the bacterial binding site.

An alternative method to examine the regulation of bacterial attachment is to examine the regulation of the expression of the three groups of genes known to be required for bacterial attachment. Although the regulation of *chvA* and *chvB* has not been examined explicitly in *A. tumefaciens*, in other organisms the production of β-1–2-D-glucan is regulated by the osmotic strength of the medium (Miller *et al.*, 1986). More of the polysaccharide is made when the bacteria are grown in low osmotic strength than in high osmotic strength medium. Regulation of the expression of *pscA* has not been examined, but since this gene is required for the production of exopolysaccharides which are generally present on the bacterial surface, it must be expressed constitutively, although the level of its expression could be altered by the presence of plant cells. When the promoter probe transposon Tn3HoHo1 was used to examine the regulation of the expression of *att* genes, these genes were found to be expressed constitutively at a low level. The expression was increased by the addition of plant extracts to the medium (Rosche & Matthysse, unpublished observations). Thus it is likely that the bacterial binding site is present on the surface of bacteria before they encounter plant cells. However, the number of binding sites per bacterium may increase in the presence of host cells.

V. The role of the host cell and the nature of the receptor to which the bacteria bind

The host cell plays a passive role in the initial attachment of *A. tumefaciens* to plant cells. Carrot cells which have been heat killed or fixed with glutaraldehyde still retain their ability to bind *A. tumefaciens* (Matthysse *et al.*, 1981; Matthysse *et al.*, 1982). The number of receptor sites on the surface of a carrot cell was estimated using the binding of a cellulose-minus mutant which is not capable of indirect binding to plant cells. A limited number of receptor sites (about 2,000) for the bacteria were found to be present on a carrot suspension culture cell (Gurlitz *et al.*, 1987).

In an attempt to elucidate the nature of the plant cell receptor to which the bacteria bind, several different types of compounds were assayed for their ability to inhibit the initial attachment of *A. tumefaciens*. The binding of *A. tumefaciens* is markedly reduced by exposing carrot cells to dilute detergents such as Triton X-100 or proteolytic enzymes such as chymotrypsin, trypsin, or proteinase K indicating the involvement of a plant protein component in the attachment mechanism (Gurlitz *et al.*, 1987). Bacterial binding to carrot cells is unaffected by chelating agents which remove divalent cations, salt, a variety of sugars, polysaccharides including pectin, and various plant lectins (Gurlitz *et al.*, 1987).

Vitronectin (also termed S-protein or serum spreading factor) is a constituent of the extracellular matrix, which promotes the attachment and spreading of a variety of animal cells to a substrate (Tomasini and Mosher, 1990). Vitronectin binds to glass and sulfated heparin, and interacts with the transmembrane vitronectin receptor – integrin (Tomasini and Mosher, 1990). Vitronectin is reported to bind to certain strains of *Staphylococcus aureus*, streptococci, and *E. coli* (Chhatwal *et al.*, 1987; Paulsson and Wadstorm, 1990); and has been proposed to play a specific role in streptococcal attachment to cultured mammalian epithelial and endothelial cells (Filippson *et al.*, 1990).

The presence of a vitronectin-like protein and its associated integrin receptor (Schindler *et al.*, 1989) has been detected in several species of higher plants including tobacco (Zhu *et al.*, 1991) tomato, soybean, broad bean, and lily (Sanders *et al.*, 1991). *A. tumefaciens* is capable of infecting all of these plants (DeCleene and DeLey, 1976; Hooykaas-Van Slogteren *et al.* 1984). A vitronectin-like protein is also found in the brown alga *Fucus* (Wagner *et al.*, 1992) and the slime mold *Physarum* (Miyazaki *et al.*, 1992).

The ability of *A. tumefaciens* to recognize and bind to the recently discovered vitronectin-like protein found on the surface of plant cells was examined. Bacterial binding to carrot cells is inhibited by either the addition of human vitronectin or antivitronectin antibodies. Wild type *A. tumefaciens* strains A6 and C58 bind radioactive human vitronectin; whereas little binding is observed to the avirulent nonattaching mutants (Att-339, Att-C43, Att-C69, PscA, and A1045 which is a ChvB mutant). Other mutants affecting cellulose (Cel-12) and lipopolysaccharide (Ivr-225) which do not affect the initial attachment to carrot cells bind radioactive vitronectin. *Rhizobium meliloti* which is closely related to *A. tumefaciens* but fails to bind to carrot cells also fails to bind vitronectin (Table 2). Thus the ability of the bacteria to bind radioactive vitronectin correlates with their ability to bind carrot cells (Wagner and Matthysse, 1992).

Table 2. Mutants of *A. tumefaciens* which affect bacterial attachment.

Bacterial Mutant	Properties and Compounds Affected	Attachment to Plant Cells	Virluence	Reference
(Type)	(primary and secondary defects)			
Cel	Fail to make cellulose.	Direct only	Unaltered or reduced $10^{-1 \text{ to } -2}$	Matthysse, 1983 Minnemeyer et al, 1991
Chv A and B	Fail to make <beta>-1,2-D glucan; reduced motility and conjugation; overproduction of EPS.	Not detectable	Avirulent	Douglas et al, 1982 Puvanesarajah et al, 1985
Att	Lack one or more minor outer membrane protein; fail to grow on Luria agar containing neomycin.	Not detectable	Avirulent	Matthysse, 1987
PscA (ExoC)	Lack of EPS; slow growth reduced motility	Not detectable	Avirulent	Thomashow et al, 1987 Cangelosi et al, 1987

The binding of bacteria to carrot suspension cells is inhibited by relatively low concentrations of polyclonal antibodies to human vitronectin, but not by antibodies to the closely-related fibronectin (Wagner and Matthysse, 1992). Furthermore, the same polyclonal antibodies that inhibit binding also recognize a vitronectin-like protein in detergent extracts of the carrot cells. The Triton X-100 detergent extraction method used was previously shown to remove the receptor to which the bacteria bind while retaining the viability of carrot cells

90

(Gurlitz *et al.*, 1987). These results suggest that a vitronectin-like protein is found on the surface of the carrot cells and that anti-human vitronectin serum inhibits binding by attaching to the exposed vitronectin-like protein on the surface of the carrot cells.

The ability of *A. tumefaciens* to attach to vitronectin which is an exterior component of a structural and functional complex that may be linked to an actin cytoskeletal network could aid in the transport of the T-DNA and associated proteins to the host cell nucleus.

VI. Conclusion

Binding of *Agrobacterium tumefaciens* to plant host cells is a two step process (Fig. 3). In the first step the bacteria bind to a receptor on the plant cell surface. The bacterial binding site pre-exists the exposure of the bacteria to host cells and the plant receptor is found on plant cells which have had no contact with bacteria. The bacterial binding site is composed of exopolysaccharides and/or outer membrane proteins. The host receptor contains a vitronectin-like protein. This initial binding of the bacteria to the host cell is a loose binding and the bacteria can be removed by shear forces. When the bacteria are in proximity to plant cells they respond to substances released from the plant cells by the elaboration of cellulose fibrils. These fibrils anchor the bacteria firmly to the

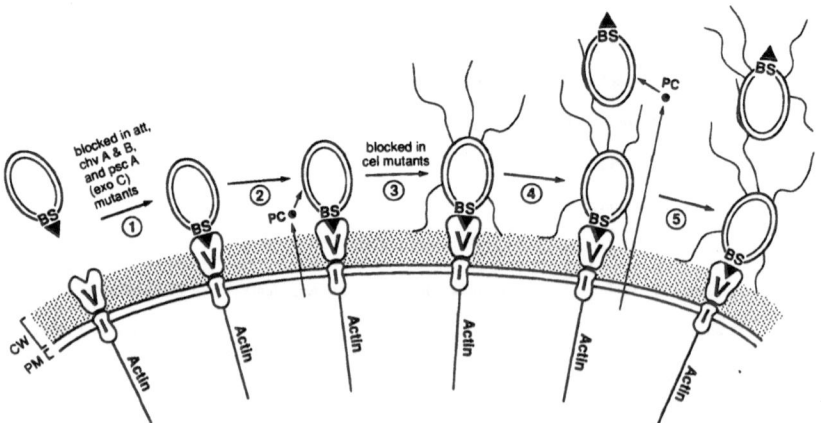

Figure 3. A model of the attachment of *Agrobacterium tumefaciens* to plant host cells. In step 1 the bacterium with a preexisting binding site (BS) binds to a vitronectin (V) receptor on the surface of the plant cell. The vitronectin is linked via integrin (I) to the actin cytoskeleton of the plant cell. CW: plant cell wall. PM: plant plasma membrane. This step is blocked in Att, ChvA and B, and PscA (ExoC) mutants. In step 2 the plant cell releases substances (PC) which (step 3) induce the bacteria to make cellulose fibrils. This step is blocked in Cel mutants. The cellulose fibrils anchor the bacteria to the plant cell. In addition some of the fibrils entrap other bacteria (step 4). These indirectly attached bacteria are induced to make cellulose fibrils by plant compounds (PC) in step 5. The repetition of steps 4 and 5 results in the formation of bacterial aggregates on the surface of the plant cell.

plant cell surface. The bacteria can no longer be removed by shear forces. In addition the cellulose fibrils result in the formation of large aggregates of bacteria on the surface of the host cells.

The initial binding of the bacteria to host cells is required for virulence and, presumably, for DNA transfer. Mutants blocked in this initial step (ChvA and ChvB, PscA, and Att) are avirulent. The second step is not required for virulence. However, mutants blocked in this step show a reduced level of virulence in the laboratory as measured by the number of bacteria required to from tumors at 50% of the inoculated sites. The synthesis of cellulose fibrils may be important in aiding bacteria to adhere to host cells under natural conditions.

Acknowledgements

This research was supported by grants from USDA (91–37303–3267)and NSF (IBN-8916586) to AGM and an NSF fellowship (DIR-9104253) to VTW.

References

Cangelosi GA, Hung I, Puvanesarajah V, Ozga AD, Leigh JA, & Nester EW (1987) Common loci for *Agrobacterium tumefaciens* and *Rhizobium meliloti* exopolysaccharide synthesis and their roles in plant interactions. J. Bacteriol. 169: 2086–2091.

Chhatwal GS, Preissner KT, Muller-Berghaus G, & Blobel H (1987) Specific binding of the human S protein (vitronectin) to Streptococci, *Staphylococcus aureus*, and *Escherichia coli*. Infect. Immun. 55: 1878–1883.

Chilton MD, Drummond MH, Merlo DJ, Sciaky D, Montoya AI, Gorden MP & Nester EW (1977) Stable incorporation of plasmid DNA into higher plant cells: the molecular basis of crown gall tumorigenesis. Cell 11: 263–271.

Deasey MC & Matthysse AG (1984) Interactions of wild-type and a cellulose-minus mutant of *Agrobacterium tumefaciens* with tobacco mesophyll and tobacco tissue culture cells. Phytopathology 74: 991–994.

DeCleene M & DeLey J (1976) The host range of crown gall. Bot. Rev. 42: 389–466.

Douglas CJ, Halperin W & Nester EW (1982) *Agrobacterium tumefaciens* mutants affected in attachment to plant cells. J. Bacteriol. 152: 1265–1275.

Douglas CJ, Staneloni RJ, Rubin RA & Nester EW (1985) Identification and genetic analysis of an *Agrobacterium tumefaciens* chromosomal virulence region. J. Bacteriol. 161: 850–860.

Filippsen LF, Valentin-Weigand P, Blobel H, Preissner KT & Chhatwal GS (1990) Role of complement S protein (vitronectin) in adherence of *Streptococcus dysgalactiae* to bovine epithelial cells. Am. J. Vet. Res. 51: 861–865.

Goldman RC, Capobiano JO, Doran CC & Matthysse AG (1992) Inhibition of lipopolysaccharide synthesis in *Agrobacterium tumefaciens* and *Aeromonas salmonicida*. J. Gen. Microbiol. 138: 1527–1533.

Gurlitz RHG, Lamb PW & Matthysse AG (1987) Involvement of carrot cell surface proteins in attachment of *Agrobacterium tumefaciens*. Plant Physiol. 83: 564–568.

Hawes MC & Pueppke SG (1987) Correlation between binding of *Agrobacterium tumefaciens* by root cap cells and susceptibility of plants to crown gall. Plant Cell Rep. 6: 287–290.

Hooykaas-Van Slogteren GMS, Hooykaas PJJ & Schilperoort RA (1984) Expression of Ti plasmid genes in monocotyledonous plants infected with *Agrobacterium tumefaciens*. Nature (London) 311: 763–764.

Kado CI (1991) Molecular mechanisms of crown gall tumorigenesis. Crit. Revs. Plant Sci. 10: 1–32.

Matthysse AG (1983) Role of bacterial cellulose fibrils in *Agrobacterium tumefaciens* infection. J. Bacteriol. 154: 906–915.

Matthysse AG (1987a) Characterization of nonattaching mutants of *Agrobacterium tumefaciens*. J. Baceriol. 169: 313–323.

Matthysse AG (1987b) Effect of plasmid pSa and of auxon on attachment of *Agrobacterium tumefaciens* to carrot cells. Appl. Environ. Microbiol. 53: 2574–2582.

Matthysse AG, Holmes KV & Gurlitz RHG (1981) Elaboration of cellulose fibrils by *Agrobacterium tumefaciens* during attachment to carrot cells. J. Bacteriol. 145: 583–595.

Matthysse AG, Wyman PM & Holmes KV (1978) Plasmid-dependent attachment of *Agrobacterium tumefaciens* to plant tissue culture cells. Infect. Immun. 22: 516–522.

Miller KJ, Kennedy EP & Reinhold VN (1986) Osmotic adaption by gram-negative bacteria: possible role for periplasmic oligosaccharides. Science 231: 48–51.

Metts J, West J, Doares SH, & Matthysse AG (1991) Characterization of three *Agrobacterium tumefaciens* avirulent mutants with chromosomal mutations that affect induction of vir genes. J. Bacteriol. 173: 1080–1087.

Minnemeyer SL, Lightfoot R & Matthysse AG (1991) A semiquantitative bioassay for relative virulence of *Agrobacterium tumefaciens* strains on *Bryophyllum daigremontiana*. J. Bacteriol. 173: 7723–7724.

Miyazaki K, Hamano T & Hayashi M (1992) *Physarum* vitronectin-like protein: an Arg-Gly-Asp-dependent cell-spreading protein with a distinct NH2-terminal sequence. Exp. Cell Res. 199: 106–110.

Paulsson M & Wadstrom T (1990) Vitronectin and type-I collagen binding by *Staphyloccus aureus* and coagulase-negative staphylococci. FEMS Microbiol. Immunol. 65: 55–62.

Puvanesarajah VF, Schell M, Stacey G, Douglas CJ & Nester EW (1985) Role for 2-linked-β-D-glucan in the virulence of *Agrobacterium tumefaciens*. J. Bacteriol. 164: 102–106.

Ream W (1989) *Agrobacterium tumefaciens* and interkingdom exchange. Annu. Rev. Phytopathol. 27: 583–618.

Robertson JL, Holiday T & Matthysse AG (1988) Mapping of *Agrobacterium tumefaciens* chromosomal genes affecting cellulose synthesis and bacterial attachment to host cells. J. Bacteriol. 170: 1408–1411.

Sanders L, Wang C-O, Walling L & Lord E (1991) A homolog of the substrate adhesion molecule vitronectin occurs in four species of flowering plants. Plant Cell 3: 629–635.

Schindler M, Meiners S & Cheresh DA (1989) RGD-dependent linkage between plant cell wall and plasma membrane: Consequences for growth. J. Cell Biol. 108: 1955–1965.

Stachel SE, An G, Carlos F & Nester EW (1985) A Tn*3lacZ* transposon for the random generation of β-galactosidase gene fusions: application to the genetic analysis of gene expression in *Agrobacterium*. EMBO J. 4: 1445–1454.

Sykes LC & Matthysse AG (1986) Time required for tumor induction by *Agrobacterium tumefaciens*. Appl. Environ. Microbiol. 52: 597–598.

Thomashow MF, Karlinsey JE, Marks JR & Hurlbert RE (1987) Identification of a new virulence locus in *Agrobacterium tumefaciens* that affects polysaccharide composition and plant cell attachement. J. Bacteriol. 169: 3209–3216.

Tomasini BR & Mosher DF (1990) Vitronectin. Prog. Hemostasis Thromb. 10: 269–305.

Wagner VT, Brian L & Quatrano RS (1992) Role of a vitronectin-like molecule in embryo adhesion of the brown alga *Fucus*. Proc. Natl. Acad. Sci. USA 89: 3644–3648.

Wagner VT & Matthysse AG (1992) Involvement of a vitronectin-like protein in attachment of *Agrobacterium tumefaciens* to carrot suspension culture cells. J. Bacteriol. 174: 5999–6003.

Whatley MH, Bodwin JS, Lippincott BB & Lippincott JA (1976) Role for *Agrobacterium* cell envelope lipopolysaccharide in infection site attachment. Infect. Immun. 13: 1080–1083.

Zhu J-K, Shi J, Sing U & Carpita NC (1991) Possible involvement of vitronectin- and fibronectin-like proteins in membrane-wall and protoplast adhesion in salt-adapted tobacco cells. Plant Physiol. (suppl.) 96: 10.

7. The function and regulation of genes required for extracellular polysaccharide synthesis and virulence in *Pseudomonas solanacearum*

P. solanacearum extracellular polysaccharide

CHRIS CHENG KAO and LUIS SEQUEIRA

Abstract. In the bacterial wilt pathogen, *Pseudomonas solanacearum*, wild-type level of virulence requires the production of extracellular polysaccharide (EPS). However, the structure of EPS and the regulation of its production in *P. solanacearum* are poorly understood. To determine the function of EPS genes, we have extensively characterized the *ops* gene cluster. Mutations in any of the seven complementation units of this cluster not only affect EPS production, but also the structure of lipopolysaccharide (LPS). Furthermore, one of the complementation units appears to encode enzymes responsible for the synthesis of rhamnose, a component of LPS and EPS. To better understand regulation of EPS production, we characterized *epsR*, a regulatory locus whose product prevents EPS production and affects virulence. The *epsR* gene encodes a polypeptide of about 26 kDa whose sequence contains conserved motifs found in several previously identified bacterial regulatory proteins.

Abbreviations: EPS, Extracellular Polysaccharide; LPS, Lipopolysaccharide; wt, Wild-Type; GalNac, N-acetyl Galactosamine

Introduction

Many cell surface molecules are important in the complex interaction between a pathogen and its host. In Gram-negative bacteria, these molecules include extracellular polysaccharide (EPS), which is secreted around the cell either as a capsule or as loose slime. EPS is required to control the exchange of charged molecules between the bacterium and its environment and can be important for virulence. For example, mutations in EPS in *Rhizobium spp.* lead to a reduced ability to nodulate host plants (Djordjevic *et al.*, 1987; Leigh *et al.*, 1985), and result in loss of virulence in species of *Erwinia* (Ayers *et al.*, 1979), *Agrobacterium* (Kamoun *et al.*, 1989), and *Haemophilus* (Hoiseth *et al.*, 1985). Despite much effort, however, we know very little about how EPS interacts with the host plant, or how specific changes in the EPS structure affect virulence.

In our laboratory, we have focused attention on EPS produced by the plant pathogen *Pseudomonas solanacearum*, which causes wilt of over one hundred plant species (for reviews, Buddenhagen and Kelman, 1964; Coplin and Cook, 1990). Interest in the role of EPS in virulence of *P. solanacearum* stemmed from the early work of Kelman (1954), who proposed that EPS contributes to wilting by occluding xylem vessels, hence reducing water movement (Husain and

C.I. Kado and J.H. Crosa (eds.), Molecular Mechanisms of Bacterial Virulence, 93–108.
© *1994 Kluwer Academic Publishers.*

Kelman, 1958; Van Alfen *et al.*, 1983 and 1984). This proposal was based on three observations: a) EPS can be detected in the xylem of infected plants, b) purified EPS can cause wilting of tomato cuttings, and c) spontaneous mutants that lack EPS are also affected in virulence. An alternate, but complementary hypothesis was that of Young and Sequeira (1986) who proposed that EPS allows survival of bacteria inside the plant. They determined that EPS protects the invading bacteria from attachment to plant cell walls, thus allowing spread of the bacterium in the xylem vessels.

In this chapter we will be concerned primarily with our current efforts in three main areas: 1) establishing unambiguously the correlation between a requirement for EPS and wild-type (wt) level of virulence in *P. solanacearum*, 2) elucidating the functions of one class of EPS biosynthetic genes, the *ops* genes, and 3) characterizing the *epsR* gene which negatively regulates EPS production. Recent reviews on the biology of *P. solanacearum* and its virulence factors (EPSI and EPSII) have been published by Hayward (1991) and Boucher (1992). Thus, in this paper, we have intentionally avoided duplicating the areas of primary concern in those reviews.

I. The structure of *P. solanacearum* EPS

A concept of the structure of EPS has evolved over the years. Akiyama *et al.* (1986) reported that EPS of *P. solanacearum* is a homopolymer of N-acetylgalactosamine (GalNac), but others reported that glucose and rhamnose are also present as minor components (1985). Consistent with the results of others, Cook and Sequeira (1991) have determined that the uridine diphosphoyl form of N-acetylgalactosamine (UDP-GalNac) is one of the major sugar nucleotides in cells of wt strains of *P. solanacearum*. More recently, A. Trigalet and collaborators (1991) reported that the EPS of *P. solanacearum* strain GM1001 is a very heterogeneous mixture that separated into at least four distinct fractions based on size and charge (A. Trigalet, personal communication). The structure of a component of the acidic fraction was elucidated and found to be a repeating unit of two unusual amino sugars, bacillosamine (2,4-diamino-2,4,6 trideoxy-glucose) and galactosaminuronic acid. The latter sugar apparently is responsible for the acidic nature of the molecule.

Much work remains to be done on the structure of EPS. The chemical and structural analyses are being complemented by genetic analyses primarily in our laboratory and those of Timothy Denny and Mark Schell at the University of Georgia (see chapter 23). In our laboratory a short-term goal is to understand the biochemical role of EPS genes in the synthesis and structure of EPS. This knowledge will better enable us to construct models of how EPS interacts with molecules in the infected plant.

II. The consistent correlation between EPS and wt virulence

Early correlations between the requirement for EPS and wt virulence in *P. solanacearum* were made with the use of spontaneously non-mucoid colonies which appeared when bacterial cultures were grown in the absence of aeration, or after passage through plants (Kleman, 1954). Kelman was the first to report that all such spontaneous mutants were defective in virulence (ibid). However, while all spontaneous avirulent mutants are phenotypically similar in that they are defective in EPS production, they may also have alterations in many other characteristics. For example, the best characterized mutant strain (B1) differs from the parental strain (K60) not only in EPS production, but in its new ability to induce a hypersensitive reaction in tobacco (Lozano and Sequeira, 1970), absence of the 0-antigen region of the lipopolysaccharide constituent of the outer membrane (Hendrick and Sequeira, 1984), increased production of indole acetic acid (Phelps and Sequeira, 1968), and polygalacturonase (Allen *et al.*, 1991), increased piliation (Stemmer and Sequeira, 1987), and the ability to attach to plant cell walls (Duvick and Sequeira, 1984; Young and Sequeira, 1986), etc. Thus the spontaneous shift from K60 to B1 is pleiotropic and apparently irreversible, facts that complicate the genetic analysis of this phenomenon.

The relationship of EPS and virulence has been confirmed with more defined transposon-generated mutants (Cook and Sequeira, 1991; Denny *et al.*, 1988 and 1989; Kao *et al.*, 1991 and 1992; Schell *et al.*, 1987; Xu *et al.*, 1998 and 1990). Complementation analyses with Tn5 insertion mutants have defined at least three EPS gene clusters named EPSI, EPSII, and *ops* (for *o*utermembrane *p*olysaccharide) (Cook and Sequeira, 1991; Denny and Baek, 1991, Kao and Sequeira, 1991). We also have several other transposon-tagged, EPS-deficient mutants that cannot be placed into these three complementation groups, suggesting the possibility of many more genes involved in EPS synthesis. The *ops* genes contain at least seven complementation units, named *opsA* to *opsG*, which are expressed *in planta*. Mutations in any of these seven complementation units reduced virulence of the bacterium to plants (Cook and Sequeira, 1991; Kao and Sequeira, 1991). The EPSI and EPSII gene clusters map to adjacent locations in the bacterial chromosome (Denny and Baek, 1991) in a position distinct from the *ops* gene cluster. Mutations in EPSI, but not in EPSII, resulted in decreased virulence probably because EPSII mutants are not affected in EPS production *in planta* (Denny and Baek, 1991). Not surprisingly, expression of the EPS genes *in planta* is required for full wt level of virulence.

There was only one biochemically defined mutant which, at first determination, disagreed with the correlation that EPS is required for wt virulence. This mutant, named KD700, produced reduced levels of GalNac in culture and *in planta*, but seemingly remained highly virulent (Xu *et al.*, 1990). These results led to the proposal that EPS was not always required for virulence (Xu *et al.*, 1990). Upon further characterization of this strain, we determined that this interpretation was incorrect. Genetic and molecular evidence suggesting that the KD700 mutation maps to the EPSI gene cluster. This result is based on the following observations: 1) KD700 can be complemented for EPS production and virulence by plasmids containing a functional EPSI region and vice versa; and 2) pL700A, which complements the mutation in KD700, hybridizes strongly to DNA from the EPSI region (kindly provided by T. Denny) and shares many restriction sites. Since EPSI mutants have been reported to be affected in virulence, we reexamined the virulence of KD700. When a range of inoculum concentrations was tested on eggplant or tobacco, KD700 was significantly less virulent than the wt strain K60 (Kao et al., 1992). Xu et al. (1990) only tested a high dose of inoculum and thus missed the decrease in virulence that was more obvious at lower inoculum doses. To be certain that the KD700 strain did not suffer additional mutations during storage, we transferred the KD700 mutation to a wt background and tested this marker-exchanged mutant for virulence to eggplants. This new mutant was also affected in virulence. Our results, therefore, confirm that in *P. solanacearum* there is a consistent association between EPS production and virulence.

III. OPS biosynthetic genes

The *ops* gene cluster was originally identified by the mutant KD500 (Xu et al., 1988). This mutation was genetically and physically distinct fom the EPSI and EPSII genes. The KD500 mutant and all other *ops* mutants are impaired in EPS production, but do not abolish it altogether. Therefore, these mutations likely affect the structure of EPS. The fragment of wt DNA corresponding to that tagged by Tn5 in KD500 was cloned in plasmid pL5001. KD500 mutants transformed with pL5001 were restored in EPS production and virulence (Xu et al., 1990). Saturation mutagenesis with Tn3-Hogus was used to determine the critical regions in pL5001 that control the virulence and EPS phenotypes (Cook and Sequeira, 1991). Several dozen derivatives of pL5001, each with a single unique Tn3 insertion, were obtained; the location of Tn3 was determined by restriction mapping. The effect of each insertion on EPS and virulence was determined by individually marker exchanging these mutations into the K60 chromosome. These chromosomal mutations allowed us to perform complementation analysis by restoring EPS production *in trans* by means of replicating plasmids, each harboring a Tn3 insertion. In all, we identified a 6.5 kb cluster in pL5001 that contains seven complementation units required for full EPS production and virulence (Fig. 1; Cook and Sequeira, 1991; Kao and Sequeira, 1991). Each unit appears to be under the control of a distinct promoter and all genes appeared to be expressed independently. Of the six complementation units in which the direction of transcription is known, five were transcribed from right to left (Fig. 1). The *opsA* complementation unit, however, is transcribed in the opposite direction.

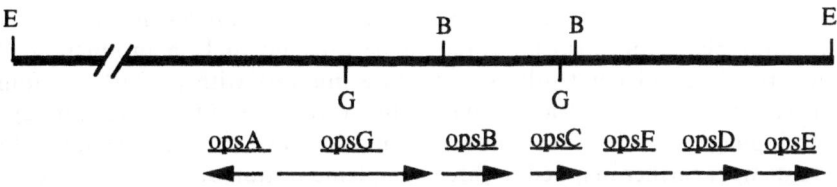

Figure 1. The seven complementation units in the *ops* gene cluster. The arrows denote the direction of transcription. For *opsA* to *opsE*, this was determined by expression from a promoterless *gus* gene (Cook and Sequeira, 1991). The *opsG* transcription direction was determined by DNA sequencing; direction of transcription of the *opsF* gene has not been determined. Tn3*gus* mutations to the right of *opsF* and to the left of *opsA* did not have an effect on EPS production in culture or on virulence (D. Cook and C. Kao, unpublished data). Abbreviations used for restriction enzyme sites are: E, EcoRI; B, BamHI; G, BglII.

a) Function of the ops complementation units

When we attempted to conduct phage-mediated transduction with the *ops* mutants, we noticed that mutations in any of the seven *ops* complementation units prevented plaque formation by the lytic bacteriophage CH154. Bacteriophage CH154 was previously characterized in our laboratory and shown to gain entry into the cell by first attaching to lipopolysaccharide (LPS) (Hendrick and Sequeira, 1984). Consistent with this observation, we determined that purified wt *P. solanacearum* LPS can protect intact cells from lysis by CH154, while LPS purified from the B1 strain cannot (Kao and Sequeira, 1991). Mutations in the EPSI and EPSII clusters do not affect the ability of the cells to propagate CH154. Thus, since mutations in the *ops* gene cluster prevented lysis by LPS-specific phage, it seemed likely that genes in this cluster were required for synthesis both of LPS and EPS.

The LPS of *P. solanacearum* plays an important role in virulence of this organism (Baker *et al.*, 1984; Hendrick and Sequeira, 1984). As with other Gram-negative bacteria, *P. solanacearum* LPS is composed of three separately assembled moieties: lipid A, an oligosaccharide core, and the O-antigen which consists of a chain of repeating oligosaccharide units. While the structure of the lipid A moiety of *P. solanacearum* LPS has not been determined, the core structure is known to be composed of rhamnose, glucose, heptose, and 2-keto-deoxy-octonate (Baker *et al.*, 1984), and the O-antigen repeating unit contains a 4:1:1 ratio of rhamnose, N-acetylglucosamine, and xylose (ibid). In *P. solanacearum* the wt LPS molecule apparently prevents agglutination by certain plant cell wall proteins (Sequeira and Graham, 1977). Most, but not all, of the mutations which affected LPS synthesis also affected virulence (Hendrick and Sequeira, 1984). Exactly how changes in *P. solanacearum* LPS structure affect virulence remains to be elucidated.

Several LPS mutants had been isolated previously in our laboratory (Hendrick and Sequeira, 1984) and the effects of these mutations on LPS structure had been analyzed. We reasoned that if the *ops* genes encoded products required for LPS biosynthesis, then pL5001 should complement these known LPS mutants back to wt structure. This hypothesis was especially attractive because several of the LPS mutants were visibly less mucoid than wt strains on plates and also appeared to be affected in EPS production. We transformed 12 independently isolated LPS mutants with pL5001 and found that seven of the 12 transformants recovered the ability to propagate bacteriophage CH154. The mutations in the seven complementable LPS mutants were further mapped to three complementation units, *opsC*, *opsD*, and *opsG*. Consistent with the idea that these *ops* genes are involved in LPS synthesis, we determined that LPS extracted from all the *ops* mutants appeared to lack the O-antigen portion, based on LPS mobility in polyacrylamide gel electrophoresis. Furthermore, these defective LPS molecules reverted to wt when the mutants harbored pL5001.

b) The ops *complementation units probably encode enzymes for nucleotide sugar synthesis*
Since the effects of *ops*-complementable mutations on LPS structure had been elucidated previously by Hendrick and Sequeira (1984), we could make reasonable predictions about the possible function of *ops* genes. For example, mutations at the *opsG* locus lacked the sugar rhamnose which is a component of both EPS and LPS. Thus, it is likely that *opsG* is involved either in the synthesis or assembly of the rhamnose residue in LPS and EPS (Fig. 2). We explored this possibility by sequencing the *opsG* complementation unit and by comparing this sequence with entries in GenBank.

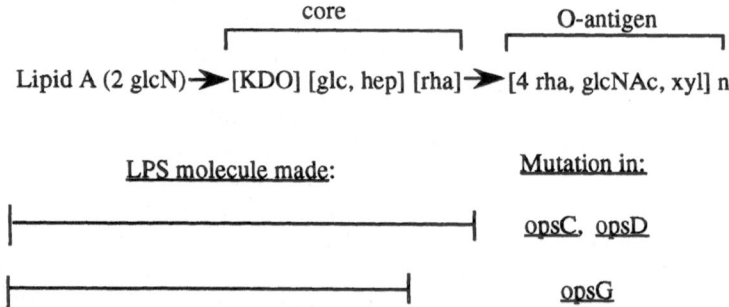

Figure 2. Diagram of the *P. solanacearum* LPS and the defects caused by mutations in the *opsC*, *opsD*, and *opsG* complementation units. Abbreviation: glcN, glucosamine; KDO, ketodeoxyoctonate; glc, glusoce; hep, heptose; rha, rhamnose; glcNAc, N-acetyl glucosamine; xyl, xylose. The 'n' denotes a repeating unit of the six sugars in the O-antigen.

In enteric bacteria, many of the genes involved in LPS synthesis have been identified and the specific enzymatic function of several of these gene products has been determined (for example: Collins and Hackett, 1991; Crowell *et al.*, 1987; Verma *et al.*, 1989), including the rhamnose synthetase gene, *rfb* (Jiang *et al.*, 1991). In these bacteria, the genes required for the synthesis of each moiety of the LPS molecule are located in different areas of the chromosome. However, the genes required for the synthesis of the core moiety or the O-antigen portion

Rhamnose Synthetase

Salmonella typhimurium rfb operon

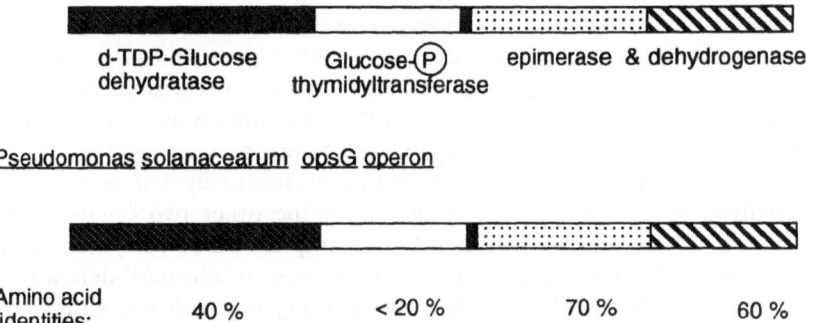

Figure 3. The *opsG* complementation unit has sequence similarity to the *rfb* operon of *S. typhimurium*, which encodes polypeptides required for the synthesis of rhamnose.

tend to be found in operons, and/or clusters (Crowell *et al.*, 1987; Makela *et al.*, 1980 and 1984).

In our laboratory, sequencing of *opsG* is nearly complete, and we have strong evidence that this gene is involved in rhamnose biosynthesis. The *opsG* complementation unit of *P. solanacearum* encodes at least four open reading frames (ORFs), whose translated sequences bear striking resemblance to the *rfb* operon of *Salmonella typhimurium* (Jiang *et al.*, 1991; Fig. 3). The *rfb* gene cluster encodes four enzymes responsible for the synthesis of rhamnose from glucose-1-phosphate (ibid): dTDP-D-glucose 4,6-dehydratase, glucose-1-phosphate thymidyltransferase, dTDP-6-deoxy-D-glucose 3,5, epimerase and dTDP-6-deoxy-L-mannose dehydrogenase. The latter two enzymes together constitute rhamnose synthetase (ibid) that converts TDP-activated glucose to TDP-rhamnose. The third and fourth ORFs in *opsG* have more than 60% identity – at the amino acid level – to the two polypeptides of the rhamnose synthetase. In these two rhamnose synthetase-like polypeptides of *opsG*, many of the divergencies in sequence to *rfb* are conserved in terms of the charge of the amino acid. We have also found a high degree of homology (approximately 40% at the amino acid level) between the first *opsG* open reading frame to the *rfb*-encoded dTDP-D-glucose 4,6-dehydratase. Interestingly, we have not found sequence similarities between the *rfb* glucose-1-phosphate thymidyltransferase and the sequence encoded by the corresponding open reading frame in *opsG*. The neighboring sugar residues present on *S. typhimurium* and *P. solanacearum* LPS are different and the divergence in the transferase function may reflect the accommodation to these differences. However, based upon the high degree of sequence conservation in three of the *opsG* open reading frames to *rfb* and the overall similarity in the position of the open reading frames, it is highly likely that *opsG* encodes enzymes responsible for the biosynthesis of TDP-rhamnose and for its assembly on the LPS molecule. Presumably, these enzymes play the same roles in assembly of rhamnose on the EPS molecule.

c) *Predicted function of the* opsC *and* opsD *Genes*

The same line of reasoning we used to derive valuable clues as to the function of the *opsG* complementation unit can be applied to *opsC* and *opsD*. Mutations in these two complementation units also lead to defects in the structure of the O-antigen moiety of the *P. solanacearum* LPS. This moiety is composed of the sugars: rhamnose, N-acetyl glucosamine (glcNAc), and xylose (Fig. 2). Since *opsG* evidently is involved in rhamnose synthesis, it is likely that *opsC* and *opsD* are involved in the synthesis or transport of the other two sugars. Neither glcNAc or xylose, however, have yet been implicated as components of *P. solanacearum* EPS. We expect that these two sugar, or chemical derivatives of them, will be found in EPS and that the inability to synthesize either sugars would result in defects in EPS synthesis and in a reduction in virulence.

IV. Regulation of EPS genes in *P. solanacearum*

There is substantial evidence that EPS production in plant-associated bacteria is subject to both negative and positive regulation. This appears to be the case in species of *Rhizobium*, for example, where the balance between activities of both types of regulators determines the amount of EPS that is produced (Gray and Rolfe, 1990; Gray *et al.*, 1990; Zhan and Leigh, 1990). In *P. solanacearum*, a potential candidate for a positive regulator of EPS expression is the product of the *phcA* gene which, as previously indicated, can complement EPS production in some but not in all spontaneous EPS mutants of *P. solanacearum* (Brumbley and Denny, 1990). This phenotype is expected since many genetic changes are associated with these phenotype shifts, even though they share the common EPS deficiency. For example, we have noticed that several independently isolated spontaneous mutants of strain K60 have incurred large deletions in areas of the chromosome that contain EPS biosynthetic genes (C. Kao, unpublished data). In addition, with other strains, Boucher and collaborators (1987, 1988) have reported that chemically mutagenized cells that simultaneously lose EPS and virulence also have a large DNA deletion in a megaplasmid that is commonly carried by virulent strains. These deletion mutants should not be reversible by *phcA*.

A potential negative regulator of EPS synthesis in *P. solanacearum*, named *epsR*, has also been described (Huang and Sequeira, 1990; Kao *et al.*, submitted). The activity of the *epsR* gene was first noticed when plasmid pBE6C containing genomic DNA from strain B1, a spontaneous derivative of strain K60, mimicked the phenotype of spontaneous mutants. In addition to shutting off EPS production, cells bearing pBE6C showed increased polygalacturonase and tyrosinase activity, but decreased virulence. Since pBE6C appeared to encode product(s) involved in the regulation of EPS production and virulence, it was of interest in terms of the control of pathogenicity and in terms of gene regulation in *P. solanacearum*. In the next section, we describe data which led us to believe that the *epsR* gene, which is contained in pBE6C, is responsible for these phenotypes. Furthermore, *epsR* appears to encode a protein that is a negative transcriptional regulator of EPS production (Kao *et al.*, submitted).

a) The epsR *gene*
Mutagenesis of pBE6C revealed a DNA region in the *P. solanacearum* genome that is responsible for the EpsR phenotype, i.e. the shut-off of EPS production in the first 4 days of growth on nutrient plates. In order to be certain that the plasmid containing B1 DNA encoded a protein that is functional in wt strains and is not a mutant product, we used the active locus in pBE6C to clone the homologous gene from the wt K60 strain. For this purpose, a subclone of pBE6C was used to screen a library of K60 genomic DNA and led to the identification of a plasmid named pKL44 which contains a 1.4 kb DNA region that shares common restriction sites with the insert from pBE6C. When electroporated into K60 or other wt strains, pKL44 and pBE6C were able to shut off EPS production, while the plasmid vector alone had no effect. Furthermore, approximately 4 days after electroporation, colonies containing either plasmid began to produce detectable amounts of EPS. This suggests that wt strains of *P. solanacearum* possess a functional *epsR* gene.

Saturation mutagenesis of pBE6 with Tn3-*gus* and subclones of the cosmids suggested that a single transcriptional unit of about 1.3 kb, named *epsR*, is responsible for the phenotype shift. Nucleotide sequencing of both the mutant and wt copies of the 1.3 kb sequence encompassing the *epsR* gene revealed that the sequences were identical, confirming results from the phenotypic assays described above.

Results from nucleotide sequencing and mutagenesis studies suggest that the putative EpsR protein consists of 236 amino acid residues encoded by an ORF present in a 1.4 kb restriction fragment (Kao *et al.*, submitted). The promoter for the *epsR* gene is present in this restriction fragment since subclones containing the fragment were functional even when we used a vector (pLAFR6) that contains transcriptional termination signals on both sides of the cloning site. Furthermore, deletion of the first 287 nucleotides of the 1.3 kb fragment (to an internal *Bam*HI site) eliminated EpsR activity, while an 8 bp insertion at the same *Bam*HI did not. At nucleotide 365 is a GUG codon which may serve as the translational start site of EpsR. Translation initiating at this position would result in a polypeptide of ca 26 kDa, in good agreement with the ca. 25 kDa protein observed in maxi-cell analysis (Huang and Sequeira, 1990).

To determine whether the GUG codon starting at nucleotide 365 is the initiation codon of the *epsR* gene, we made fusions with the promoter to the *P. solanacearum opsG* operon, which, as previously described, directs the synthesis of the rhamnose residue that is required for both EPS and LPS synthesis in *P.*

solanacearum. Prior experiments indicated that the *opsG* fragment was able to drive expression of a promoterless *gus* gene (C. Kao, unpublished data). By fusing *epsR* fragments with defined 5′ end points to the *opsG* promoter, we wanted to determine whether the EpsR phenotype requires the sequences at nucleotide 365 at the start of the GUG codon. The *opsG* promoter was synthesized by PCR as a fragment with a *Nde*I site (CATATG) at its 3′ end. A PCR fragment of the *epsR* gene from nucleotide 365 to 1170 was generated so that an *Nde*I site was added to the 5′ end, replacing the original GUG codon with an AUG codon from the *Nde*I restriction site. When cloned in pLAFR3 (pepsRA) this PCR fragment of the *epsR* sequence had no EpsR activity. However, when the same fragment was fused to the *opsG* promoter via the respective *Nde*I sites and cloned in pLAFR3 (Fig. 3), the resultant fusion construct, pG-epsRA, was able to shut off EPS production in strain K60. Colonies transformed with pG-epsRA were deficient in EPS production for approximately four days following electroporation. A second construct, pG-epsRB, which contains nucleotide 461 to 1170 of *epsR* fused to the *opsG* promoter, and cloned in pLAFR3, was not able to shutoff EPS expression. Therefore, the sequence between bp 365 and 461 was necessary for EpsR activity. Apparently, translation cannot start at or downstream of nucleotide 461 and result in a functional *epsR* product.

b) Possible function of the EpsR protein

The amino acid sequence translated from the putative EpsR polypeptide that presumably initiated at nucleotide 365 bears striking resemblance to a class of

```
BvgA       STLISVLSNR ELTVLQLLAQ GMSNKDIADS MFLSNKTVST YKTRLLQKLN ATSLVELIDL AKRNN
Dna5-ORF   SSTVTVLSNR EVTILRYLVS GLSNKEIADK LLLSNKTVSA HKSNIYGKLG LHSIVELIDY AKLYE
UhpA       DDANDILTKR ERQVAEKLAQ GMAVKEIAAE LGLSPKTVHV HRANLMEKLG VSNDVELARR MF.DG
UvrC-ORF2  ESPFASLSER ELQIMLMITK GQKVNEISEQ LNLSPKTVNS YRYRMFSKLN IHGDVELTHL AIRHG
GerE       FQSKPSLTKR EREVFELLVQ DKTTKEIASE LFISEKTVRN HISNAMQKLG VKGRSQAVVE LLRMG
DegU       RRPLHILTRR ECEVLQMLAD GKSNRGIGES LFISEKTVKN HVSNILQKMN VNDRTQAVVV AIKNG
NarL       ERDVNQLTPR ERDILKLIAQ GLPNKMIARR LDITESTVKV HVKHMLKKMK LKSRVEAAVW VHQER
FixJ       RARLQTLSER ERQVLSAVVA GLPNKSIAYD LDISPRTVEV HRANVMAKMK AKSLPHLVRM ALAGG
RcsB       GYGDKRLSPK ESEVLRLFAE GFLVTEIAKK LNRSIKTISS QKKSAMMKLG VENDIALLNY LSSVT
ComA       QKEQDVLTPR ECLILQEVEK GFTNQEIADA LHLSKRSIEY SLTSIFNKLN VGSRTEAVLI AKSDG
MalT       LIRTSPLTQR EWQVLGLIYS GYSNEQIAGE LEVAATTIKT HIRNLYQKLG VAHRQDAVQH AQQLL
AlkT       NKADALLTRK QIAVLRLVKE GCSNKQIATN MHVTEDAIKW HMRKIFATLN VVNRTQATIE AERQG
LuxR       NKSNNDLTKR EKECLAWACE GKSSWDISKI LGCSERTVTF HLTNAQMKLN TTNRCQSISK AILTG


Consensus  ------LT-R E--VL-L--- G-----IA-- L--S--TV-- H------KL- V-----AV-- A---G
EpsR       DCAMLGLTQR QYEILVLLSR GHPVKTISRM LGISEATTKA HINALYRRLE VRSRTEAIFV ATQRG
```

Figure 4. Comparison of the C-terminal sequence of EpsR to sequences of other bacterial regulatory proteins. Alignment of *P. solanacearum* EpsR residues 181–236 with the *E. coli*: DnaY-ORF (Maramatsu and Mixuno, 1990), UhpA (Friedrich and Kadner, 1987), UvrC-ORF2 (Moolenaar *et al.*, 1987), NarL (Gunzalus *et al.*, 1989), RcsB (Stout and Gottesman, 1990), MalT (Cole and Ribaud, 1986), the *B. pertussis*: BvgA (Ario *et al.*, 1989), the *B. subtilis*: GerE (Henner *et al.*, 1988), DegU (ibid., ComA (Weinrauch *et al.*, 1989), the *R. meliloti*: FixJ (David *et al.*, 1988), the *P. oleovorans*: AltK (Eggink *et al.*, 1990) and the *V. fischeri*: LuxR (Engebrecht and Silverman, 1987). A consensus sequence is defined by the residues conserved in identity or charge at more than nine residues out of 14 for each position. Computer alignment was performed using the Genetic Computer group sequence analysis software package (Devereux *et al.*, 1984). Fast A search (Pearson and Lipman, 1988) was conducted through the Genbank bionet services provided by the European Molecular Biology Laboratory.

bacterial regulatory proteins. Several of these polypeptides have been identified as members of the LuxR family of bacterial signal-transducing systems (reviewed in Gross *et al.*, 1989) that share extended homology, especially at their C-terminal end (Deteric *et al.*, 1986b, Gross *et al.*, 1989; Stock *et al.*, 1989). An alignment of all these polypeptides revealed 22 residues, mostly at the C-terminal end of each polypeptide, that are invariant in at least 9 of the 13 polypeptides. (Fig. 4). The C-terminal segment of the putative EpsR protein sequence has sequence identities at 15 residues to the highly conserved 22 residues. Of the remaining seven residues in EpsR which were different from the consensus sequence, five were conserved in terms of charge. One of the regulatory polypeptides to which EpsR bears striking homology is RcsB, which positively regulates capsule production in enteric bacteria (Stout *et al.*, 1990 and 1991). Thus, we suggest that the EpsR protein is a member of these DNA-binding regulatory proteins.

c) Mechanism of EpsR activity
The EpsR phenotype probably is mediated by the EpsR protein and is not a nucleotide sequence that merely titrates out factors that positively regulate EPS synthesis. Several lines of reasoning led to this conclusion. First, the intact ORF encoding the putative EpsR polypeptide is required in order to disrupt EPS biosynthesis; mutations at the 5', 3' ends or internal to the ORF abolished EpsR activity. Second, for activity, the *epsR* gene must be transcribed, either from its own promoter or from the *opsG* promoter. Third, the homology of the putative EpsR polypeptide sequence to other regulatory proteins is consistent with the concept that EpsR plays a regulatory role in EPS synthesis.

In plate culture the EpsR phenotype is maintained for about four days after electroporation. Afterwards, colony appearance changes as cells begin to produce EPS. We do not know whether the structure of the EPS produced by the transformants is chemically the same as that produced by wt *P. solanacearum*. It is also possible that the effect of EpsR is modulated by growth conditions or other cellular regulatory mechanisms. However, since EpsR expressed from the *opsG* promoter also eliminated EPS production for approximately four days, it is likely that modulation of EpsR activity occurs by a post-transcriptional mechanism. In *E. coli*, the regulation of capsule polysaccharide (CPS) production can be mediated through the Lon-dependent degradation of the RcsA protein (Stout *et al.*, 1991). Whether this type of regulation exists in *P. solanacearum* remains to be explored.

The concept that EPS production is regulated by environmental stimuli has been well established for *E. coli* capsular synthesis (Stout *et al.*, 1991), *P. aeruginosa* alginate synthesis (Deteric *et al.*, 1989a and 1989b), and *X. campestris* EPS synthesis (Daniel *et al.*, 1989; Ferris and Beveridge, 1985; Osbourn *et al.*, 1990; Tang *et al.*, 1990). If we assume that *epsR* is a member of an antagonistic plus-minus regulatory switch, there must be a positive regulator which has not yet been identified. The *phcA* gene of *P. solanacearum*, which encodes a potential positive regulator of EPS production (Brumbley and

Denny, 1990), cannot overcome the shut-off of EPS expression caused by *epsR* (C. Allen, personal communication). In addition, the EPS genes regulated by EpsR remain to be identified. The expression of the *ops* genes appears to be constitutive and unaffected by overexpression of EpsR. We are hopeful that the genes that are regulated by EpsR can be identified by standard bacterial genetic techniques and the use of the EpsR protein to fish out its cognate recognition site in the same way that the binding site for the mammalian p53 polypeptide was elucidated (Kern *et al.*, 1991). Once the EpsR binding site is determined, we will be able to use it to identify novel genes involved in EPS synthesis.

V. Summary

We are at a very exciting stage in these studies on the relationship of EPS biosynthesis to virulence of *P. solanacearum*. The requirement of wt EPS synthesis for full virulence is consistent and the biochemical function of at least one of the EPS biosynthetic genes can be predicted with reasonable accuracy. The characterization of the *ops* gene cluster has led to several testable hypotheses: 1) the *opsG* gene is responsible for rhamnose biosynthesis; 2) the *opsC* and *opsD* complementation units are involved in the biosynthesis of N-acetyl glucosamine and/or xylose; 3) some of the sugars affected by mutations in *ops* genes are found in EPS and LPS and their absence causes reduced virulence in *P. solanacearum*; 4) the polypeptide encoded by *epsR* is probably a DNA binding protein that regulates EPS expression. Further characterizations of the EPS biosynthetic genes and their regulation will be useful in deciphering EPS structure and in understanding the physiological interaction between the wilt pathogen *P. solanacearum* and its hosts.

Acknowledgements

We thank F. Gosti, A. Trigalet and C. Allen for sharing unpublished results and V. King for secretarial assistance. C. K. is supported by a National Science Foundation Plant Biology Postdoctoral fellowship DIR-9104366. This research was funded by NSF grant PCBB87-18310.

References

1. Akiyama, Y., Eda, S., Nishikamaji, S., Tanaka, H., and Ohnishi, A. 1986. Extracellular polysaccharide produced by a virulent strain (U-7) of *Pseudomonas solanacearum*. Ann. Phytopathol. Soc. Japan. 52: 741–744.
2. Allen, C., Huang, Y., and Sequeira, L. 1991. Cloning of genes affecting polygalacturonase production in *Pseudomonas solanacearum* polygalacturonase. Mol. Plant. Microbe Interact. 4: 147–154.

3. Arico, O.B., Miller, J.F., Roy, C., Stibitz, S., Monack, D., Falkow, S., Gross, S., and Rappouli, R. 1989. Sequences required for the expression of *Bordetella pertussis* virulence factors share homology with procaryotic signal transduction proteins. Proc. Nat. Acad. Sci. USA 86: 6671–6675.

4. Ayers, A.R., Ayers, S.B., and Goodman, R.N. 1979. Extracellular polysaccharides of *Erwinia amylovora:* a correlation with virulence. Appl. Environ. Microbiol., 138: 659–666.

5. Baker, J.M., Neilson, J., Sequeira, L., and Keegstra, K. G. 1984. Chemical characterization of the lipopolysaccharide of *Pseudomonas solancearum*. Appl. Environ. Microbiol. 47: 1096–1100.

6. Boucher, C.A., Gough, C.L., and Arlat, M. 1992. Molecular genetics of pathogenicity determinants of *Pseudomonas solanacearum* with special emphasis on *HRP* genes. Annu. Rev. Phytopathol. 30: 443–461.

7. Boucher, C.A., Barberis, P.A., and Arlet, M. 1988. Acridine orange selects for deletions of *hrp* genes in all races of *Pseudomonas solanacearum*. Mol. Plant Microbe Interact. 1: 232–238.

8. Boucher, C.A., Van Gijsegem, F., Barberis, P.A., Arlat, M., and Zischek, C. 1987. *Pseudomonas solanacearum* genes controlling both pathogenicity on tomato and hypersensitivity on tobacco are clustered. J. Bacteriol. 169: 5626–5632.

9. Brumbley, S.M. and Denny, T.P. 1990. Cloning of wild-type *Pseudomonas solanacearum phcA*, a gene that when mutated alters expression of multiple traits that contribute to virulence. J. Bacteriol. 172: 5677–5685.

10. Buddenhagen, I. and Kelman, A. 1964. Biological and physiological aspects of bacterial wilt caused by *Pseudomonas solanacearum*. Annu. Review Phytopathol. 2: 203–230.

11. Cole, S.T. and Raibaud, O. 1986. The nucleotide sequence of the *malT* gene encoding the positive regulator of the *Escherichia coli* maltose regulon. Gene 42: 201–208.

12. Collins, L.V. and Hackett, J. 1991. Molecular cloning, characterization, and nucleotide sequence of the *rfc* gene, which encodes an O-antigen polymerase of *Salmonella typhimurium*. J. Bacteriol., 173: 2521–2529.

13. Cook, D. and Sequeira, L. 1991. Genetic and biochemical characterization of a gene cluster from *Pseudomonas solanacearum* required for extracellular polysaccharide production and for virulence. J. Bacteriol. 173: 1654–1662.

14. Coplin, D.L. and Cook, D. 1990. Molecular genetics of extracellular polysaccharide biosynthesis in vascular phytopathogenic bacteria. Mol. Plant-Microbe Interact. 3: 271–279.

15. Crowell, D.N., Reznikoff, W.S., and Raetz, C.R. 1987. Nucleotide sequence of the *Escherichia coli* gene for lipid A disaccharide synthase. J. Bacteriol., 169: 5727–5734.

16. Daniels, M.J., Osbourn, A.E., and Tang, J.L. 1989. Regulation in *Xanthomonas*-plant interactions. p. 189–196. *In*: B. J. J. Lugtenberg (ed.) Signal Molecules in Plants and Plant-Microbe Interactions. NATO ASI Series, Vol. H36. Springer-Verlag, Berlin.

17. David, M., *et al.* 1988. Cascade regulation of *nif* gene expression in *Rhizobium meliloti*. Cell 54: 671–683.

18. Denny, T.D., Makini, F.W., and Brumbley, S.M. 1988. Characterization of *Pseudomonas solanacearum* Tn5 mutants deficient in extracellular polysaccharide. Mol. Plant-Microbe Interact. 1: 215–223.

19. Denny, T.P. and Baek, S.R. 1991. Genetic evidence that extracellular polysaccharide is a virulence factor of *Pseudomonas solanacearum*. Mol. Plant-Microbe Interact. 4: 198–206.

20. Deteric, V., Konyecsni, W.M., Mohr, C.D., Martin, D.W., and Hibler, N.S. 1989a. Common denominators of promoter control in *Pseudomonas* and other bacteria. Biotechnology 7: 1249–1254.

21. Deteric, V., Dikshit, R., Konyecsni, W.M., Chakrabarty, A.M., and Mista, T.K. 1989b. The *algR* gene, which regulates mucoidy in *Pseudomonas aeruginosa*, belongs to a class of environmentally responsive genes. J. Bacteriol. 171: 1278–1283.

22. Devereux, J., Haeberli, P., and Smithies, O. 1984. A comprehensive set of sequence analysis programs for the VAX. Nucl. Acids Res. 12: 387–395.

23. Djordjevic, S.P., Chen, H., Redmond, J.W., and Rolfe, B. 1987. Nitrogen-fixing ability of exopolysaccharide synthesis mutants of *Rhizobium* sp. strain NGR234 and *R. trifolii* by the

addition of homologous exopolysaccharide. J. Bacteriol. 169: 53–60.

24. Drigues, P., Demmery-Laffergue, D., Trigalet, A., Dupin, P., Samain, D., and Asselineau, J. 1985. Comparative studies of lipopolysaccharide and exopolysaccharide from a virulent strain of *Pseudomonas solanacearum* and from three avirulent mutants. J. Bacteriol. 172: 504–509.

25. Duvick, J.P. and Sequeira, L. 1984. Interaction of *Pseudomonas solanacearum* lipopolysaccharide with an agglutinin from potato tubers. Appl. Environ. Microbiol. 48: 192–198.

26. Eggink G., Engel, H., Vriend, G., Terpstra, P., and Witholt, B. 1990. Rubredoxin reductase of *Pseudomonas oleovorans*. Structural relationship to other flavoprotein oxidoreductases based on one NAD and two FAD fingerprints. J. Mol. Biol. 212: 135–142.

27. Engebrecht, J. and Silverman, M. 1987. Nucleotide sequence of the regulatory locus controlling expression of bacterial genes for bioluminescence. Nucl. Acids Res. 15: 10455–10467.

28. Ferris, F.G. and Beveridge, T.J. 1985. Functions of bacterial cell surface structures. BioScience 35: 172–177.

29. Friedrich, M.J. and Kadner, R.J. 1987. Nucleotide sequence of the *uhp* region of *Escherichia coli*. J. Bacteriol.169: 3556–3563.

30. Gray, J.X. and Rolfe, B.G. 1990. Exopolysaccharide production in *Rhizobium* and its role in invasion. Mol. Microbiol. 4: 1425–1431.

31. Gray, J.X., Djordjevic, M.A., and Rolfe, B.G. 1990. Two genes that regulate exopolysaccharide production in *Rhizobium* sp. strain NGR234; DNA sequences and resultant phenotypes. J. Bacteriol. 172: 193–203.

32. Gross, R., Arico, B., and Pappuoli, R. 1989. Families of bacterial signal-transducing proteins. Mol. Microbiol. 3: 1661–1667.

33. Gunzalus, R.P., Kalman, L.V., and Stemart, R.R. 1989. Nucleotide sequence of the *narL* gene that is involved in global regulation of nitrate controlled respiratory genes of *Escherichia coli*. Nucl. Acids Res. 17: 1965–1975.

34. Hayward, A.C. 1991. Biology and epidemiology of bacterial wilt caused by *Pseudomonas solanacearum*. Annu. Rev. Phytopathol. 29: 65–87.

35. Hendrick, C.A. and Sequeira, L. 1984. Lipopolysaccharide-defective mutants of the wilt pathogen *Pseudomonas solanacearum*. Appl. Environ. Microbiol. 48: 94–101.

36. Henner, D.J., Yang, M., and Ferrari, E. 1988. Localization of *Bacilus subtilis sacU* (Hy) mutations to two linked genes with similarities to the conserved procaryotic family of two-component signaling systems. J. Bacteriol. 170: 5102–5109.

37. Hoiseth, S., Connelly, C.J., and Moxon, R. 1985. Genetics of high frequency loss of β capsule expression in *Haemophilus influenza*. Infect. Immun., 49: 389–395.

38. Huang, Y. and Sequeira, L. 1990. Identification of a locus that regulates multiple functions in *Pseudomonas solanacearum*. J. Bacteriol. 172: 4728–4731.

39. Husain, A. and Kelman, A. 1958. Relation of slime production to mechanism of wilting and pathogenicity of *Pseudomonas solanacearum*. Phytopathol. 48: 155–164.

40. Jiang, X.M., Neal, B., Santiago, F., Lee, S.J., Romana, L. K., and Reeves, P. R. 1991. Structure and sequence of the *rfb* (O antigen) gene cluster of *Salmonella* serovar *typhimurium* (strain LT2). Molec. Microbiol. 5: 695–713.

41. Kamoun, C., Cooley, M.B., Rogowsky, P.M., and Kado. C. I. 1989. Two chromosomal loci involved in the production of exopolysaccharide in *Agrobacterium tumefaciens*. J. Bacteriol., 171: 1755–1759.

42. Kao, C., Gosti, F., Huang, Y., and Sequeira, L. Characterization of a negative regulator of exopolysaccharide production in the plant pathogenic bacterium, *Pseudomonas solanacearum*. (submitted to Mol. Plant-Microbe Interact.).

43. Kao, C., Barlow, E., and Sequeira, L. 1992. Extracellular polysaccharide is required for wildtype virulence of *Pseudomonas solanacearum*. J. Bacteriol. 174: 1068–1071.

44. Kao, C. and Sequeira, L. 1991. A gene cluster required for the coordinated biosynthesis of both lipopolysaccharides and exopolysaccharide also affects virulence of *Pseudomonas solanacearum*. J. Bacteriol. 173: 7841–7848.

45. Kelman, A. 1954. The relationship of pathogenicity of *Pseudomonas solanacearum* to colony

appearance on a tetrazolium medium. Phytopathology 44: 693–695.

46. Kern, S.E., Kinzler, K.W., Bruskin, A., Jarosz, D., Friedman, P., Prives, C., and Vogelstein, B. 1991. Identification of p53 as a specific DNA binding protein. Science 252: 1708–1711.

47. Leigh, J.A., Signer, E.R., and Walker, G.C. 1985. Exopolysaccharide deficient mutants of *Rhizobium meliloti* that form ineffective nodules. Proc. Natl. Acad. Sci. USA 82: 6231–6235.

48. Lozano, J.C. and Sequeira, L. 1970. Differentiation of races of *Pseudomonas solanacearum* by a leaf infiltration technique. Phytopathol. 60: 833–838.

49. Makela, P.H., Bradley, D.J., Brandis, H., Frank, M.M., Hahn, H., Henke, L.W., Jann, K., Morse, S.A., Robbins, J.B., Rosenstreich, D.L., Smith, H., and Timmis, K. 1980. Evasion of host-defenses. Group report. *In:* Smith, H., Skehel, J.J., and Turner, M.J. (ed). The Molecular Basis of Pathogenicity. Verlag Chenie, Weinheim.

50. Makela, P.H. and Stocker, B.A.D. 1984. Genetics of Lipopolysaccharide. *In* E.T. Rietschel (ed). The chemistry of endotoxin. p. 59–137. Elsevier Biomedical Press, New York.

51. Maramatsu, S. and Mixuno, T. 1990. Nucleotide sequence of the region encompassing the *int* gene of a cryptic prophage and the *dna Y* gene flanked by a curved DNA sequence of *Escherichia coli* K12. Mol. Gen. Genet. 220: 325–328.

52. Moolenaar, G.F., Van Sluis, C.A., Backendorf, C., and Van de Putte, P. 1987. Regulation of the *Escherichia coli* excision repair gene *uvrC*. Overlap between the *uvrC* structural gene and the region coding for a 24 kDa protein. Nucl. Acids Res. 15: 4273–4289.

53. Nohno, T., Noji, S., Taniguchi, S., and Saito, T. 1989. The *narX* and *narL* genes encoding the nitrate-sensing regulator of *Escherichia coli* are homologous to a family of procaryotic two-component regulatory genes. Nucl. Acids Res. 17: 2947–2957.

54. Orgambide, G., Montrozier, H., Servin, P., Roussel, J., Demery, D., and Trigalet, A. 1991. High heterogenity of the exopolysaccharides of *Pseudomonas solanacearum* and the complete structure of the major polysaccharide. J. Biol. Chem. 266: 8312–8321.

55. Osbourn, A.E., Clarke, B.R., Stevens, B.J.B., and Daniels, M.J. 1990. Use of oligonucleotide probes to identify members of two-component regulatory systems in *Xanthomonas campestris* pathovar *campestris*. Mol. Gen. Genet. 222: 145–151.

56. Pearson, W.R. and Lipman, D.J. 1988. Improved tools for biological sequence comparison. Proc. Natl. Acad. Sci. USA 85: 2444–2448.

57. Phelps, R.H. and Sequeria, L. 1968. Auxin biosynthesis in a host-parasite complex. pp. 197–212. *In:* F. Wightman and G.D. Setterfield (ed). "Biochemistry and Physiology of Plant Growth Substances". Runge Press, Ottawa.

58. Schell, M.A., Roberts, D.P., and Denny, T.P. 1987. Analysis of the spontaneous mutation to avirulence by *Pseudomonas solanacearum*. p. 61–66. *In:* D.P.S. Verma and N. Brisson (ed.). "Molecular Genetics of Plant Microbe Interaction", Proc. 3rd. Intl. Symp., Martinus Nijhoff, The Netherlands.

59. Sequeira, L. and Graham, T.L. 1977. Agglutination of avirulent strains of *Pseudomonas solanacearum* by potato lectin. Physiol. Plant Pathol. 11: 43–54.

60. Stemmer, W.P.C. and Sequeria. L. 1987. Fimbriae of phytopathogenic and symbiotic bacteria. Phytopath. 77: 1633–1639.

61. Stock, J.B., Ninfa, A.J., and Stock, A. 1989. Protein phosphorylation and regulation of adaptive responses in bacteria. Microb. Revs. 53: 450–490.

62. Stout, V., Torres-Cabassa, A., Maurizi, M.R., Gutnick, D., and Gottesman, S. 1991. RcsA, an unstable positive regulator of capsular polysaccharide synthesis. J. Bacteriol. 173: 1738–1747.

63. Stout, V. and Gottesman, S. 1990. RcsB and RcsC: a two-component regulator of capsule synthesis in *Escherichia coli*. J. Bacteriol. 172: 659–669.

64. Tang, J.L., Gough, C.L., and Daniels, M.J. 1990. Cloning of genes involved in negative regulation of production of extracellular enzymes and polysaccharides of *Xanthomonas campestris* pathovar *campestris*. Mol. Gen. Genet. 222: 157–160.

65. Van Alfen, N.K. 1982. Wilts: concepts and mechanisms. p. 459–474. *In:* M.S. Mount and G.H. Lacy (ed.). Phytopathogenic Procaryotes. Vol. 1. Academic Press. Inc. (London). Ltd., London.

66. Van Alfen, N.K., McMillan, B.D., Turner, V., and Hess, W.M. 1983. Role of pit membranes

in macromolecular-induced wilt of plants. Plant Physiol., 73: 1020–1023.

67. Verma, N.K. and Reeves, P.R. 1989. Identification and sequence of *rfbS* and *rfbE* which determines antigenic specificity of group A and group D *Salmonella*. J. Bacteriol. 171: 5694–5701.

68. Weinrauch, Y., Guillen, N., and Dubnau, D.A. 1989. Sequence and transcription mapping of *Bacillus subtilis* competence genes *comA* and *comB*, one of which is related to a family of bacterial regulatory determinants. J. Bacteriol. 171: 5362–5375.

69. Xu, P., Leong, S, and Sequeira, L. 1988. Molecular cloning of genes that specify virulence in *Pseudomonas solanacearum*. J. Bacteriol. 170: 617–622.

70. Xu, P., Iwata, M., Leong, S., and Sequeira, L. 1990. Highly virulent strains of *Pseudomonas solanacearum* that are defective in extracellular polysaccharide production. J. Bacteriol.172: 3946–3951.

71. Young, D.H. and Sequeira, L. 1986. Binding of *Pseudomonas solanacearum* fimbriae to tobacco leaf cell walls and its inhibition by bacterial extracellular polysaccharides. Physiol. Mol. Plant Pathol. 28: 393–402.

72. Zhan, H. and Leigh, J.A. 1990. Two genes that regulate exopolysaccharide production in *Rhizobium meliloti*. J. Bacteriol. 172: 5254–5259.

8. Effects of Opa proteins and lipooligosaccharides on surface charge and biological behavior of gonococci

JOHN SWANSON

Abstract. Three classes of the gonococcus' surface-exposed components – pili, outer membrane opacity-associated proteins, and lipooligosaccharides – vary extensively *in vitro* and *in vivo*. Pili are required for virulence. Opacity-associated proteins and lipooligosaccharides have poorly-defined pathogenetic roles, but they affect several facets of gonococci behavior: clumping; epithelial cell adherence; phagocytosis by neutrophils; and killing by complement.

The electrokinetic characteristics of gonococci with desired opacity-associated proteins, and lipooligosaccharides phenotypes were evaluated by electrophoresis of whole, viable cells. Both of these classes of outer membrane components contribute to cell surface charge. Opacity-associated (Opa) proteins influence charge of gonococci depending on their binding exogenous polyanions (such as DNA, heparin); when they express certain opacity-associated proteins, gonococci acquire a "polyanion capsule" which diminishes their adherence to tissue culture cells.

Abbreviations: EPM, Electrophoretic Mobility (of Whole, Intact Cells); Gc, Gonococcus; KDO, Ketodeoxyoctanate; LOS, Lipooligosaccharides; LPS, Lipopolysaccharide; Opa, Opacity-Associated Protein; SDS-PAGE, Sodium Dodecyl Sulfate-Polyacrylamide Gel Electrophoresis

Introduction

Bacteria interact with the 'outside world' mainly via their surface components. Many interactions, such as the binding and transport of sugars, amino acids, or ions, utilize specific outer membrane proteins. Others are mainly dictated by overall physicochemical properties of the bacterial cells including surface charge and hydrophobicity (Krekeler *et al.*, 1989); these include many interactions of bacteria with one another (aggregation) and with heterologous surfaces (colonization, 'fouling') as well as their permeabilities and susceptibilities to detergents and to antibiotics such as polymixin B.

A priori, the physical properties of bacterial cells are generated by their surface-exposed lipid, carbohydrate, and protein molecules, either alone or in combination. In some instances a single component, such as a polysaccharide capsule, has a dominant effect on surface properties which differ depending on the capsule's composition (Bayer & Sloyer, 1990). Noncapsulated Gram-negative bacteria can differ significantly in charge and hydrophobicity depending on whether they synthesize smooth versus rough lipopolysaccharide (LPS) molecules (James, 1957). Virtually nothing is known about the possible contributions of outer membrane proteins to surface charge of Gram-negative

C.I. Kado and J.H. Crosa (eds.), Molecular Mechanisms of Bacterial Virulence, 109–125.
© *1994 Kluwer Academic Publishers.*

bacteria. Secondary structure algorithms will predict which regions of polypeptide are exposed on the molecule's surface or are buried in its interior, and these algorithms are often applied to predict the portions of membrane proteins that are exposed on the membrane surface. These predictions generally conform to assessments of bacterial outer membrane protein surface exposure by antibody binding, protease susceptibility, and radiolabeling of whole cells. Porins and other outer membrane proteins typically carry charged amino acids in their surface-exposed regions and have slightly alkaline pI's, but whether their charged residues contribute to overall cell charge has not been defined.

The pathogenic personalities of gonococci (Gc) correlate with particular surface-exposed components including pili, opacity-associated (Opa) proteins, and lipooligosaccharide (LOS). Pili are necessary for pathogenicity, apparently by enhancing adherence of Gc to host mucosal surfaces and promoting their colonization (Swanson et al., 1987). The pathogenetic roles of the Opa protein family remain incompletely-defined. Various Opa proteins confer strikingly different degrees of opacity on Gc colonies, apparently correlating with the amount of intercellular aggregation that they induce (Swanson, 1982). The majority of Gc isolated from an infected male's urethra express one, another, or several Opa proteins even when infection was initiated with Opa$^-$ cells (James & Swanson, 1978; Swanson et al., 1988). But Gc recovered from the Fallopian tubes of infected females or from systemic sites typically are Opa$^-$, even when Opa$^+$ variants abound in other locales (cervix, rectum) of the same individual (Draper et al., 1980). Gc that express certain Opa proteins exhibit striking adherence to tissue culture cells and neutrophils, in vitro (Draper et al., 1985; Fischer & Rest, 1988; Belland et al., 1992), but whether these functions have relevance in vivo is unsettled. The LOS that is expressed by Gc and whether or not it is substituted with sialic acid groups dictates a number of their properties, including their sensitivity to killing by serum + complement (Rice, 1989; Smith, 1990). The molecular details for these modified behaviors of Gc are not clearly elucidated, and it is not at all clear whether 'specific' or 'nonspecific' cell surface attributes are involved.

On the assumption that Gc surface charge may contribute to their biological behaviors, the electrophoretic characteristics of intact, whole organisms that express selected outer membrane components were examined. The results indicate that both LOS and outer membrane Opa proteins – individually and together – influence cell charge. Polyanions including DNA and sulfated polysaccharides accumulate on Gc expressing those Opa proteins with the largest net excesses of positively-charged amino acids in their predicted surface-exposed regions; this polyanion accretion confers highly negative charges on the Opa$^+$ cells and markedly diminishes their adherence to tissue culture cells in vitro.

Results

EPM of LOS variants

All organisms examined were from strain MS11$_{MK}$ (Swanson *et al.*, 1985). Three LOS phenotypes (LOSa, LOSa/b, LOSb) are distinguishable in this strain by their profiles after SDS-PAGE coupled with silver staining and by immunoblotting with monoclonal antibodies αLOSa and αLOSb. LOSa has smaller apparent size and reacts only with the former monoclonal either on whole cells or by immunoblotting after SDS-PAGE. LOSb has a major band of larger apparent size that reacts with αLOSb. The third variant (LOSa/b) seems to make both LOSa and LOSb molecules, apparently in equal numbers. When Gc are incubated with cytidine-monophosphate-N-acetyl neuraminic acid (CMP-NA) (Parsons *et al.*, 1988; Parsons *et al.*, 1989), LOSb and LOSa/b add sialic acid groups; LOSa does not.

The contribution of LOS to EPM of Gc is evident on comparing pilus⁻ Opa⁻ cells of the three LOS phenotypes (LOSa = −0.8, LOSa/b = −0.5, LOSb = −0.2μM-cm/V-s). Unlike Opa⁺ cells (see below), Opa⁻ variants exhibited no changes in EPM when grown on different media. Sialylation of LOSa/b and LOSb variants markedly increases EPM negativity of their Opa⁻ cells to approximately equivalent levels (−1.0 μM-cm/V-s); accordingly, the latter show approximately twice the enhanced negativity as the LOSa/b 'mosaic.' Sialylated cells show a single, sharp EPM peak.

Table 1. Opa expression.

	HV1			HV2			Total Net	pI	DNA-binding	
	K, R, (H)	D, E	net	K, R, (H)	D, E	net			WC	blot
OpaA	6 (8)	1	+5 (7)	7 (8)	3	+ 4 (5)	+9 (12)	9.836	++++	++++
OpaB	5	3	+ 2	6 (8)	4	+2 (4)	+4 (6)	9.715	+	+
OpaC	6 (7)	4	+2 (3)	10 (11)	4	+ 6 (7)	+8 (10)	9.702	++	+++
OpaD	5	5	0	6 (8)	5	+1 (3)	+1 (3)	9.310	-	+/-
OpaE	5	5	0	5 (7)	3	+2 (4)	+2 (4)	9.340	-	-
OpaF	7 (9)	3	+4 (6)	4 (6)	2	+2 (4)	+6 (10)	9.808	++	-
OpaG	6	3	+ 3	6 (8)	5	+1 (3)	+4 (6)	9.495	+	+/-
OpaH	6 (8)	3	+3 (5)	4 (6)	2	+2 (4)	+5 (9)	9.700	++	+
OpaI	6	4	+ 2	5	4	+1	+3	9.500	+	+/-

HV1 and **HV2** denote the two major 'hypervariable' and predicted surface-exposed domains for Opa proteins, as reported by Bhat, et al., 1991.

K, R, H, D, and E denote lysine, arginine, hystidine, aspartic acid and glutamic acic, respectively that reside in HV1 and HV2. K, R, and (H) have positively-charged groups while D, and E are negatively charged. Calculations of net charge are shown with and without inclusion of H.

Calculations of **pI** are with MacVector program on deduced Opa sequences.

Opa⁺ variants in strain MS11
Nine Opa proteins were identified in strain MS11. Their sequences deduced from the respective, cloned *opa* genes predict alkaline pI's and extensive surface-exposed regions that contain abundant charged residues (Bhat *et al.*, 1991) (Table 1). On that basis, it was anticipated that Opa expression might affect gonococcal surface charge.

EPMs of agar-grown Opa⁺ variants
Initially, the Opa⁺ Gc whose EPMs were measured came from cultures on agar-containing solid medium or in liquid. The influence of each Opa on EPM was defined by comparing Opa⁺ versus Opa⁻ cells of the same LOS phenotype.

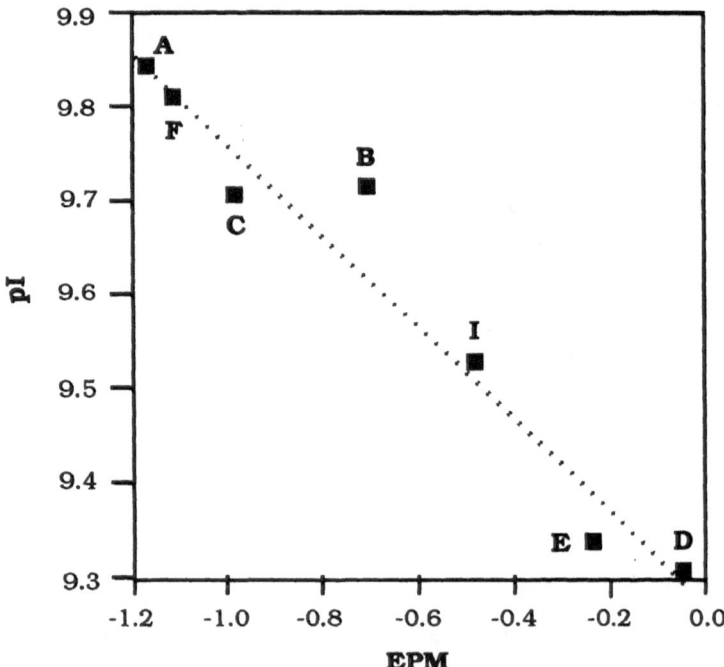

Figure 1. EPM of Opa⁺ LOSa cells versus Opa pI. Gc were swabbed from phosphate-buffered, agar containing solid medium (Swanson, 1978; Swanson, 1981) at 37°C in 5% CO_2 for 16 – 18 h and suspended (OD540 = 0.5 – 1) in 4 ml phosphate-buffered saline with the following composition: Na_2HPO_4 = 6.82 g/L; KH_2PO_4 = 2.54 g/L; NaCl = 4.24g/L; pH = 7.0). The Gc suspension was introduced into a DELSA 440 (Coulter Corp.) and electrophoretic mobility (EPM) was measured. The outputs from all four detectors that are set at different angles were averaged, and the results expressed in μM-cm/V-s (or velocity in μM/s per voltage in V/cm). At least three different EPM assessments were made on each preparation and the different values were averaged.

For Opa⁺ Gc (OpaA⁺ = A, etc.) propagated on agar-containing medium, their EPMs are inversely related to the pI's calculated for their respective Opa proteins as shown also in Table 1.

Although the values for some Opa⁺ cells (particularly OpaA⁺ and OpaC⁺) varied somewhat and sometimes yielded several small highly-negative peaks, the EPMs of Opa⁺ variant generally correlated quite well, but *inversely,* with the charge predicted for their particular Opa polypeptide (Swanson, 1991) (Fig. 1).

This inverse relationship between EPM and predicted charge for each Opa was found for Opa⁺ variants regardless of their LOS phenotype; but each Opa endowed LOSb cells with a more pronounced negative EPM change than LOSa cells expressing the same Opa (Fig. 2).

Opa influence on EPM [(Opa⁺) - (Opa⁻)]

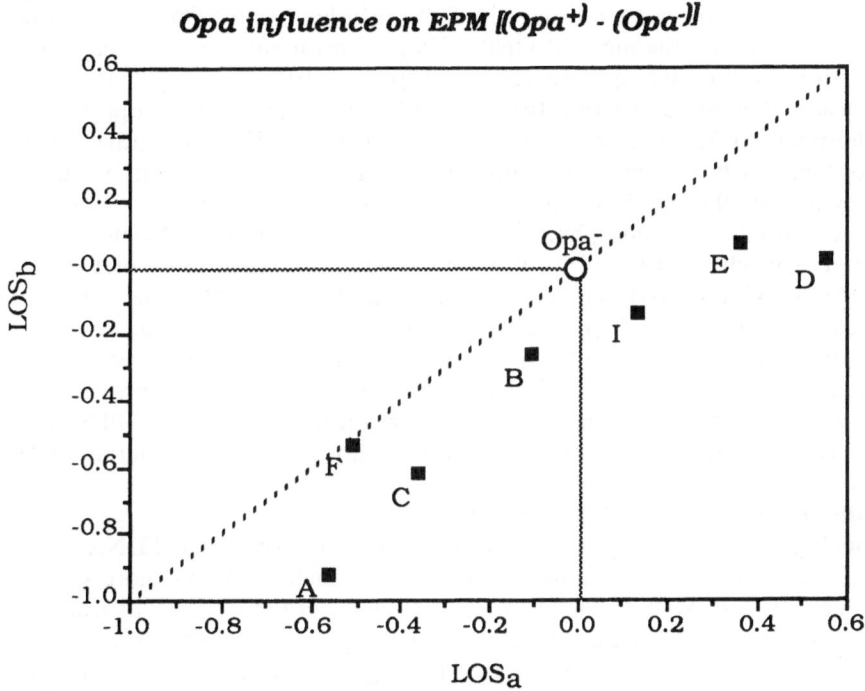

Figure 2. Opa influence on EPM of LOSa versus LOSb cells. EPMs were determined for Opa⁻ and Opa⁺ variants (OpaA⁺ = A, etc.) of both LOSa and LOSb cells grown on agar-containing medium. The values for Opa⁺ Gc were then 'corrected' by subtracting the EPM of Opa⁻ variants of the respective LOS phenotypes, and the differences are shown. These EPM differences are consistently larger for Opa⁺ variant expressing LOSb than for LOSa (e.g. LOSb OpaA = −0.92 μM-cm/V-s, LOSa OpaA = −0.055). This suggests that surface exposure of these Opa proteins is greater on LOSb than LOSa cells.

OpaA has the largest predicted net excess of positively-charged amino acids (K, R, H) in its exposed portions (Table 1), but it endows cells with the most negative EPM when compared to Opa⁻ variants of the corresponding LOS phenotype (Figs. 1, 2). When OpaA⁺ and OpaC⁺ cells were examined from 'old' (>18 h, 37°C) cultures from solid medium, as suspensions that were maintained in PBS for >15 min, or as liquid-grown organisms, they typically displayed a complex series of peaks with very negative EPMs in the $1.0 - 2.0$ μM-cm/V-s range. These negative peaks of agar-grown OpaA$^+$ cells largely disappeared upon treatment with DNase which altered their EPM to slightly less positive values than those of control Opa$^-$ cells. When agar-propagated OpaA$^+$ cells were incubated with proteinase K, their EPM changed to resemble that for Opa$^-$ cells of the same LOS phenotype.

EPMs of agarose-grown Opa$^+$ variants

The results obtained above suggested that polyanions from the medium might be responsible for the highly negative EPM profiles of some Opa$^+$ Gc. This was corroborated by finding that Opa$^+$ cells grown on solid medium containing HEPES buffer and agarose (H/As) displayed EPMs that generally were considerably more positive than those for the same cells propagated on phosphate-buffered, agar-containing medium (P/Ar) (Fig. 3). Neither OpaD$^+$ nor OpaE$^+$ cells exhibited this difference and resembled Opa$^-$ variants in that they had similar EPMs when grown on the two media. Growth of the other Opa$^+$ variants on phosphate-buffered medium containing Noble agar or agarose produced EPMs that were intermediate between those of P/Ar and H/As. Striking EPM differences were also found for cells incubated in biphasic media consisting of HEPES-buffered liquid that overlay either an agar or an agarose plug. The relative EPMs of Opa$^+$ cells from H/As displayed less clear relationships to the predicted charges of their particular Opa proteins than found for cells grown on P/Ar. Opa$^-$ organisms grown in HEPES versus phosphate-buffered liquid media did not display significantly differing EPMs.

Influence of DNA on EPM of Opa$^+$ Gc

Incubation of pilus$^-$ OpaA$^+$ cells with plasmid or chromosomal DNA or with synthetic oligonucleotides markedly changed their EPM to more negative values ($\geq$ 1.9 μM-cm/V-s). These EPM changes of OpaA$^+$ cells depended on the amount of DNA added (range 0.01 – 10 μg/ml). Some Opa$^+$ variants (OpaD, E, I) showed no such EPM change regardless of the amount of DNA to which they were exposed, and in this regard they resembled Opa$^-$ Gc; others (OpaC, F, G, H) had altered EPM at the highest concentrations of DNA (10 μg/ml) but no change at lower DNA concentrations (e.g. 1 μg/ml).

DNA binding by Gc

DNA binding to Gc and their components was assessed by two ways. In the first, whole Gc were incubated with pGEM, were washed several times, and were then subjected to SDS and proteinase K digestions followed by phenol extractions identical to those employed for recovery of total genomic DNA from Gc. These extracts were used for slot-blot Southern hybridizations with [^{32}P]pGEM to define the amount of this plasmid that had accreted on the intact organisms. OpaA$^+$ and OpaC$^+$ variants bound much more pGEM than the other Opa$^+$ variants, with OpaA$^+$ > OpaC$^+$. OpaB$^+$, F$^+$, G$^+$, and H$^+$ variants bound less plasmid, but discernibly more than Opa$^-$ cells; OpaD$^+$, E$^+$, and I$^+$

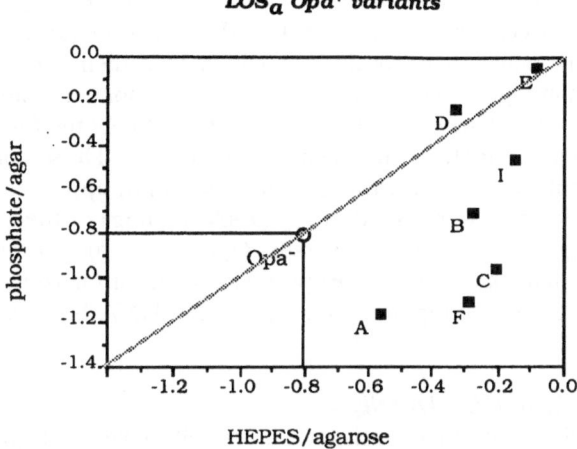

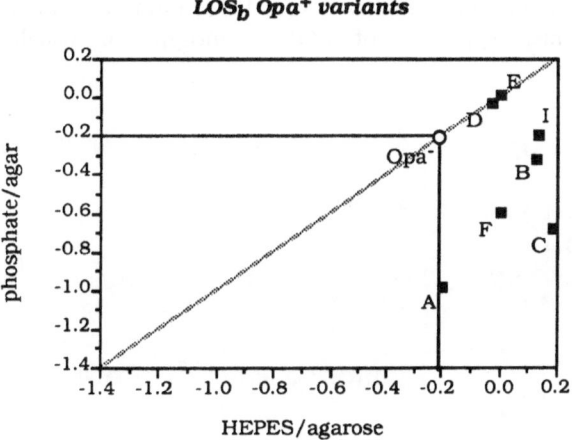

Figure 3. Comparison of EPMs for LOSa and LOSb Gc propagated on phosphate-buffered agar medium and on HEPES-buffered agarose medium. The HEPES buffered medium with agarose had the following formulation: Proteose peptone #3 (BBL) 15 g/L; soluble starch (BBL) 0.5 g/L; HEPES salt [N-(2-hydroxyethyl) piperazine-N'-(2-ethanesulfonic acid, sodium salt] (Sigma Chem. Co.), 2.15 g/L; HEPES acid (United States Biochemical Corp.), 5 g/L; NaCl, 5 g/L; agarose (SeaKem, GTG), 0.8 g/L. The pH was adjusted to 8.0 prior to autoclaving. IsoVitaleX (BBL) was added to cooled medium before plates were poured. Several batches of agarose were tried, but several did not support the growth of OpaA$^+$ cells; all other variants were successfully propagated on all agarose preparations.

The EPMs for Opa-cells of either LOS phenotype are identical on the two media (-0.8 μM-cm/V-s for LOSa, -0.2 for LOSb). Most Opa$^+$ variants exhibit much more negative EPM when grown on agar than on agarose (e.g. LOSa OpaA$^+$ = -1.2 μM-cm/V-s on agar and -0.58 on agarose). This is attributed to their adsorbing polyanions (sulfated and pyruvated polysaccharides) that 'contaminate' agar but are absent from agarose. The EPMs of OpaD$^+$ and OpaE$^+$ cells are unaltered by growth on the two media.

cells resembled Opa⁻ variants and bound little plasmid (Fig. 4). In the second assay, whole Gc were subjected to SDS-PAGE with subsequent transfer of the separated components to nitrocellulose which was then probed with [^{32}P]-pGEM, as seen in Fig. 5. Several unidentified Gc components bind DNA in this experimental format. One of these ($M_r = 32 - 33$) is likely the Rmp (*reduction-modified protein* or P.III) outer membrane protein. Those components of smaller apparent size are unidentified, but they do not appear to be LOS (see Fig. 5). No differences were noted in the DNA bindings of these unidentified, non-Opa components for Opa⁻ versus Opa⁺ cells or for different LOS phenotypes. Striking labeling by [^{32}P]-pGEM occurs for OpaA and OpaC and, at much lower levels, for OpaB, F, G, and H. No DNA binding was seen for OpaD, E, or I.

Attachment of Gc to HEC-1B cells
The levels to which Gc adhere to tissue culture cells *in vitro* is highly dependant on their pilus and Opa constitutions. Pilus⁺ Gc, regardless of their Opa phenotype, exhibit much greater adherence to a wide variety of eukaryotic cells in culture than pilus⁻ variants (Swanson, 1973; Lambden *et al.*, 1980). Pilus⁻ Opa⁻ organisms, regardless of LOS phenotype or whether they were

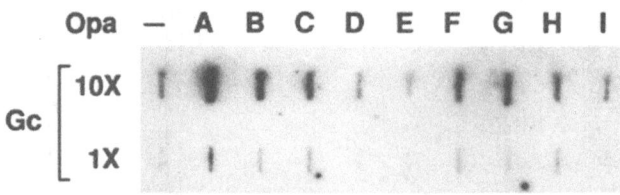

Hybridized with [^{32}P] - pGEM

Figure 4. Binding of pGEM by whole Gc. To evaluate binding of DNA by whole, intact Gc, organisms were grown overnight on HEPES/agarose medium and suspended in liquid HEPES medium OD$_{540}$ = 0.2. 2 ml of this Gc suspension was incubated with pGEM (0.0063 ng) for 1h at room temperature after which the cells were pelleted by centrifugation, washed in 1 ml HEPES liquid medium and resuspended in 150 μL STE. The cells were then lysed with SDS, digested with proteinase K, and DNA was extracted by published methods (Swanson, Bergstrom *et al.*, 1985). Two amounts (10X, 1X) of the resulting DNA preparation (Gc genomic DNA + any adsorbed pGEM) were applied on nitrocellulose in a slot-blot apparatus and washed with 10X SSC. After UV-cross linking of DNA to the membrane, it was soaked in 0.5 N NaOH, neutralized with 1M Tris, pH 7.5, and dried. After prehybridization in 1M NaCl, 1% SDS, and 10% PEG, the blot was probed with [32P]-pGEM that had been boiled along with herring sperm DNA. The blot was probed with [32P]-pGEM overnight at room temperature and was then washed twice with 2X SSC at room temperature followed by a wash with 0.1X SSC containing 1% SDS at 65°C. It was then covered with plastic wrap and exposed at room temperature to film. The results, seen most clearly in the '1X' slots, show OpaA cells to have adsorbed differing amounts of pGEM, as follows: A+ > C+ > F+, H+, B+ > G+ > D+, E+, I+ ≥ Opa⁻.

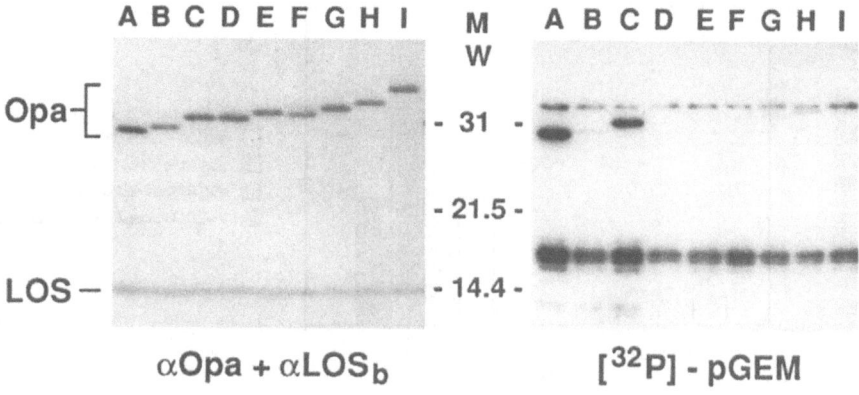

Figure 5. DNA-binding by SDS-PAGE separated Gc components. Whole Gc lysates in duplicate were separated by SDS-PAGE and transferred to membranes by electroblotting, as described previously (Swanson, 1978; Swanson, 1982). One membrane was probed with two monoclonal antibodies that recognize all Opa and LOSb, respectively, and then with [^{125}I]-protein A (left panel). The duplicate nitrocellulose sheet was incubated in PBS that contains 0.5% Tween 20 and pGEM that had been radiolabeled with [^{32}P]-dCTP by random priming (BRL, Random Primers DNA labeling system). Approximately 20 ng of [^{32}P]-pGEM was used per blot. This was incubated in a closed bag overnight at room temperature and was washed 2X, each for 15 min, with PBS-Tween containing 0.2 M NaCl. The washed blot was covered with Saran Wrap and exposed to Kodak AR2 film with or without an enhancing screen. The radiolabeled pGEM binds to several fast-migrating bands ($M_r \approx 16 - 18$ kD) that are different in size from LOS and also to a larger moiety ($M_r = 32 - 33$ kD) that likely represents the Rmp outer membrane protein. These banding profiles are identical in all Gc regardless of Opa phenotype. Both OpaA and OpaC are intensely stained by the radiolabeled pGEM while OpaB and OpaH are faintly stained. Additional, longer exposures revealed low level staining of OpaF and OpaG, but virtually no label on OpaD, OpaE, or OpaI.

sialylated, exhibit negligible levels of attachment to HEC-1B and Chang cells. Pilus$^-$ OpaA$^+$ Gc adhered best to these tissue culture cells; OpaA$^+$ organisms adhered better when propagated on H/As compared to those from P/Ar medium. When OpaA$^+$ organisms were pretreated with DNA and washed or were exposed to HEC-1B cells in the presence of DNA, their adherence was markedly reduced; this was more marked for OpaA$^+$ cells grown on H/As medium (Fig. 6). Heparin had similar adherence-reducing effects on OpaA$^+$ cells. An influence of added DNA was also evident, but less marked than with OpaA$^+$ variants, on the attachment of OpaC$^+$ variants to HEC-1B cells.

Other Opa$^+$ variants except OpaC$^+$ (OpaB, D, E, F, G, H, I) showed adherence levels that were negligibly higher than Opa$^-$ variants (Fig. 7).

118

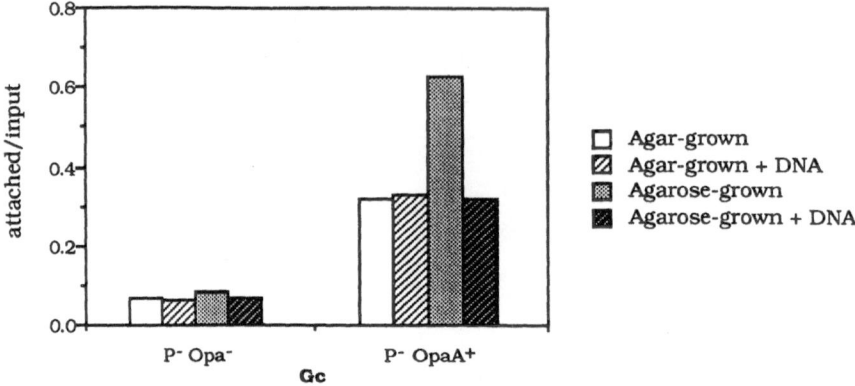

Figure 6. Influence of growth medium on Gc attachment to HEC-1B cells. Pilus⁻ cells that expressed either no Opa (P⁻ Opa⁻) or OpaA (P⁻ OpaA⁺) were grown on either agar or agarose-media to which was added containing 100 μCi [35S]-methionine (10 μCi/ml) (New England Nuclear). The [35S]-methionine-labeled Gc were then incubated with HEC-1B cells in the presence (pGEM, 10 μg/ml) or absence of DNA and their levels of attachment to these tissue cultures cells were determined, as shown. P⁻ Opa⁻ Gc exhibit little if any attachment regardless of medium or DNA presence. P⁻ OpaA⁺ cells attach much better when grown on agarose compared to agar. Addition of DNA to the agarose-grown OpaA⁺ variants markedly reduces their attachment to the HEC-1B cells while similar treatment of agar-grown OpaA⁺ variants has little effect, presumably because they are already 'coated' with polyanions that contaminate the agar.

OpaA versus OpaC

OpaA⁺ and OpaC⁺ cells bind DNA and consequently display negative shifts in their respective EPMs. When coated with DNA, OpaA⁺ variants have markedly reduced attachment to HEC-1B cells (Fig. 8). OpaC⁺ cells exhibit similar behavior in some experiments. Their behaviors differ, however, in the context of their LOS being sialylated. OpaA⁺ cells exhibit the same EPM and continue to bind DNA when they are sialylated; and sialylation does not modify the ablation of their attachment when incubated with DNA. In contrast, sialylation of OpaC⁺ cells results in a negative shift in their EPM; and when incubated with DNA, these sialylated OpaC⁺ variants neither bind DNA nor display modified attachment to HEC-1B cells. These results indicate that positively-charged portions of OpaA, more than those of OpaC, remain surface-exposed on Gc whose LOS has been substituted by sialic acid groups.

Discussion and speculations

The surface charge on bacteria has been investigated for several decades by determining the electrophoretic mobilities of whole cells, their partitioning

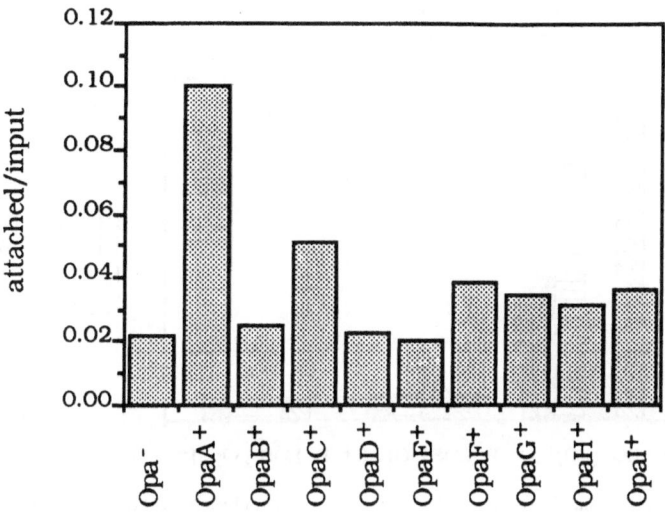

Figure 7. Attachment of pilus⁻ LOSb Gc to HEC-1B cells. LOSb Gc that expressed no Opa (Opa⁻) or a single Opa protein (OpaA⁺, OpaB⁺, etc.) were radiolabeled with [35S]-methionine and incubated with HEC-1B cells. [35S]-labeled Gc were suspended in liquid HEPES medium to OD540 = 0.1. One half ml of this Gc suspension was added to each well of a 24-well plate containing confluent HEC-1B cells covered with 0.5 ml of serum-free RPMI 1640 (Gibco Corp.); these tissue culture cells had previously been grown in the same medium containing 10% fetal bovine serum (Hyclone Corp.) in 8% CO_2 and then washed in the serum-free medium. The Gc + HEC-1B cells were incubated together at 37°C in air for 20 min with constant agitation on an orbital shaker. Unattached Gc were removed by washing 2X, 1 min each, with 1ml serum-free RPMI on the orbital shaker. The remainder was solubilized by adding 1 ml of a 1% solution of saponin (Calbiochem Corp.) in phosphate-buffered saline pH 7.0 for 30 min at room temperature. The saponin-solubilized material was mixed with scintillation cocktail (BioSafe II, Research Products) and counted in a scintillation spectrometer. All determinations were in triplicate and were compared with aliquots of the suspensions of [35S]-labeled Gc used to inoculate the tissue culture wells.

As shown, Opa⁻ variants exhibit little or no attachment to the tissue culture cells as do variants that express OpaB, D, or E. Low levels of attachment are seen for OpaF⁺, G⁺, H⁺, and I⁺ cells, but each of these cultures contained OpaA⁺ contaminants. Only OpaA⁺ variants attach at high levels followed by OpaC⁺ cells whose attachment is considerably less.

between charged aqueous phases, or their elution from charged support matrices (Brown, 1932; James, 1957; Richmond & Fisher, 1973; Krekeler, Ziehr *et al.*, 1989). This interest stems partly from a theoretical importance for net charge of bacteria in determining their interactions toward one another and toward heterologous surfaces. Bacteria, especially smaller, coccoid-shaped cells, have sizes in the range of typical colloidal particles. Surface charge, along with hydrophobicity/hydrophilicity, is the dominant determiner of attraction/repulsion of colloidal particles with other surfaces. Adsorption of diverse bacteria to inanimate (marine, dental, prosthetic) and animate substrates (epithelial, phagocytic host cells) is influenced by the charges and

120

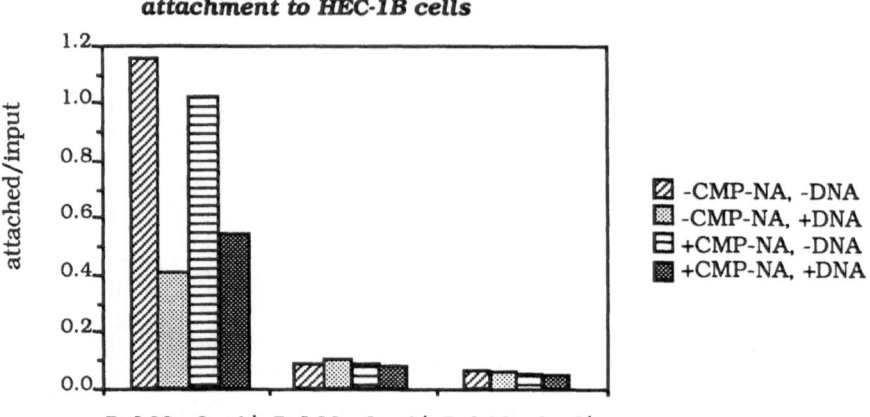

Figure 8. Effects of sialylation and DNA on Opa⁺ Gc attachment to HEC-1B cells. Pilus⁻ LOSb Gc that are OpaA⁺, OpaC⁺, or OpaI⁺ were grown overnight on solid HEPES/agarose medium, were suspended in 4 ml HEPES liquid medium (OD_{540} = 0.1) with [^{35}S]-methionine added (50 μCi/ml), and incubated for 3 – 4 h at 37°C either lacking or containing CMP-NANA to affect their sialylation. Their attachment to HEC-1B cells was then examined. OpaA⁺ variants attach at high levels to the tissue culture cells, regardless of whether they are sialylated (+CMP – NA) or not (– CMP – NA). The attachment of OpaA⁺ cells, both sialylated and not, is markedly diminished by DNA. The attachment of OpaC⁺ and OpaI⁺ cells is little influenced either by sialylation or DNA or the combination of both.

hydrophobic characteristics of the participating surfaces (van Loosdrecht *et al.*, 1987).

In spite of prolonged interest in the topic of cell surface charge, relatively little is known about its genesis. Eukaryotic and prokaryotic cells are all said to possess net negative charge at pHs near neutrality. The negativity of eukaryotic cells is thought to arise mainly from sialic or other ionized acid-containing glyoproteins of their plasma membranes. When bacteria synthesize polysaccharide capsules, they generally exhibit enhanced negativity attributable to carboxyl moieties of the capsular sugars. This is clearly-demonstrated in a recent report on *E. coli* capsule mutants whose negative charge correlated with their capsule composition: sialic acid (K1) > pyruvic acid (K29) > glucuronic acid (K30) (Bayer and Sloyer, 1990). The sources of negative charge on noncapsulated Gram-negative bacteria may come mainly from phosphate groups, such as those on ketodeoxyoctanate (KDO) in the of LOS core. Cells with 'smooth' LPS exhibit less negative charge than 'rough' LPS mutants whose side chains are much more truncated (James, 1957). Cells with long O-side chains likely exhibit less negative charge because their KDO phosphate groups are 'shielded' and are less exposed on the cell surface; this seems analogous to affects of long O-side chains on the surface exposures of outer membrane proteins (van der Ley *et al.*, 1986). But amino sugars in the O-side chains could also reduce net negativity.

Two groups have previously studied surface charge of gonococci. Heckels *et al.* found acidic pI's (5.3, 5.6) for pilus⁺ and pilus⁻ Gc to predict that both bear net negative charges at physiological pHs and that pili endow cells with slightly enhanced negativity (Heckels *et al.*, 1976). Gc with blocked amino groups (by formaldehyde) exhibited pI = 4.0 near the pK of carboxyl groups which, rather than acidic phosphates, are thereby suggested as major contributors to surface negativity. Magnusson *et al.* also found enhanced negativity of pilus⁺ cells but noted that Opa expression had 'no clear effect on the surface charge' (Magnusson *et al.*, 1979).

Our studies indicate that LOS and Opa proteins both clearly influence Gc surface charge. Although the structures of LOSa, LOSa/b, and LOSb variants that we examined are not known in detail, existing information suggests that their oligosaccharide side chains are of differing lengths (LOSb > LOSa) (Mandrell *et al.*, 1988; Schneider *et al.*, 1991), and that might alter surface charge by the 'shielding' mentioned above. But until their structures are determined, it is unclear whether these EPM distinctions might be due to differences in LOS side chain amino sugars, etc. The EPM differences among LOS variants have no apparent correlate in their interactions with host cells; the three LOS variants, if pilus⁻ and Opa⁻, show no discernible interaction (adherence, ingestion) with tissue culture cells or neutrophils, *in vitro*. Nor does sialylation alter the adherence of pilus⁻ Opa⁻ LOSb or LOSa/b cells that attain greatly enhanced negativity.

Some Opa proteins markedly alter EPM of Gc, but the nature of these effects is incompletely understood. Those Opa proteins having the largest net excesses of positively charged amino acids in their (predicted) exposed regions impart marked negative charge on Gc if the cells are exposed to abundant polyanions such as DNA and sulfated polysaccharides. The same Opa proteins would probably endow Gc with highly positive EPM if they could be examined in the absence of such polyanions. *In vitro* that is technically unattainable because of the polyanions in agar and other medium constituents and the DNA that is released from lysed sibling cells and continuously 'contaminates' liquid cultures; it seems unlikely *in vivo* where the infectious milieu of Gc includes mucus with its abundant sulfated polysaccharides and purulent exudate which also contains additional polyanions in membranes, nucleic acids, and other internal constituents of lysed eukaryotic cells.

Expression of OpaA enables pilus⁻ Gc to adhere to tissue culture cells, regardless of whether their LOS is sialylated or not; but if OpaA⁺ cells become coated with polyanions (DNA, sulfated polysaccharides, etc.), their adherence is ablated. Because OpaA⁺ Gc can stick avidly to mucosal sites bathed in the absence or low concentrations of polyanions but do not adhere when polyanions are abundant, these variants may possess 'special abilities' relevant to both initiation of a mucosal infection and in being transmitted to the next host. Similar roles for other Opa proteins are difficult to imagine. OpaC has somewhat fewer positively-charged amino acids in its predicted exposed portions than OpaA and only very high polyanion concentrations alter the

EPM of OpaC⁺ cells. Several other Opa proteins (OpaD, E, I) do not endow Gc with enhanced adherence nor do they bind exogenous polyanions. These Opa polypeptides have relatively few or no net excess of positively-charged amino acids in their exposed portions.

My current working idea is that the adherence-promoting capacities of Opa proteins derive from their positively-charged exposed regions which participate in charge-charge interactions with negatively charged groups on the surfaces of eukaryotic cells. Such interactions are compromised when those positively-charged Opa portions become 'coated' with polyanions. Any of several polyanions can be involved, including nucleotides and nucleoproteins released by lysed cells in the culture, sulfate or pyruvate-containing polysaccharides that 'contaminate' low purity agar (Duckworth, 1971), and perhaps compounds such as polyphosphate that are synthesized and released by Gc in large amounts (Noegel & Gotschlich, 1983). The polyanions that accumulate on the surfaces of OpaA⁺ and OpaC⁺ cells effectively 'encapsulate' them. The adsorbed polyanionic 'garbage' on certain Opa⁺ variants may account for capsules being observed microscopically on Gc in clinical specimens and *in vitro*, but also contributes to the failure in purifying any unique polysaccharide that might constitute such a capsule. Difficult to incorporate into this notion, however, is the abundance of sulfated polysaccharide and other polyanions in mucus that likely bathes any mucosal surface. Further complicating this hypothetical scheme is the apparent small effect that Opa expression and DNA coating of Opa⁺ Gc have on the EPM and attachment properties of pilus⁺ organisms. My preliminary observations indicate that pilus⁺ Gc have more positive EPM than pilus⁻ cells that are otherwise identical; and small EPM distinctions occur among pilus⁺ cells that synthesize structurally-different pilins. This embryonic pilus/positive EPM story currently evokes a 'deja vu' reaction 'all over again' inasmuch as pilin polypeptides typically bear excess numbers of negatively-charged amino acids in their predicted surface-exposed regions but pilus⁺ cells have reduced negativity. It is interesting to speculate whether the effects of piliation on EPM might resemble the inverse relationships between Opa predicted charge and EPM; this requires further study.

What role might Opa mediated attachment of Gc play in pathogenesis of gonorrhea? It is clear from preliminary experiments that Opa phenotype has relatively little influence on the adherence of pilus⁺ Gc to tissue culture cells. However, studies by Tie Chen and Bob Belland (Rocky Mountain Labs) strongly suggest that, although pilus⁺ cells are superior in their attachment propensities, it is mainly pilus⁻ OpaA⁺ Gc that are endocytosed by tissue culture cells. These findings suggest that pilus⁺ Gc, after initially adhering to mucosal cells, spawn pilus⁻ OpaA⁺ variants that promote their own endocytosis and perhaps transcytosis by epithelial cells. Such pilus⁻ OpaA⁺ variants might also be well-suited for emigration from one infected host to the next when the inflammation they incite supplies sufficient DNA or other polyanions to render them nonadherent to mucosal cells. Certainly OpaA⁺

variants are 'popular' among Gc shed in urines of men experimentally infected with predominantly Opa⁻ organisms.

My attempts to relate biological behaviors of Gc to their surface charge may be overly simplistic and quite naive. Mine is certainly not the first attempt to understand such relationships. Watt and Ward have discussed the issue extensively (Watt & Ward, 1980), but their thinking is based on the premise that all cells bear net negative charge. My studies on Gc summarized above plus my recent examination of campylobacteria and salmonella EPM and attachment in collaboration with Mike Konkel (Rocky Mountain Labs) all contain indications that bacteria can possess net positive charge. Although measuring EPM of intact cells is facilitated greatly by new technology, there are few benchmarks for evaluation of the results. The EPM differences reported above were easily and reproducibly obtained with the DELSA apparatus. The organism remains in running PBS buffer for only 5 – 10 min at room temperature and effluent from the running cell contains viable Gc. Explanations for the observed EPM shifts are tentative. But it should be clear that 1) surface-exposed outer membrane components markedly influence EPM, 2) some surface-exposed moieties may dominate EPM and eclipse any contribution by others, 3) but several outer membrane components can simultaneously contribute to EPM, and 4) polyions of the environment can bind to bacteria and alter both their EPM and their surface behavior.

Both well-defined mutants and *E. coli* or other bacterial hosts expressing recombinant molecules should be appropriate objects for study by this technique; *a priori*, their examination should yield new understanding of the surface properties imparted by particular components. Indeed, *E. coli* that express recombinant *opa* genes exhibit an enhanced negative EPM resembling that of Gc with the same Opa⁺ phenotype. But I have been unable to define conditions under which Opa⁺ *E. coli* will accrete polyanions from the medium or, conversely, will exhibit positive EPM shift like that seen for Opa⁺ Gc propagated on agarose containing medium.

Acknowledgements

I thank Susan Smaus for secretarial help, Gary Hettrick for photography, and Jeanne Wilson for technical help.

References

Bayer, ME & Sloyer JL Jr. (1990) The electrophoretic mobility of gram-negative and gram-positive bacteria: an electrokinetic analysis. J Gen Microbiol 136: 867–874.

Belland, RJ, Chen T, Swanson J & Fischer SH (1992) Human neutrophil response to recombinant neisserial Opa proteins. Mol Microbiol 6: 1729–1737.

Bhat, KS *et al.* (1991) The opacity proteins of *Neisseria gonorrhoeae* strain MS11 are encoded by a family of 11 complete genes. Mol Microbiol 5: 1889–1901.

124

Brown, HC, & Broom JC (1932) Further observations on bacterial cataphoresis. British J Exp Path 13: 334–342.

Draper, DL, James JF, Brooks GF & Sweet RL (1980) Comparison of virulence markers of peritoneal and fallopian tube isolates with endocervical *Neisseria gonorrhoeae* isolates from women with acute salpingitis. Infect Immun 27: 882–888.

Draper, DL, Lammel CJ, Sweet RL & Brooks GF (1985) Attachment of gonococcal outer membranes containing protein II variants to HeLa 229 cells. p. 271–275. *In:* GK Schoolnik, Brooks GF, Falkow S *et al.* (ed.) The Pathogenic Neisseriae. American Society for Microbiology, Washington, DC.

Duckworth, M, & Yaphe W (1971) The structure of agar. Part I. Fractionation of a complex mixture of polysaccharides. Carbohdr. Res 16: 189–197.

Fischer, SH & Rest RF (1988) Gonococci possessing only certain P.II outer membrane proteins interact with human neutrophils. Infect Immun 56: 1574–1579.

Heckels, JE, Blackett B, Everson JS & Ward ME (1976) The influence of surface charge on the attachment of *Neisseria gonorrhoeae* to human cells. J Gen Microbiol 96: 359–364.

James, AM (1957) The electrochemistry of the bacterial surface. Prog Biophys Biophys Chem 8: 95–142.

James, JF & Swanson J (1978) Color/opacity colony variants of *Neisseria gonorrhoeae* and their relationship to the menstrual cycle. p. 338–345. *In:* GF Brooks, Gotschlich EC, Holmes KK, Sawyer WD & Young FE (ed.) Immunobiology of *Neisseria gonorrhoeae*. American Society for Microbiology, Washington, DC.

Krekeler, C, Ziehr H & Klein J (1989) Physical methods for characterization of microbial cell surfaces. Experientia 45: 1047–1055.

Lambden, PR, Robertson JN & Watt PJ (1980) Biological properties of two distinct pilus types produced by isogenic variants of *Neisseria gonorrhoeae* strain P9. J Bacteriol 141: 393–396.

Magnusson, KE *et al.* (1979) Effect of colony type and pH on the surface charge and hydrophobicity of *Neisseria gonorrhoeae*. Infect Immun 26: 397–401.

Mandrell, RE, Griffis JM & Macher BA (1988) Lipooligosaccharides (LOS) of *Neisseria gonorrhoeae* and *Neisseria meningitidis* have components that are immunochemically similar to precursors of human blood group antigens. J Exp Med 168: 107–126.

Noegel, A & Gotschlich EC (1983) Isolation of a high molecular weight polyphosphate from *Neisseria gonorrhoeae*. J Exp Med 157: 2049–2060.

Parsons, NJ *et al.* (1989) Sialylation of lipopolysaccharide and loss of absorption of bactericidal antibody during conversion of gonococci to serum resistance by cytidine 5′-monophospho-N-acetyl-neuraminic acid. Microb Pathog 7: 63–72.

Parsons, NJ *et al.* (1988) Cytidine 5′-monophospho-N-acetyl neuraminic acid and a low level molecular weight factor from human blood cells induce lipopolysaccharide alteration in gonococci when conferring resistance to killing by human serum. Microb Pathog 5: 303–309.

Rice, PA (1989) Molecular basis for serum resistance in *Neisseria gonorrhoeae*. Clin Microbiol Rev 2: S112–S117.

Richmond, DV & Fisher DJ (1973) The electrophoretic mobility of microorganisms. p. 129. *In:* AH Rose & Tempest DW (ed.) Advances in microbial physiology. Academic Press, London, England.

Schneider, H *et al.* (1991) Expression of paragloboside-like lipooligosaccharides may be a necessary component of gonococcal pathogenesis in men. J Exp Med 174: 1601–1605.

Smith, H (1990) Pathogenicity and the microbe *in vivo*. The 1989 Fred Griffith Lecture. J Gen Microbiol 136: 377–383.

Swanson, J (1973) Studies on gonococcus infection. IV. Pili: their role in attachment of gonococci to tissue culture cells. J Exp Med 137: 571–589.

Swanson, J (1978) Studies on gonococcus infection. XIV. Cell wall protein differences among color/opacity colony variants of *Neisseria gonorrhoeae*. Infect Immun 21: 292–302.

Swanson, J (1981) Surface-exposed protein antigens of the gonococcal outer membrane. Infect Immun 34: 804–816.

Swanson, J (1982) Colony opacity and protein II compositions of gonococci. Infect Immun 37: 359–368.

Swanson, J (1982) Protein II variants of gonococci. p. 353–356. *In:* D Schlessinger (ed.) Microbiology-1982. American Society for Microbiology, Washington, DC.

Swanson, J (1991) Some affects of LOS and Opa on surface properties of gonococci. Proceedings of the Seventh International Pathogenic Neisseria Conference, Berlin, Walter de Gruyter & Co.

Swanson, J, Barrera O, Sola J & Boslego J (1988) Expression of outer membrane protein II by gonococci in experimental gonorrhea. J Exp Med 168: 2121–2129.

Swanson, J *et al.* (1985) Pilus⁻ gonococcal variants. Evidence for multiple forms of piliation control. J Exp Med 162: 729–744.

Swanson, J *et al.* (1987) Gonococcal pilin variants in experimental gonorrhea. J Exp Med 165: 1344–1357.

Van der Ley, P, Kuipers O, Tommassen J & Lugtenberg B (1986) O-antigenic chains of lipopolysaccharide prevent binding of antibody molecules to an outer membrane pore protein in *Enterobacteriaceae.* Microb Pathog 1: 43–49.

Van Loosdrecht, MCM *et al.* (1987) Electrophoretic mobility and hydrophobicity as a measure to predict the initial steps of bacterial adhesion. Appl Environ Microbiol 53: 1898–1901.

Watt, PJ & Ward ME (1980) Adherence of *Neisseria gonorrhoeae* and other *Neisseria* species to mammalian cells. p. 253–288. *In:* EH Beachey (ed.) Bacterial adherence. Series B, Volume 6. Chapman & Hall, Ltd, New York, London.

9. Anthrax toxin mechanisms of receptor binding and internalization

STEPHEN H. LEPPLA, KURT R. KLIMPEL and NAVEEN ARORA

Abstract. The anthrax toxin complex contains two catalytic proteins, edema factor (EF) and lethal factor (LF), which must be translocated to the cytosol of eukaryotic cells to cause toxicity. The third protein component of the toxin, protective antigen (PA), achieves this internalization. PA binds to cell surface receptors, is cleaved after the sequence Arg164-Lys165-Lys166-Arg167 by a cell surface protease, and thereby exposes a binding site to which EF or LF bind with high affinity. The complexes enter endosomes, which then become acidified, causing transfer of LF and EF to the cytosol.

To identify the cell surface protease causing PA activation, PA residues 164–167 were altered by mutagenesis. The minimum sequence needed to retain toxicity is RXXR, where X is any residue. The requirement for an Arg at position −4 relative to the bond cleaved suggested that the cellular protease cleaving PA is furin, a subtilisin-like protease involved in processing many protein precursors. Purified furin was shown to cleave the PA mutants at rates proportional to their toxicity. Furthermore, cleavage of PA bound by cells was blocked by the same protease inhibitors which inhibit purified furin. This data strongly suggests that furin is the protease which activates anthrax toxin.

To study the translocation and intracellular trafficking of LF and EF, fusions of LF to the ADP-ribosylation domain of *Pseudomonas* exotoxin A were prepared. These fusions were highly potent for eukaryotic cells when combined with PA, indicating that LF efficiently crosses endosomal membranes and can carry fused polypeptides to the cytosol. This system may be exploited in the future for delivery of other polypeptides to the cytosol.

Abbreviations: PA, protective Antigen; PA63, 63-kDa, Carboxyl-Terminal Fragment of PA; LF, Lethal Factor; EF, Edema Factor; PE, *Pseudomonas* Exotoxin A; DTT, Dithiothreitol; pCMB, *p*-Chloromercuribenzoate; PCR, Polymerase Chain Reaction

Genetics of virulence factors of *Bacillus anthracis*

The ability of *Bacillus anthracis* to infect a wide variety of animal species and to cause rapidly progressive disease and death has been unambiguously attributed to the presence of two major virulence factors. These are the poly-D-glutamic acid capsule and the three-component protein exotoxin (Keppie, 1963; Thorne, 1960a; Leppla, 1991a). The capsule acts to prevent phagocytosis of the vegetative cells, and is probably essential during the early stages of an infection when host defense systems are intact. The toxin may play some role in depressing the host defenses early in the infection, but its action is more apparent at later stages in infection as the cause of the extensive tissue edema

C.I. Kado and J.H. Crosa (eds.), Molecular Mechanisms of Bacterial Virulence, 127–139.
© 1994 *Kluwer Academic Publishers.*

that is a major cause of death. The toxin is either not highly potent or is slow to act on cells, as indicated by the fact that bacteria often grow to very high density in the blood (10^8 bacteria per ml) prior to death of an infected animal (Turnbull, 1992).

Our understanding of the role and significance of the two major *B. anthracis* virulence factors evolved rapidly in the 1980s once the almost classical design of the genetic elements encoding them was discovered (Mikesell, 1983; Robillard, 1983; Thorne, 1985; Green, 1985). The toxin and capsule are each encoded by a separate large plasmid, designated pXO1 and pXO2, respectively (Green, 1985; Uchida, 1985; Uchida, 1986). For a strain to be virulent, it must possess both plasmids. Virulent strains can be cured of each plasmid individually, so that strains synthesizing either the toxin or capsule are readily available. Furthermore, methods for conjugation and transduction have been developed that allow return of plasmids to cured strains (Battisti, 1985; Heemskerk, 1990; Makino, 1987). This has made it possible to attribute moderate differences in the virulence of certain natural strains possessing both plasmids to additional unidentified virulence factors located on pXO2 (Welkos, 1991).

The genes encoding each of the three protein components of the toxin as well as three genes involved in capsule synthesis have been cloned and sequenced (see Leppla, 1991a, for review). In the case of the toxin genes, this allowed targeted mutation of pXO1 to inactivate each of the toxin components individually (Cataldi, 1990; Pezard, 1991). Strains lacking any single toxin component were at least 10-fold less virulent than wild type, proving that the toxin is a major determinant of *B. anthracis* virulence.

Overview of anthrax toxin structure and internalization

Early studies of anthrax toxin established the presence of three separate proteins, and the requirement that these be combined in pairs to obtain toxic effects (Keppie, 1963; Thorne, 1960b; Smith, 1955; Smith, 1962). The three proteins (Fig. 1) are designated protective antigen (PA), lethal factor (LF), and edema factor (EF). A more recent nomenclature emphasizes the requirement for combinations of components and the fact that two different types of toxic effect are produced (Friedlander, 1986; Leppla, 1991a). This nomenclature defines the combination of PA and LF as lethal toxin, and the combination of PA with EF as edema toxin. Lethal toxin injected intravenously kills experimental animals, with rats being especially susceptible to rapid death. Rat and mouse macrophages are rapidly lysed, providing a convenient and more humane assay for lethal toxin. Edema toxin causes tissue edema when injected intradermally, but is not lethal for animals or cultured cells.

As its name implies, PA is the material in culture supernatants that induces protective immunity to infection. PA was actually first identified during attempts to develop anthrax vaccines (Gladstone, 1948; Turnbull, 1992), many

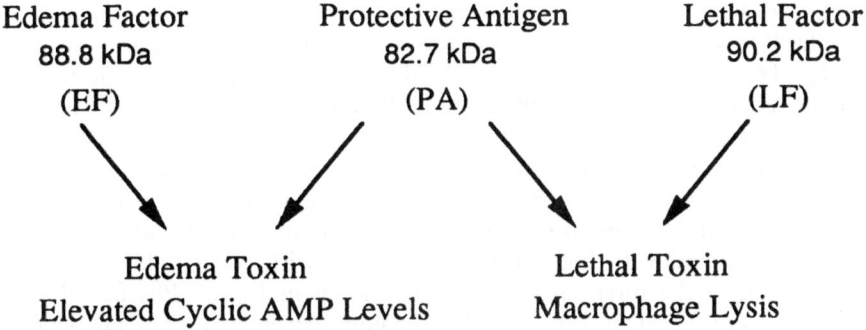

Figure 1. The anthrax toxin complex. Anthrax toxin consists of three separate proteins which combine to produce two distinct toxic activities, as shown by the converging arrows.

years before existence of the toxin was recognized. After the protein components were separated (although still not purified), studies in animals suggested that PA becomes bound to tissues as the first step in toxin action (Molnar, 1963). Taken together with concepts derived from study of other toxins, these data led to the suggestion that PA was binding to receptors on cells and causing internalization of LF and EF. This view was a logical extension of a model formulated to describe the structures of several other toxins that act intracellularly (Gill, 1978). This "AB" model holds that toxins which act enzymatically in the cytosol of eukaryotic cells typically have two domains with clearly distinct functions. One domain (the "B" chain or domain) has a ligand binding ability that directs the toxin to target cell surface structures. The other domain ("A" chain or domain) has a catalytic activity that disrupts cellular metabolism, often leading to cell death. In this view, PA was the common "B" moiety that was causing uptake of two alternate "A" moieties, LF and EF, each of which would have a different catalytic activity. The first attempts to find enzymatic activities associated with the toxin failed (Stanley, 1963), but recognition that the vascular permeability enhancing activity of edema toxin resembles that caused by cholera toxin led to the discovery that EF is an adenylate cyclase (Leppla, 1982; Leppla, 1991b). Almost simultaneously it was found that *Bordetella pertussis* produces a similar "invasive" adenylate cyclase (Confer, 1982; Hanski, 1991). It is assumed by persons working in this area that LF also has a catalytic activity, but none has yet been identified.

The major functional domains of PA, LF, and EF were deduced by a combination of sequence analysis, characterization of proteolytic fragments, and mutagenesis (for review, see Leppla, 1991a). The ability to purify large amounts of the proteins was helpful in the early stages of this work (Leppla, 1988a; Quinn, 1988; Leppla, 1991b). Instrumental in understanding the role of PA was the finding that the protein is exceptionally sensitive to cleavage at single sites by low concentrations of trypsin and chymotrypsin (Leppla, 1988a; Leppla, 1988b). Trypsin cleaves specifically at a single site to produce fragments of 19 and 63 kDa. These remain associated but can be separated and purified by anion exchange chromatography at pH 9.0. The larger fragment (PA63) was found to have all the toxic activity of the intact protein. Subsequently it was found that PA bound on the surface of cells incubated at low temperature was specifically cleaved at Arg-167 by a cell surface protease. Separate studies established that PA63 but not intact PA had a site to which LF and EF bound competitively and with high affinity. Together with other data, this led to formulation of the internalization model shown in Fig. 2. Pharmacological evidence showed that the PA-LF or PA-EF complex enters cells by endocytosis and passes to an acidic vesicle from which LF or EF escapes to the cytosol (Friedlander, 1986; Gordon, 1988). The adenylate cyclase activity of EF then causes large and unregulated increases in intracellular cAMP concentrations. LF is presumed to be transferred in the same way to the cytosol where its putative catalytic activity damages physiological processes, and in the case of macrophages, causes cell lysis.

Our knowledge of the functional domains of LF and EF is derived largely from sequence analysis and comparison. The amino-terminal 300 residues of EF and LF share substantial sequence homology, indicating it is this region that binds to PA (Robertson, 1988a; Robertson, 1988b; Mock, 1988; Escuyer, 1988). The carboxyl-terminal 470 amino acids of EF include the catalytic domain, as deduced from homology of this region to the more fully characterized adenylate cyclase of *B. pertussis* (Labruyere, 1990). Amino acid residues 278–383 of LF contain five imperfect repeats of 19 amino acids. No function is known for this region. The sequence homology to EF and the fact that the region containing the repeats is unlikely to be part of a enzyme active center, lead to the proposal that residues 384–776 contain the putative enzymatic domain of LF.

One of the unique features of the anthrax toxins is the requirement that the three proteins interact to gain access to the cytosol of target cells. Although the studies described above provide the outlines of how the proteins act, at least three important questions remain. First, it will be important to identify the PA receptor. This protein is present on all cell types examined to date, although amounts vary somewhat. The PA receptor certainly is needed to transport the toxin to endosomes, but it may play a more active role by contributing to formation of a membrane pore required for LF and EF translocation. This can only be determined after the receptor has been identified. The recent identification of the receptors for *Pseudomonas* exotoxin A (Kounnas, 1992) and diphtheria toxin (Naglich, 1992) illustrate ways this might be achieved. Second, understanding the cell specificity and uptake mechanism requires identification of the cell surface protease that activates PA. Because the cleavage of PA has parallels in the activation of several other toxins by proteolysis, identification of the eukaryotic protease has been a major goal of our work. This is discussed in detail below. Finally, and perhaps most difficult, is the question of how a protein can cross a lipid bilayer membrane. Our approach to this

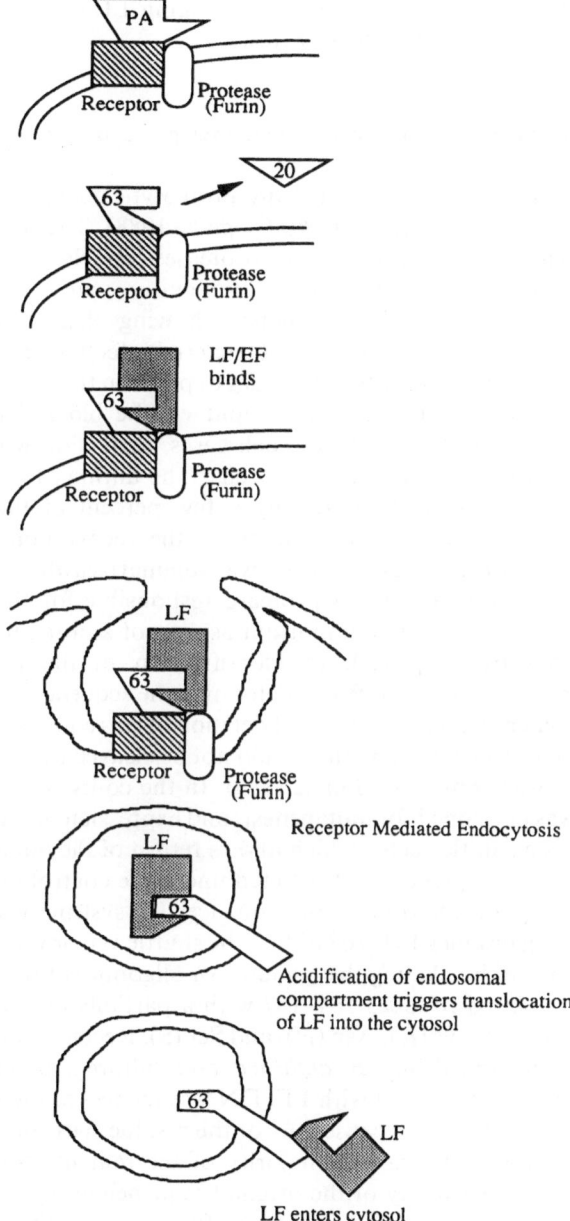

Figure 2. Anthrax toxin binding and internalization. After binding to a receptor on sensitive animal cells, PA is cleaved by the cellular protease furin. EF or LF binds to the newly exposed site on the receptor-bound PA 63-kDa fragment. The complex is internalized via receptor-mediated endocytosis. Acidification of the endosome induces translocation of EF, LF, or LF fusion proteins across the endosomal membrane into the cytosol.

question has been to prepare fusion proteins linking LF sequences to other proteins. This too is discussed in detail below.

Identification of furin as the protease that activates protective antigen

A number of bacterial protein toxins require proteolytic activation to acquire toxic activity (Ohishi, 1987, Ogata, 1990, Cieplak, 1988). The bacteria which produce these toxins typically also produce proteases capable of activating the toxins, so that the toxins purified from bacterial cultures are at least partially activated. However, there is little evidence showing that these bacterial proteases activate the toxins *in vivo* during a bacterial infection or intoxication. Indeed, it would seem difficult for the proteases to perform this activation, since they will be greatly diluted in body fluids and will be blocked by protease inhibitors such as α_2-macroglobulin. In such cases, activation would require proteases present in the sensitive tissue or cells. The anthrax PA is obtained from the usual growth medium with only a few percent of the molecules proteolytically nicked. The evidence leading to the recognition that PA is cleaved by a eukaryotic cell-surface protease was summarized above. We sought to identify the eukaryotic activating protease responsible for the activation because this would help to explain important aspects of anthrax toxin action.

The most productive approach to identification of the protease was mutagenesis of the cleavage site of PA to determine the sequence recognized by the protease. We had previously shown that deletion of six residues spanning the cleavage site produced a PA protein that could not be cleaved by the cell surface protease and that was non-toxic (Singh, 1989). In the course of that work an efficient system was developed for mutagenesis and expression of the altered PA proteins. This uses a shuttle vector which allows return of the mutated gene to *B. anthracis* where it is expressed and secreted under the control of its original *B. anthracis* transcription and translation sequences. To systematically examine the effects of varying residues 164–167 of PA, the shuttle vector was modified to include restriction sites bracketing this region. An oligonucleotide cassette was synthesized that replaced residues 164–167 with a partially random sequence containing only Arg (R), Lys (K), Asn (N), and Ser (S). Recombinants obtained in *E. coli* were transformed into *B. anthracis* and culture supernatants were tested for toxicity when combined with LF. DNA sequences of the randomized regions were also determined. Because 50% of the residues introduced were the basic amino acids Lys and Arg, the majority of the mutant sequences were expected to retain the sensitivity of the original sequence to trypsin. Nineteen different sequences were obtained, and of these, fifteen were active after trypsin treatment (Klimpel, 1992). Three of the other mutants contained a single Arg or Lys but were not activated by a low concentration of trypsin, for unknown reasons. In contrast to the sensitivity of nearly all of the mutant sequences to trypsin, only four of the mutants were toxic without trypsin treatment, indicating that they were substrates for the cell surface protease. These four all had the sequence Arg-X-(Arg/Lys)-Arg, with X being Ser, Lys, or Arg. To verify the sequence requirements of the protease, three additional PA mutants were prepared, this time using defined oligonucleotides for cassette mutagenesis. Of the three mutants prepared, AKKR, RKKA, and RAAR, only the latter was cleaved by the cell surface protease. This proves that three basic residues are not sufficient for cleavage. Instead, the protease requires Arg at positions 164 and 167, and is largely indifferent to the identity of residues 165 and 166 (Klimpel, 1992).

When this consensus substrate sequence data was compared to published information about cellular proteases, an obvious correlation was found with the specificity of the recently characterized protease furin. Discovery of this subtilisin-like serine protease ended a search extending over many years to identify the proteases which process proproteins and prohormones (Barr, 1991). The original goal of those efforts was to identify the proteases that recognize paired basic residues in hormone precursors. After identification and cloning of the yeast enzyme KEX2, which cleaves after paired basic residues in the killer toxin precursor, it became possible to probe DNA libraries of higher eukaryotes for related proteases. This search identified the protease furin and then other members of a family of proteases having sequence homology to the prokaryote enzymes of the subtilisin family (van de Ven, 1990; van den Ouweland, 1990). Recent work indicates that furin is expressed in nearly all cell types, where it achieves processing of a number of essential cellular proteins. In contrast, expression of the related enzymes, PC1/PC3 and PC2, is restricted to neuroendocrine cells, where it is required for processing of peptide hormone precursors at paired basic residues. A comparison of the sequences of many different precursors cleaved in cells expressing only furin suggested that furin requires the presence of the -4 Arg, while PC3/PC1 requires only the paired basic residues at positions -2 and -1 (Hosaka, 1991). For furin, some substrates lack a basic residue at -2, so that the furin consensus sequence can be represented as RXXR (Barr, 1991).

The fact that furin is present in all types of cells and is anchored to membranes, combined with the specificity data described above, make it is an excellent candidate for the enzyme that activates PA. To test this idea further, we asked whether purified human furin can cleave PA and the mutant PA proteins. Both PA and the RAAR mutant were cleaved rapidly, while the AKKR and RKKA mutants were not cleaved (Klimpel, 1992).

The final type of evidence that furin is the activating protease comes from inhibitor profiles. Subtilisin and the related proteases furin and PC3/PC1 are classified as a subgroup of the serine proteases because they share with trypsin the catalytic triad of Ser, His, Asp. In spite of this similarity, furin is quite resistant to the classical serine protease inhibitors. Only very high concentrations of diisopropylflurophosphate and phenylmethylsulfonyl fluoride inhibit these enzymes (Stieneke-Grober, 1992). To compare the sensitivity to inhibitors of furin with that of the cell surface protease, we took advantage a gentle method of chemical fixation that was found to preserve activity of the PA receptor and the cell surface protease (Klimpel, 1992). Cells could then be incubated with radiolabelled PA and inhibitors at room temperature and under conditions that would detach unfixed cells. An excellent correlation was found between the set of inhibitors which inhibit the cleavage by purified furin of a synthetic peptide substrate and those which blocked cleavage of PA bound to fixed cells (Klimpel, 1992). The only agents that inhibited both processes were calcium chelators, certain metal ions, dithiothreitol (DTT), and p-chloromercuribenzoate (pCMB). The effects of the chelators and metal ions can be attributed to the requirement of furin for calcium ion, presumably for substrate binding. The inhibition by DTT and pCMB cannot be attributed to reaction at a sulfhydryl group at the catalytic center, because furin has no such essential cysteine residue. Furthermore, other sulfhydryl reagents such as iodoacetamide are without effect. A cysteine is located near the active site serine, and it may be that the bulky pCMB molecule bound there would block the active site. There is no apparent explanation for the inhibition by DTT.

Although the evidence presented above argues strongly that furin is the protease which is principally responsible for activating PA, a more extensive kinetic analysis of the rate of macrophage lysis by the PA mutants suggests that other proteases may substitute for furin in some cases. For example, when assays were extended beyond 4 hr, PA mutants that were barely active in shorter assays eventually equalled PA in potency. This suggests that other proteases are able to activate PA, but that they act more slowly. The presence of four basic residues at the PA cleavage site should make it an excellent substrate for other serine proteases that recognize basic residues. Because these slower acting proteases are likely to be intracellular, it appears that PA is able to recycle to the surface after being activated intracellularly. Others have shown that PA can form ion-conductive channels (Blaustein, 1989), so it is possible that PA becomes inserted in membranes and recycles to the cell surface where it would bind LF and cause its internalization. Quantitative kinetic studies could be used to determine whether the number of LF molecules internalized exceeds the number of cell-associated PA molecules.

Analysis of the toxin internalization route using protein fusions

The other major question that must be answered to fully understand anthrax toxin action is how the catalytic moieties, LF and EF, reach the cytosol. Studies modelled on work with diphtheria toxin showed that cells are protected against lethal toxin by amines, implying that the toxin passes through an acidified compartment (Friedlander, 1986; Gordon, 1988). Further work showed that the receptor-bound toxin could be artificially internalized by exposure to low pH. In these and other pharmacological studies, the anthrax lethal toxin behaved exactly like diphtheria toxin. Furthermore, the active proteolytic fragment, PA63, was shown to form ion-conductive channels in artificial lipid membranes (Blaustein, 1989). Our own studies (Leppla, 1988b; Leppla, 1991a) showed that

PA63 could form a stable oligomer, reminiscent of multimeric pore forming membrane proteins such as porins. With the recent determination of the structure of diphtheria toxin (Choe, 1992) it becomes possible to visualize how exposure to low pH can allow a normally soluble protein to insert in membranes and cause translocation to the cytosol. However, even with the analogy to diphtheria toxin, it is not apparent how the anthrax toxin proteins insert into and cross membranes. Unlike diphtheria toxin, none of the three anthrax toxin proteins contain strongly hydrophobic domains that are obvious candidates for membrane insertion.

To characterize the translocation mechanism, and identify the protein domains and sequences that are essential parts of this uptake system, we created fusion proteins of LF linked to an easily assayed marker enzyme, the ADP-ribosylation domain of *Pseudomonas* exotoxin A (PE). A large number of fusion proteins containing PE have previously been made (Pastan, 1991), and this has led to a detailed understanding of the roles of the PE domains (Fig. 3). The three domains of PE recognized from the X-ray structure of PE are proven to function in receptor binding (domain I), membrane translocation (II), and ADP-ribosylation activity of elongation factor 2 (III). Delivery of a single molecule of domain III to the cytosol is considered sufficient to kill a eukaryotic cell by inhibition of protein synthesis. Therefore a fusion of LF to PE domain III was expected to kill cells if it were translocated efficiently to the cytosol. Another goal of preparing LF-PE fusions was the hope that this might yield new types of therapeutic chimeric toxins to add to the array of those already available.

The desired sequences from LF and PE were obtained in PCR reactions that also added new restriction sites to their ends. These fragments were then ligated into a T7-based expression vector, and the expressed proteins were purified by gel filtration and anion-exchange chromatography to near homogeneity. High purity was considered essential to obtain accurate measurements of the potencies of the proteins. Fusions of LF to PE domains II+III and to domain III (Fig. 3) were found to inhibit protein synthesis and kill CHO cells (Arora, 1992). The most potent fusions were more toxic for CHO cells than was native PE. This cannot be attributed to differences in receptor number, because the receptor for PE, now known to be the α_2-macroglobulin receptor (Kounnas, 1992), and the (unidentified) receptor for PA are both present at 5,000–20,000 per cell. Toxicity of the LF-PE fusions required addition of PA, showing that uptake of the fusion protein was by the normal, PA-dependent route. Inclusion of PE domain II sequences decreased the potency of the fusions, indicating that the fusions were trafficked differently than PE. These data show that LF is capable of completely crossing endosomal membranes and of carrying with it other polypeptides as passengers (Arora, 1992).

The design of the LF-PE fusions also was varied to examine the role of the carboxyl terminus of PE. The native terminal sequence, REDLK, fits the consensus for signals present on proteins which are retained in or actively retrieved to the endoplasmic reticulum. Mutagenesis of the terminus of PE had previously demonstrated that an active ER retention signal was needed to make

136

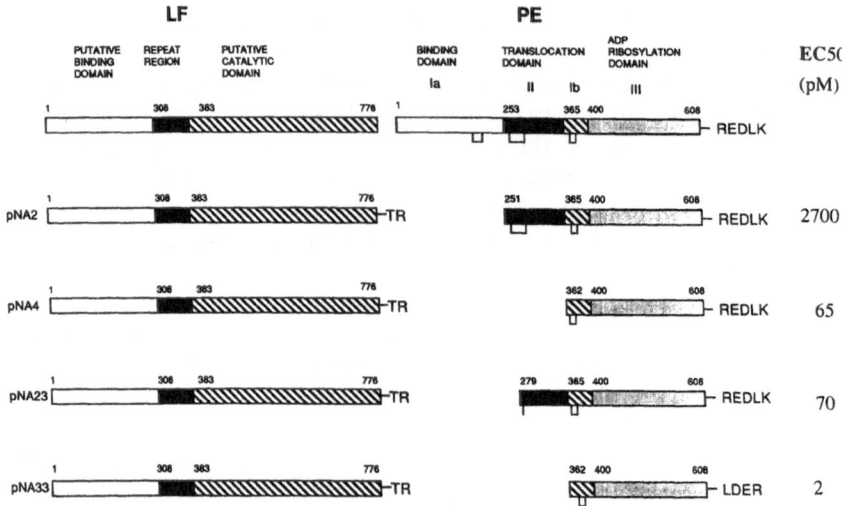

Figure 3. Composition and toxicity of fusion proteins of LF and *Pseudomonas* exotoxin A (PE). Four fusion proteins containing LF residues 1–776 fused to varying portions of PE were produced and tested for toxicity on Chinese hamster ovary cells in the presence of 1 ug/ml PA. The concentration of fusion protein that caused 50% inhibition of protein synthesis is shown at the right. The fusion protein shown at the bottom differed from the others in having the native carboxyl terminus REDLK changed to LDER (see text). (Figure adapted from Arora (1992) with permission of the publisher).

PE toxic (Chaudhary, 1990). This was explained by suggesting that translocation to the cytosol of a proteolytic fragment, residues 280–613 of PE, occurs in the ER. Because the evidence discussed above argues that LF and LF fusions are translocated directly from endosomes, passage through the ER would not be part of the required trafficking, and indeed any signal causing such trafficking could be inhibitory. Our studies showed that the presence of a functional ER retention signal actually decreases the potency of the LF fusions, as might be expected if the ER retrieval system efficiently cleared the cytosol of polypeptides having the KDEL sequence.

In an extension of the work described above, we defined the minimum LF sequence needed to bind to PA and cause internalization. Residues beyond about 270 in LF and EF are probably involved in catalysis, which suggested that only the amino-terminal region was needed for binding to PA. This was confirmed by showing that a fusion of LF residues 1–254 with PE domain III was as potent as an analogous fusion containing all 776 residues of LF.

As discussed above, three different polypeptides (the catalytic domains of LF, EF, and PE) fused to the amino-terminal domains of LF and EF are efficiently internalized into the cytosol of cells through binding to PA. This suggests that the uptake system is quite tolerant of diverse sequences and therefore has substantial promise as a tool for delivery of polypeptides into the cytosol of cells. Ongoing studies are designed to determine what constraints exist on the peptides and proteins that can be internalized by attachment to the LF sequence.

Acknowledgements

We thank Yogendra Singh, Vijay Chaudhary, and David FitzGerald for reagents and advice, Ray Fields, Sheila Haley, Deborah White, and Stuart Cohn for technical assistance, and Jerry M. Keith for his continuing encouragement and support.

References

Arora N, Klimpel KR, Singh Y & Leppla SH (1992) Fusions of anthrax toxin lethal factor to the ADP-ribosylation domain of *Pseudomonas* exotoxin A are potent cytotoxins which are translocated to the cytosol of mammalian cells. J. Biol. Chem. 267: 15542–15548.

Barr PJ (1991) Mammalian subtilisins: The long-sought dibasic processing endoproteases. Cell 66: 1–3.

Battisti L, Green BD & Thorne CB (1985) Mating system for transfer of plasmids among *Bacillus anthracis, Bacillus cereus*, and *Bacillus thuringiensis*. J. Bacteriol. 162: 543–550.

Blaustein RO, Koehler TM, Collier RJ & Finkelstein A (1989) Anthrax toxin: channel-forming activity of protective antigen in planar phospholipid bilayers. Proc. Natl. Acad. Sci. USA. 86: 2209–2213.

Cataldi A, Labruyere E & Mock M (1990) Construction and characterization of a protective antigen-deficient *Bacillus anthracis* strain. Mol. Microbiol. 4: 1111–1117.

Chaudhary VK, Jinno Y, FitzGerald D & Pastan I (1990) *Pseudomonas* exotoxin contains a specific sequence at the carboxyl terminus that is required for cytotoxicity. Proc. Natl. Acad. Sci. USA. 87: 308–312.

Choe S, Bennett MJ, Fujii G, Curmi PM, Kantardjieff KA, Collier RJ & Eisenberg D (1992) The crystal structure of diphtheria toxin. Nature (London) 357: 216–222.

Cieplak W, Hasemann C & Eidels L (1988) Specific cleavage of diphtheria toxin by human urokinase. Biochem. Biophys. Res. Commun. 157: 747–754.

Confer DL & Eaton JW (1982) Phagocyte impotence caused by the invasive bacterial adenylate cyclase. Science 217: 948–950.

Escuyer V, Duflot E, Sezer O, Danchin A & Mock M (1988) Structural homology between virulence-associated bacterial adenylate cyclases. Gene 71: 293–298.

Friedlander AM (1986) Macrophages are sensitive to anthrax lethal toxin through an acid-dependent process. J. Biol. Chem. 261: 7123–7126.

Gill DM (1978) Seven toxin peptides that cross cell membranes. p. 291–332. *In*: Jeljaszewicz J & Wadstrom T (ed.) *Bacterial Toxins and Cell Membranes*. Academic Press, New York.

Gladstone GP (1948) Immunity to anthrax. Production of the cell-free protein antigen in cellophane sacs. Brit. J. Exp. Path. 29: 379.

Gordon VM, Leppla SH & Hewlett EL (1988) Inhibitors of receptor-mediated endocytosis block the entry of *Bacillus anthracis* adenylate cyclase toxin but not that of *Bordetella pertussis* adenylate cyclase toxin. Infect. Immun. 56: 1066–1069.

Green BD, Battisti L, Koehler TM, Thorne CB & Ivins BE (1985) Demonstration of a capsule plasmid in *Bacillus anthracis*. Infect. Immun. 49: 291–297.

Hanski E & Coote JG (1991) *Bordetella pertussis* adenylate cyclase toxin. p. 349–366. *In*: Alouf JE & Freer JH (ed.) *Sourcebook of Bacterial Protein Toxins*. Academic Press, London.

Heemskerk DD & Thorne CB (1990) Genetic exchange and transposon mutagenesis in *Bacillus anthracis*. Salisbury Med. Bull. 68, Spec. suppl.: 63–67.

Hosaka M, Nagahama M, Kim WS, Watanabe T, Hatsuzawa K, Ikemizu J, Murakami K & Nakayama K (1991) Arg-X-Lys/Arg-Arg motif as a signal for precursor cleavage catalyzed by furin within the constitutive secretory pathway. J. Biol. Chem. 266: 12127–12130.

Keppie J, Harris-Smith PW & Smith H (1963) The chemical basis of the virulence of *Bacillus anthracis*. IX. Its aggressins and their mode of action. Brit. J. Exp. Path. 44: 446–453.

Klimpel KR, Molloy SS, Thomas G & Leppla SH (1992) Anthrax toxin protective antigen is activated by a cell surface protease with the sequence specificity and catalytic properties of furin. Proc. Natl. Acad. Sci. USA. 89:10277–10281.

Kounnas MZ, Morris RE, Thompson MR, FitzGerald DJ, Strickland DK & Saelinger CB (1992) The α_2-macroglobulin receptor/low density lipoprotein receptor-related protein binds and internalizes *Pseudomonas* exotoxin A. J. Biol. Chem. 267: 12420–12423.

Labruyere E, Mock M, Ladant D, Michelson S, Gilles AM, Laoide B & Barzu O (1990) Characterization of ATP and calmodulin-binding properties of a truncated form of *Bacillus anthracis* adenylate cyclase. Biochemistry 29: 4922–4928.

Leppla SH (1982) Anthrax toxin edema factor: a bacterial adenylate cyclase that increases cyclic AMP concentrations of eukaryotic cells. Proc. Natl. Acad. Sci. USA. 79: 3162–3166.

Leppla SH (1988a) Production and purification of anthrax toxin. *In*: Harshman S (ed.) *Methods in Enzymology, Vol. 165* (pp. 103–116) Academic Press, San Diego.

Leppla SH, Friedlander AM & Cora E (1988b) Proteolytic activation of anthrax toxin bound to cellular receptors. p. 111–112. *In*: Fehrenbach F, Alouf JE, Falmagne P, Goebel W, Jeljaszewicz J, Jurgen D & Rappouli R (ed.) *Bacterial Protein Toxins*. Gustav Fischer, New York.

Leppla SH (1991a) The anthrax toxin complex. p. 277–302. *In*: Alouf JE & Freer JH (ed.) *Sourcebook of Bacterial Protein Toxins*. Academic Press, London.

Leppla SH (1991b) Purification and characterization of adenylyl cyclase from *Bacillus anthracis*. p. 153–168. *In*: Johnson RA & Corbin JD (ed.) *Methods in Enzymology, Vol. 195* Academic Press, San Diego.

Makino S, Sasakawa C, Uchida I, Terakado N & Yoshikawa M (1987) Transformation of a cloning vector, pUB110, into *Bacillus anthracis*. FEMS Microbiol. Lett. 44: 45–48.

Mikesell P, Ivins BE, Ristroph JD & Dreier TM (1983) Evidence for plasmid-mediated toxin production in *Bacillus anthracis*. Infect. Immun. 39: 371–376.

Mock M, Labruyere E, Glaser P, Danchin A & Ullmann A (1988) Cloning and expression of the calmodulin-sensitive *Bacillus anthracis* adenylate cyclase in *Escherichia coli*. Gene 64: 277–284.

Molnar DM & Altenbern RA (1963) Alterations in the biological activity of protective antigen of *Bacillus anthracis* toxin. Proc. Soc. Exp. Biol. Med. 114: 294–297.

Naglich JG, Metherall JE, Russell DW & Eidels L (1992) Expression cloning of a diphtheria toxin receptor: identity with a heparin-binding EGF-like growth factor precursor. Cell 69: 1051–1061.

Ogata M, Chaudhary VK, Pastan I & FitzGerald DJ (1990) Processing of *Pseudomonas* exotoxin by a cellular protease results in the generation of a 37,000-Da toxin fragment that is translocated to the cytosol. J. Biol. Chem. 265:20678–20685.

Ohishi I (1987) Activation of botulinum C2 toxin by trypsin. Infect. Immun. 55:1461–1465.

Pastan I & FitzGerald D (1991) Recombinant toxins for cancer treatment. Science 254: 1173–1177.

Pezard C, Berche P & Mock M (1991) Contribution of individual toxin components to virulence of *Bacillus anthracis*. Infect. Immun. 59: 3472–3477.

Quinn CP, Shone CC, Turnbull PC & Melling J (1988) Purification of anthrax-toxin components by high-performance anion-exchange, gel-filtration and hydrophobic-interaction chromatography. Biochem. J. 252: 753–758.

Robertson DL (1988a) Relationships between the calmodulin-dependent adenylate cyclases produced by *Bacillus anthracis* and *Bordetella pertussis*. Biochem. Biophys. Res. Commun. 157: 1027–1032.

Robertson DL, Tippetts MT & Leppla SH (1988b) Nucleotide sequence of the *Bacillus anthracis* edema factor gene (cya): a calmodulin-dependent adenylate cyclase. Gene 73: 363–371.

Robillard NJ, Koehler TM, Murray R & Thorne CB (1983) Effects of plasmid loss on the physiology of *Bacillus anthracis*. Annu. Mtg. Amer. Soc. Microbiol. 115 (Abstract).

Singh Y, Chaudhary VK & Leppla SH (1989) A deleted variant of *Bacillus anthracis* protective antigen is non-toxic and blocks anthrax toxin action *in vivo*. J. Biol. Chem. 264: 19103–19107.

Smith H, Keppie J & Stanley JL (1955) The chemical basis of the virulence of *Bacillus anthracis*. V. the specific toxin produced by *B. anthracis in vivo*. Brit. J. Exp. Path. 36: 460–472.

Smith H & Stanley JL (1962) Purification of the third factor of anthrax toxin. J. Gen. Microbiol. 29: 517–521.

Stanley JL & Smith H (1963) The three factors of anthrax toxin: their immunogenicity and lack of demonstrable enzymic activity. J. Gen. Microbiol. 31: 329–337.

Stieneke-Grober A, Vey M, Angliker H, Shaw E, Thomas G, Roberts C, Klenk HD & Garten W (1992) Influenza virus hemagglutinin with multibasic cleavage site is activated by furin, a subtilisin-like endoprotease. EMBO J. 11: 2407–2414.

Thorne CB (1960a) Biochemical properties of virulent and avirulent strains of *Bacillus anthracis*. Ann. N. Y. Acad. Sci. 88: 1024–1033.

Thorne CB, Molnar DM & Strange RE (1960b) Production of toxin *in vitro* by *Bacillus anthracis* and its separation into two components. J. Bacteriol. 79: 450–455.

Thorne CB (1985) Genetics of *Bacillus anthracis*. p. 56. *In*: Lieve L, Bonventre PF, Morello JA, Schlessinger S, Silver SD & Wu HC (ed.) *Microbiology-85*. American Society for Microbiology, Washington, D.C.

Turnbull PCB (1992) Anthrax vaccines: past, present and future. Vaccine 9: 533–539.

Uchida I, Hashimoto K & Terakado N (1986) Virulence and immunogenicity in experimental animals of *Bacillus anthracis* strains harbouring or lacking 110 MDa and 60 MDa plasmids. J. Gen. Microbiol. 132: 557–559.

Uchida I, Sekizaki T, Hashimoto K & Terakado N (1985) Association of the encapsulation of *Bacillus anthracis* with a 60 megadalton plasmid. J. Gen. Microbiol. 131: 363–367.

Van de Ven WJ, Voorberg J, Fontijn R, Pannekoek H, van den Ouweland AM, van Duijnhoven HL, Roebroek AJ & Siezen RJ (1990) Furin is a subtilisin-like proprotein processing enzyme in higher eukaryotes. Mol. Biol. Rep. 14: 265–275.

Van den Ouweland AM, van Duijnhoven HL, Keizer GD, Dorssers LC & van de Ven WJ (1990) Structural homology between the human fur gene product and the subtilisin-like protease encoded by yeast KEX2 [published erratum appears in Nucleic Acids Res 1990 Mar 11;18(5):1332]. Nucleic. Acids. Res. 18: 664.

Welkos SL (1991) Plasmid-associated virulence factors of non-toxigenic (pX01⁻) *Bacillus anthracis*. Microb. Pathog. 10: 183–198.

10. Host-specific virulence genes of *Xanthomonas*

D.W. GABRIEL, M.T. KINGSLEY, Y. YANG, J. CHEN and
P. ROBERTS

Abstract. At least some *Xanthomonas* virulence genes are not *hrp* genes, but instead are host-specific determinants of disease and/or host range. We have been interested in altering the host range and pathovar status of several different xanthomonads causing economically serious diseases. Several genes affecting host range were identified and complementing DNA clones isolated from *X. campestris* pvs. citrumelo and translucens. These genes affect growth in some hosts, but not in other hosts. One of these genes, *opsX* of *X.campestris* pv. citrumelo, appears to be involved in lipopolysaccharide (LPS) core assembly; a specifically modified LPS may be needed by the strain as a barrier against citrus, but not bean, defense compounds. In addition to genes which affect host range, genes were identified and complementing clones isolated which elicit host-specific disease symptoms, and in some cases, may determine pathovar status. For example, a 5 kb DNA fragment was cloned from *X. campestris* pv. campestris that conferred to *X.campestris* pv. armoraceae the ability to induce systemic blight and black vein symptoms when inoculated onto Brassica plants. Since the clone conferred ability to elicit symptoms diagnostic of blight disease to *X.campestris* pv. armoraceae (which does not cause blight), the 5 kb cloned insert carried all information necessary to alter the apparent pathovar status of the recipient strains. Similarly, gene *pthA* of *X. citri* confers citrus-specific mitogenic ability to many different xanthomonads, enabling them to elicit symptoms diagnostic of citrus canker disease. Gene *pthA* belongs to a host-specific virulence/ avirulence gene family from *Xanthomonas*; about half of all microbial avirulence (*avr*) genes reported cloned to date from any source are members of this gene family. Most members of this gene family have no known virulence function. However, another member of the gene family, *avrb6* of *X.campestris* pv. malvacearum, is essential for cotton-specific watersoaking ability. Members of this family reported to date are > 97% identical in DNA sequence, even when derived from phylogenetically distinct strains. Nearly identical inverted terminal repeats mark the boundaries of homology of many members. The presence of multiple, nearly identical members of this gene family bounded by inverted terminal repeats in phylogenetically distinct strains provides evidence for horizontal gene transfer and transposition. Horizontal gene transfer of a few host-specific virulence genes might explain why phylogenetically distinct strains can cause nearly identical disease symptoms and be placed together in the same pathovar. Perhaps the most striking observation of this work is that both host range and pathovar status of *Xanthomonas* may be determined in some cases by single, host-specific, virulence genes.

Abbreviations: pv., pathovar; LPS, Lipopolysaccharide

Introduction

Microbial genes directly involved in conditioning the symbiotic interactions of bacteria with plants may be functionally categorized into four groups: those

C.I. Kado and J.H. Crosa (eds.), Molecular Mechanisms of Bacterial Virulence, 141–158.
© 1994 *Kluwer Academic Publishers.*

affecting parasitism (growth *in planta* on all hosts), pathogenicity (pathogenic symptoms), host range (growth *in planta* on specific hosts), and avirulence (eliciting a host defense) (Gabriel, 1986). We have been interested in investigating the genetics of pathogenicity and host range of the genus *Xanthomonas*, with a view to understanding its population structure, its genes that determine host range and the physiological mechanisms by which the bacteria signal plant cells to condition acceptance and avoid rejection. In the course of this work, a number of parallels were noticed among various plant pathogenic bacterial genera in terms of the genetics of virulence and the molecular biology of virulence. Some similarities with animal pathogens were also identified, raising the possibility of convergent evolution of virulence mechanisms by prokaryotes in symbioses with eukaryotes generally.

One of the main advantages of utilizing plant-microbe interactions to study bacterial virulence mechanisms is that large numbers of variants in both parasite and host can be engineered and experimentally manipulated. One very applied goal of our work is to be able to deliberately manipulate host range and pathogenicity genes in the construction of genetically engineered biological control agents of weeds and perhaps other microbes. A second applied goal is to engineer plants to be unresponsive to virulence signals. A third goal is to identify mechanisms whereby pathogens become tolerant to plant defense compounds.

Xanthomonas *populations*

Members of the genus *Xanthomonas* are all plant-associated, and a relatively few of them cause economically serious diseases. The genus as a whole has a wide host range, including a large number of plant species in many different plant families. However, individual strains are quite limited in host range and are grouped into pathovars based primarily on the host plants from which they were first isolated. Phylogenetic analyses of several different xanthomonads reveal a clonal population structure, which correlates generally with the pathovar status of the strains (Gabriel *et al.*, 1988, 1989). That is, a given pathovar might contain one or more clonal groups, and the clonal groups can generally be distinguished on the basis of pathogenicity. A clonal population structure is typical of several pathogenic enteric bacterial species, and is thought to be due to 1) low rates of recombination and 2) selective host amplification of strains with specific virulence factors (Selander, 1985). Selective host amplification of strains with specific virulence factors has also been hypothesized to account for the clonality observed among strains of some plant pathogens (Gabriel, 1989). In this hypothesis, host-specific virulence genes with high selective value on a given host are thought to cause strains to be clonally amplified in direct proportion to the distribution and quantity of the host. Some of these genes may be transferred horizontally among xanthomonads colonizing the same host niche. Since most major crop plants have a very narrow genetic base (and some crops are clonally propagated), those crops which are widespread in geographical distribution (cereal grains, citrus, beans, etc.) might be selecting a proportionally widespread distribution of a genetically corresponding pathogen population. If so, clonal pathogen groups, and therefore, some pathovars, might differ from one another by only a few virulence genes having a high selective value on specific crop genotypes. Such genes would necessarily be host-specific.

Host-specific virulence

Members of the genus *Rhizobium* are all capable of causing cortical hypertrophies, leading to nodules, on leguminous plants. The genes responsible for nodulation have been identified and cloned. The common nodulation (*nod*) genes of *Rhizobium* are conserved at the genus level, are functionally interchangeable (Djordjevic *et al.*, 1987), and are detected by hybridization in all *Rhizobium* and *Bradyrhizobium* species tested (Kondorosi *et al.*, 1986). Analogously, some virulence genes in the Pseudomonadaceae (e.g., the hypersensitive response and pathogenicity (*hrp*) genes (Willis, *et al.*, 1991)) may be common to the family, and are found by hybridization in all plant pathogenic strains tested of both *Pseudomonas* and *Xanthomonas*. By contrast with common *nod* genes, the host specific nodulation (*hsn*) genes of *Rhizobium* determine host range, and may be manipulated to alter the host range of some strains (Long, 1984). Perhaps analogously, some host specific virulence genes have been reported cloned from *P. solanacearum* (Ma *et al.*, 1988), *P. syringae* (Salch & Shaw, 1988), *X. citri* (Swarup *et al.*, 1991), from *X.campestris* pv. translucens (Waney *et al.*, 1991), from *X.campestris* pv citrumelo (Kingsley *et al.*, 1993). Recent work in our lab, some of it published only as abstracts and presented below, provides evidence for several additional *host-specific virulence* (*hsv*) genes from *Xanthomonas*, and for the horizontal transfer of members of a host specific virulence/avirulence gene family. As with *hsn* genes of *Rhizobium*, these *Xanthomonas hsv* genes may be manipulated to alter host range, and/or pathovar status (refer below).

Host-specific virulence effectors

Recently, a family of lipo-oligosaccharides, products of the *nodABC* and *nodFEL* operons (the latter are *hsn* genes), have been demonstrated to mediate host specificity in *Rhizobium* (Spaink *et al.*, 1991; Schultze *et al.*, 1992). These lipo-oligosaccharides are mitogenic to cells of host plants, and induce both root hair curling and cortical hyperplasia, both of which are required developmental steps leading to successful symbioses. More recently, the lipopolysaccharide (LPS) of *X.campestris* was implicated in determining host-specific virulence; The *Xanthomonas* LPS may serve as a barrier against specific host defense compounds (Kingsley *et al.*, 1993; and discussed below). To our knowledge, the LPS is the only molecule implicated as playing a determinative role in *Xanthomonas* host specific virulence. Nevertheless, a growing number of genes have been identified and cloned which determine host specific virulence, and for which there is no known biochemical function. The remainder of this paper is devoted to a description of the pathological phenotypes affected by these *Xanthomonas* genes.

Pathovar names and Xanthomonas *taxonomy*

A phenotype-based taxonomic revision of the genus *Xanthomonas* was proposed in 1972 in which over 100 different pathogenic species were lumped into five species, with *X. campestris* containing most strains of the old species (Lelliot, 1972). Despite the strong evidence that the proposal was artificial (the phenetic data were known to be unrepresentative of the genetic diversity (Murata & Starr, 1973)), the taxonomy became widely accepted. It is also widely accepted among bacterial taxonomists that phylogenetic relationships should determine taxonomy, and that nomenclature should agree with (and reflect) genomic information (Wayne *et al.*, 1987). The commonly accepted standard is that strains with 70% or greater relatedness by DNA-DNA reassociation belong in the same species (Wayne *et al.*, 1987). Nevertheless, the artificial 1972 taxonomy of *Xanthomonas* remains largely in place, and most strains of the genus remain lumped into one species, *X. campestris*.

The pathovar (short for *patho*genic *var*iant) designation is used to indicate pathogenic information (Dye, *et al.*, 1980). Since pathogenic information is based – by definition – on the host response phenotype, and not on the basis of intrinsic properties of the bacterium, pathovar names may or may not reflect phylogenetic relatedness. Since editors of journals generally require that an organism be named, strains are often given pathovar names as a matter of convenience, most often on the basis of the host from which they were first isolated. Strains within a given pathovar may not even cause similar diseases, since inclusion of a strain can be incidental and arbitrary. For example, *X. campestris* pv. alfalfae attacks both beans and alfalfa plants, and could therefore be named after either host. Similar observations have been made with some *Rhizobium* strains (Keyser et al., 1982). Even though pathovar names have no standing in nomenclature, many journals require pathovar names be italicized, resulting in a trinomial designation in which the pathovar name appears to be elevated to species status (e.g., *X.c. alfalfae* or *P.s. syringae*).

An artificial taxonomy can be scientifically misleading and economically costly. Leaf-spotting xanthomonads (now known as *X. campestris* pv. citrumelo (Gabriel *et al.*, 1989)) were misidentified, at a cost of well over $25 million, as agents of citrus canker disease (Schoulties *et al.*, 1987; Longman, 1989). Since pathogenic *Xanthomonas* cells were isolated from citrus, they were placed in species *campestris,* were designated pathovar "citri", and were then automatically subject to federal quarantine and eradication laws. This costly misdiagnosis is not an isolated incident. The USDA-APHIS-PPQ imposed a nine-year embargo on Mexican citrus, based upon the isolation of non pathogenic *Xanthomonas* bacteria from citrus trees that were infected by the fungus (!), *Alternaria limicola* (Federal Register, Jan 11, 1991, Vol 56, #8, pp. 1122–3; Simmons, 1990; Stapleton & Garza-Lopez, 1988). With *Xanthomonas* toxonomy in its current state, relationships among strains with identical names must be experimentally verified. Experiments designed to alter host range or ability to elicit pathogenic symptoms may depend on knowledge of the taxonomic distance between strains. Only one or a few genes appear needed to interconvert the host ranges of phylogenetically similar strains, while multiple genes may be needed to interconvert more phylogenetically distant strains.

Avirulence genes and host range

Avirulence (*avr*) genes are bacterial genes that act as negative factors to limit the growth of strains within a pathovar or species to a subset of *hosts* within the range of the pathovar or species. Avirulence genes determine pathogenic races, and in no case have *avr* genes been shown to determine pathovar status of a strain. It appears to be generally true that *avr* genes confer avirulence to strains only on hosts that carry specific resistance (*R*) genes; this genetic requirement is often termed gene-for-gene (*avr*-for-*R*) specificity (for reviews, refer Keen, 1992; Gabriel & Rolfe, 1990). Resistance in hosts to avirulent pathogens is usually accompanied by a hypersensitive response (HR). Some *avr* genes have been shown to confer ability to elicit a non-host HR, but no *avr* gene or group of *avr* genes has been shown to determine non-host status; the non-host HR in at least some cases may therefore be a gratuitous plant response (Swarup et al., 1992).

At least 28 *avr* genes have been cloned, defined sufficiently to be named, and published (refer Table 1). Twenty-seven of these are from bacterial pathogens and 1 from a fungus. Of these 28 genes, almost half (17–28 in Table 1) are members of a multi-gene family, all of which were found in *Xanthomonas*.

Table 1. Cloned *avr* genes.

Gene	Cloning source	Reference
1. *avrA*	*Pseudomonas syringae* pv. glycinea	Staskawicz *et al.*, 1984
2. *avrB0*	*Pseudomonas syringae* pv. glycinea	Staskawicz *et al.*, 1987
3. *avrB1*	*Pseudomonas syringae* pv. glycinea	Staskawicz *et al.*, 1987
4. *avrC*	*Pseudomonas syringae* pv. glycinea	Staskawicz *et al.*, 1987
5. *avrPm1*	*P. syringae* pv. maculicola	Dangl *et al.*, 1992
6. *avrPpiA1*	*P. syringae* pv. pisi	Dangl *et al.*, 1992
7. *avrPpi2*	*P. syringae* pv. pisi	Fillingham *et al.*, 1992
8. *avrPph3*	*P. syringae* pv. phaseolicola	Fillingham *et al.*, 1992
9. *avrD*	*P. syringae* pv. tomato	Kobayashi *et al.*, 1990
10. *avrPto*	*P. syringae* pv. tomato	Ronald *et al.*, 1992
11. *avrRpt2*	*P. syringae* pv. tomato	Whalen *et al.*, 1991
12. *avrBs1*	*X. campestris* pv. vesicatoria	Swanson *et al.*, 1988
13. *avrBs2*	*X. campestris* pv. vesicatoria	Minsavage *et al.*, 1990
14. *avrBsT*	*X. campestris* pv. vesicatoria	Minsavage *et al.*, 1990
15. *avrRxv*	*X. campestris* pv. vesicatoria	Whalen *et al.*, 1988
16. *avr9*	*Cladosporium fulvum*	van Kan *et al.*, 1988

146

Table 1. Continued.

Gene	Cloning source	Reference
17. *avrBn*	*X. campestris* pv. malvacearum	Gabriel *et al.*, 1986
18. *avrB4*	*X. campestris* pv. malvacearum	De Feyter and Gabriel, 1991
19. *avrb6*	*X. campestris* pv. malvacearum	De Feyter and Gabriel, 1991
20. *avrb7*	*X. campestris* pv. malvacearum	De Feyter and Gabriel, 1991
21. *avrBIn*	*X. campestris* pv. malvacearum	De Feyter and Gabriel, 1991
22. *avrB101*	*X. campestris* pv. malvacearum	De Feyter and Gabriel, 1991
23. *avrB102*	*X. campestris* pv. malvacearum	De Feyter and Gabriel, 1991
24. *avrxa5*	*X. campestris* pv. oryzae	Hopkins *et al.*, 1992
25. *avrXa7*	*X. campestris* pv. oryzae	Hopkins *et al.*, 1992
26. *avrXa10*	*X. campestris* pv. oryzae	Hopkins *et al.*, 1992
27. *avrBs3*	*X. campestris* pv. vesicatoria	Bonas *et al.*, 1989
28. *avrBsP*	*X. campestris* pv. vesicatoria	Canteros *et al.*, 1991

In addition to those members of the *Xanthomonas* gene family which function for avirulence, at least one additional gene, *pthA* of *X. citri*, is a member of this gene family based upon DNA sequence data. All of these genes have been determined to be members of a single family of *avr* genes by Southern hybridization, restriction analyses and DNA sequence similarity (refer Fig. 1). DNA sequence comparisons of *avrBs3* (Bonas *et al.* 1989) and *avrBsP* (Canteros *et al.* 1991) of *X. campestris* pv. vesicatoria, *avrB4* and *avrb6* of *X. campestris* pv. malvacearum (DeFeyter *et al.* 1993) and *pthA* (Swarup *et al.*, 1992) revealed > 97% sequence identity among all these genes. Gene *pthA* is not included in Table 1 because it functions for virulence in the source strain and is not known to function for avirulence in the source strain on any host (Swarup *et al.*, 1991). (Gene *pthA* can function to elicit a non-host HR, but that function is gratuitous in terms of avirulence on at least some nonhosts (Swarup *et al.*, 1992)). Other members of this gene family have been cloned but have not yet been published (see below).

Much published work on bacterial and fungal avirulence genes and host specificity led Gabriel (1989) to hypothesize that *avr* genes could not determine host range at the pathovar level. According to the hypothesis, if *avr* genes, which determine race specificity, also determined pathovar specificity, then pathovar-change mutants should be isolated at frequencies comparable to race change mutants. That is, induced or natural pathovar-change mutations affecting host range should have been reported. They have not. Experiments designed to select such mutants have been reported (for example, see (Schnathorst, 1966)), but without success. Of the 28 *avr* genes in Table 1 and *pthA*, only three genes (*avrBs2*, *avrb6* and *pthA*) have a known or suspected pleiotropic function. Although most of these genes have been mutagenized by marker-exchange, there is no known pleiotropic function for these genes. Of these 29 genes, therefore, 90% have no known selective value to the microbe.

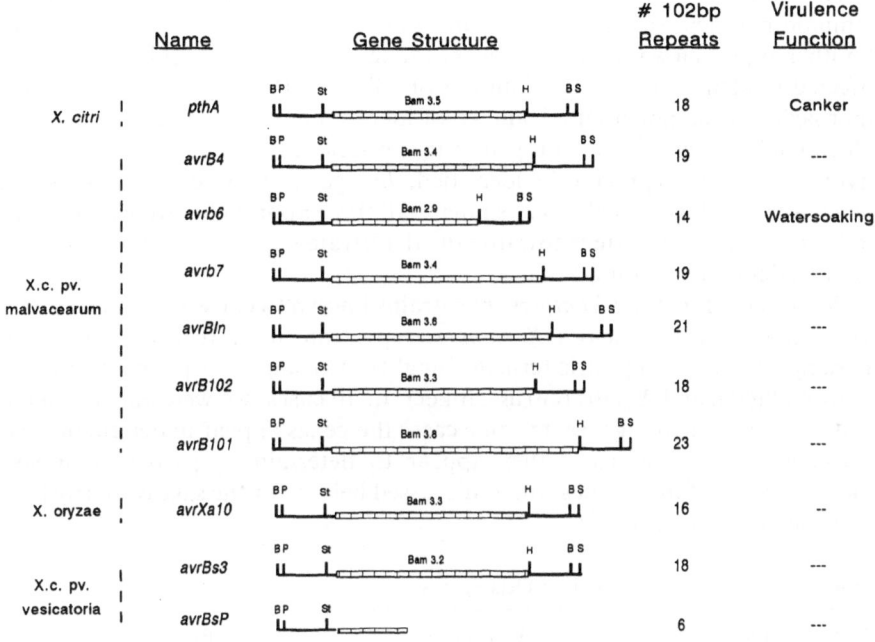

Name	Gene Structure	# 102bp Repeats	Virulence Function
X. citri pthA	BP St Bam 3.5 H B S	18	Canker
avrB4	BP St Bam 3.4 H B S	19	---
avrb6	BP St Bam 2.9 H B S	14	Watersoaking
X.c. pv. avrb7 *malvacearum*	BP St Bam 3.4 H B S	19	---
avrBln	BP St Bam 3.8 H B S	21	---
avrB102	BP St Bam 3.3 H B S	18	---
avrB101	BP St Bam 3.8 H B S	23	---
X. oryzae avrXa10	BP St Bam 3.3 H B S	16	--
X.c. pv. avrBs3 *vesicatoria*	BP St Bam 3.2 H B S	18	---
avrBsP	BP St	6	---

Figure 1. A *Xanthomonas hsv/avr* gene family. The structure of members of the gene family reported to date is illustrated. Characteristics shared by members of this gene family are: 1) a relatively large size (> 3kb); 2) nearly identical 5′ and 3′ ends, and 3) a 102bp repeated DNA motif in the central portion of all members. The hatched region in the middle represents the 102 bp direct repeat region. Restriction sites indicated are: B, *Bam*HI; P, *Pst*I; St, *Stu*I; H, *Hinc*II; S, *Sst*I.

Genes determining host range and pathovar status

Based on the above considerations, we hypothesized that host-specific virulence (*hsv*) genes, not *avr* genes, were the primary determinants of host range and pathovar status and were responsible for the clonal population structure observed among xanthomonads. A corollary hypothesis is that relatively rare horizontal gene transfer events (involving *hsv* genes) might create phylogenetically distinct clones with a similar host range. In some cases, one or two *hsv* genes might determine host range, and in these cases host range should be readily manipulable by transfer or swapping of single genes. One unsuccessful approach to altering host range involved running a gene library from *X. campestris* pv. translucens into *X. campestris* pv. campestris, and vice versa (Sawczyc *et al.*, 1989). These experiments may have failed because the *X. campestris* pv. translucens strain used was likely phylogenetically distant from the *X. campestris* pv. campestris strain used. We took a variety of different approaches. One approach used, the "virulence enhancement" approach (Swarup *et al.*, 1991), was to run DNA libraries between strains of different pathovars which shared a common host and screen for symptoms diagnostic of the DNA donor strain. Another approach was to utilize transposon mutagenesis to disable genes which affected virulence on one host, but not on all hosts (Waney *et al.*, 1991). This approach has the potential advantage of identifying multiple, unlinked genes, which could never be cloned together in a cosmid library. Another approach involved swapping previously identified, *hsv* genes between xanthomonads which were phylogenetically very similar. All three approaches were successful, and represent the first steps towards the deliberate engineering of strains with specific, defined host ranges.

We selected for study *Xanthomonas* strains known to cause some of the most economically serious diseases: *X. campestris* pv. translucens (cereal black chaff), *X. campestris* pv. campestris (crucifer blight), *X. campestris* pv. malvacearum (cotton blight), and *X. citri* (citrus canker). In all cases, we were able to isolate host-specific virulence genes; in some cases the genes appear to determine host range, while in other cases they appear to determine a particular disease phenotype. Six of these genes will be discussed below; for the sake of clarity they are summarized in Table 2:

Table 2. Xanthomonas host-specific virulence genes.

Gene	Source	Function	Reference
hsvB	*X.c.* pv. translucens	Host-range, barley	Waney *et al.*, 1991
hsvW	*X.c.* pv. translucens	Host-range, wheat	Waney *et al.*, 1991
opsX	*X.c.* pv. citrumelo	Host-range, citrus	Kingsley *et al.*, 1993
sysC	*X.c.* pv. campestris	Blight disease, Brassicas	Chen *et al.*, 1992
avrb6	*X.c.* pv. malvacearum	Leaf-spot disease, cotton	De Feyter *et al.*, 1991
pthA	*X. citri*	Canker disease, citrus	Swarup *et al.*, 1991

Host range genes can affect *in planta* growth

X. campestris *pv. translucens*
X. campestris pv. translucens strain Xt-216 causes black chaff on barley, wheat, oats, rye and triticale. Transposon induced mutations were isolated affecting virulence on all hosts generally (i.e., Hrp⁻), and mutations were isolated affecting host range on only one or more of these cereal hosts, but not all (i.e., Hsv⁻) (Waney *et al.*, 1991). Furthermore, DNA fragments complementing two of the host range mutations, one affecting host range on wheat (HsvW⁻), and another affecting host range on barley (HsvB⁻) were isolated in that study. Inactivation of *hsvB* in Xt-216 (*hsvB*::Tn5-*gus*) resulted in a significant loss of *in planta* growth on barley, but not on wheat. Transfer of *hsvB* into a wild-type *X. campestris* pv. translucens strain increased the host range of the strain to include barley, based on water-soaking symptoms. Growth of transconjugants of the same wild-type *X. campestris* pv. translucens strain with *hsvB* in barley are

increased by 2–3 logs. Since these strains are from the same pathovar, this extension of host range did not alter pathovar status.

X. campestris *pv. citrumelo*

X. campestris pv. citrumelo strain 3048 causes bacterial spot of citrus and also attacks some legumes (Gabriel *et al.*, 1989). Spontaneous mutants of 3048 which lost virulence on citrus, but not common bean, were recovered at surprisingly high frequency (0.1%) (Kingsley *et al.*, 1993). None of the mutants exhibited a hypersensitive response and pathogenicity (*hrp*) phenotype. The loss of virulence on one host, but not another, presumably affected a gene essential for host-specific virulence. Two of these mutants, M27 and M28, were examined in some detail. Studies of the pathogenicity phenotype and kinetics of growth on both hosts were performed. In Fig. 2 are growth curves on both bean and citrus for the mutant M28, the wild-type 3048, and *X. campestris* pv. malvacearum XcmN (a pathogen of cotton) as a control. As expected, XcmN failed to grow significantly on either nonhost, while 3048 grew well on both bean and citrus. Both mutants M27 and M28 were slightly affected in growth on bean, but were rapidly killed on citrus. This rapid loss of viable cell counts was unusual, since most wild-type xanthomonads neither grow well nor are killed off rapidly when inoculated on nonhosts. In our experience, most xanthomonads are recoverable many days after inoculations into nonhost leaf tissue at roughly the same cell counts as when they were first inoculated. Since we had observed that both M27 and M28 exhibited reduced gum production on some media, we hypothesized that perhaps

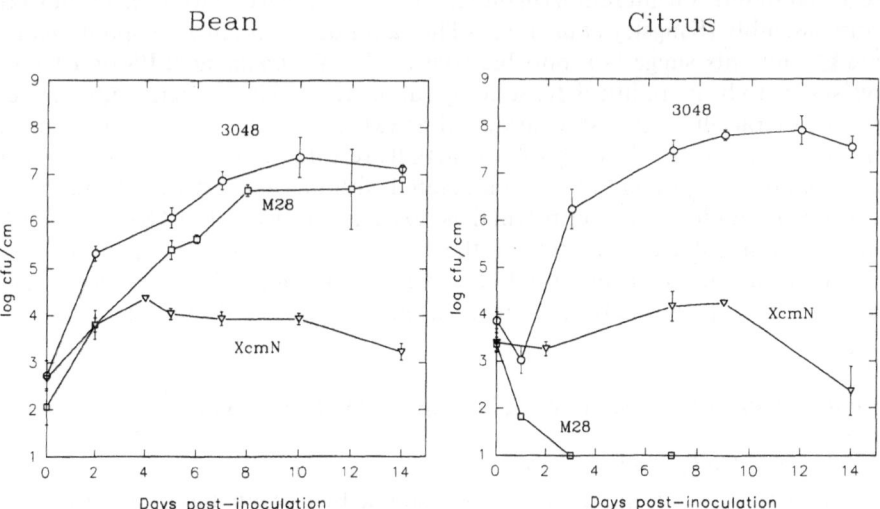

Figure 2. Comparative growth of *Xanthomonas campestris* pv. citrumelo 3048 and host-specific virulence mutants in bean and in citrus. Open circles represent data points of the wild type strain 3048. Open squares represent data points of the mutant strain M28. Open triangles represent data points of *X. campestris* pv. malvacearum strain XcmN (bean and citrus are not hosts for XcmN).

these mutants were more sensitive to citrus-specific host defense compounds, such as phenolics or phytoalexins.

A single clone was isolated which complemented both M27 and M28, restoring growth on citrus and gum production (not shown). Subcloning and DNA sequencing revealed a single 800 bp gene *ops X⁻* (oligo-*poly*-saccharide from *Xanthomonas*) is responsible (Kingsley *et al.*, 1993). Gene *ops X⁻* was specifically disrupted by marker-interruption, using the method of Kamoun *et al.*, (1992). The phenotype of these gene disruption mutants was identical with the spontaneous mutants. Southern hybridization analyses using a DNA fragment internal to *ops X⁻* as a probe revealed that this gene is conserved among all xanthomonads tested, but with some polymorphisms. Marker-disruption mutants were also made in *X. campestris* pv. campestris strain 528; these also exhibited reduced gum production on some media and greatly attenuated virulence in mature cabbage plants inoculated by stem wounding. An *ops X⁻* hybridizing clone from a 528 library complemented the 528 mutant in both gum production and virulence on cabbage. However, the 528 clone failed to complement the marker-disrupted *X. campestris* pv. citrumelo mutants to virulence on citrus, and *ops X⁻* carrying clones failed to complement the 528 mutants to virulence on cabbage, providing strong evidence that analogues of *opsX* are not functionally interchangeable.

Interestingly, *ops-X* has high sequence similarity to lsi-I of *Neisseria gonorrhoeae* and to RfaQ of *Escherichia coli* [Kingsley *et al.*, 1993]. Both the *lsi-1* and *rfaQ* genes are involved in lipopolysaccharide (LPS) core assembly (Petricoin *et al.*, 1991; Parker *et al.*, 1992). Extracts of LPS from *ops X⁻* mutants revealed significant alterations in the LPS of 3048, consistent with defects in LPS core assembly (Kingsley *et al.* 1993). The rapid death in citrus of spontaneous *opsX⁻* mutants suggests a possible role of the *Xanthomonas* LPS or EPS in resistance to host-specific defense compounds. In repeated experiments, crude, sterile, extracellular polysaccharide (EPS) extracts of the culture supernatant from wild type 3048 cells were able to partially restore water-soaking and growth *in planta* by M28, when M28 was inoculated with the 3048 extracts (Kingsley & Gabriel, unpublished). (Controls in these experiments were the EPS extracts and M28 cells inoculated separately on the same citrus leaves). It is apparent that *ops X⁻* is an enzyme involved in LPS and EPS biosynthesis in *Xanthomonas*, and in its absence, a highly effective citrus defense response kills the pathogen.

Alteration of pathovar status using host-specific virulence genes

Leaf spot to blight and black vein
X. campestris pv. campestris causes a systemic black vein and rot of crucifers. Some strains, such as *X. campestris* pv. campestris 528, cause severe blight symptoms on crucifers. *X. campestris* pv. armoraceae causes a non-systemic leaf spot of crucifers. Both pathovars have the same or similar host range. In an effort to determine whether or not the *X. campestris* pv. armoraceae strains carry some

avirulence function which limits systemic infection, or whether or not *X. campestris* pv. campestris strains carried genes which enable systemic movement and/or blight symptoms, total DNA libraries of *X. campestris* pvs. campestris and armoraceae were constructed using cosmid vectors in *E. coli*. The *X. campestris* pv. campestris library was transferred by conjugation into *X. campestris* pv. armoraceae, and vice versa. A seedling assay was used in initial screens. No effects on pathogenicity of *X. campestris* pv. campestris were observed by any of the DNA clones from *X. campestris* pv. armoraceae. However, a cosmid clone from *X. campestris* pv. campestris significantly altered the phenotype of *X. campestris* pv. armoraceae in the seedling assays (Chen, 1992). A subcloned 5 kb fragment from this cosmid carried all of the activity observed in the seedling assays, and in mature plant assays, conferred to *X. campestris* pv. armoraceae the ability to cause blight and black vein symptoms. Interestingly, a vascular hypersensitive response, characteristic of *X. campestris* pv. armoraceae (refer Kamoun *et al.*, 1992) was also observed on mature crucifers inoculated with these transconjugants. This hypersensitive response appeared to be tissue specific and localized to the vascular region, but was insufficient to stop the spread of blight symptoms induced by these transconjugants. We do not yet have growth data to know whether or not growth was affected. Significantly, Southern hybridization analyses of 18 different *Xanthomonas* species and pathovars revealed that only *X. campestris* pv. campestris strains capable of causing blight symptoms hybridized with the 5 kb fragment (Chen, 1992; Chen *et al.*, 1992). The gene(s) conferring blight symptoms may therefore be unique to *X. campestris* pv. campestris strains capable of causing blight. Since *X. campestris* pv. armoraceae causes only a leaf spot on crucifers, and since the clone conferred symptoms diagnostic of *X. campestris* pv. campestris, the 5 kb fragment carried all information necessary to alter pathovar status of the recipient strains.

Leaf spot to hyperplastic cankers

Gene *pthA* is required by *X. citri* to elicit the hyperplastic cankers on citrus that make *X.citri* the subject of U.S. federal and state quarantine and eradication laws. Furthermore, *pthA* confers the ability to elicit hyperplastic cankers on citrus (and only on citrus) to several xanthomonads, including *X. campestris* pvs. alfalfae and citrumelo (Swarup *et al.*, 1991). The reinstatement of *X. citri* (ex Hasse) to species and creation of a new pathovar, *X. campestris* pv. citrumelo (Gabriel) was based on restriction fragment length polymorphism analyses showing a large phylogenetic distance between strains of *X. citri* and strains of *X. campestris*, including *X. campestris* pv. citrumelo strains (Gabriel *et al.*, 1989). Subsequent work using DNA-DNA hybridizations have confirmed that *X. citri* is distinct from *X. campestris* 528[T] using a different phylogenetically based method (Egel *et al.*, 1991). Further work using restriction fragment length polymorphism analyses, DNA-DNA hybridizations, polyacrylamide gel electrophoresis of proteins and fatty acid analyses also support separation of *X. citri* from *X. campestris* pv. citrumelo (Vauterin *et al.*, 1991). Gene *pthA* is therefore not only capable of altering the pathovar status of recipient strains, but on citrus, may provide a selective advantage that leads to clonal amplification of strains carrying the gene. *X. citri* may have diverged from other xanthomonads virulent on citrus a relatively long time ago.

A Xanthomonas *host-specific virulence/avirulence* (hsv/avr) *gene family*
Surprisingly, gene *pthA* of *X. citri* (Swarup *et al.*, 1992) is a member of a large gene family that includes 12 *avr* genes (published) to date (refer Table 1, above, and Fig. 1). DNA fragments which hybridize to internal probes derived from members of this gene family were found in 9 of 12 *Xanthomonas* species or pathovars examined, including *X. citri*, *X. phaseoli*, and *X. campestris* pvs. alfalfae, aurantifolii, cyamopsidis, glycines, malvacearum, translucens and vignicola, and almost always with multiple members per strain (Bonas *et al.*, 1989; Swarup *et al.*, 1992). All strains tested of *X. citri*, *X. phaseoli* and *X. campestris* pv. malvacearum carried members of this gene family (Fig. 1 and Swarup *et al.*, 1992), suggesting that these genes may be needed in some strains. We recently cloned the three other DNA fragments from *X. citri* that hybridize to *pthA* from *X. citri* 3213; only *pthA* conferred ability to cause cankers on citrus (Swarup, Yang, Kingsley & Gabriel, unpublished). We hypothesize that only one member of this gene family, *pthA*, is needed for virulence of *X. citri* strain 3213 on citrus. However, since not all *Xanthomonas* strains carry DNA fragments which hybridize to members of this gene family, these genes do not appear to be needed in the genus generally.

The largest number of published members of this gene family (seven) derived from one strain have been cloned from *X. campestris* pv. malvacearum strain XcmH (Tabel 1). (Gabriel *et al.*, 1986). All strains tested of *X. campestris* pv. malvacearum originating in North America, such as XcmH, had at least 8–11 potential members of the gene family, including 4–6 hybridizing fragments on large plasmids (De Feyter *et al.*, 1993). Strains originating in West Africa (such as XcmN) had 4–5 potential members of the family, including at least one on a plasmid. Yet strain XcmN exhibits no known avirulence activity, and most of the other African strains are virulent on a wide range of cotton cultivars. The presence of potential members of the gene family in all cotton blight inducing *X. campestris* pv. malvacearum strains – some with no known avirulence activity – was unexpected, and indicated that at least one or more members may be needed for virulence on cotton.

To study possible pleiotropic functions of members of the gene family found in XcmH, marker exchange-eviction mutagenesis (Reid & Collmer, 1987) was carried out (Yang and Gabriel, 1992). A suicide vector carrying an internal fragment cloned from a member of the gene family, and carrying a *nptI-sacB-sacR* cartridge cloned in the middle of the internal fragment was constructed. The *nptI-sacB-sacR* cartridge confers kanamycin resistance and sucrose sensitivity (due to the production of levansucrase by *sacB*). The marker exchange mutants of XcmH were tested for loss of virulence on susceptible cotton plants, and for loss of avirulence on congenic cotton lines differing by single resistance *R* genes.

Mutants causing altered phenotype in comparison to XcmH were analyzed by Southern hybridization. Some mutants were subjected to eviction of the *nptI-sacB-sacR* marker in the presence of 5% sucrose and further rounds of marker exchange-eviction mutagenesis. More than 120 marker exchange mutants were generated and 40 of them were tested. As predicted by gene-for-gene theory, a marker exchange mutation of *avrBIn, avrB4, avrb7, avrB101, avrB102* or *avrb6* in XcmH resulted in the loss of *R* gene specific avirulence. However, a marker exchange mutation in *avrb6* also resulted in reduced watersoaking ability of XcmH on the susceptible cotton line Ac44 (Yang *et al.*, 1991). We have previously shown that *avrb6* affects virulence as indicated by enhancing watersoaking symptoms on the susceptible cotton line Ac44 (De Feyter & Gabriel, 1991). As with *pthA*, we hypothesize that only one member of this gene family, *avrb6*, is needed for virulence of XcmH on cotton. The remaining members may function for avirulence only.

Alteration of both host range and pathovar status
Since *pthA* of *X. citri* and *avrb6* of *X. campestris* pv. malvacearum appeared to confer host specific virulence and were structurally similar, we attempted gene swapping experiments to determine whether or not the host range of XcmH could be modified to include citrus, and whether or not the host range of an *X. citri* strain could be modified to include cotton. Surprisingly, *X. citri* and cotton blight strains of *X. campestris* pv. malvacearum are closely related (90%) by total DNA-DNA reassociation (Egel *et al.*, 1991), and therefore the number of gene differences between them that determine host range may be few. To test this idea, we mobilized *pthA* into a mutant strain of XcmH which had lost seven members of this *avr* gene family, including *avrb6*. Although the resulting strain induced a visible HR on citrus, it also elicited cankers on citrus, when inoculated at high (10^8 cfu ml^{-1}) concentrations. At lower concentrations, small cankers were also induced. We have not yet tested growth *in planta*. Gene *pthA* did not confer to *X. campestris* pv. malvacearum strains the ability to elicit cankers on cotton, but instead these transconjugants exhibited cultivar-specific avirulence on cotton. Similarly, *avrb6* did not confer to *X. citri* strains the ability to elicit water soaking symptoms on citrus, but instead these transconjugants elicited an HR on citrus. The virulence function(s) of these genes is clearly host specific. We therefore refer to this gene family as a *hsv/avr* gene family.

Horizontal transfer of *xanthomonas hsv/avr* genes

The nucleotide sequence of *avrb6* revealed the presence of 62 bp nearly perfect, inverted repeats, which are also found in *avrB4, avrBs3* and *pthA* (De Feyter *et al.*, 1993). From the beginning of the left inverted repeat to the end of the right inverted repeat, *avrb6* is 98% identical in DNA sequence along its entire length (3.4 kilobases) to the complete nucleotide sequence of *avrBs3*. The presence of inverted terminal repeat sequences flanking members of this gene family sequenced to date was unexpected, as was the high level of sequence homology. Since strains of *X. campestris* pv. malvacearum and *X. campestris* pv. vesicatoria are only 34–42% similar by DNA-DNA hybridizations (Kingsley and Gabriel, unpublished), these genes have obviously moved horizontally among genetically dissimilar strains. The wide distribution of multiple copies of hybridizing fragments among natural strains of the genus *Xanthomonas* may indicate that the horizontal transfer of these genes is not rare. The presence of terminal inverted repeats, the evidence for horizontal gene transfer and the presence of multiple hybridizing bands in nearly all strains examined suggested transposition as a possible mechanism of genetic exchange. Transposable elements with 50 ± 18 bp inverted repeats have been isolated from *X. citri* (Tu *et al.*, 1989).

It has long been puzzling as to why microbial plant pathogens carry genes which function for avirulence. The neo-Darwinian assumption that "*avr* genes (generally) must encode important functions for the pathogen" (de Wit, 1992) is a widely held dogma for which there is no evidence. Indeed, the evidence from 29 cloned *avr* genes demonstrates that less than 10% of the genes studied exhibit any detectable function other than avirulence. The *Xanthomonas avr* genes belonging to the *hsv/avr* gene family may transfer horizontally within the genus and may transpose. The evidence from XcmH, the cloning source of seven members of this gene family, indicates that recombination among duplicated members is the basis for spontaneous race-change mutations (De Feyter *et al.*, 1993). Deletion of those members of the gene family not involved in virulence appears to have no effect on virulence. There is no compelling evidence that alleles must have selective value because they exist; in fact, the vast majority of alleles appear to be selectively neutral (Nei, 1987). A neutral theory view holds that some *avr* alleles may have pleiotropic selective value, but there is no need for it *a priori* (Gabriel, 1989).

In summary, it is useful to distinguish between genes which confer host range and those which confer disease-specific phenotypes, since pathovar status can be determined by either. Pathovars are defined as pathogenic variants, and phylogenetically similar strains with the same host range may be in different pathovars. For example, both *X. campestris* pv. armoraceae (leaf spot) and *X. campestris* pv. campestris (black rot) have the same host range on Brassicas. Alteration of pathovar status can be a matter of simply adding disease specific gene(s), such as the one(s) on the 5 kb fragment from *X. campestris* pv. campestris to *X. campestris* pv. armoraceae. Similarly, phylogenetically diverse strains with the same host range may be altered in disease phenotype by adding a single gene, as was done with *pthA* in *X. campestris* pv. citrumelo. The alteration of host range may be more complicated. On the one hand, alteration of host range among phylogenetically similar strains may be easy, as in the case of *hsvB*, moved from one *X. campestris* pv. translucens strain to another. On the other hand, alteration of host range that affects pathovar status may be more complicated, and require the deletion of one or more *avr* genes, and perhaps *hsv* genes, such as *avrb6*. Perhaps the most striking observation of this work is that both host range and pathovar status of *Xanthomonas* may be determined in some cases by single, host-specific, virulence genes. By experiments designed to uncover the basis of pathovar-level specificity, it seems clear that host specific virulence genes can be identified and manipulated to alter the host range and/or pathovar status of *Xanthomonas spp.* strains. This is a relatively new idea, and the protocols used and perhaps the genes themselves may be useful to investigators desiring to create biological weed control agents.

References

Bonas U, Stall RE, Staskawicz B (1989) Genetic and structural characterization of the avirulence gene, *avrBs3* from *Xanthomonas campestris* pv. *vesicatoria*. Mol Gen Genet 218: 127–136

Canteros B, Minsavage G, Bonas U, Pring D, Stall R (1991) A gene from *Xanthomonas campestris* pv. vesicatoria that determines avirulence in tomato is related to *avrBs3*. Molec Plant-Microbe Interact 4: 628–632

Chen J (1992) Isolation of disease-specific virulence genes from *Xanthomonas campestris*. MS Thesis, U. Florida, Gainesville, 77 pp.

Chen J, Roberts PD, Gabriel DW (1992) Isolation and cloning of a potential systemic movement factor from *Xanthomonas campestris* pv. campestris. Phytopathology 82: 1118 (abstract)

Dangl JL, Ritter C, Gibbon MJ, Mur LAJ, Wood JR, Goss S, Mansfield J & Vivian A. (1992) Functional homologs of the Arabidopsis *RPM1* disease resistance gene in bean and pea. Plant Cell 4: 1359–69

DeFeyter R, Yang Y, Gabriel DW (1993) Gene-for-genes interactions between cotton *R* genes and *X. campestris* pv. malvacearum *avr* genes. Molec. Plant-Microbe Interact 6: 225–237.

De Feyter R & Gabriel DW (1991) At least six avirulence genes are clustered on a 90-kilobase plasmid in *Xanthomonas campestris* pv. malvacearum. Mol. Plant-Microbe Interact. 4: 423–432

de Wit PJGM (1992) Molecular characterization of gene-for-gene systems in plant-fungs interactions and the application of avirulence genes in control of plant pathogens. Annu Rev Phytopathol 30: 391–418

Djordjevic MA, Gabriel DW & Rolfe BG (1987) *Rhizobium* the refined parasite of legumes. Ann. Rev. Phytopath. 25: 145–168

Djordjevic MA, Schofield PR & Rolfe BG (1985) Tn5 mutagenesis of Rhizobium trifolii host-specific nodulation genes result in mutants with altered host-range ability. Mol. Gen. Genet. 200: 463–471

Dye DW, Bradbury JF, Goto AC, Hayward AC, Lelliott RA, Schroth MN (1980) International standards for naming pathovars of phytopathogenic bacteria and a list of pathovar names and pathotype strains. Rev. Plant Pathol. 59: 153–168

Egel DS, Graham JH & Stall RE (1991) Genomic relatedness of *Xanthomonas campestris* strains causing diseases of citrus. Appl. Environ. Microbiol. 57: 2724–30

Fillingham AJ, Wood J, Bevan JR, Crute IR, Mansfield JW, Taylor JD & Vivian A (1992) Avirulence genes from *Pseudomonas syringae* pathovars phaseolicola and pisi confer specificity towards both host and non-host species. Physiol. Mol. Plant Pathol. 40: 1–15

Gabriel DW (1986) Specificity and gene function in plant/ pathogen interactions. ASM News 52: 19–25

Gabriel DW (1989) The genetics of plant pathogen population structure and host-parasite specificity. p. 343–379. *In:* Kosuge T & Nester EW (eds.) Plant-Microbe Interactions: Molecular and Genetic Perspectives, Vol 3 Macmillan Publishing Co., New York

Gabriel DW, Burges A & Lazo GR (1986) Gene-for-gene recognition of five cloned avirulence genes from *Xanthomonas campestris* pv. *malvacearum* by specific resistance genes in cotton.

Proc. Natl. Acad. Sci. USA. 83: 6415–6419

Gabriel DW, Hunter J, Kingsley M, Miller J & Lazo G (1988) Clonal population structure of *Xanthomonas campestris* and genetic diversity among citrus canker strains. Mol. Plant-Microbe Interact. 1: 59–65

Gabriel DW, Kingsley MT, Hunter JE & Gottwald TR (1989) Reinstatement of *Xanthomonas citri* (*ex* Hasse) and *X. phaseoli* (*ex* Smith) to species and reclassification of all *X. campestris* pv. *citri* strains. Int. J. System. Bacteriol. 59: 14–22

Gabriel DW & Rolfe BG (1990) Working models of specific recognition in plant-microbe interactions. Ann. Rev. Phytopathol. 28: 365–391

Hopkins CM, White FF, Choi S-H, Guo A & Leach JE (1992) Identification of a family of avirulence genes from *Xanthomonas oryzae* pv. oryzae. Mol Plant-Microbe Interact 5: 451–459

Kamoun S, Kamdar HV, Tola E & Kado CI (1992) Incompatible interactions between crucifers and *Xanthomonas campestris* involve a vascular hypersensitive response: role of the *hrpX* locus. Molec. Plant-Microbe Interact. 5: 22–33

Keen NT (1992) The molecular biology of disease resistance. Plant Mol. Biol. 19: 109–122

Keyser HH, van Berkum P, and Weber DF (1982) A comparative study of the physiology of symbioses formed by *Rhizobium japonicum* with *Glycine max, Vigna unguiculata,* and *Macroptilium atropurpurem*. Plant Physiol. 70: 1626–1631

Kingsley MT, Gabriel DW, Marlow GC & Roberts PD (1993) The *opsX* locus of *Xanthomonas campestris* affects host range and biosynthesis of lipopolysaccharide and extracellular polysaccharide. J. Bacteriol. 175: 5839–5850

Kobayashi, DY, Tamaki SJ, Keen NT (1990) Molecular characterization of avirulence gene D from *Pseudmonas syringae* pv. tomato. Mol. Plant-Microbe Interact. 3: 94–102

Kondorosi A, Horvath B, Rostas K, Gottfert M, Putnoky P, Rodriguez-Quinones F, Banfalvi Z & Kondorosi E (1986) Common and host-specific nodulation genes of *Rhizobium meliloti*. 88. Third International Symposium on the molecular genetics of plant-microbe interactions, July 27-31 McGill University, Montreal

Lelliot RA (1972) *Proc. Third Intern. Conf. Plant Pathogenic Bacteria* (ed. Geesteranus), p269, Centre for Agricultural Publication and Documentation, Wageningen

Long SR (1984) Genetics of *Rhizobium* nodulation. p. 265–306. *In:* Kosuge T & Nester EW (eds.) Plant-microbe interactions, Vol.1 (pp 265–306) Macmillan, New York

Longman P (1989) The big lie. Florida Trend 32: 40–46

Ma QS, Chang MF, Tang JL, Feng JX, Fan MJ, Han B & Liu T (1988) Identification of DNA sequences involved in host specificity in the pathogenesis of *Pseudomonas solanacearum* strain T2005. Molecular Plant Microbe Interactions 1: 169–174

Minsavage GV, Dahlbeck D, Whalen MC, Kearney B, Bonas U, Staskawicz BJ and Stall RE (1990) Gene-for-gene relationships specifying disease resistance in *Xanthomonas campestris* pv. vesicatoria-pepper interactions. Mol. Plant-Microbe Interact. 3: 41–47

Murata N & Starr MP (1973) A concept of the genus *Xanthomonas* and its species in the light of segmental homology of deoxyribonucleic acids. Phytopath.Z. 77: 285

Nei M (1987) Molecular evolutionary genetics, Columbia University Press, New York

Parker CT, E Pradel and Schnaitman CA (1992). Identification and sequence of the lipopolysaccharide core biosynthetic genes *rfaQ, rfaP,* and *rfaG* of *Escherichia coli* K-12. J. Bacteriol. 174: 930–934.

Petricoin III EF, Danaher RJ, and Stein DC (1991) Analysis of the *lsi* region involved in lipo-oligosaccharide biosynthesis in *Neisseria gonorrhoeae*. J. Bacteriol. 173: 7896–7902

Reid JL & Collmer A (1987) An *nptI-sacB-sacR* cartridge for constructing directed, unmarked mutations in Gram-negative bacteria by marker exchange-eviction mutagenesis. Gene 57: 239–246

Ronald PC, Salmeron JM, Carland FM & Staskawicz BJ (1992) The cloned avirulence gene *avrPto* induces disease resistance in tomato cultivars containing the *Pto* resistance gene. J. Bacteriol 174: 1604–1611

Sawczyc MK, Barber CE & Daniels MJ (1989) The role of pathogenicity of some related genes

In *Xanthomonas campestris* pathovars campestris and translucens: a shuttle strategy for cloning genes required for pathogenicity. Molec. Plant-Microbe Interact 2: 249–255

Salch YP & Shaw PD (1988) Isolation and characterization of pathogenicity genes of *Pseudomonas syringae* pv. *tabaci*. J. Bacteriol. 170: 2584–2591

Schnathorst WC (1966) Unaltered specificity in several xanthomonads after repeated passage through *Phaseolus vulgaris*. Phytopathology 56: 58–60

Schoulties CL, Civerolo EL, Miller JW, Stall RE, Krass CJ, & Poe SR (1987) Citrus canker in Florida. Plant Disease 71: 388–394

Schultze M, Quiclet-Sire B, Kondorosi E, Virelizier H, Glushka JN, Endre G, Gero SD, Kondorosi A (1992) *Rhizobium meliloti* produces a family of sulfated lipo-oligosaccharides exhibiting different degrees of plant host specificity. Proc. Natl. Acad. Sci. USA 89: 192–196

Selander RK (1985) Protein polymorphism and the genetic structure of natural populations of bacteria. p. 85–106. *In:* Ohta T & Oaki K, (eds.), Population genetics and molecular evolution Japan Scientific Societies Press, Tokyo

Simmons EG (1990) Alternaria themes and variations. Mycotaxon 37: 79–119

Spaink HP, Sheeley DM, van Brussel AAN, Glushka J, York WS, Tak T, Geiger O, Kennedy EP, Reinhold VN, Lugtenberg BJJ (1991) A novel highly unsaturated fatty acid moiety of lipo-oligosaccharide signals determines host specificity of *Rhizobium*. Nature 354: 125–130

Stapleton JJ & Garza-Lopez JG (1988) Epidemiology of a citrus leaf-spot disease in Colima, Mexico. Phytopathology 78: 440–443

Staskawicz B, Dahlbeck D, Keen N & Napoli C (1987) Molecular characterization of cloned avirulence genes from race 0 and race 1 of *Pseudomonas syringae* pv. *glycinea*. J. Bacteriol. 169: 5789–5794

Staskawicz BJ, Dahlbeck D & Keen NT (1984) Cloned avirulence gene of *Pseudomonas syringae* pv. *glycinea* determines race-specific incompatibility on *Glycine max* (L.) Merr. Proc. Natl. Acad. Sci. USA 81: 6024–6028

Swanson J, Kearney B, Dahlbeck D & Staskawicz B (1988) Cloned avirulence gene of *Xanthomonas campestris* pv. *vesicatoria* complements spontaneous race change mutants. Mol. Plant-Microbe Interact. 1: 5–9

Swarup S, DeFeyter R, Brlansky RH, Gabriel DW (1991) A pathogenicity locus from *Xanthomonas citri* enables strains from several pathovars of *X. campestris* to elicit cankerlike lesions on citrus. Phytopathology 81: 802–809

Swarup S, Yang Y, Kingsley MT, Gabriel DW (1992) A *Xanthomonas citri* pathogenicity gene, *pthA*, pleiotropically encodes gratuitous avirulence on nonhosts. Molec Plant-Microbe Interact 5: 204–213

Trevors JT, Barkay T, and Bourquin AW (1986) Gene transfer among bacteria in soil and aquatic environments: a review. Can. J. Microbiol. 33: 191–198

Tu J, Wang HR, Chang SF, Charng, YC, Lurz R, Dobrinski B & Wu WC (1989) Transposable elements of *Xanthomonas campestris* pv. citri originating from indiginous plasmids. Mol. Gen. Genet. 216: 505–510

Van Kan JAL, van den Ackerveken GFJM & de Wit PJGM (1991) Cloning and characterization of cDNA of avirulence *avr9* of the fungal pathogen *Cladosporium fulvum*, causal agent of tomato leaf mold. Molec. Plant-Microbe Interact. 4: 52–59

Vauterin L, Yang P, Hoste B, Vancanneyt M, Civerolo EL, Swings J, Kersters K. (1991) Differentiation of *Xanthomonas campestris* pv. citri strains by sodium dodecyl sulfate-polyacrylamide gel electrophoresis of proteins, fatty acid analysis, and DNA-DNA hybridization Inter.J.System.Bacteriol. 41: 535–542

Wayne LG, Brenner DJ, Colwell RR, Grimont PAD, Kandler O, Krichevsky MI, Moore LH, Moore WEC, Murray RGE, Stackebrandt E, Starr MP & Truper, HG (1987) Report of the ad hoc committee on reconciliation of approaches to bacterial systematics. Int. J. System. Bacteriol. 37: 463–464

Waney VR, Kingsley MT & Gabriel DW (1991) *Xanthomonas campestris* pv. translucens genes determining host-specific virulence on cereals identified by Tn5-*gusA* insertion mutagenesis.

Mol. Plant-Microbe Interact. 4: 623–627

Whalen MC, Innes RW, Bent AF & Staskawicz BJ (1991) Identification of *Pseudomonas syringae* pathogens of Arabidopsis and a bacterial locus determining avirulence on both Arabidopsis and soybean. Plant Cell 3: 49–59

Whalen MC, Stall RE & Staskawicz BJ (1988) Characterization of a gene from a tomato pathogen determining hypersensitive resistance in non-host species and genetic analysis of this resistance in bean. Proc. Natl. Acad. Sci. USA 85: 6743–6747

Willis DK, Rich JJ & Hrabak EM (1991) *hrp* genes of phytopathogenic bacteria. Molec Plant-Microbe Interact 4: 132–138

Yang Y, De Feyter R, Swarup S & Gabriel DW (1991) The avirulence gene *avrb6* from *Xanthomonas campestris* pv. malvacearum affects pathogenicity in compatible interactions with cotton, but not bacterial growth *in planta*. Phytopathology 81: 1226 (abstract)

Yang Y & Gabriel DW (1992) Functional analysis of avirulence genes in *Xanthomonas campestris* pv. malvacearum. Phytopathology 82: 1167 (abstract)

SECTION II

Pathogen ingression and invasive mechanisms

11. Iron and plant pathogenesis: the systemic soft rot disease induced by *Erwinia chrysanthemi* 3937 on saintpaulia plants

DOMINIQUE EXPERT, CLAIRE NEEMA, J. PIERRE LAULHÈRE, CHRYSTÈLE SAUVAGE, CÉLINE MASCLAUX and BRUNO MAHÉ

Abstract. Competition for iron plays an important role in bacterial virulence and animal host defense. Several attempts have been made to determine the role of microbial siderophores in plant infection, but only recently it has been clarified. In *Erwinia chrysanthemi* 3937, a plant pathogenic bacterium responsible for soft-rot diseases, Enard *et al.* (1988) have shown the relevance of an iron assimilation system mediated by the siderophore chrysobactin, in the virulence of this bacterium in saintpaulia plants. The level of iron in the intercellular fluid of plant leaves proved to be low enough to restrict the growth of iron uptake mutants during early stages of infection. To investigate the problem of iron acquisition in this pathogenic interaction, we undertook a molecular study of the chrysobactin system with a special emphasis on its regulation by iron. In addition, iron was found to modulate the expression of several pectinase-encoding genes of major importance for the virulence. We identified the presence of chrysobactin as well as the induction of the chrysobactin operon in infected tissues. Competition for nutritional iron was further studied through a plant-bacterial system from which drastic changes were shown to occur in iron metabolism of infected cultured soybean cells.

Abbreviations: GUS, β-Glucuronidase

Introduction

In the course of evolution, pathogenic bacteria have acquired accurate mechanisms which allow them to specifically infect animal or plant organisms. Although they are fundamentally different in their level of cellular differentiation and organization, plants and animals are subject to the same challenges upon bacterial attack. For the bacterium, entering the host and attaching to target tissues are central issues which are tightly controlled by the production of factors promoting the establishment of the microorganism and subsequent evasion of host defense. In the later stages of infection, the bacterium may adopt new features to allow survival or transmission to a new host. However, structural biology of the host and especially the cytoskeletal anatomy (Knox 1992) are important for determining the nature of the invasive process (Isberg 1991; Falkow 1991). In animal organisms, epithelial or fibroblastic cells are often the sites of microbial multiplication (Galan 1992). In contrast, the presence of rigid walls surrounding plant cells limits the possibilities for the bacterium to develop intracellularly, unless parts of the cell walls are completely dissolved. In this regard, receptor-ligand interactions may be more relevant in animal than in plant infection. But in any case, the dynamic

C.I. Kado and J.H. Crosa (eds.), Molecular Mechanisms of Bacterial Virulence, 161–171.
© 1994 *Kluwer Academic Publishers.*

nature of the cytoskeleton reflects the diversity of ingression factors elaborated by a pathogen to move inside its host. The release of large amounts of cell wall degrading enzymes may be, for instance, considered as an advantage for a plant pathogen, in that clearance of the intercellular matrix can facilitate its movement between cells.

In addition to structural constraints, a second important aspect of pathogenesis relies on the conditions imposed on the microbe by the microenvironment (temperature, oxygen tension, ionic strength, nutrient status...), at the colonization site (Cornelis *et al.*, 1989; Parsot & Mekalanos, 1990; Dorman *et al.*, 1990; Scarlato *et al.*, 1991; Tobe *et al.*, 1991; Deretic *et al.*, 1991). Pathogenic bacteria display the remarkable ability to exploit the host conditions encountered at various stages of their parasitic life cycle for their own benefit (Stachel & Zambryski, 1986 Rogowsky *et al.*, 1987 Veluthambi *et al.*, 1989 Coote, 1991 Porat *et al.*, 1991). Iron does not escape this rule. This essential metal is not readily bioavailable. Excretion of siderophores and subsequent transport of their ferric complexes define a possible route used by pathogenic bacteria to overcome the low iron availability of body fluids in vertebrates. Iron transport proteins of the transferrin family may prevent proliferation of the pathogen by depriving it of nutritional iron (Weinberg, 1984). Furthermore, several pathogenic bacteria are armed with toxins not related to iron transport *per se*, that are specifically induced in low iron environments (Calderwood & Mekalanos, 1987; Poole & Braun, 1988). Since the intrinsic property of ferric iron to be insoluble at physiological pH must also prevail in plant fluids, a similar nutritional problem may occur during the systemic development of a phytopathogen.

We focused a particular attention to this question in the case of the soft rot disease triggered by *E. chrysanthemi* 3937 on saintpaulia plants (Expert & Toussaint, 1985). The bacterial cells invade the intercellular spaces of leaf parenchymatous tissues and may further move to other aerial parts of the plant, through xylem vessels. The symptom, trivially named maceration, is a disorganization of parenchyma and results from several secreted enzymes including a set of pectinases which work in concert to degrade the plant cell walls (Garibaldi & Bateman, 1973). In addition, a functional genetic system responsible for the production of the siderophore chrysobactin and the utilization of its ferric complex is required for the spreading phase of the disease (Enard *et al.*, 1988). Interestingly, this enterobacterium cannot use ferric citrate, i.e., the major iron carrier in plant vessels (Brown, 1978), as an iron source (Expert & Gill, 1991). To investigate the problem of iron acquisition in this pathogenic interaction, we undertook a molecular study of the chrysobactin system with a special emphasis on its regulation by iron. In addition, iron was found to modulate the transcription of several pectinase-encoding genes. We identified the presence of chrysobactin as well as the induction of the system in infected tissues. Competition for nutritional iron was further studied through a plant-bacterial system: drastic changes were shown to occur in plant iron metabolism upon infection of cultured soybean cells. A review of these different findings is presented.

Genetic analysis of the *E. chrysanthemi* 3937 chrysobactin-mediated iron assimilation system

Like other high affinity iron transport systems, the chrysobactin-mediated pathway involves a clustered organization of genes (Franza *et al.*, 1991). They encode the functions required for (i) the biosynthesis of the siderophore which roughly consists of the condensation of a D-lysyl-L-serine peptide to a catechol moiety (Persmark *et al.*, 1988); (ii) the release of the molecule into the external medium and (iii) its subsequent transport inside the cell, once it has been loaded with iron (Persmark *et al.*, 1992). We found that the steps subsequent to the passage of the chrysobactin ferric complex across the outer-membrane are the same as those required for the ferric enterobactin complex which can be used as an exogenous iron source by *Erwinia* cells. We identified an operon (*fct, cbsCEBA*) extending beyond ca. 8 kb which contains the ferri-siderophore outer membrane receptor (Fct, 80-kDa) gene located upstream of the genes involved in the catechol pathway (Franza & Expert, 1991). By Northern blot hybridization analysis, using a radioactive probe containing *cbsCE*, a cognate polycistronic RNA transcript was identified in cells grown only under iron limitation but, so far, we ignore if it is the only representative transcript to be expressed from the operon (Fig. 1). Indeed, as a first gene of the operon, *fct* might be controlled differently than the downstream *cbs* genes. The promoter region is currently being analysed.

Iron transport is negatively regulated in *E. chrysanthemi* by an iron-responsive two factor system encoded by the *cbr* operon

Insertional mutagenesis of strain 3937 allowed isolation of a regulatory mutant which expressed chrysobactin and the outer membrane receptor functions for the two exogenous siderophores used by *Erwinia* cells (enterobactin and ferrichrome), regardless of the iron level. Cloning and analysis of the responsible locus led to identification of an operon (*cbrAB*) encoding two polypeptides CbrA and CbrB, with molecular weights of 34,000 and 55,000, respectively (Expert *et al.*, 1992) (Fig. 1). Any insertion in the locus confers on the cell a constitutive phenotype with regard to chrysobactin production. Accordingly, a chrysobactin polycistronic RNA transcript was observed in a *cbrB* mutant. Regulation by iron was further analysed, using chromosomal *lac* chrysobactin operon fusions (*fct::lacZ*) in diverse *cbr* backgrounds (Table 1). In *cbrA+B+* cells, low iron stimulates *fct::lacZ* expression while a constitutive level, although lower than the induced level, was observed in *cbrA+B* cells. Extra copies of *cbrA* allowed repression of the fusion whereas no repression (although visualized in the presence of the whole *cbrAB* locus) was shown on the catechol production (Expert *et al.*, 1992). In the absence of iron, no repression occurred, regardless of the *cbr* background. The *cbr* locus expression was studied through chromosomal *lac* gene fusions in the presence of plasmid-

164

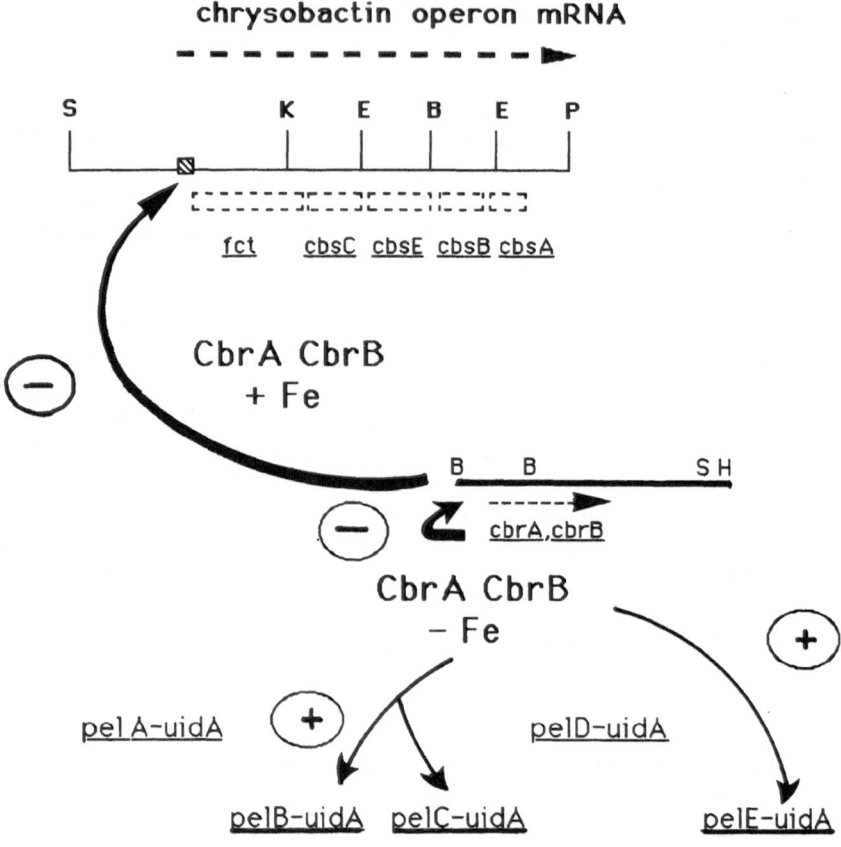

chrysobactin operon mRNA

CbrA CbrB + Fe

CbrA CbrB – Fe

pectate lyase gene expression

Figure 1. General scheme for iron regulation in *E. chrysanthemi* 3937. Regulation is mediated through the activity of the *cbr* operon: the encoded products CbrA and CbrB allow negative control of the chrysobactin operon, in the presence of iron. In the absence of iron, *cbr* is subject to negative autogeneous regulation and up-regulates the transcriptional activity of the pectate lyase genes *pelB*, *pelC* and *pelE*.

borne wild type constructions (Expert *et al.*, 1992): the data show that in the presence of iron, *cbr* negatively regulates the chrysobactin biosynthetic and transport genes, while under conditions of depletion, *cbr* is subject to negative autogenous regulation. Sequencing data should help us to further understand how the two products CbrA and CbrB interact in their iron sensing and regulatory functions. It seems that there is a fairly constrained concentration range for the Cbr proteins over which normal regulation by iron occurred.

Table 1. β-Galactosidase activity. Expression of a *lac* chrysobactin operon fusion under different environmental conditions and *cbr* backgrounds. [a]Plasmid borne *cbr* constructions (p*cbr*) were present at 1–2 copies per cell (Expert *et al.*, 1992). [b]Highest mean β galactosidase activities (given in Miller units) recorded in time course experiments are indicated. Bacterial cells were grown in deferrated or 10μM FeCl₃ supplemented M63 medium (– Fe, +Fe) and in leaf intercellular fluid (De Witt & Spikman, 1982) supplemented with glucose (IF). *In planta* activities were measured in bacterial cells collected from IF of leaves harvested after infection of potted saintpaulia plants (the reported value refers to an infection of 18 hours). The activity observed after inoculation of the parental strain L37 is of plant origin. ND: not done.

Strain	β galactosidase activity[b]			
cbr genotype[a]	*in vitro*		*in vivo*	
	+ Fe	- Fe	IF	*in planta*
L37 Lac⁻	< 5	< 5		15
L37 *cbrA+B+fct::lac*	20	250	200	120
L37 *cbrA+B fct::lac*	120	300	ND	ND
+ p*cbrA+B+*	20	250	ND	ND
+ p*cbrA+*	40	120	ND	ND

Iron, via the *cbr* locus, exerts a positive control on pectate lyase gene transcription

The first evidence which led us to examine the role of iron as a modulating signal of pectinolytic activity was the pathogenicity of *cbr* mutants: the aspect of the maceration symptom which normally results from the activity of several enzymes, of which five pectate lyase isozymes encoded by independent cistrons (*pelA* to *pelE*) (Reverchon *et al.*, 1986) differed from the wild type phenotype. The production of total pectate lyase activities appeared to be influenced by iron levels, when bacterial cells from the wild type strain were grown in minimal medium supplemented with galacturonate, a catabolic inducer of *pel* gene expression. We thus studied the effect of iron on the expression of transcriptional GUS fusions generated in each *pel* gene, in the wild type strain and in its *cbrAB* or *cbrB* mutant derivatives (data not shown): in wild type cells, low iron concentrations roughly resulted in a three-fold increase in the transcriptional activity of *pelB*, *pelC* and *pelE*, regardless of the presence of the inducer. Both *cbr* mutants proved to be completely insensitive to iron fluctuations, and GUS activity corresponded to the basal levels produced in iron replete wild type cells. Iron regulation could be restored by the presence of a plasmid-borne *cbrAB* wild type construction, thus indicating that *cbr* acts as a positive regulator of *pelB*, *pelC* and *pelE* in the absence of iron (Fig. 1).

The problem of iron acquisition in plant infection

We had previously shown that the integrity of the chrysobactin system was required for the spreading phase of the disease. However, it was still unknown if this high affinity iron transport system was expressed during infection. The low iron content found in intercellular fluids from healthy plants and the inability of this enterobacterium to use ferric citrate supported the idea that chrysobactin was produced *in planta*. We first observed that in intercellular fluids from healthy plants, bacteria behaved like that in any iron depleted synthetic medium i.e., they induce their high affinity iron transport systems

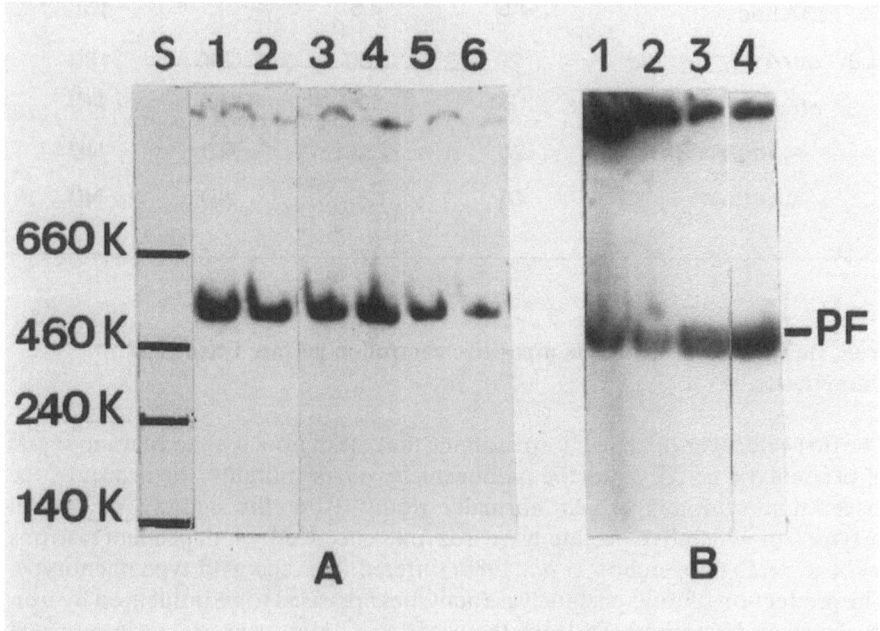

Figure 2. Incorporation of ^{59}Fe into plant ferritin (PF) in soybean cells infected under various conditions. PAGE analysis of native protein extracts and detection of ferritin radioiron were achieved as described previously (Laulhère *et al.*, 1992). Panel A: infection with *E. chrysanthemi* wild type cells. Lanes 1, 2: PF from an axenic plant cell suspension, 2 and 1 ml respectively. Lanes 3 to 6: PF from 2 ml cell suspensions infected with 2×10^6, 2×10^7, 5×10^7 bacterial cells, respectively. ^{59}Fe associated to plant cells was reduced by a 5-fold factor in the presence of bacteria. Panel B: infection with bacteria free culture supernatants. Lanes 1 to 3: infection with 1, 2, 1.5 ml of an iron starved culture (2×10^8 cells per ml) supernatant from the wild type strain. Lane 4: as in A, lane 1. Lane S: molecular weights of standard proteins are indicated in kilodaltons.

(Table 1). Second, using an approach to differentiate strong iron ligands from weak ones, we identified the presence of high amounts of a strong ligand only in intercellular fluids of infected plants. By its spectrophotometric characteristics and its biological activity, this compound proved to be chrysobactin. The following question was thus to know if the high iron chelating capacity of chrysobactin could affect the iron metabolism of the infected host. To explore this problem, we looked at the distribution of iron when supplied with ferric citrate to cultured plant cells infected with bacteria. As a relevant level of iron availability, plant cell ferritins were used because they represent one of the first molecules to be labelled in soybean cells (Laulhère et al., 1992). We found that the bacterial cells as well as their culture supernatants which contained iron-free chrysobactin, shunted the mobilization of iron into plant ferritins (Fig. 2). This indicates that chrysobactin when present in infected plants is required for bacterial iron nutrition but is deleterious to plant cells, by denying them essential iron.

Finally, we looked at the expression of β-galactosidase from *fct::lacZ* fusions in leaves of plants inoculated with the Lac⁻ mutant carrying the fusion, during the primary stages of infection. From 12 to 18 hours of infection, increasing levels of bacterial β galactosidase activity originating from expression of the *fct::lacZ* fusion were produced (Table 1). Beyond this period, the infected area was too much damaged because of the high pectinolytic activity for the assay to be significant. The plant enzyme activity thus masked the bacterial activity.

Conclusion

By controlling the functions involved in its transport and storage, living cells respond to iron in order to balance levels of an essential element which is potentially toxic. Iron has also long been recognized as a regulatory signal of microbial virulence, in animals. As a plant-pathogen, *E. chrysanthemi* does not seem to escape to this scope. Like in animal or human infections (Aumont et al.,1988; Bullen et al., 1991), it seems that the production of siderophore(s) is required when bacteria are multiplying in the extracellular spaces of the host tissues. But, when plant cells of the infected area become sufficiently damaged, there must be enough readily available iron to satisfy the needs of the pathogen. Production of siderophores is energy consuming and other mechanisms to acquire iron from host cell proteins may be more advantageous for the pathogen. In animal or human hosts, hemolytic or proteolytic activities as well as the possibilities to directly use heme or iron-loaded transferrin molecules are alternative mechanisms (Payne, 1989; also see Payne et al., this volume). Our study showing the expression of a *lacZ* chrysobactin operon fusion in infected plant leaves, indicates that the system comes into play at the initial stages of vascularisation, when bacteria are spreading beyond the commensal necrotic region. The release of chrysobactin allows the pathogen to efficiently compete for iron (III), which is probably complexed to ligands related to citrate, considering the stability constant towards the metal. Catechol type siderophores must be stronger ligands than the diverse potential plant iron chelators and the presence of chrysobactin *in planta* may represent an important deficit for the host. Indeed, the release of chrysobactin results in drastic effects on host iron metabolism: the lack of iron incorporation into plant ferritins in the presence of chrysobactin shows that the siderophore can potentially shunt the normal course of nutritional iron in infected tissues. This leads us to assume that production of the siderophore *in planta* can have a significant impact on photosynthetic and plant defense functions since both require a number of important haem and iron proteins. In addition, the chrysobactin ferric complex may interfere with the level of reduction of Fe(III) to Fe(II), the latter being more available for plant use than Fe(III). Furthermore, catechols are susceptible to oxidation and their metal complexes can undergo intramolecular electron transfer reactions (Hider, 1984) making them good catalysts for the formation of free hydroxyl radicals during the Haber-Weiss reaction.

In the light of these data, the regulatory effect of iron on three of the five pectate lyases released during infection is noteworthy. A differential role for these redundant enzymatic activities has long been addressed (Collmer & Keen 1986). The impact of each of them on the pathogenicity on saintpaulia plants has been analyzed (Boccara *et al.*, 1988; Boccara & Chatain, 1989; Temsah *et al.*, 1991). The complexity and the diversity of pectic polymers as well as their distribution inside cell walls may be related to the origin of such a panoply of enzymes. Their cytotoxic effect together with the fact that their hydrolysis products are elicitors of plant defenses (Davies *et al.*, 1984) likely explain why their production must be tightly controlled. An unbalanced production of these isozymes changes the normal course of infection. The different products of degradation generated by these enzymes were also analyzed and their diversity may support these observations (Preston *et al.*, 1992). In addition, several signals, including the same pectic oligosaccharides, have been identified as central catabolic regulatory molecules of the pectinolytic pathway (Reverchon & Robert-Baudouy 1987; Reverchon *et al.*, 1991). An interesting study performed on *E. carotovora* showed a sequential pattern for the induction of diverse pectinases *in planta* (Yang *et al.*, 1992). In this context, the low iron availability of plant intercellular fluids acts as a signal which modulates in a co-ordinated manner the production of several ingression factors which are likely called into play at some critical stages of infection.

The whole data leads us to assume that the *cbr*-encoded sensory system used in *Erwinia* cells to control functions by iron could be of biological significance. The ubiquity of the *fur* system found within enterobacteria and even more distant species suggested the existence in *E. chrysanthemi* of a similar, if not homologous system. The Fur protein acts as a global iron-responsive transcriptional repressor which turns off the structural genes in the presence of iron (Schäffer *et al.*, 1985; Bagg & Neilands, 1987). The dimeric metal-induced conformational form of the 17 kDa polypeptide interacts with its operator when bound to ferrous iron (Coy & Neilands, 1991) probably via the 12 histidine residues typical of the molecule. A more sophisticated regulatory system based on a functional interaction between two components encoded by an operon is perhaps more sensitive to minor and/or qualitative iron fluctuations. This allows us to speculate about the possibility in this case of an adaptive response to the plant environment.

Acknowledgements

We thank Nicole Hugouvieux Cotte Pattat for providing us with *pel::uidA* constructions, Dominique Fréchon and Thierry Franza for critical reading of the manuscript.

References

Aumont P, Enard C, Expert D, Pieddeloup C, Tancrède C & Andremont A (1988) Production of hemolysin, aerobactin and enterobactin, and resistance to human serum in strains of *E. coli* causing bacteremia in cancer patients. Res. Microbiol. 140: 21–27.

Bagg A & Neilands JB (1987) Molecular mechanism of regulation of iron uptake systems in *Escherichia coli* K-12. Microbiol. Revs. 51: 509–518.

Boccara M & Chatain V (1989) Regulation and role in pathogenicity of *Erwinia chrysanthemi* 3937 pectin methylesterase. J. Bacteriol. 171: 4085–4087.

Boccara M, Diolez A, Rouve M & Kotoujansky A (1988) The role of individual pectate lyases of *Erwinia chrysanthemi* strain 3937 in pathogenicity on saintpaulia plants. Physiol. Mol. Plant Pathol. 3: 95–104.

Brown JC (1978) Mechanism of iron uptake by plants. Plant Cell Environ. 1: 249–257.

Bullen JJ, Ward CG & Rogers HJ (1991) The critical role of iron in some clinical infections. Eur. J. Clin. Microbiol. Infect. Dis. 10: 613–617.

Calderwood ST & Mekalanos JJ (1987) Iron regulation of shiga-like toxin expression in *Escherichia coli* is mediated by the *fur* locus. J. Bacteriol. 169: 4759–4764.

Collmer A and Keen NT (1986) The role of pectic enzymes in plant pathogenesis. Annu. Rev. Phytopathol. 24: 343–409.

Coote JG (1991) Antigenic switching and pathogenicity: environmental effects on virulence gene expression in *Bordetella pertussis*. J. Gen. Microbiol. 137: 2493–2503.

Cornelis GR, Biot T, de Rouvroit L, Michiels T, Mudler B, Sluiters M, Sory P, Van Bouchaute M & Vanooteghem JC (1989) The *Yersinia yop* regulon. Mol. Microbiol. 3: 1455–1459.

Coy M & Neilands JB (1991) Structural dynamics and functional domains of the Fur protein. Biochemistry 30: 8201–8210.

Davies KR, Lyon GD, Darvill AG & Albersheim (1984) Host-pathogen interactions XXV. Endopolygalacturonic acid lyase from *Erwinia carotovora* elicits phytoalexin accumulation by releasing plant cell walls fragments. Plant Physiol. 74: 52–60.

Deretic V, Mohr CD & Martin DW (1991) Mucoid *Pseudomonas aeruginosa* in cystic fibrosis: signal transduction and histone-like elements in the regulation of bacterial virulence. Mol. Microbiol. 5: 1577–1583.

De Witt PGJM & Spikman G (1982) Evidence for the occurance of race and cultivar-specific elicitors of necrosis in intercellular fluids of compatible interactions of *Cladosporium fulvum* and tomato. Physiol. Plant Pathol. 21: 1–11.

Dorman CJ, Bhriain NN & Higgins CF (1990) DNA supercoiling and environmental regulation of virulence gene expression in *Shigella flexneri*. Nature (London) 344: 789–792.

Enard C, Diolez A & Expert D (1988) Systemic virulence of *Erwinia chrysanthemi* 3937 requires a functional iron assimilation system. J. Bacteriol. 163: 221–227.

Expert D & Gill PG (1991) Iron: a modulator in bacterial virulence and symbiotic nitrogen-fixation. p. 229–245. *In*: Verma DPS (ed.) Mol. Signals in Plant-Microbe Communications. CRC Press Inc., Boca Raton.

Expert D, Sauvage C & Neilands JB (1992) Negative transcriptional control of iron transport in *Erwinia chrysanthemi* involves an iron-responsive two factor system. Mol. Microbiol. 6: 2009–2017.

Expert D & Toussaint A (1985) Bacteriocin-resistant mutants of *Erwinia chrysanthemi*: possible involvement of iron acquisition in phytopathogenicity. J. Bacteriol. 163: 222–227.

Falkow S (1991) Bacterial cell entry into eucaryotic cells. Cell 65: 1099–1102.

Franza T, Enard C, van Gijsegem F & Expert D (1991) Genetic analysis of the *Erwinia chrysanthemi* 3937 chrysobactin iron-transport system: characterization of a gene cluster involved in uptake and biosynthetic pathway. Mol. Microbiol. 5: 1319–1329.

Franza T & Expert D (1991) The virulence-associated chrysobactin iron uptake system of *Erwinia chrysanthemi* 3937 involves an operon encoding transport and biosynthetic functions. J. Bacteriol. 173: 6874–6881.

Galan JE, Pace J & Hayman MJ (1992) Involvement of the epidermal growth factor receptor in the invasion of cultured mammalian cells by *Salmonella typhimurium*. Nature (London) 357: 588–589.

Garibaldi A & Bateman DF (1971) Pectic enzymes produced by *Erwinia chrysanthemi* and their effects on plant tissue. Physiol. Pl. Path. 1: 25–40.

Hider RC (1984) Siderophore mediated absorption of iron. p. 25–87. *In:* Clarke MJ *et al.* (eds.) Structure and bonding 58. Springer-Verlag.

Isberg RR (1991) Discrimination between intracellular uptake and surface adhesion of bacterial pathogens. Science 252: 934–938.

Knox JP (1992) Cell adhesion, cell separation and plant morphogenesis. Plant Journal 2: 137–141.

Laulhère, JP Labouré, AM Lobraux, S Proudhon D, Briat JF (1992) Purification, characterization and function of bacterioferritin from the cyanobacterium *Synechocystis* P.C.C.6803. Biochem J. 281: 785–793.

Parsot C & Mekalanos JJ (1990) Expression of ToxR, the transcriptional activator of the virulence factors in *Vibrio cholerae*, is modulated by the heat shock response. Proc. Natl. Acad. Sci. USA 87: 9898–9902.

Payne SM (1989) Iron and virulence in *Shigella*. Mol. Microbiol. 3: 1301–1306.

Persmark M, Expert D & Neilands JB (1989) Isolation, characterization and synthesis of chrysobactin, a compound with siderophore activity from *Erwinia chrysanthemi*. J. Biol. Chem. 264: 3187–3193.

Persmark M, Expert D & Neilands JB (1992) Ferric iron uptake in *Erwinia chrysanthemi* mediated by chrysobactin and related catechol compounds. J. Bacteriol. 174: 4783–4789.

Poole K & Braun V (1988) Iron regulation of *Serratia marcescens* hemolysin gene expression. Infect. Immun. 56: 2967–2971.

Porat R, Clark BD, Wolff SM, Dinarello CA (1991) Enhancement of growth virulent strains of *Escherichia coli* by interleukin-1. Science 254: 430–432.

Preston III JF, Rice JD, Ingram LO & Keen NT (1992) Differential depolymerization mechanisms of pectate lyases secreted by *Erwinia chrysanthemi* EC16. J. Bacteriol. 174: 2039–2042.

Reverchon S, Nasser W & Robert Baudouy J (1992) Characterization of *kdgR*, a gene of *Erwinia chrysanthemi* that regulates pectin degradation. Mol. Microbiol. 5: 2203–2205.

Reverchon S & Robert-Baudouy J (1987) Regulation of expression of pectate lyase genes *pelA*, *pelD*, *pelE* in *Erwinia chrysanthemi*. J. Bacteriol. 169: 2417–2423.

Reverchon S, van Gijsegem F, Rouve M, Kotoujansky A & Robert-Baudouy J (1986) Organization of a pectate lyase gene family in *Erwinia chrysanthemi*. Gene 49: 215–224.

Rogowsky PM, Close TJ, Chimera JJ, Shaw JJ & Kado CI (1991) Regulation of the *vir* genes of *Agrobacterium tumefaciens* plasmid pTiC58. J. Bacteriol. 169: 5101–5112.

Scarlato V, Arico AP & Rappuoli R (1991) Sequencial activation and environmental regulation of virulence genes in *Bordetella pertussis*. The EMBO J. 10: 3971–3975.

Schäffer S, Handke K & Braun V (1985) Nucleotide sequence of the iron regulatory gene *fur*. Mol. Gen. Genet. 200: 110–113.

Stachel SE & Zambryski PC (1986) *virA* and *virG* control the plant-induced activation of the T-DNA transfer process of *A. tumefaciens*. Cell 46: 325–333.

Temsah M, Bertheau Y & Vian B (1991) Immunolocalisation of pectate-lyase of *Erwinia chrysanthemi* strain 3937 in infected leaves of *Saintpaulia ionantha*. Cell Biol. Intern. Reports 15: 611–620.

Tobe T, Nagai S, Okada N, Adler B, Yoshikawa M & Sasakawa C (1991) Temperature-regulated expression of invasion genes in *Shigella flexneri* is controlled through the transcriptional activation of the *virB* gene on the large plasmid. Mol. Microbiol. 5: 887–893.

Veluthambi K, Krishman M, Gould JH, Smith RH & Gelvin SB (1989) Opines stimulate induction of the vir genes of *Agrobacterium tumefaciens* Ti plasmid . J. Bacteriol. 171: 3696–3703.

Yang Z, Cramer Cl & Lacy (1992) *Erwinia carotovora subsp. carotovora* pectic enzymes: *in planta* gene activation and roles in soft-rot pathogenesis. Mol. Plant-Microbe Interact. 5: 104–112.

Weinberg ED (1984) Iron witholding: A defence against infection disease. Physiol. Rev. 64: 65–102.

12. Characterization of the *Vibrio cholerae* heme iron transport system and its role in pathogenesis

SHELLEY M. PAYNE and DOUGLAS P. HENDERSON

Abstract. *Vibrio cholerae* can utilize hemin or hemoglobin as its sole source of iron. The ability to use these iron sources is independent of production of the siderophore vibriobactin. The *V. cholerae* heme utilization system can be reconstituted in *Escherichia coli* 1017 by transformation with two recombinant plasmids, pHUT3 and pHUT10. These plasmids contain at least two genes required for utilization of heme iron. One of the genes encoded by pHUT10 specifies a 26 kDa inner membrane protein. The second plasmid, pHUT3, encodes a 77 kDa outer membrane protein. Tn5 mutagenesis of the plasmids indicates that loss of either the 77 kDa or the 26 kDa protein leads to loss of ability to use heme as an iron source. The effect of iron on expression of these proteins was determined by analysis of protein levels and measurement of Tn5 *lac* gene fusions. Expression of the 77 kDa protein, like the siderophore and hemolysin, is negatively regulated by iron. Regulation is mediated by the *V. cholerae* analog of the *E. coli fur* gene. *V. cholerae* mutants defective in heme transport and/or siderophore production were constructed and tested for virulence in the infant mouse model. A mutant defective in both siderophore production and heme transport was unable to cause disease in this model.

Abbreviations: EDDA, Ethylenediamine-di(*o*-hydroxyphenylacetic acid)

Introduction

Pathogenic bacteria require iron for growth. However, the element is complexed to various iron binding proteins in the human body, making iron difficult for the pathogen to obtain. Most of the iron inside the human body is intracellular, bound to hemoglobin or ferritin. Other iron binding proteins, such as lactoferrin and transferrin, bind the relatively small amounts of extracellular iron (Theil and Aisen, 1987). In order to acquire iron from the host, pathogens have evolved systems enabling them to compete with host proteins for the element or to use these host compounds directly as sources of iron. For example, certain pathogenic bacteria can utilize transferrin (Archibald and DeVoe, 1979; Mickelsen and Sparling, 1981) or hemoglobin (Stoebner and Payne, 1988; Mickelsen and Sparling, 1981; Stull, 1987; Francis *et al.*, 1985) as sources of iron. Other pathogens may synthesize and secrete siderophores, which are high affinity iron binding compounds that remove iron from proteins such as transferrin and transport the element into the cell (Konopka *et al.*, 1982).

In a variety of pathogens, the ability to compete successfully with the host for iron has been correlated with virulence. For highly invasive, septicemic strains,

C.I. Kado and J.H. Crosa (eds.), Molecular Mechanisms of Bacterial Virulence, 173–184.
© 1994 *Kluwer Academic Publishers.*

acquisition of iron *in vivo* often correlates with siderophore production. Among invasive strains of *E. coli*, production of the siderophore aerobactin is associated with virulence (Williams, 1979). Similarly, a virulent strain of the fish pathogen *Vibrio anguillarum* exhibits reduced virulence when cured of a plasmid encoding the siderophore anguibactin. Virulence is restored when iron is added to the inoculum or when siderophore synthesis and transport genes are reintroduced into the organism (Crosa, 1980).

Iron transport and virulence of *V. cholerae*

Vibriobactin synthesis and transport

The relationship between iron acquisition and virulence also has been investigated in non-invasive pathogens such as *V. cholerae*. This pathogen, which colonizes the epithelial surface of the small intestine and secretes a potent enterotoxin causing the diarrheal disease cholera, has several mechanisms for iron acquisition. Under low iron conditions, the organism synthesizes and secretes a siderophore of the phenolate class called vibriobactin (Griffiths *et al.*, 1984) which complexes with ferric iron. The vibriobactin-iron complex then binds to the ferric vibriobactin receptor, a 74 kDa iron-regulated outer membrane protein (Stoebner *et al.*, 1992) which is required for transport of vibriobactin-bound iron. The vibriobactin receptor gene (*viu A*), which was recently cloned and sequenced, is regulated at the transcriptional level by the Fur (ferric uptake regulation) repressor protein (Butterton *et al.*, 1992). The *V. cholerae fur* is analogous to the *E. coli fur* gene, which has been well characterized (Bagg and Neilands, 1987). Under conditions of iron sufficiency, iron (II) binds to the Fur protein, altering its conformation so that it binds to a region upstream of many iron regulated genes, preventing transcription. The Fur binding region is a seventeen base pair sequence containing dyad symmetry, called the Fur box (Butterton *et al.*, 1992). Two possible Fur boxes have been identified in the promoter of *vui A* (Butterton *et al.*, 1992).

Siderophore independent iron transport systems

V. cholerae also can acquire iron from ferric citrate (Sigel *et al.*, 1985) or from heme or hemoglobin (Stoebner and Payne, 1988). The ability to utilize these compounds as iron sources is independent of siderophore production (Table 1). All strains tested, including classical and El Tor biotypes and non-O1 isolates, could use hemin or hemoglobin as a source of iron. *V. cholerae* 2076–79 is a non-01 strain, Lou 15 is an El Tor strain, and CA401 is a classical strain. The siderophore transport mutant Lou 1510 and the siderophore synthesis mutant CA40130 utilized both compounds as efficiently as the parent strain, indicating that a functional siderophore-mediated transport system is not required for utilization of iron from hemin or hemoglobin. Similarly, ferric citrate utilization was found to be independent of siderophore production (Sigel *et al.*, 1985). The ferric citrate system in *V. cholerae*, however, has not been further characterized.

Table 1. Utilization of heme and hemoglobin by *Vibrio cholerae.*

Strain	Zone of stimulation (mm) with[a]	
	Hemin	Hemoglobin
2076-79	26	19
Lou 15	21	14
Lou 1510 (Iut⁻)	18	10
CA401	21	12
CA40130 (Sid⁻)	19	13

[a] Low iron medium was seeded with various strains of *V. cholerae*. Wells were cut and filled with hemin (8 μM) or hemoglobin (3 μM) and, after 18 h, the zones of growth around the wells were measured.

Virulence of iron transport mutants of V. cholerae

To determine whether the *V. cholerae* high affinity iron transport systems play a role in virulence of the pathogen, wild type and iron transport mutants of the organism were tested for the ability to cause fluid accumulation in the intestines of orally inoculated infant mice. The disease cholera is characterized by excessive watery diarrhea; fluid accumulation ratios quantitatively measure this effect (Baselski *et al.*, 1977), with higher ratios indicating more severe disease. As shown in Table 2, infection with a relativily high dose of either the parent strain (CA401) or the siderophore synthesis mutant (CA40130) resulted in high fluid accumulation levels in the mice. Both strains also caused death in the majority of the infant mice. These data indicate that siderophore synthesis is not essential for production of disease. In contrast, DHH1, a heme utilization mutant constructed from CA40130, produced fluid accumulation levels significantly lower than those of the other two strains and did not cause the death of any of the mice. Therefore, the ability of *V. cholerae* to acquire iron from hemin or hemoglobin may play a role in virulence.

The non-O1 *V. cholerae* strains are relatively invasive and may produce a bloody diarrhea, indicating the presence of heme or hemoglobin in their environment *in vivo* (McIntyre and Freeley, 1965). In contrast, the O1 strains of V. *cholerae* are not invasive or septicemic pathogens. However, they also would be likely to encounter heme compounds *in vivo*. Dietary heme would be present in the lumen of the intestine, and intestinal epithelial cells are a potential reservoir of the compound. Release of the intracellular heme from the epithelial cells could be effected by the cytolytic hemolysin produced by *V. cholerae*

Table 2. Virulence of iron transport mutants of *V. cholerae.*

Strain	Deaths [a]	FA ratio[b]
CA401	4/6	0.120
CA40130	6/6	0.121
DHH1	0/6	0.097
L broth	0/4	0.079

[a] Infant mice were orally challenged with 5×10^6 organisms and deaths recorded after 18 hours

[b] Fluid accumulation ratios [intestine weight/(body weight - intestine weight)]

(Honda and Finkelstein, 1979). Such a toxin has the potential of releasing heme from cells in the small intestine, providing a source of iron *in vivo.* Of particular interest is the observation that the *V. cholerae* hemolysin is produced in much greater amounts when the cells are starved for iron (Stoebner and Payne, 1988). Thus, like the iron transport systems, the hemolysin is synthesized in response to bacterial iron stress.

Genetics of heme iron transport

To characterize the heme utilization system on a genetic level, additional heme utilization mutants of *V. cholerae* were constructed. Fig. 1 shows the growth of one of the mutants, DHH4, in media containing hemin as the sole source of iron. The growth rate of DHH4 in this medium was approximately ten-fold lower than that of the parent strain. The strain also grew poorly in media containing hemoglobin as the iron source (data not shown).

To isolate the genes involved in heme utilization, a library of wild type *V. cholerae* DNA was constructed and transferred into one of the heme utilization mutants. The heme utilization defect was complemented by the recombinant plasmid pHUT10, which contains a 10 kB insert of *V. cholerae* DNA. As shown in Fig. 1, growth of DHH4/pHUT10 in media containing hemin as the sole iron source was comparable to that of the parent strain. The recombinant plasmid also restored the ability of DHH4 to utilize hemoglobin as an iron source (data not shown).

An attempt was made to reconstitute the heme utilization system in *E. coli* K-12, which cannot utilize hemin as an iron source. *E. coli* 1017, a siderophore synthesis mutant, was transformed with pHUT10. The transformed strain was tested for its ability to grow in media containing hemin as the iron source. As shown in Fig. 2, 1017/pHUT10 grew as poorly as the untransformed strain in hemin-containing media. This suggested that other genes in addition to those

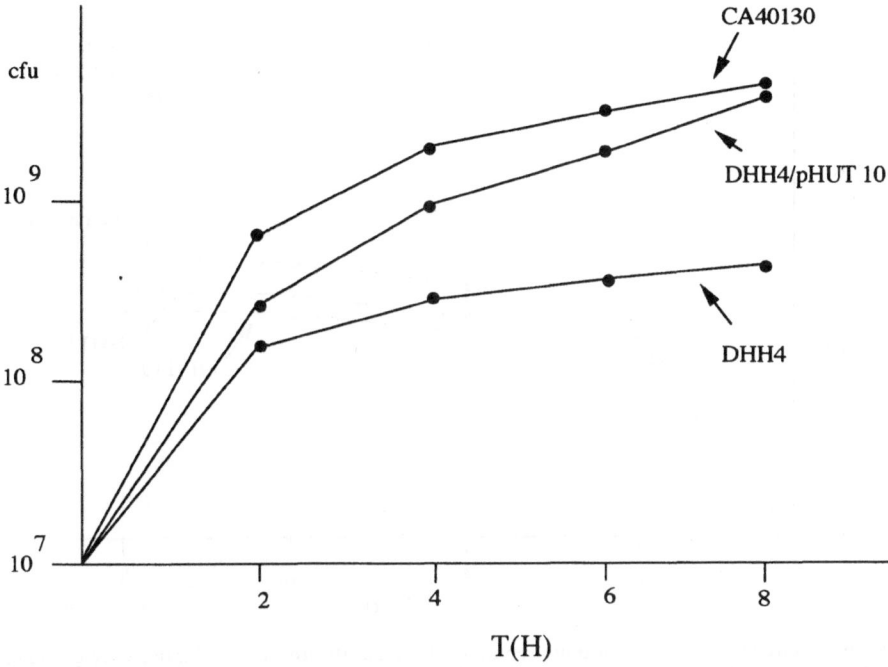

Figure 1. Growth of *V. cholerae* in medium containing hemin as the iron source. Strains were grown in L broth containing 150 μg/ml EDDA and 7.6 μM hemin. Absorbance A_{650} was measured at two hour intervals.

encoded on pHUT10 were required for heme utilization. When a *V. cholerae* DNA library was transferred to *E. coli* 1017/pHUT10, a second clone encoding genes involved in heme utilization was identified. This second plasmid, called pHUT3, contained a 3 kB insert of *V. cholerae* DNA. *E. coli* 1017/pHUT10 transformed with pHUT3 could grow in media containing hemin as a source of iron (Fig. 2). Neither plasmid alone permitted growth of the *E. coli* strain under these conditions. Therefore, at least two unlinked genes were required to reconstitute the *V. cholerae* heme iron transport system.

To identify the regions critical for heme utilization on pHUT10 and pHUT3, the recombinant plasmids were subjected to transposon mutagenesis. *E. coli* 1017 containing one of the non mutagenized plasmids was transformed with the transposon insertion derivatives of the other plasmid. The strains were then tested for growth on medium containing hemin as the sole iron source. Figure 3 shows a restriction enzyme map of the two plasmids with the sites of Tn5 insertions which disrupt heme utilization indicated by closed circles. Insertions into a 2.5–3 kB region on pHUT10 destroyed heme utilization. The right most insertion on pHUT10, indicated by the open circle, had no effect on heme utilization. Insertions into a 2 kB *EcoRV* fragment of pHUT3 disrupted heme utilization.

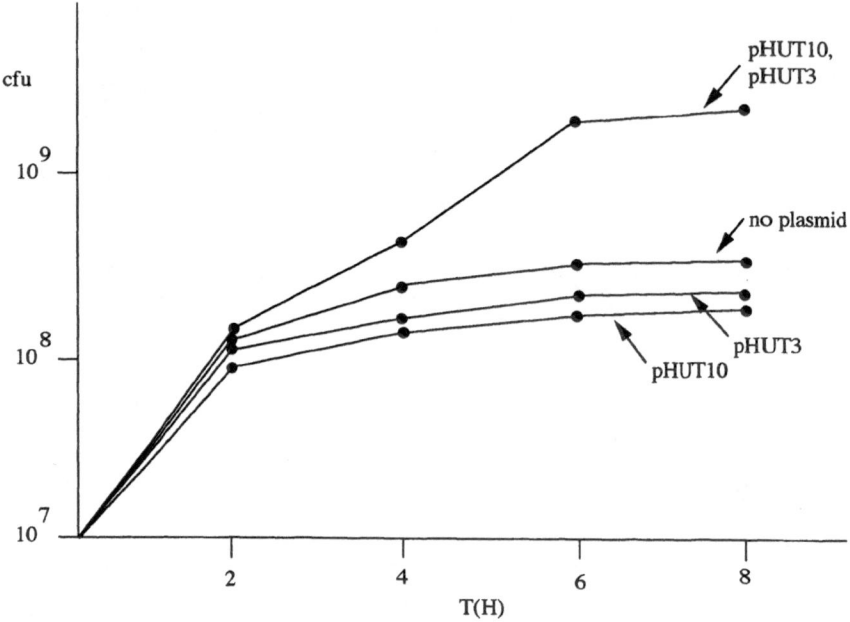

Figure 2. Growth of *E. coli* in medium containing hemin as the iron source. Strains were grown in L broth containing 75 μg/ml EDDA and 7.6 μM hemin. Absorbance A_{600} was measured at two hour intervals.

Restriction map of pHUT10

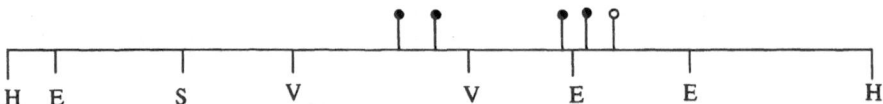

Restriction map of pHUT3

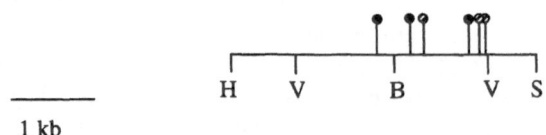

1 kb

Figure 3. Restriction enzyme and transposon insertional analysis of recombinant heme utilization plasmids. The restriction enzyme sites are as follows: **B** = *Bam*HI, **E** = *Eco*RI, **V** = *Eco*RV, **H** = *Hind*III, **S** = *Sal*I. Sites of Tn5 insertions that destroy heme utilization are marked with solid circles and those that have no effect with open circles. Tn5 *lac* insertions that disrupt heme utilization are marked with hatched circles.

Regulation of heme iron transport

Because many of the genes involved in vibriobactin synthesis and transport are iron regulated, iron regulation of the recombinant heme utilization plasmids was assessed. Tn*5 lac* (Kroos and Kaiser, 1984), which contains a promoterless *lacZ* gene, was used to mutagenize pHUT3, and derivatives were tested for iron-regulated fusions. Several iron-regulated Tn*5 lac* fusions that disrupted heme utilization were identified (Fig. 3, hatched circles).

One iron-regulated *lacZ* fusion (Fig. 3, right most hatched circle on pHUT3) was transferred to *V. cholerae* CA401 and further analyzed to determine the effect of increasing concentrations of iron and hemin on β-galactosidase expression. Fig. 4, which shows the results of β-galactosidase assays conducted on this strain, indicates that expression is highest when no iron is added to the medium. At 1 uM FeCl$_3$, β-galactosidase expression was reduced to a baseline level which did not change when the iron concentration was increased. When the cells were grown in media containing hemin as the source of iron, the baseline level of expression was approximately two-fold higher than that achieved when cells were grown in media containing equalmolar concentrations of FeCl$_3$. The two-fold increase in β-galactosidase expression was not diminished by higher concentrations of hemin. This suggests that an additional

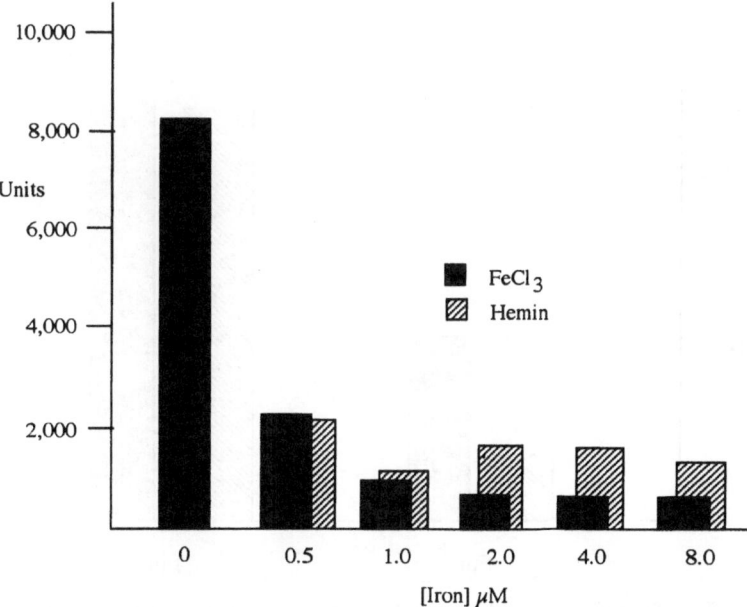

Figure 4. Effect of iron or hemin levels on β-galactosidase activity of CA401/pHUT3::Tn*5 lac*. An overnight L broth culture was washed twice in saline and inoculated 1 to 50 into T medium with varying concentrations of iron (filled bars) or hemin (hatched bars). After 4 hours of growth, β-galactosidase assays were conducted.

180

component of regulation involving heme may control transcription of the heme utilization gene(s) encoded by pHUT3. Alternatively, the cells grown in media containing hemin as an iron source may be slightly iron-starved because the pool of iron in the cell is tied up in the heme molecule. Thus, there may be insufficient iron released from heme to saturate the pool of Fur repressor protein.

Because the level of transcription of many iron-regulated genes in *V. cholerae* is controlled by the Fur protein (Stoebner and Payne, 1988; Butterton *et al.*, 1992; Goldberg *et al.*, 1990) the Tn*5lac* fusion was tested to determine if iron regulation of β-galactosidase synthesis was mediated by Fur. The fusion was transferred to MFT-5, a *fur* mutant of *E. coli* and to MFT-5/pABN203, which contains a functional plasmid-encoded *fur* gene. As shown in Fig. 5, β-galactosidase expression from pHUT3/Tn*5 lac* is iron-regulated in the Fur$^+$ strain but not in the Fur$^-$ strain. In the Fur$^+$ strain, activity of the strain grown in low iron conditions is higher than activity in iron-replete medium. In the Fur$^-$ strain, β-galactosidase activity was not related to the level of iron in the medium. This suggests that the heme utilization gene(s) on pHUT3 are controlled at the transcriptional level by Fur.

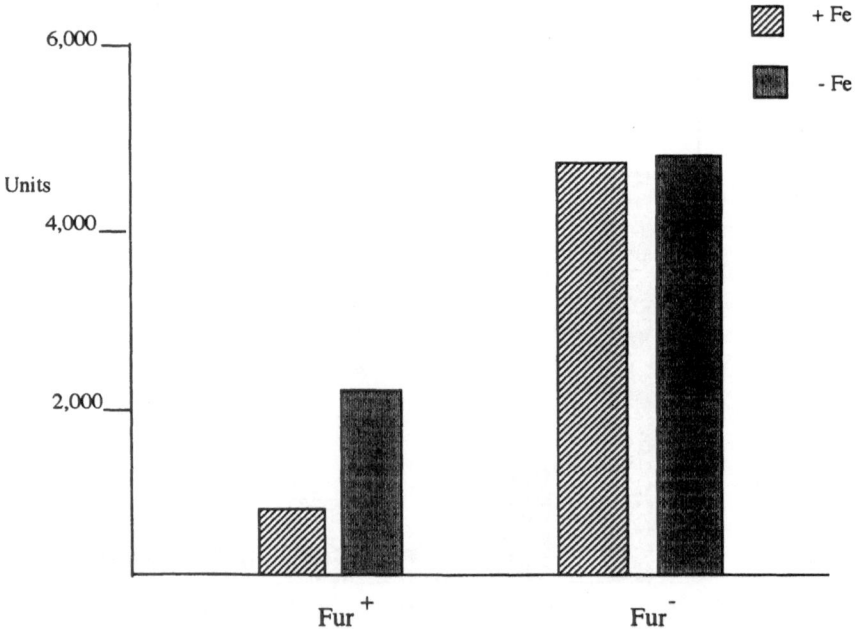

Figure 5. Effect of iron level on β-galactosidase activity of an *E. coli fur* mutant transformed with pHUT3::Tn*5 lac*. An overnight L broth culture of each strain was inoculated 1 to 100 into L broth (hatched bar) or L broth with 500 μg/ml EDDA (filled bar). After 4 hours of growth, β-galactosidase assays were conducted. The Fur$^-$ strain is MFT-5/pHUT3::Tn*5 lac* and the Fur$^+$ strain is MFT-5/pHUT3::Tn*5 lac* transformed with pABN203, a plasmid which encodes a functional *fur* gene.

Proteins associated with heme iron transport

To determine the proteins encoded by pHUT3 and pHUT10, the plasmids were transferred to a minicell strain of *E. coli*. Minicell analysis of pHUT3 and pBR322, the vector for pHUT3, indicated that pHUT3 encoded a 77 kDa protein (data not shown). To determine the cellular location of the 77 kDa protein, membrane extractions of *E. coli* containing pHUT3 or pBR322 were performed using the detergent sarkosyl, which selectively releases inner membrane proteins but leaves the outer membrane intact. No differences were detected in inner membrane preparations from *E. coli* transformed with pHUT3 or pBR322. However, outer membrane preparations of *E. coli*/pHUT3 (Fig. 6, lanes 1 and 2) contained a 77 kDa protein that was absent from preparations of *E. coli*/pBR322 (lane 3). Furthermore, the protein was absent from membrane

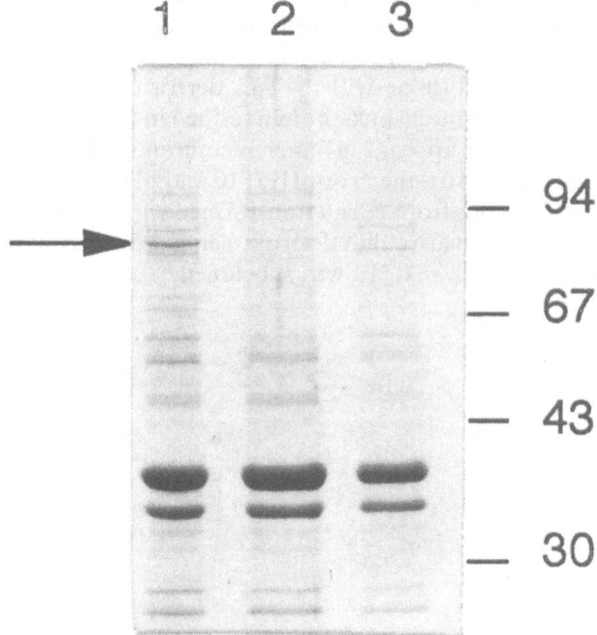

Figure 6. Outer membranes of *E. coli* P678-54 transformed with pHUT3 or pBR322. Outer membranes were isolated by the method of Filip *et al.* (1973) and the proteins were separated by SDS-PAGE and stained with Coomassie blue. Lanes 1 and 2: P678-54/pHUT3 grown in (1) low iron medium or (2) high iron medium. Lane 3: P678-54/pBR322 grown in low iron medium. The arrow marks the position of the 77 kDa protein.

preparations of *E. coli* transformed with Tn5 derivatives of pHUT3 that disrupted heme utilization (data not shown). These data suggest that pHUT3 encodes a 77 kDa protein which is required for heme utilization.

Growth of the cells in media containing different iron concentrations indicated that the 77 kDa protein encoded by pHUT3 is iron regulated. The first two lanes of Fig. 6 show outer membrane profiles of *E. coli* /pHUT3 grown in low iron medium (lane 1) and high iron medium (lane 2). EDDA was added to the medium to chelate nonheme iron. The 77 kDa protein was synthesized in higher amounts when the strain was grown in a low iron medium (lane 1) than when the strain was grown in iron replete medium (lane 2). These data support the *lacZ* fusion data which indicates the presence of an iron-regulated gene involved in heme utilization on pHUT3.

Minicell analysis also was performed on pHUT10 and its Tn5 derivatives. The results indicated that pHUT10 encoded proteins of 22 kDa, 24 kDa and 26 kDa (Henderson & Payne, 1992). The 26 kDa protein was missing in one of the Tn5 derivatives that disrupted heme utilization, suggesting that this protein is necessary for heme utilization. Analysis of membrane fractions of *E. coli* transformed with pHUT10 or with a Tn5 derivative that disrupted heme utilization indicated that the 26 kDa protein resided in the inner membrane (Fig. 7). The 26 kDa protein appears in inner membrane preparations of *E. coli* transformed with pHUT10 (lane 2) or pHUT10 and pHUT2 (lane 3), but not in inner membrane fractions from *E. coli* transformed with vector alone (lane 1) or with a Tn5 insertion derivative that destroys heme utilization (lane 4). pHUT2 is the plasmid from which pHUT3 was subcloned.

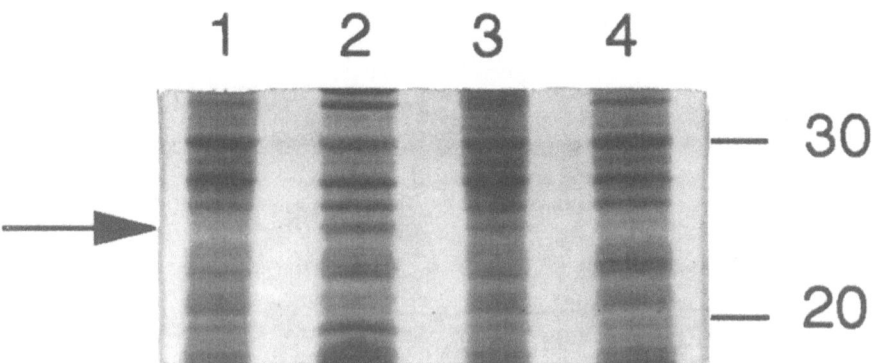

Figure 7. Inner membranes of *E. coli* 1017 transformed with recombinant heme utilization plasmids. Inner membranes were isolated by the method of Filip *et al.* (1973) and proteins were separated by SDS-PAGE and stained with Coomassie blue. Lanes 1–4 contain proteins from *E. coli* transformed with pACYC184 (vector), pHUT10, pHUT10 and pHUT2, or a Tn5 derivative of pHUT10 that disrupts heme utilization, in that order. The arrow marks the position of the 26 kDa protein.

Summary and discussion

V. cholerae acquires iron in at least three ways. 1) Under low iron conditions, the organism produces the siderophore vibriobactin which binds iron and transports it into the cell; 2) the organism also acquires iron from ferric citrate; and 3) iron is removed from heme or hemoglobin. The heme iron transport system requires at least two genes. One encodes a 77 kDa outer membrane protein and the other a 26 kDa inner membrane protein. Data obtained from a Tn*5 lac* fusion to the gene encoding the 77 kDa protein suggest that the protein is regulated by iron at the transcriptional level. When the Tn*5 lac* fusion was transferred to an *E. coli fur* mutant, β-galactosidase activity was no longer controlled by the concentration of iron in the media. Iron regulation of the Tn*5 lac* fusion was restored when the strain was transformed with a plasmid containing the *fur* gene. This indicates that the gene encoding the 77 kDa protein may be regulated by the Fur-like protein in *V. cholerae*. β-galactosidase assays of *V. cholerae* transformed with the Tn*5 lac* fusion indicated that hemin was less effective than ferric iron in repressing expression of this gene. When hemin was provided as the iron source, the baseline level of β-galactosidase activity was two-fold higher than when equal concentrations of $FeCl_3$ were provided in the media.

Because the ability of pathogens to acquire iron correlates with virulence, various iron transport mutants of *V. cholerae* were tested for virulence using the infant mouse model. A siderophore synthesis mutant was capable of causing disease. A double mutant, containing the siderophore defect and a heme utilization defect, was avirulent, suggesting that heme utilization may play a role in the ability of *V. cholerae* to cause disease. It is not yet known whether the heme iron uptake system is essential for virulence. It is possible that the presence of either the siderophore or the heme transport system is sufficient for iron acquisition *in vivo* and that virulence is only lost in the absence of both systems. Additional mutants will have to be constructed and further studies are needed to determine whether the presence of two systems represents redundancy or whether the systems serve different functions in different environments. The presence of multiple systems for acquisition of the essential element iron may provide a survival advantage to *V. cholerae*. The ability to use heme iron may promote infection and multiplication when the bacteria are in the intestine or closely associated with intestinal epithelial cells. The production of a cytolysin/hemolysin which is coordinately regulated with heme utilization could increase levels of available heme *in vivo* by lysing host cells. The high affinity, siderophore-mediated iron uptake system is not essential *in vivo* but may be crucial to the survival of the organism in the lumen of the gut or in nature. It has been noted that El Tor and non-O1 strains, which can survive and multiply in nature, typically produce greater amounts of vibriobactin than do classical *V. cholerae* strains (Stoebner and Payne, 1988).

Acknowledgements

We are indebted to Janice Stoebner and Suzanne Sigel for providing data and assistance. These studies were supported by grants DMB-8819169 and MCB-9120412 from the National Science Foundation.

References

Archibald FS & DeVoe IW (1979) Removal of iron from human transferrin by *Neisseria meningitidis*. FEMS Microbiol Lett 6: 159–162.

Bagg A & Neilands JB (1987) Ferric uptake regulation protein acts as a repressor, employing iron (II) as a cofactor to bind the operator of an iron transport operon in *Escherichia coli*. Biochemistry 26: 5471–5477.

Baselski V, Briggs R & Parker C (1977) Intestinal fluid accumulation induced by oral challenge with *Vibrio cholerae* or cholera toxin in infant mice. Infect Immun 15: 704–712.

Butterton JR, Stoebner JA, Payne SM & Calderwood SB (1992) Cloning, sequencing and transcriptional regulation of *viuA*, the gene encoding the ferric vibriobactin receptor for *Vibrio cholerae*. Infect Immun 174: 3729–3738.

Crosa JH (1980) A plasmid associated with virulence in the marine fish pathogen *Vibrio anguillarum* specifies an iron-sequestering system. Nature (London) 284: 566–568.

Filip C, Fletcher G, Wulff JL & Earhart CF (1973) Solubilization of the cytoplasmic membrane of *Escherichia coli* by the ionic detergent sodium-lauryl sarcosinate. J Bacteriol 115: 717–722.

Francis Jr RT, Booth JW & Becker RR (1985) Uptake of iron from hemoglobin and the haptoglobin-hemoglobin complex by hemolytic bacteria. Int J Biochem 17: 767–773.

Goldberg MB, Boyko SA & Calderwood SB (1990) Transcriptional regulation by iron of *Vibrio cholerae* virulence gene and homology of the gene to the *Escherichia coli* Fur system. J Bacteriol 172: 6863–6870.

Griffiths GL, Sigel SP, Payne SM & Neilands JB (1984) Vibriobactin, a siderophore from *Vibrio cholerae*. J Biol Chem 259: 383–385.

Honda T & Finkelstein RA (1979) Purification and characterization of a hemolysin produced by *Vibrio cholerae* biotype El Tor: another toxic substance produced by cholera vibrios. Infect Immun 26: 1020–1027.

Konopka K, Bindereif A & Neilands JB (1982) Aerobactin-mediated utilization of transferrin iron. Biochemistry 21: 6503–6508.

Kroos L & Kaiser D (1984) Construction of Tn5 *lac*, a transposon that fuses *lacZ* expression to the exogenous promoters, and its introduction into *Myxococcus xanthus*. Proc Natl Acad Sci USA 81: 5816–5820.

McIntyre OR & Freeley JC (1965) Characterization of non-cholera vibrios isolated from cases of human diarrhoea. Bull WHO 32: 627–632.

Mickelsen PA & Sparling PF (1981) Ability of *Neisseria gonorrhoeae*, *Neisseria meningitidis*, and commensal *Neisseria* species to obtain iron from transferrin and iron compounds. Infect Immun 33: 555–564.

Sigel SP, Stoebner JA & Payne SM (1985) Iron-vibriobactin transport system is not required for virulence of *Vibrio cholerae*. Infect Immun 47: 360–362.

Stoebner JA, Butterton JR, Calderwood SB & Payne SM (1992) Identification of the vibriobactin receptor of *Vibrio cholerae*. J Bacteriol 174: 3270–3274.

Stoebner JA & Payne SM (1988) Iron-regulated hemolysin production and utilization of heme and hemoglobin by *Vibrio cholerae*. Infect Immun 56: 2891–2895.

Stull TL (1987) Protein sources of heme for *Haemophilus influenzae*. Infect Immun 55: 148–153.

Theil EC & Aisen P (1987) The storage and transport of iron in animal cells. p. 491–520. *In*: Winkelmann G, van der Helm D & Neilands JB (eds.) Iron Transport in Microbes, Plants and Animals. VCH Publishers, Cambridge, U.K.

Williams PH (1979) Novel iron uptake system specified by ColV plasmids: an important component of virulence in invasive strains of *Escherichia coli*. Infect Immun 26: 925–932.

13. Toxin secretion in *Bordetella pertussis*: breaking the Gram-negative barrier

ALISON WEISS

Abstract. Unlike cells of any other living creature, Gram-negative bacteria have two membranes. This presents a severe barrier to moving compounds in and out of the cell. Pathogenic Gram-negative bacteria in particular have devised unique strategies to secrete the potent toxins and virulence factors they find necessary to subvert the host defenses. *Bordetella pertussis*, the causative agent of human whooping cough is no exception. This microorganism produces three potent protein toxins: dermonecrotic toxin, pertussis toxin and adenylate cyclase toxin (Weiss & Hewlett, 1986), each of which localizes to a different cellular compartment. The dermonecrotic toxin is cytoplasmic, and appears to be incapable of crossing any membranes. Both pertussis toxin and adenylate cyclase toxin, however, are found outside of the bacterial cell. Pertussis toxin is secreted into the culture supernatant, while adenylate cyclase toxin remains bound to the outer membrane. These two toxins appear to utilize entirely different secretion pathways. The adenylate cyclase toxin utilizes a protein-specific, Sec-independent pathway. In contrast, each of the five pertussis toxin subunits has a secretion signal-sequence which should allow the protein to be secreted past the first membrane and into the periplasm by the non-specific Sec-dependent pathway. It was thought that secretion outside of the second membrane would occur spontaneously once the protein made it to the periplasm. We have identified new bacterial mutants which suggest that this is incorrect, and that in addition to the generalized Sec-dependent pathway, a novel secretion apparatus is required for pertussis toxin secretion. This pertussis toxin secretion pathway shares homology with post-periplasmic secretion pathways present in other Gram-negative pathogens.

Abbreviations: *E. coli, Escherichia Coli*

The Gram-negative dilemma

The architecture of a Gram-negative bacterium is unique among all living creatures. They possess two membranes. Like all biological membranes, both membranes present a permeability barrier to large hydrophilic compounds. While this double protection is often beneficial to the microorganism, it can also present a considerable challenge if the microorganism produces large proteins which are designed to function outside of the cell. An example includes bacterial toxins, large proteins produced by pathogenic bacteria designed to damage the host and aid in the infectious process. Secretion mechanisms of Gram-negative bacteria have received extensive study in recent years, and it now appears that several different mechanisms have evolved to move proteins past both bacterial membranes. They can be roughly divided into two different classes (Fig. 1); Sec-dependent secretory mechanisms (Schatz & Beckwith, 1990; Wickner *et al.*,

C.I. Kado and J.H. Crosa (eds.), Molecular Mechanisms of Bacterial Virulence, 185–191.
© 1994 *Kluwer Academic Publishers.*

SECRETION BY
GRAM NEGATIVE BACTERIA

**Sec-
Mediated**

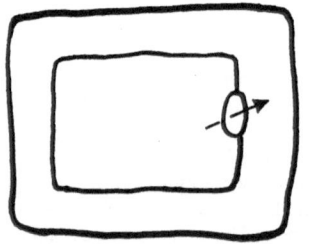 

**Not Sec-
Mediated**

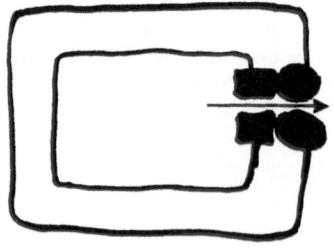

Figure 1. Schematic drawing of Sec-dependent vs Sec-independent secretion pathways. TOP. The Sec-secretion machinery recognizes a very broad category of proteins via the secretion signal peptide. This signal appears to not be dependent on the primary amino acid sequence, rather a very general structure containing positively charged and hydrophobic domains is sufficient to allow access to the Sec-secretion pathway. In Gram-negative bacteria, the Sec-secretion machinery will localize a protein into the periplasmic space (TOP-LEFT). True extracellular secretion requires additional components to move the substrate across the outer membrane (TOP-RIGHT). BOTTOM. The Sec-independent pathway, as defined by the *E. coli* hemolysin system appears to mediate secretion past both membranes without a periplasmic intermediate. These appear to be dedicated secretion pathways, in that the machinery has evolved to recognize and transport a single substrate.

1991), and Sec-independent secretory mechanisms (Fath *et al.*, 1991; Pugsley *et al.*, 1990). Both types of systems have been described in *Bordetella pertussis*. Adenylate cyclase toxin appears to be secreted by a Sec-independent pathway utilizing a "dedicated" transport system, or a secretion apparatus that transports a single protein. The protein products of three genes are required for adenylate cyclase toxin secretion (Glaser *et al.*, 1988), one of which (CyaB) is homologous to an ATP-transporter gene (Welch, 1991), known to mediate secretion in both prokaryote (example, *E. coli* hemolysin) and eukaryotic systems (examples; the P-glycoprotein that mediates drug resistance to antineoplastic agents, or the cystic fibrosis protein that is involved in chlorine ion secretion). In contrast to adenylate cyclase toxin, pertussis toxin appears to be secreted by the Sec-mediated pathway.

Sec-mediated secretory processes

The Sec-secretion pathway appears to be universally conserved in the prokaryote and eukaryote kingdoms and is responsible for moving proteins across the cytoplasmic membrane. The best characterized system is in *Escherichia coli*. Biochemical and genetic studies suggest the actual secretion apparatus involves an integral membrane complex formed from three different proteins, SecA, SecE, and SecY (Schatz & Beckwith, 1990; Wickner *et al.*, 1991). SecE and SecY co-purify and must be tightly associated. Other accessory proteins have also been described, but unlike SecA, SecE, and SecY they are only required for secretion of certain proteins. The energy to drive the secretion process appears to be derived from both ATP and energy due to either the electrical or pH gradient across the membrane (Wickner *et al.*, 1991). Proteins destined to enter this secretion pathway must possess a secretion signal, a small hydrophobic peptide (16–26 amino acids) preceded by a positively charged amino acid (Schatz & Beckwith, 1990; Wickner *et al.*, 1991). This secretion signal is apparently recognized by the SecA protein and secretion proceeds by an ill-defined mechanism.

Post-periplasmic secretion

For Gram-negative bacteria, entry into the Sec-dependent pathway will not result in secretion of a protein, but rather the protein will localize in the periplasmic space, the compartment between the two membranes.

Some post-periplasmic secretion systems have been described. The best documented are the pullulanase system of *Klebsiella oxytoca* (Kornacker *et al.*, 1988), secretion of pili and toxins by *Pseudomonas aeruginosa* (Nunn & Lory, 1992; Whitchurch *et al.*, 1990), and the *out* genes necessary for secretion of the plant cell wall-degrading enzymes of *Erwinia chrysanthemi* (He *et al.*, 1991a;1991b). These systems appear to be quite homologous at the level of both genetic organization and DNA sequence. The pullulanase is the best characterized system and 14 individual gene have been shown to be required for proper secretion of functional pullulanase (Pugsley *et al.*, 1990).

Pertussis toxin secretion

Each of the pertussis toxin subunits has a secretion signal sequence which is cleaved off prior to secretion (Nicosia *et al.*, 1986). This should allow the proteins to be secreted past the first membrane and into the periplasm where the toxin is assembled from the individual subunits. In the past it was thought that secretion outside of the second membrane would occur spontaneously once the protein made it to the periplasm, perhaps by leakage during cell division. Our data suggests that this is incorrect and a novel pathway is necessary to perform this process. By screening for bacterial mutants with reduced virulence we have discovered an interesting mutation that maps downstream from the structural gene for pertussis toxin (Weiss *et al.*, 1989a). We have shown that this mutant is deficient in the secretion, but not production of pertussis toxin, since the mutant accumulates intracellular toxin, unlike the wild type strains which secrete toxin (Johnson *et al.*, 1992).

We have named the pertussis toxin secretion system, *ptl* for pertussis toxin liberation. Mapping and sequence data suggest this operon contains seven open reading frames The predicted proteins are designated PtlB through PtlH. Initially, we speculated that we would identify a complex similar to the pullulanase system. However, with the exception of the predicted PtlH gene which is homologous to the PulE gene (Table 1), the entire Ptl operon appears to share extensive homology with a secretion complex which is required to

Table 1. Homologues of Ptl operon.

PROTEIN	% IDENTITY	AMINO ACID OVERLAP	HOMOLOGUE	
PtlB	26.4%	91	VirB3,	A. tumefaciens
PtlC	30.5%	600	VirB4,	A. tumefaciens
PtlD	17.7%	300	VirB6,	A. tumefaciens
PtlE	32.2%	180	VirB8,	A. tumefaciens
PtlF	26.6%	252	VirB9,	A. tumefaciens
PtlG	32.0%	363	VirB10,	A. tumefaciens
PtlH	34.6%	312	VirB11,	A. tumefaciens
	27.6%	214	PulE,	K. oxytoca
	26.7%	210	PilB,	P. aeruginosa
	25.8%	178	PilT,	P. aeruginosa
	51.6%	31	MDR,	P-glycoprotein
	21.5%	65	PrtD,	E. chrysanthemi

transfer DNA, not protein, out of the cell of a plant pathogen, *Agrobacterium tumefaciens* (Kuldau *et al.*, 1990; Ward *et al.*, 1990).

Why would the pertussis toxin secretion system resemble a DNA secretion apparatus rather than that of the other protein systems? Unlike pullulanase, pertussis toxin is not a single polypeptide chain. It is formed from an association of five unique proteins. Four are present in a single copy, and one protein is present in two copies, comprising a complex of six polypeptides. While the VirB operon does indeed ultimately transfer a DNA molecule past both bacterial membranes, it is not transferred as naked DNA, but rather as a protein-coated DNA complex. Similarly, the multisubunit complex of pertussis toxin may be better suited to be transported by a system evolved to handle a multisubunit protein/DNA complex, than one evolved to transport a single polypeptide.

In spite of these differences, all of the Gram-negative post-periplasmic secretion systems appear to share one protein. The PtlH protein of *B. pertussis* is homologous to the VirB11 protein necessary for transfer of the Ti DNA of *Agrobacterium tumefaciens* (Ward *et al.*, 1990), as well as the PulE protein of *Klebsiella oxytoca* necessary for secretion of pullulanase (Pugsley *et al.*, 1990), and the PilB and PilT proteins (Whitchurch *et al.*, 1990) necessary for pili secretion by *Pseudomonas aeruginosa* (Table 1). In addition, the ComG protein, necessary for development of competence in *Bacillus subtilis* is also a member of this protein family. The predicted amino acid sequence suggests these must be cytoplasmic proteins since they are very hydrophilic. In addition, these protein have a conserved ATP binding consensus sequence. Perhaps the role of these proteins is to supply cytoplasmic ATP to energize the secretion process, since the periplasm is not thought to contain ATP and the outer membrane (unlike the inner membrane) is not thought to have an electrical-chemical gradient to serve as a energy source.

Interestingly, the Sec-independent secretion pathway appears to converge with the Sec-dependent pathway by the presence of a homologue to the PtlH protein. The PrtD protein, necessary for the Sec-independent secretion of the protease genes of *Erwinia chrysanthemi* (Fath *et al.*, 1991), as well as the mammalian multidrug resistance (P-glycoprotein) also share homology to PtlH (Table 1). In the case of the P-glycoprotein, the homologous region is centered around the ATP binding site, again suggesting a role in supplying energy to the system. However, the PrtD protein homology is not within the ATP binding site, suggesting a different, but conserved function is performed by this domain.

Implications

Conventional chemotherapy against bacterial infections requires that the agent have a selective toxicity to the microorganism and not the human hosts. In most cases, antimicrobial agents are targeted against a specific biochemical pathway possessed by the bacteria and not by man. For example, penicillin and related compounds disrupt bacterial cell wall synthesis. The cells of our body do not have a cell wall, so inhibition of this process has no effect on us. However, these agents indiscriminately kill microorganisms. Often serious side effects occur because the harmless bacteria that reside on our bodies actually protect us from more dangerous bacteria by out-competing them for nutrients and other substrates. This natural protection is often lost during broad-spectrum antibiotic treatment. An antimicrobial agent targeted against only pathogenic bacteria would be an important development. We believe these observations will allow us to screen for a whole new class of antimicrobial agents that will be targeted against a subset of bacteria, Gram-negative bacteria which secrete toxic factors. We have already shown that *B. pertussis* strains with a mutational block in the *ptl* pathway are reduced in virulence (Weiss & Goodwin, 1989b). It seems likely that a therapeutic, including a vaccine, directed against this pathway would also result in less severe infection.

Acknowledgements

This work was supported in part by Public Health Service grant AI-23695 from the National Institute of Allergy and Infectious Disease and by a Pew Foundation Fellowship.

References

Fath MJ, Skvirsky RC, & Kolter R (1991). Functional complementation between bacterial MDR-like export systems: colicin V, alpha-hemolysin, and *Erwinia* protease. J. Bacteriol. 173: 7549–7556.

Glaser, P, Sakamoto, H, Bellalou, J, Ullman, A, & Danchin, A (1988). Secretion of cyclolysin, the calmodulin-sensitive adenylate cyclase-haemolysin bifunctional protein of *Bordetella pertussis*. EMBO J. 7: 399–4004.

He SY, Lindeberg M, Chatterjee AK, & Collmer A (1991a). Cloned *Erwinia chrysanthemi out* genes enable *Escherichia coli* to selectively secrete a diverse family of heterologous proteins to its milieu. Proc. Nat. Acad. Sci. USA 88: 1079–1083.

He SY, Schoedel C, Chatterjee AK, & Collmer A (1991b). Extracellular secretion of pectate lyase by the *Erwinia chrysanthemi out* pathway is dependent upon Sec-mediated export across the inner membrane. J. Bacteriol. 173: 4310–4317.

Johnson FD, Weiss AA, & Burns DL. 1992. Characterization of a mutant of *Bordetella pertussis* having reduced levels of pertussis toxin in the culture supernatant. B-15, p. 28. Abstr. 92nd Annul. Meet. Am. Soc. Microbiol. 1992.

Kornacker MG, Boyd A, Pugsley AP, & Plastow GS (1988). A new regulatory locus of the maltose regulon in *Klebsiella pneumoniae* strain K21 identified by the study of pullulanase secretion mutants. Molec. Microbiol. 3: 497–503.

Kuldau GA, De Vos G, Owen J, McCaffrey G, & Zambryski P (1990). The *virB* operon of *Agrobacterium tumefaciens* pTiC58 encodes 11 open reading frames. Mol. Gen. Genet. 221: 256–266.

Nicosia A, Perugini M, Franzini C, Casagli MC, Borri MG, Antoni G, Almoni M, Neri P, Ratti G, & Rappuoli R (1986). Cloning and sequencing of the pertussis toxin genes: operon structure and gene duplication. Proc. Natl. Acad. Sci. USA 83: 4631–4635.

Nunn DN, & Lory S (1992). Components of the protein-excretion apparatus of *Pseudomonas aeruginosa* are processed by the type IV prepilin peptidase. Proc. Nat. Acad. Sci. USA 89: 47–51.

Pugsley AP, d'Enfert C, Reyss I, & Kornacker MG (1990). Genetics of extracellular protein secretion by Gram-negative bacteria. Annu. Rev. Genet. 24: 67–90.

Schatz, PJ & Beckwith J (1990). Genetic analysis of protein export in *Escherichia coli*. Ann. Rev. Genet. 24: 215–248.

Shirasu K., Morel P., and Kado CI (1990). Characterization of the *virB* operon of *Agrobacterium tumefaciens* Ti plasmid: nucleotide sequence and protein analysis. Mol. Mirobiol. 4: 1153–1163.

Ward JE, Jr., Dale EM, Christie PJ, Nester EW, & Binns AN (1990). Complementation analysis of *Agrobacterium tumefaciens* Ti plasmid *virB* genes by use of a *vir* promoter expression vector: *virB9*, *virB10*, and *virB11* are essential virulence genes. J. Bacteriol. 172: 5187–5199.

Welch RA (1991). Pore-forming cytolysins of Gram-negative bacteria. Molec. Microbiol. 5: 521–528.

Whitchurch CB, Hobbs M, Livingston SP, Krishnapillai V, & Mattick JS (1990). Characterisation of a *Pseudomonas aeruginosa* twitching motility gene and evidence for a specialised protein export system widespread in eubacteria. Gene 101: 33–44.

Wickner W, Driessen AJM, & Hartl FU (1991). The enzymology of protein translocation across the *Escherichia coli* plasma membrane. Annu. Rev. Biochem. 60: 101–124.

Weiss AA & Hewlett E. 1986. Virulence factors of *Bordetella pertussis*. Annu. Rev. Microbiol. 40: 661–86.

Weiss AA, Melton AR, Walker KE, Andraos-Selim C & Meidl JJ. 1989a. Use of the promoter fusion transposon, Tn5 *lac* to identify *Bordetella pertussis* mutants in *vir*-regulated genes. Infect. Immun. 57: 2674–2682.

Weiss AA, & Goodwin MS. 1989b. Lethal infection by *Bordetella pertussis* mutants in the infant mouse model. Infect. Immun. 57: 3757–3764.

14. Factors affecting the virulence of soft rot *Erwinia* species: the molecular biology of an opportunistic phytopathogen

GEORGE P.C. SALMOND

Abstract. Gram-negative bacteria of the genus *Erwinia* cause soft rots and wilt diseases of a spectrum of plants. *Erwinia* species in the "soft rot" group include *E. chrysanthemi, E. carotovora* subsp. *carotovora* and *E. carotovora* subsp. *atroseptica*. The diseases caused by this group include soft rot, stem rot and blackleg. A characteristic feature of these *Erwinia* species is the ability to synthesize and actively secrete a cocktail of plant cell wall degrading enzymes, including multiple isoenzymic forms of pectinases, cellulases and proteases. The pectinases are of greatest significance in the virulence of *Erwinia*, and within the pectinase class, the endo-pectate lyases play a key role in plant cell wall dissolution. There is evidence that the different pectate lyases contribute differentially to the virulence of the pathogen. Pectinases and cellulases of soft rot *Erwinia* species are secreted by a general secretory pathway which is distinct from that used by the proteases. Homologues of the protein components of the first pathway are widely distributed in Gram-negative bacteria, including other plant and animal pathogens, and are often similarly used for the secretion of virulence factors. Regulation of extracellular enzyme synthesis is complex. In *E. chrysanthemi*, the pectinolytic pathway enzymes are co-ordinately regulated via a common repressor which binds to a defined sequence in the regulatory regions of the respective genes. In *E. carotovora* there is also a global regulatory system which co-ordinately regulates the synthesis of all of the extracellular enzymes and this global control system also responds to thermoregulation.

Abbreviations: KDG, 2-keto-3-deoxygluconate

Introduction

The enteric genus *Erwinia* contains Gram-negative, facultative anaerobes. A subdivision of the genus – the "soft rot group" – contains several species which can cause soft rots, black rots, stem rots and blackleg diseases on a wide spectrum of plants, including commercially-important crops such as potato and ornamentals such as chrysanthemum (Perombelon, 1982; Perombelon and Kelman, 1980; Perombelon and Kelman, 1987). It has been estimated that $50–100 million worth of crop damage is caused annually by the soft rot *Erwinia* spp. and there is no chemical control method currently available for the treatment of these diseases (Perombelon and Kelman, 1980). The major soft rot *Erwinia* spp are *E. chrysanthemi* (Echr) *E. carotovora* subsp. *carotovora* (Ecc) and *E. carotovora* subsp. *atroseptica* (Eca) and their ecological distribution varies considerably. Eca is environmentally rare and is generally only found in cool climates – usually associated with potatoes (Perombelon and Hyman, 1989). Ecc is found in temperate and tropical regions and in association with a broad

C.I. Kado and J.H. Crosa (eds.), Molecular Mechanisms of Bacterial Virulence, 193–206.
© *1994 Kluwer Academic Publishers. Printed in the Netherlands.*

spectrum of plant hosts. Echr is only important in tropical and subtropical regions or as a pathogen of ornamental greenhouse crops in cooler climates (Perombelon and Kelman, 1980). Ecc strains show no host specificity in pathogenesis. Although certain isolates of Echr have been reported to show enhanced host specificity towards maize or ornamentals such as Saintpaulia or Dieffenbachia, none of these *Erwinia* spp exhibit host specificity similar to that described for some other bacterial plant pathogens such as *Pseudomonas* and *Xanthomonas*. Some pseudomonad and xanthomonad phytopathogens show a tightly regulated "gene for gene" relationship with their respective hosts in which only certain genetically-defined races of pathogen can infect correspondingly-specific cultivars of the host (Daniels *et al.*, 1988). There is no evidence for such a phenomenon in the soft rot *Erwinia* spp. and so these bacteria should really be viewed as opportunistic phytopathogens. The strongest argument in favour of this view is the fact that some host plants can be infected by these organisms in significant numbers without any pathological symptoms. For example, the blackleg pathogen Eca can often be found in lenticels of asymptomatic potato tubers in a state of "latency" (Perombelon, 1982; Perombelon and Kelman, 1980; Perombelon and Kelman, 1987.). As with opportunistic animal pathogens, this state is stable until environmental or physiological factors encourage bacterial proliferation in a way that is still not understood. The consequence is the aggressive attack of the host plant in blackleg rot disease. In general, soft rot is due to Ecc either as a postharvest disease of stored crops, or in the field. The organism usually gains entry to the inside of the plant after trauma but, as with Eca in potato blackleg, the environmental conditions (particularly temperature, anaerobiosis and humidity) determine the outcome of the interaction of the plant and the pathogen (Perombelon, 1982; Perombelon and Kelman, 1980; Perombelon and Kelman, 1987; Perombelon and Hyman, 1989). Anaerobiosis is thought to encourage disease progression simply by decreasing the ability of the plant host to mount an effective host response to the infection. Although it is also theoretically possible that anaerobiosis induces virulence determinants in the pathogen, there is currently no strong evidence for this. Temperature has a dramatic effect on the progression of soft rot diseases. One explanation for this is that there is a tight thermoregulation system for the synthesis of the extracellular enzymes which are the key virulence factors of the pathogen (see below).

Virulence factors in the soft rot *Erwinia* spp.

1. Strategies for defining virulence determinants

Two basic approaches have been attempted to define the main virulence factors of *Erwinia*. The first is a directed approach in which determinants thought likely to affect virulence were investigated. The obvious candidates for analysis were the extracellular enzymes (pectinases, cellulases and proteases) which, by their very catalytic nature, would be expected to degrade plant cell walls composed of pectins, celluloses and proteins. Consequently there has been considerable effort directed at the analysis of the structure, regulation, secretion and function of many of these enzymes. Another aspect of the directed approach has been to specifically address the question of whether or not the cell surface of the pathogen plays a significant role in the virulence of the organism. A third type of directed approach has been to identify genes in the pathogen which are only induced in the presence of host plant extracts, then determine whether or not these genes encode virulence factors. Of course the directed approaches suffer from the limitation that such strategies may well lead to the exclusion of important traits which could play significant roles in virulence.

To try to identify all features important in virulence, a "black box" approach can also be taken in which large numbers of randomly generated mutants are tested individually to try to identify rare mutants with altered virulence phenotypes. In this way, in principle at least, it should be possible to identify all genes in the pathogen which play a role in the disease. The latter argument only applies of course if the pathogenicity assay is performed in a way that is truly representative of the natural pathogenesis – something which is very difficult to do with large numbers of plant assays. Both random mutagenesis/black box approaches and the directed approaches have been taken by various workers in the genetic analysis of *Erwinia* soft rot diseases. The major result of all of these studies is confirmation of the major importance of the pectinases in pathogenicity, although other virulence factors have also been identified.

2. The extracellular enzymes of Erwinia

Pectinases, cellulases and proteases are made by Echr, Ecc and Eca. Some strains also make DNases and lipases, but there has been little research effort on the latter.

2.1. The pectinases

The pectinases made by soft rot *Erwinia* spp. include pectin lyase (Pnl), pectin methylesterase (Pme) pectate lyases (Pel), of both endo- and exo- activities, and polygalacturonases (Peh) of both endo- and exo- activities (Kotoujansky, 1987) (see Table 1). Most of the soft rot *Erwinia* spp make Pnl, which cleaves plant cell wall pectin – a highly methoxylated form of polygalacturonic acid (PGA). The regulation of this enzyme is interesting because it is induced by DNA damaging agents such as UV radiation and mitomycin C (Zink *et al.*, 1985). Plant hosts produce DNA damaging molecules such as phytoalexins in response to bacterial and fungal attack and these could play a role in the SOS response induction of Pnl in *Erwinia*. The Pel enzymes are made in multiple isoenzymic forms which can be differentiated by isoelectrofocussing. The number and type made is species and (to a limited extent) strain dependent (Kotoujansky, 1987). In general Echr makes five Pels (ABCDE) all of which are secreted, and Ecc and Eca makes four Pels (ABCD) although only Pels C and D of Ecc are secreted

Table 1. Summary of the major pectinases of the soft rot *Erwinia* species.

ENZYME	HOST	LOCATION	SIZE (Da)	pI (FAMILY)
Pel:				
PLa	Eca-EC	secreted	38,720	9.4 (1)
PLb	Eca-EC	secreted	38,720	9.4 (1)
PLc	Echr-EC16	secreted	39,923	9 (1)
PLb	Echr-EC16	secreted	40,213	8.8 (1)
PLc	Ecc-193	secreted	40,571	10.3 (1)
PLa	Echr-3937	secreted	41,555	4.6 (2)
PLa	Echr-EC16	secreted	42,077	4.2 (2)
PLe	Echr-3937	secreted	43,095	9.8 (2)
PLe	Echr-EC16	secreted	41,115	10 (2)
PLd	Echr-B374	secreted	42,019	9.2 (2)
PLe	Echr-B374	secreted	42,995	9.3 (2)
PLb	Ecc-193	periplasm	63,532	11 (3)
PL153	Ecc-EC153	periplasm	63,528	8.8 (3)
Pme:				
Pme	Echr-B374	secreted	39,318	9.9
Peh:				
Peh	Ecc-193	secreted	42,639	~10
PehA	Ecc-3193	secreted	42,849	>10
PehX	Echr-EC16	secreted	64,608	8.3
Pnl:				
Pnl	Ecc-71	"secreted"	32,100	9.92

The Pel isoenzymes have been assigned to a specific family (1 or 2 -- secreted; or 3 -- periplasmic) based on sequence homology alignments (see text and Hinton et al. 1989a).

with the A and B isozymes being retained in the periplasm (Hinton *et al.*, 1989a; Kotoujansky, 1987). Although the Echr *pel* genes are arranged in two clusters (*pelAED* and *pelBC*) they are all transcribed individually from independent promoters (Boccara and Chatain, 1989; Kotoujansky, 1987; Reverchon *et al.*, 1986). In Ecc the *pelC* and *pelD* genes are together but *pelA* and *pelB* are unlinked to each other or the *pelCD* cluster (Hinton *et al.*, 1989a).

There is now a large number of Pel sequences available and this shows that there are at least three families of Pel enzyme as determined by homology analysis (Hinton *et al.*, 1989a). The secreted enzymes fall into two distinct groups (Table 1) although it is possible to identify some common residues between the two distinct families and it seems likely that these residues could be important in catalysis (Hinton *et al.*, 1989a). However it is not immediately clear why *Erwinia* should make such an arsenal of pectinases. One explanation is that the different isoenzymes play different roles in pathogenesis or they could have different functions in survival in various ecological habitats outside of plants. The hypothesis that the isozymes have a differential importance in virulence has been tested by the directed approach. Defined deletion and insertion mutants of Echr were made lacking each of the Pel enzymes and these mutants were tested in plant assays to determine whether or not the plant responded differently to the respective mutants (Boccara *et al.*, 1988). The results suggested that Pels D,E and A are more important than Pels B and C in virulence and systemic spread (Boccara *et al.*, 1988). However the basis of this is not known.

The *peh* gene of Ecc has been cloned and sequenced to reveal that the product has homology with the polygalacturonase of tomato (Hinton *et al.*, 1990; Saarilahati et el, 1990). Also, the Pme of Echr shows homology with plant enzymes (Hinton *et al.*, 1990). Echr mutants defective in Pme show reduced virulence *in planta* (Boccara and Chatain, 1989).

Interestingly, when a strain of Echr was constructed which lacked all of the known Pels and the exo-Peh, the resultant mutant was still capable of plant maceration and this led to the discovery of novel plant-inducible Pels – although the inducing molecule has not been defined (Collmer *et al.*, 1991). All of this information begs the question as to the precise functions of each of the pectinases of the soft rot *Erwinia* spp., and although, as described above, there is some evidence that different isozymes play different functions in virulence and invasiveness, it is still not possible to define any of these roles. It may be that one set of enzymes is needed during saprophytic growth and another set is used *in planta* – or in specific tissues of the plant. The fact that these are opportunistic pathogens may mean that they need a spectrum of pectinases to degrade the variable cell wall substrates that they encounter in the diverse host plants which they infect i.e. the ability to elaborate a cocktail of pectinases may enhance their host range.

2.2. Other extracellular enzymes

In Echr two Cellulases have been identified (CelZ and CelY) (Boyer *et al.*, 1987a,b). CelZ is the major secreted enzyme and CelY is periplasmic. Ecc also produces two cellulases (CelV and CelS) (Cooper and Salmond, submitted; Saarilahti *et al.*, 1990). The latter enzyme is not expressed *in vitro* by Ecc but it could be plant-inducible (E.T. Palva, pers. comm.). Of these four Cels only the CelZ and CelV show any homology, and that sequence identity lies only in the N-terminal 2/3 of the proteins with totally divergent C-terminal cellulose-binding domains (Cooper and Salmond, submitted). Interestingly, homologues of *celV* are widely distributed in the *Erwinia* genus (F. Ellard, pers. comm.).

The soft rot *Erwinia* spp. also make multiple proteases and some of these genes have been cloned and sequenced to reveal diversity, yet some intruiging similarities, of structure (Allen *et al.*, 1986; Wandersman *et al.*, 1987, Dahler *et al.*, 1990; Kyostio *et al.*, 1991). Most is known about the proteases of Echr than the other soft rot species, and this is largely because of the comparative ease of cloning of the Echr enzyme genes. Echr strains secrete three proteases (Prt A, B and C) which are members of the metalloprotease family which includes proteases from opportunistic animal pathogens such as *Serratia marcescens* (PrtSM) and *Pseudomonas aeruginosa* (AprA) (Wandersman, 1987; Ghigo and Wandersman, 1992). Although there are strain variations, the enzyme structural genes and the protease secretory apparatus genes (PrtD, E and F) tend to be clustered and the secretory "signal" for the proteases lies in the C-terminal region of each protein – although this signal is not cleaved on translocation to the extracellular medium (Wandersman, 1987; Dahler *et al.*, 1990). The Prt A, B and C proteins share sequence homology and could have different substrate specificities, presumably allowing a broad proteolytic spectrum (Ghigo and Wandersman, 1992) which might be an advantage in virulence. However, Dahler *et al.* (1990) showed that Prt⁻ mutants of Echr remained virulent in plant tests.

An Echr phospholipase gene (*plcA*) has also been characterised (Keen *et al.*, 1992). Marker exchange mutants of Echr lacking the *plcA* gene failed to make a secreted phospholipase, but, as found with the Prt⁻ mutants, remained virulent (Dahler *et al.*, 1990). These results suggest that exoenzymes such as the proteases and phospholipase play little, or no, role in pathogen ingression. However, it is very important to bear in mind that the interpretation of the phytopathological significance of these enzymes is always dependent on the assay used – and to date very crude plant test sytems have been used with the soft rot *Erwinia* spp. which could miss key elements in the initial plant/pathogen interaction where these enzymes might have a role.

3. Extracellular enzyme secretion

To be effective in macerating plant cell walls, the pectinase enzymes (the major virulence factors) have to be efficiently secreted from the pathogen. Thus, by definition, the secretory machinery of the pathogen is a virulence determinant. This is confirmed by the observation that pectinase non-secreting mutants show significantly reduced virulence on host plants (Murata *et al.*, 1990; Pirhonen *et al.*, 1991) presumably because the ingression of the pathogen is retarded by an inability to dissolve the host cell walls. These mutants (Out mutants) continue to synthesize the Pels and Peh enzymes but they accumulate in the periplasm of the pathogen. The cellulases are also retained in the periplasm of Out mutants, showing that all of these enzymes are targeted through the same pathway – referred to as the "general secretory pathway" or "type II pathway" (Pugsley, 1993; Salmond and Reeves, 1993). In contrast, the protease enzymes are secreted via an independent route (the "type I pathway") which is functionally identical to that used by the haemolysins of *E. coli* and the proteases of other animal and plant pathogens – including Echr (Salmond and Reeves, 1993; Wandersman, 1992).

The *out* genes which encode the type II secretory apparatus proteins (which have been referred to as "Membrane Traffic Wardens") of Echr and Ecc lie in large clusters of at least 13 genes (*outCDEFGHIJKLMNO*); although curiously, the *outN* gene is absent from Echr (Salmond and Reeves, 1993). These have now been sequenced and their products characterised in some detail. It is now known that the Out proteins have homologues in other bacteria, including other plant and animal pathogens such as *Pseudomonas, Xanthomonas, Aeromonas,* and *Vibrio* (Salmond and Reeves, 1993; Wandersman, 1992). In some of these other pathogens, virulence determinants – especially enzymes aggressive to the plant or animal host – are similarly targeted through the homologous secretory pathway. One of the Out homologues (OutO) functions as a specific peptidase which processes the so-called MePhe signal sequence found in the N-terminus of homologues of OutGHI and J proteins. The homologous enzyme (PilD/XcpA) in the opportunistic animal pathogen *Pseudomonas aeruginosa* processes type IV (MePhe) pilins which are involved in "twitching motility" (Pugsley, 1993; Salmond and Reeves, 1993; Wandersman, 1992). It is possible that when the OutGHI and J proteins are processed by the OutO peptidase they may assemble into a structure or channel which spans the inner and outer membranes. An interesting recent observation is that two of the Out membrane traffic wardens (OutD and OutE) seem to be highly conserved in other bacteria – some of which do not have this particular secretory pathway e.g. YscC (OutD homologue) in *Yersinia enterocolitica* Yop protein secretion (Salmond and Reeves, 1993). It is possible that the Out D homologues from diverse sources could act as outer membrane "pores" in the final translocation of the corresponding macromolecules across the outer membrane (Salmond and Reeves, 1993). The OutE homologues probably function as "Traffic ATPases" (Pugsley, 1993) which provide energy for either the translocation process or for assembly of the secretory machinery itself. Little is known about the functions of the other secretory apparatus proteins, although they are located in the inner membrane. Possible functions are that they could also be involved in the formation of a secretory structure or may interact with the secreted enzymes as chaperones which control the folding of the enzymes such that they gain access to this general secretory pathway (Pugsley, 1993; Salmond and Reeves, 1993). Irrespective of the precise functions of these Out proteins, what is clear is that this general secretory pathway is widely dispersed throughout plant and animal pathogens providing similar routes for the transmembrane targeting of the major virulence determinants in diverse bacterial pathogens.

4. *Extracellular enzyme regulation*

In addition to the regulation of the Pnl enzyme (see 2.1) the regulation of the extracellular enzymes of Echr and Ecc is complex. In Echr the complete pectinolytic pathway (including synthesis of extracellular and intracellular enzymes) is co-ordinately negatively regulated by the KdgR repressor (Reverchon *et al.*, 1989). This is achieved by binding to a consensus sequence (the KdgR "box") which lies in the 5′ regions of genes encoding the enzymes of the catabolic pathway (Reverchon *et al.*, 1989). The inducer for the pathway enzymes is not the actual substrate (pectin or polygalacturonate) but is 2-keto-3-deoxygluconate (KDG) which is an intermediate in the catabolic pathway (Nasser *et al.*, 1991). In addition to the negative regulation via the KdgR repressor there are other forms of regulation in operation, at least in Echr, involving other positive and negative regulators and including catabolite repression (Kotoujansky, 1987; Reverchon *et al.*, 1989) growth phase-dependent regulation and regulation by environmental conditions such as oxygen and temperature (Perombelon and Kelman, 1980).

There have been comparatively few regulatory studies done on Pel or Peh enzymes in Ecc or Eca (Hinton *et al.*, 1989b; Saarilahti *et al.*, 1992). However, of considerable interest is the observation that a global regulatory system is in operation, and possibly in all three soft rot *Erwinia* spp.. This global regulatory system co-ordinately controls the synthesis of the extracellular enzymes (Pel, Peh, Cel and Prt) and this system may be superimposed on any of the other regulatory circuits previously described. In addition, this global system is responsive to thermoregulation in such a way that these pathogens will down-regulate the synthesis of the secreted enzymes at several degrees below their maximal growth temperatures. So, the pathogen somehow senses temperature and shuts down enzyme production above a specific threshold. When the corresponding enzyme genes are cloned and expressed in *E. coli* the enzymes are synthesised and are catalytic at temperatures well in excess of the maximal growth temperatures for any of the *Erwinia* spp. (e.g. 42°C) (Hinton *et al.*, 1989a). Currently the mechanism of this thermoregulation is unknown but it seems likely that it is only one aspect of the global regulation cascade rather than an independent system. The relationship of the thermoregulation system to the ecological distribution and host range of the various species is still not clear, although it is likely that they are connected.

The main studies on global regulation of the extracelluar enzymes have been in Ecc. Three independently-isolated sets of mutants in different strains have been identified which show global regulatory mutant phenotypes. The mutant phenotypes are either co-ordinate down-regulation of the enzymes or co-ordinate up-regulation leading to a hyperproduction phenotype. The down-regulated mutants in the three strains are phenotypically similar and have been called Aep (*A*ctivator of *e*xtracellular protein *p*roduction) Exp (*Ex*oenzyme *p*roduction) and Rex (*R*egulation of *ex*oenzymes) respectively (Murata *et al.*, 1991; Pirhonen *et al.*, 1991; Jones *et al.*, 1993.). Mutants of this general phenotype can be identified by enzyme plate assay but they also arise in plant screening for reduced virulence mutants (Pirhonen *et al.*, 1991). So, by definition, the relevant genes are extremely important in the virulence of the pathogen – although whether this is simply due to the decreased pectinase production or other factors is not easy to determine. Several genes have been cloned by direct complementation of these Aep/Exp/Rex phenotypes but, interestingly, some of the resultant clones restore the wild type phenotype by suppression rather then by *bona fide* allelic complementation (Pirhonen *et al.*, 1991) and to date the predicted protein products of some of the suppressors (P. Golby, *et al.*, in preparation) or the allelic complementers (Chatterjee *et al.*, 1993) show no strong sequence identities with known DNA-binding proteins. One possibility is that the global regulatory system operates as a cascade and there are several positive and negative regulatory inputs to this – hence the suppression could be due to copy effects of gene products and target binding sites causing imbalance in the titration of activators and repressors. This is obviously a regulatory system of central importance for Ecc because it appears to integrate multiple physiological and environmental signals into global control of the major virulence factors of the pathogen. Given this importance to the pathogen it seems highly likely that the synthesis of other important, but currently unidentified, proteins will also fall within this global control system. Finally, the recent discovery that a small molecule [*N*-(3-Oxohexanoyl)-L-homoserine lactone] plays a central role in this global regulatory system now suggests a testable model for the molecular basis of exoenzyme regulation in Ecc (Jones *et al.*, 1993; Pirhonen *et al.*, 1993). That model predicts that co-ordinate synthesis of the Ecc virulence factors is controlled by a system analogous to the bioluminescence (Lux) system of the marine bacterium, *Vibrio fischeri* (Jones *et al.*, 1993; Pirhonen *et al.*, 1993).

5. Is there a role for the bacterial cell surface in virulence?

Bacteriophage and bacteriocins have been used to select for cell surface mutants of *Erwinia*. Some phage and bacteriocin resistant mutants of Echr showed reduced virulence on plants (Enard *et al.*, 1988). Some of these mutants were affected in lipopolysaccharide (LPS) profiles and may have had a reduced viability in the face of host plant defence mechanisms (Schoonejans *et al.*, 1987). Some other reduced-virulence mutants were affected in the ability to sequester iron via the siderophore chrysobactin, perhaps implying that the host plant competes for the available iron with the pathogen (Enard *et al.*, 1988). In such circumstances the reduced virulence can be viewed as a result of *in planta* iron auxotrophy leading to poor systemic spread – a situation also encountered in some animal pathogens. These results show that the cell surface of Echr and the chrysobactin siderophore are virulence factors to some limited extent. However, attempts to identify similar siderophores in Ecc and Eca have been fruitless. Unlike the phage resistant mutants of Echr, T4-resistant Ecc mutants were unaffected in virulence on tobacco, even though they had LPS alterations (Pirhonen *et al.*, 1988). In contrast, we have found that some Eca mutants isolated as multiply phage resistant show reduced virulence on potato when inoculated into stems (Toth, 1991). In addition to reduced virulence and multiple phage resistance, some of these mutants are pleiotropic and show defects in extracellular enzyme synthesis, LPS profiles, and sensitivity to surface active agents. The relevant genes have been cloned and sequencing is in progress. It is too early to say whether or not it is strictly the cell surface alterations which are causally connected with the reduced virulence. An alternative explanation is that these highly pleiotropic mutations cause other effects which are responsible for the phenotype in plants.

6. Identification of plant-inducible genes in the pathogen

Beaulieu and Van Gijsegem (1990) have identified Echr mutants which have insertions in plant-inducible genes. Most of those mutants analysed showed a reduced virulence in plants and some were characterised as being defective in cation transport, enzyme synthesis or pectin catabolism. Interestingly, however, some of these plant-inducible mutations which caused reduced virulence showed no other identifiable phenotype and are presumably in previously-unidentified virulence genes. The relationship of these genes to the previously-cryptic, plant-inducible Pels discovered by Collmer *et al.* (see 2.1) is currently unknown. This approach of identifying pathogen genes which are only activated in the presence of plant extracts (or *in planta*) should lead to the identification of novel chemical signals which pass between plant and pathogen at different stages of pathogenesis. It must be the case that there is significant chemical dialogue between pathogen and host – both before pathogenesis is initiated, and throughout the infection. To date there has been little effort directed at identifying the chemical communication signals which occur in these interactions.

7. Identification of virulence determinants by random transposon mutagenesis

Random transposon mutagenesis and plant tests have been used to identify reduced virulence mutants of Ecc and Eca (Hinton *et al.*, 1989b; Pirhonen *et al.*, 1991). In Ecc, the major class of reduced-virulence mutants on tobacco were affected in motility, whereas the other main category of mutants were avirulent and defective in exoenzyme synthesis or secretion (Pirhonen *et al.*, 1991). Similarly, in Eca the main reduced virulence mutants were affected in enzyme production or secretion (Hinton *et al.*, 1989b). However, recent studies with Eca has shown another class of reduced virulence (Rvi⁻) mutants which is capable of synthesizing the exoenzymes but remains Rvi⁻. This class can be divided into two subgroups, one of which is also affected in motility (Mot⁻) like some of the Ecc mutants. These mutants again show pleiotropic phenotypes because they have also become resistant to a set of bacteriophages (although a different set from those used for the positive selection of surface mutants – see 5). These mutants have been complemented and the genes on the complementing fragment of DNA appear to encode proteins which are homologous with several flagellar assembly proteins and pathogenicity proteins of *Bacillus, E. coli, Salmonella, Shigella* and *Yersinia* (Mulholland *et al.*, 1993) implying, once again, common proteins involved in plant and animal pathogenesis.

8. Summary of virulence factors, and similarities with other pathogens

The net outcome from the searches for virulence genes of soft rot *Erwinia* spp. is confirmation of the central importance of the extracellular enzymes – especially pectate lyases – in pathogenesis. All strategies employed to identify mutants with reduced virulence have turned up isolates with defects in either the structure, regulation or secretion of these enzymes. However, the fact that other reduced-virulence mutants have been isolated which appear to have wild type exoenzyme levels confirms the presumption that factors other than cell wall degrading enzymes are important in soft rot diseases.

Two common themes seem to be emerging in the molecular study of plant and animal bacterial pathogens. The first is that the secretory systems of many pathogens are similar. In some cases extensive homologies exist between most of the secretory machinery proteins of taxonomically-unrelated pathogens, and this has to imply evolution from common ancestry. Even in the cases where there are only one or two homologous secretory proteins the sheer fact that they are conserved suggests that they are critically important in delivering virulence factors to the site of host attack (Salmond and Reeves, 1993).

The second common theme that is arising in plant and animal pathogens is global regulatory control systems for the major virulence determinants. These pathogens, although taxonomically diverse and parasitising totally different hosts, are faced with analogous microecological problems. Most plant and animal pathogens have to exist in at least two, totally different, niches. One is on, or inside, the host which will be attacked and the other is usually in the natural environment. Survival in such diverse microcosms necessitates considerable physiological adaptability and, to achieve this, subtle regulatory systems are required to modulate virulence factor expression appropriate to the local environment of the pathogen. This can be done by globally regulating virulence determinants such that they are co-ordinately expressed in response to major environmental insults e.g. starvation and temperature fluctuations. There are many ways of achieving this e.g. via osmoregulated DNA supercoiling changes or by two component systems which sense specific environmental factors and translate that information into intracellular action (e.g. see Mekalanos, 1992). However, another way of responding to environmental signals is simply via small, freely-diffusible signalling molecules, which can be made by the pathogen and/or by the host, and which cause major (global) shifts in the physiology of the pathogen – ultimately leading to virulent pathogenesis. These freely-diffusible signalling molecules could act by binding to intracellular proteins thereby allowing the latter to act as transcriptional activators of virulence regulons in the pathogen. The search for such molecules is now in progress in several laboratories and recent results (Jones *et al.*, 1993; Pirhonen *et al.*, 1993) confirm a key role for the small signalling molecule *N*-(3-oxohexanoyl)-L-homoserine lactone in the global regulation of virulence determinant production in the plant pathogen Ecc (and in the opportunistic animal pathogen *Pseudomonas aeruginosa*). These recent discoveries now mean that research workers in the field of bacterial pathogenicity are about to embark on an exciting, new era of study of the chemistry of small molecules – and their roles in biological regulation.

Acknowledgements

I acknowledge generous support from the AFRC, UK (Awards PG88/501; PG88/503; PG88/511 and PG88/513).

References

Allen C, Stromberg VK, Smith FD, Lacy GH and Mount MS (1986) Complementation of an *Erwinia carotovora* protease mutant with a protease-encoding cosmid. Mol. Gen. Genet., 202: 276–279

Beaulieu C and Van Gijsegem F (1990) Identification of plant-inducible genes in *Erwinia chrysanthemi* 3937. J. Bacteriol. 172: 1569–1575

Boccara M, Diolez A, Rouve M and Kotoujansky A (1988) The role of individual pectate lyases of *Erwinia chrysanthemi* strain 3937 in pathogenicity on saintpaulia plants. Physiol. and Mol. Plant Pathol, 33: 95–104.

Boyer M-H, Cami B, Chambost J-P, Magnan M and Cattaneo J (1987a) Characterisation of a new endoglucanase from *Erwinia chrysanthemi*. Eur. J. Biochem., 162: 311–316

Chatterjee A, Liu Y, Murata H, Souissi T and Chatterjee AK (1993) Physiological and genetic regulation of a pectate lyase structural gene, *pel-1* of *Erwinia carotovora* subsp. *carotovora* strain 71. p. 241–251. *In*: Nester, EW and Verma, DPS (eds), Advances in Molecular Genetics of Plant-Microbe Interactions, Kluwer Academic Publishers, Dordrecht.

Boyer M-H, Cami B, Kotoujansky A, Chambost J-P, Frixon C and Cattaneo J (1987b) Isolation of the gene encoding the major endoglucanase of *Erwinia chrysanthemi*. Homology between *cel* genes of two strains of *Erwinia chrysanthemi*. FEMS Microbiol. Letts., 41: 351–356

Collmer A, Bauer DW, He SY, Lindberg M, Kelemu SK, Rodriguez-Palenzuela P, Burr TJ and Chatterjee A (1991) Pectic enzyme production and bacterial plant pathogenicity. p. 65–72. *In*: Hennecke H and Verma DPS (eds), Advances in Molecular Genetics of Plant-Microbe Interactions, Kluwer Academic Publishers, Dordrecht.

Cooper V and Salmond GPC (1993) Cloning and sequencing of the major cellulase (CelV) of *Erwinia carotovora*: evidence for an evolutionary "mix-and-match" of enzyme domains. Mol. Gen. Genet. (submitted for publication)

Dahler GS, Barras F and Keen NT (1990) Cloning of genes encoding metalloproteases from *Erwinia chrysanthemi* EC16. J. Bacteriol., 172: 5803–5815

Daniels MJ, Dow JM and Osborn AE (1988) Molecular genetics of pathogenicity in phytopathogenic bacteria. Ann. Rev. Phytopathol., 26: 285–312

Enard C, Diolez A and Expert D (1988) Systemic virulence of *Erwinia chrysanthemi* 3937 requires a functional iron assimilation system. J. Bacteriol., 170: 2419–2456

Ghigo J-M and Wandersman C (1992) Cloning, nucleotide sequence and characterization of the gene encoding the *Erwinia chrysanthemi* B374 PrtA metalloprotease: a third metalloprotease secreted via a C-terminal signal. Molec. Gen. Genet. 236: 135–144

Hinton JCD, Gill DR, Lalo D, Plastow GS and Salmond GPC (1990) Sequence of the *peh* gene of *Erwinia carotovora*: homology between *Erwinia* and plant enzymes. Molec. Microbiol., 4: 1029–1036

Hinton JCD, Sidebotham JM, Gill DR and Salmond GPC (1989) Extracellular and periplasmic isoenzymes of pectate lyase from *Erwinia carotovora* belong to different gene families. Molec. Microbiol., 3: 1785–1795

Hinton JCD, Sidebotham JM, Hyman LJ, Perombelon MCM and Salmond GPC (1989) Isolation and characterisation of transposon-induced mutants of *Erwinia carotovora* subsp. *atroseptica* exhibiting reduced virulence. Molec. Gen. Genet., 217: 141–148

Jones SE, Yu B, Bainton NJ, Birdsall M, Bycroft BW, Chhabra SR, Cox AJR, Golby P, Reeves PJ, Stephens SKS, Winson MK, Salmond GPC, Stewart GSAB and Williams P (1993) The Lux autoinducer regulates the production of exoenzyme virulence determinants in *Erwinia carotovora* and *Pseudomonas aeruginosa*. EMBO J. 12: 2467–2476.

Keen NT, Ridgway D, Boyd C (1992) Cloning and characterisation of a phospholipase gene from *Erwinia chrysanthemi* EC16. Molec. Microbiol., 6: 179–187

Kotoujansky A (1987) Molecular genetics of pathogenesis by soft-rot erwinias. Ann. Rev. Phytopathol., 25: 405–430

Kyostio SRM, Cramer CL and Lacy GH (1991) *Erwinia carotovora* subsp. *carotovora* extracellular protease: characterisation and nucleotide sequence of the gene. J. Bacteriol., 173: 6537–6546

Mekalanos JJ (1992) Environmental signals controlling expression of virulence determinants in bacteria. J. Bacteriol., 174: 1–7

Mulholland V, Hinton JCD, Sidebotham JM, Toth I, Hyman LJ, Perombelon MCM, Reeves PJ and Salmond GPC (1993) Pleiotropic reduced virulence (Rvi⁻) mutants of *Erwinia carotovora* subspecies *atroseptica* are defective in flagella assembly proteins which are conserved in plant and animal pathogens. Molec. Microbiol. 9: 343–356.

Murata H, Fons M, Chatterjee A, Collmer A and Chatterjee AK (1990) Characterisation of transposon insertion Out- mutants of *Erwinia carotovora* subsp. *carotovora* defective in enzyme export and of a DNA segment that complements out mutations in *E. carotovora* subsp. *carotovora*, *E. carotovora* subsp. *atroseptica*, and *E. chrysanthemi*. J. Bacteriol., 172: 2970–2978

Murata H, McEvoy JL, Chatterjee A, Collmer A and Chatterjee AK (1991) Molecular cloning of an *aepA* gene that activates production of extracellular pectolytic, cellulolytic and proteolytic enzymes in *Erwinia carotovora* subsp. *carotovora*. Molec. Plant-Microbe Interact., 4: 239–246

Nasser W, Condemine G, Plantier R, Anker D and Robert-Baudouy, J (1991) Inducing properties of analogs of 2-keto-3-deoxygluconate on the expression of pectinase genes of *Erwinia chrysanthemi*. FEMS Microbiol. Letts., 81: 73–78

Perombelon MCM and Hyman LJ (1989) Survival of soft rot coliforms, *Erwinia carotovora* subsp. *carotovora* and *E. carotovora* subsp. *atroseptica* in soil in Scotland. J. Appl. Bacteriol. 66: 95–106

Perombelon MCM (1982) The impaired host and soft rot bacteria. p. 55–68. *In:* G. Lacy and M. Mount (ed.) Phytopathogenic Prokaryotes, Vol2, Academic Press, New York.

Perombelon MCM and Kelman A (1980) Ecology of the soft rot erwinias. Ann. Rev. Phytopathol., 18: 361–387

Perombelon MCM and Kelman A (1987) Blackleg and other potato diseases caused by soft rot erwinias: proposal for revision of terminology. Plant Disease, 71: 283–285

Pirhonen M, Heino P, Helander I, Harju P and Palva ET (1988) Bacteriophage T4 resistant mutants of the plant pathogen *Erwinia carotovora*. Microbial. Pathogen., 4: 359–367

Pirhonen M, Saarilahti H, Karlsson M-B and Palva ET (1991) Identification of pathogenicity determinants of *Erwinia carotovora* subsp. *carotovora* by transposon mutagenesis. Molec. Plant-Microbe Interact., 4: 276–283

206

Pirhonen M (1992) Characterisation of pathogenicity determinants of the plant pathogen *Erwinia carotovora*. Ph.D thesis, Swedish University of Agricultural Sciences, Uppsala, Sweden

Pirhonen M, Flego D, Heikinheimo R and Palva ET (1993) A small diffusible signal molecule is responsible for the global control of virulence and exoenzyme production in the plant pathogen *Erwinia carotovora*. EMBO J.

Pugsley A (1993) The complete general secretory pathway in Gram-negative bacteria. Microbiological Reviews.

Reverchon S, Huang Y, Bourson C and Robert-Baudouy J (1989) Nucleotide sequence of the *Erwinia chrysanthemi ogl* and *pelE* genes, negatively regulated by the *kdgR* gene product. Gene, 85: 125–134

Reverchon S, Van Gijsegem F, Rouve M, Kotoujansky A and Robert-Baudouy J (1986) Organisation of a pectate lyase gene family in *Erwinia chrysanthemi*. Gene. 49: 215–224

Salmond GPC and Reeves PJ (1993) Membrane traffic wardens and protein secretion in Gram-negative bacteria. Trends in Biochem Sci. 18: 7–12

Saarilahti HT, Heino P, Pakkenen R, Kalillinen N, Palva I and Palva ET (1990) Structural analysis of the *pehA* gene and the characterisation of its protein product, endopolygalacturonase, of *Erwinia carotovora* subsp. *carotovora*. Molec. Microbiol., 4: 1037–1044

Schoonejans E, Expert D and Toussaint A (1987) Characterisation and virulence properties of *Erwinia chrysanthemi* lipopolysaccharide-defective, phiEC2-resistant mutants. J. Bacteriol., 169: 4001–4007

Toth IK (1991) The isolation of novel *Erwinia* phages and their use in the study of bacterial phytopathogenicity. Ph.D Thesis, University of Warwick, Coventry, UK

Wandersman C, Delapelaire P, Letoffe S and Schwartz M (1987) Characterisation of *Erwinia chrysanthemi* extracellular protease genes in *Escherichia coli*. J. Bacteriol., 169: 5046–5053

Wandersman C (1992) Secretion across the bacterial outer membrane. Trends in Genet., 8: 317–322

Zink RT, Engwall JK, McEvoy JL and Chatterjee AK (1985) *recA* is required in the induction of pectin lyase and carotovoricin in *Erwinia carotovora* subsp. *carotovora*. J. Bacteriol., 164: 390–396

15. Processing of the T-DNA from the *Agrobacterium tumefaciens* Ti-plasmid

STANTON B. GELVIN and SERGEI A. FILICHKIN

Abstract. During the initial stages of crown gall tumorigenesis, the T-DNA of the *Agrobacterium tumefaciens* Ti-plasmid is processed, forming single-stranded T-DNA molecules (T-strands) that most likely are the form of the T-DNA that is transferred to plant cells. This processing is effected by a T-DNA border-specific endonuclease that is encoded by the *virD1* and *virD2* genes. The VirD2 protein nicks the T-DNA between bases 3 and 4 of the 25 bp T-DNA border, following which it attaches to the 5' end of the T-strand. This nicking reaction is aided by the VirD1 protein with topoisomerase-like activity and possibly by the proteins encoded by the *virC1* and *virC2* genes. This process has been likened to the initial stages of transfer of bacterial conjugal plasmids. In the case of the plasmid RP4, the TraI protein nicks and attaches to the plasmid DNA at the origin of transfer (*oriT*). This process is aided by the TraJ and TraH proteins to form the relaxosome complex. To detect relaxosome-like complexes, we have initiated a detailed study of T-DNA processing in both *A. tumefaciens* and in *E. coli*. Using high copy-number plasmids that constitutively express VirC1, VirC2, VirD1, and VirD2 proteins and process the T-DNA, we have identified a putative intermediate in T-DNA processing. This intermediate consists of a double-stranded plasmid containing VirD2 protein attached to a nick at the T-DNA border.

Abbreviations: BSA, Bovine Serum Albumin; EDTA, Ethylene Diamine Tetraacetic Acid; GST, Glutathione-*S*-Transferase; IPTG, Isopropyl β-D-Thioga-lactopyranoside; MOPS, 3-[N-Morpholino] Propanesulfonic Acid; *nptII*, Neomycin Phosphotransferase II; *Pat*, Phosphinothricin Acetyl Transferase; PMSF, Phenylmethanesulfonyl Fluoride; PAGE, Polyacrylamide Gel Electrophoresis; PBS-150 mM NaCl, 16 mM Na_2HPO_4, 4 mM NaH_2PO_4, pH 7.3; SDS-Sodium Dodecyl Sulfate; SSC-0.15 M NaCl, 0.015 M Na Citrate; TE-10 mM Tris-HCl, pH 8.0, 1mM EDTA

Introduction

During the process of crown gall tumorigenesis, the T-(transferred) DNA region is processed from the Ti-(tumor inducing) plasmid of *Agrobacterium tumefaciens* and is transferred to plant cells. Processing and transfer of the T-DNA is initiated when *A. tumefaciens* perceives phenolic and sugar signal molecules released from plant wounds. The phenolic chemical signal is detected by the VirA protein, encoded by a gene located on the Ti-plasmid (Stachel and Nester, 1986; Stachel and Zambryski, 1986a; Rogowsky *et al.*, 1987). The protein encoded by the *chvE* (glucose/galactose transporter) gene (Cangelosi *et al.*, 1990), located on the *A. tumefaciens* chromosome, potentiates this signal transduction process. The VirA protein subsequently phosphorylates the VirG protein. In octopine-type Ti-plasmids, the phosphorylated VirG protein stimulates the transcription of the genes *virA*, *virB*, *virC*, *virD*, *virE*, *virF*, *virG*, and *virH* (Stachel and Nester, 1986;

C.I. Kado and J.H. Crosa (eds.), Molecular Mechanisms of Bacterial Virulence, 207–222.
© 1994 *Kluwer Academic Publishers. Printed in the Netherlands.*

Stachel and Zambryski, 1986a; Rogowsky *et al.*, 1987; Winans *et al.*, 1988). The *virC* and *virD* operons are additionally under the negative control of the product of the chromosomal *ros* gene (Close *et al.*, 1985, 1987; Tait and Kado, 1988; Cooley *et al.*, 1991).

Processing of the T-DNA initiates with nicking (Albright *et al.*, 1987; Wang *et al.*, 1987) or double-stranded cleavage (Veluthambi *et al.*, 1987, 1988; Jayaswal *et al.*, 1987; Steck *et al.*, 1989) at the T-DNA borders, which are 25 bp directly repeated sequences that flank and delimit the T-DNA (Yadav *et al.*, 1982; Wang *et al.*, 1984). This process is mediated by a T-DNA border-specific endonuclease encoded by the *virD1* and *virD2* genes (Yanofsky *et al.*, 1986; Jayaswal *et al.*, 1987; Stachel *et al.*, 1987; Veluthambi *et al.*, 1987). The products of the *virC1* and *virC2* genes may enhance T-DNA processing by binding to a sequence, called overdrive, that is present near the right T-DNA border in octopine Ti-plasmids (Peralta and Ream, 1985; Peralta *et al.*, 1986; Van Haaren *et al.*, 1987; Toro *et al.*, 1988, 1989; DeVos and Zambryski, 1989). The result of this cleavage at the T-DNA borders is the accumulation within the bacterium of both single-stranded (T-strands; Stachel *et al.*, 1986, 1987; Albright *et al.*, 1987; Jayaswal *et al.*, 1987; Veluthambi *et al.*, 1988) and double-stranded (Durrenberger *et al.*, 1989; Steck *et al.*, 1989; Gelvin, 1992) T-DNA molecules. Both of these forms of the processed T-DNA contain the VirD2 protein tightly associated with the 5' end (Durrenberger *et al.*, 1989; Herrera-Estrella *et al.*, 1988; Ward and Barnes, 1988; Young and Nester, 1988; Howard *et al.*, 1989). It has been proposed that the single-stranded form of the T-DNA is transferred to the plant in a process related to that of plasmid conjugation between bacteria (Stachel and Zambryski, 1986b; Zambryski, 1988). The early molecular T-DNA processing events preceding T-strand formation, however, remain unknown.

During conjugal transfer of plasmids between bacterial cells a nick is introduced at the origin of transfer (*oriT*) and single-stranded DNA molecules, capped at the 5' end by a "pilot protein", are transferred unidirectionally, beginning from the 5' end, into the recipient bacterium (Willets and Wilkins, 1984; Ippen-Ihler and Minkley, 1986). In the RP4 conjugation system the TraI protein, in the presence of TraJ and TraH proteins, attaches to the supercoiled plasmid DNA at *oriT* to form a structure called the "relaxasome" (Furste *et al.*, 1989; Ziegelin *et al.*, 1989; Pansegrau *et al.*, 1990a). This complex was isolated under mild lysis conditions or assembled *in vitro* (Pansegrau *et al.*, 1990a). Upon treatment with SDS, the supercoiled plasmid was converted to the relaxed form with a "pilot protein" attached covalently through the nick at oriT (Pansegrau *et al.*, 1990b). During T-DNA processing in *A. tumefaciens*, the VirD2 protein covalently attaches to the 5' end of the T-strand or double-stranded cleavage sites in the T-DNA border repeat sequences. Several workers (Durrenberger *et al.*, 1989; Steck *et al.*, 1989; Howard and Citovsky, 1990) have hypothesized that, following nicking at the right T-DNA border, the VirD2 protein bound to the 5' end of the T-strand acts as a pilot protein to guide the T-DNA from the bacterial cell into the plant cell. The analogy between the conjugal transfer of plasmids between bacteria and T-DNA transfer to plants has recently been reinforced by the identification of homologous DNA sequences preceding the nick sites in the T-DNA borders and in the *oriT* region of IncP plasmids (Waters *et al.*, 1991). In this paper we describe the isolation of a nucleoprotein complex formed by the tight, possibly covalent, association of the VirD2 protein with the T-DNA right border region of a plasmid. This complex may be analogous to the relaxation intermediate involved in the conjugal transfer of plasmids between bacteria. In addition, we show that both in *E. coli* and in *A. tumefaciens*, proteins encoded by the *virD* genes of the nopaline-type Ti-plasmid pTiC58 cleave T-DNA borders derived both from octopine-type and nopaline-type Ti-plasmids with approximately equal frequency.

Results

Isolated double-stranded plasmid-protein complexes bind with high affinity to nitrocellulose

To prove that plasmid molecules isolated as described in Materials and Methods contain bound proteins, they were subjected to a nitrocellulose filter binding assay. Double-stranded DNA binds to a nitrocellulose membrane only when complexed with proteins. Double-stranded plasmid DNA was isolated from *E. coli* cells harboring pTRIvir, a plasmid containing a right T-DNA border sequence from the nopaline-type Ti-plasmid pTiT37 and the genes *virD1* and *virD2*. Plasmid DNA isolated as described in Materials and Methods was digested with *Sau*3AI, the ends labelled with ^{32}P, and the DNA filtered through nitrocellulose in the presence of increasing quantitities of BSA. Some of the labelled restriction endonuclease fragments revealed high binding affinity to the nitrocellulose. Increasing concentrations of BSA, a protein that non-specifically competes for binding sites on the nitrocellulose membrane, decreased the ability of the restriction endonuclease fragments to bind to the nitrocellulose (Fig. 1, A and B). Binding was also inhibited when the restriction endonuclease fragments were treated with Proteinase K prior to filtration through the nitrocellulose membrane. Proteins tightly bound to the intact isolated complex were labelled *in vitro* with ^{125}I and the DNA molecules subjected to electrophoresis through an agarose gel containing urea and SDS. The molecules were subsequently blotted onto a nitrocellulose membrane by a capillary transfer method without prior denaturation. The form of the plasmid with the lowest mobility through the gel contained more than 95% of the radioactivity (Fig. 1C), although it represented less than 10% of the DNA. This form of the plasmid most likely represents relaxed plasmid molecules complexed with protein. Protease treatment prior to electrophoresis prevented binding of these molecules to the nitrocellulose membrane, demonstrating that the double-stranded DNA was retained on the membrane by the associated proteins (Fig. 1C). Because the plasmid isolation procedure involved the use of SDS, high salt, and urea, these protein molecules are most likely covalently bound to the DNA.

210

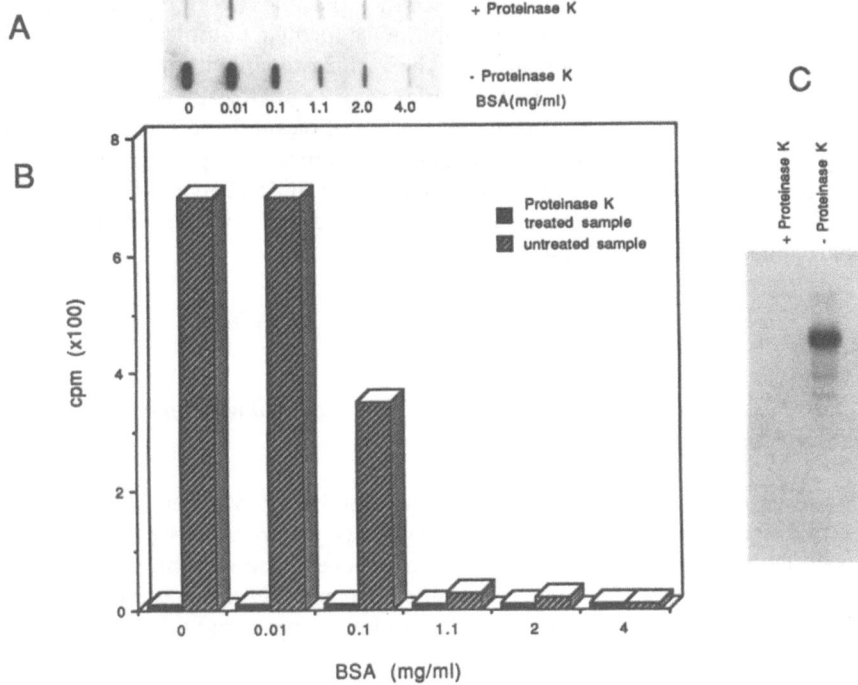

Figure 1. Binding of relaxed plasmid nucleoprotein molecules to a nitrocellulose membrane. A. Nitrocellulose slot blot binding assay of the plasmid pTRIvir. The plasmid nucleoprotein complex isolated as described as follows: *E.coli* cells harvested from 3 liters of 2 × YT medium were washed with 1 l of ice-cold TE buffer, frozen in liquid nitrogen, mixed with 10 g of coarse sand, and ground with a pestle and mortar. The mixture was resuspended in 250 ml of ice-cold buffer A (50 mM MOPS, pH 7.0; 2 mM EDTA; 0.2 mM PMSF; 0.01 U/ml aprotinin [Sigma Chemical Co., St. Louis, MO], 1% Triton X-100 and 100 μg/ml RNase A) and incubated at 4°C for 30 min NaCl was added to a final concentration 400 mM and the slurry was centrifuged at 18000 × g at 4°C for 40 min to remove the cell debris and the bulk of the chromosomal DNA. The supernatant fraction was passed twice through a 2.5 × 4 cm Qiagen-500 column equilibrated with buffer A. The column was washed twice with 100 ml of buffer E (1M NaCl, 50 mM MOPS, pH 7.0, 4 M urea) to remove unbound proteins, RNA and single-stranded DNA. Double-stranded DNA was eluted with 20 ml of buffer F (1.5 M NaCl, 50 mM MOPS, pH 8.0). The DNA was concentrated using a Centricon 30 device (Amicon, Beverly, MA) and the protein-DNA complex was precipitated for 1hr at 4°C by the addition of SDS and KCl to final concentrations of 1.0% and 100 mM, respectively. Pellets were collected by centrifugation at 5000 × g for 30 min and solubilized in a minimal volume of TE buffer. Approximately 5 ml of the sample was loaded onto a 2.5 × 40 cm Sephacryl S-1000 column equilibrated with buffer S (0.5 M NaCl, 50 mM Tris-HCl, pH 8.0, 0.1% Triton X-100, 0.1 mM PMSF). The first peak (monitored at a wavelength of 260 nm) was collected. Buffer S was substituted with TE buffer containing 0.1 mM PMSF by the repeated concentration of the sample at 4°C using an Amicon 30 microconcentrator. The complex was cleaved with *Sau*3AI and the DNA labelled with ^{32}P by filling in the ends with Klenow fragment of DNA polymerase (Maniatis *et al.*,1982). Half of the sample was digested for 30 min at 37°C with 100 mg/ml Proteinase K (+) and the other half was left intact (−). BSA was added to the labelled DNA as a competing protein at concentrations of 0.01–

Full length and truncated VirD2 proteins are bound to the plasmid through a T-DNA border-containing fragment

To determine whether proteins are tightly bound to the T-DNA border region, three plasmids were constructed and introduced into *E. coli*. pTRIvir, already described above, contains a nopaline-type right T-DNA border and the genes *virD1* and *virD2*. pTRI is similar to pTRIvir except that it lacks the *virD* genes. pTRIvirΔ is also similar to pTRIvir except that the *virD2* gene is deleted at the 3′ end, resulting in the production of a truncated VirD2 protein containing only the first 226 amino acids. Several laboratories have previously shown that a similarly truncated VirD2 protein retains the ability to process the T-DNA (Yanofsky *et al.*, 1986; Jayaswal *et al.*, 1987). Plasmid DNA was isolated from *E. coli* cells as described in Materials and Methods and digested with *Not*I and *Bam*HI. These restriction endonucleases cleave a 1.1 kbp fragment containing the T-DNA border from the plasmids (Fig. 2, panel I). Proteins bound to these restriction endonuclease fragments were labelled with ^{125}I and the fragments analyzed by agarose gel electrophoresis. Fig. 2 (panel IIA, lane 1) indicates the mobility of the deproteinized 1.1 kbp *Not*I-*Bam*HI restriction endonuclease

---

4.0 mg/ml (the concentration of BSA for each slot is indicated on the top of the autoradiograph), the aliquots were diluted 1:10 with buffer (20 mM Tris-HCl, pH 7.5, 100 mM NaCl) and passed through a nitrocellulose membrane using a filtration manifold. After exposure of the membrane to X-ray film, the individual spots were excised and the amount of radioactivity determined using a scintillation counter. B. Specific binding of the fragments shown in panel A was determined as the difference between the binding of samples treated and not treated with Proteinase K. C. Filter-binding assays were performed essentially as described in Ausubel *et al.* (1987) using a slot blot manifold. For DNA-protein capillary transfer, the ^{125}I-labelled complex was separated by electrophoresis through a 0.8% agarose gel in the presence of 0.1 mM PMSF and transferred to a nitrocellulose membrane in 10 × SSC. The membrane was dried and exposed to X-ray film at −70°C. ^{125}I-labelled relaxed plasmid nucleoprotein complexes were labelled with ^{125}I (Amersham, Arlington Heights, IL) using chloramine-T (Parker, 1990). Five ml (1mg/ml) of chloramine-T were added to 50 ml of the sample in 50 mM phosphate buffer, pH 7.0. After 30 min the reaction was stopped by the addition of 2-mercaptoethanol and NaI to final concentrations of 1.0% and 20 mM, respectively, and unreacted ^{125}I was removed by filtration of the sample through a Sephadex G-50 spin-column. pTRIvir was labelled with ^{125}I, separated by electrophoresis through a 0.8% agarose gel containing 0.1% SDS and 4 M urea, and blotted onto a nitrocellulose membrane using capillary transfer in 10 × SSC. A portion of the sample was treated with Proteinase K prior to electrophoresis (+). The remainder of the sample was left intact (−).

Bacterial strains and growth conditions: *E. coli* HB101 (Boyer and Roulland-Dessoix, 1969), DH5α (Hanahan, 1983), LE392 (Murray *et al.*, 1977), and MC1061 (Casabadan and Cohen, 1980) were grown at 37°C in LB medium (Maniatis *et al.*, 1982). *E. coli* LE392 containing pGEX-VD2 was grown in 500 ml of 2 × YT medium (Ausubel *et al.*, 1987) containing 50 μg ml^{-1} of ampicillin with vigorous shaking until an O.D. of one at 550 nm was reached. IPTG was added to a final concentration 0.5 mM, the culture was incubated for 4 hours at 30°C, and cooled on ice before harvesting the cells. 30 ml of an overnight culture of *E. coli* MC1061 containing pTDCNvir was inoculated into 3 l of 2 × YT medium and incubated vigorously at 37°C until an O.D. of 2.0 at 550 nm was reached. All other *E. coli* strains were grown on LB medium according to standard procedures (Maniatis *et al.*, 1982).

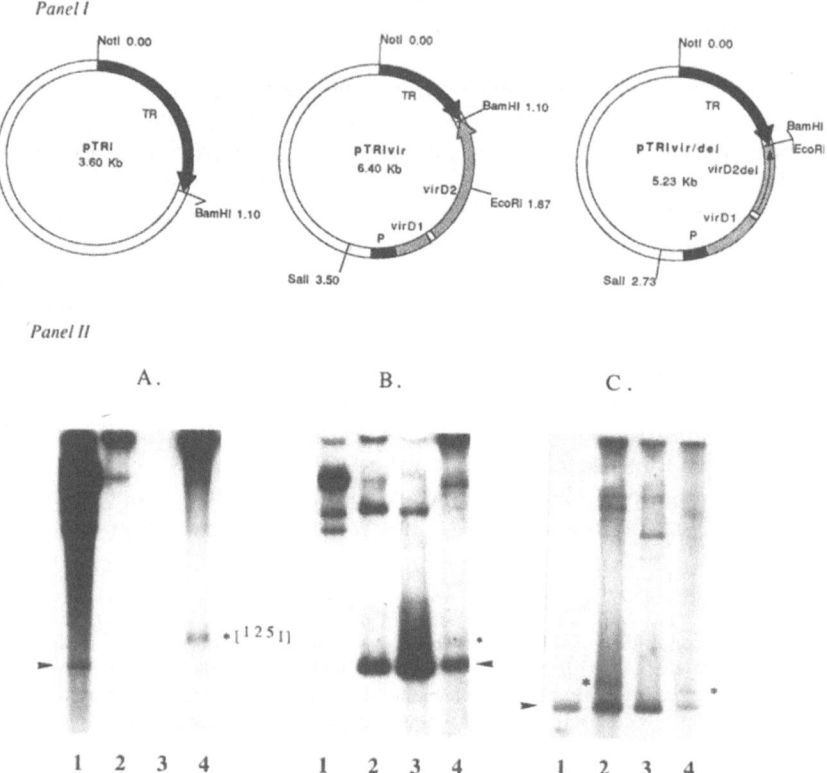

Figure 2. Gel retardation analysis of a restriction endonuclease fragment containing the right T-DNA border. A set of three plasmids was constructed for the gel retardation experiments. The base plasmid, pTRI, carries a 1.1 kbp *Not*I-*Bam*HI restriction endonuclease fragment with a nopaline right border repeat (TR) from pTiT37 cloned into the vector pBluescript KS$^+$ (Stratagene, La Jolla, CA). A 2.4 kbp *Sal*I-*Bam*HI fragment of pTiC58 containing the *virD1* and *virD2* genes was cloned into pTRI. The resulting plasmid, pTRIvir, contains sequences required both for the expression of the VirD1,2 endonuclease and its binding to the substrate DNA. pTRIvirΔ is a derivative of pTRIvir encoding a truncated VirD2 protein that lacks 221 amino acids from the C-terminus (VirD2Δ).

Plasmid DNA containing ^{125}I-labelled nucleoprotein complex was digested with the restriction endonucleases *Not*I and *Bam*HI in the presence of 0.01 U/ml aprotinin and 0.1mM PMSF. Following digestion, urea and SDS (final concentrations of 4 M and 0.1%, respectively) were added to the sample and the DNA was subjected to electrophoresis through a 0.9% agarose gel containing 0.1% SDS and 4 M urea. The gels were dried and exposed to X-ray film at −70°C for two days. After a further 70 days, when the labelled bands could not be detected by three days exposure to X-ray film, dried gels were directly hybridized with a 1.1 kbp ^{32}P-labelled *Not*I-*Bam*HI fragment of pTRI (containing the nopaline T-DNA border sequence) as follows: The gel was soaked in 0.5 M NaOH, 1.5 M NaCl for 30 min, rinsed in water, and neutralized in 0.5 M Tris-HCl, pH 8.0 for 30 min. Hybridization was carried out for 16 hours at 42°C in a hybridization solution containing 50% of formamide (Maniatis *et al.*, 1982). The gel was washed twice for 15 min in 6 × SSC at room temperature, once for 15 min in 0.2 × SSC, 0.2% SDS at 50°C, and exposed to X-ray film at −70°C.

fragment containing a T-DNA border repeat. When the two fragments from the isolated pTRI complex were labelled with ^{125}I instead of ^{32}P, no label was detected in the region of the 1.1 kbp T-DNA border fragment. A small amount of radioactiviy associated with pTRI was detected near start of the gel lane, but not in the vicinity of the T-DNA border fragment (Fig. 2, panel IIA, lane 2). After protease treatment of the sample, no radioactivity was detected in the lane (Fig. 2, panel IIA, lane 3). This result can be explained by the presence of residual amounts of proteins non-specifically bound to the plasmid. When the restriction endonuclease fragments from the isolated pTRIvir complex were labelled with ^{125}I, a fragment whose mobility resembled that of the 1.1 kbp fragment was detected (Fig. 2, panel IIA, lane 4). The mobility of this fragment was slightly retarded relative to that of the 1.1 kbp fragment, however. To detect both retarded and non-retarded T-DNA border fragments, DNA fragments in the agarose gel shown in Fig. 2, panel IIA were denatured and hybridized directly in the gel with a ^{32}P-labelled right T-DNA border probe (Fig. 2, panel IIB). The retarded band hybridized with this probe, as did the 1.1 kbp fragments from pTRI and pTRIvir that were not labelled with ^{125}I. The extent of hybridization of the retarded band from pTRIvir was approximately 5–10% that of the non-retarded 1.1 kbp restriction endonuclease fragment, suggesting that not all border-containing DNA fragments from pTRIvir were complexed with protein.

The mobility of the retarded fragment from pTRIvirΔ, which directed the synthesis of a truncated VirD2 protein, was greater than the mobility of the retarded fragment from pTRIvir. This differential mobility suggests that VirD2 protein (either full-length or truncated) binds to the restriction endonuclease fragment containing the T-DNA border sequence in these plasmids. The detection of other bands with lower mobilities than the 1.1 kbp fragment in

---

Panel I. Genetic and restriction endonuclease maps of the plasmids used for the gel retardation experiments. Each plasmid carries a 1.1 kbp *Not*I-*Bam*HI restriction endonuclease fragment containing a nopaline right T-DNA border from pTiT37 (TR). A 2.4 kbp fragment from pTiC58 containing the *virD1* and *virD2* genes under the control of their native promoter (P) was cloned into pTRI. pTRIvirΔ is a derivative of pTRIvir that expresses a truncated form of the VirD2 protein lacking 221 amino acids from the C-terminus. *Panel II.* Gel-retardation of a right T-DNA border-containing fragment.. In Panel II, the DNA in all lanes was digested with *Not*I and *Bam*HI to release the T-DNA border-containing fragment. Panel II, A and B: Lane 1, deproteinized pTRI digested with *Not*I and *Bam*HI. The ends of the restriction endonuclease fragments were labelled by filling in using Klenow fragment of DNA polymerase and ^{32}P-dCTP; Lanes 2 and 3, DNA from pTRI lacking the *virD* genes. pTRI plasmid preparations were loaded onto the gel without (lane 2) or after (lane 3) Proteinase K treatment. Lane 4, pTRIvir plasmid preparation without Proteinase K treatment. In panel A, Lanes 2, 3, and 4 contain ^{125}I-labelled material. In panel B (after a period of time when the initial ^{125}I labeling could no longer be detected), the DNA from the gel shown in panel A was denatured and hybridized in the gel with a ^{32}P-labelled T-DNA right border probe. C. Lane 1, purified 1.1 kbp *Not*I-*Bam*HI fragment; Lane 2, plasmid preparation from an *E. coli* strain containing pTRIvir; Lane 3, plasmid preparation from an *E. coli* strain containing pTRI; Lane 4, plasmid preparation from an *E. coli* strain containing pTRIvirΔ . The 1.1 kbp restriction endonuclease fragments are shown by arrows. The retarded 1.1 kbp restriction endonuclease fragments are shown by *.

Fig. 2, panels IIB and IIC may represent hybridization of the probe to incompletely cleaved plasmid DNA forms.

The retarded ^{125}I-labelled fragments from pTRIvir and pTRIvirΔ, and the equivalent region of the gel lane containing DNA from pTRI, were isolated and the fragments treated with DNAseI. Proteins were fractionated by electrophoresis through a SDS polyacrylamide gel, and labelled proteins detected by autoradiography. Although several protein bands were detected from preparations of total plasmid DNA isolated by non-denaturing conditions (data not shown), only a single protein bound to the T-DNA border fragment was detected from pTRIvir or pTRIvirΔ isolated using denaturing conditions (Fig. 3). These proteins had mobilities similar to those observed for VirD2 (58 kDa; Fig. 3, lane 1) and for the truncated VirD2 lacking 221 amino acids from the carboxy terminus (32 kDa; Fig. 3, lane 2) (The mobility of VirD2 through SDS polyacrylamide gels is anomalous because the protein is highly charged; Yanofsky *et al.*, 1986; Jayaswal *et al.*, 1987; DeVos and Zambryski, 1989). No protein was detected bound to the T-DNA border fragment from pTRI (Fig. 3, lane 3).

The protein bound to pTRIvir reacts with anti-VirD2 antibodies

A portion of the VirD2 coding sequence (the amino terminal 286 amino acids) was fused in frame to the sequence encoding glutathione-S-transferase (Fig. 4, panel I) and the fusion protein expressed in *E. coli*. Antibodies against the fusion protein were raised in chickens and purified by affinity to the VirD2 portion of the fusion protein that was cleaved with thrombin. The Western blot in Fig. 4, panel IIA shows the specificity of this antibody when reacted with the purified fusion protein (Lane 1), purified VirD2 fragment cleaved from the fusion protein (Lane 2), and proteins produced in *E. coli* harboring pTDCNvir (Lane 4) and pTRIvir (Lane 6). No protein with a molecular weight of 58,000 was detected in extracts from *E. coli* lacking the *virD2* gene (Lanes 3 and 5). Plasmid DNA from *E. coli* harboring pTRIvir was isolated according to Materials and Methods and digested with DNAseI. Western blot analysis of proteins tightly bound to the plasmid revealed the presence of a protein with the mobility of VirD2 (Fig. 4, panel IIB, lanes 1 and 3). Thus, VirD2 protein is tightly bound to pTRI, most likely at the T-DNA border.

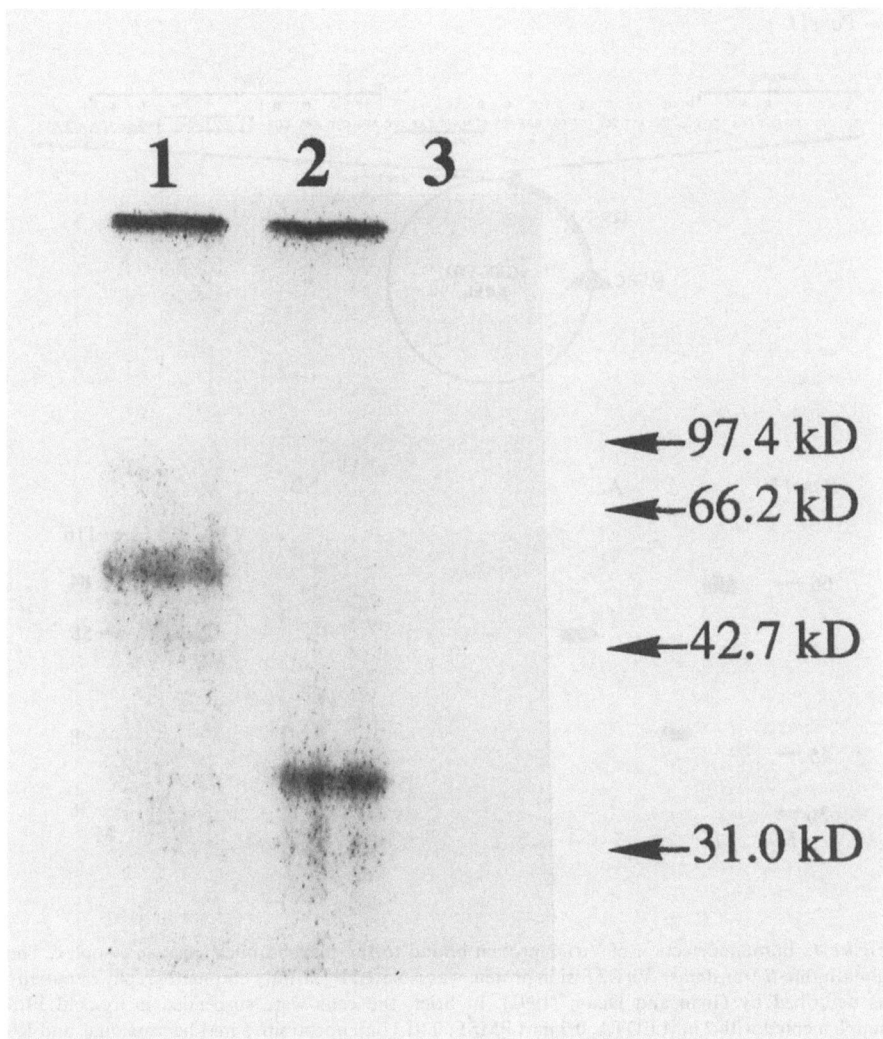

Figure 3. Analysis of ^{125}I-labelled proteins bound to the right T-DNA border restriction endonuclease fragment. The restriction endonuclease fragment containing the ^{125}I-labelled protein-DNA complex was excised from a low melting point agarose gel, the DNA purified by ion-exchange chromatography, and subsequently digested with DNase I. The proteins were analyzed by electrophoresis through 10% SDS polyacrylamide gels. Lane 1, protein bound to the retarded band from pTRIvir. Lane 2, protein bound to the retarded band from pTRIvirΔ. Lane 3, protein bound to pTRI. The positions of the protein molecular weight markers are indicated by the arrows.

216

Panel I

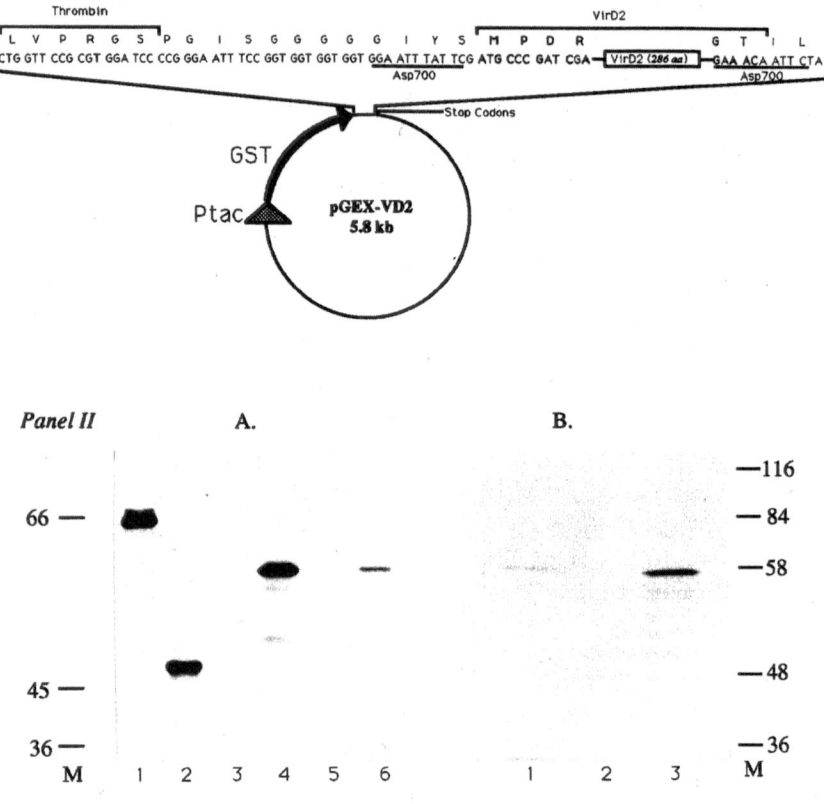

Figure 4. Immunodetection of VirD2 protein bound to the plasmid nucleoprotein complex. The glutathione-*S*-transferase-VirD2 fusion protein was isolated by affinity chromatography essentially as described by Guan and Dixon (1991). In brief, the cells were suspended in ice-cold PBS supplemented with 2 mM EDTA, 0.2 mM PMSF, 0.01 U/ml aprotinin, 5 mM benzamidine, and 1% Triton X-100 and lysed by passage through a French pressure cell three times at 14000 lb/in[2]. The lysate was passed twice at 4°C through a 2 × 2 cm column packed with glutathione-agarose (Sigma Chemical Co., St. Louis, MO). The fusion protein was eluted from the column with 10 mM glutathione, concentrated in a Centricon-30 concentrator, and cleaved with human thrombin. The cleaved proteins were separated by electrophoresis through a 10% SDS polyacrylamide gel (Laemmli, 1970), stained with Coomassie Brilliant Blue R-250, and the band containing VirD2 was excised for immunization. Alternatively, proteins from the gel were transferred onto a PVDF membrane (ProBlott[TM], Applied Biosystems, Inc.) and the amino-terminus of the membrane-bound VirD2 was microsequenced using an Applied Biosystems protein sequenator at the Purdue University Protein Sequencing facility. About 1 mg of VirD2 cleaved from the fusion protein was used for the immunization of laying hens as described (Song *et al.*, 1985). The IgG fraction of serum was precipitated with 0.7 volumes of saturated ammonium sulfate and dialyzed against PBS buffer. VirD2 band-specific antibodies were additionally purified by elution with glycine (pH 2.5) from preparative Western blots of cleaved GST-VirD2 fusion protein as described (Olmsted, 1981). VirD2-primary

Discussion

We present here direct evidence that during T-DNA processing in *E. coli* VirD2 protein is tightly associated with the relaxed substrate plasmid, most likely through a nick at the right T-DNA border. Both native and truncated forms of ^{125}I-labelled VirD2 protein were detected bound to a double-stranded restriction endonuclease fragment containing a T-DNA border. The association of these proteins with the border-containing fragment resulted in the retarded mobility of this fragment during electrophoresis through an agarose gel containing SDS and urea. In addition, the presence of VirD2 protein tightly associated with the relaxed substrate plasmid was shown by Western blot analysis using anti-VirD2 antibodies. Primer extension analysis has shown that the nick in this relaxed plasmid occurs between nucleotides 3 and 4 of the T-DNA border (ZhenWu Lin, unpublished). Proposed models for the production of T-strands (or double-stranded processed T-DNA molecules) assume that *virD*-encoded proteins associate with the nicked border sequences during T-DNA processing (Stachel *et al.*, 1986; Albright *et al.*, 1987; Durrenberger *et al.*, 1989; Herrera-Estrella *et al.*, 1988; Ward and Barnes, 1988; Young and Nester, 1988; Howard *et al.*, 1989). The association of VirD2 protein with the 5' ends of processed double-stranded T-DNA molecules and the 5' end of the cleaved double-stranded vector has been reported (Durrenberger *et al.*, 1989; Steck *et al.*, 1989). This is the first report, however, of the tight association of VirD2 protein with a T-DNA border fragment that has not been fully cleaved.

The tight association of VirD2 protein with a relaxed double-stranded substrate plasmid suggests that this structure is an intermediate in the formation of T-strands from such plasmids. This tight, most likely covalent, association of VirD2 protein with a T-DNA border may facilitate DNA duplex opening and T-strand replication and displacement. In these experiments, ^{125}I-labeling of these

---

antibody comlexes were detected on immunoblots by using affinity purified anti-chicken IgG-alkaline phosphatase conjugates (Jackson ImmunoResearch Laboratories, Inc., West Grove, PA). Panel I. Structure of the glutathione-*S*-transferase-VirD2 fusion in the plasmid pGEX-VD2. A 800 bp *Asp*700 fragment of *virD2* was ligated into the filled-in *Eco*RI site of pGEX-KG. The resulting fusion protein contains a 26 kD portion of GST at the amino-terminus of VirD2. This fusion protein migrates as 66 kD through a SDS polyacrylamide gel. After affinity purification on a glutathione agarose column followed by thrombin treatment, the cleaved VirD2 portion migrates as a band with a molecular weight of 47,000. An excess of a thrombin during digestion resulted in the appearance of an additional 37 kD band. The filled arrow represents the glutathione-*S*-transferase coding sequence. Thrombin, thrombin cleavage site; Ptac, *tac* promoter. Nucleotides and amino acids originating from *virD2* are shown in boldface. Panel II, A. and B. Immunodetection of VirD2 protein by Western blot analysis. A. Lanes 1 and 2, GST-VirD2 fusion protein affinity purified on a glutathione-agarose column before (Lane 1) and after (Lane 2) thrombin cleavage; Lanes 3 and 5, total proteins from *E. coli* lacking *virD* genes; Lane 4, total proteins from *E. coli* containing pTDCNvir; Lane 6, total proteins from *E. coli* containing pTRIvir. B. Lane 1, pTRIvir plasmid nucleoprotein complex was digested with DNase I for 1 hour at 37°C, concentrated in an Amicon 30 microconcentrator, and boiled in gel loading buffer for 5 min Lanes 2 and 3, total proteins from *E. coli* containing pTRI (lane 2) and pTRIvir (lane 3). M, molecular weight markers.

relaxed nucleoprotein molecules revealed the presence of only VirD2. Because our plasmid isolation conditions used high salt, SDS, and urea, proteins non-covalently bound to the DNA would most likely have been removed. It is likely, however, that other proteins associate with the relaxed substrate plasmid because the radioiodination of proteins associated with such molecules isolated under non-denaturing conditions revealed the presence of a few other proteins in addition to VirD2 (Filichkin and Avramova, unpublished). We do not yet know, however, whether these additional proteins bind specifically to the T-DNA border region of the substrate plasmid, or whether they might even be involved in protein-protein interactions with VirD2 and not interact with the plasmid DNA directly. One may expect that VirD1 protein, that has topoisomerase-like activity (Ghai and Das, 1989), might associate with the border during T-DNA processing because VirD1 protein is essential for the functioning of VirD2 protein during T-DNA border cleavage (Jayaswal et al., 1987). VirC1 protein binds to overdrive, a DNA sequence near the right T-DNA border of many Ti- and Ri-plasmids, and may facilitate nicking at the border (Toro et al., 1988, 1989). This protein may also associate with the T-DNA border region during T-DNA processing.

We chose E. coli as a system in which to investigate the early events of T-DNA processing for a number of reasons. Both the nicking and double-stranded cleavage of the T-DNA borders, as well as the production of T-strands, have been demonstrated in E. coli (Yanofsky et al., 1986; Jayaswal et al., 1987; Herrera-Estrella et al., 1988; DeVos and Zambryski, 1989). In A. tumefaciens, the virC and virD promoters are under both the positive regulation of the VirA/VirG signal transduction pathway (Stachel and Nester, 1986) and the negative control of the ros gene product (Close et al., 1985,1987). In E. coli (DeVos and Zambryski, 1989; Avramova, Filichkin, and Gelvin, unpublished) and in A. tumefaciens ros mutants (Close et al., 1987; Avramova, Filichkin, and Gelvin, unpublished) the virC and virD genes function constitutively. Thus, the analysis of T-DNA processing is not complicated by varying levels of induction of the vir genes by acetosyringone. In addition, we were able to obtain relatively large amounts of the putative T-DNA processing intermediates by using high copy number substrate plasmids containing both the vir genes and T-DNA border sequences in a defined (one-to-one) stoichiometry. In previous studies, separate plasmids containing either the vir genes or the T-DNA borders were used, and thus the vir gene to border ratio could not be controlled (Yanofsky et al., 1986; Jayaswal et al., 1987; DeVos and Zambryski, 1989).

T-DNA transfer from A. tumefaciens to plant cells has many features in common with the conjugal transfer of plasmids between bacteria (Stachel and Zambryski, 1986b; Zambryski, 1988). In F plasmid conjugal transfer, a transient single-stranded copy of the plasmid located at the membrane is capped at the 5' end with a pilot protein that may lead the strand through the bacterial membranes and into the recipient bacterium (Willetts and Wilkins, 1984; Ippen-Ihler and Minkley, 1986). Although the transfer of a single-stranded form of the T-DNA from A. tumefaciens to plant cells has not yet been demonstrated, single-stranded T-DNA molecules (T-strands) accumulate in *A. tumefaciens* cells in which the *vir* genes have been induced. T-strands are tightly associated at the 5' end with VirD2 protein which may protect them from exonucleolytic degradation within the plant cell (Herrera-Estrella *et al.*, 1988; Ward and Barnes, 1988; Young and Nester, 1988; Durrenberger *et al.*, 1989; Howard *et al.*, 1989). VirD2 protein also has nuclear targeting domains that may guide the T-strand from the bacterial cell to the plant nucleus (Herrera-Estrella *et al.*, 1990; Howard *et al.*, 1992). In addition, the product of the *virE2* gene, a single-stranded DNA binding protein, may coat the T-strand and thus protect it from endonucleolytic degradation (Gietl *et al.*, 1987; Christie *et al.*, 1988; Citovsky *et al.*, 1988,1989; Das, 1988; Sen *et al.*, 1989).

The analogy between the early events of T-DNA processing and conjugal plasmid transfer has recently been reinforced by the discovery that the origins of transfer (*oriT*) of IncP plasmids and the T-DNA borders contain homologous DNA sequences (Waters *et al.*, 1991). Three proteins (TraH, TraI, and TraJ) appear to be required for the formation of a stable nucleoprotein conjugation complex, the relaxasome, at the origin of transfer of IncP plasmids (Furste *et al.*, 1989; Ziegelin *et al.*, 1989; Pansegrau *et al.*, 1990a,b). Only TraI protein, however, remains covalently attached to the nick at *oriT* upon relaxation of the supercoiled DNA substrate (Pansegrau *et al.*, 1990b). In T-DNA processing, the VirD2 protein may play a similar functional role as does TraI in IncP plasmid conjugal transfer. In this work we have shown that *E. coli* cells containing the appropriate substrate plasmids contain a population of relaxed substrate plasmid molecules containing VirD2 protein tightly associated with a T-DNA right border fragment. In IncP plasmid conjugal transfer, the TraJ protein binds to inverted repeat sequences near *oriT* (Ziegelin *et al.*, 1989). VirD1 protein may play a similar role in T-DNA processing, although no sequence specificity has yet been demonstrated for VirD1 topoisomerase activity (Ghai and Das, 1989). Finally, the TraH protein may stabilize the relaxation complex during the initial stages of IncP plasmid conjugal transfer. A similar role in T-DNA processing may be played by the overdrive-binding protein VirC1. In the absence of VirC1 protein, lower levels of T-strands accumulated in *E. coli* cells in which the level of VirD2 protein was limiting (DeVos and Zambryski, 1989).

We are currently attempting to identify other proteins bound at or near the T-DNA border sequences during the early stages of T-DNA processing. The isolation of such complexes from *E. coli* cells or their reconstitution *in vitro* should provide us with useful information regarding the nature of the protein-DNA and protein-protein interactions in the T-DNA border region during early stages of T-DNA processing.

Acknowledgements

The authors thank Drs. Susan Karcher and Vidadi Yusabov for critical reading of the manuscript, and Dr. Zoya Avramova for help in the iodination experiments. This work was funded by the DuPont Co., with matching funds from the Midwest Plant Biotechnology Consortium.

References

Albright LM, Yanofsky MF, Leroux B, Ma D, & Nester, EW (1987) Processing of the T-DNA of *Agrobacterium tumefaciens* generates border nicks and linear, single-stranded T-DNA. J. Bacteriol. 169: 1046–1055.

Ausubel FM, Brent R, Kingston RE, Moore DD, Seidman JG, Smith JA, & Strugh K (eds). (1987) Current protocols in molecular biology. Green Publishing Associates, N.Y.

Boyer HW, & Roulland-Dussoix D (1969) A complementation analysis of the restriction and modification of DNA in *Escherichia coli*. J. Mol. Biol. 41: 459–472.

Cangelosi GA, Ankenbauer RB, & Nester EW (1990) Sugars induce the *Agrobacterium* virulence genes through a periplasmic binding protein and a transmembrane signal protein. Proc. Natl. Acad. Sci. USA 87: 6708–6712.

Casabadan MJ, & Cohen SN (1980) Analysis of gene control signals by DNA fusion and cloning in *Escherichia coli*. J. Mol. Biol. 138: 179–207.

Christie PJ, Ward JE, Winans SC, & Nester EW (1988) The *Agrobacterium tumefaciens virE2* gene product is a single-stranded-DNA-binding protein that associates with T-DNA. J. Bacteriol. 170: 2659–2667.

Citovsky V, DeVos G, & Zambryski P (1988) Single-stranded DNA binding protein encoded by the *virE* locus of *Agrobacterium tumefaciens*. Science 240: 501–504.

Citovsky V, Wong ML, & Zambryski P (1989) Cooperative interaction of *Agrobacterium* VirE2 protein with single-stranded DNA: Implications for the T-DNA transfer process. Proc. Natl. Acad. Sci. USA 86: 1193–1197.

Close TJ, Rogowsky PM, Kado CI, Winans SC, Yanofsky MF, Nester EW (1987) Dual control of *Agrobacterium tumefaciens* Ti plasmid virulence genes. J. Bacteriol. 169: 5113–5118.

Close TJ, Tait RC, & Kado CI (1985) Regulation of Ti plasmid virulence genes by a chromosomal locus of *Agrobacterium tumefaciens*. J. Bacteriol. 164: 774–781.

Cooley MB, D'Souza MR, & Kado CI (1991) The *virC* and *virD* operons of the *Agrobacterium* Ti plasmid are regulated by the *ros* chromosomal gene: Analysis of the cloned *ros* gene. J.Bacteriol. 173: 2608–2616.

Das, A. (1988) *Agrobacterium tumefaciens virE* operon encodes a single-stranded DNA-binding protein. Proc. Natl. Acad. Sci. USA 85: 2909–2913.

DeVos G, & Zambryski P (1989) Expression of *Agrobacterium* nopaline-specific VirD1, VirD2, and VirC1 proteins and their requirement for T-strand production in *E. coli*. Mol. Plant-Microbe Interact. 2: 43–52.

Durrenberger F, Crameri A, Hohn B, & Koukolikova-Nicola Z (1989) Covalently bound exonucleolytic degradation. Proc. Natl. Acad. Sci. USA 86: 9154–9158.

Furste JP, Pansegrau W, Ziegelin G, Kroger M, & Lanka E (1989) Conjugative transfer of promiscuous IncP plasmids: Interaction of plasmid-encoded products with the transfer origin. Proc. Natl. Acad. Sci. USA 86: 1771–1775.

Gelvin SB (1992) Chemical signaling between *Agrobacterium* and its plant host. *In:* Verma, D.P.S. (ed.) Molecular signals in plant-microbe communications. CRC Press (Boca Raton). pp.137–167.

Ghai J, & Das A (1989) The *virD* operon of *Agrobacterium tumefaciens* Ti plasmid encodes a DNA-relaxing enzyme. Proc. Natl. Acad. Sci. USA 86: 3109–3113.

Gietl C, Koukolíková-Nicola Z, & Hohn B (1987) Mobilization of T-DNA from *Agrobacterium* to plant cells involves a protein that binds single-stranded DNA. Proc. Natl. Acad. Sci. USA 84: 9006–9010.

Guan K-L, & Dixon JE (1991) Eukariotic proteins expressed in *Escherichia coli*: An improved thrombin cleavage and purification procedure of fusion proteins with glutathione S-transferase. Anal. Biochem. 192: 262–267.

Hanahan D (1983) Studies on transformation of *Escherichia coli* with plasmids. J. Mol. Biol. 166: 557–580.

Herrera-Estrella A, Chen Z-M, Van Montagu M, & Wang K (1988) VirD proteins of *Agrobacterium tumefaciens* are required for the formation of a covalent DNA-protein complex at the 5′ terminus of T-strand molecules. EMBO J. 7: 4055–4062.

Herrera-Estrella A, Van Montagu M, & Wang K (1990) A bacterial peptide acting as a plant nuclear targeting signal: The amino-terminal portion of *Agrobacterium* VirD2 protein directs a β-galactosidase fusion protein into tobacco nuclei. Proc. Natl. Acad. Sci. USA 87: 9534–9537.

Howard E, & Citovsky V (1990) The emerging structure of the *Agrobacterium* T-DNA transfer complex. BioEssays 12: 103–108.

Howard EA, Winsor BA, DeVos G, & Zambryski P (1989) Activation of the T-DNA transfer process in *Agrobacterium* results in the generation of a T-strand-protein complex: Tight association of VirD2 with the 5′ ends of T-strands. Proc. Natl. Acad. Sci. USA 86: 4017–4021.

Howard EA, Zupan JR, Citovsky V, & Zambryski PC (1992) The VirD2 protein of *A. tumefaciens* contains a C-terminal bipartite nuclear localization signal: Implications for nuclear uptake of DNA in plant cells. Cell 68: 109–118.

Ippen-Ihler KA, & Minkley EGJ (1986) The conjugation system of F, the fertility factor of *Escherichia coli*. Annu. Rev. Genet. 20: 593–624.

Jayaswal RK, Veluthambi K, Gelvin SB & Slightom JL (1987) Double-stranded cleavage of T-DNA and generation of single-stranded T-DNA molecules in *Escherichia coli* by a *virD*-encoded border-specific endonuclease from *Agrobacterium tumefaciens*. J. Bacteriol. 169: 5035–5045.

Laemmli UK (1970) Cleavage of structural proteins during the assembly of the head of bacteriophage T4. Nature (London) 227: 680–685.

Maniatis T, Fritsch EF, & Sambrook J (1982) Molecular cloning: A laboratory manual. Cold Spring Harbor Laboratory, Cold Spring Harbor, N.Y.

Murray NE, Brammer WJ, & Murray K (1977) Lambdoid phages that simplify the recovery of *in vitro* recombinants. Mol. Gen. Genet. 150: 53–61.

Olmsted JB (1981) Affinity purification of antibodies from diazotized paper blots of heterogeneous protein samples. J. Biol. Chem. 256: 11955–11957.

Pansegrau W, Balzer D, Kruft V, Lurz R, & Lanka E (1990a) *In vitro* assembly of relaxosomes at the transfer origin of plasmid RP4. Proc. Natl. Acad. Sci. USA 87: 6555–6559.

Pansegrau W, Ziegelin G, & Lanka E (1990b) Covalent association of the *traI* gene product of plasmid RP4 with the 5′-terminal nucleotide at the relaxation nick site. J. Biol. Chem. 265: 10637–10644.

Parker CW (1990) Radiolabeling of proteins. Meth. Enzym. 182: 721–737.

Peralta EG, Hellmiss R, & Ream W (1986) Overdrive, a T-DNA transmission enhancer on the *A. tumefaciens* tumour-inducing plasmid. EMBO J. 5: 1137–1142.

Peralta EG & Ream LW (1985) T-DNA border sequences required for crown gall tumorigenesis. Proc. Natl. Acad. Sci. USA 82: 5112–5116.

Rogowsky PM, Close TJ, Chimera JA, Shaw JJ, & Kado CI (1987) Regulation of the *vir* genes of *Agrobacterium tumefaciens* plasmid pTiC58. J. Bacteriol. 169: 5101–5112.

Sen P, Pazour GJ, Anderson D, & Das A (1989) Cooperative binding of *Agrobacterium tumefaciens* VirE2 protein to single-stranded DNA. J. Bacteriol. 171: 2573–2580.

Song C-S, Yu J-H, Bai DH, Hester PY, & Kim K-H (1985) Antibodies to the α-subunit of insulin receptor from eggs of immunuzed hens. J. Immun. 135: 3354–3359.

Stachel SE, & Nester EW (1986) The genetic and transcriptional organization of the *vir* region of the A6 Ti plasmid of *Agrobacterium tumefaciens*. EMBO J. 5: 1445–1454.

222

Stachel SE, Timmerman B, & Zambryski P (1986) Generation of single-stranded T-DNA molecules during the initial stages of T-DNA transfer from *Agrobacterium tumefaciens* to plant cells. Nature 322: 706–712.

Stachel SE, Timmerman B, & Zambryski P (1987.) Activation of *Agrobacterium tumefaciens vir* gene expression generates multiple single-stranded T-strand molecules from the pTiA6 T-region: Requirement for 5' *virD* gene products. EMBO J. 6: 857–863.

Stachel SE, & Zambryski P (1986a.) *virA* and *virG* control the plant-induced activation of the T-DNA transfer process of *Agrobacterium tumefaciens*. Cell 46: 325–333.

Stachel SE, & Zambryski PC (1986b) *Agrobacterium tumefaciens* and the susceptible plant cell: A novel adaptation of extracellular recognition and DNA conjugation. Cell 47: 155–157.

Steck TR, Close TJ, & Kado CI (1989) High levels of double-stranded transferred DNA (T-DNA) processing from an intact nopaline Ti plasmid. Proc. Natl. Acad. Sci. USA 86: 2133–2137.

Tait RC, & Kado CI (1988) Regulation of the *virC* and *virD* promoters of pTiC58 by the *ros* chromosomal mutation of *Agrobacterium tumefaciens*. Mol. Microbiol. 2: 385–392.

Toro N, Datta A, Carmi OA, Young C, Prusti RK, & Nester EW (1989) The *Agrobacterium tumefaciens virC1* gene product binds to overdrive, a T-DNA transfer enhancer. J. Bacteriol. 171: 6845–6849.

Toro N, Datta N, Yanofsky M, & Nester E (1988) Role of the overdrive sequence in T-DNA border cleavage in *Agrobacterium*. Proc. Natl. Acad. Sci. USA 85: 8558–8562.

Van Haaren MJJ, Sedee NJA, Schilperoort RA, & Hooykaas PJJ (1987) Overdrive is a T-region transfer enhancer which stimulates T-strand production in *Agrobacterium tumefaciens*. Nuc. Acids Res. 15: 8983–8997.

Veluthambi K, Jayaswal RK, & Gelvin SB (1987) Virulence genes *A*, *G*, and *D* mediate the double-stranded border cleavage of T-DNA from the *Agrobacterium* Ti plasmid. Proc. Natl. Acad. Sci. USA 84: 1881–1885.

Veluthambi K, Ream W, & Gelvin SB (1988) Virulence genes, borders, and overdrive generate single-stranded T-DNA molecules from the A6 Ti plasmid of *Agrobacterium tumefaciens*. J. Bacteriol. 170: 1523–1532.

Wang K, Herrera-Estrella L, Van Montagu M, & Zambryski P (1984) Right 25 bp terminus sequence of the nopaline T-DNA is essential for and determines direction of DNA transfer from *Agrobacterium* to the plant genome. Cell 38: 455–462.

Wang K, Stachel SE, Timmerman B, Van Montagu M, & Zambryski PC (1987) Site-specific nick in the T-DNA border sequence as a result of *Agrobacterium vir* gene expression. Science 235: 587–591.

Ward ER, & Barnes WM (1988) VirD2 protein of *Agrobacterium tumefaciens* very tightly linked to the 5' end of T-strand DNA. Science 242: 927–930.

Waters VL, Hirata KH, Pansegrau W, Lanka E, & Guiney G (1991) Sequence identity in the nick regions of IncP plasmid transfer origins and T-DNA borders of *Agrobacterium* Ti plasmids. Proc. Natl. Acad. Sci. USA 88: 1456–1460.

Willets N, & Wilkins B (1984) Processing of plasmid DNA during bacterial conjugation. Microbiol. Rev. 48: 24–41.

Winans SC, Kerstetter RA, & Nester EW (1988) Transcriptional regulation of the *virA* and *virG* genes of *Agrobacterium tumefaciens*. J. Bacteriol. 170: 4047–4054.

Yadav NS, Vanderleyden J, Bennett DR, Barnes WM, & Chilton M-D (1982) Short direct repeats flank the T-DNA on a nopaline Ti plasmid. Proc. Natl. Acad. Sci. USA 79: 6322–6326.

Yanofsky MF, Porter SG, Young C, Albright LM, Gordon MP, & Nester, EW (1986) The *virD* operon of *Agrobacterium tumefaciens* encodes a site-specific endonuclease. Cell 47: 471–477.

Young C, & Nester EW (1988) Association of the VirD2 protein with the 5' end of T-strands in *Agrobacterium tumefaciens*. J. Bacteriol. 170: 3367–3374.

Zambryski P (1988) Basic processes underlying *Agrobacterium*-mediated DNA transfer to plant cells Annu. Rev. Genet. 22: 1–30.

Ziegelin G, Furste JP, & Lanka E (1989) TraJ protein of plasmid RP4 binds to a 19-base pair inverted sequence repetition within the transfer origin. J. Biol. Chem. 264: 11989–11994.

16. T-DNA transfer from *Agrobacterium* to the plant cell nucleus

BRUNO TINLAND, LUCA ROSSI and BARBARA HOHN

Abstract. The T-DNA (transferred DNA) of *Agrobacterium tumefaciens* is the only non-viral nucleic acid known to genetically transform a higher eukaryotic cell in nature. The bacterium has to produce plant-infectious DNA which then has to penetrate the bacterial and plant membranes, find the plant cell nucleus, and finally has to integrate into the chromosomal DNA.

An adaptation of the bacterium to the eukaryotic condition of the recipient must have been the nuclear targeting of the T-DNA. Here we address the question of how the T-DNA is directed to the plant nucleus. Virulence protein D2, covalently attached to T-DNA in the bacterium, is a prime candidate for the function of nuclear targeting. We show that VirD2 protein contains two nuclear localization signals (NLS) which are able to target an otherwise cytoplasmatically localized β-galactosidase protein to the nuclei of both yeast and plant cells. Experiments employing a transient assay for T-DNA transfer reveal that deletion of the C-terminal NLS of VirD2 drastically reduces (but does not abolish) the efficiency of T-DNA transfer, whereas mutations in the N-terminal NLS coding sequence seem to have no effect on T-DNA transfer. This result complements the *in vitro* nuclear localization data.

Abbreviations: GUS, ß-glucuronidase; NLS, Nuclear Localization Signals

Introduction

Transfer of genetic information from the phytopathogenic bacterium *Agrobacterium tumefaciens* to plant cells resembles bacterial conjugation in many respects (Zambryski *et al.*, 1989). However, several adaptations had to be made in the course of evolution. The recipient cell in the case of *Agrobacterium*-plant interaction belongs to the eukaryotic kingdom. This implies that 1) the transferred DNA (T-DNA) carries genetic information which is recognized in a eukaryotic cell, and that 2) this genetic information reaches the location in the eukaryotic organism which is uniquely equipped for deciphering this information – the nucleus.

In the bacterium, the T-DNA is localized on the Ti (tumor inducing) plasmid. In the course of induction of *Agrobacterium* by the plant, the T-DNA becomes separated from the Ti plasmid. The genes localized on the T-DNA, however, are designed to be active in the recipient cell only, and therefore carry eukaryotic expression signals. It is not clear how the bacterial T-DNA evolved to be equipped with these plant-specific regulation sequences.

It is not known how nucleic acids enter nuclei of eukaryotic organisms.

C.I. Kado and J.H. Crosa (eds.), Molecular Mechanisms of Bacterial Virulence, 223–230.
© 1994 *Kluwer Academic Publishers. Printed in the Netherlands.*

However, in the case of T-DNA, which most probably does not travel to the recipient cell as pure DNA, the specific role of nuclear targeting is likely to be taken over by a protein. This protein, the virulence protein VirD2, is covalently attached to the T-DNA liberated from the Ti plasmid upon induction by plant compounds.

Entry of proteins into the nucleus is a selective process which requires the activation of the nuclear pore complex (Newmeyer & Forbes, 1988; Richardson *et al.*, 1988; and for review see Garcia Bustos *et al*, 1991; and Silver, 1991). This activation is mediated by a *nuclear localization signal* (NLS) carried either by the transported protein itself or by a helper protein (for review, see Goldfarb & Michaud, 1991). Two types of nuclear localization signals have been described. The first consists of a single cluster of positively charged amino acids, the consensus sequence being K-R/K-X-R/K (Chelsky *et al.*, 1989). The second type is a bipartite signal in which two necessary sequence elements made up of basic amino acids are separated by about ten undefined amino acids (Robbins *et al.*, 1991; Dingwall & Laskey, 1991). This latter motif is present in only 4% of non-nuclear eukaryotic proteins but in 56% of nuclear proteins (Dingwall & Laskey, 1991). The SV40 large T-antigen and the nucleoplasmin NLSs are examples of these two types of signals. Inspection of the 424 amino acids of the VirD2 protein reveals two regions with similarities to these sequences (Wang *et al.*, 1990). One is located in the N-terminal part of the protein and resembles the SV40 large-T antigen or single cluster type signal. The other is in the C-terminal part of the protein and belongs to the nucleoplasmin or bipartite type signal. Interestingly, the sequence corresponding to the C-terminal putative NLS is perfectly maintained in different *Agrobacterium* strains although it belongs to the C-terminal half of VirD2 which is less than 20% conserved, whereas the N-terminal half is more than 80% conserved (Wang *et al.*, 1990; Steck *et al.*, 1990). The C-terminal signal was reported to be the only one having nuclear targeting properties (Howard *et al.*, 1992), whereas the N-terminal 292 amino-acids of the VirD2 protein have also been shown to confer such a property upon linked proteins (Herrera-Estrella *et al.*, 1990). Plant nuclear proteins contain sequences which are similar to the NLSs described for other eukaryotic organisms (Restrepo *et al.* 1990; Varagona *et al.*, 1991). Defined NLS sequences from non plant-proteins are correctly recognized in different organisms; the SV40 large-T-antigen NLS has been succesfully tested in mammalian cells (Kalderon *et al.*, 1984), in yeast cells (Nelson & Silver, 1989; Benton *et al.*, 1990) and more recently also in plant cells (van der Krol & Chua, 1991), and a yeast nuclear protein has been shown to enter the nucleus of a higher eukaryotic organism (Wagner & Hall, 1993). Thus the recognition system involved in nuclear import is most likely universal.

Results

Nuclear targeting of proteins by NLSs of VirD2

The nuclear localization potential of the VirD2 protein was tested in yeast cells and in plant protoplasts. As an antigenic tag β-galactosidase was chosen because its large size prevents it from passing through the nuclear pore and because it can be fused to other proteins without loss of enzymatic activity (Silhavy & Beckwith, 1985). VirD2 protein, deletion derivatives, and short segments coding for the isolated N-and C-terminal nuclear targeting signals were tested by indirect immunofluorescence for their ability to direct β-galactosidase entry into the nucleus. In yeast cells and in plant protoplasts both NLS sequences were shown to target β-galactosidase to nuclei (Tinland *et al.*, 1992). This observation confirms the conservation of nuclear targeting sequences throughout the eukaryotic kingdom. Moreover, these results confirm and extend the finding of Herrera-Estrella *et al.* (1990), who showed that the N-terminal 292 amino acids of the VirD2 protein (70% of VirD2 protein) are able to target β-galactosidase to the plant nucleus. The N-terminal NLS alone was not directly tested by these authors. However, Howard *et al.* (1992) tested peptides containing either the N-terminal or the C-terminal NLS of VirD2 fused to the C-terminus of *E. coli* β-glucuronidase and found that only the C-terminal NLS had nuclear targeting properties. Our results demonstrate that both the C-terminal and N-terminal sequences are efficient in targeting.

Nuclear targeting of the VirD2 protein containing T-DNA complex

In T-DNA transfer not only has a protein to be targeted to the plant nucleus but a big complex composed of DNA and proteins. In this complex the NLSs of VirD2 may or may not play the same nuclear targeting functions as they have in the protein in its free form. To elucidate the biological relevance of these putative NLS sequences in the T-DNA transfer process we studied T-DNA import into the nucleus. Our strategy was to construct a series of mutations affecting one or both NLSs in the *virD2* sequence of *Agrobacterium tumefaciens* and to test whether these mutated strains were still able to transfer the T-DNA to the plant cell nucleus. The efficiency of each mutant to perform nuclear entrance was determined by performing a transient expression assay using the β-glucuronidase (GUS) gene as reporter. We established a new assay system using very young tobacco seedlings as extremely competent and homogenous recipients for T-DNA transfer. Transfer could be quantified by extracting the seedlings and measuring enzymatic activity fluorometrically. Efficiency of T-DNA transfer could be measured over a range of three orders of magnitude. The assay, as used here, most probably measures nuclear entry because 1) it is independent of integration (Jansen and Gardner, 1989; Castle and Morris, 1990) because 2) gene expression requires nuclear entry of the T-DNA, and because 3) mutations in sequences shown to be involved in nuclear targeting of the isolated protein are being analyzed. It should be pointed out, however, that a lack of nuclear entry could not be distinguished from lack of entry into the plant cell.

To test whether VirD2 is able to pilot the T-DNA to the nucleus of the plant cell, sequences of virD2 containing different mutations in the N- and C-terminal NLS, shown before to be able to target foreign proteins to the plant cell nucleus, (Herrera-Estrella et al., 1990; Tinland et al., 1992; Howard, et al., 1992) were analyzed for their ability to transfer the T-DNA (Rossi et al., 1993). As general recipient for vir D2 mutants a Ti-plasmid was constructed in which 70% of the virD2 coding sequence was deleted. Plasmids carrying the mutated virD2 gene, placed under the control of the virD promoter, were transferred by conjugation into the Agrobacterium strain carrying the virD2 deletion, where they integrated into the Ti plasmid.

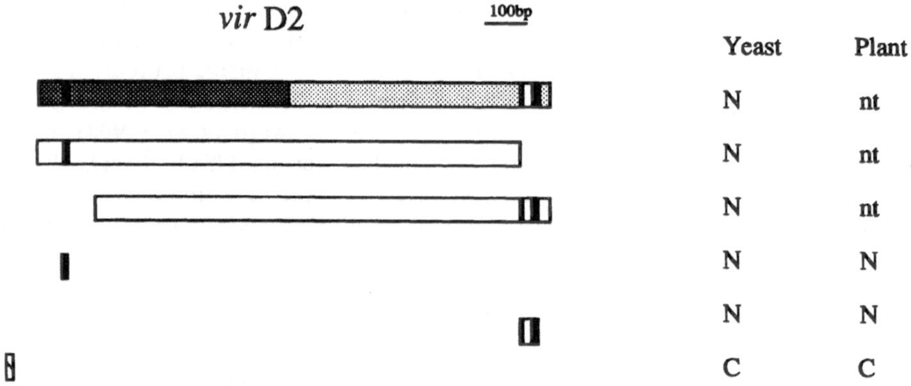

Figure 1. Nuclear targeting of VirD2 and derivatives in yeast and plant cells. The shadowed N-terminal part of VirD2 corresponds to the highly conserved part, the dotted C-terminal part to the less well conserved part; black bars indicate the NLSs. The entire virD2 gene, deletion derivatives lacking the N- or C-terminal NLS, the NLS sequences and a control sequence alone were fused to the lacZ gene at the 5′ end of the latter and tested for the intracellular localization of the VirD2-β-galactosidase fusion derivatives. N, nuclear localization; C, cytoplasmic localization; nt, not tested. Data are from Tinland et al., (1992).

To evaluate the importance of the C-terminal NLS for T-DNA transfer, a derivative of VirD2 in which the signal was deleted was analyzed. The transfer efficiency was reduced to 4% (see Table 1A), compared to the activity given by the wild-type. This result indicates that the C-terminal NLS is essential for the transfer process. To analyze the two parts of the C-terminal NLS precisely, small deletions encompassing the N-terminal or the C-terminal part of this NLS were introduced by PCR. The T-DNA transfer activity of the three strains deleted in different segments of the bipartite C-terminal NLS was tested. Deletion of both parts of the signal resulted in a reduction of the transfer efficiency to 4–6%. Deletion of only the first block of basic amino-acids resulted

Table 1. Involvement of C-terminal and N-terminal NLSs of VirD2 in T-DNA transfer.

| N-terminal NLS | C Terminal NLS | | GUS activity |
	C ter 1	C ter 2	
A RKGK	+	+	100
RKGK[a]	-	-	4
RKGK[b]	+	-	0.43
RKGR[c]	-	+	40
B RKGK	+	+	100
RKGK	-	-	4.6
RKG*N*	+	+	51.5
RKG*N*	-	-	4.5
*N*KGK	+	+	111
*N*KGK	-	-	3.4
*T*KGK	+	+	16.5
*T*KGK	-	-	0.31
R*N*GK	+	+	92
R*N*GK	-	-	4.5

a) The deletions spanned aminoacids 396 to 404 and 409 to 413 of the VirD2 protein

b) The deletion spanned aminoacids 409 to 413

c) The deletion spanned aminoacids 396 to 404

in a small decrease of efficiency whereas deletion of the second block abolished T-DNA transfer completely.

A series of point mutations affecting different positively charged aminoacids of the N-terminal NLS were tested. Out of the four tested mutants only one (arginine to threonine) gave rise to a reduced level of measured GUS activity whereas the three other showed a level of transfer efficiency comparable to the one of wild type VirD2, although the N-terminal NLS was likely to be abolished (see Table 1B). In order to assess a possible N-terminal NLS in the T-DNA transfer more conclusively, the mutations described above were tested in combination with a deletion of the C-terminal NLS. Introduction of the deletion covering the C-terminal NLS into the N-terminal NLS point mutations resulted in a drop to 2–8% of T-DNA transfer efficiency, in comparison to the corresponding point mutation construct in the N-terminal NLS. This drop is comparable to the one obtained when the C-terminal NLS sequence is deleted from an otherwise wildtype VirD2 protein. This result therefore indicates that the N-terminal NLS has no effect on T-DNA transfer.

Discussion

We tested the nuclear localization potential of VirD2 protein in stably transformed yeast cells. Different parts of VirD2 were fused to β-galactosidase. Each construct containing either one or both nuclear localization signals were expressed in yeast. Their products were localized by indirect immuno-fluorescence in the yeast nucleus, whereas in the absence of a nuclear localization signal, β-galactosidase was identified throughout the cell (Tinland *et al.*, 1992). The nuclear targeting function was precisely defined in yeast cells as two peptides of 11 and 20 amino-acids belonging to the N-terminal and to the C-terminal part of the VirD2 protein, respectively. Each of the peptides, tested individually, was then shown to target β-galactosidase into the nucleus of plant cell protoplasts.

Whereas these data clearly show that the VirD2 protein contains properties to target itself and other proteins to the nucleus, they do not automatically imply that the VirD2 protein can also transport the linked T-DNA to the plant nucleus. Therefore the relevance of these nuclear localization sequences for T-DNA transfer *in vivo* was elucidated. Using a new transfer assay with very young tobacco seedlings, the efficiency of T-DNA mediated β-glucuronidase expression could precisely and reproducibly be established.

Mutagenesis of the sequences shown to exhibit nuclear localization functions in the targeting experiments revealed the following: deletions spanning the C-terminal NLS caused a decrease in T-DNA transfer to about 4% of an intact VirD2 protein, suggesting the involvement of this NLS for the nuclear entrance of T-DNA. The involvement of the C-terminal NLS of VirD2 in T-DNA transfer has also been shown by Shurvinton *et al.*, (1992) and Koukolíková-Nicola *et al.*, 1993.

To test whether this residual transfer could be mediated by the N-terminal NLS, we mutagenized each of the three positively charged amino acids belonging to the N-terminal NLS (RKgK). These mutations alone were not expected to interfere with nuclear targeting because the wild-type N-terminal NLS was not able to compensate the effect of the C-terminal NLS deletion. However, a combination of the N-terminal NLS mutation with a C-terminal deletion was expected to reveal to which extent the T-DNA transfer is dependent on the two VirD2 NLSs. The N-terminal NLS is very close to a tyrosine proposed to be involved in the T-DNA-VirD2 linkage (Vogel & Das, 1992). Therefore, in order to find at least one mutation which would not interfere with the T-DNA processing four different point mutations were introduced separately into the N-terminal NLS. The replacing amino acids, threonine and asparagine, were chosen on the basis of a mutation analysis performed on the NLS of the SV40 large-T-antigen (Kalderon *et al.*, 1984; Landford & Butel, 1984). Only one of the four mutated constructs showed a drop in the T-DNA transfer efficiency (to 16.5% of the wild-type level) whereas the activity of the others resembled that of non-mutated VirD2. The mutation giving rise to the low transfer efficiency can possibly be interpreted as interfering with T-DNA processing. The combination of each mutation in the N-terminal NLS with the deletion in the C-terminal NLS resulted in a decrease in transfer efficiency, which was comparable to the decrease when the C-terminal NLS was deleted from the wild-type VirD2. This result indicates that the N-terminal NLS has no nuclear localization function for T-DNA.

Because it is not due to a partial compensation by the N-terminal NLS, the residual activity of the C-terminal mutation could be due to targeting by VirE2 (Citovsky *et al.*, 1992). An alternative explanation could be cryptic NLS sequences in VirD2 or free entry into the nucleus during its desintegrated phase in the cell cycle.

Acknowledgements

We thank M. Hall for help in designing and executing targeting experiments in yeast, H. Puchta for critical reading of the manuscript, Z. Koukolíková-Nicola and W. Ream for communication of unpublished information, and K. Asfar for excellent technical help. B. Tinland was recipient of a long term EMBO fellowship.

References

Benton BM, Eng WK, Dunn JJ, Studier FW, Sternglantz R & Fisher PP (1990) Signal-mediated import of bacteriophage T7 RNA polymerase into the *Saccharomyces cerevisiae* nucleus and specific transcription of target genes. Mol. Cell. Biol. 10: 353–360.

Castle LA & Morris RO (1990) A method for early detection of T-DNA transfer. Pl. Mol. Biol. Reporter 8: 28–39.

Chelsky D, Ralph R & Jonak G (1989) Sequence requirements for synthetic peptide-mediated translocation to the nucleus. Mol. Cell. Biol. 9: 2487–2492.

Citovsky V, Zupan J, Warnick D & Zambryski P (1992) Nuclear localization of *Agrobacterium VirE2* protein in plant cells. Science 256: 1802–1804.

Dingwall C & Laskey RA (1991) Nuclear targeting sequences – a consensus? TIBS 16: 478–481.

Garcia Bustos J, Heitman J & Hall MN (1991) Nuclear protein localization. Biochim. Biophys. Acta Rev. 1071: 83–101.

Goldfarb D & Michaud N (1991) Pathways for the nuclear transport of proteins. Trends Cell Biology 1: 20–24.

230

Herrera-Estrella A, Van Montagu M & Wang K (1990) A bacterial peptide acting as a plant nuclear targeting signal: the amino-terminal portion of *Agrobacterium VirD2* protein directs a β-galactosidase fusion protein into tobacco nuclei. Proc. Natl. Acad. Sci. USA 87: 9534–9537.

Howard EH, Zupan JR, Citovsky V & Zambryski PC (1992) The VirD2 protein of A. tumefaciens contains a C-terminal bipartite nuclear localization signal: implication for nuclear uptake of DNA in plant cells. Cell 68: 109–118.

Jansen BJ & Gardner RC (1989) Localized transient expression of GUS in leaf discs following cocultivation with *Agrobacterium*. Plant Mol. Biol. 14: 61–62.

Kalderon D, Roberts BL, Richardson WD & Smith AE (1984) A short amino acid sequence able to specify nuclear location. Cell 39: 499–509.

Koukolíková-Nicola Z, Raineri D, Stephens K, Ramos C, Tinland B, Nester Ew & Hohn B (1993) Genetic analysis of the *vir*D operon of *Agrobacterium tumefaciens* : a search for functions involved in transport of T-DNA into the plant cell nucleus and in T-DNA integration. 1. Bacteriol. 175: 723–731.

Landford RE & Butel JS (1984) Construction and characterization of an SV40 mutant defective in nuclear transpot of T antigen. Cell 37: 801–813.

Nelson M & Silver P (1989) Context affects nuclear protein localization in *Saccharomyces cerevisiae*. Mol. Cell. Biol. 9: 384–389.

Newmeyer DD & Forbes DJ (1988) Nuclear import can be separated into distinct steps in vitro: nuclear pore binding and translocation. Cell 52: 641–653.

Restrepo MA, Freed DD & Carrington JC (1990) Nuclear transport of plant potyviral proteins. The Plant Cell 2: 987–998.

Richardson WD, Mills AD, Dilworth SM, Laskey RA & Dingwall C (1988) Nuclear protein migration involves two steps: rapid binding at the nuclear envelope followed by slower translocation through the nuclear pores. Cell 52: 655–664.

Robbins J, Dilworth SM, Laskey RA & Dingwald C (1991) Two interdependent basic domains in nucleoplasmin nuclear targeting sequence: identification of a class of bipartite nuclear targeting sequence. Cell 64: 615–623.

Rossi L, Hohn B & Tinland B (1993). The *vir*D$_2$ Protein of *Agrobacterium tumefaciens* carries nuclear localization signals important for transfer of T-DNA to plants. Mol. Gen. Genet 239: 345–353.

Shurvinton CE, Hodges L & Ream W (1992) A nuclear localization signal in the *Agrobacterium tumefaciens vir* D2 endonuclease is important for tumor formation. Proc. Natl. Acad. Sci. USA, 89: 11837–11841.

Silhavy TJ & Beckwith JR (1985) Microbiol. Rev. 49: 398–418.

Silver PA (1991) How proteins enter the nucleus. Cell 64: 489–497.

Steck TR, Lin TS & Kado CI (1990) VirD2 gene product from the nopaline plasmid pTiC58 has at least two activities required for virulence. Nucleic Acids Res. 18: 6953–6958.

Tinland B, Koukolíková-Nicola Z, Hall MN & Hohn B (1992) The T-DNA linked VirD2 contains two distinct functional nuclear localization signals. Proc. Natl. Acad. Sci. USA 49: 7442–7746.

Van der Krol AR & Chua NH (1991) The basic domain of plant B-ZIP proteins facilitates import of a reporter protein into the plant nuclei. Plant Cell 3: 667–675.

Varagona JM, Schmidt RJ & Raikhel NV (1991) Monocot regulatory protein opaque-2 is localized in the nucleus of maize endosperm and transformed tobacco plants. The Plant Cell 3: 105–113.

Vogel AM & Das A (1992) Mutational analysis of *Agrobacterium tumefaciens vir*D2: tyrosine-29 is essential for endonuclease activity. J. Bacteriol. 174: 303–312.

Wagner P & Hall MN (1993) Nuclear protein transport is functionally conserved between yeast and higher eukaryotes. Febs. 321: 261–266.

Wang K, Herrera-Estrella A & Van Montagu M (1990) Overexpression of *vir*D1 and *vir*D2 genes in *Agrobacterium tumefaciens* enhances T-complex formation and plant transformation. J. Bacteriol. 172: 4432–4440.

Zambryski P, Tempé J & Schell J (1989) Transfer and function of T-DNA genes from *Agrobacterium* Ti and Ri plasmids in plants. Cell 56: 193–201.

17. Mechanisms of T-DNA transfer and integration into plant chromosomes: role of *virB*, *virD4* and *virE2* and a short interspersed repetitive element (SINE) from tobacco

YASUSHI YOSHIOKA, YOSHITO TAKAHASHI, SHOGO MATSUMOTO, SHOKO KOJIMA, KEN MATSUOKA, KENZO NAKAMURA, KAZUHIKO OHSHIMA, NORIHIRO OKADA and YASUNORI MACHIDA

Abstract. The present paper describes two topics on the early events during the formation of crown gall. (1) To measure an efficiency of T-DNA transfer from *Agrobacterium* cells to plant nuclei, we have developed a simple procedure which relies on *Agrobacterium*-mediated transient expression of the intron-GUS gene in plant cells. The results of experiments by this procedure indicate that products encoded by the *virB* locus and by the *virD4* gene are necessary for transfer of T-DNA. VirE2 protein is also required for efficient transfer of T-DNA, although it is not absolutely essential. (2) We found a new family of short interspersed repetitive element (SINE) around the T-DNA integration target sites in the tobacco genome. SINEs are one of retroposons that is thought to originate from a tRNA or its gene. The SINE designated here as the TS family is the first example of a SINE family of plant bearing the significant homology to specific tRNAs including a vertebrate tRNA^{Lys} which is thought to be a cognate molecule of most animal SINEs. The TS family occurs in approximately 5.0×10^4 copies per tobacco genome at least. Correlation between the presence of the TS family in tobacco chromosomes and integration of T-DNA will be discussed.

Abbreviations: SINE, Short Interspersed Repetitive Element; AS, Acetosyringone; GUS, β-Glucuronidase

Introduction

Agrobacterium tumefaciens harboring the Ti plasmid incites crown gall tumors on a wide variety of dicotyledonous plants. Upon infection of plants, transferred DNA (T-DNA), which is flanked by 25 base pair direct repeats on Ti plasmid, is transferred to plant cells and integrated into plant nuclear DNA (reviewed in Zambryski, 1988, Kado, 1991). The T-DNA processing in *Agrobacterium* cells and its transfer require products encoded in the *vir* region of Ti plasmid. Mutations in this region abolish or lower the virulence such as tumorigenicity or transformation ability of *Agrobacterium*. The *vir* region of octopine Ti plasmid contains at least seven complementation loci (*virA, virB, virG, virC, virD, virE* and *virF*) (Stachel & Nester, 1986; Melchers *et al.*, 1990). While *virA* and *virG* are expressed constitutively, the expression of *virB, C, D* and *E* is positively regulated at the transcriptional level by plant phenolic signal molecules (Stachel *et al.*, 1986a; Rogowsky *et al.*, 1987). The plant signal is transduced into *Agrobacterium* cells through functions of the VirA and VirG

C.I. Kado and J.H. Crosa (eds.), Molecular Mechanisms of Bacterial Virulence, 231–248.
© 1994 *Kluwer Academic Publishers. Printed in the Netherlands.*

proteins (Stachel & Zambryski, 1986). By the functions of VirD1 and VirD2 proteins, transferable T-DNA molecules such as single-stranded and double-stranded T-DNA were generated (Albright *et al.*, 1987; Stachel *et al.*, 1986b; Yanofsky *et al.*, 1986). VirD2 protein has been shown to attach to 5' ends of both single- and double-stranded T-DNA molecules (Herrera-Estrella *et al.*, 1988; Ward & Barnes, 1988; Young & Nester, 1988; Dürrenberger *et al.*, 1989). VirE2 protein has the ability to bind non-specifically to single-stranded DNA (Gietl *et al.*, 1987; Das, 1988; Christie *et al.*, 1988; Citovsky *et al.*, 1988), although the *virE* locus is not absolutely essential for the virulency (Stachel & Nester, 1986). Thus, such protein(s)/T-DNA complexes are thought to migrate to plant cells and to be targeted to the nuclei (Hohn *et al.*, 1989; Zambryski 1992). The discovery that both VirD2 and VirE2 have nuclear localization signals and localized in plant nuclei when they are synthesized in plant cells support the above idea (Herrera-Estrella *et al.*, 1990; Howard *et al.*, 1992; Citovsky *et al.*, 1992). Since some of VirB proteins (Christie *et al.*, 1989; Ward *et al.*, 1990) and VirD4 protein (Okamoto *et al.*, 1991) are anchored in the inner membrane, they are thought to participate in the T-DNA transfer event at the bacterial cell surface. Functions of *virB* and *virD* loci which can encode a number of proteins, however, remain to be defined.

In contrast to the progress made in understanding the molecular events in *Agrobacterium* cells, little is known about the events involved in transfer of the T-DNA into the plant cell, mobilization of T-DNA into the plant nucleus, and its integration into the plant chromosomes. Recently, several numbers of studies have been made on T-DNA integration into the plant chromosomes. Genetic mapping indicates that T-DNA can be randomly inserted in plant chromosomes (Chyi *et al.*, 1986; Wallroth *et al.*, 1986). T-DNA tagging experiments have revealed preferential integration into transcriptionally active plant DNA loci (Koncz *et al.*, 1989; Herman *et al.*, 1990). More recently, it was shown that short sequence similarity is present between break points of the T-DNA and integration target site(s) on plant chromosomal DNA (Matsumoto *et al.*, 1990; Gheysen *et al.*, 1991; Mayerhofere *et al.*, 1991). In addition, repetitive sequences were found in the DNA regions adjacent to integration sites of T-DNA in two independently isolated tobacco transformants (Matsumoto *et al.*, 1990 and this work). The presence of a repetitive sequence near the integration target in the unique chromosome region was also reported by Gheysen *et al.* (1987). These observations allow us to examine molecular characteristic of the repetitive sequences and correlation between the presence of these sequences in the plant genome and the events of T-DNA integration.

In the first section of this paper, we report the transient gene expression system mediated by *Agrobacterium* which is useful for studying processes of T-DNA transfer from *Agrobacterium* to plant cells. The results of experiments by this system indicated that *virB*, *virD4* and *virE2* are required for efficient transfer of T-DNA into plant nuclei, although *virE2* is not absolutely essential. In the latter section, we report the presence of the short interspersed repetitive element (SINE) around the T-DNA integration target sites of two independently isolated tobacco transformants. SINEs are a repetitive sequence family that belong to a retroposon (Singer, 1982), which can be transposed and amplified through transcription by RNA polymerase and reverse transcription (Weiner *et al.*, 1986). Because most of them share high sequence and structural homologies with tRNA, it is believed to be derived from a tRNA or its gene (Okada, 1991a, b). The SINE family found here is the first example of a SINE family of plant bearing the significant homology to a vertebrate tRNA[Lys]. This SINE is at least made up of 5.0×10^4 copies in the tobacco genome. The sequences that are highly homologous to this family are commonly present in Solanaceae plants.

Results and discussion

Agrobacterium-*mediated transient gene expression in plant cell is useful for studying processes of T-DNA transfer*

A procedure for detecting transfer of T-DNA from Agrobacterium *to plant nuclei*

A procedure for assaying expression levels of genes that are transiently introduced into plant cells is currently used for estimating activity of a promoter of interest since results can be rapidly obtained. It has been shown that a certain type of tobacco suspension cultured cell lines are able to be directly transformed by co-culture with *Agrobacterium* harboring binary vector plasmid (An, 1985). We here examined whether transient expression of the GUS reporter gene can be detected by using the co-culture procedure.

To detect expression of the reporter gene specifically in plant cells, we used binary plasmid pIG121–HM (Ohta *et al.*, 1990), which carries the cauliflower mosaic virus 35S promoter-linked GUS gene with the intron sequence in the GUS coding region (it was referred to as intron-GUS) so that normal GUS protein from this construct can be synthesized only in plant cells. BY-2 tobacco suspension cultured cells were co-cultured with various strains of *Agrobacterium* harboring pIG121–HM and expressed GUS activity was assayed. As shown in Fig. 1A, activity of GUS was detected in all the samples at 48 hours after co-culture, although the level of activity differed from one strain to another. They scarcely increased by further co-culture. Subsequently, we examined whether the activities of GUS we observed were due to transient expression of the intron-GUS before it is integrated into tobacco nuclear DNA. We measured the rate of synthesis of GUS protein by pulse-labeling aliquots of the co-cultured sample of the EHA101 strain with [35S] protein labeling mix (NEN). Fig. 1B shows that the rate of GUS synthesis was maximum between 36 and 42 hours after co-culture, then decreased rapidly. When carbenicillin was added into the co-culture at 48 hours to inhibit growth of *Agrobacterium*, the degree of decrease was exaggerated. Such transient synthesis cannot be explained by assuming expression of intron-GUS stably resident in tobacco chromosomes. It is most likely that the GUS protein observed was synthesized

234

from the intron-GUS molecules that had been transferred from *Agrobacterium* cells to tobacco nuclei but not yet integrated into the chromosomal DNA. In addition, the present results suggest that a large number of intron-GUS

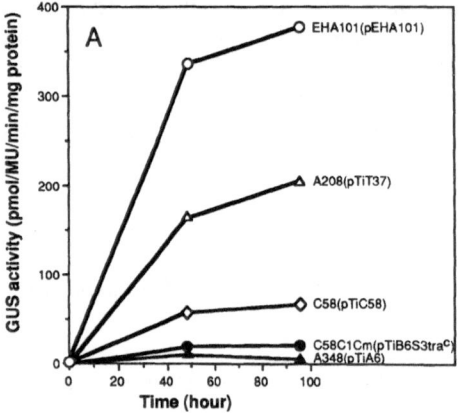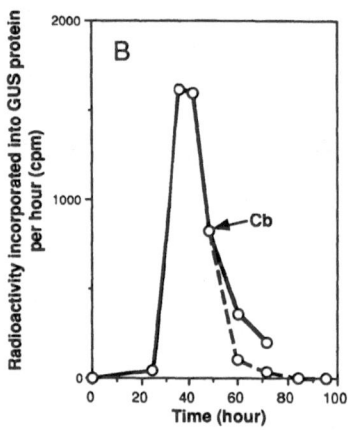

Figure 1. (A) Kinetics of GUS activity after co-culture of BY-2 cells with the *Agrobacterium* strains carrying pIG121–HM (intron-GUS). Co-culture and measurement of GUS activity were carried out as described below. (B) Kinetics of the rate of GUS protein synthesis after co-culture of BY-2 cells with EHA101 carrying pIG121–HM. Dashed line shows kinetics of GUS protein synthesis rate after addition of 200 μg ml[-1] of carbenicillin (Cb) to the co-culture at 48 hr. *Agrobacterium tumefaciens* C58C1Cm harboring pTiB6S3*tra*[c], EHA101 carrying pEHA101, A208 carrying pTiT37 and C58 carrying pTiC58 were described previously (Petit *et al.*, 1978; Hood *et al.*, 1986; Sciaky *et al.*, 1978). *A. tumefaciens* A348mx226, A348mx238, A348mx355 and A348mx341 harbor pTiA6 with insertions of transposon Tn3-HoHo1 in the *virA, virB, virD4* and *virE2*, respectively (Stachel & Nester 1986). *Escherichia coli* JM109 (Yanisch-Perron *et al.*, 1985) was used for plasmid transformation, *E. coli* strain NM538 was used for growing phage λEMBL-3 (Frischauf *et al.*, 1983).

pIG121–HM having the intron-containing GUS gene was previously described (Ohta ·t al. 1990). pAO416VG is constructed by inserting the 5-kilobase-pairs *Sal*I DNA fragment of pTOK9 (Jin *et al.*, 1987) covering the region from *virC* to *virG* of pEHA101 into *Sal*I site of mini Ri plasmid which was constructed by Nishiguchi *et al.* (1987).

BY-2 cells were co-cultured with *Agrobacterium tumefaciens* as described by An (1985). Stable transformants were selected for resistance to 200 μg ml^{-1} kanamycin. *Agrobacterium* cells were cultured until the absorbance at 600 nm was 1.5 and the culture was concentrated ten-fold in experiments with *vir* mutants. One hundred μl of this culture was mixed with 4 ml of a suspension culture of BY-2 cells and the mixture was incubated with Linsmaier and Skoog (LS) liquid medium (Linsmaier & Skoog, 1965) medium with 0.2 mg l^{-1} 2,4-dichlorophenoxyacetic acid (2,4-D) at 26°C in the dark for various periods.

Fluorometric assay for β-glucuronidase (GUS) activity was performed by the procedures of Jefferson *et al.* (1986). BY-2 cells were collected various periods after co-culture and disrupted by sonication (three 30-sec pulses). After removal of cell debris by centrifugation, the supernatants were recovered and used for determination of protein concentration and GUS activity by a fluorometric assay. An aliquot of the co-culture was washed with LS medium and incubated for 30 min in LS medium containing 50 μg ml^{-1} chloramphenicol. Then, cells were labeled with 75 μCi of [^{35}S] protein labeling mix (NEN) for 30 min and the reaction was quenched by chilling. Cells were lysed by sonication and GUS protein was immunoprecipitated with antisera against β-glucuronidase. Precipitated proteins were fractionated on SDS-polyacrylamide gel. Radioactivity of ^{35}S in the region corresponding to the GUS protein band was measured.

molecules can be first transferred to plant nuclei, then only a limited portion of them can be integrated into nuclear DNA. It has been reported that when *Agrobacterium* harboring the GUS gene in a binary vector was inoculated on leaf discs, a similar pattern of transient expression is observed (Janssen & Gardner, 1989; Vancanneyt *et al.*, 1990).

As shown in Fig. 1A, the level of GUS expression varied from one *Agrobacterium* strain to another: the highest GUS activity was found in the co-culture with EHA101, which is known to carry the disarmed supervirulent Ti plasmid. There seems to be a parallel correlation between the level of GUS activity and the frequency of transformation of BY-2 by *Agrobacterium* strains used (Table 1). These results imply that the level of GUS activity reflects amounts of intron-GUS molecules in tobacco nuclei which must have been transferred from *Agrobacterium*. Therefore, the efficiency of T-DNA transfer can be roughly measured by assaying GUS activity after co-culture BY-2 tobacco cells.

Table 1. Comparison of transient activity of GUS to the frequency of transformation by various *Agrobacterium* strains

Strain	Ti plasmid	Transformation frequency (number of Km^r calli 0.1ml culture^{-1})		GUS activity[1] (pmol MU min^{-1} mg protein^{-1})	
		+pIG121-HM	-pIG121-HM	+pIG121-HM	-pIG121-HM
EHA101	pEHA101	494 ±203	0	338 ±202	1.5 ± 0.70
A208	pTiT37	229 ± 19	0	164 ± 55	1.1 ± 1.3
C58	pTiC58	86 ± 75	0	58 ± 42	0.80± 0.10
C58C1Cm	pTiB6S3tra^c	25 ± 8	0	19 ± 9.3	0.73± 0.16
A348	pTiA6	15 ± 1	0	12 ± 2.6	1.6 ± 0.74

[1] Assayed at 48 hr after co-culture.

VirB, VirD4 and VirE2 proteins are responsible for transfer of T-DNA

We examined mutants of *vir* genes for abilities to transfer intron-GUS to plant nuclei. A series of avirulent mutants were created with the A348 strain harboring pTiA6 (Stachel & Nester, 1986), however, A348 was found to be inefficient in T-DNA transfer in the present study (Table 1). To increase the efficiency of T-DNA transfer, we introduced pAO416VG into the A348 strain. pAO416VG contains the DNA region from *virG* to *virC* from supervirulent Ti plasmid pTiBo542 which is shown to be responsible for the supervirulent characteristics and the high level of AS-inducible expression of *vir* genes (Jin *et al.*, 1987). As shown in Table 2, the efficiency of intron-GUS transfer increased 10-fold, when the A348 strain harboring pAO416VG and pIG121–HM was used for co-culture.

We introduced pAO416VG and pIG121–HM into A348 derivative strains with Tn*3*-HoHo1 insertions in *virA*, *virB*, *virD4* and *virE2*, respectively, and co-cultured BY-2 cells with each of these *vir* mutants. Table 2 summarizes activities of GUS detected at 72 hours after co-culture. When co-cultured with mutants of *virA*, *virB* and *virD4*, GUS activities were in the background level. Significant

Table 2. Effects of *vir* mutations on transient activity of GUS

Strain	Mutation	GUS activity[1] (pmol MU min[-1] mg protein[-1])
A348 / pAO416VG , pIG121-HM	Wild type	171
A348 / pIG121-HM	Wild type	15.0
A348 / pAO416VG	Wild type	1.0
A348mx226 / pAO416VG , pIG121-HM	*virA*	0.59
A348mx238 / pAO416VG , pIG121-HM	*virB*	0.31
A348mx355 / pAO416VG , pIG121-HM	*virD4*	0.79
A348mx341 / pAO416VG , pIG121-HM	*virE2*	8.6

[1] Assayed at 72 hr after co-culture.

activity was detected in the co-culture with the *virE2* mutant, although the level of activity decreased 20-fold.

We have demonstrated that transient gene expression is observed when tobacco BY-2 cells are co-cultured with *Agrobacterium* cells and that this transient expression system is useful for studying the process of T-DNA transfer. When BY-2 cells were co-cultured with *Agrobacterium* having a mutation in either *virA, virB, virD4* or *virE2*, transient expression was abolished or lowered (Table 2). T-DNA processing normally takes place in cells of *virB* and *virD4* mutants when they are incubated with AS or plant exudates (Yanofsky *et al.*, 1986; Yamamoto *et al.*, 1987; Veluthambi *et al.*, 1987; Stachel *et al.*, 1987), although they are not able to induce crown gall tumor. Therefore, the results described above suggest that products of *virB* and *virD4* are involved in the process of T-DNA transfer from the bacterial cells to plant nuclei. Previous observations that some of VirB proteins and VirD4 protein are localized in inner membranes of *Agrobacterium* cells (Christie *et al.*, 1989; Ward *et al.*, 1990; Okamoto *et al.*, 1991) support this suggestion.

Since only the low level of GUS activity was found in the co-culture with the *virE2* mutant, VirE2 protein is required for efficient transfer of T-DNA, but it seems not to be essential. This result is consistent with the observation that *virE* mutants incite attenuated tumor on tobacco plants (Stachel & Nester, 1986). VirE2 protein binds T-strand and is localized in plant nuclei when it is synthesized in plant cells (Citovsky *et al.*, 1992). Based on these results, it is proposed that VirE2 protein could facilitate nuclear transport of T-strand (Citovsky *et al.*, 1992). Our observation is in line with this idea. Occurrence of the reduced level of transient expression (Table 2), however, suggests that T-DNA is transported into plant nuclei without VirE2 protein, which may be mediated by other protein(s) that may have the function similar to that of VirE2. The most likely candidate for that is VirD2 protein that also binds to T-DNA molecules and is localized to plant nuclei (Herrera-Estrella *et al.*, 1990; Howard *et al.*, 1992).

The present results suggest that a large number of T-DNA molecules are transiently transferred to plant nuclei during co-culture. Our previous estimation has shown that T-DNA can be transferred into at least 5% of BY-2 cells co-cultured with EHA101 cells (Onouchi *et al.*, 1991). Therefore, the experimental system presented here can be applicable for analysis of structures of T-DNA molecules after transferred to plant cells.

A short interspersed repetitive element (SINE) found around the T-DNA integration sites in tobacco genome

A repetitive sequence present around T-DNA integration target sites in tobacco genome is a SINE

We previously reported that an interspersed repetitive sequence was present around the T-DNA integration target site of tobacco transformant designated as D-2T (Matsumoto *et al.*, 1990). Lower panel of Fig. 2A shows a restriction map of the tobacco genomic DNA fragment which contained the target site of

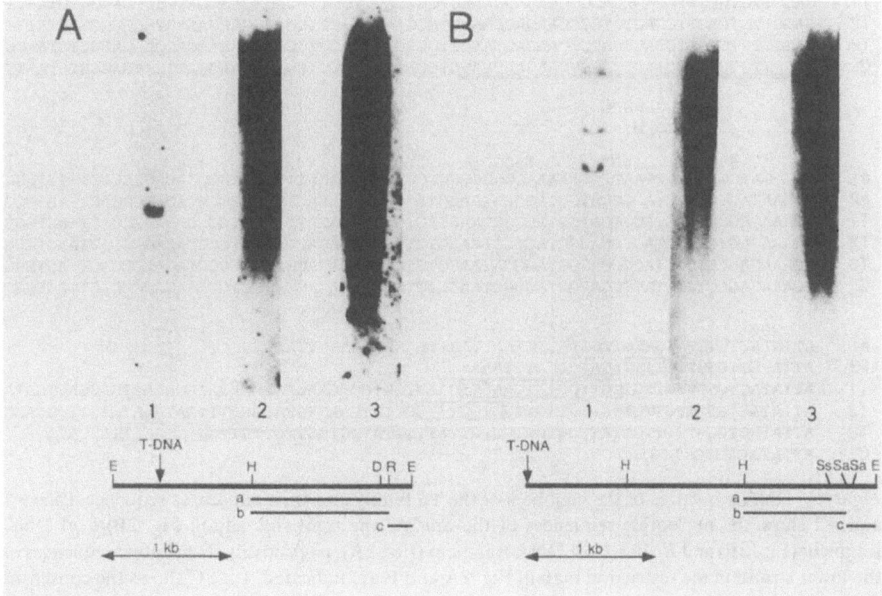

Figure 2. Determination of the presence of repetitive sequences in the tobacco genome near the target sites of T-DNA integration. (A) Southern hybridization with the DNA fragments around the target site of T-DNA integration in the transformant D-2T (Matsumoto *et al.*, 1990). Probes used: Lane 1, fragment a; lane 2, fragment b; lane 3, fragment c. Arrow indicates the integration site of T-DNA. (B) Southern hybridization with the DNA fragments around the target site of T-DNA integration in the tobacco transformant B-4-3. Probes used: Lane 1, fragment a; lane 2, fragment b; lane 3, fragment c. Arrow indicates the T-DNA integration site. E, *Eco*RI; H, *Hin*dIII; D, *Dra*I; R, *Rsa*I; Ss, *Ssp*I; Sa, *Sau*3AI.

238

D-2T transformant. It has been revealed that the repetitive sequence in D-2T transformant was present within the right hand *Hind*III-*Eco*RI fragment (Matsumoto *et al.*, 1990). To identify the region containing the repetitive sequence more precisely, we hybridized the DNA fragments a, b and c (Fig. 2A) to *Eco*RI digests of genomic DNA of non-transformant tobacco leaves. As shown in the upper panel of Fig. 2A, fragments b and c hybridized with many heterogeneous bands, whereas fragment a hybridized with a single genomic fragment. This result showed that the *Dra*I-*Eco*RI region which was 1.7 kb away from the target site contained the repetitive sequence. The nucleotide sequence of this repetitive sequence was determined and shown in line #2 of Fig. 3.

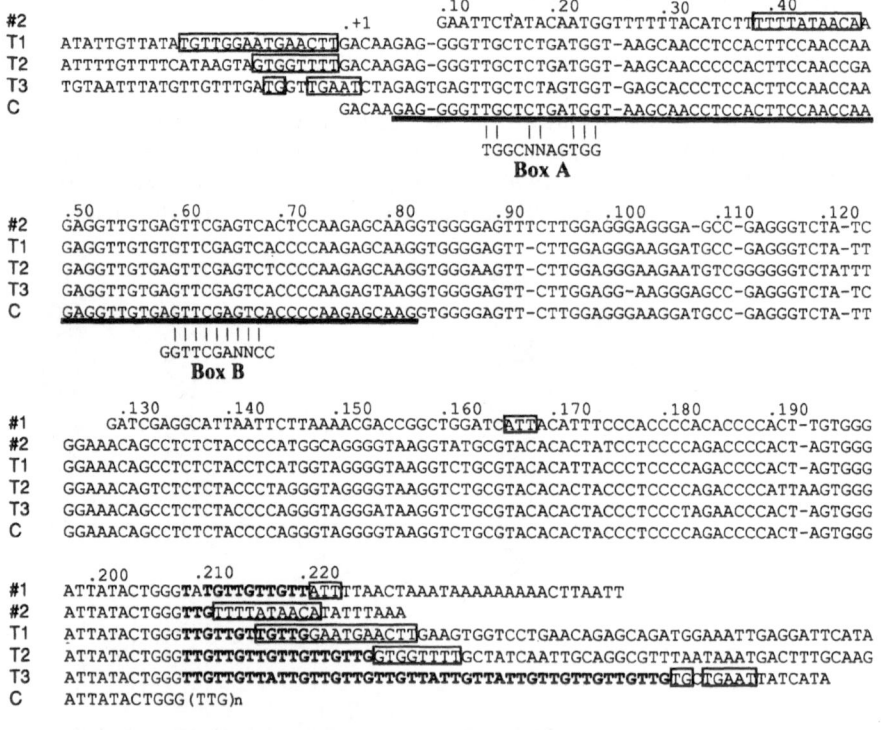

Figure 3. DNA sequences of the members of the TS family and their consensus sequence. Line #1 and #2 show the nucleotide sequences of the *Sau*3AI(the right side site in Fig. 2B)-*Ssp*I DNA fragment (Fig. 2B) and *Eco*RI-*Dra*I DNA fragment (Fig. 2A), respectively. Nucleotide sequences of the lower strand in the restriction map in Fig. 2A and B are indicated. Line C shows the consensus nucleotide sequence deduced from those of clone T1, T2, and T3. Deletions are shown by bars. Direct repeats are boxed. The tRNA-related region is underlined. Box A and Box B show consensus sequences for the promoter of RNA polymerase III (Galli *et al.*, 1981). We isolated three additional members from the tobacco genome, and comparison of nucleotide sequences of six members showed that in five out of six the sequence identity begins with the G residue marked with +1, suggesting the this G residue is the 5′ end of the TS family.

A tobacco genomic library of *Nicotiana tabacum* L. var. NK326 (purchased from CLONTECH Laboratories, Inc.) was screened using the ^{32}P-labled *Eco*RI-*Dra*I DNA fragment (nucleotide position 11 to 228 of Lane #2 in Fig. 3) as a probe which contains the repetitive sequence. The

As shown in Fig. 2B, we found that a repetitive sequence was also present around the T-DNA integration target site of another transformant of tobacco designated as B-4-3. When we hybridized the DNA fragments a, b and c (Fig. 2B) to EcoRI digests of genomic DNA of non-transformant tobacco leaves, the DNA fragment b and c hybridized with many heterogeneous bands, whereas DNA fragment a hybridized with a few genomic fragments. This result showed that the SspI-EcoRI region which was 2.2 kb away from the target site contained the repetitive sequence. The nucleotide sequence of this fragment was determined. As shown in line #1 of Fig. 3, the region between SspI and Sau3AI sites (the left side site in Fig. 2B) of this fragment contained a highly homologous sequence to a part of the #2 repetitive sequence present in the DraI-EcoRI fragment (from nucleotide position 169 to 211 in Fig. 3).

Using the DNA fragment containing the #2 repetitive sequence (DraI-EcoRI fragment in Fig. 2A) as a probe, we isolated from a tobacco genomic library three independent phage clones which contained other members of this repetitive sequence family (designated as T1, T2, and T3). Nucleotide sequence analyses of these DNA fragments showed that they contained the sequences which are almost identical to the sequence present in the probe DNA. Therefore, these nearly identical sequences were expected to be members of the repetitive sequence family. We here assumed their boundary sequences as described in the legend for Fig. 3. The lengths of these members were found to be 215, 227, and 247 bp, respectively. Each of them was composed of three distinct regions: a region homologous to tRNA (tRNA-related region, thick line in Fig. 3) (discussed in detail in a latter section), a region non-homologous to tRNA (tRNA-unrelated region), and a TTG repeat with variable length from 12 to 45 nucleotides at the 3' end. Any pair of T1, T2, and T3 shared sequence homologies more than 86.8 %. Sequences were flanked by perfect or nearly perfect short direct repeats with length of 15 bp for T1, 8 bp for T2, and 7 bp for T3, which seem to have been generated upon their insertion. Comparison of the #1 and #2 repetitive sequences to other members suggests that #1 and #2 are

---

EcoRI-DraI DNA fragment was isolated from the T-DNA integration target site of clone D-2T (Fig. 1A. Matsumoto et al., 1990). The probe DNA was labeled with [γ − ^{32}P]ATP (222 TBq mmol^{-1}, Amersham) using T4 polynucleotide kinase. Phage clones were plated on E. coli NM538 cells and phage plaques were blotted on a nylon membrane (Gene Screen Plus, New England Nuclear) as recommended by the supplier. Filters were hybridized with ^{32}P-labeled EcoRI-DraI DNA fragment in 6 × SSC, 1% SDS at 65°C and then washed with 2 × SSC (1 × SSC is 0.15 M NaCl, 0.015 M sodium citrate), 1% SDS at room temperature (Feinberg and Vogelstein, 1983). We obtained three independent positive clones and designated them as T1, T2, and T3 respectively. A 0.7 kb Sau3AI fragment of T1, a 0.78 kb Sau3AI-EcoRI fragment of T2, and a 0.56 kb Sau3AI-EcoRI fragment of T3 were detected by Southern hybridization (Southern, 1975) using ^{32}P-labeled EcoRI-DraI DNA fragment as a probe. These fragments were subcloned in pUC18 vector and sequenced. DNA sequencing was carried out by the dideoxy-chain-termination method (Sanger et al., 1977) using a 7-deaza sequencing kit (Toyobo). We also carried out DNA sequencing using an A. L. F. DNA sequencer (Pharmacia LKB Biotechnology) and an AutoRead sequencing kit (Pharmacia LKB Biotechnology).

truncated ones. They were also flanked by a perfect direct repeat of the three-base-pair sequence and the ten-base-pair sequence, respectively (Fig. 3). These structural characteristics suggest that this repetitive sequence belongs to SINE (short interspersed repetitive element) which is thought to be one of retroposons (Rogers, 1985; Weiner *et al.*, 1986). Many SINEs are transcribed by RNA polymerase III and supposed to be amplified *via* cDNA intermediates (Van Arsdell *et al.*, 1981; Jagadeeswaran *et al.*, 1981). In the case of the TS family, a member of the family (T1 in Fig. 2) was also shown to be transcribed by the RNA polymerase III in HeLa cell extract (data not shown), and the initiation site of transcription was assigned to the G residue at position 1 in Fig. 3 (data not shown). This result also confirms the TS family as a member of SINEs.

TS family is at least made up of 50,000 copies in the tobacco genome
We determined copy number of the TS family in the tobacco genome by the quantitative dot-blot hybridization (Cullis *et al.*, 1984) using the synthetic oligonucleotide homologous to the TS family as a probe (Fig. 4). We used the DNA fragment which contains an almost entire portion of the TS sequence (from 3 to 191 of clone T3 in Fig. 3) as a copy number standard. As shown in Fig. 4C, the amount of radioactivity of tobacco genomic DNA was 9.80 PSL ng^{-1} whereas that of the standard was 1.95×10^2 PSL $fmol^{-1}$ (1 PSL mm^{-2} is 1.5×10^{-6} roentgen). Therefore, tobacco gemonic DNA must contain 5.02×10^{-2} fmol of TS sequences per ng DNA. Assuming that a tobacco cell contains 1.6×10^9 bp of DNA per haploid genome and thus 1.67×10^{-3} ng DNA per haploid genome (Meyerowitz & Pruitt, 1985), we estimated that 5.0×10^4 copies of TS family are present in the haploid genome. This, however, may be an underestimation of the copy number, because the TS family does not have identical sequences but has sequence diversities such as base substitutions, insertions, and deletions among its members. Therefore, there is at least one member of the TS family every 32 kb in the tobacco genome on the average.

As described above, in D-2T and B-4-3 transformants, T-DNAs were integrated into the DNA regions which were 1.7 kb and 2.2 kb away from the chromosomal locations of #2 and #1 of the TS family, respectively. These results, taken together with the observation that there are preferential integration areas on a genome to transposable elements (Machida & Machida, 1987; Tenzen & Ohtsubo, 1991; Chalker & Sandmeyer, 1992), suggest that T-DNA is not randomly integrated into plant chromosomes: it may prefer chromosomal regions near the TS family for its integration to other regions. To judge whether this suggestion is reasonable, we must analyse target sites and their adjacent regions in more T-DNA-transformed cell lines. It is interesting to know how close correlation there is between integration events and chromosomal locations of the TS family.

A

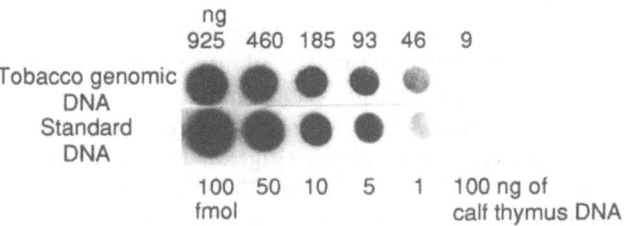

	ng					
	925	460	185	93	46	9
Tobacco genomic DNA						
Standard DNA						
	100	50	10	5	1	100 ng of
	fmol					calf thymus DNA

B

Genomic DNA (ng)	Amount of radioactivity (PSL)	Standard DNA (fmol)	Amount of radioactivity (PSL)
925	8,828	100	17,543
460	4,936	50	10,537
185	2,082	10	2,583
93	1,000	5	1,290
46	465	1	269
9	103		
100[1]	41		

[1]Calf thymus DNA

C

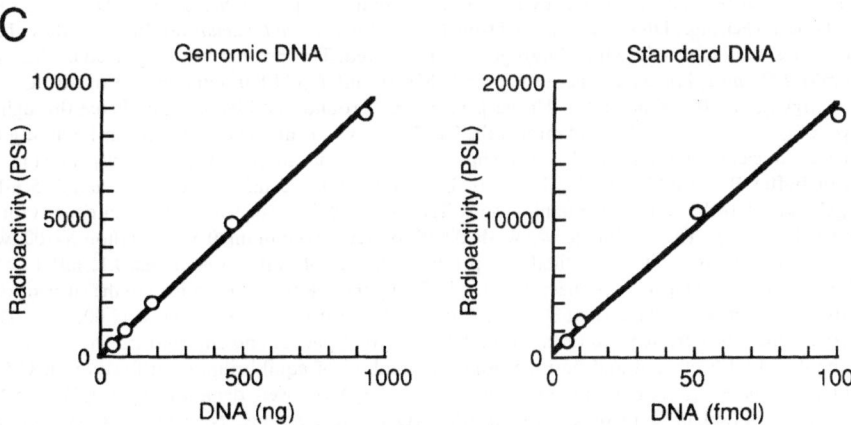

Figure 4. Estimation of the copy number of the TS family. (A) Dot blot hybridization of tobacco genomic DNA by using the ^{32}P-labeled 40-mer synthetic oligonucleotide homologous to a portion of the TS sequence (complementary sequence from position 176 to 215 in Fig. 3) as a probe. A series of different amount of tobacco genomic DNA was spotted on a nylon membrane filter (Gene Screen Plus, New England Nuclear). The DNA fragment containing the region from position 3 to 191 of clone T3 (Fig. 3) was similarly spotted as a copy number standard DNA. (B) Amount of radioactivity of each spots measured by Fujix Bio-imageanalyzer BAS2000. One PSL mm^{-2} is 1.5 ×

TS family homologous sequences are present in the genomes of other Solanaceae plants

Southern-blot analysis showed that the sequences that are homologous to the TS family are detected in Solanaceae plants such as *Datura stramonium, Hyosyamus niger, Lycopersicon esculentum, Petunia hybrida, Solanum tuberosum, Nicotiana glauca,* and *Nicotiana langsdorfii,* but not in *Arabidopsis thaliana* and monocotyledonus plants (Fig. 5). In the rice genome, the SINE family designated as p-SINE *1* was found (Mochizuki *et al.,* 1992), but it does not have significant sequences similarity to a tRNA and to the TS family. We found members of the TS family in plant genes or their adjacent regions which have been published: in the promoter adjacent region of the gene for pathogenesis-related 1c protein (PR1c) of tobacco (Ohshima *et al.,* 1990), in that of the auxin-inducible gene GNT35 of tobacco (Van der Zaal *et al.,* 1991) and in the second intron of the gene for ATP-dependent protease CD4A of tomato (Gottesman *et al.,* 1990). These members are 5′ truncated forms like the #1 and #2 sequences shown in Fig. 3 (data not shown). Thus, truncated forms as well as the full length of the TS family scatter all over the genome of Solanaceae plants and may generate structural diversities of their genomes.

tRNA-related region of TS family is similar to vertebrate tRNA^{Lys}

As described in the previous section, the TS family contains the tRNA-related sequence. Among the tRNA species of angiospermosida that have been

---

10^{-6} roentgen. (C) The relationship between the radioactivity and the amount of genomic DNA (left) and that between the radioactivity and the amount of copy number standard DNA (right).

Tobacco genomic DNA was prepared from leaves of *Nicotiana tabacum* var. Bright Yellow. Ten g of leaves were soaked into liquid nitrogen and powdered. This powder was suspended in 40 ml of 25 mM 2-(N-morpholine) ethanesulfonic acid (MES) buffer (pH 6.0) containing 1 mM $CaCl_2$, 5 mM $MgCl_2$, and 0.3 M mannitol. The suspension was ground in a blender and filtered through 4 layer of gauze and then 2 layer of miracloth. The filtrate was centrifuged for 5 min at 1,500 × *g* to separate nuclei from most cytoplasmic components. The resulting precipitate was suspended in 30 ml of buffer B (10 mM Tris-HCl buffer (pH 8.0) containing 5 mM β-mercaptoethanol, 5 mM $MgCl_2$, and 1.14 M sucrose) containing 0.5 % Triton X-100. The suspension was centrifuged for 5 min at 1,500 × *g*. The precipitate was washed with buffer B containing 0.5 % of Triton X-100 two more times. The precipitate was finally suspended in 0.9 ml of buffer B, and then 0.15 ml of 20 % SDS, 0.18 ml of 10 mg ml^{-1} pronase, and 15 μl of 5 mg ml^{-1} RNase were added to the suspension. This suspension was incubated at 37°C for 4 hr and then centrifuged for 5 min at 12,000 × *g*. The supernatant was extracted two times with a 1:1 mixture of phenol and chloroform. The DNA was precipitated with ethanol and purified using the method of equilibrium centrifugation in CsCl-ethidium bromide gradient (Sambrook *et al.,* 1989). The DNAs were denatured by NaOH (0.25 N) and then spotted on a nylon membrane filter (Gene Screen Plus, New England Nuclear) as recommended by the supplier. This filter was hybridized with the ^{32}P-labeled 40-mer synthetic oligonucleotide (5′-AAC ATA CCC AGT ATA ATC CCA CAA GTG GGG TGT GGG GAG G-3′) which is similar to a portion of the TS sequence (complementary sequence from position 174 to 213 in Fig. 3) in 6 × SSC, 1% SDS at 55°C and then washed with 2 × SSC, 1% SDS at 55°C. The probe DNAs were labeled with [γ−^{32}P]ATP (222 TBq mmol^{-1}, Amersham) using T4 polynucleotide kinase. The synthetic oligonucleotide probe was chemically synthesized using a Gene Assembler Plus (Pharmacia LKB Biotechnology).

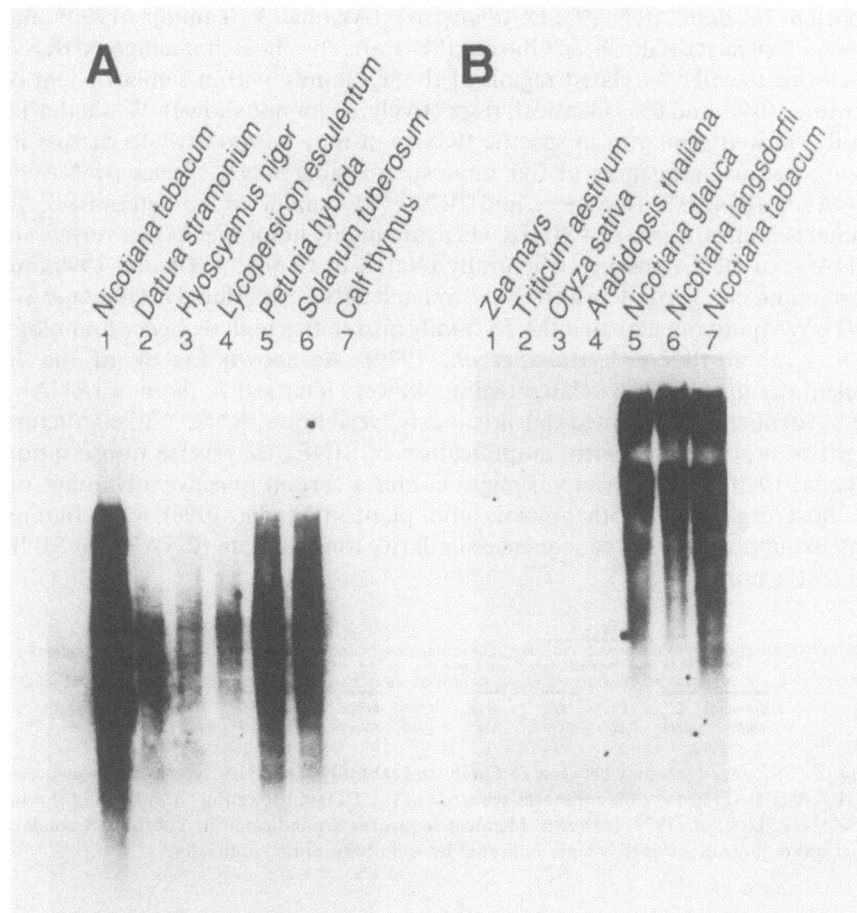

Figure 5. Southern hybridization of genomic DNAs from various plants with DNA fragments containing the TS family. (A) *Eco*RI-digested genomic DNA (5 μg) of *Nicotiana tabacum* (lane 1), *Datura stramonium* (lane 2), *Hyoscyamus niger* (lane 3), *Lycopersicon esculentum* (lane 4), *Petunia hybrida* (lane 5), *Solanum tuberosum* (lane 6), and calf thymus (lane 7) were hybridized with *Dra*I-*Eco*RI fragment shown in Lane #2 in Fig. 3. (B) *Eco*RI-digested genomic DNA (10 μg) of *Zea mays* (lane 1), *Triticum aestivum* (lane 2), *Oryza sativa* (lane 3), *Arabidopsis thaliana* (lane 4), *Nicotiana glauca* (lane 5), *Nicotiana langsdorfii* (lane 6), *Nicotiana tabacum* (lane 7) were hybridized with the *Dra*I-*Eco*RI fragment of clone #2 (Fig. 2).

Plant total genomic DNAs were prepared by the method previously described (Matsumoto *et al.*, 1990). These DNAs were digested with restriction endonuclease *Eco*RI and subjected to electrophoresis on a 0.7% agarose gel containing Tris-acetate/EDTA buffer (40 mM Tris-acetate (pH 8.0), 1 mM EDTA). DNA was blotted on a nylon membrane filter (Gene Screen Plus, New England Nuclear) using 1.5 M NaCl, 0.5 M NaOH, and filter was washed with 2 × SSC. The filter was incubated at 65 °C for 1 hr in 6 × SSC, 1% SDS. The treated filter was hybridized with ^{32}P-labeled DNA fragment (Feinberg and Vogelstein, 1983) in 6 × SSC, 1% SDS at 65 °C and then washed with 2 × SSC, 1% SDS at 65 °C. The probe DNA was labeled with [α–^{32}P]dCTP (222 TBq mmol^{-1}, Amersham) using the Random primed DNA labeling kit (Boehringer Mannheim).

reported to date, tRNA^{Phe} (*Arabidopsis*) (Akama & Tanifuji, 1990) and tRNA^{Trp} (wheat) (Ghosh & Ghosh, 1984) are the most homologous tRNA species to the tRNA-related region of the TS family with a similar extent of identities (64% and 63% identical, respectively. Data not shown). While the TS family showed similarity to specific tRNAs, it may be too early to discuss its origin, because sequences of five angiospermosida tRNA species (tRNA^{Asp}, tRNA^{Lys}, tRNA^{Glu}, tRNA^{Leu}, and tRNA^{Thr}) remain to be determined. In animal genomes, a group of SINEs which are highly homologous to a vertebrate tRNA^{Lys} or tRNA species structurally related to tRNA^{Lys} (Okada, 1990) are present and categorized as a tRNA^{Lys}-related SINE superfamily (Okada *et al.*, 1991). We point out here that the TS family also shares high sequence homology with a rabbit tRNA^{Lys} (Raba *et al.*, 1979). As shown Fig. 6, of the 76 nucleotides in the tRNA-related region, 48 were identical to those of tRNA^{Lys} (63 % identical). As pointed out previously, vertebrate tRNA^{Lys} like structure might be crucial for efficient amplification of SINEs *via* reverse transcription (Okada, 1990). Alternatively, it might confer a certain selective advantage on the host organisms, both animals and plants (Okada, 1990). Our finding suggests importance of the sequence similarity to vertebrate tRNA^{Lys} for SINE in plant genome.

	Box A							Box B			

TS family 5'-GAGGGGUU-GCUCUGAU-GGUA-AGCAACCUCC-ACU-UCCAA-CCAAGAGGUUGUGAGUUCGAGUCACCCCAAGAGCAAG-3'
        * ** * **** * * **** ****   *   *** * ** ** ** *** *** ******* * * * **
tRNA^{Lys} 5'-GCCCGGCUAGCUCAG-UCGGUAGAGCA---UGGGACUCUU-AAUCCCAG-GGUCGUGGGUUCGAGCCCCACGUUGGGCGCCA-3'

| aminoacyl | D | D | D | antic. | antic. | antic. | TΨ | TΨ | TΨ | aminoacyl |
| stem | stem | loop | stem | stem | loop | stem | stem | loop | stem | stem |

Figure 6. Sequence homology between TS family and rabbit lysine tRNA. Nucleotide sequence of the tRNA-related region of the consensus sequence of the TS family (see Fig. 3) and that of Rabbit tRNA^{Lys} (Raba *et al.*, 1979) is shown. Identical sequences are indicated by asterisks. Secondary structural organizations of tRNA are indicated by solid bars. antic., anticodon.

Acknowledgements

We are grateful to Dr. A. Oka for the gift of pTOK9 and mini Ri plasmid. This research was supported in part by a Grant-in-Aid for General Science and Culture of Japan and a grant from the Ministry of Agriculture, Forestry, and Fisheries of Japan.

References

Akama K & Tanifuji S (1990) Sequence analysis of three tRNA^{Phe} nuclear genes and a mutated gene, and one genes for tRNA^{Ala} from *Arabidopsis thaliana*. Plant Mol. Biol. 15: 337–346.

Albright LM, Yanofsky MF, Leroux B, Ma D & Nester EW (1987) Processing of the T-DNA of *Agrobacterium tumefaciens* generates border nicks and linear, single-stranded T-DNA. J. Bacteriol. 169: 1046–1055.

An G (1985) High efficiency transformation of cultured tobacco cells. Plant Physiol. 79: 568–570.

Chyi Y-S, Jorgensen RA, Goldstein D, Tanksley SD & Loaiza-Figueroa F (1986) Locations and stability of *Agrobacterium*-mediated T-DNA insertions in *Lycopersicon* genome. Mol. Gen. Genet. 204: 64–69.

Chalker DL & Sandmeyer SB (1992) Ty3 integrates within the region of RNA polymerase III transcription initiation. Genes & Dev. 6: 117–128.

Christie PJ, Ward JE, Winans SC & Nester EW (1988) The *Agrobacterium tumefaciens virE2* gene product is a single-stranded-DNA-binding protein that associates with T-DNA. J. Bacteriol. 170: 2659–2667.

Christie PJ, Ward Jr. JE, Gordon MP & Nester EW (1989) A gene required for transfer of T-DNA to plants encodes an ATPase with autophosphorylating activity. Proc. Natl. Acad. Sci. USA 86: 9677–9681.

Citovsky V, De Vos G & Zambryski P (1988) Single-stranded DNA binding protein encoded by the *virE* locus of *Agrobacterium tumefaciens*. Science 240: 501–504.

Citovsky V, Zupan J, Warnick D & Zambryski P (1992) Nuclear localization of *Agrobacterium* VirE2 protein in plant cells. Science 256: 1802–1805.

Cullis CA, Rivin CJ & Walbot V (1984) A rapid procedure for the determination of the copy number of repetitive sequences in eukaryotic genomes. Plant Mol. Biol. Reptr. Vol 2 (4): 24–31.

Das A (1988) *Agrobacterium tumefaciens virE* operon encodes a single-stranded DNA-binding protein. Proc. Natl. Acad. Sci. USA 85: 2909–2913.

Dürrenberger F, Crameri A, Hohn B & Koukolíková-Nicola Z (1989) Covalently bound VirD2 protein of *Agrobacterium tumefaciens* protects the T-DNA from exonucleolytic degradation. Proc. Natl. Acad. Sci. USA 86: 9154–9158.

Feinberg AP & Vogelstein B (1983) A technique for radiolabeling DNA restriction endonuclease fragments to high specific activity. Anal. Biochem. 132: 6–13.

Frischauf A-M, Lehrach H, Poustka A & Murray N (1983) Lambda replacement vectors carrying polylinker sequences. J. Mol. Biol. 170: 827–842.

Galli G, Hofstetter H & Birnstiel ML (1981) Two conserved sequence blocks within eukaryotic tRNA genes are major promoter elements. Nature (London) 294: 626–631.

Gheysen G, Van Montagu M & Zambryski P (1987) Integration of *Agrobacterium tumefaciens* transfer DNA (T-DNA) involves rearrangements of target plant DNA sequences. Proc. Natl. Acad. Sci. USA 84: 6169–6173.

Gheysen G, Villarroel R & Van Montagu M (1991) Illegitimate recombination in plants: a model for T-DNA integration. Genes & Dev. 5: 287–297.

Ghosh K & Ghosh HP (1984) Structure and function of tryptophan tRNA from wheat germ. Nucleic Acids Res. 12: 4997–5003.

Gietl C, Koukolíková-Nicola Z & Hohn B (1987) Mobilization of T-DNA from *Agrobacterium* to plant cells involves a protein that binds single-stranded DNA. Proc. Natl. Acad. Sci. USA 84: 9006–9010.

Gottesman S, Squires C, Pichersky E, Carrington M, Hobbs M, Mattick JS, Dalrymple B, Kuramitsu H, Shiroza T, Foster T , Clark WP, Ross B, Squires CL & Maurizi MR (1990) Conservation of the regulatory subunit for the Clp ATP-dependent protease in prokaryotes and eukaryotes. Proc. Natl. Acad. Sci. USA 87: 3513–3517.

Herman L, Jacobs A, Van Montagu M & Depicker A (1990) Plant chromosome/marker gene fusion assay for study of normal and truncated T-DNA integration events. Mol. Gen. Genet. 224: 248–256.

Herrera-Estrella A, Chen Z-m, Van Montagu M & Wang K (1988) VirD proteins of *Agrobacterium tumefaciens* are required for the formation of a covalent DNA-protein complex at the 5′ terminus of T-strand molecules. EMBO J. 7: 4055–4062.

Herrera-Estrella A, Van Montagu M & Wang K (1990) A bacterial peptide acting as a plant nuclear targeting signal: The amino-terminal portion of *Agrobacterium* VirD2 protein directs a β-galactosidase fusion protein into tobacco nuclei. Proc. Natl. Acad. Sci. USA 87: 9534–9537.

Hohn B, Koukolíková-Nicola Z, Bakkeren G & Grimsley N (1989) *Agrobacterium*-mediated gene transfer to monocots and dicots. Genome 31: 987–993.

246

Hood EE, Helmer GL, Fraley RT & Chilton M-D (1986) The hypervirulence of *Agrobacterium tumefaciens* A281 is encoded in a region of pTiBo542 outside of T-DNA. J. Bacteriol. 168: 1291–1301.

Howard EA, Zupan JR, Citovsky V & Zambryski PC (1992) The VirD2 protein of *A. tumefaciens* contains a C-terminal bipartite nuclear localization signal: implications for nuclear uptake of DNA in plant cells. Cell 68: 109–118.

Jagadeeswaran P, Forget BG & Weissman SM (1981) Short interspersed repetitive DNA elements in eukaryotes: transposable DNA elements generated by reverse transcription of RNA pol III transcripts? Cell 26: 141–142.

Janssen B-J & Gardner RC (1989) Localized transient expression of GUS in leaf discs following cocultivation with *Agrobacterium*. Plant Mol. Biol. 14: 61–72.

Jefferson RA, Burgess SM & Hirsh D (1986) β-Glucuronidase from *Escherichia coli* as a gene-fusion marker. Proc. Natl. Acad. Sci. USA 83: 8447–8451.

Jin S, Komari T, Gordon MP & Nester EW (1987) Genes responsible for the supervirulence phenotype of *Agrobacterium tumefaciens* A281. J. Bacteriol. 169: 4417–4425.

Kado CI (1991) Molecular mechanisms of crown gall tumorigenesis. Crit. Revs. Plant Sci. 10: 1–32.

Koncz C, Martini N, Mayerhofer R, Koncz-Kalman Z, Körber H, Redei GP & Schell J (1989) High-frequency T-DNA-mediated gene tagging in plants. Proc. Natl. Acad. Sci. USA 86: 8467–8471.

Linsmaier EM & Skoog F (1965) Organic growth factor requirements of tobacco tissue cultures. Physiol. Plant 18: 100–127.

Machida C & Machida Y (1987) Base substitutions in transposable element IS*1* cause DNA duplication of variable length at the target site for plamid co-integration. EMBO J. 6: 1799–1803.

Matsumoto S, Ito Y, Hosoi T, Takahashi Y & Machida Y (1990) Integration of *Agrobacterium* T-DNA into a tobacco chromosome: possible involvement of DNA homology between T-DNA and plant DNA. Mol. Gen. Genet. 224: 309–316.

Mayerhofer R, Koncz-Kalman Z, Nawrath C, Bakkeren G, Crameri A, Angelis K, Redei GP, Schell J, Hohn B & Koncz C (1991) T-DNA integration: a model of illegitimate recombination in plants. EMBO J. 10: 697–704.

Melchers LS, Maroney MJ, den Dulk-Ras A, Thompson DV, van Vuuren HAJ, Schilperoort RA & Hooykaas PJJ (1990) Octopine and nopaline strains of *Agrobacterium tumefaciens* differ in virulence; molecular characterization of the *virF* locus. Plant Mol. Biol. 14: 249–259.

Meyerowitz EM & Pruitt RE (1985) *Arabidopsis thaliana* and plant molecular genetics. Science 229: 1214–1218.

Mochizuki K, Umeda M, Ohtsubo H & Ohtsubo E (1992) Characterization of a plant SINE, p-SINE*1*, in rice genomes. Japan J. Genet. 67: 155–166.

Nishiguchi R, Takanami M & Oka A (1987) Characterization and sequence determination of the replicator region in the hairy-root-inducing plasmid pRiA4b. Mol. Gen. Genet. 206:1–8.

Ohshima M, Harada N, Matsuoka M & Ohashi Y (1990) The nucleotide sequence of pathogenesis-related (PR) 1c protein gene of tobacco. Nucleic Acids Res. 18: 182.

Ohta S, Mita S, Hattori T & Nakamura K (1990) Construction and expression in tobacco of a β-glucuronidase (GUS) reporter gene containing an intron within the coding sequence. Plant Cell Physiol. 31: 805–813.

Okada N (1990) Transfer RNA-like structure of the human Alu family: implications of its generation mechanism and possible functions. J. Mol. Evol. 31: 500–510.

Okada N (1991a) SINEs: short interspersed repeated elements of the eukaryotic genome. Trends Ecol. Evol. 6: 358–361.

Okada N (1991b) SINEs. Curr. Opin. in Genet. & Dev. 1: 498–504.

Okada N, Aono M, Endoh H, Kido Y, Koishi R, Matsumoto K-I, Matsuo M, Murata S, Nagahashi S & Yamaki T (1991) Evolution of repetitive sequences. p. 175–186. *In:* Osawa S & Honjo T (eds) Evolution of Life. Springer-Verlag Tokyo.

Okamoto S, Toyoda-Yamamoto A, Ito K, Takebe I & Machida Y (1991) Localization and orientation of the VirD4 protein of *Agrobacterium tumefaciens* in the cell membrane. Mol. Gen. Genet. 228: 24–32.

Onouchi H, Yokoi K, Machida C, Matsuzaki H, Oshima Y, Matsuoka K, Nakamura K & Machida Y (1991) Operation of an efficient site-specific recombination system of *Zygosaccharomyces rouxii* in tobacco cells. Nucleic Acids Res. 19: 6373–6378.

Petit A, Tempe J, Kerr A, Holsters M, Van Montagu M & Schell J (1978) Substrate induction of conjugative activity of *Agrobacterium tumefaciens* Ti plasmids. Nature (London) 271: 570–571.

Raba M, Limburg K, Burghagen M, Katze JR, Simsek M, Heckman JE, Rajbhandary UL & Gross HJ (1979) Nucleotide sequence of three isoaccepting lysine tRNAs from rabbit liver and SV40-transformed mouse fibroblasts. Eur. J. Biochem. 97: 305–318.

Rogers JH (1985) The origin and evolution of retroposon. Int. Rev. Cytol. 93: 187–279.

Rogowsky PM, Close TJ, Chimera JA, Shaw JJ & Kado CI (1987) Regulation of the *vir* genes of *Agrobacterium tumefaciens* plasmid pTiC58. J. Bacteriol. 169: 5101–5112.

Sambrook J, Fritsch EF & Maniatis T (1989) Molecular cloning: A laboratory manual, second edition. Cold Spring Harbor Laboratory, Cold Spring Harbor, N.Y. pp 1.42–1.43.

Sanger F, Nicklen S & Coulson AR (1977) DNA sequencing with chain-terminating inhibitors. Proc. Natl. Acad. Sci. USA 74: 5463–5467.

Sciaky D, Montoya AL & Chilton MD (1978) Fingerprints of *Agrobacterium* Ti plasmids. Plasmid 1: 238–253.

Singer MF (1982) SINEs and LINEs: highly repeated short and long interspersed sequences in mammalian genomes. Cell 28: 433–434.

Southern EM (1985) Detection of specific sequences among DNA fragments separated by gel electrophoresis. J. Mol. Biol. 98: 503–517.

Stachel SE & Nester EW (1986) The genetic and transcriptional organization of the *vir* region of the A6 Ti plasmid of *Agrobacterium tumefaciens*. EMBO J. 5: 1445–1454.

Stachel SE & Zambryski PC (1986) *virA* and *virG* control the plant-induced activation of the T-DNA transfer process of A. tumefaciens. Cell 46: 325–333.

Stachel SE, Nester EW & Zambryski PC (1986a) A plant cell factor induces *Agrobacterium tumefaciens vir* gene expression. Proc. Natl. Acad. Sci. USA 83: 379–383.

Stachel SE, Timmerman B & Zambryski P (1986b) Generation of single-stranded T-DNA molecules during the initial stages of T-DNA transfer from *Agrobacterium tumefaciens* to plant cells. Nature (London) 322: 706–712.

Stachel SE, Timmerman B & Zambryski P (1987) Activation of *Agrobacterium tumefaciens vir* gene expression generates multiple single-stranded T-strand molecules from the pTiA6 T-region: requirement for 5′ *virD* gene products. EMBO J. 6: 857–863.

Tenzen T & Ohtsubo E (1991) Preferential transposition of an IS*630*-associated composite transposon to TA in the 5′-CTAG-3′ sequence. J. Bacteriol. 173: 6207–6212.

Van Arsdell SW, Denison RA, Bernstein LB, Weiner AM, Manser T & Gesteland RF (1981) Direct repeats flank three small nuclear RNA pseudogenes in the human genome. Cell 26: 11–17.

Van der Zaal EJ, Droog FNJ, Boot CJM, Hensgens LAM, Hoge JHC, Schilperoort RA & Libbenga KR (1991) Promoters of auxin-induced genes from tobacco can lead to auxin-inducible and root tip-specific expression. Plant Mol. Biol. 16: 983–998.

Vancanneyt G, Schmidt R, O'Connor-Sanchez & A, Willmitzer L & Rocha-Sosa M (1990) Construction of an intron-containing marker gene: Splicing of the intron in transgenic plants and its use in monitoring early events in *Agrobacterium*-mediated plant transformation. Mol. Gen. Genet. 220: 245–250.

Veluthambi K, Jayaswal RK & Gelvin SB (1987) Virulence genes *A, G* and *D* mediate the double-stranded border cleavage of T-DNA from the *Agrobacterium* Ti plasmid. Proc. Natl. Acad. Sci. USA 84: 1881–1885.

Wallroth M, Gerats AGM, Rogers SG, Fraley RT & Horsch RB (1986) Chromosomal localization of foreign genes in *Petunia hybrida*. Mol. Gen. Genet. 202: 6–15.

Ward ER & Barnes WM (1988) VirD2 protein of *Agrobacterium tumefaciens* very tightly linked to the 5′ end of T-strand DNA. Science 242: 927–930.

Ward Jr. JE, Dale EM, Nester EW & Binns AN (1990) Identification of a VirB10 protein aggregate in the inner membrane of *Agrobacterium tumefaciens*. J. Bacteriol. 172: 5200–5210.

Weiner AM, Deininger PL & Efstratiadis A (1986) Nonviral retroposons: genes, pseudogenes, and transposable elements generated by the reverse flow of genetic information. Ann. Rev. Biochem. 55: 631–661.

Yamamoto A, Iwahashi M, Yanofsky MF, Nester EW, Takebe I & Machida Y (1987) The promoter proximal region in the *virD* locus of *Agrobacterium tumefaciens* is necessary for the plant-inducible circularization of T-DNA. Mol. Gen. Genet. 206: 174–177.

Yanisch-Perron C, Vieira J & Messing J (1985) Improved M13 phage cloning vectors and host strains: nucleotide sequences of the M13mp18 and pUC19 vectors. Gene 33: 103–119.

Yanofsky MF, Porter SG, Young C, Albright LM, Gordon MP & Nester EW (1986) The *virD* operon of *Agrobacterium tumefaciens* encodes a site-specific endonuclease. Cell 47: 471–477.

Young C & Nester EW (1988) Association of the VirD2 protein with the 5' end of T strands in *Agrobacterium tumefaciens*. J. Bacteriol. 170: 3367–3374.

Zambryski P (1988) Basic processes underlying *Agrobacterium*-mediated DNA transfer to plant cells. Annu. Rev. Genet. 22:1–30.

Zambryski PC (1992) Chronicles from the *Agrobacterium*-plant cell DNA transfer story. Ann. Rev. Plant Mol. Biol. 43: 465–490.

18. Molecular pathogenesis of viridans streptococcal endocarditis

CINDY MUNRO and FRANCIS L. MACRINA

Abstract. The viridans streptococci comprise a heterologous group of organisms which are part of the normal oral flora of humans. These organisms are the most common etiology of native valve endocarditis in humans. Binding to platelets, binding to fibrin, exopolysaccharide production, and binding to fibronectin have been identified as factors associated with virulence of viridans streptococci. Allelic exchange mutagenesis has been used to construct isogenic strains of *S. mutans* which differ from wild type only in defined genes; these mutants have been tested in model systems of human infectious diseases. Mutants unable to synthesize exopolymers from sucrose have been found to be less virulent in a rat endocarditis model. Our data suggest that the presence of sucrose-derived polymers plays a role in fibrin binding and in reducing phagocytic killing.

Abbreviations: PTS, phosphotransterase system

Introduction

Viridans streptococci are part of the normal oral flora of humans, and are important human pathogens. For example, *Streptococcus mutans* is the etiologic agent of smooth surface dental caries, the most prevalent human infectious disease. Streptococci of the viridans group are also the most common etiology of native valve endocarditis in humans, accounting for 45 to 80% of cases (Bayliss *et al.*, 1983; Dall *et al.*, 1987; Freeman *et al.*, 1990; van der Meer *et al.*, 1991).

Bacterial endocarditis, the most common cardiac infection (Bayer & Norman, 1990), is life threatening. This disease was uniformly fatal prior to the development of antibiotic therapy. Despite recommendations for prophylaxis prior to scheduled procedures which place individuals at risk for endocarditis (Dajani *et al.*, 1990), the incidence of endocarditis remains essentially unchanged and mortality is estimated at 10–20%. In the United States, an estimated 15,000 to 30,000 new cases occur each year.

The development of effective strategies to combat viridans streptococcal infections is important to maintenance of optimal human health. Reduction of oral diseases could potentially yield a bonus effect; healthy oral tissues are less likely to predispose individuals to bacteremias which precede the development of endocarditis. Several strategies to reduce the virulence of viridans streptococci, and in particular *S. mutans*, are being pursued. Any effective

C.I. Kado and J.H. Crosa (eds.), Molecular Mechanisms of Bacterial Virulence, 249–265.
© 1994 *Kluwer Academic Publishers. Printed in the Netherlands.*

strategy, however, must be based upon a thorough understanding of the genetic determinants which contribute to the organism's virulence.

The viridans streptococci comprise a heterologous group of organisms with several characteristics common to most members (Coykendall, 1989; Facklam, 1977). The groups of viridans streptococci most commonly associated with human infections are summarized in Table 1. Viridans streptococci are generally susceptible to penicillin, and do not demonstrate β-hemolysis on blood agar plates. Viridans streptococci usually do not have carbohydrate group antigens which can be distinguished by Lancefield typing. These species are able to ferment a wide variety of carbohydrates and many species produce exopolysaccharides from the hydrolysis of sucrose. The mutans streptococci are distinct from other viridans streptococci in that only the mutans streptococci ferment mannitol and sorbitol and produce water-insoluble glucans from sucrose.

Table 1. Viridans streptococci.

Group	Species
"*S. mutans* group"	*S. cricetus*
	S. downei
	S. ferus
	S. macace
	S. mutans
	S. rattus
	S. sobrinus
"*S. salivarius* group"	*S. intestinalis*
	S. salivarius
	S. vestibularis
"*S. sanguis* group"	*S. gordonii*
	S. sanguis group H
	S. sanguis group W
"*S. mitis* group"	*S. mitis/ S. mitior*
	S. oralis
	S. sanguis biotype II
"*S. anginosus* group"	*S. anginosus*
	S. intermedius
	S. constellatus
	S. milleri

Sucrose metabolism in *Streptococcus mutans*

S. mutans uses host dietary sucrose to supply itself with glucose and fructose for cellular energy needs as well as for a substrate for extracellular polymers (Hamada & Slade, 1980). *S. mutans* V403, the strain used in our laboratory, produces several enzymes involved in sucrose metabolism. Approximately 95% of sucrose available to *S. mutans* is transported into the cell and used as a substrate for glycolysis. Two determinants, *scrA* and *scrB*, are important in sucrose-specific transport via a phosphoenolpyruvate-dependent phosphotransferase system (PTS) (Macrina *et al.*, 1991). *scrA* encodes a sucrose-specific enzyme II component of the PTS which is responsible for formation of the transmembrane channel and phosphorylation of the sucrose. The *scrB* product is a sucrose-6-phosphate hydrolase; it cleaves phosphorylated sucrose which has entered the cell into metabolically useful forms of carbon. Sucrose may also enter the cell via the PTS which transports trehalose although this is believed to be a minor pathway (Jacobson *et al.*, 1989).

Glucosyltransferases (Gtf enzymes, EC 2.4.1.5) and fructosyltransferase (Ftf enzyme, EC 2.4.1.10) are important in formation of sucrose-dependent exopolysaccharides. In *S. mutans*, the glucosyltransferase B (*gtfB*) gene directs production of an enzyme which converts sucrose to an insoluble glucan polymer and fructose (Shiroza *et al.*, 1987). The product of glucosyltransferase C (*gtfC*) converts sucrose to insoluble glucan, soluble glucan, and fructose (Hanada & Kuramitsu, 1988). The glucosyltransferase D (*gtfD*) gene product converts sucrose to a soluble glucan polymer and fructose (Hanada & Kuramitsu, 1989). Fructosyltransferase produces fructan polymer and glucose from sucrose (Shiroza & Kuramitsu, 1988). All four genes have been sequenced (Honda *et al.*, 1990; Shiroza & Kuramitsu, 1988; Shiroza *et al.*, 1987; Ueda *et al.*, 1988), and the three glucosyltransferase genes exhibit high levels of nucleotide and amino acid sequence homology. The carboxyl terminus of glucosyltransferases and related enzymes has several direct repeats each consisting of 65 amino acids. In work with GtfB, Kato and Kuramitsu (1990) demonstrated that deletion of these repeating units resulted in abolition of both sucrase activity and glucosyltransferase activity; if one repeat unit remained partially intact, insignificant glucosyltransferase activity was observed, but substantial sucrase activity remained. Mooser and colleagues (1991) defined the active site by peptide fragment mapping. An aspartic acid residue at position 451 is important in sucrose binding of GtfB. When this residue was changed to threonine, no enzymatic activity was detected (Kato & Kuramitsu, 1991). *gtfB* and *gtfC* are tightly linked, but neither *gtfD* nor *ftf* are linked to the other known genes involved in exopolysaccharide production (Perry & Kuramitsu, 1990).

The role of exopolysaccharides in development of tooth decay has been the subject of intensive research, and many studies have implicated insoluble glucan polymers of *S. mutans* in tenacious attachment to and accumulation on the tooth surface (de Stoppelaar *et al.*, 1971; Hamada & Slade, 1980; Koga *et al.*, 1986; Munro *et al.*, 1991; Schroeder *et al.*, 1989; Tanzer *et al.*, 1974). In addition, some studies have suggested that exopolysaccharides are important factors in the development of streptococcal endocarditis (Pulliam *et al.*, 1985; Ramirez-Ronda, 1978; Ramirez-Ronda, 1980; Scheld *et al.*, 1978).

Allelic exchange mutagenesis

Many streptococcal genes are readily expressed in *E. coli*, although few streptococcal proteins are exported into the periplasm. Expression of cloned genes provides the means to overproduce and study the gene product. It is possible to create a defective copy of the gene in question using standard recombinant DNA methods in *E. coli*, by insertional inactivation of the cloned gene using a directly selectable marker such as an antibiotic resistance gene (Zealey *et al.*, 1990). This defective gene then can be introduced into the streptococcal cell by transformation or electroporation. A chromosomal double recombinational event results in replacement of the wild-type gene by the defective copy which is easily detected by the presence of the accompanying antibiotic resistance gene. Recombination into the chromosome can be definitively evaluated using Southern blot hybridization analysis to confirm the predicted recombinational products. Although the genetic lesion is well defined, it is possible for untoward effects of the insertion (such as polarity on downstream genes) to occur. Therefore, verification of insertion effects is essential. Properly constructed and verified allelic exchange mutants provide powerful tools for the analysis of virulence factors in bacteria. Such mutants can be evaluated *in vitro* to assess biochemical and physiological implications of the loss of the gene in question. The importance of specific genes and their gene products as colonization and virulence factors can be tested in appropriate animal models. Allelic exchange mutants can provide information that can lead to the design of rational intervention strategies, including the identification of potential immunogens for antibacterial vaccines. Indeed, the method of allelic exchange itself can be use to create strains of reduced virulence which, in turn, can be considered for use as whole cell vaccine candidates, or which could be the basis of replacement therapy, in which the virulent normal flora would be supplanted by nonvirulent mutants of the same strain (Zealey *et al.*, 1990).

Development of allelic exchange mutants in *S. mutans*

Isogenic strains of *S. mutans* V403, which carry single and multiple mutations of the *gtfB*, *gtfC*, *gtfD*, and *ftf* genes, have been constructed by allelic exchange in our laboratory (Munro *et al.*, 1991). With this panel of strains, it was possible to examine the importance of water soluble glucan, water insoluble glucan, and fructan production in virulence while controlling for the effects of strain and species variability. Genetic and biochemical characterization and assays of glucosyltransferase and fructosyltransferase activity have been performed. No obvious changes in phenotype of mutant strains were detected other than the expected growth on antibiotics and colonial morphology changes on mitis salivarius (MS) agar. MS agar contains sucrose allowing production of polymers which impart a unique colonial morphology. Protein preparations from those strains in which both *gtfB* and *gtfC* were inactivated lost the ability to produce water-insoluble glucan in the *in vitro* assay of glucosyltransferase activity. In strains where all three glucosyltransferase enzymes were insertionally inactivated, ability to make water-soluble and water-insoluble glucan was lost. Strains which had an insertional inactivation of *ftf* lost the ability to form fructan in the *in vitro* fructosyltransferase activity assay. When tested for the ability to liberate reducing sugars from sucrose, V1996, in which all four genes of exopolysaccharide synthesis had been insertionally inactivated, demonstrated a complete loss of the ability to hydrolyze sucrose. These phenotypes were consistent with the appearance of the extracellular protein preparations of the strains on SDS-PAGE treated with PAS, and remained stable following passage in the rat.

Sequence data indicate that inverted repeat sequences characteristic of rho-independent termination sequences are located 200 bp downstream from the *gtfD* stop codon (Honda *et al.*, 1990), 78 bp downstream from the *ftf* stop codon (Shiroza & Kuramitsu, 1988), and 621 bp downstream (following a 185 bp open reading frame) from the *gtfC* stop codon (Ueda *et al.*, 1988). Presence of these sequences makes alteration of downstream genes in these mutants unlikely.

The ability of these organisms to cause caries in the gnotobiotic rat model has been described (Munro *et al.*, 1991). Mutant strains were less virulent than the wild type in almost every location and level of involvement. Inactivation of either *gtfB* and *gtfC* or *ftf* dramatically reduced virulence; the subsequent inactivation of *gtfD* did not enhance the effect of reduced cariogenicity.

The technique of allelic exchange has been used by several groups to construct isogenic mutants which have been comparatively studied *in vitro* and *in vivo*. Table 2 is a compilation of some of the results obtained when *in vivo* virulence testing has been used to evaluate the cariogenicity of mutant strains.

The virulence of the wild type strain and an isogenic mutant which produced no exopolysaccharide (V1996) were compared in a rodent endocarditis model and in an *in vitro* phagocytosis assay (Munro & Macrina, 1993). In the rat model of endocarditis, animals inoculated with the wild type *S. mutans* V403 developed endocarditis more frequently than animals inoculated with the polymer-deficient mutant (58% versus 12%, $p < 0.01$). Interestingly, if the wild type organism was grown without sucrose added, it produced significantly fewer cases of endocarditis than when grown in sucrose. In an *in vitro* phagocytosis assay using ^{3}H-labeled bacteria, both sucrose-grown strains were found to be associated with human granulocytes. However, colony counts of the cell pellets following lysis of granulocytes indicated that a greater proportion of wild type organisms than of mutant organisms were alive at the conclusion of assays. The ability of the wild type and polymer-defective mutants to bind to fibrin-coated

Table 2. Properties of isogenic mutants constructed by allelic exchange.

Mutated Gene	Product/Activity	Evaluation	Virulence	Reference
gtfB/C	insoluble and soluble glucan	rodent caries model	reduced	Munro et al. (1991)
ftf	fructan	rodent caries model	reduced	Schroeder et al. (1989)
gtfA	sucrose phosphorylase	rodent caries model	wild type	Barletta et al. (1988)
scrA/B	sucrose transport	rodent caries model	wild type	Macrina et al. (1991)
spaP1	surface protein antigen	rodent caries model	wild type[A]	Bowen et al. (1991)
glgR	regulation of intracellular glycogen synthesis	rodent caries model	reduced	Spatafora Harris et al. (1992)
fruA	extracellular fructanase	rodent caries model	wild type	Wexler et al. (1992)

[A] Although smooth surface caries scores were not reduced in an isogenic *spaP1*-deficient mutant relative to wild type, the mutant did show a significant reduction in salivary induced agglutination and adherence to saliva-coated hydroxylapatite (Bleiweis et al., 1990).

plates was examined. Colony counts recovered from fibrin plates incubated with the mutant were lower than those incubated with sucrose-grown wild type *S. mutans*. These experiments indicated that exopolysaccharides produced by *Streptococcus mutans* contribute to its infectivity in endocarditis.

Pathophysiology of streptococcal endocarditis

Streptococcal endocarditis is generally associated with native valves of the left side of the heart (Bayer & Norman, 1990; Freeman & Hall, 1989). Viridans streptococci may escape from their normal habitat and enter the bloodstream during any number of seemingly innocuous procedures. Induction of bacteremia following dental extractions has been demonstrated experimentally in rats (Moreillon *et al.*, 1988; Overholser *et al.*, 1987). Transient bacteremia has been demonstrated in humans following dental extractions, toothbrushing, and chewing (Fekete, 1990; Freeman & Hall, 1989). A recent history of dental manipulation is a common finding in patients with bacterial endocarditis (Bayliss, *et al.*, 1983). Although transient bacteremia does not generally result in disease in otherwise healthy persons, those with underlying cardiac disease and pre-existing valvular damage are particularly at risk. Damage to the valve results in formation of a sterile vegetation consisting largely of fibrin and platelets (Freedman, 1987). In experimental animal models, a sterile vegetation on a cardiac valve can serve as a nidus for infection. In human endocarditis, vegetations containing microorganisms are a hallmark of the disease.

In general, bacterial densities tend to be higher in left-sided endocarditis (Bayer & Norman, 1990). Infected vegetations may become large and segments may dislodge to become systemic emboli. Persons who have endocarditis caused by viridans streptococci are at particular risk for embolic events (Steckelberg *et al.*, 1991). Emboli from the left side of the heart most commonly are found in the brain, kidney, spleen, or eye (Freeman & Hall, 1989). Embolic events may cause serious consequences. Neurologic complications occur in 27 to 39% of all cases of bacterial endocarditis (Salgado, 1991); the most frequently occurring central nervous system complication is cerebrovascular accident.

Several investigators have investigated the phenomenon that rabbit sera raised against *S. mutans* contains antibodies cross-reactive with heart and skeletal muscle tissue of human and rabbit origin (Ayakawa *et al.*, 1985; Choi & Stinson, 1989; Doyle *et al.*, 1986; Wu & Russell, 1990). The explanation of cross-reactivity is not yet entirely clear. It has been hypothesized that streptococcal components may cause injury of muscle fibers during immunization of rabbits, resulting in exposure of cryptic self-antigens and stimulation of self-reactive B cells (Choi & Stinson, 1989). Membrane-associated streptococcal proteins have been implicated by some researchers (Ayakawa *et al.*, 1985), although no consensus has been reached regarding which streptococcal components are responsible. Whatever its basis, this cross-reactivity warrants serious consideration in development of vaccine strategies.

Factors associated with virulence of viridans streptococci

The problem of relating particular endocarditis virulence factors to viridans streptococci is complicated by the diverse composition of the group, and satisfactory differentiation of species within the group is difficult (Coykendall, 1989; Facklam, 1977). In addition, since viridans streptococci are generally susceptible to penicillin, it is usually not essential to identify an infecting organism to the species level in order to initiate appropriate antibiotic therapy. For these reasons, epidemiologic data related to incidence of infection by particular viridans streptococci is difficult to interpret. Potential virulence traits are frequently studied across species despite the possibility that different virulence traits may be employed to a greater or lesser extent in different species. For example, previous studies have compared the adherence of bacterial species which naturally produce glucan to species which do not produce glucan (Crawford & Russell, 1986), or compared clinical isolates of viridans streptococci from a variety of sources to each other (Herzberg *et al.*, 1990b).

Binding to platelets
Binding of viridans streptococci, in particular *S. sanguis* and *S. salivarius*, to platelets has been hypothesized to influence the binding of viridans streptococci to sterile vegetations on heart valves and to stimulate platelet aggregation and subsequent increased size of the vegetation (Herzberg *et al.*, 1983; Herzberg *et*

al., 1990b; Sullam *et al.*, 1990). In *S. sanguis*, interaction with platelets appears to be mediated by a bacterial cell wall-bound protein (Herzberg *et al.*, 1990a; Erickson & Herzberg, 1990). Sullam *et al.* (1990) suggested that the interaction of platelets and *S. sanguis* is mediated by a specific receptor-ligand interaction, since they observed rapid, reversible, saturable binding. However, it has been noted that *S. mutans* interacts minimally with human platelets (Herzberg *et al.*, 1983). Additionally, in a study of 18 endocarditis and oral isolates of *S. sanguis*, interactivity of the bacterial cells with platelets was not associated with levels of dextran or fructan produced by the strain.

Exopolysaccharide production

In addition to its importance in caries formation, glucan polymer synthesis also has been implicated in the development of streptococcal endocarditis (Ramirez-Ronda, 1978; Scheld *et al.*, 1978). It has been suggested that glucan formation, which promotes tenacious adherence to the tooth surface in the oral cavity, may also enhance adherence to the vegetations present on the valves of the heart (Ramirez-Ronda, 1980). Pulliam *et al.* (1985) hypothesized that production of exopolysaccharide may protect viridans streptococci from opsonization and phagocytosis. Research with glucan producing and non-glucan producing non-isogenic strains of *S. sanguis* indicated that formation of glucan did increase virulence (Meddens *et al.*, 1984), but no published studies have compared isogenic mutants which differed only in the production of specific exopolysaccharides. Noteworthy is that *S. sanguis* produces a glucan which is structurally different from that produced by *S. mutans* (Coykendall, 1989). Prior to our work, reported here and elsewhere (Munro & Macrina, 1993), there had been no published reports which used isogenic strains to specifically address the role of exopolymer production in cardiac virulence.

Binding to fibrin

Since fibrin deposition is common when valvular damage occurs, the ability to bind fibrin is thought to reflect the ability to bind to damaged cardiac valves. Two types of fibrin binding assays are commonly described. A large scale method was developed by Scheld *et al.* (Scheld *et al.*, 1978). We used a microtiter method, adapted from Chhatwal *et al.* (1990), to investigate the binding characteristics of *S. mutans* V403.

The microtiter fibrin binding assays required two parallel sets of wells. Bacteria were added to the first set of wells at the outset of the assay, and provided information about adherent bacteria. The parallel set of wells, treated identically throughout the assay, had bacteria added following final washing of the plate, and reflected the total number of bacteria initially added to the first set of wells. Fibrin coated plates were prepared by the addition of 50 μl fibrinogen (10 mg/ml) and 25 μl thrombin(25 U/ml) to each well of a 96 well microtiter plate (Falcon 3027, Becton-Dickinson, Lincoln Park, NJ). Plates were incubated at 37°C for 30 min and washed twice with phosphate buffered saline (PBS, pH 7.4). Overnight cultures of bacteria grown in brain-heart infusion broth (BHI) were diluted 1:10 in fresh BHI supplemented with 50 μCi/ml [methyl-^{3}H]-thymidine (NEN Research Products, Boston, Mass.; 20.0 Ci/mmol) and 250 μg/ml deoxyadenosine, and grown anaerobically at 37°C to an $OD_{660} \approx 0.600$. ^{3}H-labeled bacteria were serially diluted in PBS to yield 1 × 10^8 cells/ml, and 100 μl were added to wells of the microtiter plate. For each set of wells containing bacteria, 100 μl PBS were added to a parallel set of wells. Microtiter plates were incubated for 30 min at 37°C unless otherwise stated and washed twice with PBS to remove nonadherent cells. 100 μl of bacteria were added to the wells which had previously contained PBS, 100 μl of PBS were added to wells containing bacteria during incubation, and fibrin clots in wells were dissolved by the addition of 50 μl 2.5% trypsin to each well. Three 50 μl samples of each well were spotted onto number 3 qualitative filter paper disks (2.3 cm circles, Whatman International, Maidstone, England) and acid insoluble counts were precipitated with 5% trichloroacetic acid. Disks were air dried and radioactivity of disks was determined by liquid scintillation counting. The percent adherence was calculated as: (mean cpm of samples from wells initially containing bacteria ÷ mean cpm of samples from wells to which bacteria were added following final washes) × 100.

The microtiter assays provided information about the characteristics of binding to fibrin clots *in vitro* (Fig. 1 and Fig. 2). Since other investigators have noted that *S. mutans* interacts minimally with human platelets (Herzberg *et al.*, 1983) and that in *S. sanguis* interaction with platelets is mediated by a bacterial cell wall-bound protein (Erickson & Herzberg, 1990; Herzberg *et al.*, 1990a), we focused on the interaction of bacterial cells and fibrin matrices which did not include platelets.

The linear relationship between number of bacteria added to wells and cpm recovered from clots indicated a nonspecific interaction (Fig. 1), although percent adherence of *S. mutans* V403 was not affected by the addition of an unrelated protein (albumin). In some groups, 10, 100, or 1000 μg of albumin were added to the wells containing fibrin clots 15 min before addition of ^{3}H-labeled bacteria. In other groups, 10, 100, or 1000 μg of albumin were added to wells which had been incubated with bacteria for 30 min; these plates were incubated an additional 15 min following addition of albumin. None of the groups displayed adherence which differed from untreated controls.

Effects of altering the length of time bacteria were in contact with fibrin clots in microtiter wells was examined. When bacteria were in contact with wells for less than 1 minute, cpm of wells was equivalent to that of background control wells. Little difference was observed between the percentages of cells adherent to the fibrin coated wells at 15, 30, 45, or 60 min (adherences of 9.5, 9.5, 10.3, and 10.8%, respectively).

Variation in the temperature at which interaction between *S. mutans* V403 and the fibrin clots took place indicated that binding was best at 37°C (18.7%), and was reduced at 4°C (9%), 20°C (12.3%), and 42°C (13.7%).

In order to investigate the possibility that adherence to fibrin clots *in vitro* was protein-mediated, the microtiter fibrin binding assay was performed with

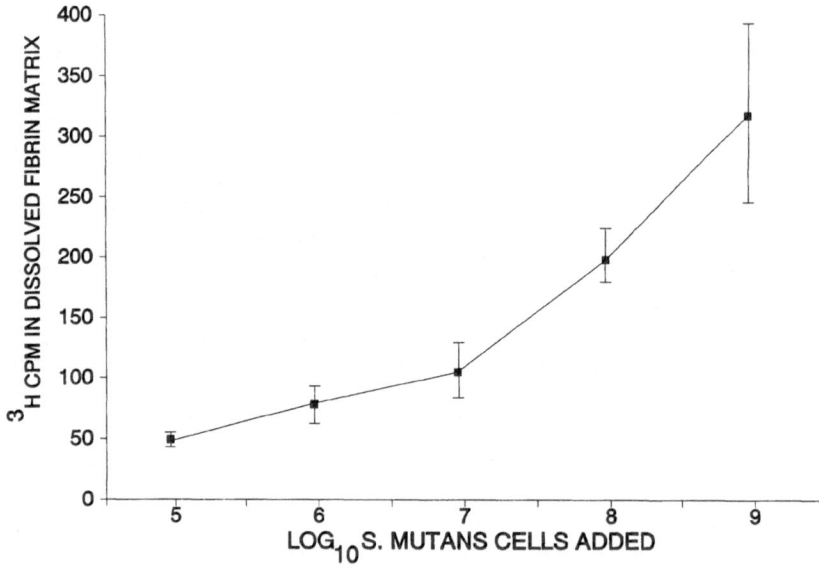

Figure 1. Cpm recovered from dissolved fibrin matrices following incubation of *S. mutans* V403 on fibrin coated plates. Varying numbers of ^{3}H-labeled bacterial cells were incubated on fibrin-coated 60 × 15 mm Petri dishes for 30 min at 37°C. Following washes of plates, the fibrin matrices were dissolved with trypsin. Material present in the dissolved clot fluid was precipitated with tricholoacetic acid and radioactivity of samples of the precipitated material was determined in a liquid scintillation counter. Error bars indicate standard deviation.

cells which had been preincubated with 2 mg of trypsin per 10^8 bacteria for 10 min. Phenylmethylsulfonyl fluoride (PMSF, Sigma) was added to bacteria to a concentration of 10 mM following the incubation with trypsin in order to avoid disintegration of the fibrin clot, and bacteria were centrifuged and suspended in fresh PBS. Adherence of *S. mutans* V403 was similar in untreated groups (13.0%), in groups treated with trypsin (14.8%), and in groups treated with PMSF alone (13.3%). In experiments performed with *S. sanguis* V213, adherence was decreased from 4.9% with untreated bacteria to 2.6% with trypsin treatment; adherence of V213 treated with PMSF alone was 5.7%. The finding that adherence was not affected by treatment of *S. mutans* V403 with trypsin was consistent with a non-protein binding mediator such as exopolysaccharides. Our data suggest further that bacterial proteins may influence binding of *S. sanguis* V213 to fibrin.

Addition of unlabeled *S. mutans* V403 cells to microtiter wells containing fibrin clots reduced the cpm recovered, demonstrating that these cells were able to compete for binding sites on fibrin clots with ^{3}H-labeled V403 cells. When 10^5 or 10^6 unlabeled V403 were incubated on the fibrin clots prior to the addition of 10^6 labeled V403, recovery of cpm was reduced by 57%. When 10^5 or 10^6 unlabeled V403 were added to microtiter wells with fibrin clots following the incubation of labeled V403 in the wells, recovery of cpm was reduced by 28%

and 48%, respectively. This may indicate that individual bacteria are rather loosely adherent, as one might expect for an interaction mediated by material which is not tightly associated with individual bacterial cells. The observations fit a model in which the presence of exopolysaccharides enhances binding of cells in the aggregate, but in which individual bacteria are alternately adherent and free.

Fibronectin binding

Fibronectin is a large (425 kDa) glycoprotein produced by endothelial cells. A soluble form of the protein is present in plasma. A bound immobilized form can be demonstrated at sites of tissue injury and it is thought that immobilized fibronectin may be associated with sterile vegetations. Since fibronectin can bind a variety of Gram-positive bacteria and result in bacterial aggregation (Chhatwal & Blobel, 1987; Vercellotti *et al.*, 1985), it has been postulated that binding to fibronectin is one mechanism of establishment of endocarditis.

It has been reported that *S. sanguis* adherence was affected by fibronectin, one component of sterile vegetations *in vivo*, incorporated into a fibrin clot (Lowrance *et al.*, 1988) or immobilized in a tissue culture well (Vercellotti *et al.*, 1985), but that *S. sanguis* adherence was unaffected by soluble fibronectin (Lowrance *et al.*, 1988). The ability of four strains of *S. mutans* to bind fibronectin was tested by Babu *et al.* (1983). Marked heterogeneity in both ability to bind and subsequent aggregation was observed by these workers.

We tested the effects of addition of varying amounts of human fibronectin (Sigma Chemical Co, St. Louis, MO) to bacterial cultures or incorporated into fibrin matrices on the ability of *S. mutans* V403 and *S. sanguis* V213 to bind to fibrin in the microtiter assay described above (Fig. 2). Fibronectin was incorporated into the clot by adding 0, 2, 10, 50, 100, or 200 μg of fibronectin to the microtiter wells prior to the addition of thrombin (Fig. 2, Panel A). The percent adherence of V403 remained unchanged despite increasing concentrations of fibronectin. Low levels of binding were demonstrated by *S. sanguis*, but the pattern of increasing adherence in the presence of increasing amounts of fibronectin incorporated into the clot was consistent with findings of other groups. Soluble fibronectin, preincubated with V403 or V213 at concentrations of 1, 10, 35, or 50 μg per 10^8 bacteria for 30 min, did not affect adherence of either strain (Fig. 2, Panel B).

Rat model of endocarditis

The rat is an excellent model for the study of endocarditis. Rat cardiovascular anatomy is very similar to that of humans, and the pathophysiology and course of infection closely mimics human disease. Like humans, rats are generally susceptible to endocarditis only if the cardiac valve has been damaged. Damage of the aortic valve, which is the valve most frequently involved in human native valve endocarditis (Bayliss *et al.*, 1983), is easily accomplished in the rat through introduction of a plastic catheter.

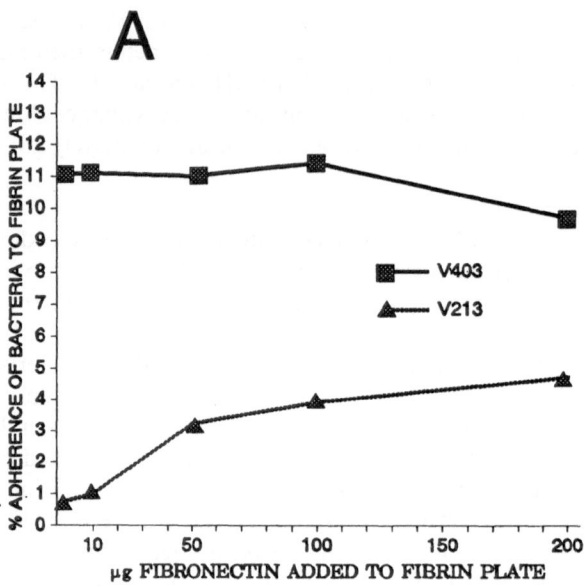

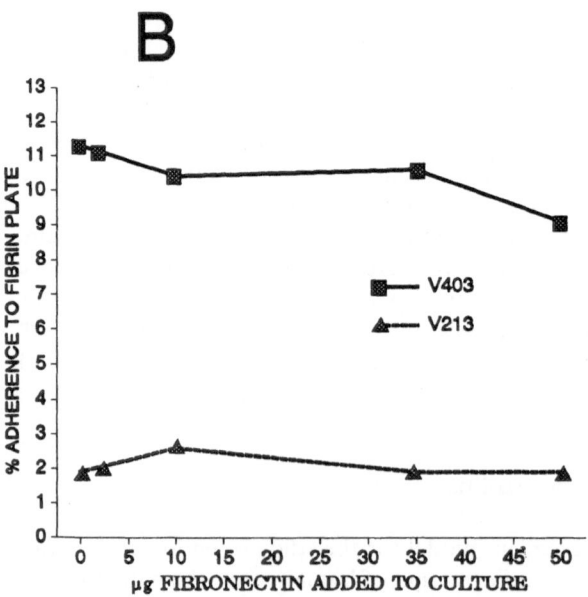

Figure 2. Adherence of *S. mutans* and *S. sanguis* to fibrin plates in the presence of fibronectin. Assays were performed in wells of 96 well microtiter plates. Panel A, Fibronectin (0 to 200 µg) incorporated into fibrin clot prior to addition of bacteria. Panel B, Bacterial cells incubated with 0 to 50 µg soluble fibronectin prior to addition of bacteria to fibrin plate.

We use variation of the rat models of endocarditis described by Santoro and Levinson (1978) and Heraief *et al.* (1982). Approval for animal use was obtained from the Virginia Commonwealth University Institutional Animal Care and Use Committee (Protocol #9009–1291) prior to initiation of experiments. Male Sprague-Dawley rats, weighing 200–250 grams, were anesthetized with 100 μg ketamine per gram body weight. Anesthesia was maintained with whiffs of methoxyflurane. A 2 cm midline incision was made in the neck and the right carotid artery was sutured closed at the cephalic end. A 7 cm length of PE 10 intramedic polyethylene tubing (Clay Adams, Parsippany. N.J.), occluded at one end, was inserted through a small incision in the carotid and advanced toward the chest until resistance was met and pulsation of the catheter was observed. The carotid artery with catheter in place was tied with suture at the base of the neck in order to prevent exsanguination and maintain catheter placement in the left ventricle. Following approximation of the neck tissues, closure of the neck incision, and recovery from anesthesia, animals were permitted to resume their previous levels of activity. The catheter remained in place throughout the course of the experiment. Twenty four or 48 hours following catheter placement, rats were randomly assigned to groups and received an inoculation of either V403 or V1996 in 0.5 ml PBS via tail vein. Three days later, animals were sacrificed by CO_2 inhalation. The heart was removed and opened. Proper catheter placement and the presence or absence of vegetations on the aortic valve were scored by visual inspection. The aortic valve and associated vegetations were removed, homogenized in a sterile disposable tissue grinder (No. 3505, Sage Products, Cary, Il.), serially diluted in PBS, and plated on BHI to determine colony forming units per valve. Histologic examinations of sections of heart from representative animals were performed, and vegetations from those animals were Gram-stained.

In our experience with the rat model, production of polymer appeared to play a major role in development of *S. mutans* endocarditis Animals inoculated with V403 grown in sucrose had a much higher incidence of infected vegetations than did animals inoculated with V1996, which lacks functional enzymes of exopolysaccharide production. Animals inoculated with V403 grown in sucrose also had a much higher incidence of infected vegetations than did animals inoculated with V403 grown in glucose, which lacked substrate for polymer formation.

The differences in groups could be explained by differences in the ability of the organisms to bind to the sterile vegetation on the aortic valve, or by differences in interaction with phagocytic cells. Variations between wild type and polymer-deficient strains were found in *in vitro* assessments of both phagocytosis and fibrin binding (Munro & Macrina, 1993). In *in vitro* assays of phagocytosis using ^{3}H-labeled bacteria, both strains were found to be associated with human granulocytes. However, colony counts of cell pellets following lysis of granulocytes indicated that a greater proportion of sucrose-grown wild type organisms than of mutant organisms were alive at the conclusion of assays ($p < 0.05$). When granulocytes which had been incubated with either sucrose-grown V403 or V1996 were visually compared, a smaller percentage of granulocytes had ingested the wild type bacteria, and granulocytes incubated with V403 contained a smaller mean number of bacteria. In addition, a larger chemiluminescent burst was generated by granulocytes incubated with V1996 than by granulocytes incubated with V403. Compared to adherence of the wild type organism grown in sucrose, adherence to fibrin coated plates was significantly reduced (p < 0.05) for the wild type strain grown in glucose and for the polymer-deficient mutant grown in either sucrose or glucose.

Concluding remarks

Endocarditis in humans caused by viridans streptococci is an important health problem (Freeman & Hall, 1989). Allelic exchange mutants provide a powerful tool to study the effects of specific gene products on virulence while controlling for strain and species variability. We have used isogenic strains derived from *S. mutans* V403 by allelic exchange, and differing from the parental wild type only by insertional inactivation of *gtfB*, *gtfC*, *gtfD*, and/or *ftf*, to demonstrate the influence of extracellular polysaccharides on virulence in endocarditis in several model systems, including infectivity in the rat and *in vitro* investigations of binding to fibrin and avoidance of phagocytic killing. In each model, the wild-type organism grown in sucrose, which produced exopolysaccharides, behaved differently from the wild type organism grown in glucose (unable to produce polymer due to lack of substrate) and the mutant organism (unable to produce polymer due to allelic exchange mutagenesis). Sucrose-grown V403 demonstrated greater infectivity in the rat model, enhanced binding to fibrin, and reduced mortality in the presence of granulocytes. Taken together, our data indicate that exopolysaccharides are an important component of virulence in *S. mutans* endocarditis.

Acknowledgements

We thank James Desiderio, Cindy Lamb, and Jeff Hibbard of Bristol-Myers Squibb, Wallingford, CT, for assistance with the rat model of endocarditis. Helpful discussions with Gordon Archer are gratefully acknowledged. This work was supported by U.S.P.H.S. grants R37 DE04224 (F.L.M.) and F31 NR06498 (C.M.).

References

Ayakawa GY, Siegel JL, Crowley PJ & Bleiweis AS (1985) Immunochemistry of the *Streptococcus mutans* BHT cell membrane: detection of determinants cross-reactive with human heart tissue. Infect. Immun. 48: 280–286.

Babu JP, Simpson WA, Courtney HS & Beachey EH (1983) Interaction of human plasma fibronectin with cariogenic and non-cariogenic oral streptococci. Infect. Immun. 41: 162–168.

Barletta RG, Michalek SM & Curtiss R 3d (1988) Analysis of the virulence of *Streptococcus mutans* serotype c *gtfA* mutants in the rat model system. Infect. Immun. 56: 322–330.

Bayer AS & Norman DC (1990) Valve site-specific pathogenetic differences between right-sided and left-sided bacterial endocarditis. Chest 98: 200–205.

Bayliss R, Clarke C, Oakley CM, Somerville W, Whitfield AGW & Young SEJ (1983) The microbiology and pathogenesis of infective endocarditis. British Heart J. 50: 513–519.

Bleiweis AS, Lee SF, Brady LJ, Progulske-Fox A & Crowley PJ (1990) Cloning and inactivation of the gene responsible for a major surface antigen on *Streptococcus mutans*. Arch. Oral Biol. 35 Suppl: 15S–23S.

Bowen WH, Schilling K, Giertsen E, Pearson S, Lee SF, Bleiweis AS & Beeman D (1991) Role of a cell surface associated protein in adherence and dental caries. Infect. Immun. 59: 4606–4609.

Chhatwal GS & Blobel H (1987) Heterogeneity of fibronectin reactivity among streptococci as revealed by binding of fibronectin fragments. Comp. Immun. Microbiol. and Infect. Dis. 10: 99–108.

Chhatwal GS, Valentin Weigand P & Timmis KN (1990) Bacterial infection of wounds: fibronectin-mediated adherence group A and C streptococci to fibrin thrombi in vitro. Infect. Immun. 58: 3015–3019.

Choi SH & Stinson MW (1989) Purification of a *Streptococcus mutans* protein that binds to heart tissue and glycosaminoglycans. Infect. Immun. 57: 3834–3840.

Coykendall A (1989) Classification and identification of the viridans streptococci. Clin. Microbiol. Rev. 2: 315–328.

Crawford I & Russell C (1986) Comparative adhesion of seven species of streptococci isolated the blood of patients with sub-acute bacterial endocarditis to fibrin-platelet clots in vitro. J. Appl. Bacteriol. 60: 127–133.

Dajani AS, Bisno AL, Chung KJ, Durack DT, Freed M, Gerber MA, Karchmer AW, Millard HD, Rahimtoola S, Shulman ST, Watanakunakorn C & Taubert KA (1990) Prevention of bacterial endocarditis: Recommendations by the American Heart Association. J. A. M. A. 264: 2919–2922.

Dall L, Barnes WG, Lane JW & Mills J (1987) Enzymatic modification of glycocalyx in the treatment of experimental endocarditis due to viridans streptococci. J. Infect. Dis. 156: 736–740.

De Stoppellaar JD, Konig K, Plasschaert A & van der Hoeven J (1971) Decreased cariogenicity of a mutant of *Streptococcus mutans*. Arch. Oral Biol. 16: 971–975.

Doyle G, Everhard D, Mallett C, Ayakawa G & Bleiweis AS (1986) Demonstration of shared antigenic determinants between *Streptococcus mutans* BHT cell membrane human heart tissue and myosin using monoclonal antibodies to *S mutans*. J. Gen. Microbiol. 132: 2885–2892.

Erickson PR & Herzberg MC (1990) Purification and partial characterization of a 65-kDa platelet aggregation-associated protein antigen from the surface of *Streptococcus sanguis*. J. Biol. Chem. 265: 14080–14087.

Facklam RR (1977) Physiological differentiation of viridans streptococci. J. Clin. Microbiol. 5: 184–201.

Fekete T (1990) Controversies in the prevention of infective endocarditis related to dental procedures. Dental Clinics of North America 34: 79–90.

Freedman LR (1987) The pathogenesis of infective endocarditis. J. Antimicrob. Chemotherapeutics 20 suppl A: 1–6.

Freeman J, Epstein MF, Smith NE, Platt R, Sidebottom DG & Goldmann DA (1990) Extra hospital stay and antibiotic usage with nosocomial coagulase-negative staphylococcal bacteremia in two neonatal intensive care unit populations. Amer. J. Diseases of Children 144: 324–329.

Freeman R & Hall R (1989) Infective Endocarditis. p. 853–876. *In:* Julian DG, Camm AJ, Fox KM, Hall RJC & Poole-Wilson PA (ed.) Diseases of the heart Bailliere Tindall, Philadelphia.

Hamada S & Slade HD (1980) Biology immunology and cariogenicity of *Streptococcus mutans*. Microbiol. Rev. 44: 331–384.

Hanada N & Kuramitsu H (1989) Isolation and characterization of the *Streptococcus mutans gtfD* gene coding for primer-dependent soluble glucan synthesis. Infect. Immun. 57: 2079–2085.

Hanada N & Kuramitsu HK (1988) Isolation and characterization of the *Streptococcus mutans gtfC* gene coding for synthesis of both soluble and insoluble glucan. Infect. Immun. 56: 1999–2005.

Heraief E, Glauser MP & Freedman LR (1982) Natural history of aortic valve endocarditis in rats. Infect. Immun. 37: 127–131.

Herzberg MC, Brintzenhofe KL & Clawson CC (1983) Aggregation of human platelets and adhesion of *Streptococcus sanguis*. Infect. Immun. 39: 1457–1469.

Herzberg MC, Erickson PR, Kane PK, Clawson DJ, Clawson CC & Hoff FA (1990a) Platelet-interactive products of *Streptococcus sanguis* protoplasts. Infect. Immun. 58: 4117–4125.

Herzberg MC, Gong K, MacFarlane GD, Erickson PR, Soberay AH, Krebsbach PH, Manjula G, Schilling K & Bowen WH (1990b) Phenotypic characterization of *Streptococcus sanguis* virulence factors associated with bacterial endocarditis. Infect. Immun. 58: 515–522.

Honda O, Kato C & Kuramitsu HK (1990) Nucleotide sequence of the *Streptococcus mutans gtfD* gene encoding the glucosyltransferase-S enzyme. J. Gen. Microbiol. 136: 2099–2105.

Jacobson GR, Lodge J & Poy F (1989) Carbohydrate uptake in the oral pathogen *Streptococcus mutans*: Mechanisms and regulation by protein phosphorylation. Biochimie 71: 997–1004.

Kato C & Kuramitsu HK (1990) Carboxyl-terminal deletion analysis of the *Streptococcus mutans* glucosyltransferase-I enzyme. FEMS Microbiol. Lett. 72: 299–302.

Kato C & Kuramitsu HK (1991) The carboxyl terminal domain of *Streptococcus mutans* glucosyltransferases prevents secretion in *Escherichia coli*. FEMS Microbiol. Lett. 81: 107–110.

Koga T, Asakawa H, Okahashi N & Hamada S (1986) Sucrose-dependent cell adherence and cariogenicity of serotype c *Streptococcus mutans*. J. Gen. Microbiol. 132: 2873–2883.

Lowrance JH, Hasty DL & Simpson WA (1988) Adherence of *Streptococcus sanguis* to conformationally specific determinants in fibronectin. Infect. Immun. 56: 2279–2285.

Macrina FL, Jones KR, Alpert C-A, Chassy BM & Michalek SM (1991) Repeated DNA sequence involved in mutations affecting transport of sucrose into *Streptococcus mutans* V403 via the phosphoenolpyruvate phosphotransferase system. Infect. Immun. 59: 1535–1543.

Meddens MJ, Thompson J, Leijh PC & Van Furth R (1984) Role of granulocytes in the induction of an experimental endocarditis with a dextran-producing *Streptococcus sanguis* and its dextran-negative mutant. Brit. J. Exp. Pathol. 65: 257–265.

Mooser G, Hefta SA, Paxton RJ, Shively JE & Lee TD (1991) Isolation and sequence of an active-site peptide containing a catalytic aspartic acid from two *Streptococcus sobrinus* alpha-glycosyltransferases. J. Biol. Chem. 266: 8916–8922.

Moreillon P, Overholser CD, Malinverni, Bille J & Glauser MP (1988) Predictors of endocarditis in isolates from cultures of blood following dental extractions in rats with periodontal disease. J. Infect. Dis. 157: 990–995.

Munro C & Macrina FL (1993) Sucrose-derived exoploysaccharides of *Streptococcus mutans* V403 contribute to infectivity in endocarditis. Mol Microbiol. 8: 133–142.

Munro C, Michalek SM & Macrina FL (1991) Cariogenicity of *Streptococcus mutans* V403 glucosyltransferase and fructosyltransferase mutants constructed by allelic exchange. Infect. Immun. 59: 2316–2323.

Overholser CD, Moreillon P & Glauser MP (1987) Experimental bacterial endocarditis after dental extractions in rats with periodontitis. J. Infect. Dis. 155: 107–112.

Perry D & Kuramitsu HK (1990) Linkage of sucrose-metabolizing genes in *Streptococcus mutans*. Infect. Immun. 58: 3462–3464.

Pulliam L, Dall L, Inokuchi S, Wilson W, Hadley WK & Mills J (1985) Effects of exopolysaccharide production by viridans streptococci on penicillin therapy of experimental endocarditis. J. Infect. Dis. 151: 153–156.

Ramirez-Ronda CH (1978) Adherence of glucan-positive and glucan-negative streptococcal strains to normal and damaged heart valves. J. Clin. Invest. 62: 805–814.

Ramirez-Ronda CH (1980) Effects of molecular weight of dextran on the adherence of *Streptococcus sanguis* to damaged heart valves. Infect. Immun. 29: 1–7.

Russell RRB, Aduse-Opoku J, Tao L & Ferretti JJ (1991) Binding protein-dependent transport system in *Streptococcus mutans*. p. 244–247. *In:* Dunny GM, Cleary PP & McKay LL (eds.) Genetics and molecular biology of streptococci lactococci and enterococci, American Society for Microbiology, Washington.

Salgado AV (1991) Central nervous system complications of infective endocarditis. Curr. Concepts of Cerebrovascular Dis. Stroke 26: 19–22.

Santoro J & Levison ME (1978) Rat model of experimental endocarditis. Infect. Immun. 19: 915–918.

Scheld WM, Valone JA & Sande MA (1978) Bacterial adherence in the pathogenesis of endocarditis: Interaction of bacterial dextran platelets and fibrin. J. Clin. Invest. 61: 1394–1404.

Schroeder VA, Michalek SM & Macrina FL (1989) Biochemical characterization and evaluation of virulence of a fructosyltransferase-deficient mutant of *Streptococcus mutans* V403. Infect. Immun. 57: 3560–3569.

Shiroza T & Kuramitsu HK (1988) Sequence analysis of the *Streptococcus mutans* fructosyltransferase gene and flanking regions. J. Bacteriol. 170: 810–816.

Shiroza T, Ueda S & Kuramitsu HK (1987) Sequence analysis of the *gtfB* gene from *Streptococcus mutans*. J. Bacteriol. 169: 4263–4270.

Spatafora Harris G, Michalek SM & Curtis R 3rd (1992) Cloning of a locus involved in *Streptococcus mutans* intracellular polysaccharide accumulation and virulence testing of an intracellular polysaccharide-deficient mutant. Infect. Immun. 60: 3175–3185.

Steckelberg JM, Murphy JG, Ballard D, Bailey K, Tajik AJ, Taliercio CP, Guiliani ER & Wilson WR (1991) Emboli in infective endocarditis: The prognostic value of echocardiography. Annals Int. Med. 114: 635–640.

Sullam PM, Payan DG, Dazin PF & Valone FH (1990) Binding of viridans streptococci to human platelets: A quantitative analysis. Infect. Immun. 58: 3802–3806.

Tanzer JM, Freedman ML, Fitzgerald RJ & Larson RH (1974) Diminished virulence of glucan synthesis defective mutants of *Streptococcus mutans*. Infect. Immun. 10: 197–203.

Ueda S, Shiroza T & Kuramitsu HK (1988) Sequence analysis of the *gtfC* gene from *Streptococcus mutans*. Gene 69: 101–109.

Van der Meer JTM, van Vianen W, Hu E, van Leeuwen WB, Valkenburg HA, Thompson J & Michel MF (1991) Distribution antibiotic susceptibility and tolerance of bacterial isolates in culture-positive cases of endocarditis in the Netherlands. European J. Clin. Microbiol. Infect. Dis. 10: 728–734.

Vercellotti GM, McCarthy JB, Lindholm P, Peterson PK, Jacob HS & Furcht LT (1985) Extracellular matrix proteins (fibronectin laminin and type IV collagen) bind and aggregate bacteria. Amer. J. Pathol. 120: 13–21.

Wexler DL, Penders JEC, Bowen WH & Burne RA (1992) Characteristics and cariogenicity of a fructanase-defective *Streptococcus mutans* strain. Infect. Immun. 60: 3673–3681.

Wu H & Russell MW (1990) Immunological cross-reactivity between *Streptococcus mutans* and human heart tissue examined by cross-immunization experiments. Infect. Immun. 58: 3545–3552.

Zealey GR, Loosmore SM, Yacoob RK, Cockle SA, Boux LJ, Miller LD & Klein MH (1990) Gene replacement in *Bordetella pertussis* by transformation with linear DNA. Bio Technology 8: 1025–1029.

19. Method for detecting gene expression of internalized *Salmonella typhimurium* in macrophages

STEPHEN J. GRACHECK and SARA A. WOLD

Abstract. Expression of virulence factors is essential for microbial pathogenesis. We examined the interaction of *Salmonella typhimurium* and macrophages in an *in vitro* assay to assess the effects of antibiotics on bacterial viability and gene expression. The murine macrophage cell line J774A.1 and *S. typhimurium* strains TT12308 with a *pur*G::*lacZ* (*Mud*J) fusion and $_x$3181(pGTRO90) with a *spv*A::*lacZ* fusion were used. The TT12308 fusion is a constitutively expressed purine biosynthetic gene and the $_x$3181(pGTRO90) fusion is an inducible plasmid virulence gene necessary for systemic infection. The bacteria were exposed to ciprofloxacin and sparfloxacin, quinolone DNA gyrase inhibitors, and tetracycline either simultaneously to the macrophages, or following adherence and invasion. Exposure of the *pur*G strain to ciprofloxacin at the time of adherence resulted in a dose dependent decrease in viable bacterial counts recovered internally from macrophages and ranged from a 0.5 to 2.2 log decrease at 1/2 to 16 × MIC (MIC = 0.003 μg/ml), respectively. Similarly *pur*G expression, as measured by β-galactosidase activity, decreased in a dose dependent manner, ranging from 25% to 75% of the control at 1/8 to 1/2 × MIC. No β-galactosidase activity was seen above 1/2 × MIC. Ciprofloxacin, when added after adherence and invasion, did not decrease either bacterial viability or *pur*G expression. Probenecid, a carrier-mediated transport inhibitor, added simultaneously with ciprofloxacin, had an additive effect on the decrease in recovery of viable bacteria. When sparfloxacin was added at the time of infection, a 2.5 log decrease occurred, and, when added 3 h post infection, only a 1.3 log decrease in viability of the *pur*G strain was seen. In studies with the *spv*A strain treated in macrophages at the time of infection, sparfloxacin reduced the viability 4.5 logs and tetracycline had a bacteriostatic effect. When the drugs were added 3 h post infection, sparfloxacin reduced the number of internalized organisms 3.3 logs, whereas tetracycline had little effect. In both treatment regimens, subinhibitory concentrations of tetracycline induced a β-galactosidase response in the *spv*A strain, whereas sparfloxacin had no effect at any concentration. Induction was due to a tetracycline resistance promoter on the vector. This model can be utilized to evaluate the effect of antibiotics on the viability and gene expression of bacteria in macrophages.

Abbreviations: MIC, Minimal Inhibitory Concentration; CPRG, Chlorophenal redß-D-galactopyranoside

Introduction

Expression of virulence factors by pathogenic bacteria is well established as essential for infection (Falkow, 1991). Traditional chemotherapy relies on the outright killing ability of antibiotics, or on bacteriostatic effects followed by immune intervention (Goldberg, 1988, Pratt, 1986). Many medically significant pathogenic bacteria reside in an intracellular environment where they are

C.I. Kado and J.H. Crosa (eds.), Molecular Mechanisms of Bacterial Virulence, 267–280.
© 1994 *Kluwer Academic Publishers. Printed in the Netherlands.*

protected from both high extracellular concentrations of antibiotics, and from the host immune response. The ability of intracellular pathogens to evade conventional therapy makes them more intractable to elimination. Antibiotics which are quite effective *in vitro* can be rendered useless because of their inability to penetrate intracellularly and affect the resident bacteria. Other antibiotics are capable of penetrating intracellularly, but cellular efflux and detoxification mechanisms prevent them from accummulating to an effective concentration (Cao, 1992, Noumi, 1990, Une, 1988, Pascual, 1992, Rastogi, 1991). As an adjunct to traditional chemotherapy, modulation of either virulence factors or host immunity represent a potential means of therapy to eliminate infections caused by intracellular pathogens.

Salmonella typhimurium is a medically significant pathogen which invades and multiplies intracellularly during an infection (Falkow, 1991, Finlay, 1989, Portnoy,1992, Goldberg, 1988). The organisms can invade host cells either by receptor mediated invasion or by engulfment by phagocytic cells. Expression of virulence factors by *Salmonella typhimurium* which direct attachment to host cell receptors and mediate invasion are required to penetrate non-phagocytic host cells and establish an infection (Galan, 1992, Ginnochio, 1992). *Salmonella typhimurium* harbor a large 90-kb virulence plasmid which encodes a number of factors necessary for intracellular infection (Gulig, 1990). Elimination of this plasmid, or transposon mutagenesis into the virulence encoding genes, renders *S. typhimurium* avirulent in murine models (Gulig, 1992). It is known that virulence genes are expressed differentially during the infection process, however their regulation is poorly understood (Finlay, 1989, Caldwell, 1991). Bacterial surface proteins specific for host cell receptors are expressed initially, and following adherence, invasion factors are sequentially expressed to gain entry into the target cell (Finlay, 1988, Buchmeier, 1990). The potential to modulate the expression of bacterial adhesion or invasion gene expression could prevent the subsequent infection of other cells. Since most chemotherapy is initiated during an active infection, it is necessary for antibiotics to penetrate host cells and affect internalized bacteria.

The purpose of this investigation was the development of methodologies for testing the ability of traditional chemotherapeutic agents to penetrate and affect

Table 1. S. typhimurium strains.

Strain[a]	Genotype	Characteristics
TT12308[b]	*pur*G2149::*lacZ*(*mudJ*)	constitutive, non-virulent[d]
SR-11,3181(pGTR090)[c]	pStSR100*spv*A-*lacZ*	inducible, virulent[d]

[a] **Methods.** The bacteria were grown in Oxoid Nutrient Broth or Agar #2 at 35^0C in 5% CO_2. Overnight broth cultures gave an OD_{600} of 0.18 with a corresponding CFU/ml of 10^8.

[b] Source: T. Guo-Min, 1988.

[c] Source: P. Gulig, 1990.

[d] **Methods.** CD-1 female mice, 18-20 g, were injected interperitoneally with 0.5 ml of log phase broth cultures diluted in 20% hog gastric mucin and observed for 5-7 days.

Salmonella typhimurium in an intracellular macrophage model. Antibiotics were added to the macrophages at the time of infection or 6 hours post infection. The model ideally would be able to independently evaluate the effects of antibiotics on bacterial cell viability and on gene expression of virulence factors. The intracellular activity of most classes of antibiotics has been well characterized (Peterson, 1988, Pratt, 1986). Currently, little information is available on the ability of antibiotics to modulate gene expression of a pathogenic bacteria intracellularly. An earlier study examined the expression of an outer membrane protein of *Yersinia pestis* by means of a *lacZ* fusion to the *yopK* gene (Pollack, 1986). This study found that Ca^+ dependent *in vitro* gene expression was also observed in *Y. pestis* resident in macrophage phagolysosomes.

Studies on the effects of antibiotics on the expression of two *Salmonella*

Table 2. Minimal inhibitory concentrations.

Antibiotics[b]	MIC (μg/ml)	
	TT12308[c]	SR-11,3181(pGTRO90)[c]
ampicillin	1.0	1.0
aztreonam	$\leq$0.06	0.1
cefaclor	0.5	3.0
chloramphenicol	>100	>100
ciprofloxacin	0.003	0.006
clindamycin	>10	>10
imipenem	0.3	0.3
isoniazid	>100	>100
metronidazole	>100	>100
mitomycin C	2.5	1.25
novobiocin	>100	>100
penicillin G	6.0	12
rifampin	12	10
sparfloxacin	0.012	0.012
spectinomycin	25	>10
streptomycin	12.5	25
tetracycline	0.06	100
vancomycin	>10	>10

[a] MICs were determined after 24 h incubation at 35^{0}C in 5% CO_2.

[b] Antibiotic dilutions were made in Oxoid Broth #2 in microtitre trays.

[c] An inoculum size of 2-5x10^5 CFU/ml in Oxoid Broth #2 was added to each well.

typhimurium genes, *purG* and *spvA* are presented here. The bacterial strains containing either a *purG::lacZ* fusion, a constitutively expressed chromosomal purine metabolism gene, or *spvA*, an inducible gene essential for invasive infecton found on the 90-kb virulence plasimd (spv), were used to examine the expression of these genes. Both intracellular viability and gene expression were characterized *in vitro*. The results show that these features can be studied in an intracellular environment.

Results

Minimal inhibitory concentrations

The MIC of various antibiotics was determined (Table 2). Ciprofloxacin and sparfloxacin, quinolone antibiotics, were the most potent inhibitors of both strains. The MIC of ciprofloxacin was 0.003 μg/ml and 0.006 μg/ml against *S. typhimurium* TT12308 and $_x$3181(pGTRO90), respectively. Sparfloxacin inhibited both strains at 0.012 μg/ml, while the remaining antibiotics had MIC values ranging from 0.06 to > 100 μg/ml. Strain $_x$3181(pGTRO90) was markedly resistant to tetracycline due to the presence of a tetracycline resistance marker.

In vitro *viability and β-galactosidase activity*

In vitro bacterial viability was correlated to β-galactosidase activity in both strains (Table 3 and 4). β-Galactosidase was qualitatively assessed and

Table 3. *In vitro* comparison of viability and β-galactosidase activity.

Strain	CFU/ml log$_{10}$[a]	β-galactosidase activity[b]	Miller Units[c]
TT12308	8.7	4	1591
	8.4	4	
	8.1	4	
	7.8	4	
	7.5	3	
	7.2	2	
	6.9	±	
	6.6	0	
x3181 (pGTRO90)	8.8	0	146

[a] **Methods.** Bacterial dilutions were made in microtitre plates. Samples were removed after 6 h incubation at 35°C in 5% CO$_2$ and CFU/ml determined.

[b] **Methods.** The β-galactosidase activity was determined by adding a 100 μl aliquot of the sample to a microtitre plate. The bacteria were lysed by the addition of 10 μl of 0.1% solution of hexadecyltrimethyl ammonium bromide. 100 μl of a 1 mg/ml solution of chlorophenol red-β-D-galactopyranoside (CPRG) in Z buffer was added to detect β-galactosidase activity. The plates were incubated at 37°C for one hour, the time required for a deep red color to develop in the positive control. To stop color development, 50 μl of a 1 M NaCO$_3$ solution was added to each well. Color development was qualitatively rated on a scale of 0 to 4 with 4 being a strong response.

[c] **Methods.** Miller Units of a 6 h log phase culture were determined using CPRG as the substrate (Miller, 1972, Eustice, 1991).

Table 4. In vitro comparison of antibiotic concentration and β-galactosidase activity.

Strain	Drug	Concentrations ug/ml	β-galactosidase activity[a] [b]
TT12308	Ciprofloxacin[c]	0.0002	4
		0.0004	3
		0.0008	2
		0.0016	1
		0.003	0
	Sparfloxacin[d]	0.006	4
		0.012	2
		0.025	0
	Tetracycline[e]	0.0006--0.03	4
		0.06	0
x 3181 (pGTRO90)	Ciprofloxacin[f]	0.000006--10	0
	Sparfloxacin[g]	0.000006--10	0
	Tetracycline[h]	0.0001	0
		0.0003	0
		0.0006	4
		0.006	4
		0.06	4
		0.5	4
		1.0	4
		25	4
		50	2
		100	0

[a] 1 h β-galactosidase induction period in MIC microtiter plates.

[b] Rated on a scale of 0 to 4 with 4 being a strong response.

[c] MIC = 0.003 μg/ml
[d] MIC = 0.012 μg/ml
[e] MIC = 0.06 μg/ml
[f] MIC = 0.006 μg/ml
[g] MIC = 0.012 μg/ml
[h] MIC = 100 μg/ml

compared to the control for relative production. Strain TT12308 required a minimum cell density of 10^7 cells/ml to detect enzyme activity. Strain $_x$3181(pGTRO90) with the *spv*A::*lacZ* fusion is not constitutively expressed. Subinhibitory concentrations of all antibiotics tested against strain TT12308 resulted in low enzyme activity which was a function of reduced cell viability. Concentrations at or above the MIC resulted in no β-galactosidase activity. The spvA::lacZ fusion in $_x$3181(pGTRO90) was strongly induced by subinhibitory concentrations of tetracycline from 0.006 μg/ml to 50 μg/ml due to a tetracycline resistance promoter on the vector. No other antibiotic induced the spvA::lacZ fusion at any concentration tested.

Mouse virulence studies
Only the *S. typhimurium* $_x$3181(pGTRO90) strain was mouse virulent with an LD$_{50}$ of $10^{1.9}$. The LD$_{50}$ for TT12308 was $10^{5.3}$ which indicates a significant attenuation of virulence. It has been shown that other strains with partial purine defects also have reduced virulence (McFarland, 1987).

Effect of simultaneous exposure
The effect of the simultaneous addition of the antibiotics and bacteria to the macrophage J774A.1 monolayers was measured. Based on *in vitro* potency, ciprofloxacin and sparfloxacin were chosen for this evaluation in *S. typhimurium* TT12308. The organism was internalized by macrophages within one hour and grew to 6.6 log$_{10}$ CFU/ml, about a 1.5 log$_{10}$ increase over the 6

Table 5. Viability and β-galactosidase activity of *Salmonella* TT12308 in macrophages after treatment with antibiotics.[a]

Drug	Concentrations	CFU/ml log$_{10}$	β-galactosidase[e] activity
Control	-------	6.6	4
probenecid[b]	2mM	6.4	4
ciprofloxacin	1/8 x MIC[c]	6.2	3
	1/2 x MIC	5.6	1
	2 x MIC	5.3	0
	16 x MIC	4.4	0
ciprofloxacin	16 x MIC		
+ probenecid		3.8	0
sparfloxacin	16 x MIC[d]	4.1	2

[a] **Methods.** Mouse monocyte-macrophage-like cell line J774A.1 was grown at 37°C in 5%CO$_2$ in 89% Dulbecco's modified Eagle's medium with 10% fetal bovine serum. Macrophage monolayers were prepared by adding 500 μl of 2x10^5 J774A.1 cells to wells of a 24-well tissue culture tray and incubating overnight. Each well was inoculated with 10 μl of an overnight *Salmonella* broth culture (about 10^6 CFU/well). Antibiotic solutions were added at the time of infection in 10 μl aliquots. The trays were incubated and one tray removed for sampling every hour for six hours. The wells were washed 3 X with PBS and 500 μl of a 200 μg/ml gentamicin solution in media was added to each well. After an additional 2 h incubation period, the wells were washed 3 X with PBS and 500 μl of a 0.1% triton X-100 solution was added to disrupt the J774A.1 cells and release the bacteria. The plates were refrigerated at 4°C for 20 min. 100 μl aliquots from each well were removed to determine *Salmonella* viability in CFU/ml (Vladoianu, 1990). Duplicate samples were pooled and concentrated 10 X by centrifugation to look for β-galactosidase activity.

[b] 20 μl of 2 mM probenecid was added to the macrophages at the time of infection along with the antibiotic and incubated for 6 h.

[c] MIC = 0.003 μg/ml

[d] MIC = 0.012 μg/ml

[e] Rated on a scale of 0 - 4 with 4 being maximum activity.

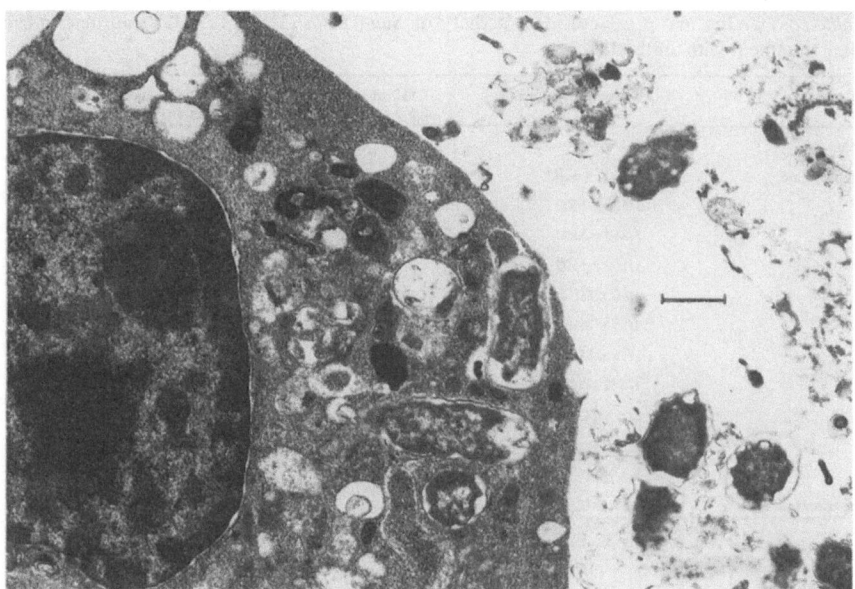

Figure 1. Transmission electron micrograph of a J774A.1 mouse macrophage infected with *Salmonella typhimurium* TT12308. (Bar = 1.0 μm). Infected monolayers were fixed in buffered 1.5% gluteraldehyde and postfixed in OsO$_4$. The samples were dehydrated with ethanol, embedded in Polybed 812, sectioned, poststained with uranyl acetate and lead citrate, and examined on a Philips CM-10 transmission electron microscope.

hour incubation period (Fig. 1). Bacterial viability and β-galactosidase activity were determined at the end of the 6 hour incubation period. Cell viability (CFU/ml) for strain TT12308 was reduced from $\log_{10}$ 6.6 to 4.4 at 16 × MIC for ciprofloxacin and to 4.1 for sparfloxacin (Table 5). β-Galactosidase activity was seen at concentrations up to 1/2 × MIC for ciprofloxacin but not at higher concentrations, again reflecting the correlation with cell density (Table 5). Sparfloxacin treatment of strain TT12308, at up to 16 × MIC, consistently retained β-galactosidase activity at a low cell density, $\log_{10}$ 4.1 vs 7, when compared to the *in vitro* results. Sparfloxacin was the only compound to be non-inhibitory for β-galactosidase activity in strain TT12308 at 16 × MIC.

The effect of sparfloxacin, tetracycline, and streptomycin on *S. typhimurium* $_x$3181(pGTRO90) was measured when the antibiotics were added at the same time the macrophages were infected. Sparfloxacin was chosen for testing based on potency while tetracycline was used because it was the only agent to induce the *spv*A::*lac*Z fusion, as expected. Streptomycin was tested to determine if other protein synthesis inhibitors induced activity. Sparfloxacin at 16 × MIC reduced intracellular bacterial viability, measured in CFU/ml, 4.7 logs from $\log_{10}$ 6.9 to 2.2, and streptomycin at 8 × MIC decreased viability 2.1 logs, from $\log_{10}$ 6.9 to 4.8. Neither antibiotic induced β-galactosidase activity at any concentration tested. Tetracycline reduced bacterial viability over a MIC range

Table 6. Viability and β-galactosidase activity of *Salmonella* [x]3181(pGTRO90) in macrophages after treatment with antibiotics.[a]

Drug	Concentrations	CFU/ml log$_{10}$	β-galactosidase[d] activity
Control	-------	6.9	0
tetracycline	1/1000 x MIC[b]	7.0	0
	1/500 x MIC	6.8	±
	1/250 x MIC	7.1	1
	1/100 x MIC	6.7	3
	1/50 x MIC	6.3	2
	1/25 x MIC	6.2	2
	1/10 x MIC	6.7	2
	1/5 X MIC	6.0	2
	1/2 X MIC	5.8	0
	1 x MIC	5.1	0
	2 x MIC	4.8	0
sparfloxacin	16 x MIC[c]	2.2	0

[a] Methods. Same as Table 5.

[b] MIC = 100 μg/ml

[c] MIC = 0.012 μg/ml

[d] Rated on a scale from 0 - 4 with 4 being maximum activity.

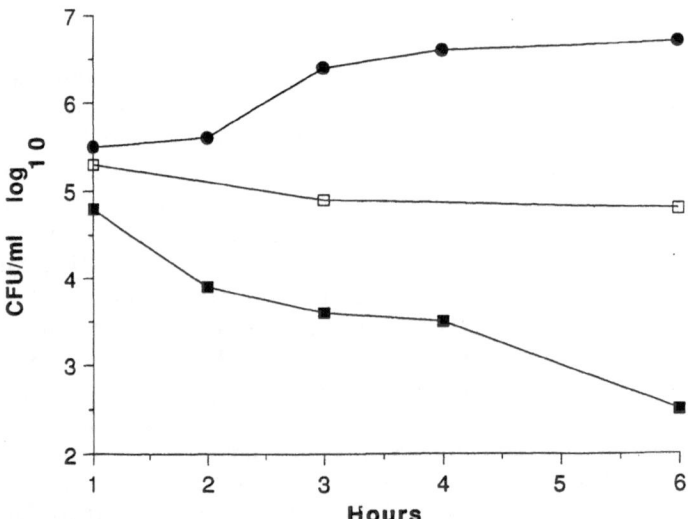

Figure 2. Effect of sparfloxacin and tetracycline on the growth of *Salmonella* [x]3181(pGTRO90) in macrophages. Antibiotics were added at the time of infection. [x]3181(pGTRO90) growth control (●), tetracycline 2 x MIC (□), and sparfloxacin 16 x MIC (■).

of 1/1000 to 2 × MIC. from a CFU/ml log$_{10}$ 7.0 to 4.8 (Table 6). Tetracyline at 2 × MIC was bacteriostatic following a six hour intracellular exposure with the infected macrophages (Fig. 2). β-Galactosidase activity was strongly induced at tetracycline concentrations of 1/100 to 1/5 × MIC. At 1/2 × MIC, activity was 50% of control values while at the MIC, no β-galactosidase activity was detected, indicative of a lack of *de novo* protein synthesis.

Active efflux and detoxification mechanisms have been shown to be active in macrophages (Cao, 1992). In order to test whether diminished antibiotic effects on cell viability or gene expression was due to one of these mechanisms, probenecid was tested in combination with the above system. Probenecid is an inhibitor of carrier-mediated transport of anionic compounds (Buisman, 1991, Steinberg, 1987, Henderson, 1989). Probenecid, when added at the time of simultaneous infection and treatment with ciprofloxacin, further reduced bacterial viability (CFU/ml) by log$_{10}$ 0.6 for strain TT12308. This effect appears to be additive in that viability decreased arithmatically when the combination of probenecid and ciprofloxacin was compared to either agent alone (Fig. 3).

Effects on internalized infection

To study the effects of antibiotics on the viability and β-galactosidase expression of internalized *S. typhimurium*, test compounds were added to

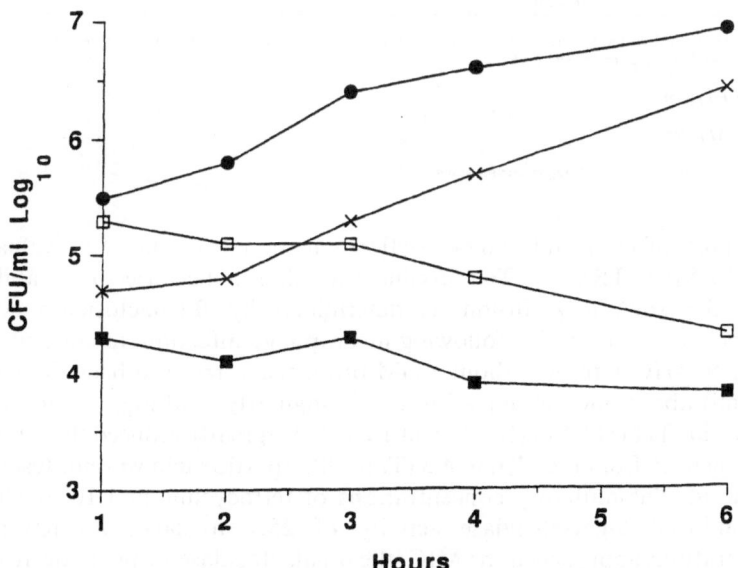

Figure 3. Effect of ciprofloxacin and probenecid on the viability of *Salmonella* TT12308 in macrophages. Ciprofloxacin and probenecid were added at the time of infection. TT12308 growth control (●), probenecid 2mM (X), ciprofloxacin 16 × MIC (□), ciprofloxacin 16 × MIC and probenecid 2mM (■).

macrophage monolayers 3 hours post infection with the bacteria and following gentamicin treatment. The cells were incubated in the presence of antibiotics another 6 hours, and intracellular viability measured by CFU/ml and β-galactosidase activity were determined. Ciprofloxacin with and without probenecid and sparfloxacin singularly were tested against strain TT12308. Ciprofloxacin at 16 × MIC with and without probenecid did not reduce bacterial viability. CFU/ml of $\log_{10}$ 5.7 to 5.6 were obtained. Sparfloxacin at 16 × MIC did reduce the number of viable internalized bacteria by 1.3 logs from CFU/ml of $\log_{10}$ 5.7 to 4.4. Ciprofloxacin did not affect β-galactosidase activity, while the combination of ciprofloxacin and probenecid reduced enzyme activity by 25%. Sparfloxacin also reduced β-galactosidase activity by 25% (Table 7).

Table 7. Viability and β-galatosidase activity of *Salmonella* TT12308 in macrophages treated with antibiotics 3 h post infection.[a]

Drug	Concentrations	CFU/ml $\log_{10}$	β-galactosidase[d] activity
Control	-------	5.7	4
probenecid	2mM	5.7	4
ciprofloxacin	16 x MIC[b]	5.6	4
ciprofloxacin	16 x MIC		
+ probenecid		5.6	3
sparfloxacin	16 x MIC[c]	4.4	3

[a] Macrophages were infected with bacteria for 3 h, extracellular bacteria were eradicated, and the macrophages were treated with antibiotics and probenecid for 6 h.

[b] MIC = 0.003 μg/ml

[c] MIC = 0.012 μg/ml

[d] Rated on a scale of 0 - 4 with 4 being maximum activity.

The post infection addition of sparfloxacin and tetracycline was evaluated in Strain $_x$3181(pGTRO90). Tetracycline was chosen because of its ability to induce the *spv*A::*lac*Z fusion as determined by β-galactosidase activity. Tetracycline treatment 3 h following macrophage infection, at concentrations up to 2 × MIC with or without 2 mM probenecid, reduced bacterial viability (CFU/ml) about one log, $\log_{10}$ 7.6 to 6.3 singularly, and $\log_{10}$ 7.6 to 6.2 with probenecid (Table 8). Sparfloxacin at 16 × MIC greatly reduced the number of viable bacteria from $\log_{10}$ 7.6 to 4.3 (Table 8). Sparfloxacin was not tested with probenecid. Subinhibitory concentrations of tetracycline at 1/10 to 1/100 × MIC induced β-galactosidase activity of 25% to 50%. As tetracycline concentrations approached the MIC, the β-galactosidase response decreased to 0. Tetracycline with probenecid abolished β-galactosidase acitvity. Sparfloxacin treatment resulted in a lack of detectable β-galactosidase activity at all concentrations tested.

Table 8. Viability and β-galactosidase activity of *Salmonella* $_x$3181(pGTRO90) in macrophages treated with antibiotics 3 h post infection.[a]

Drug	Concentrations	CFU/ml log$_{10}$	β-galactosidase[d] activity
Control	-------	7.6	0
probenecid	2mM	7.2	0
tetracycline	1/500 x MIC[b]	7.6	±
	1/100 x MIC	7.0	3
	1/50 x MIC	6.9	2
	1/10 x MIC	6.8	2
	2 x MIC	6.3	0
probenecid			
+ tetracycline	1/500 x MIC	7.3	0
	1/100 x MIC	6.4	0
	1/50 x MIC	6.2	0
	1/10 x MIC	6.5	0
	2 x MIC	6.2	0
sparfloxacin	16 x MIC[c]	4.3	0

[a] Macrophages were infected with bacteria for 3 h, extracellular bacteria were eradicated, and the macrophages were treated with antibiotics and probenecid for 6 h.

[b] MIC = 100 μg/ml

[c] MIC = 0.012 μg/ml

[d] Rated on a scale of 0 - 4 with 4 being maximum activity.

Discussion

The effects of antibiotics on the internalization of *Salmonella* into macrophages and on the intracelllular gene expression were evaluated. Three measures were used to determine these effects. First, bacterial viability was determined by measuring the number of colony forming units recovered from internalized bacteria in macrophages. Secondly, the metabolic state of the intracellular bacteria was determined by assessing the activity of a purine biosynthetic pathway. Thirdly, the ability of antibiotics to modulate the expression of a virulence factor was also assessed by use of an inducible *spvA::lacZ* fusion.

The ability of antibiotics to affect engulfment of *S. typhimurium* by macrophages was evaluated. Antibiotics, when added simultaneously with bacteria at the time of macrophage infection, should mimic a potential prophylactic clinical situation. The results from this model indicate that cellular viability and metabolic activity were directly related. Antibiotics which reduced viability also reduced metabolic activity, as expected. Ciprofloxacin and sparfloxacin, quinolone antibiotics, were evaluated because of their potency, rapid bactericidal activity, and their ability to coordinately inhibit all major metabolic pathways. As seen for strain TT12308, ciprofloxacin at ≥ 1/2 × MIC reduces both viability and metabolic activity in a concentration dependent manner. Recovery of viable bacteria post treatment indicates that either length of exposure or internal ciprofloxacin concentration was insufficient to kill all bacteria at the concentrations tested. Other studies report similar results with ciprofloxacin and suggest that uptake is a simple diffusion process dependent on internal vs external equilibrium (Easmon, 1985). A reduced intracellular concentration effect is supported by the $\log_{10}$ 0.6 further reduction in CFU/ml when probenecid was added to block efflux. Sparfloxacin, at 16 × MIC, reduced viability by 2.5 logs while only reducing *pur*G::*lac*Z expression by 50%. Sparfloxacin also was shown to be active in other intracellular models studying intracellular viability only (Rastogi, 1991). Sparfloxacin was unique for its ability to potentiate a greater than two log reduction in CFU/ml while still maintaining a qualitatively higher metabolic state as indicated by β-galactosidase activity.

The expression of the *spv*A gene in strain ₓ3181(pGTRO90) was induced by sub-MIC levels of tetracycline because of the presence of a tetracycline resistance promoter upstream from the *spv* region of the virulence plasmid. Bacterial viability decreased as the MIC of tetracycline was approached. Sparfloxacin, as well as streptomycin, caused a greater than 2 log reduction in CFU/ml while not inducing the virulence gene fusion. Streptomycin was used to test whether protein synthesis inhibition in general would induce the *spv*A::*lac*Z fusion. The lack of induction by a separately targeted protein synthesis inhibitor would suggest that protein synthesis inhibition alone is not responsible for the induction. Other targeted protein synthesis inhibitors were not evaluated.

The data from the simultaneous addition experiment indicate that both viability and gene expression can be assessed. However, it cannot separate the effects of initial antibiotic exposure at the time of addition and infection, as compared to later intracellular effects, where the antibiotic must effectively concentrate in the macrophage.

The internalized infection experiments should simulate the clinical situation of active infection where chemotherapy is initiated post infection, and the antibiotic must effectively penetrate and concentrate in an intracellular environment. Ciprofloxacin was ineffective in causing either a reduction in viability or change in metabolic status in strain TT12308. The addition of probenecid did not affect the antibacterial activity of ciprofloxacin but did cause a reduction in metabolic activity. Sparfloxacin demonstrated greater antibacterial activity and caused a similar reduction in metabolic state, consistent with earlier observations.

When strain ₓ3181(pGTRO90) was used in the internalized experiments, less reduction in bacterial viability was seen for most of the antibiotics tested even at high MIC multiples. Sparfloxacin still produced about a 3 log decrease in viability. Modulation of *spv*A expression by tetracycline was still apparent. Addition of probenecid resulted in no enhancement of antibacterial activity as measured by viability, but did block the ability of tetracycline to induce *spv*A expression. The reason for this is unknown.

In summary, the effects of antibiotics on the intracellular pathogen,

Salmonella typhimurium, are evaluated. The experiments simulate two separate clinical presentations and the results indicate that both bacterial viability and gene expression can be studied independently. Differential antibacterial efficacy was seen in these models and did correlate to standard *in vitro* inhibition measurements. Gene expression of both metabolic status and a *Salmonella typhimurium* virulence factor was modulated as a function of antibiotic concentration.

Acknowledgements

We would like to acknowledge the assistance of our colleagues at Parke Davis, Jyoti Bhatia, Jeffery Gage, Paul Miller, Maureen Dazer, James Saunders, Kenny Lee, Heather Lewandowski, and Jacqui Ficaj. We would also like to thank Bruce Donohoe from the University of Michigan for the electron micrograph and Paul Gulig from the University of Florida for the $_x$3181(pGTRO90) strain.

References

Buchmeier NA & Hefron F (1990) Induction of *Salmonella* stress proteins upon infection of macrophages. Science 248: 730–732.

Buisman HP, Buys LFM, Langermans JAM, Van Den Broek PJ & Van Furth R (1991) Effect of probenecid on phagocytosis and intracellular killing of *Staphylococcus aureus* and *Escherichia coli* by human monocytes and granulocytes. Immunol. 74: 338–341.

Caldwell AL & Gulig PA (1991) The *Salmonella typhimurium* virulence plasmid encodes a positive regulator of a plasmid-encoded virulence gene. J. Bacteriol. 173: 7176–7185.

Cao CX, Silverstein SC, Neu HC & Steinberg TH (1992) J774 macrophages secrete antibiotics via organic anion transporters. J. Infect. Dis. 165: 322–328.

Easmon CSF, & Crane JP (1985) Uptake of ciprofloxacin by macrophages. J. Clin. Pathol. 38: 442–444.

Eustice DC, Feldman PA, Colberg-Poley AM, Buckery RM & Neubauer RH (1991) A sensitive method for the detection of β-galactosidase in transfected mammalian cells. Biotech. 11: 739–742.

Falkow S (1991) Bacterial entry into eukaryotic cells. Cell 65: 1099–1102.

Finlay BB & Falkow S (1988) Virulence factors associated with *Salmanella* species. Microbiol. Sci. 5: 324–328.

Finlay BB & Falkow S (1989) Common themes in microbial pathogenicity. Microbiol. Rev. 53: 210–230.

Galan JE, Ginocchio C & Costeas P (1992) Molecular and functional characterization of the *Salmonella* invasion gene *inv*A: homology of invA to members of a new protein family. J. Bacteriol. 174: 4338–4349.

Ginocchio C, Pace J, & Galan JE (1992) Identification and molecular characterization of a *Salmonella typhimurium* gene involved in triggering the internalization of salmonellae into cultured epithelial cells. Proc. Natl. Acad. Sci. USA 89: 5976–5980.

Goldberg MB & Rubin RH (1988) The spectrum of *Salmonella* infection. Infect. Dis. Clin. North Am. 2: 571–598.

Gulig PA (1990) Virulence plasmids of *Salmonella typhimurium* and other salmonellae. Microb. Pathog. 8: 3–11.

Gulig PA, Caldwell AL & Chiodo VA (1992) Identification, genetic analysis, and DNA sequence of a 7.8 kb virulence region of the *Salmonella typhimurium* virulence plasmid. Mol. Microbiol. 8: 1395–1411.

Guo-Min T (1988) A repressor gene for the purine biosynthetic pathway of *Salmonella typhimurium*. Abstract H-163, Am. Soc. Microbiol. 88th General Meeting.

Henderson GB (1989) Mediation of cellular anion detoxification in leukemic cells by unidirectional efflux pumps. Adv. Enzyme Regul. 29: 61–72.

McFarland WC & Stocker BA (1987) Effect of different purine auxotrophic mutations on mouse virulence of a vi-positive strain of *Salmonella dublin* and two strains of *Salmonella typhimurium*. Microb. Pathog. 3: 129–141.

Miller JH (1972) Assay of β-galactosidase, p. 352–360. Experiments in Molecular Genetics. Cold Spring Harbor Laboratory, Cold Spring Harbor, NY.

Noumi T, Nishida N, Minami S, Watanabe Y & Yasuda T (1990) Intracellular activity of tosufloxacin (T-3262) against *Salmonella enteritidis* and ability to penetrate into tissue culture cells of human origin. Antimicrob. Agents Chemother. 34: 949–953.

Pascual A, Garcia I & Perea EJ (1992) Entry of lomefloxacin and temafloxacin into human neutrophils, peritoneal macrophages, and tissue culture cells. Diagn. Microbiol. Infect. Dis. 15: 393–398.

Peterson PK & Verhoef J (1988) The Antimicrobial Agents Annual 3. Elsevier, Amsterdam.

Pollack C, Straley SC, & Klempner MS (1986) Probing the phagolysosomal environment of human macrophages with a Ca_2+-responsive operon fusion in *Yersinia pestis*. Nature (London) 322: 834–836.

Portnoy DA & Smith GA (1992) Devious devices of *Salmonella*. Nature (London) 357: 536–537.

Pratt WB & Fekety R (1986) The Antimicrobial Drugs. Oxford University Press, New York.

Rastogi N, Labrousse V, Goh KS & De Sousa JPC (1991) Antimycobacterial spectrum of sparfloxacin and its activities alone and in association with other drugs against *Mycobacterium avium* complex growing extracellularly and intracellularly in murine and human macrophages. Antimicrob. Agents Chemother. 35: 2473–2480.

Steinberg TH, Newman AS, Swanson JA & Silverstein SC (1987) Macrophages possess probenecid-inheritable organic anion transporters that remove fluorescent dyes from the cytoplasmic matrix. J. Cell Biol. 105: 2695–2702.

Une T & Osada Y (1988) Penetrability of ofloxacin into cultured epithelial cells and macrophages. Arzneim.-Forsch./Drug Res. 38 no. 9: 1265–1267.

Vladoianu IR, Chang HR & Pechere JC (1990) Expression of host resistance to *Salmonella typhimurium*: bacterial survival within macrophages of murine and human origin. Microb. Pathog. 8: 83–90.

Elaboration of pathogenic factors

20. Structural and functional analysis of HpmA hemolysin of *Proteus mirabilis*

TIMOTHY S. UPHOFF and RODNEY A. WELCH

Abstract. The calcium-independent hemolysins of *Proteus mirabilis* (HpmA) and *Serratia marcescens* (ShlA) have significant amino acid sequence similarity. These hemolysins are secreted extracellularly when expressed in *E. coli*. This secretion requires an amino-terminal leader sequence and the transport proteins HpmB or ShlB. We have examined truncated forms of HpmA with carboxyl-terminal deletions of various lengths on the basis of hemolytic activity and extracellular secretion. A domain necessary for stable extracellular hemolytic activity was identified between amino acids 1,034 and 1,165 of HpmA. An HpmA truncated form possessing the amino-terminal 872 amino acids of HpmA was the smallest peptide which exhibited detectable cell-bound hemolytic activity. In addition, six HpmA::PhoA protein fusions resulting from TnphoA mutagenesis of pWPM100 were examined. All of these fusion proteins were secreted to the periplasm and exhibited alkaline phosphatase activity. Despite having all of the domains necessary for secretion of HpmA, none of these protein fusion were secreted extracellularly. Finally, a comparison between these findings and the results of previously reported ShlA structure/function analyses demonstrated conservation of functional domains between these two hemolysins.

Abbreviations: EDTA, Ethylene Diamine Tetraacetic Acid

Introduction

The amino acid sequences of the *Proteus mirabilis* HpmA and HpmB hemolysin proteins are very similar to the ShlA and ShlB pore forming hemolysin proteins, respectively, of *Serratia marcescens* (Uphoff and Welch, 1990). Predicted structural domains are also highly conserved among these proteins (Uphoff and Welch, 1990). HpmA and ShlA possess leader peptides of 29 and 30 amino acids, respectively (Poole *et al.*, 1988; Uphoff and Welch, 1990). The carboxyl-terminus of ShlA is very hydrophobic; however, this degree of hydrophobicity is not conserved in the carboxyl-terminus of HpmA. Amino acids 100 to 300 in both proteins are also predominantly hydrophobic but to a lesser extent. Based on a Goldman-Engleman-Steiz hydrophobicity scale, no stretches of amino acids (aside from the leader peptides) appear long enough to span a lipid membrane (Devereux *et al.*, 1984). Amphipathic domains may be involved in formation of HpmA and ShlA pores. Cytolysins often possess bundles of amphipathic α-helices or β-sheets which can form a hydrophilic face on the interior of the pore and interact with the hydrophobic membrane (Ojicius and Young, 1991).

C.I. Kado and J.H. Crosa (eds.), Molecular Mechanisms of Bacterial Virulence, 283–292.
© *1994 Kluwer Academic Publishers. Printed in the Netherlands.*

Early studies of ShlA expression in *E. coli* suggested that the hemolysin was cell-bound and not secreted extracellularly (Braun *et al.*, 1987, Poole *et al.*, 1988). A set of carboxyl-terminal truncated ShlA polypeptides were studied. A polypeptide containing the amino-terminal 1,030 amino acids of ShlA (a deletion of 37%) retains 3% of the full-length ShlA hemolytic activity (Poole *et al.*, 1988). A polypeptide consisting of the amino-terminal 696 amino acids can still insert into red blood cell membranes in a ShlB-dependent manner, but displays no hemolytic activity (Poole *et al.*, 1988). It was hypothesized that ShlB is necessary for activation but not secretion of ShlA (Poole *et al.*, 1988).

These results contrasted with our findings regarding HpmA which is secreted extracellularly from *E. coli* by an HpmB-dependent mechanism (Uphoff and Welch, 1990). Swihart and Welch (1990a) constructed a *Cla*I deletion of *hpmA* (pWPM99) which produces a nonhemolytic peptide containing the amino-terminal 356 amino acids of HpmA. This truncated form of HpmA is still secreted extracellularly from *E. coli* (Swihart and Welch, 1990a). Another construct, pWPM97, fused these 356 amino acids to amino acids 953 to 1,220 of HpmA. This form is also secreted from *E. coli* and does not exhibit any hemolytic activity (Swihart and Welch, 1990a).

In the present study, we have constructed additional carboxyl-terminal truncated forms of HpmA and have examined HpmA::PhoA protein fusions to further define functional domains necessary for extracellular secretion and hemolytic activity. In addition, we compare these results with recent studies of ShlA which demonstrated that this protein is also secreted extracellularly from *E. coli* (Schiebel *et al.*, 1989).

Results

HpmA *deletion mutants*

Thirty-five *hpmA* deletion mutants generated by exonuclease III treatment of pWPM100 were chosen on the basis of varying hemolytic phenotypes on a 5% sheep blood agar plate. Based on restriction endonuclease, immunoblot, and DNA sequence analyses, eight clones were selected for further study (WPM145, 147, 149, 152, 154,155, 156, and 165). The structural and phenotypic characteristics of the truncated forms of HpmA produced by these clones were compared to those of previously described clones in Table 1. Only WPM140, WPM100 and WPM147 exhibited any hemolytic activity in the culture supernatant.

Smaller truncated forms of HpmA produced by WPM149, WPM152, WPM154 and WPM165 demonstrated cell-associated hemolytic activity. No extracellular activity was detected among these clones, although the HpmA antigen is secreted into the supernatant (data not shown). WPM155 and WPM165 exhibit different hemolytic phenotypes yet produce HpmA truncated forms that include the same length of HpmA. The phenotypic differences are probably the result of differences in the number of amino acids fused at their carboxyl termini derived from different vector openreading frames.

Table 1. Phenotypes of strains producing truncates of hpmA

Strain	Amino acids from HpmA[a]	Amino acids from vector[b]	Secretion	%Extracellular Hemolytic Activity[c] Cell	Sup
WPM140	1577	none	yes	100	100
WPM100	1220	49	yes	10	10
WPM147	1165	55	yes	11	10
WPM149	1034	IF lacZ'	yes	10	<0.1
WPM152	977	IF lacZ'	yes	5	<0.1
WPM154	897	IF lacZ'	yes	1	<0.1
WPM155	872	55	yes	<0.1	<0.1
WPM165	872	5	yes	1	<0.1
WPM287	862	IF PhoA	no	<0.1	<0.1
WPM156	449	15	yes	<0.1	<0.1
WPM97	623	49	yes	<0.1	<0.1
WPM99	356	none	yes	<0.1	<0.1
WPM282	142	?	no	<0.1	<0.1

a Refers to the amino-terminal amino acids of HpmA except pWPM97 which is an in-frame deletion of amino acids 357 to 952 from pWPM100.
b pWPM97, pWPM99 and pWPM282 have not been confirmed by DNA sequence analysis. IF lacZ', in frame protein fusion with the ß-galactosidase α-fragment. IF PhoA, in frame protein fusion with alkaline phosphatase.
c Expressed as the percentage of the hemolytic activity of full length HpmA (from pWPM140). Cell; whole culture sample: Sup; cell-free culture supernatant sample.

Growth media, reagents and chemicals. The reagents and conditions used for bacterial growth and recombinant DNA methods were identical to those described in previous publications (Welch, 1987; Uphoff and Welch, 1990).

Bacteria and Bacteriophage Strains. The construction of WPM 100 and TnphoA-mediated mutagenesis of that same plasmid were previously described (Welch, 1987). Construction of WPM97, WPM99, and WPM140 have also been described (Swihart and Welch, 1990a, Uphoff and Welch, 1990). A 1.2 kb *Afl*III fragment was isolated from pWPM140, the cohesive ends were made blunt by T7 DNA plymerase treatment, and the fragment was cloned into the *Sma*I site of pUC19 to produce pWPM282.

The construction of pWPM100 deletion mutants was performed by isolating a set of 3' nested deletions of *hpmA* using a commercial kit (Eras-a-base; Promega, Madison, WI) that is based on the exonuclease III digestion method described by Henikoff (Henikoff, 1984). Ten micrograms of pWPM100 DNA was digested with *Bam*HI, and this site was endfilled with alpha-phosphorothioate dNTPs, according to the manufacturer's protocol. The *Bam*HI-digested plasmid was then digested with *Sal*I. During the exonuclease III treatment, samples were removed at 30 sec. intervals and held on ice to stop digestion. SI nuclease treatment and ligations were carried out according to the manufacturer's protocol. Ligation mixtures from various time points were transformed into *E. coli* DH1 and plated on blood agar plates containing ampicillin and supplemented with 5% sheep blood. Plasmids pWPM147, 149, 152, 154, 155, and 156 were obtained by this procedure. pWPM165 is a derivative of pWPM155 which has an *Xba*I linker (CTAGTCTAGACTAG, New England Biolabs) inserted at the 3' end of the deletion to provide a translational stop codon in all three reading frames.

Hemolysis assays. Liquid hemolysis assays were performed on supernatant and whole culture samples as previously described (Uphoff and Welch, 1990) with the following exceptions. A 200 µl sample (5 µl for WPM140 whole culture and cell-free supernatants) was mixed with 800 µl of 0.85% saline containing a suspension of washed sheep red blood cells at a final concentration of 2.5 or 5.0%. This mixture was incubated at 37 °C for 10 or 30 min, pelleted in a microcentrifuge for 30 sec and the absorbance of the released hemoglobin in the supernatant was measured spectrophotometrically at a wavelength of 540 nm. Hemolytic activity was reported as a percentage of hemolytic activity compared to the full length HpmA (WPM140) hemolysin.

WPM99 produces the smallest truncated HpmA we have identified that is still secreted extracellularly. A smaller, nonhemolytic, HpmA truncated form consisting of the amino-terminal 142 amino acids of HpmA is produced by WPM282. This truncate is not secreted extracellularly from *E. coli*, even in the presence of HpmB (Fig. 1, lanes 1 and 2).

HpmA::PhoA fusion proteins

TnphoA insertions into pWPM100 were previously constructed (Welch, 1987). A library of TnphoA insertion mutants was grown on LB agar plates supplemented with 5-bromo-4-chloro-indolyl phosphate. From these plates, we selected blue colonies producing active PhoA fusion proteins. By restriction endonuclease and immunoblot analyses, six clones appeared to be producing enzymatically active HpmA::PhoA fusion proteins of various sizes. Our objective was to determine whether any of these fusion proteins were secreted

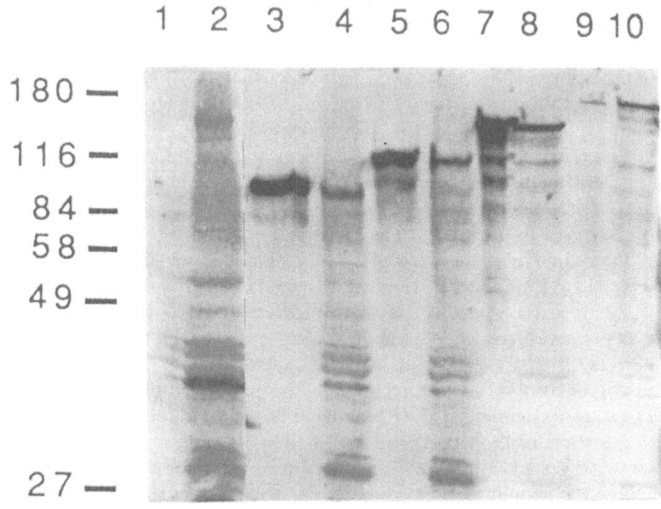

Figure 1. Secretion of HpmA truncated forms. Immunoblot analysis using rabbit polyclonal anti-HpmA antiserum. Lanes: 1 and 2, WPM282; 3 and 4, WPM165; 5 and 6, WPM147; 7 and 8, WPM100; 9 and 10, WPM140. Lanes 1, 3, 5, 7, and 9 contain cell-free supernatant samples; lanes 2, 4, 6, 8, and 10 contain washed cells. Ten µl of sample representing 100 µl of original culture material was loaded in each lane. Positions of prestained molecular mass markers (in kilodaltons) are shown on the left.

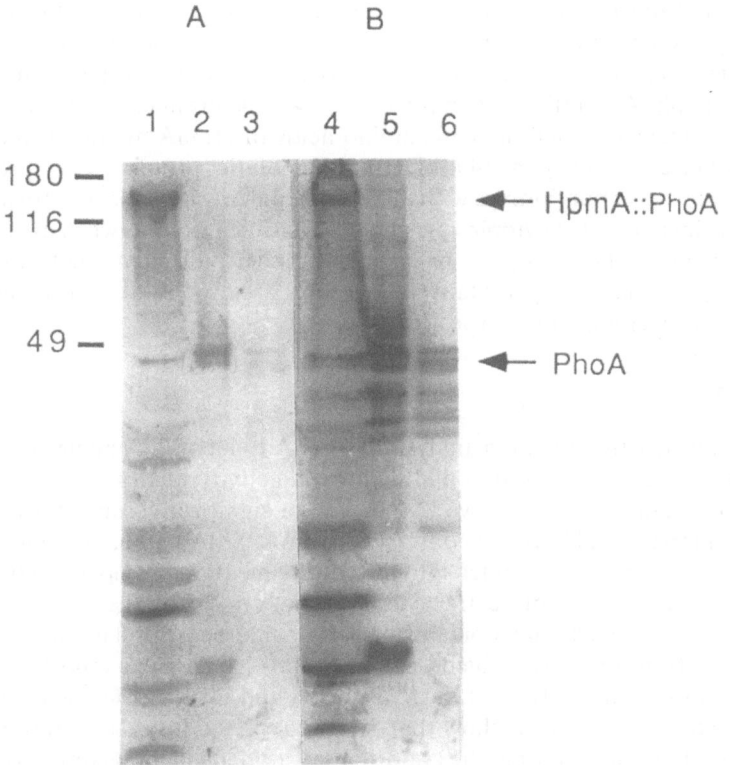

Figure 2. Cellular localization of a representative HpmA::PhoA protein fusion. Immunoblot analysis of WPM287 cellular fractions using rabbit polyclonal anti-HpmA (lanes 1–3) and rabbit anti-alkaline phosphatase (lanes 4–6) antisera. Lanes: 1 and 4 contain spheroplast fractions; 2 and 5 are periplasmic fractions; 3 and 6 are cell-free supernatant fractions. Ten µl of sample representing 200 µl of original culture material was loaded in each lane. Positions of prestained molecular mass markers (in kilodaltons) are shown on the left.

Growth and preparation of *E. coli* strains: Cultures of *E. coli* DH1 or CC118 harboring designated plasmids were grown to an OD600 of 0.9 in LB broth with constant aeration at 37°C and antibiotic selection. Culture samples were centrifuged at 7,000 × g for 10 min and the supernatant was filtered through a 0.2 µm Acrodisc (Gelman Sciences, Ann Arbor, MI). Whole culture samples were taken directly from the culture flask. Strains containing HpmA::PhoA protein fusions were also separated into periplasmic and spheroplast fractions by the following protocol. Five mls of culture was centrifuged and filtered as described above. The cell pellet was resuspended in 1 ml of 0.85% saline. A 400 µl aliquot of these cells was removed, washed once with saline and resuspended in 1 ml of saline. This represented the washed cell fraction. The remaining 600 µl of resuspended cells were centrifuged and the pellet was resuspended in 0.75 ml of spheroplast buffer (100 mM Tris acetate [pH8.2], 0.5 M sucrose, 0.5 mM EDTA), 60 µl of 2 mg/ml lysozyme was added and the cells held at 0–4°C for 1min, 750 uls of distilled water was then added and the cells were held at 0–4°C. After 5 min, 30 µl of 1 M MgSO₄ was added and the spheroplasts were centrifuged (10,000 × g, 3 min). The supernatant was removed as the periplasmic fraction. Spheroplasts were then resuspended in 750 µl of saline. To insure complete lysis, the spheroplasts were disrupted by twenty 0.2 sec bursts using a Branson sonifer at setting 2. The spheroplast fraction was then pelleted by centrifugation at 3,000 × g for 5 min to remove unlysed cells. The supernatant was centrifuged at

into the culture medium from *E. coli*. Immunoblot analysis was performed using anti-HpmA and anti-alkaline phosphatase antisera. None of these TnphoA mutants secreted extracellular HpmA::PhoA fusion proteins (data not shown).

The TnphoA insertion site in pWPM287 was determined by DNA sequence analysis. The amino-terminal 862 amino acids of HpmA are fused to alkaline phosphatase to create a 140 kDa nonhemolytic protein. This protein is representative of our findings with all of the HpmA::PhoA fusions, because it is found primarily in the cytoplasmic and periplasmic fractions when expressed in *E. coli* (Fig. 2). There appears to be some smaller molecular size forms of the PhoA antigen in the supernatant (lane 6, Fig. 2) however, there is no evidence of HpmA or HpmA::PhoA species.

Discussion

We began structure/function analysis of HpmA to identify domains responsible for the extracellular secretion and hemolytic activity of this molecule. At the time we began these studies, we were aware of significant sequence similarities between HpmA and ShlA. In contrast to HpmA, ShlA had been reported to be cell-associated and not secreted into the culture medium. Comparison of HpmA domains necessary for extracellular secretion with the corresponding, apparently nonfunctional, domains of ShlA could be helpful in understanding this secretion process. Schiebel and Braun have since reported further structure/function analyses of ShlA (Schiebel and Braun, 1989; Schiebel *et al.*, 1989). These authors now show that ShlA is secreted into the culture medium via a mechanism requiring ShlB and the leader dependent secretion pathway. Thus, HpmA and ShlA appear to be secreted in an analogous fashion.

Fig. 3 is a representation of a gapped alignment of ShlA and HpmA for comparison of structure/function analyses of these proteins. While we have not further defined the minimum region necessary for secretion, our findings

---

15,000 × g for 30 min. Following centrifugation, the supernatant (cytoplasmic fraction) was removed and the pellet (membrane fraction) was resuspended in 750 μl saline. These samples were used for hemolysis assays and immunoblotting. Assays for β-galactosidase and β-lactamase activities were performed on all cell fractions to assure proper localization of cell compartments (Maniatis 1982).

Immunoblot analysis: Immunoblotting was performed as previously described (Welch, 1987). Trichloroacetic acid (TCA) was added to each supernatant sample to a final concentration of 10%. Following overnight incubation at 4°C, samples were centrifuged for 30 min at 15,000 × g, the pellets dried and resuspended in equal volumes of 2X electrophoresis sample buffer (Laemmli, 1970) and 1M Tris (pH 9.0) to yield a 10-fold concentration of the culture sample. Sample volume loaded on SDS-polyacrylamide gels was 10 μl. Rabbit anti-HpmA antiserum was produced against an electrophoretically purified HpmA carboxyl-terminal truncate encoded by pWPM100 (Swihart and Welch, 1990b). The anti-HpmA antiserum was diluted 1:10,000 in 0.5% Tween 20-phosphate buffered saline and was added to the blots. Anti-alkaline phosphatase antiserum was provided by Steven Lory (University of Washington, Seattle). This antiserum was diluted 1:1,000 for use in immunoblots. The presence of bound primary antibodies was detected using alkaline phosphatase-conjugated goat anti-rabbit antiserum (Sigma Chemical).

regarding the secretion of these truncated forms of HpmA agree with those of ShlA reported by Schiebel and Braun (1989). It was surprising that none of the HpmA::PhoA fusion proteins which we examined were secreted extracellularly. As an example, the fusion protein encoded by pWPM287 should contain all the HpmA domains necessary for extracellular secretion. The full length fusion appears to be localized in the cytoplasm while smaller forms reactive with anti-alkaline phosphatase antiserum are principally localized to the periplasm. Since none of the fusion proteins were secreted extracellularly *in toto*, it appears that the functional block in extracellular secretion is not specific to the the site of the gene fusion. There may be some portion of alkaline phosphatase which is not compatible with this HpmB protein-dependent system of extracellular protein secretion. β-galactosidase is an example of a protein that has several domains which inhibit its secretion past the *E. coli* inner membrane (Lee *et al.*, 1989). Since alkaline phosphatase is found in the periplasm, this inhibition appears to involve the interaction of the fusion protein with HpmB and not the leader dependent secretion system. A study of additional HpmA::PhoA fusion proteins, or different target proteins in the fusion system, is necessary to confirm this hypothesis.

We have defined the minimal amino-terminal domain necessary for HpmA hemolytic activity (pWPM165, 872 aa, 54% of HpmA). This protein is 168 amino acids shorter than the smallest active ShlA truncate reported (pES20, 1030 aa, 63% of ShlA). The region from aa 860 to 874 contains the strongest predicted amphipathic stretch of amino acids in HpmA. This domain is also strongly conserved in ShlA. The predicted secondary structure in this area is a β-sheet, which may interact with other amphipathic β-sheet domains as an integral part of the functional pore. Truncated forms produced by WPM155 and WPM165 contain the same amino acids from HpmA; however, WPM155 has an additional 55 vector-encoded amino acids fused to the end of the HpmA truncate while the WPM165 form contains only an additional 5 amino acids. These carboxyl-terminal 50 amino acids completely abolish detectable hemolytic activity in WPM155. The additional amino acids may be directly interfering with pore formation by inhibiting β-sheet formation in this critical domain. We recognize that such evidence does not prove this domain is part of the pore, since these additional amino acids may be affecting the structural conformation elsewhere in the peptide.

All of the hemolytic activities reported in Fig. 3 represent the whole cell activity. Schiebel and Braun(1989) used urea extracts of whole cells and we used whole culture samples. In addition, we performed hemolysis assays of the culture supernatants of *E. coli* producing HpmA truncated polypeptides. The extracellular hemolytic activities of HpmA and ShlA are very unstable (Schiebel *et al.*, 1989; Uphoff and Welch, 1990). Only two truncated peptides, produced by WPM100 and WPM147, demonstrated extracellular hemolytic activity. The HpmA truncated form produced by WPM149 exhibited the same whole cell hemolytic activity as WPM100 and WPM147, but no detectable cell-free activity, although the antigen was detected in the supernatant. This suggests

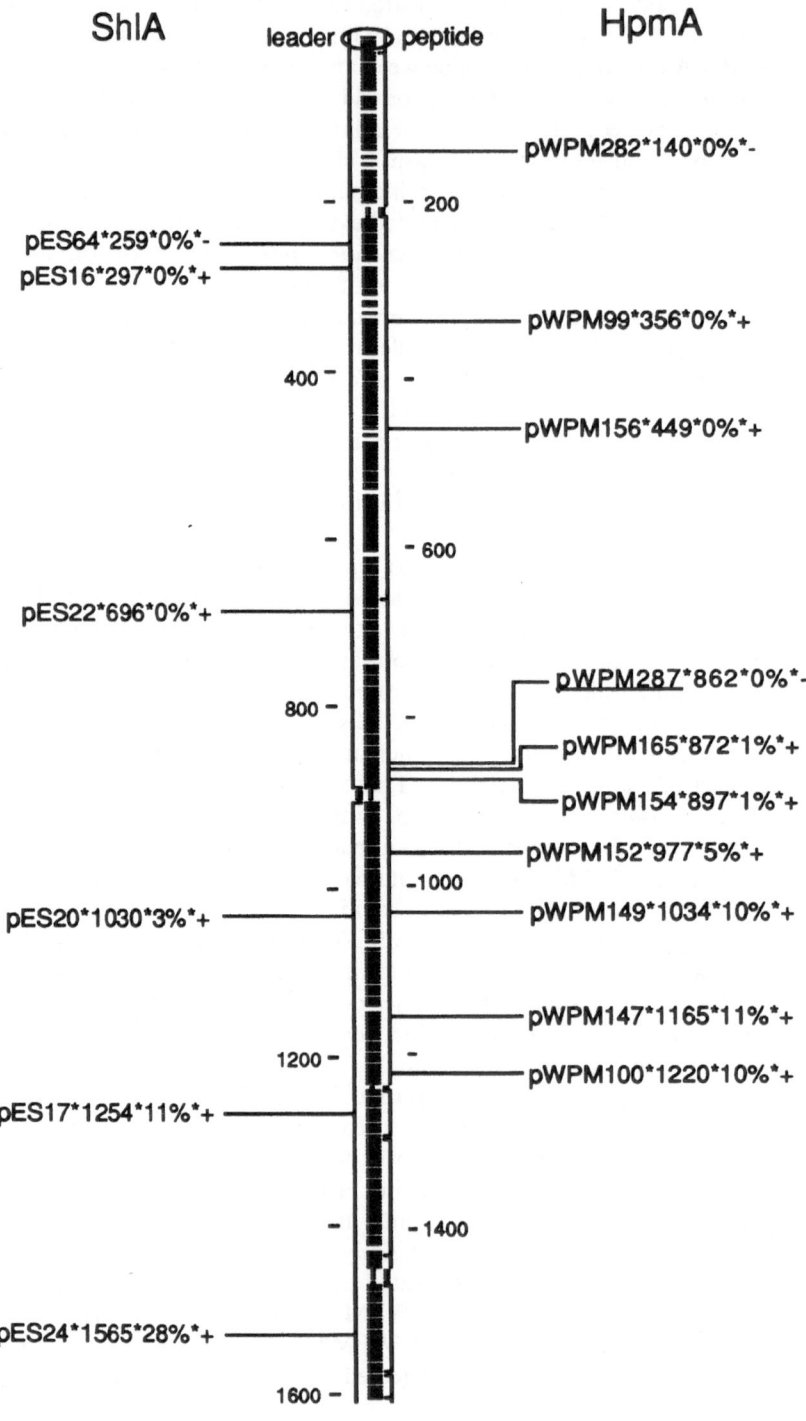

ShlA

HpmA

leader peptide

pWPM282*140*0%*-

- 200

pES64*259*0%*-
pES16*297*0%*+

pWPM99*356*0%*+

400

pWPM156*449*0%*+

- 600

pES22*696*0%*+

pWPM287*862*0%*-

800

pWPM165*872*1%*+

pWPM154*897*1%*+

pWPM152*977*5%*+

-1000

pES20*1030*3%*+

pWPM149*1034*10%*+

pWPM147*1165*11%*+

1200

pWPM100*1220*10%*+

pES17*1254*11%*+

- 1400

pES24*1565*28%*+

1600

Figure 3. Comparison of hpmA and shlA structure/function analyses. Comparison of amino-terminal truncated forms of HpmA and ShlA (10,12). The vertical broken lines represent the gapped alignment of HpmA and ShlA proteins. Short horizontal lines in the center represent matches between the amino acid sequences. The short horizontal lines on each side of the center and the breaks in the vertical lines designate gaps inserted to conserve the alignment between the two sequences. The scale of amino acids is marked by tics which are alternately numbered. The leader peptide domain is encircled by an ellipse at the amino-terminal end of the proteins. The site of each truncate is marked along the length of the gapped alignment. Each is labeled as: the encoding plasmid*, the number of HpmA or ShlA amino acids*, cell associated hemolytic activity (% of full length HpmA or ShlA)*, extracellular secretion of the truncate (+ or −). WPM155 and WPM165 contain the same amino acids from HpmA; however, WPM155 has an additional 55 vector-encoded amino acids fused to the end of the HpmA truncate while the WPM165 form contains only an additional 5 amino acids. WPM155 is not hemolytic but does secrete the HpmA truncate. pWPM287 is underlined to denote that it encodes a TnphoA fusion protein.

DNA sequencing: The dideoxy-sequencing reactions were performed on double stranded plasmid DNA templates using $(\alpha^{-32}P)$ dATP as a radiolabel. The protocols used were described by Kraft *et al.* (1988) and by the the commercial supplier of the Sequenase enzyme (United States Biochemical) based on the method of Sanger *et al.* (1977). Oligonucleotides complementary to pUC18 (AACAGCTATGACCATG) and TnphoA (AATATCGCCCTGAGC) were produced for use as sequencing primers by using an Applied Biosystems DNA synthesizer model 381A according to the manufacturer's protocol. The labeled reaction mixtures were separated by electrophoresis on 8M urea, 6 or 8% polyacrylamide gels. Following electrophoresis, the gels were dried and autoradiograms made using X-ray film (XAR-5 Kodak film).

that a domain between amino acids 1,034 and 1,165 of HpmA is necessary for stabilization of extracellular hemolytic activity. Immunoblot analysis did not reveal any differences in the stability of these proteins, as very little breakdown was detected.

In summary, we have demonstrated a conservation of the functional domains between these related hemolysins and more precisely described domains responsible for hemolytic activity for HpmA.

Acknowledgements

This work was supported by a Pew Scholarship and a Romnes Fellowship to RAW. We thank Steven Lory for the gift of anti-alkaline phosphatase antiserum and Janine Brander for her helpful advice. We also thank the University of Wisconsin Genetics Computer Group technical staff for help in the DNA sequence analysis.

References

1. Braun, V., B. Neuss, Y. Ruan, E. Schiebel, H. Schoffler and G. Jander. 1987. Identification of the *Serratia marcescens* hemolysin determinant by cloning into *Escherichia coli*. J. Bacteriol. 169: 2113–2120.
2. Devereux, J., P. Haeberli and O. Smithies. 1984. A comprehensive set of sequence analysis programs for the VAX. Nucleic Acids Res. 12: 387–395.

3. Henikoff, S. 1984. Unidirectional digestion with exonuclease III creates targeted breakpoints for DNA sequencing. Gene 28: 351–359.

4. Kraft, R., T.J.K. Krauter and L. Leinwand. 1988. Using mini-prep plasmid DNA for sequencing double stranded templates with Sequenase. BioTechniques. 6: 544–546.

5. Laemmli, U.K. 1970. Cleavage of structural proteins during the assembly of the head of bacteriophage T4. Nature (London). 227: 680–685.

6. Lee, C., L. Ping, H. Inouye, E. Brickman and J. Beckwith. 1989. Genetic studies on the inability of β-galactosidase to be translocated across *Escherichia coli* cytoplasmic membrane. J. Bacteriol. 171: 4609–4616.

7. Maniatis, T., E.F. Fritsch and J. Sambrook. 1982. Molecular Cloning. A Laboratory Manual. Cold Spring Harbor Laboratory. Cold Spring Harbor, NY.

8. Ojcius, D.M. and J. D. Young. 1991. Cytolytic pore-forming proteins and peptides: is there a common structural motif? TIBS. 16: 225–229.

9. Poole, K., E. Schiebel and V. Braun. 1988. Molecular characterization of the hemolysin determinant of *Serratia marcescens*. J. Bacteriol. 170: 3177–3188.

10. Sanger, F., S. Nicklen and A.R. Coulson. 1977. DNA sequencing with chain-terminating inhibitors. Proc. Natl. Acad. Sci. USA. 74: 5463–5467.

11. Schiebel, E. and V. Braun. 1989. Integration of the *Serratia marcescens* haemolysin into human erythrocyte membranes. Mol. Microbiol. 3: 445–453.

12. Schiebel, E., H. Schwarz and V. Braun. 1989. Subcellular location and unique secretion of the hemolysin of *Serratia marcescens*. J. Biol. Chem. 264: 16311–16320.

13. Swihart, K.G. and R.A. Welch. 1990. Cytotoxic activity of the *Proteus hemolysin* HpmA. Infect. Immun. 58: 1861–1869.

14. Swihart, K.G. and R.A. Welch. 1990. The HpmA hemolysin is more common than HlyA among *Proteus* isolates. Infect. Immun. 58: 1853–1860.

15. Uphoff, T.S. and R.A. Welch. 1990. Nucleotide sequencing of the *Proteus mirabilis* calcium independent hemolysin genes (*hpmA* and *hpmB*) reveals sequence similarity with the *Serratia marcescens* hemolysin genes (*shlA* and *shlB*). J. Bacteriol. 172: 1206–1216.

16. Welch, R.A. 1987. Identification of two different hemolysin determinants in uropathogenic *Proteus* isolates. Infect. Immun. 55: 2183–2190.

21. Assembly of *Escherichia coli* heat-labile enterotoxin and its secretion from *Vibrio cholerae*

MARIA SANDKVIST, LINDA J. OVERBYE, TITIA K. SIXMA, WIM G.J. HOL and MICHAEL BAGDASARIAN

Abstract. Subunits of the heat-labile enterotoxin of *Escherichia coli* (LT) assemble in the periplasm and are secreted through the outer membrane in *Vibrio cholerae*. Deletions or substitutions of residues at the carboxyl terminus of the B subunit (EtxB) result in mutant polypeptides that assemble into normal pentamers at 30°C but cannot assemble at 42°C *in vivo*. This defect may be suppressed by substitutions of single amino acid residues in regions that interact directly with the modified carboxyl terminus. Carboxyl terminal residues of EtxB thus appear to be required for formation or stabilization of an assembly intermediate of B subunit pentamerization but are not essential for the stability of the final pentamer.

Secretion of the cholera toxin (CT) or of EtxB through the outer membrane of *V. cholerae* requires the functions of several genes that display extensive similarities to genes required for macromolecular translocation in other Gram-negative bacteria. One of the gene products required seems to be a cytoplasmic protein containing ATP-binding domains. It may be a protein involved in the regulatory signal transduction.

Abbreviations: CT, Cholera Toxin; C-Terminal, Carboxyl-Terminal; ELISA, Enzyme-Linked Immunosorbent Assay; EtxA, A Subunit of Heat-Labile Enterotoxin; EtxB, B Subunit of Heat-Labile Enterotoxin; *etxA*, Gene Encoding EtxA; *etxB*, Gene Encoding EtxB; G_{M1}, Galactosyl-N-Acetylgalactosaminyl-(N-Acetylneuraminyl)-Galactosylglucosylceramede; LT, *Escherichia coli* Heat-Labile Enterotoxin; IPTG, Isopropyl-β-D-Thiogalactopyranoside; Kb, Kilobase(s) or 1000 Bp; LT, Heat-Labile Enterotoxin of *E. coli*; PAGE, Polyacrylamide Gel Electrophoresis; SDS-Sodium Dodecyl Sulfate

Introduction

Escherichia coli and *Vibrio cholerae* are, along with rotavirus, responsible for most of the annually recorded one billion cases of acute diarrhea, which result in nearly four million deaths, mostly of children under age five (Holmgren, 1981; Hirschhorn and Greenough, 1991). Enterotoxigenic strains of *E. coli* (ETEC) are also the etiological agent of severe diarrheal diseases in farm animals. In piglets and calfs ETEC induced diarrhea is one of the economically most important diseases and one of the most common infectious cause of mortality (Sussman, 1985; Morris & Sojka, 1985; Fairbrother *et al.*, 1989; Cieslicki, 1989; Duchet-Suchaux, 1991).

Heat-labile enterotoxin (LT) is one of the main pathogenicity factors of enterotoxigenic *E. coli*. It is a multimeric protein composed of five identical B subunits (EtxB) of 11.6 kDa each and a single A subunit (EtxA) of 28 kDa (Gill

C.I. Kado and J.H. Crosa (eds.), Molecular Mechanisms of Bacterial Virulence, 293–309.
© 1994 *Kluwer Academic Publishers*.

et al., 1981). Its biological activity, immunological properties and structure resemble closely those of cholera toxin (CT), produced and secreted by *V. cholerae* (Holmgren, 1973; Moss & Richardson, 1978; Clements & Finkelstein, 1978; Dallas & Falkow, 1980; Gill *et al.*, 1981; Dallas, 1983; Yamamoto *et al.*, 1987). The three-dimensional structure of LT has recently been determined at 2.3 Å resolution (Sixma *et al.*, 1991; 1992). It revealed a doughnut-shaped B subunit pentagonal ring to which the A subunit is attached via interactions between its extended carboxyl terminus and amino acid residues within the highly charged central pore of the B subunit pentamer. The B subunit is responsible for binding the toxin to the G_{M1}-ganglioside receptor of intestinal epithelial cells (Cuatrecasas, 1973; Holmgren, 1973, Sixma *et al.*, 1992). A portion of the A subunit, the A1 peptide, which is formed by proteolytic cleavage of the A subunit in the intestine, is then transferred across the membrane of the epithelial cells. Inside the cell, the A1 peptide activates adenylate cyclase by ADP-ribosylation of the stimulatory GTP-binding protein (Moss and Richardson, 1978). This leads to loss of ions and fluids which results in diarrhea and dehydration (Sack, 1975). If not treated, the dehydration rapidly leads to death.

The genes specifying the subunits of LT, *etxA* and *etxB*, were found on large conjugative plasmids and have been cloned from *E. coli* isolated from both humans and pigs (So *et al.*, 1978; Dallas *et al.*, 1979; Dallas and Falkow, 1980; Yamamoto and Yokota, 1983; Yamamoto *et al.*, 1984; 1987; Leong *et al.*, 1985; Spicer and Noble, 1982). Nucleotide sequence analysis revealed that the genes are part of a contiguous operon; the *etxA* gene is promoter-proximal and its termination codon overlaps the *etxB* initiation codon. Both the A and B subunits are synthesized as precursor proteins with 18 and 21 amino acid N-terminal signal peptides, respectively, and are translocated across the cytoplasmic membrane (Hofstra and Witholt, 1984; Palva *et al.*, 1981). Following removal of the leader peptides, the mature A and B subunits are released into the periplasm, where they assemble noncovalently into AB_5 complexes or into B subunit pentamers if the B subunits are produced alone (Hofstra and Witholt, 1984; 1985; Hirst *et al.*, 1983; 1984a). In contrast to *E. coli*, which retains the LT in the periplasm, *V. cholerae* secrets its toxin into the extracellular medium. This is not due to differences between LT and CT, since LT was successfully secreted into the medium when produced in *V. cholerae* (Neill *et al.*, 1983; Hirst *et al.*, 1984) and CT remained in the periplasm when synthesized in *E. coli* (Pearson and Mekalanos, 1982). Production of B subunits in the absence of A subunits resulted in secretion of the B subunit pentamers through the outer membrane of *V. cholerae*. On the other hand, when only A subunits were produced, they remained cell-associated. It was therefore concluded that the B subunit determines the extracellular location of the toxin and the association between the A and B subunits prior to outer membrane translocation is essential for the successful secretion of the A subunit (Hirst *et al.*, 1984b). It was also suggested that the outer membrane of *V. cholerae* must contain a specific translocation apparatus that provides for the secretion of certain proteins, such as CT, LT (Neill *et al.*, 1983; Hirst and Holmgren, 1987) or the hemagglutinin/protease (HA/protease) (Hanne & Finkelstein, 1982), while other proteins, such as β-lactamase or alkaline phosphatase, remain in the periplasm. This suggestion was recently strengthened by the indication that specific genes are required for the translocation of proteins across the outer membrane into the extracellular medium in several Gram-negative species (for review see Pugsley, 1992).

LT is a convenient model for studying the assembly of multimeric proteins *in vivo* and their translocation through the outer membrane. The structure of the mature B subunit pentamers is so stable that it remains pentameric in the presence of the ionic detergent, SDS, at temperatures up to 70°C. These pentamers migrate as a 45 kDa band on SDS-polyacrylamide gel electrophoresis clearly distinguishable from the 11.6 kDa band of the monomers (Hirst *et al*, 1983; Sandkvist *et al.*, 1987; 1990). It is possible, therefore, to analyze the steps of toxin biogenesis, presented schematically in Fig. 1, by radioactive pulse-chase labelling, fractionation of subcellular components and electrophoresis.

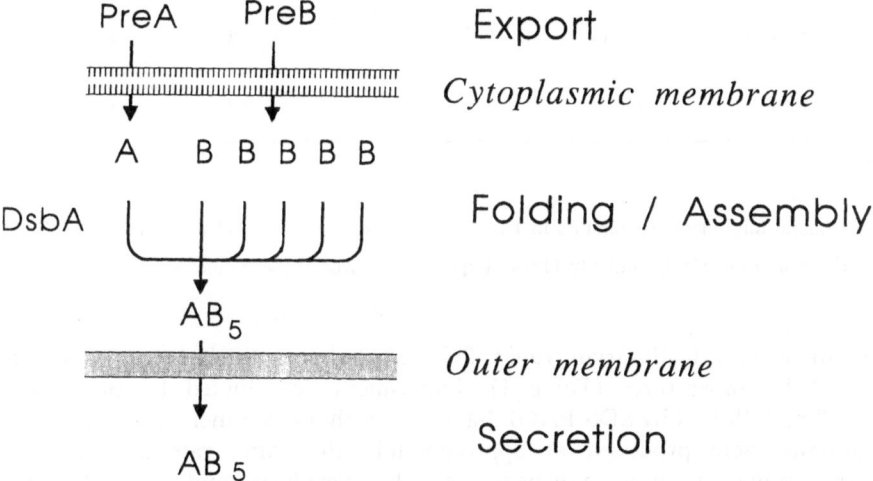

Figure 1. Pathway of biogenesis of the heat-labile enterotoxin in *Vibrio cholerae*. PreA and PreB, precursors of the A and B subunit polypeptides, respectively, AB₅, assembled holotoxin; DsbA, disulfide bond-forming oxidoreductase (Sandkvist 1992; Peek & Taylor, 1992; Yu *et al.*, 1992).

The role of carboxyl terminus in the assembly of the B subunits into pentamers

Mapping of functional domains in the B subunit monomers revealed that the carboxyl terminal amino acids are essential for the assembly into pentameric structure. Deletion or substitution of just a few of the terminal residues by site-directed mutagenesis of the 3′ end of the *etxB* gene resulted in lower yields of B subunit oligomers if assayed by G_{MI}-ELISA (Sandkvist *et al.*, 1987; 1990;

Table 1. Assembly of wild type and mutant B subunits at different temperatures.

B subunit	COOH-terminal residues	B subunit oligomer concentration (ng/ml)[a]	
		30°C	42°C
wt EtxB	...Ala-Ile-Ser-Met-Glu-Asn	3290	2500
EtxB214	...Ala-Ile-Ser-Met-Glu	2990	200
EtxB215	...Ala-Ile-Ser-Met	3000	860
EtxB191.5	...Ala-Ile-**Gly-Leu-Asn**	1910	10
EtxB216	...Ala-Ile-Ser	20	<3
EtxB217	...Ala-Ile	<3	<3
EtxB218	...Ala	<3	<3

[a] B subunit oligomer concentration in 1 ml of cell suspension of A_{650}=1.0 determined by G_{MI}-ELISA with monoclonal antibody 118-8. Experimental error of these determinations is 5-10%.

Sandkvist and Bagdasarian, 1993). This defect was particularly pronounced at elevated temperatures (Table 1). Experiments conducted by pulse-chase labelling followed by SDS-PAGE have shown that monomers of the mutant B subunits were produced at approximately the same rate as wild type polypeptides. They were exported to the periplasm and assembled into pentamers at 30°C, but their assembly into pentamers was inhibited at 42°C (Sandkvist and Bagdasarian, 1993; see also Fig. 2).

Once formed at permissive temperature, the pentamers of mutant B subunits remained as stable as those of the wild type subunits *in vivo* when the temperature was raised to 42°C and *in vitro* at temperatures up to 65°C. Also, monomers of the mutant B subunits produced at 42°C were able to associate into stable pentamers when the temperature was lowered to 30°C (Sandkvist and Bagdasarian, 1993). This indicated that carboxyl-terminal domain is particularly important in the early stages of subunit assembly but once other subunit-subunit interactions have been established, carboxyl-terminal residues no longer play a decisive role in the overall stability of the pentamer.

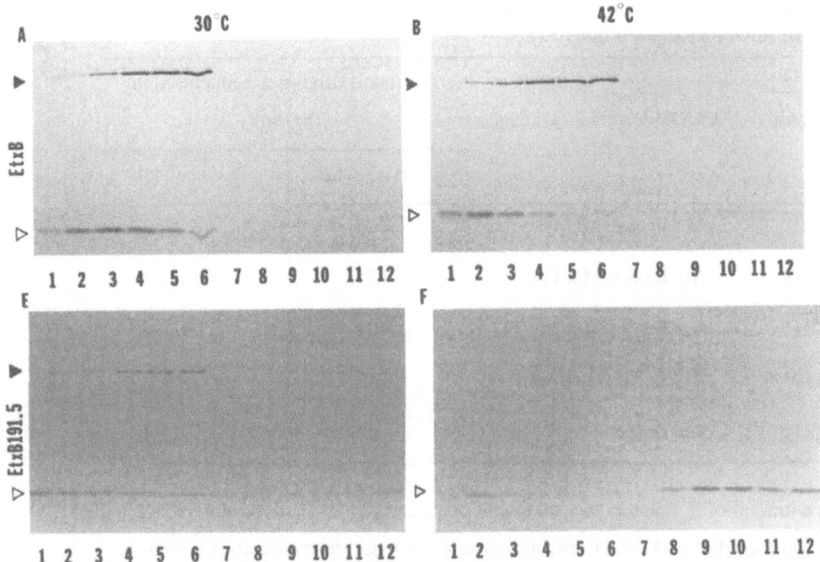

Figure 2. Effect of temperature on the B subunit assembly. Cultures of *E. coli* strain KS476 producing wild type or mutant B subunits were grown at 30°C or 42°C, pulse-labeled with [^{35}S]-methionine for 20 s (lanes 1 and 7) and chased with unlabeled methionine for 0.5 min (lanes 2,8); 2 min (lanes 3,9); 5 min (lanes 4,10); 10 min (lanes 5,11); and 20 min (lanes 6,12). The cells were fractionated into periplasm (lanes 1–6) and spheroplasts (lanes 7–12), the B subunits were immunoprecipitated and analyzed without heat treatment by SDS-PAGE and autoradiography. Monomers and pentamers are indicated by open and filled arrowheads respectively.

Suppression of B subunit assembly defect

Although wild type B subunit monomers can efficiently associate into pentameric structures in the absence of the A subunit, the presence of A polypeptides accelerates the assembly (Hardy *et al.*, 1988). This effect of the A subunit was attributed not only to the ability of its carboxyl-terminal residues to "anchor" the A subunit in the central pore of the B pentamer, but also to a number of charged and polar interactions between the A_2 domain and the B pentamer (Sixma *et al.*, 1991; Streatfield *et al.*, 1992).

The presence of A subunit in the cells producing mutant B subunits suppressed the effect of mutations for those mutant B subunits that were still capable of assembly at permissive temperatures. The assembly of EtxB216, however, could not be rescued by the presence of the wild type A subunits (Table 2). Since mutant B subunits formed at permissive temperatures are almost as stable as those of the wild type (Sandkvist and Bagdasarian, 1993) the suppression effect of the A subunit may be attributed to the ability to stabilize an assembly intermediate.

Table 2. B subunit oligomers detected in the absence or presence of EtxA, determined by G_{M1}-ELISA with monoclonal antibody 118–8.

B subunit	COOH-terminal residues	B subunit oligomer concentration (ng/ml)[a]	
		- A subunit	+ A subunit
wt EtxB	...Ala-Ile-Ser-Met-Glu-Asn	725	1071
EtxB214	...Ala-Ile-Ser-Met-Glu	270	1340
EtxB215	...Ala-Ile-Ser-Met	530	1370
EtxB191.5	...Ala-Ile-**Gly-Leu-Asn**	40	690
EtxB216	...Ala-Ile-Ser	20	<3

[a] Concentration of B subunits in 1 ml of cell extract of $A_{650}=1.0$, determined by G_{M1}-ELISA with monoclonal antibody 118-8. Experimental error of these determinations is 5-10%.

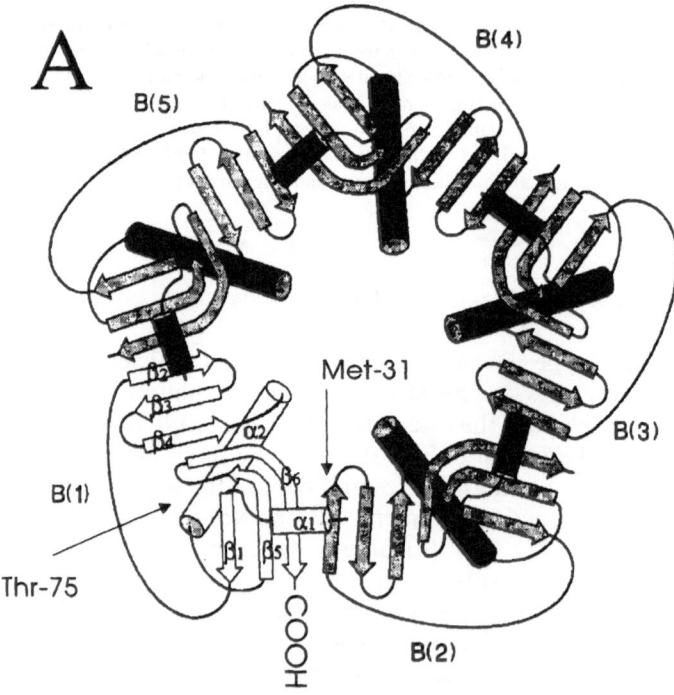

Figure 3A. Schematic secondary structure (projection) of the B subunit pentamer. Modified from (Sixma *et al.*, 1991). Individual B subunits are indicated as B(1), B(2) etc. α-helices and β-strands are numbered in the B(1) subunit. Carboxyl terminus is indicated by COOH in the B(1) subunit. Approximate positions of Met-31 and Thr-75 are indicated by arrows.

In vitro mutagenesis of the *etxB191.5* gene has permitted the isolation of intragenic suppresors of the assembly defect. The effects of two such suppressors is presented in Table 3. Although the last four C-terminal amino acid residues have been replaced by three different residues in EtxB191.5, a single replacement at a different site, distant from the original mutation, could suppress the assembly defect. In the X-ray structure of the B subunit of LT (Sixma *et al.*, 1991) residue Thr-75 is positioned towards the end of the α_2-helix (cosisting of residues 59–78; Fig. 3). This residue is in van der Waals contact with the C^δ of Met-101 in the C-terminal (β_6-strand) of the same B monomer (Fig. 3B and Fig. 4). Substitution of Met-101 by Leu in the mutant EtxB191.5 subunit shortens the residue 101 by one atom resulting in the loss of this contact. The subsequent substitution of Thr-75 by Ile in the suppressor mutant extends the residue in position 75 by one atom and thus presumably restores the lost van der Waals contact. The fact that substitution of Thr-75 restores the ability of EtxB191.5 to assemble at elevated temperatures indicates that this interaction between α_2-helix and the β_6-strand of the same subunit plays an important role

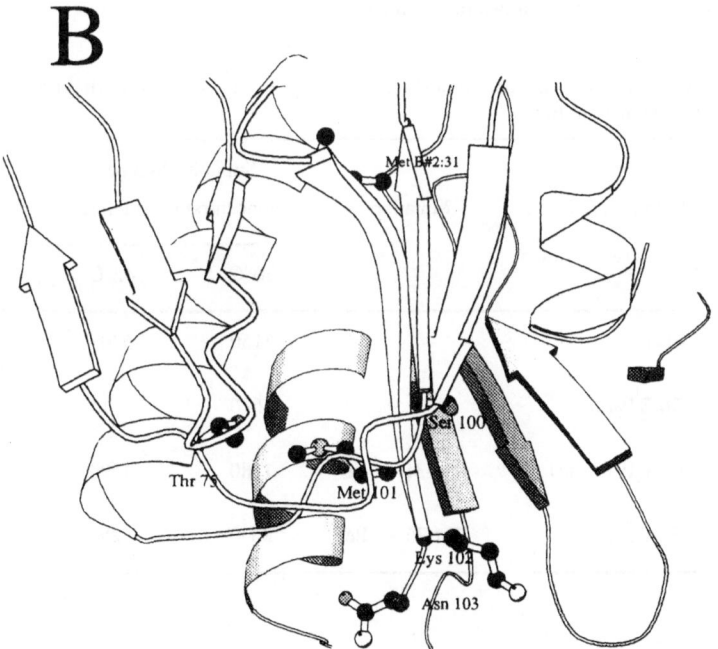

Figure 3B. Schematic MOLSCRIPT (Kraulis, 1991) representation, showing the surroundings of the carboxyl-terminal residues of the EtxB (Sixma *et al.*, 1991). Please note that the porcine variant of LT has a Lys-102 instead of Glu, present in the human LT. Subunit B#1 is drawn in clear and the neighboring subunits in shaded bands. Also shown are two residues, Thr-75 (in B#1) and Met-31 (in the neighboring monomer) which are substituted in suppressor mutants. Note the long distance between Met-31 and the carboxyl-terminal residues in strand β_6.

300

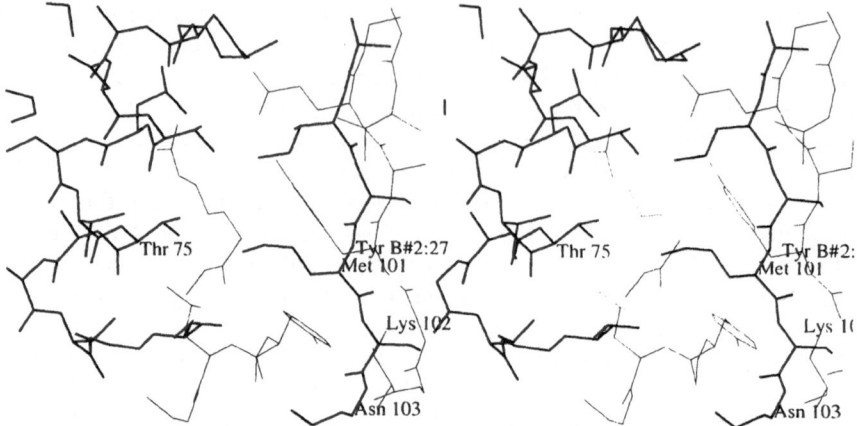

Figure 4. Stereo figure of the surroundings of Thr-75 in the X-ray structure of the wild type porcine LT (Sixma *et al.*, 1991) (the porcine variant has a Lys in position 102 instead of Glu, as one of the four changes in the amino acid sequence as compared to the human B subunit). Note the close proximity to Met-101 in the same subunit. Residues in B#1 are shown in bold lines, residues in the neighboring B#2 subunit are drawn in thin lines.

Table 3. Suppression of the thermosensitive assembly in the mutant EtxB191.5 subunit by secondary intragenic mutations.

Subunit produced	Mutation	Subunit oligomer concentration (ng/ml)[c]	
		30°C	42°C
EtxB	[a]	3150	5300
EtxB191.5	[b]	1520	15
EtxB191.5sup1	Met-31 → Ile	1780	1590
EtxB191.5sup2	Thr-75 → Ile	2640	250

[a] Wild type

[b] Alteration at the carboxyl terminus is shown in Table 1.

[c] Amount of B subunits in cell suspension of $A_{650}=1.0$ determined by G_{M1}=ELISA and monoclonal antibody 118-8. Experimental error of these determinations is 5-10%.

in the stabilization of either folding conformation, essential for assembly, or of an assembly intermediate.

The other identified suppressor mutation, *sup1* (Table 3), is located at codon 31 at the end of the β_2-strand consisting of residues 26–30 (Fig. 3 and Fig. 5). This strand interacts with the β_6-strand of the neighboring subunit (Sixma *et al.*, 1991; see also Fig. 5). The side chain of Met-31 points into a pocket created by the neighboring subunit (Fig. 6) and upon assembly a large hydrophobic area is buried at this location. Mutation to an Ile at this position may change either the monomer stability or the subunit interaction at this site and, therefore, it is difficult to say at this moment why this mutation improves the assembly of the carboxy-terminally altered LT-B polypeptide. Although the effect may be kinetic it could well be that three-dimensional structure of the mutant subunit could give an explanation of the suppressor effect.

The importance of domains or individual amino acid residues in the stabilization of assembly intermediates can not be predicted solely from the structure of the final mature protein. Such information can, however, be obtained from the combination of structural data with the comparison of

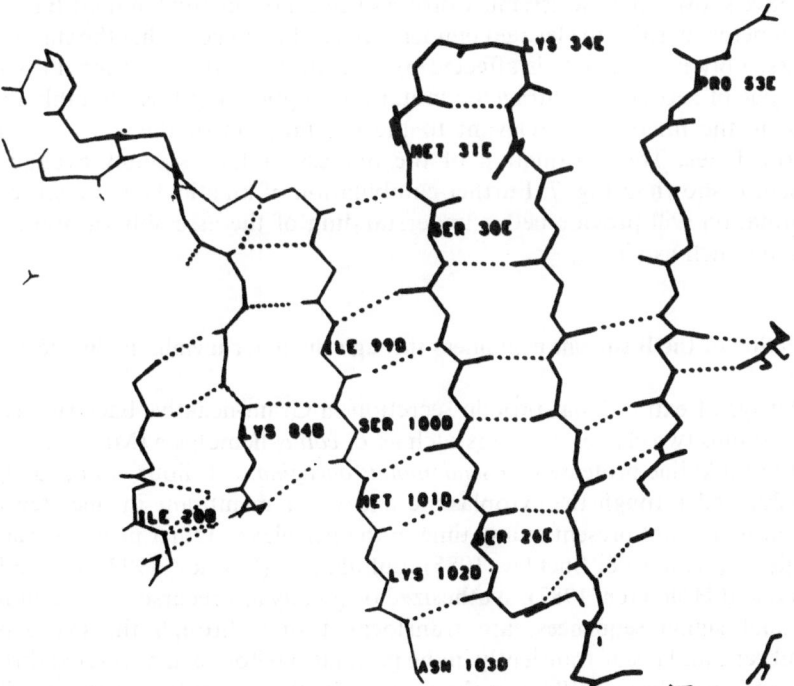

Figure 5. Main-chain hydrogen bonding in the inter-subunit β-sheet of the LT B-subunit pentamer showing the location of Met-31 (in the β_2-strand) with respect to the carboxyl-terminal residues (in the β_6-strand) of the neighboring monomer. Subunit labelling: D, for B#1 and E, for B#2.

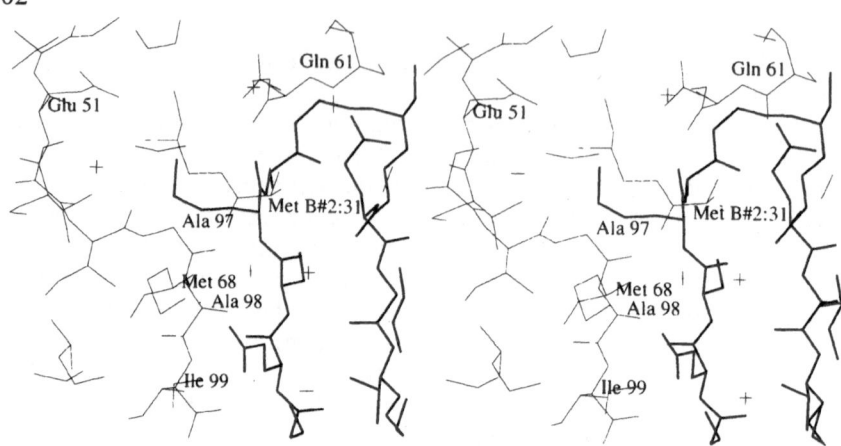

Figure 6. Stereo figure of the X-ray structure showing the surroundings of Met-31 at the end of strand β_2 in the subunit B#2 of the wild type porcine LT. Residues in B#2 are shown in bold lines and residues in the neighboring B#1 monomer in thin lines. Note the way in which the side chain points into a pocket made by neighboring residues.

stability and assembly kinetics of mutant polypeptides and their suppressors. We have shown that the assembly process itself, not the final mutant B subunit pentamer is sensitive to elevated temperatures. This suggests that the stability of an assembly intermediate is affected by alterations of the C-terminal domain. Analysis of second site intragenic mutations suppressing this assembly defect indicate the interactions relevant to the stabilization of this early assembly intermediates. The information of the process of LT assembly available at present is shown in Fig. 7. Further combination of kinetic data and structural information will provide better understanding of the assembly of multimeric proteins such as LT.

Secretion of the B subunit pentamers through the outer membrane in *Vibrio*

Pathways of extracellular protein secretion in Gram-negative bacteria may be divided into two classes. Proteins such as *E. coli* α-hemolysin (Mackman *et al.*, 1986) or alkaline protease of *Pseudomonas aeruginosa* (Filloux *et al.*, 1990) are translocated through the cytoplasmic and outer membrane in one step and, therefore, are not present at any time in the periplasm. Other proteins, such as aerolysin (Howard & Buckley, 1985), pullulanase (Pugsley, 1992), CT and LT (Hirst and Holmgren, 1987), synthesized originally as precursors containing N-terminal signal sequences, are translocated first through the cytoplasmic membrane and reside transiently in the periplasm before being secreted through the outer membrane. While in the periplasm, their periplasmic intermediates may undergo structural modifications that involve folding and assembly as indicated for the LT biogenesis pathway in Fig. 1.

In the past, only one mutant defective in the translocation of CT or EtxB

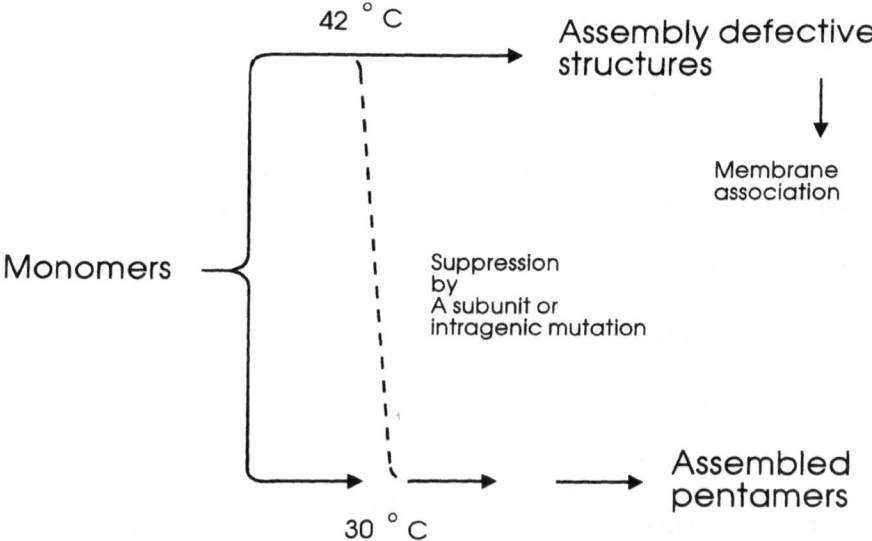

Figure 7. Fate of the mutant B subunits with alterations at the C-terminus, such as EtxB191.5, under different conditions.

pentamers through the outer membrane has been described. This mutant, M14, was isolated from the classical strain of *V. cholerae*, 569B (Holmes *et al.*, 1975; Neill *et al.*, 1983). We have isolated two other classes of secretion mutants (PU-3 and –5 and PU-4 and –6) from the strain TRH7000, a *thy* derivative of the strain JBK70 (Hirst *et al.*, 1984a) carrying a deletion of the *ctxA* and *ctxB* genes. Each of these mutants accumulates EtxB pentamers in the periplasm, but does not secrete them through the outer membrane (unpublished).

We have identified genes that rescue the defect in M14, PU-3 and PU-5 mutants on a 14 kb DNA fragment isolated from a gene library of the strain TRH7000. Location of these genes on the physical and genetic map of this fragment are presented in Fig. 8 and the results of complementation in Table 4. Determination of the nucleotide sequence of the *epsE* gene, which rescues the defect in M14 mutant, indicated that it shares extensive homology to genes essential for protein secretion or DNA translocation through the outer membrane in several different species of bacteria. The summary of homology determination is presented in Table 5. A common feature of all these proteins are the "Walker motifs", considered to represent ATP binding domains (Walker *et al.*, 1982). Fractionation of the complemented M14 cells into periplasmic, membrane and cytoplasmic fractions by differential centrifugation has shown that the EpsE protein, induced with IPTG from an expression vector, was present in the cytoplasmic compartment (Sandkvist *et al.*, 1993). Since the predicted amino acid sequence of the protein did not reveal a recognizable leader peptide sequence, we assume at present that the protein is normally

Table 4. Extracellular secretion of CT or EtxB pentamers from wild type and mutant *V. cholerae* strains.

Strain	Plasmids[a]	Secretion of protease[b]	Secretion of CT or EtxB (fraction of total)[c]
569B	None	+++	0.78
M14	None	-	0.08
M14	pMMB384	++	0.60
TRH7000	pWD615	+++	0.77
PU-3	pWD615	-	0.10
PU-4	pWD615	-	0.09
PU-5	pWD615	_	0.09
PU-6	pWD615	-	0.13
PU-3	pWD615, pMMB347	+++	0.72
PU-5	pWD615, pMMB347	+++	0.68

[a] pMMB384 carries *epsE* gene (Fig. 8); pWD615 carries the *etxB* gene (Dallas 1983); pMMB347 carries the *epsM* gene (Fig. 8).

[b] secretion of extracellular protease was estimated on skim milk agar plates.

[c] concentration of CT or EtxB was determined in the medium and in the cells by G_{M1}-ELISA (Svennerholm & Holmgren, 1978).

located in the cytoplasm, either as a completely soluble protein or loosely associated with the secretion machinery in the cytoplasmic membrane. Further confirmation of this must await purification of this protein, determination of its N-terminal amino acid sequence and development of a procedure for its detection without overexpression. The DNA fragment able to complement PU-3 and PU-5 mutants contains an open reading frame with a predicted amino acid sequence homologous to the sequence of PulM protein from *Klebsiella*, required for the secretion of pullulanase (Possot *et al.*, 1992). Nucleotide sequences flanking the the *epsE* gene revealed reading frames exhibiting homology to the PulD and PulF proteins, whereas those flanking the *epsM* gene had homologies

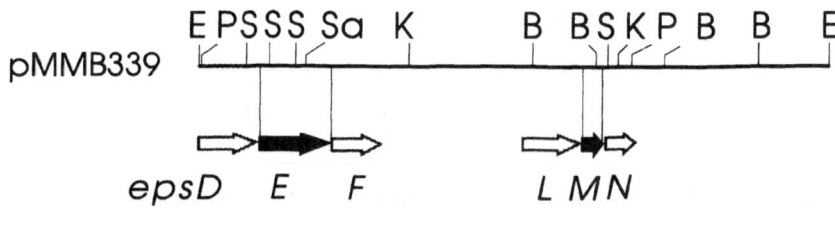

Figure 8. Physical and genetic map of the *Vibrio cholerae* chromosomal fragment containing *eps* genes, essential for extracellular secretion of proteins. Shaded arrows indicate the positions and functional orientation of the genes identified by complementation of mutants and nucleotide sequence determination. Unshaded arrows, the genes identified by sequence homology to *pul* genes. Restriction sites: B, *Bam*HI; E, *Eco*RI; K, *Kpn*I; P, *Pst*I; S, *Sac*I; Sa, *Sal*I.

to PulL and PulM proteins, all part of the secretion operon of pullulanase in *Klebsiella oxytoca* (Possot *et al.*, 1992). It appears thus that there are extensive genetic and biochemical similarities in the secretion apparatus of *V. cholerae* and *Klebsiella*.

It is difficult at present to understand the function of cytoplasmic proteins in the secretion of proteins from the periplasm into the outer medium. They may supply the energy for a step (or steps) in the transport pathway or they may regulate the activity of proteins involved in translocation. Although the EpsE protein appears to be located in the cytoplasm, as are its homologues PulE (Possot *et al.*, 1992) and XpsE (Dums *et al.*, 1991), it is essential for the translocation of toxin from the periplasm through the outer membrane. We must assume, therefore, that if it performs a regulatory function it presumably does so by trans-membrane signal transduction, perhaps similarly to OmpR of *E. coli*, VirG of *Agrobacterium tumefaciens*, or ToxR of *V. cholerae* (for review see Miller *et al.*, 1989; Gross *et al.*, 1989).

Similarities of primary structure have been found between genes essential for extracellular secretion in different Gram-negative bacteria (He *et al.*, 1991; Dums *et al.*, 1991; Possot *et al.*, 1992; Pugsley, 1992). This indicates that there are common mechanisms for translocation of different proteins through the outer membrane in a variety of Gram-negative bacteria and it prompted some authors to call this system of translocation the "general secretion pathway" (Pugsley 1992). It is interesting to note, however, that steps involving specificity and discrimination between individual proteins must be part of the secretion pathways since EtxB can not be secreted by either *Klebsiella*, *Pseudomonas* or *Xanthomonas* that harbor secretion genes highly homologous to those of *Vibrio cholerae*.

Table 5. Homology of EpsE protein to other proteins involved in macromolecular translocation through the outer membrane.

Protein	Organism	% Identity	Reference
ExeE	*Aeromonas*	67.5	Jiang and Howard (1992)
OutE	*Erwinia*	65.1	Lindeberg and Collmer (1992)
XcpR	*Pseudomonas*	62.5	Bally et al. (1992)
PulE	*Klebsiella*	60.8	Possot et al. (1992)
XpsE	*Xanthomonas*	44.6	Dums et al. (1991)
PilB	*Pseudomonas*	44.1	Nunn et al. (1990)
ComG-1	*Bacillus*	35.3	Albano et al. (1989)
KilB	RK-2 plasmid	25.8	Motallebi-Veshareh et al. (1992)
VirB	*Agrobacterium*	25.7	Thompson et al. (1988)

Acknowledgements

This work was supported by grants from US Department of Agriculture (MICLO 6874), NSF Center for Microbial Ecology (BIR 9120006), Biotechnology Research Center at Michigan State University and Research Excellence Fund from the State of Michigan.

References

Albano M, Breitling R & Dubnau DA (1989) Nucleotide sequence and genetic organization of the *Bacillus subtilis comG* operon. J Bacteriol. 171: 5386–5404.

Bally M, Filloux A, Akrim M, Ball G, Lazdunski A & Tommassen J (1992) Protein secretion in *Pseudomonas aeruginosa*: characterization of seven *xcp* genes and processing of secretory apparatus components by prepilin peptidase. Mol. Microbiol. 6: 1121–1131.

Cieslicki M (1989) Einsatz von porcimune/coli zur Kontrolle der *E. coli* Diarrhoe in Ferkelerzeugerbetrieben. Prakt. Tierarzt 70:29–34.

Clements JD & Finkelstein RA (1979) Isolation and characterization of homogeneous heat-labile enterotoxins with high specific activity from *Escherichia coli* cultures. Infect. Immun. 24: 760–769.

Cuatrecasas P (1973) Gangliosides and membrane receptors for cholera toxin. Biochemistry 12: 3558–3566.

Dallas WS, Gill DM & Falkow S (1979) Cistrons encoding *Escherichia coli* heat-labile toxin. J. Bacteriol. 139: 850–858.

Dallas WS & Falkow S (1980) Amino acid sequence homology between cholera toxin and *Escherichia coli* heat-labile toxin. Nature (London) 288: 499–501.

Dallas WS (1983) Conformity between heat-labile toxin genes from human and porcine enterotoxigenic *Escherichia coli*. Infect. Immun. 40: 647–652.

Duchet-Suchaux MF, Bertin AM & Menanteau PS (1991) Susceptibility of Chinese Meishan and European Large White pigs to enterotoxigenic *Escherichia coli* strains bearing colonization factors K88, 987P, K99 or F41. Am. J. Vet. Res. 52: 40–44.

Dums F, Dow JM & Daniels MJ (1991) Structural characterization of protein export genes of the bacterial phytopathogen *Xanthomonas compestris* pathovar *compestris*: relatedness to export systems of other Gram-negative bacteria. Mol. Gen. Genet. 229: 357–364.

Fairbrother JM, Broes A, Jacques M & Larivière S (1989) Pathogenicity of *Escherichia coli* O115:K"V165" strains isolated from pigs with diarrhea. Am. J. Vet Res. 50: 1029–1036.

Filloux A, Bally M, Ball G, Akrim M, Tomassen J. & Lazdunski, A (1990) Protein secretion in Gram-negative bacteria: transport across the outer membrane involves common mechanisms in different bacteria. EMBO J. 9: 4323–4329.

Gill DM, Clements JD, Robertson DC & Finkelstein RA (1981) Subunit number and arrangement in *Escherichia coli* heat-labile enterotoxin. Infect. Immun. 33: 677–682.

Gross R, Aricò B & Rappuoli R (1989) Families of bacterial transducing proteins. Mol. Microbiol. 3: 1661–166.

Hanne LF and Finkelstein RA (1982) Characterization and distribution of the hemagglutinins produced by *Vibrio cholerae*. Infect. Immun. 36: 209–214.

Hardy SJS, Holmgren J, Johansson S, Sanchez J & Hirst TR (1988) Coordinated assembly of multisubunit proteins: Oligomerization of bacterial enterotoxins *in vivo* and *in vitro*. Proc. Natl. Acad. Sci. USA. 85: 7109–7113.

He SY, Lindeberg M, Chatterjee AK & Collmer A (1991) Cloned *Erwinia chrysanthemi out* genes enable *Escherichia coli* to secrete a diverse family of heterologous proteins into its milieu. Proc. Natl. Acad. Sci. USA 88: 1079–1083.

Hirschhorn N & Greenough III WB (1991) Progress in oral rehydration therapy. Scientific Am. 264: 50–56.

Hirst TR, Hardy SJS & Randall LL (1983) Assembly *in vivo* of enterotoxin from *Escherichia coli*: formation of the B subunit oligomer. J. Bacteriol. 153: 21–26.

Hirst TR, Randall LL & Hardy SJS (1984a) Cellular location of heat-labile enterotoxin in *Escherichia coli*. J. Bacteriol. 157: 637–642.

Hirst TR, Sanchez J, Kaper JB, Hardy SJS & Holmgren J (1984b) Mechanism of toxin secretion by *Vibrio cholerae* investigated in strains harboring plasmids that encode heat-labile enterotoxins of *Escherichia coli*. Proc. Natl. Acad. Sci. USA 81: 7752–7756.

Hirst TR & Holmgren J (1987) Transient entry of enterotoxin subunits into the periplasm during their secretion from *Vibrio cholerae*. J. Bacteriol. 169: 1037– 1045.

Hofstra H & Witholt B (1984) Kinetics of synthesis, processing and membrane transport of heat-labile enterotoxin, a periplasmic protein in *Escherichia coli*. J. Biol. Chem. 259: 15182–15187.

Hofstra H & Witholt B (1985) Heat-labile enterotoxin in *Escherichia coli*. J.Biol. Chem. 260: 16037–16044.

Holmgren J (1973) Comparison of the tissue receptors for *Vibrio cholerae* and *Echerichia coli* enterotoxins by means of gangliosides and natural cholera toxoid. Infect. Immun. 8: 851–859.

Holmgren J (1981) Actions of cholera toxin and the prevention and treatment of cholera. Nature (London) 292: 413–417.

Holmes RK, Vasil ML, Finkelstein RA (1975) Studies on toxinogenesis in *Vibrio cholerae*. III. Characterization of non-toxinogenic mutants *in vitro* and in experimental animals. J. Clin. Invest. 55: 551–560.

Howard SP & Buckley JT (1985) Protein export by a Gram-negative bacterium: production of aerolysin by *Aeromonas hydrophila*. J. Bacteriol. 161: 1118–1124.

Jiang B & Howard SP (1992) The *Aeromonas hydrophila exeE* gene, required both for protein secretion and normal outer membrane biogenesis, is a member of a general secretion pathway. Mol. Microbiol. 6: 1351–1361.

Kraulis P (1991) MOLSCRIPT: a programto produce both detailed and schematic plots of proteins. J. Appl. Cryst. 24: 946–950.

Leong J, Vinal AC & Dallas WS (1985) Nucleotide sequence comparison between heat-labile toxin B subunit cistrons from *Escherichia coli* of human and porcine origin. Infect. Immun. 48: 73–77.

Lindeberg M & Collmer A (1992) Ananlysis of eight *out* genes in a cluster required for pectic enzyme secretion by *Erwinia chrysanthemi*: sequence comparison with secretion genes from other Gram-negative bacteria. J. Bacteriol. 174: 7385–7397.

Mackman N, Nicaud JM, Gray L & Holland IB (1986) Secretion of hemolysin by *E. coli*. Curr. Top. Microbiol. Immunol. 125: 159–181.

Miller JF, Mekalanos JJ & Falkow S (1989) Coordinate regulation and sensory transduction in the control of bacterial virulence. Science 243: 916–922.

Morris JA & Sojka WJ (1985) *Escherichia coli* as a pathogen in animals. p. 44–77. *In*: Susmann M (ed.) The virulence of *Escherichia coli*: Reviews and methods. Academic Press, London.

Moss J & Richardson SH (1978) Activation of adenylate cyclase by heat-labile enterotoxin. Evidence for ADP-ribosyl transferase activity similar to that of choleragen. J. Clin. Invest. 62: 281–285.

Motallebi-Veshareh M, Balzer D, Lanka, E, Jagura-Burdzy G, & Thomas C (1992) Conjugative transfer functions of broad-host-range plasmid RK2 are coregulated with vegetative replication. Mol. Microbiol. 6: 907–920.

Neill RJ, Ivins BE & Holmes RK (1983) Synthesis and secretion of the plasmid-coded heat-labile enterotoxin of *Escherichia coli* in *Vibrio cholerae*. Science. 221: 289–291.

Nunn D, Bergman S & Lory S (1990) Products of three accessory genes, *pilB*, *pilC* and *pilD* are required for biogenesis of *Pseudomonas aeruginosa* pili. J. Bacteriol. 172: 2911–2919.

Palva ET, Hirst TR, Hardy SJS, Holmgren J & Randall LL (1981) Synthesis of a precursor to the B subunit of heat-labile enterotoxin in *Escherichia coli*. J. Bacteriol. 146: 325–330.

Pearson GDN & Mekalanos JJ (1982) Molecular cloning of *Vibrio cholerae* enterotoxin genes in *Escherichia coi* K-12. Proc. Natl. Acad. Sci. USA. 79: 2976–2980.

Peek JA & Taylor RK (1992) Characterization of a periplasmic thiol:disulfide interchange protein required for the functional maturation of secreted virulence factors of *Vibrio cholerae*. Proc. Natl. Acad. Sci. USA. 89: 6210–6214.

Possot O, d'Enfert C, Reyss I & Pugsley AP (1992) Pullulanase secretion in *Escherichia coli* K-12 requires a cytoplasmic protein and a putative polytopic cytoplasmic membrane protein. Mol. Microbiol. 6: 95–105.

Pugsley AP (1992) Superfamilies of bacterial Transport systems with nucleotide binding components. p. 223–248. *In*: Mohan S, Dow C, and Coles JA (eds.) Prokaryotic structure and function: A new perspective. Cambridge University Press, UK.

Pugsley AP (1992) Translocation of folded protein across the outer membrane in *Escherichia coli*. Proc. Natl. Acad. Sci. USA. 89: 12058–12062.

Sack, BR (1975) Human diarrheal disease caused by enterotoxigenic *Escherichia coli*. Annu. Rev. Microbiol. 29: 333–353.

Sandkvist M (1992) Assembly and secretion of *E. coli* heat-labile enterotoxin. Doktors thesis. University of Umeå, Umeå, Sweden.

Sandkvist M, Hirst TR & Bagdasarian M (1987) Alterations at the carboxyl terminus change assembly and secretion properties of the B subunit of *Escherichia coli* heat-labile enterotoxin. J. Bacteriol. 169: 4570–4576.

Sandkvist M, Hirst TR & Bagdasarian M (1990) Minimal deletion of amino acids from the carboxyl terminus of the B subunit of heat-labile enterotoxin causes defects in its assembly and release from the cytoplasmic membrane of *Escherichia coli*. J. Biol. Chem. 265: 15239–15244.

Sandkvist M & Bagdasarian M (1993) Suppression of temperature-sensitive assembly mutants of heat-labile enterotoxin B subunits. Mol. Microbiol. (submitted).

Sandkvist M, Morales V & Bagdasarian M (1993) A soluble protein required for secretion of cholera toxin through the outer membrane of *Vibrio cholerae*. Gene. 123: 81–86.

Sixma TK, Pronk SE, Kalk KH, Wartna ES, van Zanten BAM, Witholt B & Hol WGJ (1991) Crystal structure of a cholera toxin-related heat-labile enterotoxin from *E. coli*. Nature (London). 351: 371–377.

Sixma TK, Pronk SE, Kalk KH, van Zanten BAM, Berghuis AM & Hol WGJ (1992) Lactose binding to heat-labile enterotoxin revealed by X-ray crystallography. Nature (London). 355: 561–564.

So M, Dallas WS & Falkow S (1978) Characterization of an *Escherichia coli* plasmid encoding for synthesis of heat-labile toxin: molecular cloning of the toxin determinant. Infect. Immun. 21:405–411.

Spicer EK & Noble JA (1982) *Escherichia coli* heat-labile enterotoxin. Nucleotide sequence of the A subunit gene. J. Biol. Chem. 257: 5716–5721.

Streatfield, SJ, Sandkvist M, Sixma TK, Bagdasarian M, Hol WGJ & Hirst TR (1992) Intermolecular interactions between the A- and B-subunits of heat-labile enterotoxin (LT) from *E. coli* which promote holotoxin assembly and stability *in vivo*. Proc. Natl. Acad. Sci. USA. 89: 12140–12144.

Sussman M (1985) *Escherichia coli* in human and animal disease. p. 7–45. *In*: Susmann M (ed.) The virulence of *Escherichia coli*: Reviews and methods. Academic Press, London.

Svennerholm AM & Holmgren J (1978) Identification of *Escherichia coli* heat-labile enterotoxin by means of ganglioside immunosorbent assay (G_{M1}-ELISA) procedure. Curr. Microbiol. 1: 19–27.

Thompson DV, Melchers LS, Idler KB, Schilperoort RA and Hooykaas PJJ (1988) Analysis of the complete nucleotide sequence of the *Agrobacterium tumefaciens* *vir*B operon. Nucleic Acids Research 16: 4621–4636.

Walker JE, Saraste M, Runswick MJ & Gay NJ (1982) Distantly related sequences in the α- and β-subunits of ATP synthase, myosin, kinases and other ATP-requiring enzymes and a common nucleotide binding fold. EMBO. J. 1: 945–951.

Yamamoto T & Yokota T (1983) Sequence of heat-labile enterotoxin of *Escherichia coli* pathogenic for humans. J. Bacteriol. 155: 728–733.

Yamamoto T, Tamura T & Yokota T (1984) Primary structure of heat-labile enterotoxin produced by *Escherichia coli* pathogenic for humans. J. Biol. Chem. 259: 5037–5044.

Yamamoto T, Gojobori T & Yokota T (1987) Evolutionary origin of pathogenic determinants in enterotoxigenic *Escherichia coli* and *Vibrio cholerae*. J. Bacteriol. 169: 1352–1357.

Yu J, Webb H & Hirst TR (1992) A homologue of the *Escherichia coli* DsbA protein in volved in disulpide bond formation is required for enterotoxin biogenesis in *Vibrio cholerae*. Mol. Microbiol. 6: 1949–1958.

22. Extracellular virulence factors of *Pseudomonas solanacearum*: role in disease and regulation of expression

MARK A. SCHELL, TIMOTHY P. DENNY and JIANZHONG HUANG

Abstract. The wilt pathogen *Pseudomonas solanacearum* infects and kills a wide variety of plants worldwide. *P. solanacearum* produces many extracellular molecules, which include several plant cell wall-degrading enzymes, plant growth hormones, and extracellular polysaccharides. Characterization of the genes and gene products responsible for synthesis of these extracellular molecules show that they play important roles in development of disease symptoms in solanaceous plant hosts, and that production of some of them involves unusual extracellular export systems. Coordinated synthesis of these virulence factors is controlled by an interacting, environmentally-sensitive regulatory network comprised of at least six unlinked regulatory loci that may encode as many as eight different regulatory proteins. Inactivation of any one of the genes in the network causes dramatic, but different, effects on virulence and production of extracellular polysaccharides and enzymes. In this chapter we summarize the biochemical properties and regulation of the many extracellular virulence factors of *P. solanacearum* and their role in disease.

Abbreviations: CWDE, Cell-Wall-Degrading Enzyme; Egl, Endoglucanase; EPS, Extracellular Polysaccharide; EXP, Extracellular Protein; EXVF, Extracellular Virulence Factor; IM, Inner Membrane; OM, Outer Membrane; PG, Polygalacturonase; SDS-PAGE, Sodium Dodecyl Sulfate Polyacrylamide Gel Electrophoresis

Introduction

Pseudomonas solanacearum is a major phytopathogenic bacterium that causes a lethal wilting disease of several hundred diverse species of plants worldwide, including the important crops peanut, potato, tomato, and banana (Hayward, 1991; Buddenhagen & Kelman, 1964). The disease is especially a problem in tropical and subtropical environments where the severity and incidence may be on the rise. *P. solanacearum* is a soil-borne pathogen that gains access to the vascular system of susceptible hosts via the root system (Wallis & Truter, 1978). Once infected, plants rapidly wilt and die, and large numbers of bacteria (> 10^{10}) can be found inside the stem where they have caused extensive damage to the vascular system and pith (Husain & Kelman, 1958b). Early studies of *P. solanacearum* (Husain & Kelman; 1958b; Kelman & Cowling, 1965) showed that its culture supernatants contained pectinolytic and cellulolytic enzyme activities that *in vivo* or *in vitro* could damage plant cell walls and vascular tissues, although they did not cause wilting. Virulent *P. solanacearum* strains also produce large amounts of a noncapsular extracellular polysaccharide (EPS)

C.I. Kado and J.H. Crosa (eds.), Molecular Mechanisms of Bacterial Virulence, 311–324.
© 1994 *Kluwer Academic Publishers.*

that is released as a slime; in partially purified form it was shown to cause wilt symptoms *in vitro* (Husain & Kelman, 1958a). More detailed information about the nature and role of individual putative extracellular virulence factors (EXVF) in wilt disease by *P. solanacearum* was not obtained until nearly 30 years later with the application of molecular genetic and biochemical tools. In this chapter we will summarize past and present data from both our group and others about the biochemical properties of the variety of extracellular macromolecules produced by *P. solanacearum*, their role in disease, and their complex regulation.

Biochemistry of extracellular macromolecules

Extracellular proteins (EXP)

The diversity of extracellular proteins (EXP) found in culture supernatants of *P. solanacearum* grown in a rich culture medium are illustrated by the electrophoretic analysis shown in Fig. 1. Positions of polypeptides whose functions have been partially or completely identified are labelled and discussed below. In general, the qualitative aspects of the EXP profiles do not change dramatically when cells are grown in minimal medium (not shown). The EXP profile produced *in planta* is unknown and could vary qualitatively or quantitatively from that observed *in vitro*. Egl, PglA, and PehB (PglB) have been shown to be produced *in planta*, but the levels were not quantified (Schell *et al.*, 1988; Roberts *et al.*, 1988; Allen *et al.*, 1991).

Endoglucanase (cellulase)

The first EXP of *P. solanacearum* to be purified and characterized was the cellulolytic 43-kDa β-1,4-endoglucanase, Egl (Schell, 1987). Egl is the major, if not the only, cellulolytic enzyme produced by *P. solanacearum in vitro*, since inactivation of the *egl* gene encoding it reduces cellulase activity by 99% (Roberts *et al.*, 1988). Immunologically-based assays showed that diverse isolates of *P. solanacearum* produce a very similar or identical enzyme (Schell, 1987). Egl is exclusively found in culture supernatants and the purified enzyme releases cellobiose from soluble, but not crystalline, cellulose. The *in planta* substrate(s) is unknown (likely candidates would be β-glucans of the plant cell wall), but attempts to demonstrate activity of Egl on purified plant cell walls have failed (Kelman & Cowling, 1965; Schell, 1987). The amino acid sequence of Egl (derived from the DNA sequence of cloned *egl*) shows significant similarity to cellulase proteins from *Bacillus* and *Erwinia* spp. (Huang & Schell, 1992), particularly residues 195–206 (i.e., PRVILGLNEPN) of Egl. Most striking was the finding of 33% identity (46% similarity) between the last 250 residues of Egl and endoglucanase III of the fungus *Trichoderma reesei*.

Egl is exported across the envelope of *P. solanacearum* by a novel two-step

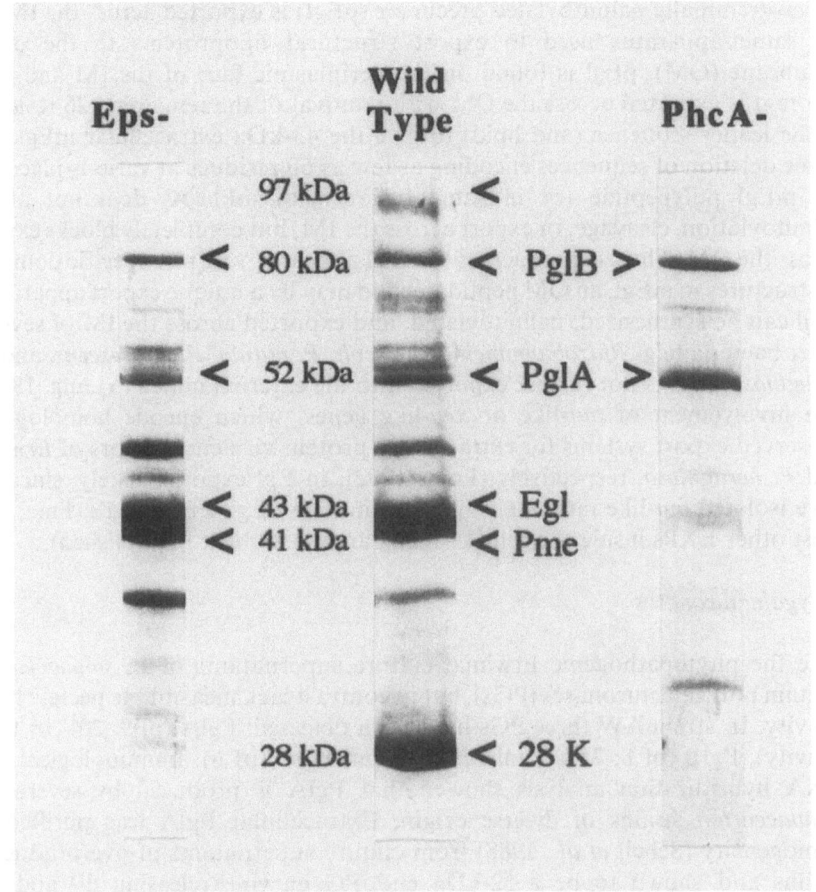

Figure 1. Extracellular protein profiles of *P. solanacearum* strains. One ml culture supernatants from stationary phase cultures [5×10^9 cells, grown in EG medium with glycerol (Schell, 1987)] of the following strains: wild type, strain AW (Schell, 1987); PhcA-, strain AW1-80 (Brumbley & Denny 1990); Eps-, strain AW1-1 (Denny & Baek, 1991) were prepared and analyzed by SDS-PAGE and staining with Coomassie Blue as previously (Huang & Schell, 1990a). Pgl, polygalacturonase; Egl, endoglucanase; Pme, pectinmethylesterase; 28 K, 28-kDa EXP.

process (Huang *et al.*, 1989; Huang & Schell, 1990a; Huang & Schell, 1992), which in part appears to be distinct from the reported mechanisms used by other pathogens (Lory, 1992). The *egl* gene is initially translated into a 48-kDa primary precursor that has a 45-residue, two-part leader sequence preceding the amino-terminus of the mature extracellular protein (mEgl). The first 19 residues of this leader are a lipoprotein signal sequence that directs modification, processing, and export of Egl (or Egl-PhoA hybrids) across the inner membrane (IM). Residue –26 (Cys) of the leader is covalently modified with a diglyceride of palmitate prior to cleavage at –27 by signal peptidase II; the resultant 46-kDa amino-terminally palmitoylated precursor (pEgl) is exported across the IM by the same apparatus used to export structural lipoproteins to the outer membrane (OM). pEgl is found on the periplasmic face of the IM and OM before it is exported across the OM with removal of the remaining 26 residues of the leader sequence (and lipid) to give the 43-kDa extracellular mEgl. In-frame deletion of sequences encoding as few as 60 residues at various places in the mEgl polypeptide (or in-frame fusion with mPhoA) does not affect palmitoylation, cleavage, or export across the IM, but completely blocks export across the OM. Thus, export across the OM appears to require specific domains or structures in mEgl, an OM peptidase, and may be a unique export apparatus. pEgl can be synthesized, palmitoylated, and exported across the IM of several other bacteria (e.g. *Rhizobium meliloti*, *E. coli*, *P. putida*, *P. fluorescens*, and *P. aeruginosa*), but is not further exported into the external milieu (Yeung, 1991). The involvement of *out*-like or *xcp*-like genes, which encode homologous, conserved export systems for extracellular protein virulence factors of *Erwinia* and *P. aeruginosa*, respectively (Lory, 1992), in Egl export is likely, since we have isolated *out*-like mutants which accumulate pEgl, PglA, PglB, Pme, and most other EXPs inside the cell (Y. Kang and M. Schell, unpublished).

Polygalacturonases

Like the phytopathogenic Erwinia, culture supernatants of *P. solanacearum* contain polygalacturonases (PGs), but in contrast lack measurable pectate lyase activity. In strain AW three PGs have been detected: PglA (pI 9; 20% of total activity), PglB (pI 8; 75% total activity) and PglC (pI 6). Immunological and DNA hybridization analysis showed that PglA is produced by several *P. solanacearum* strains of diverse origin. Extracellular PglA was purified to homogeneity (Schell *et al.*, 1988) from culture supernatants of overproducing strains and shown to be a 52-kDa endoPG enzyme releasing di- and tri-galacturonides from polygalacturonic acid (Yeung, 1991). Purified PglA has been crystallized and its tertiary structure is being determined by X-ray crystallography (F. Jurnak, personal communication). The *pglA* gene was cloned and sequenced (Huang & Schell, 1990b); residues 309–371 of its derived amino acid sequence have major (55%) similarity with sequences in PGs from tomato and *Erwinia carotovora* (Hinton *et al.*, 1990). Within this region are two 8-residue sequences showing nearly 90% conservation; these residues are likely to be important for enzyme function (Huang & Schell, 1990b). In contrast to PglA, PglB (and likely PglC) appears to be an exoPG enzyme with a molecular mass of 80 kDa (Denny *et al.*, unpublished). *P. solanacearum* strain K60 produces three extracellular PGs: PehA, PehB, and PehC, which are homologous (or likely identical) to PglA, PglB, and PglC, respectively (Allen *et al.*, 1991; C. Allen, personal communication).

The mechanism for export of PglA from *P. solanacearum* is very different from that used for Egl (Huang & Schell, 1990b). *PglA* is translated as a precursor with a 21-residue signal peptide similar to those of periplasmic proteins. This signal sequence directs export of PglA (or PglA-PhoA hybrids) across the IM into the periplasm of *P. solanacearum, E. coli, P. putida,* or *P. aeruginosa,* and is removed during the process. It is likely that this occurs via the general export pathway encoded by *sec* genes and widely used in many prokaryotes to export proteins into the periplasm (Randall *et al.,* 1987). Export of PglA across the OM (which occurs only in *P. solanacearum*) probably involves a specific apparatus and mechanism requiring specific structural motifs of mature PglA, since deletion of the last 13 residues of PglA blocks its export across the OM, but not the IM. Export of several extracellular pectate lyase enzymes of *Erwinia* also uses the common signal peptide and the Sec system, in addition to an "outer membrane" export apparatus encoded by the *out* genes, which are similar to *pul* genes required for pullulanase export from *Klebsiella pneumoniae* and *xcp* genes of *P. aeruginosa* (He *et al.,* 1991). Involvement of *out*-like genes in PglA export has not been demonstrated, but is likely, since, by means of insertional inactivation we identified a gene(s) required for export of all known extracellular enzymes of *P. solanacearum* (Kang *et al.,* submitted), a phenotype expected of an *out* mutant.

Pectin methylesterase

Culture supernatants of *P. solanacearum* contain pectin methylesterase activity that removes methoxy groups from pectin to produce polygalacturonic acid, which is a substrate for polygalacturonase. Spok *et al.* (1991) cloned, sequenced, and analyzed the 1.2-kb *pme* gene of *P. solanacearum* DSM 50905. The Pme enzyme produced in *E. coli,* was not extracellular and had a molecular mass of 42 kDa. The amino acid sequence of the *P. solanacearum* enzyme is very (30%) similar to pectin methylesterases of *E. chrysanthemi* and tomato; several highly conserved domains were identified in a three-way comparison (Spok *et al.,* 1991). The amino-terminal sequence of the *pme* translation product appears to possibly have a lipoprotein signal sequence similar to that found on Egl, and thus Pme may be exported by the same system and/or mechanism as Egl. Consistent with this hypothesis, Pme synthesized from cloned *pme* in *E. coli* is 70% membrane-associated (Schell, unpublished), identical to what is found for Egl (Huang & Schell, 1990).

28-kDa EXP

The predominant EXP in culture supernatants of several diverse isolates of *P. solanacearum* is a 28-kDa polypeptide representing > 25% of the total EXPs. This protein was purified and partially characterized (Schell & Vinson, 1985; Schell *et al.,* unpublished). It co-purifies in a high-molecular-weight complex with EPS, but the functional significance of this association is unclear. Pulse chase-immunoprecipitation analyses suggest the 28-kDa EXP is initially synthesized as a 59-kDa precursor that is exported out of the cell as 56-kDa preprotein, which is then cleaved extracellularly to give one (or possibly two) 28-

kDa polypeptides. The reason for the synthesis of a preprotein is not clear, although one possibility is that the 28-kDa polypeptide is toxic if inside the cell. Whereas the function of the 28-kDa EXP is also unclear, circumstantial evidence suggests an involvement with EPS and virulence, since SDS-PAGE analysis of culture supernatants of several different types of EPS-deficient, reduced-virulence mutants (including insertions in *eps* structual genes; Fig. 1) shows they all produce dramatically less 28-kDa EXP. We are at present cloning the gene encoding the 28-kDa EXP to investigate its function and role in wilt disease caused by *P. solanacearum*.

Other EXPs

Recent experiments show that an EXP of *Erwinia amylovora*, called harpin, plays a central role in its phytopathogenicity and ability to induce the hypersensitive defense response (HR) (Wei *et al.*, 1992). Preliminary data strongly suggest that a protein with similar properties and function is also produced by *P. solanacearum* (Arlat *et al.*, 1992). Moreover, new data from several laboratories suggests that the products of *hrp* genes (some of which are highly conserved in all plant pathogenic bacteria and required for pathogenicity, growth *in planta*, and HR [Arlat *et al.*, 1991]) encode an extracellular export apparatus similar to the one used by the animal pathogen *Yersinia pestis* to export Yops, its proteinaceous extracellular virulence factors. By analogy it has been suggested that *hrp* gene products of *P. solanacearum* (and other plant pathogens) are involved in export of certain extracellular molecules required for production of plant disease.

Huang *et al.* (1989) purified a 60-kDa, pI 9.1 EXP from *P. solanacearum* K60 that induced a rapid "browning response" in tissue culture calli of incompatible, but not compatible cultivars of potato, implicating it in elicitation of HR. Initially it appeared that this 60-kDa EXP was PehA (PglA), since *pehA*-deficient mutants failed to produce the browning response (Allen *et al.*, 1991). However *pehA*-mutants still give HR, and the amino acid composition of the browning-factor is quite different from PglA making any conclusions about it difficult. Although protease activity has been reported in culture supernatants of *P. solanacearum*, the levels produced by strain AW *in vitro* are very low in comparison to other plant pathogens such as *Xanthomonas campestris* pv. campestris.

Extracellular polysaccharides of P. solanacearum

Many, if not all, virulent *P. solanacearum* strains produce an abundance of noncapsular extracellular polysaccharide (EPS) in amounts nearly equivalent to their cell mass. Analysis of EPS from one strain (Orgambide *et al.*, 1991) showed it is composed of four fractions. The major carbohydrate fraction (we call EPS I) represents 40% of the total mass of EPS and is a high-molecular-weight (> 10^6 Da) acidic polysaccharide composed of a repeating unit of three amino sugars: N-acetyl galactosamine, 2-N-acetyl-2-deoxy-L-galacturonic acid, and 2-N-acetyl-4-N-(3-hydroxybutanoyl)-2,4,6-trideoxy-D-glucose (Fig. 2). Another 40% of the total EPS mass is an uncharged, primarily non-carbohydrate fraction of unknown structure. The two remaining minor fractions are composed primarily of rhamnose and mannose, respectively (Orgambide *et al.*, 1991). The high nitrogen content and structure of EPS I is very unusual for a prokaryotic EPS (Sutherland, 1991). The nutritional expense of making large amounts of such a nitrogen-rich EPS, and the exceedingly complex regulation of its synthesis (described later), suggests that EPS I is very important to *P. solanacearum* and may have multiple functions.

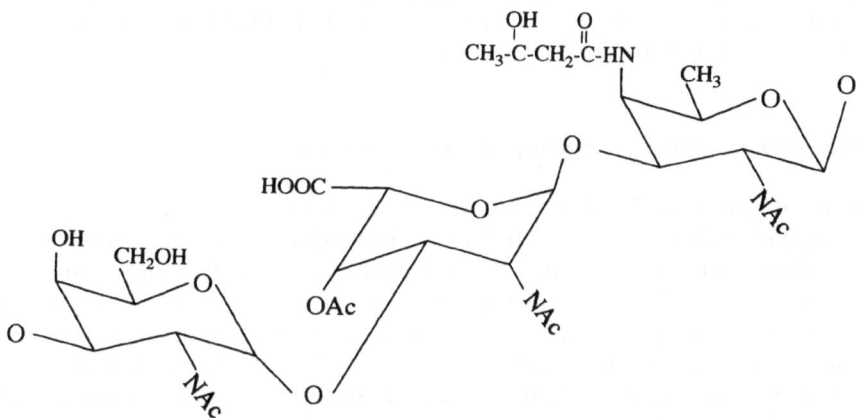

Figure 2. Structure of EPS I. Repeat unit of the major extracellular polysaccharide produced by *P. solanacearum* GMI 1000 (after Orgambide *et al.*, 1992). NAc, N-acetyl; OAc, O-acetyl.

It is likely that synthesis of the acidic galactosamine-rich EPS I is in part encoded by the 18-kb *eps* locus, since transposon insertions throughout *eps* block production of EPS I without affecting production of other EPSs (Denny & Baek, 1991; Schell *et al.*, 1993). The *eps* locus has been cloned and shown to encode at least ten individual polypeptides (some membrane-associated), but specific enzymatic activities or functions have not yet been assigned. Insertions throughout *eps* also cause a loss of production of the 28-kDa EXP and a minor 97-kDa EXP (Fig. 1); while it is possible these EXPs are encoded by *eps*, definitive evidence for this is lacking.

Cook & Sequeira (1991) and later Kao & Sequeira (1991) characterized the 6.5-kb *ops* gene cluster of *P. solanacearum* K60 in which some Tn*3* insertions caused loss of fluidal mucoid colony morphology and virulence. However, chemical measurement of EPS I (i.e., amount of extracellular polymeric N-acetyl galactosamine) showed that these mutants still produced substantial amounts of EPS (from 19 to 95% of wild type), in contrast to *eps*::Tn*5* mutants which always produce less than 3% of wild type EPS I levels (Schell *et al.*, 1993; Denny & Baek, 1991). A subsequent study (Kao & Sequeira, 1991) showed this region contains 7 genes (*opsA-F*) and that *ops*::Tn*3* mutants also had an altered lipopolysaccharide (LPS) structure. This implies that the *ops* genes encode for synthesis of precursors for both EPS and LPS. Consistent with this hypothesis, DNA sequence data (Kao & Sequeira, 1991 and this book) suggest that this region may be involved in biosynthesis of rhamnose, a component of both EPS and LPS of *P. solanacearum*.

The *rgnII* locus (7 kb downstream of *eps* in the *P. solanacearum* genome) is required for EPS I production only if cells are grown in the presence of peptone, but not in minimal medium or *in planta* (Denny & Baek, 1991). Several insertions just upstream of *rgnII* also cause EPS I production to be affected by medium composition in a manner different from those in *rgnII* (Schell *et al.*, 1993). Alteration of EPS I production in response to medium conditions is not surprising given its high nitrogen content.

Role of extracellular macromolecules in wilt disease

Many of the aforementioned enzymes and polysaccharides are important extracellular virulence factors (EXVF). Most information about their role in wilt disease comes from studies of the virulence of *P. solanacearum* strains that are specifically defective in production of one or two putative EXVF. Mutant strains are usually constructed by replacing the wild type gene with one inactivated by insertion mutagenesis *in vitro* or by transposon mutagenesis *in vivo* (Roberts *et al.*, 1988; Denny & Baek, 1991). Virulence of strains is usually analyzed in growth chambers by monitoring wilt disease development after injection of 10^4 to 10^6 bacteria into 2- to 8-week-old host plants such as tomato (Denny *et al.*, 1990) or eggplant (Kao & Sequeira, 1991; Allen *et al.*, 1991). Examples of the effects of loss of different types of EXVF on wilt disease development is shown in Fig. 3. The figure is hypothetical, but is based on a variety of published and unpublished research; specific results are discussed below.

Plant-cell-wall-degrading enzymes

Six-week-old tomato plants stem-inoculated with wild type strains at 10^5 cfu begin to show wilting after 3 days and are completely wilted and dead by 7 to 10 days. The final number of viable cells per plant sometimes reaches 10^{11}. Mutants unable to produce any one of the individual cell-wall-degrading enzymes (CWDE): Egl, PglA (also called PehA), or PglB (also called PehB), still grow normally *in planta*, produce wilt disease symptoms, and ultimately kill the infected plant. However, the onset of symptoms and time before death are delayed by anywhere from 25% to 80% depending on which enzyme is missing, initial inoculation density, and age of plants. In general, the lack of Egl causes a greater delay than lack of either PglA (Denny *et al.*, 1990) or PehC (C. Allen, personal communication). Lack of PehB (PglB) appears to cause greater delays

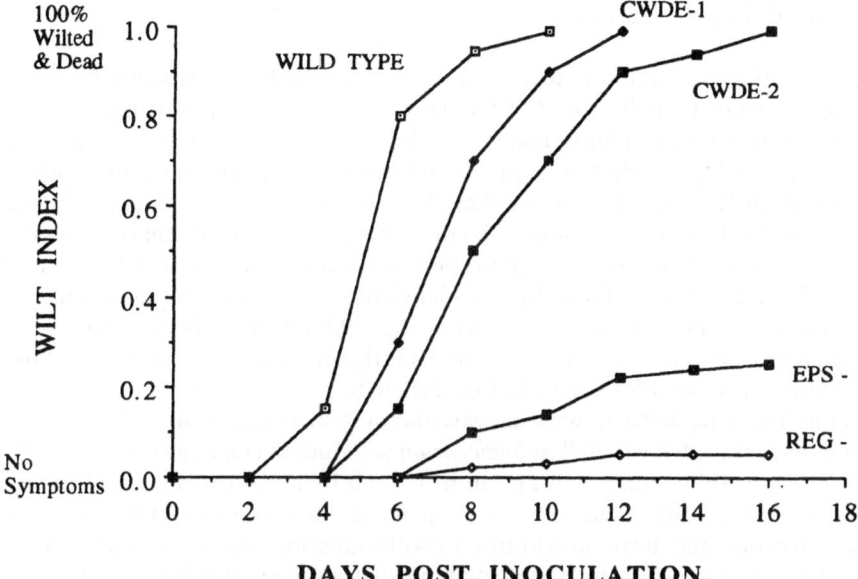

Figure 3. Effect of loss of various extracellular macromolecules on virulence of *P. solanacearum.* Graphic representation based on data from: Denny *et al.*, 1990; Denny and Baek, 1991; Schell *et al.*, 1988. Roberts *et al.*, 1988; Allen *et al.*, 1991, and several other unpublished studies. Wilt index, fraction of wilted leaves; CWDE-1, lacking one plant cell wall-degrading enzyme (e.g Egl); CWDE-2, lacking two plant cell-wall degrading enzymes; EPS-, deficient in production of EPS I; REG-, missing a component of virulence regulatory network.

than lack of PglA or PehC (C. Allen, personal communication). Strains missing two CWDE (e.g. Egl + PglA or PehB + PehC) show at best additive, rather than synergistic, delays.

The majority of evidence suggests that after *P. solanacearum* enters the xylem vessels of the stem, the absence of one or two plant-cell-wall-degrading enzymes does not prevent it from wilting and killing the plant, but rather only slows disease development (i.e., reduces its aggressiveness). However, nearly all these studies used stem-injection or leaf-cutting to inoculate, which bypasses the natural infection process occurring at the roots. It is believed that in the natural situation in the rhizosphere (Schmit, 1978) *P. solanacearum* infects plants by penetrating at emergence points of secondary roots, moving through the root cortex, and finally entering the vascular system (Wallis & Truter, 1978). It is possible that some CWDE may be more important or even required during these early stages of infection, where they may facilitate penetration of the pectin-rich cortical tissues of the root or entrance into the xylem elements. One soil-inoculation study suggested that Egl was important for root infection (Denny *et al.*, 1990). CWDE may be more important for disease in plants other than the few hosts studied to date, or when growing in soil after release from a killed plant.

Extracellular polysaccharide

Loss of EPS production reduces wilt disease caused by *P. solanacearum* much more severely than loss of CWDE (Fig. 3). When stem-inoculated at low numbers into tomato plants, many EPS-deficient mutants can grow to the same extent as wild type, albeit sometimes a bit more slowly, but most often they do not kill. In the rare cases when they do wilt and/or kill, it occurs only after prolonged infection (> 15 days), and may partly result from stem rot caused by CWDE; in comparison, wild type rapidly wilts and kills completely in 7 to 10 days. We recently found that the EPS-deficient mutants we have studied (caused by transposon insertion in *eps* [Denny *et al.*, 1990; Denny & Baek, 1991]) do not appear to be pleiotropic, and most importantly are defective in synthesis of only the nitrogen-rich EPS I shown in Fig. 2 (Schell *et al.*, 1993). EPS I likely causes wilt by blocking water flow in the vascular system and appears to be the single most important EXVF of *P. solanacearum* yet studied (Denny *et al.*, 1990; Kao & Sequeira, 1991). Since wilting can be caused by non-nitrogenous EPSs (Van Alfen, 1989), the metabolically expensive nitrogen-rich EPS I of *P. solanacearum* may have, in addition to wilt induction, undiscovered functions, such as a storage polymer or protectant against some plant defense mechanisms. The role of other EPSs in wilt disease is unknown.

Regulation of virulence factors by a complex network

Only one other type of *P. solanacearum* mutant shows greater reductions in virulence than EPS-deficient strains; these mutants (REG⁻; Fig. 3) have insertions in regulatory genes which are required for production of multiple virulence factors (e.g. *phcA*, Brumbley and Denny, 1990; *pehR*, Allen *et al.*, 1991). We have identified and isolated four more of this type of locus (*vsrA*, *xpsR*, *vsrB*, and *phcB*), each of which when inactivated give mutants that essentially lose the ability to produce wilt disease in stem-inoculated tomato plants (Schell *et al.*, in preparation). Except for *vsrA* and *xpsR* mutants, all strains seem to grow *in planta* or *in vitro* like wild type. When grown in rich medium, all are deficient in production of EPS I, probably as a result of reduced *eps* transcription, since inactivation of any of the loci causes at least a 25-fold reduction of expression of *eps::lacZ* reporters. In addition, SDS-PAGE analysis of EXP production by each mutant shows that inactivation of any of these loci causes allele-specific, distinct changes in EXP profiles; the only common feature is that all mutants produce at least 20-fold less 28-kDa EXP (see above). Cellular protein profiles of all mutants were largely identical to wild type and exhibited no evidence of accumulation of any affected EXPs or their precursors, suggesting that affects are not the result of an export block. Thus, most evidence suggests that these genes are part of a regulatory network that controls production of multiple EXVF of *P. solanacearum*.

In both qualitative and quantitative aspects, inactivation of *phcA* has the most dramatic effect on EXVF production, reducing levels of EPSI, Egl, Pme, and the 28-kDa EXP by at least 15-fold (Schell, 1987; Brumbley and Denny, 1990; Table 1 and Fig. 1). This reduction of Egl and EPSI probably results from reduced transcription of the genes encoding them (Huang *et al.*, 1989), and it is likely that the same will be found for the other affected EXVF. Levels of several other uncharacterized EXPs are also drastically reduced by the *phcA* mutation, while levels of two other EXPs appear to increase (Fig. 1). One of the EXPs that increases has a 52-kDa size, which is the same size as PglA; since we have found that *phcA* mutants produce 10-fold more PglA than wild type (without affecting PglB and PglC) (Table 1), we conclude that *phcA* negatively controls expression of *pglA*. In addition, *phcA* mutants are dramatically more motile than wild type (Brumbley and Denny, 1990). Thus, *phcA* negatively regulates expression of some virulence genes (*pglA*, motility) and positively regulates others (*eps, egl, pme*). PhcA has recently been shown to encode a LysR-type transcriptional regulator (Brumbley *et al.*, I. Bacteriol. 175: 5477).

Table 1. Levels of extracellular cell-wall degrading enzymes produced by *P. solanacearum* strains.

Enzyme	Specific Activity[a]	
	Wild Type	*phcA::Tn5*[b]
Egl	1.20	0.05
Total PG	0.80	1.88
PglA[c]	0.11	1.16
PglB+C[d]	0.69	0.72
Pme	0.31	0.02

[a] Units per ml of culture supernatant from 10^9 cells with carboxymethylcellulose (Egl), Na-polygalacturonate (PG), or pectin (Pme) as substrate using previously described assay methods (Schell *et al.*, 1988; Roberts *et al.*, 1988). 1 unit = μmol ml^{-1} min^{-1} for PG and Egl; 1 unit = nmol ml^{-1} min^{-1} for Pme.
[b] *phcA::Tn5* = strain AW 1–80 (Brumbley and Denny, 1990).
[c] Determined by immunoadsorption analysis with antiserum against PglA (Schell 1988).
[d] Remaining PG activity after immunoadsorption of PglA.

Inactivating any of the other regulators (*vsrB, xpsR,* or *vsrA*) affects production of only some EXVF that are regulated by *phcA*. From this and other data (Huang *et al.*, unpublished) we propose that these loci encode an underlayer of intermediate virulence regulators working below the global level of PhcA. Although PhcA appears to be a master switch that decides if expression of a variety of virulence genes is warranted, it probably does this indirectly by regulating the expression of components in the underlayer. In support of this, we found that PhcA positively regulates transcription of at least one of the intermediate regulators, *xpsR*, and that constitutive expression of *xpsR* partially restores *eps* expression in *phcA* mutants. Most likely, XpsR protein controls *eps* transcription, while its own transcription is in turn controlled by *phcA*. Other experiments (Huang *et al.* unpublished) suggest that XpsR protein requires VsrA for maximal activity, a behavior expected of a two-component regulatory pair (Stock *et al.*,1989). Tn*phoA* and other localization analyses show that VsrA (and also VsrB) are membrane-associated proteins, and may be autokinase sensors of the two-component type. Similar to *phcA*, *vsrB* appears to negatively regulate PglA polygalacturonase, implying that *phcA* may also regulate a *vsrB*-related component. *PhcA*- and *vsrB*-mediated control of *pglA* may occur via regulation of, or crosstalk with, the *pehR* locus. Data of Allen *et al.* (1991) suggests *pehR* encodes a two-component regulator required for virulence and high level transcription of *pehA* (*pglA*) in strain K60. The activity of PhcA may be affected by a volatile signal molecule which is self-produced by *P. solanacearum* and requires the *phcB* locus. This may be a manifestation of a cell density-sensing regulatory system that controls the PhcA virulence switch (Clough, 1991).

Perspective

As many of the chapters in this book show, host-pathogen interactions are dynamic and complex, and it is not suprising that a pathogen would rely on a variety of different molecules to accomplish its goals. The phytopathogen *P. solanacearum* is no exception, as it makes a wide variety of EXVF for use in wilt disease. Producing these EXVF requires the support of at least two distinct extracellular export systems. The importance of some EXVF such as the plant cell wall-degrading enzymes in disease appears to be minor, but we have only looked at their role for a few selected conditions and hosts. Examining EXVF involvement at various stages of the disease may provide a new perspective. Our understanding of the disease process and important molecules is still in the early stages, and many aspects of the plant-bacterial interaction remain to be investigated. While EPS appears to be the major EXVF, we still don't know why it is produced in such large quantities, why it has such an unusual structure, or what advantage is conferred by its apparent ability to wilt a plant. Studies to address the purpose and function of the complex regulation system for EPS and other EXVF may help to answer some of these questions. While our understanding of the regulation of EXVF is rudimentary, it is clear that it involves a sophisticated multicomponent system that is interacting and environmentally-sensitive. Elucidation of the multiple signals that affect the network, the nature of all the network regulated genes, and how diverse signal input is sorted into concerted transcription are important questions to investigate. Knowledge of the types of signals may give insight into the conditions and environments that are perceived as favorable for pathogenic behavior.

Acknowledgements

Work from our labs presented here was supported by grants from NSF (DMB 89–4472 and IBN 91–17544) and the University of Georgia Biotechnology Program. We thank Rosemary Wood for typing the manuscript and Caitlyn Allen for providing unpublished data.

References

Allen C, Huang Y & Sequeira L (1991) Cloning of genes affecting polygalacturonase production in *Pseudomonas solanacearum*. Mol Plant-Microbe Interact 4: 147–154.

Arlat M, Gough CL, Barber CE, Boucher C & Daniels MJ (1991) *Xanthomonas campestris* contains a cluster of *hrp* genes related to the larger *hrp* cluster of *Pseudomonas solanacearum*. Mol Plant-Microbe Interact 4: 593–602.

Arlat M, Van Gijsegem F & Boucher C (1992) Proc. Sixth Intl. Symp. on Mol. Plant Microbe Interactions, Seattle, WA, Abstract #225.

Brumbley SM & Denny TP (1990) Cloning of *phcA* from wild-type *Pseudomonas solanacearum*, a gene that when mutated alters expression of multiple traits that contribute to virulence. J Bacteriol 172: 5677–5685.

Buddenhagen IW & Kelman A (1964) Biological and physiological aspects of bacterial wilt caused by *Pseudomonas solanacearum*. Annu Rev Phytopathol 2: 203–230.

Clough SJ (1991) Regulation of virulence in *Pseudomonas solanacearum* by an endogenous volatile compound. M.S. Thesis Univ. of Georgia, Athens.

Cook D & Sequeira L (1991) Genetic and biochemical characterization of a *Pseudomonas solanacearum* gene cluster required for extracellular polysaccharide production and virulence. J Bacteriol 173: 1654–1662.

Denny TP & Baek SR (1991) Genetic evidence that extracellular polysaccharide is a virulence factor of *Pseudomonas solanacearum*. Mol Plant-Microbe Interact 4: 198–206.

Denny TP. Carney BF & Schell MA (1990) Inactivation of multiple virulence genes reduces the ability of *Pseudomonas solanacearum* to cause wilt symptoms. Mol Plant-Microbe Interact 3: 293–300.

Hayward AC (1991) Biology and epidemiology of bacterial wilt caused by *Pseudomonas solanacearum*. Annu Rev Phytopathol 29: 65–87.

He SY, Lindeberg M, Chatterjee AK & Collmer A (1991) Cloned *Erwinia chrysanthemi out* genes enable *Escherichia coli* to selectively secrete a diverse family of heterologous proteins to its milieu. Proc Natl Acad Sci USA 88: 1079–1084.

Hinton JC, Gill DR, Lalo D, Plastow GS & Salmond GP (1990) Sequence of the *peh* gene of *Erwinia carotovora*: homology between *Erwinia* and plant enzymes. Mol Microbiol 4: 1029–1034.

Huang J, Sukordhaman M & Schell MA (1989) Excretion of the *egl* gene product of *Pseudomonas solanacearum*. J Bacteriol 171: 3767–3774.

Huang J & Schell MA (1990a) Evidence that extracellular export of the endoglucanase encoded by *egl* of *Pseudomonas solanacearum* occurs by a two-step process involving a lipoprotein intermediate. J Biol Chem 265: 11628–11632.

Huang J & Schell MA (1990b) DNA sequence analysis of *pglA* and mechanism of export of its polygalacturonase product from *Pseudomonas solanacearum*. J Bacteriol 172: 3879.

Huang J & Schell MA (1992) Role of the two-component leader sequence and amino acid sequences of the mature endoglucanase *egl* in extracellular export from *Pseudomonas solanacearum*. J Bacteriol 174: 1314–1323.

Huang Y, Helgeson JP & Sequeira L (1989) Isolation and purification of a factor from *Pseudomonas solanacearum* that induces a hypersensitive-like response in potato cells. Mol Plant-Microbe Interact 2: 132–138.

Husain A & Kelman A (1958a) Relation of slime production to mechanisms of wilting and pathogenicity of *Pseudomonas solanacearum*. Phytopathology 48:155–165.

Husain A & Kelman A (1958b) The role of pectic and cellulolytic enzymes in pathogenesis by *Pseudomonas solanacearum*. Phytopathology 48: 376–385.

Kao CC & Sequeira L (1991) A gene cluster required for coordinated biosynthesis of lipopolysaccharide and extracellular polysaccharide also affects virulence of *Pseudomonas solanacearum*. J Bacteriol 173:7841–7847.

Kelman A & Cowling EB (1965) Cellulase of *Pseudomonas solanacearum* in relation to pathogenesis. Phytopathology 55: 148–155.

Lory S (1992) Determinants of extracellular protein secretion in Gram negative bacteria. J Bacteriol 174: 3423–3428.

Orgambide G, Montrozier H, Servin P, Roussel J, Trigalet-Demery D & Trigalet A (1991) High heterogeneity of the exopolysaccharides of *Pseudomonas solanacearum* strain GMI1000 and the complete structure of the major polysaccharide. J Biol Chem 266: 8312–8321.

Randall LL, Hardy SJ & Thom JR (1987) Export of protein: a biochemical view. Annu Rev Microbiol 41:507–541.

Roberts DP, Denny TP & Schell MA (1988) Cloning of the *egl* gene of *Pseudomonas solanacearum* and analysis of its role in phytopathogenicity. J Bacteriol 170: 1445–1451.

Schell MA & Vinson DJ (1985) Molecular basis of virulence of *Pseudomonas solanacearum*. In: A. Szalay and R. Legocki (eds.) Advances in Molecular Genetics of the Bacteria-Plant Interaction. Proceedings of the 2nd Intl. Symp. on Molecular Platn Microbe Interactions, p. 182–185. Ithaca, New York, June 1984. Media Services, Cornell University, Ithaca.

Schell MA (1987) Purification and characterization of an endoglucanase from *Pseudomonas solanacearum*. Appl Environ Microbiol 53: 2237–2241.

Schell MA, Roberts DP & Denny TP (1988) Analysis of the *Pseudomonas solanacearum* polygalacturonase encoded by *pglA* and its involvement in phytopathogenicity. J Bacteriol 170: 4501–4508.

Schell MA, Denny TP, Clough SJ & Huang J (1993) Further characterization of genes encoding extracellular polysaccharide of *Pseudomonas solanacearum* and their regulation. p. 231–239. In: E. Nester and D.P.S. Verma (eds) Advances in Molecular Genetics of Plant Microbe Interactions, Kluwer Academic Publishers, Dordrecht.

Schmit J (1978) Microscopic study of early stages of infection by *Pseudomonas solanacearum* E.F.S. on *"in vitro"* grown tomato seedlings. p. 184. In: Proceedings of the Fourth International Conference on Plant Pathogenic Bacteria. Station de Pathologie Vegetale et Phytobacteriologie, Anger, France, 27 Aug. – 2 Sept 1978" INRA, Angers, France.

Stock JB, Ninfa A & Stock A (1989) Protein phosphorylation and regulation of adaptive responses in bacteria. Microbiol Rev 53:450–490.

Spök A, Stubenrauch G, Schorgendorfer K & Schwab H (1991) Molecular cloning and sequencing of a pectinesterase gene from *Pseudomonas solanacearum*. J Gen Microbiol 137: 131–140.

Sutherland IW (1991) Microbial Polysaccharides. Marcel Dekker, NY, NY.

Van Alfen NK (1989) Reassessment of plant wilt toxins. Annu Rev Phytopathol 27: 533–550.

Wallis FM & Truter SJ (1978) Histopathology of tomato plants infected with *Pseudomonas solanacearum*, with emphasis on ultrastructure. Physiol Plant Pathol 13: 307–316.

Wei, Z-M, Laby RJ, Zumoff, JH, Baller DW, He SY, Collmer A & Beer SV (1992) Harpin, elicitor of the hypersensitive response produced by the plant pathogen *Erwinia amylovora*. Science 257:85–88.

Yeung, K-H A (1991) Expression of *pglA* and *egl* genes encoding polygalacturonase and endoglucanase in heterologous soil bacteria. M.S. Thesis, Univ. Georgia, Athens.

23. RTX-toxins in *Actinobacillus pleuropneumoniae* and their potential role in virulence

JOACHIM FREY

Abstract. *Actinobacillus pleuropneumoniae*, the causative agent of swine pleuropneumonia, secretes hemolysins which are considered to play an important role in virulence. They are proteins with apparent molecular weights of 105 kDa. They are strongly immunogenic in naturally or experimentally infected pigs. The strongly hemolytic and cytolytic hemolysin I (HlyI) is produced by serotypes or strains which are particularly virulent, while the less hemolytic and cytolytic hemolysin II (HlyII) is produced by all serotypes except type 10. Several serotypes secrete both HlyI and HlyII which have the same apparent molecular weight, but which can be distinguished by monoclonal and polyclonal antibodies. The hemolysin I operon consists of a structural *hlyIA* gene encoding the prohemolysin, an activator gene *hlyIC* necessary for the activation of prohemolysin to active hemolysin and two genes *hlyIB* and *hlyID* encoding proteins for specific hemolysin secretion. This is a common feature of the group of RTX-toxins (repeats in the structural *t*oxin) which are widely spread among human and animal pathogenic Gram-negative bacteria. The amino acid sequence of HlyI, which resembles the *E. coli* α-hemolysin, shows a membrane-active amphipathic helix at its N-terminus, followed by three strongly hydrophobic domains forming a typical transmembrane structure, and a segment of 13 repeated glycine-rich nonapeptides at the C-terminus of the protein which is known from other related RTX toxins to bind Ca^{2+}. The *hlyII* operon only contains the *hlyIIC* and *hlyIIA* genes. The proteins involved in the secretion of HlyII are provided by the *hlyI* operon. In those serotypes which produce HlyII but not HlyI, only part of the *hlyI* operon, containing the promoter, the *hlyIB* and *hlyID* secretion genes and a truncated *hlyIA* gene is found. The amino acid sequence of HlyII is more similar to the *Pasteurella haemolytica* leukotoxin than to HlyI and *E. coli* α-hemolysin. In contrast to HlyI, it only contains 8 glycine-rich nonapeptide repeats suggesting a lower Ca^{2+}-binding capacity for HlyII than for HlyI. So far *A. pleuropneumoniae* is the first bacterium known to contain and express two different RTX-hemolysins. From structural data we assume that the two hemolysins recognize different target cells and have different tasks in bacteria-host interactions during infection. A third RTX-toxin named pleurotoxin with an apparent molecular weight of 120 kDa has no hemolytic, but strong cytotoxic activity and has been detected in some serotypes which are devoid of HlyI. The following new and uniform designations for *A. pleuropneumoniae* RTX-toxins are proposed: ApxI for HlyI (ClyI); Apx II for HlyII (ClyII, App); and ApxIII for Ptx (ClyIII).

Abbreviations: RTX, Repeats in the structural Toxin

Introduction

Actinobacillus pleuropneumoniae is the etiological agent of swine pleuropneumonia, a severe disease causing great economic losses in industrialized swine production. The disease is either acute extensive and fibrinohemorrhagic,

C.I. Kado and J.H. Crosa (eds.), Molecular Mechanisms of Bacterial Virulence, 325–340.
© 1994 *Kluwer Academic Publishers. Printed in the Netherlands.*

or chronic localized and necrotizing associated with pleuritis (Shope, 1964). Transmission of *A. pleuropneumoniae* occurs from pig to pig, which is the main host. Survival of the organism in the environment is possible but is assumed to be of short duration (Nicolet, 1992). *A. pleuropneumoniae* is a Gram-negative bacterium belonging to the family of *Pasteurellaceae* (Pohl *et al.*, 1983). Twelve serotypes of *A. pleuropneumoniae* have been described on the basis of the serological typing of capsular polysaccharides. Serotype 5 can be divided in subtypes 5a and 5b (Nielsen, 1986a,b). It has been frequently reported that serotypes 1 and 5 and to a lesser extent also 9 and 11 are isolated from severe outbreaks characterized by strong pulmonary lesions and high mortality. The other serotypes are much less virulent, causing low mortality, but are frequently found in outbreaks in certain countries (Martineau *et al.*, 1984; Rapp *et al.*, 1985., Fales *et al.*, 1989). In experimental infections of pigs, the lethal dose of the strong hemolytic serotype 1 reference strain was 100 times lower than that of the low hemolytic serotype 2 indicating that the reference strain of serotype 1 is much more virulent than serotype 2 (Nicolet, 1970). In intranasal or intraperitoneal infections of mice, the strongly hemolytic reference strains of serotypes 1, 5a, 5b, 9, 10 and 11 were pathogenic, while the other less hemolytic serotypes were not lethal (Komal and Mittal, 1990). In spite of the strict host specificity of *A. pleuropneumoniae*, it is interesting to note that the results obtained in the mouse model are in agreement with the observations from naturally or experimentally infected pigs (Table 1).

Table 1. Hemolysins and pleurotoxin produced by the different *Actinobacillus pleuropneumoniae* serotype reference strains and their virulence properties.

Serotype	1	2	3	4	5a	5b	6	7	8	9	10	11	12
Reference strain	4074	S1536	S1421	M62	K17	L20	femø	WF83	405	CVI13261	13039	56153	8329
Hemolysin I	HlyI				HlyIc)	HlyIc)				HlyI	HlyIc)	HlyI	
Hemolysin II	HlyII	HlyII	HlyII	HlyII	HlyII	HlyII	HlyII	HlyII	HlyII	HlyII		HlyII	HlyII
Pleurotoxin		Ptxd)	Ptx	Ptx			Ptx		Ptx				
Lethality in mouse model a)	+++	-	-	-	+++	+++	-	-	-	+++	+++	+++	-
Virulence in pigs b)	+++	+	+		+++	+++		+		+++		++	

a) The mouse challenge (Komal and Mittal 1990) was made with the reference strains and can therefore directly be compared with the results of the hemolysin and pleurotoxin analysis made in the same strains.

b) Estimated virulence computed from published results (see introduction) and unpublished results are based on mortality and lesions observed in pigs infected with a given serotype (and do not correspond with the frequency with which a given serotype is isolated).

c) HlyI of serotypes 5a, 5b and 10 is expected to be slightly different from HlyI of serotypes 1,9 and 10 as deduced from analysis of their *hlyIA* genes (this paper).

d) Ptx of serotype 2 is slightly different from Ptx in the other serotypes as shown by monoclonal antibodies (van den Bosch, 1990; van den Bosch et al., 1992)

It is likely that several virulence factors are involved in pathogenicity of *A. pleuropneumoniae* including capsular polysaccharides (Jacques *et al.*, 1988;

Jensen and Bertram, 1986), lipopolysaccharides (Fenwick *et al.*, 1986; Bertram, 1988) and exotoxins (Nakai *et al.*, 1984; Frey and Nicolet 1988b, Rosendal *et al.*, 1988). In addition, fimbriae have been detected in freshly isolated strains, which might be involved in adhesion to the host tissue (Tomcik *et al.*, 1988; Utrera and Pijoan, 1991). Characteristic for all *A. pleuropneumoniae* serotypes is a secreted hemolytic activity which seems to play a central role in pathogenesis, since the strength of hemolytic activity correlates with the degree of virulence of the different serotypes (Frey and Nicolet 1988b; Frey and Nicolet, 1990). The data presented below are obtained from serotype reference strains unless otherwise indicated.

Biochemical analysis of hemolysins in *A. Pleuropneumoniae*

The first isolation of hemolysin was made from culture supernatants of *A. pleuropneumoniae* type strain 4074, serotype 1 and was characterized as a 105 kDa protein with an isoelectric point pI = 4.3 (Frey and Nicolet, 1988a). The protein was hemolytically active as a monomer. Although the protein fraction seemed to be homogeneous on SDS polyacrylamide gel electrophoresis, it was later shown that the preparation contained mainly hemolysin I (HlyI) with a calculated molecular weight of 110.1 kDa and to a lesser extent hemolysin II (HlyII) with a calculated molecular weight of 102.5 kDa (Frey *et al.*, 1992). Due to their same apparent molecular weight of 105 kDa on SDS polyacrylamide gels and the fact that the two hemolysins are co-purified, the two proteins could only be distinguished on immunoblots with specific antibodies (Kamp *et al.*, 1991; Frey *et al.*, 1992). The specific hemolytic activity of the purified hemolysin fraction from serotype 1 was 210000 HU/mg protein (hemolytic units per mg protein), and the strong hemolysin was termed HlyI (Frey and Nicolet, 1988a; Frey and Nicolet, 1988b). The hemolytic activity of HlyI required very low amounts of Ca^{2+} as cofactor for erythrocyte lysis. The production of HlyI protein in serotype 1 strain 4074 was strongly enhanced by free Ca^{2+} ions in the growth medium. Transcription and translation inhibition experiments indicated that Ca^{2+} acts as an inducer of HlyI biosynthesis at the transcriptional level (Frey and Nicolet, 1988b). The addition of 1.5 mM free Ca^{2+}, a concentration found in the interstitial fluid of vertebrates, into the growth medium resulted in optimal expression of HlyI. These results were confirmed later on by Northern blot analysis of *hlyI* mRNA (Gygi *et al.*, 1992). Serotypes producing HlyI are generally found to be virulent types and are frequently isolated from pigs in severe pleuropneumonia outbreaks (Table 1). The hemolysin of the weakly hemolytic serotypes which was termed HlyII (Table 1) was also identified as a protein of 105 kDa. Its hemolytic activity required the presence of 5 mM Ca^{2+} as cofactor but its biosynthesis was independent of the presence of Ca^{2+} in the growth medium (Frey and Nicolet, 1998b; Frey and Nicolet, 1990). In some serotype reference strains producing HlyII but not HlyI a protein with an apparent molecular weight of 120 kDa

Table 2. New designation and synonyms used for the various RTX toxins found in *A. pleuropneumoniae*

The 3 RTX toxins of *A.pleuropneumoniae* were initially named Hemolysin I [1] (HlyI), Hemolysin II [1] (HlyII), and Pleurotoxin (Ptx) [10]. Various groups have later on used different designations which are listed below. Unless specific enzymatic or toxic activities have been found for these toxins, we suggest to use in the future the new and uniform designations ApxI, ApxII and ApxIII in order to avoid confusions.

new designation		Synonyms							
Name Toxin	abbreviation protein/ genes	Name Toxin	abbreviation protein/ genes	Name Toxin	abbreviation protein genes	Name Toxin	abbreviatio protein genes	Name Toxin	abbreviation protein/ genes
A.pleuropneumoniae RTX-toxin I	**ApxI** apxIC apxIA apxIB apxID	Hemolysin I	HlyI [1] hlyIC [2,3] hlyIA [2,3] hlyIB [2,3] hlyID [2,3]	Cytolysin I	ClyI [5] clyIC [6] clyIA [6] clyIB [6] clyID [6]		appB [8] appD [8]		
A.pleuropneumoniae RTX-toxin II	**ApxII** apxIIC apxIIA	Hemolysin II	HlyII [1] hlyIIC [4] hlyIIA [4]	Cytolysin II	ClyII [5] clyIIC [6] clyIIA [6]	Hemolysin	App [7] appC [7] appA [7]	Cytolysin	Cyt [9] cytC [9] cytA [9]
A.pleuropneumoniae RTX-toxin III	**ApxIII** apxIIIC apxIIIA apxIIIB apxIIID	Pleurotoxin	Ptx [10] ptxA [11]	Cytolysin III	ClyIII [5] clyIIIC [13,14] clyIIIA [13,14] clyIIIB [13] clyIIID [13]			Macrophage toxin-	Mat [12]

References (see list) : 1) Frey and Nicolet, 1988b; 2) Gygi et al,. 1990; 3) Gygi et al., 1992; 4) Frey et al., 1992; 5) Kamp et al., 1991; 6) Jansen et al,. 1992; 7) Chang et al., 1989; 8) Chang et al,. 1991; 9) Anderson et al., 1991) 10) Rycroft et al., 1991; 11) Macdonald and Rycroft 1992; 12) van den Bosch et al., 1992; 13) Jansen et al., 1992; 14) Jansen et al., 1993.

was co-purified with the 105 kDa HlyII. This 120 kDa protein had no hemolytic activity and was shown to be different from HlyII (Frey *et al.*, 1991b). Other groups had found that it was cytotoxic on porcine alveolar macrophages and was termed pleurotoxin (Ptx) (Rycroft *et al.*, 1991) or cytolysin III (ClyIII) (Kamp *et al.*, 1991), or macrophage toxin (Mat) (van den Bosch, 1990) (Table 1 and 2).

Hemolysin I (HlyI, new designation Apx I)

Hemolysin I (HlyI) is produced by the reference strains of serotypes 1, 5a, 5b, 9, 10 and 11 as determined by biochemical analysis (Frey and Nicolet 1990) or by HlyI-specific monoclonal antibodies (Frey *et al.*, 1992) (Table 1). HlyI is identical to cytolysin I (ClyI) (Table 2), which was detected in the same serotype reference strains by monoclonal antibodies (Kamp *et al.*, 1991).

The structural gene *hlyIA* encoding inactive 105 kDa prohemolysin I of *A. pleuropneumoniae* serotype 1, has been cloned and expressed in *Escherichia coli* (Gygi *et al.*, 1990). *Trans*-activation of prohemolysin I (HlyIA) was achieved with the HlyC activator proteins from *E. coli* and *Proteus vulgaris*, and HlyIA secretion was effected by the *E. coli* HlyB and HlyD proteins supplied *in trans* (Gygi *et al.* 1990). Sequence analysis of the *hlyIA* gene (Frey *et al.*, 1991a) confirmed that HlyI belongs to the RTX (repeats in the structural *toxin*) family of pore-forming hemolysins (Strathdee and Lo, 1989) and shows strong similarities to the *E. coli* α-hemolysin (Felmlee *et al.*, 1985) and to a lesser extent also to the *Pasteurella haemolytica* leukotoxin (Lo *et al.*, 1987) (Fig. 1). The deduced amino acid sequence showed that HlyIA is a protein of 1024 amino acids with a molecular weight of 110.1 kDa. It has a membrane-active amphipathic helix at its N-terminal, followed by three strongly hydrophobic domains forming a typical trans-membrane structure, and a segment of 13 repeated glycine-rich nonapeptides (Fig. 1). These repeats bind Ca^{2+} in other related RTX toxins and are necessary for specific recognition of target cells (Welch, 1991). Using $^{45}Ca^{2+}$ and nitrocellulose blots we and others showed that HlyI and prohemolysin HlyIA have a strong Ca^{2+} binding capacity (Frey *et al.*, 1991a; Devenish and Rosendal, 1991). Cloning the full genetic determinant of HlyI permitted the expression and secretion of active HlyI in *E. coli* (Gygi *et al.*, 1992). Four contiguous genes, *hlyIC* encoding the activator, *hlyIA* encoding the structural protein, and *hlyIB* and *hlyID* encoding the proteins involved in secretion of HlyI were identified by Southern blot hybridization with corresponding *E. coli* hemolysin genes *hlyC*, *hlyA*, *hlyB* and *hlyD* (Fig. 2). Their functions were confirmed by analysis of deletion mutants and complementation experiments (Gygi *et al.*, 1992). DNA sequence analysis of the complete determinant (unpublished results) confirms the presence of these four individual genes and revealed a *rho*-independent transcriptional termination site between *hlyIA* and *hlyIB* (Frey *et al.*, 1991).

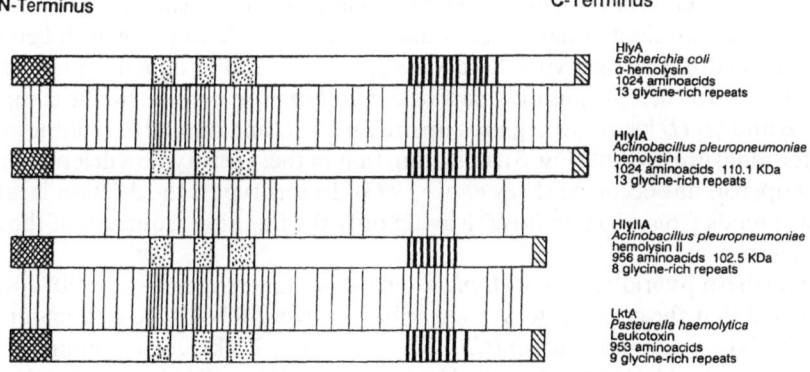

Figure 1. Schematic representation and comparison of the prohemolysin proteins HlyIA and HlyIIA of *Actinobacillus pleuropneumoniae* with *Escherichia coli* α-hemolysin (HlyA) and *Pasteurella haemolytica* leukotoxin LktA. Checkered boxes represent amphiphilic helix structures. Dotted boxes show the hydrophobic domains. The bold-face vertical lines represent the glycine-rich repeated nonapeptides. Thin vertical lines show regions of high sequence identity, the density of the lines represent the extent of similarity between two adjacent proteins.

Southern blot analysis of genomic DNA from all serotype reference strains with probes for the individual genes of the *hlyI* operon (Fig. 2) revealed the presence of the complete *hlyI* operon in the HlyI producing serotypes 1, 5a, 5b, 9, 10 and 11 (Fig. 3). Further analysis of the *hlyIA* genes by cloning and digestion with frequently cutting restriction enzymes showed, that the *hlyIA*

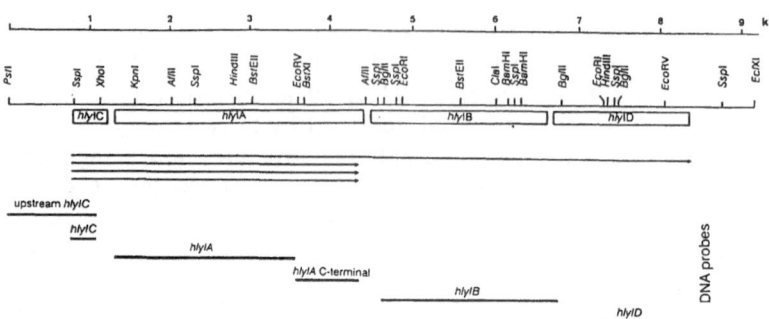

Figure 2. Map of the *hlyI*-operon and location of the probes used for the hybridization experiments. The upper part represents a physical and genetic map of the *hlyI* operon as described (Gygi *et al.*, 1992). The scale is given in kilobasepairs (kbp). Arrows show the two transcripts of this operon. The lower part shows the location of the various DNA probes used in the hybridization experiments.

gene in serotypes 5a, 5b and 10, is different from the *hlyIA* gene in serotypes 1, 9 and 11 (Fig. 3) (Frey *et al.*, 1993). In this respect, it is interesting to note that strains of serotypes 1, 9 and 11 are serologically closely related and show strong immunological cross reactions (Nicolet, 1988; Mittal, 1990). The sequence divergence between *hlyIA* of the group with the serotypes 1, 9 and 11, and that of serotypes 5a, 5b and 10 is estimated to be 5%. The HlyI proteins from both groups of serotypes seem to be biochemically very similar. So far, no monoclonal antibodies have been found which are able to distinguish between HlyI from different serotypes. Serotypes 2, 4, 6, 7, 8 and 12 do not contain *hlyIC* and *hlyIA*. In these strains only part of the *hlyI* operon containing the complete *hlyIB* and *hlyID* locus, the segment upstream *hlyIC* and a small part of the *hlyIA* N-terminal are present (Fig. 3) indicating that in these serotypes a deletion in the *hlyI* operon has occurred (Frey *et al.*, 1993). In serotype 3 the deletion is larger and extends from *hlyIC* to *hlyID* leaving only the flanking sequences of the *hlyI* operon.

Northern hybridization of *A. pleuropneumoniae* serotype 1 strain 4074 RNA revealed that the gene cluster is transcribed as two mRNA species, a major one of 3.5 kb corresponding to *hlyICA* and a second of 7.5 kb corresponding to the whole operon *hlyICABD* (Fig. 2). The level of *hlyICA* mRNA was substantially higher in *A. pleuropneumoniae* serotype 1 cells grown in the presence of Ca^{2+} (Gygi *et al.*, 1992). Control RNA hybridization with random gene probes showed that this Ca^{2+} regulation is specific for the *hlyI* determinant (Gygi *et al.* 1992). These experiments support previous evidence that the biosynthesis of HlyI is regulated by Ca^{2+} (Frey and Nicolet, 1988). So far, the only procaryotic genes known to be regulated by Ca^{2+} ions are genes encoding virulence factors of *Yersinia pestis* (Cornelis *et al.*, 1989; Straley and Browmer, 1986) and of *Pseudomonas aeruginosa* (Blumentals *et al.*, 1987), indicating that some pathogenic bacteria may use the strong difference between intracellular and extracellular Ca^{2+} concentrations in host tissues as signal for regulation of virulence factors.

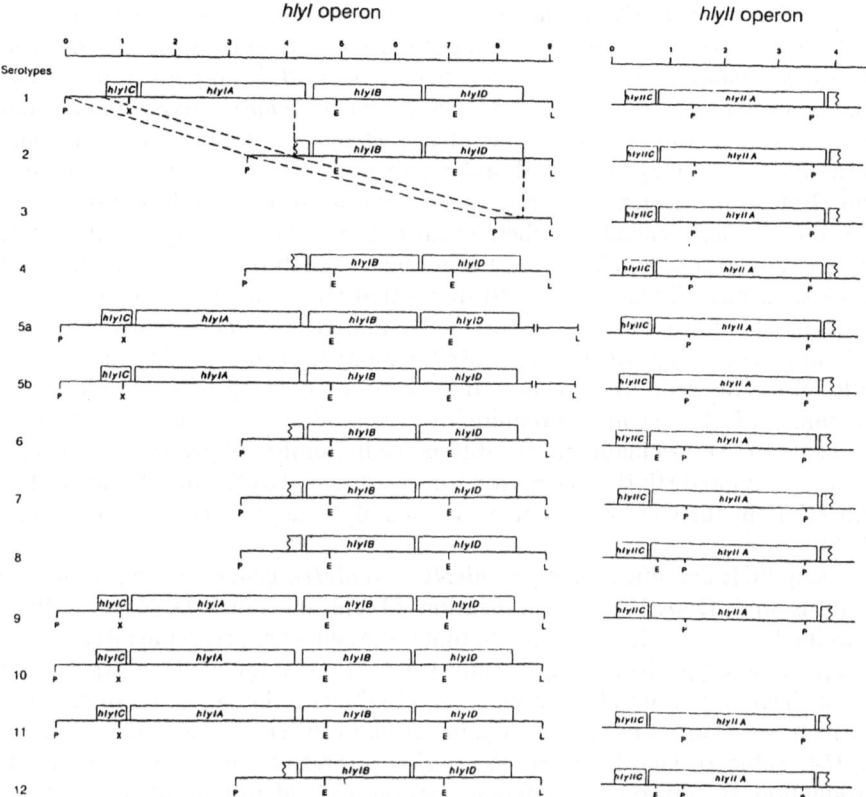

Figure 3. Genetic organization of the *hlyI* and *hlyII* operons in the different serotype reference strains. Open boxes represent the location of the various *hlyI* and *hlyII* genes. Broken boxes represent the remaining part of *hlyIA* in the *hlyI* operon or *hlyIIB* in the *hlyII* operon. Broken lines show the possible deletion formations. The restriction enzymes indicated are P: *Pst*I, E: *Eco*RI, L: *Ecl*XI. The scale on the top is given in kbp.

Hemolysin II (HlyII, new designation: ApxII)

The production of hemolysin II (HlyII) was identified in all serotypes except in serotype 10 by means of HlyII-specific monoclonal and polyclonal antibodies (Frey *et al.*, 1992). HlyII is identical to Cytolysin II (ClyII) (Table 2), which was detected in the same strains by monoclonal antibodies (Kamp *et al.*, 1991). It has been shown by means of gene expression in *E. coli* and polyclonal and monoclonal antibodies directed against HlyII that the genes *appC* and *appA* (Chang *et al.*, 1989) encode HlyII.

The operon specifying HlyII containing the activator gene *appC* and the structural gene *appA* were cloned from an *A. pleuropneumoniae* serotype 5 strain using the *P.haemolytica* leukotoxin gene (*lkt*) as a probe for screening a gene library (Chang *et al.*, 1989). The DNA sequence of the *appA* gene (Chang *et al.*, 1989) shows more similarity to the *lkt* gene of *P.haemolytica* than to *hlyIA* of *A. pleuropneumoniae* or the *E. coli hlyA* gene. Using PCR amplification, a gene analogous to *appA* was cloned from *A. pleuropneumoniae* serotypes 1 and 2 and expressed in *E. coli*. The gene was identified by polyclonal and monoclonal antibodies directed against HlyII and capable of differentiation between HlyII and HlyI, to be the gene encoding the structural protein of HlyII (Frey *et al.*, 1992). Gene *appA* encodes prohemolysin II (HlyIIA) while *appC* is the HlyII specific activator gene. The amino acid sequence of HlyII derived from the *appA* DNA sequence (Chang *et al.*, 1989) shows that HlyII, like HlyI has the typical domains of RTX toxins with the exception that HlyII only contains 8 glycine-rich nonapeptide repeats (Fig. 1). Since the number of such glycine-rich repeats in RTX-toxins seems to be related to the capacity of the proteins to bind Ca^{2+} (Boehm *et al.*, 1990) which is required for the binding to erythrocytes (Ludwig *et al.*, 1988), we speculate that the difference in number of glycine-rich repeats between HlyI and HlyII is the reason for the observed differences between HlyI and HlyII in the Ca^{2+} requirement for hemolytic activity (Frey and Nicolet, 1988b).

Using PCR amplification of the *hlyIIC* and *hlyIIA* genes, restriction enzyme analysis and DNA:DNA hybridization of chromosomal DNA with PCR amplified probes, we have confirmed that these genes are present in all serotypes except in serotype 10 and show only very little divergence (Fig. 3). Similar results have been found by Jansen *et al.*, (1992) with the exception of serotype 6 which was claimed to have a large insertion in the structural gene *clyIIA* (≡ *hlyIIA*, Table 2). This, however, contradicts our results and the fact that HlyII production by serotype 6 reference strain was identified by monoclonal antibodies (Kamp *et al.*, 1991; Frey *et al.*, 1992).

The operon encoding HlyII only contains *appC*, *appA* and some sequences resembling a 5'-end of a *hlyB* gene, while genes *(appB* and *appD)* encoding the proteins involved in secretion of hemolysin were reported to be separated in a particular serotype 5 strain (Chang *et al.*, 1991). Our sequencing results of the complete *hlyI* operon from serotype 1 show that the sequences of *appB* and *appD* are identical to *hlyIB* and *hlyID* of the *hlyI* operon (Gygi *et al.*, 1992). The fact that no full *hlyIA* gene was found upstream *appB* signifies that the particular serotype 5 strain used for the study (Chang *et al.*, 1989; Chang *et al.*, 1991) must be different from the serotype 5a and 5b reference strains which were found to produce both HlyI and HlyII (Frey *et al.*, 1992; Kamp *et al.*, 1991). The serotype 5 strain used by Chang *et al.*, (1989) must have a deletion in its *hlyI* operon similar to serotypes 2, 4, 6, 7, 8 and 12 (Fig. 3). The sequences found upstream *appB* and which were called hemolysin-pseudogenes (Chang *et al.*, 1991) are indeed identical to the N-terminal part of the *hlyIA* sequence (Frey *et al.*, 1991). The same genes were found in serotype 9 where the activator and structural genes *clyIIC* and *clyIIA* (≡ *hlyIIC* and *hlyIIA* respectively) were also reported to be unlinked to the secretion genes *clyIB* and *clyID* (≡ *hlyIB* and *hlyID*; Table 2) (Smits *et al.*, 1991).

Pleurotoxin (Ptx, new designation: ApxIII)

A secreted protein with an apparent molecular weight of 120 kDa which does not have hemolytic activity, but which is strongly cytotoxic for porcine alveolar macrophages was found in *A. pleuropneumoniae* serotypes 2 (Rycroft *et al.*, 1991; Frey *et al.*, 1991b) and was first named pleurotoxin (Ptx) (Rycroft *et al.*, 1991). The same toxin, also named Cytolysin III (ClyIII) (Table 2), was identified in serotypes 2, 3, 4 and 8 and was shown to have a strong cytotoxic activity on alveolar macrophages (Kamp *et al.*, 1991). The same toxin was also named macrophage toxin (Mat), expressed by serotype reference strains 2, 3, 4, 6 and 8 as identified with neutralizing monoclonal antibodies (van den Bosch, 1990; van den Bosch *et al.*, 1992). Variants of Ptx (Mat) exist, since besides crossreacting monoclonal antibodies also monoclonal antibodies were found that reacted only with the 120 kDa protein expressed by serotype 2 reference strain (van den Bosch *et al.*, 1992). Using immunoblot analysis with polyclonal antibodies directed against the Ptx protein purified from serotype 2 (data not shown), we demonstrated the presence of this protein in supernatants of serotypes 2, 3, 4, 6 and 8. It is interesting to note that in the serotype 3 reference strain which lacks the *hlyIBD* secretion genes, the 120 kDa Ptx protein is found in large amounts in the culture supernatant where as only faint amounts of HlyII can be detected (data not shown), predicting a separate secretion system for Ptx. The gene(s) encoding Ptx were cloned and expressed in *E. coli*. Active cytotoxic activity was detected in culture supernatants of recombinant strain indicating that the cloned DNA also contained, besides the structural gene for the Ptx protein called *ptxA*, also genes involved in secretion (Macdonald and Rycroft, 1992). Hybridization experiments with the cloned *ptxA* gene revealed the presence of this gene in serotypes 2, 3, 4, 6 and 8 (Macdonald and Rycroft, 1992). DNA sequencing data of the *clyIII* ($\equiv$ *ptx*) genes confirm that the 120 kDa pleurotoxin (cytolysin III) protein belongs to the group of RTX toxins (Jansen *et al.*, 1993). In addition, this protein was shown to be secreted in *E. coli* when the *E. coli hlyBD* genes were provided *in trans* (Jansen *et al.*, 1991).

Evolution of hemolysin genes in the *A. pleuropneumoniae* serotypes

Our hybridization results (Fig. 2) show that all serotypes of *A. pleuropneumoniae* contain two different hemolysin operons, although frequently truncated (Frey *et al.*, 1993). In addition, a third operon for an RTX-toxin containing the genes for the non-hemolytic Pleurotoxin is expected to be present in certain serotypes. The operon encoding HlyII only contains the activator gene *hlyIIC* ($\equiv$ *appC* or *clyC*) and the structural gene *hlyIIA* ($\equiv$ *appA* or *clyC*). Sequence analysis (Chang *et al.*, 1991; Smits *et al.*, 1991) and hybridization experiments (Jansen *et al.*, 1991) revealed a short gene fragment resembling the N-terminal part of a secretion protein B next to the structural *appA* gene (Fig. 3). One can therefore speculate that ancestors of *A. pleuropneumoniae* have contained two full hemolysin operons with all four genes (CABD). During evolution, the *hlyII* operon probably has lost its genes encoding the secretion proteins B and D. The function for the secretion of HlyII was therefore complemented by the *hlyIBD* gene products. It is generally known for RTX-toxins, that the gene encoding the B-secretion protein (e.g. *hlyB*, *lktB*) is highly conserved and is expressed in very low amounts (Koronakis *et al.*, 1987). Probably due to high similarity, the *B and D* genes belonging to the *hlyII* operon were deleted. The presence of only one copy of *B and D* genes was most likely sufficient or even beneficial to the bacteria. Subsequently, certain serotypes might have lost the structural and activator genes *hlyICA* from the *hlyI* operon. However, they retained the *hlyIB* and *hlyID* genes as well as the segment containing the promoter of this operon to allow expression of *hlyIB* and *hlyID* (Fig. 3). Serotype 3 seems to have lost the whole *hlyI* operon except some sequences upstream the *hlyIC* gene (Fig. 3). It is, therefore, not surprising that only faint amounts of HlyII are found in culture supernatants of serotype 3. This extracellular HlyII could originate from lysed cells or could be co-secreted by different systems.

The data presented above were obtained with *A. pleuropneumoniae* serotype reference strains (Table 1). Our preliminary results from investigations of *A. pleuropneumoniae* field strains of various serotypes isolated worldwide from different geographical areas indicate that the hemolysin patterns for the reference strains seem in general to be conserved in field strains.

The guanine + cytosine (G+C) composition of RTX-toxin genes is generally between 39% and 41% which corresponds to the G+C composition of the *A. pleuropneumoniae* chromosome. In *E. coli*, in contrast, the G+C composition of *hlyA* is 41.2% and differs from that of the chromosome which is 50%, indicating that the hemolysin genes originate from a different species. The fact that three different RTX operons have been found in *A. pleuropneumoniae* and that several strains contain up to two different RTX operons together with the finding that the chromosome of *A. pleuropneumoniae* has the same G+C content as the RTX operons, leads to the hypothesis that RTX genes might have originated from this species or from closely related species of the Pasteurellaceae family.

Immunological properties of *A. pleuropneumoniae* hemolysins

The hemolysin proteins of *A. pleuropneumoniae* were shown to be the predominant immunogenic proteins if sera from field convalescent, or sera from experimentally infected pigs with *A. pleuropneumoniae* were investigated (Devenish *et al.*, 1990a; Frey and Nicolet, 1991). Strong immunological reactions with the 105 kDa hemolysin proteins were observed in the sera of pigs that were experimentally infected with each of the 12 serotypes. Some immunological reactions with the 105 kDa hemolysins were also found in sera from pigs that were confirmed to be negative for *A. pleuropneumoniae*. These cross reactions could originate from related proteins of *Actinobacillus rossii* or

Actinobacillus suis, since these species showed also proteins in the 105 kDa range which reacted immunologically with antibodies against *A. pleuropneumoniae* hemolysins (Frey and Nicolet, 1991). The apparent immunological cross reactions between HlyI and HlyII (Frey and Nicolet, 1991), which raised doubts about the existence of two different hemolysins in *A. pleuropneumoniae* (Devenish *et al.*, 1989), are due to the fact that *A. pleuropneumoniae* serotype 1 produces both HlyI and HlyII (Frey *et al.*, 1992) and not only HlyI as initially believed.

Because of high antibody levels in convalescent sera from pigs after *A. pleuropneumoniae* infection, hemolysins were considered as useful candidates for vaccine components. Purified hemolysin from a serotype 1 strain, supposedly containing HlyI and HlyII, induced protective immunity in pigs. Vaccinated pigs showed no mortality and reduced lung lesions compared to non-vaccinated pigs after challenge with a virulent serotype 1 strain (Devenish *et al.*, 1990b). In general, purified hemolysins induced protection against mortality, but only limited or no protection against the development of typical lung lesions (Fedorka-Cray *et al.*, 1990; Devenish *et al.*, 1990b; Inzana *et al.*, 1991; Rossi-Campos *et al.*, 1992). The hemolysins (HlyI + HlyII) purified from serotypes 1 or 5b in combination with a 42 kDa outer membrane protein from serotype 1 showed full protection against death and very high protection against lung lesions without concomitant adverse side effects of vaccination (van den Bosch, 1990). A similar synergistic effect was seen with a vaccine containing HlyII and a 60 kDa transferrin binding protein using challenge with the homologous serotype 7 strain (Rossi-Campos *et al.*, 1992). Since most experiments were done before it was known that the purified hemolysin fraction of serotype 1 contained both HlyI and HlyII, no results are available to discern whether HlyI or/and HlyII are needed to induce protective immunity. However, protection induced only by HlyII antibodies showed protection only against those serotypes which produce HlyII and no other RTX toxins. Vaccines containing HlyI and HlyII were only protective against challenge with HlyI + HlyII expressing strains and not with Ptx + HlyII expressing strains, whereas vaccines containing Ptx and HlyII were only protective against challenge with Ptx + HlyII expressing strains and not with HlyI + HlyII expressing strains (van den Bosch *et al.*, 1992). Furthermore, cloned HlyII from a serotype 7 strain induced protection against mortality after homologous challenge with serotype 7 (expressing HlyII only) but not after heterologous challenge with a serotype 1 strain (expressing HlyI + HlyII) (Rossi-Campos *et al.*, 1992). Bhatia *et al.*, (1991) showed that purified hemolysins from serotype 1 (HlyI+HlyII) are important for protective immunity against challenge with *A. pleuropneumoniae* in a mouse model. In this model, intranasal or intraperitoneal challenge of mice with *A. pleuropneumoniae* serotype 1 or serotype 5 was shown to be lethal. Active immunization with purified hemolysin (HlyI+HlyII) in combination with washed formalinized whole *A. pleuropneumoniae* serotype 1 or serotype 5 cells induced full protection against challenge with both serotypes 1 and 5. The hemolysin fraction alone or formalinized cells alone gave significantly lower protection. In the same experiment, virtually no protection was obtained with purified capsular polysaccharide or lipopolysaccharide (Bhatia *et al.*, 1991). Preliminary results using a mouse model indicate that HlyII alone does not induce protective immunity against serotype 1 or serotype 5 (which both produce HlyI and HlyII).

New and uniform designations of *A. pleuropneumoniae* RTX toxins

Due to intense research efforts on *A. pleuropneumoniae* RTX toxins which were made simultaneously by various groups during the last few years, and also due to the uncommon genetic structure of some or these RTX operons, several different toxin names and gene designations have been used and led to confusions. The researchers working on *A. pleuropneumoniae* RTX toxins therefore made the effort to agree for a common, uniform nomenclature for these toxins which was proposed at this Fallen Leaf Lake Conference 1992 on "Bacterial Virulence Mechanisms". These new designations propose the use of ApxI for the strong hemolytic and cytotoxic hemolysin I (HlyI, ClyI), the designation of ApxII for the weak hemolytic and cytotoxic hemolysin II (HlyII, ClyII), and the designation ApxIII for the non-hemolytic and strongly cytotoxic pleurotoxin Ptx (ClyIII). The designations for the toxins and for the corresponding genes with their respective synonyms are listed in Table 2.

Concluding remarks

The hemolysins of *A. pleuropneumoniae* belong to the group of RTX toxins and are considered important virulence factors. *A. pleuropneumoniae* is the first bacterium known to simultaneously produce two different RTX-hemolysins. A third secreted protein, the pleurotoxin, with cytotoxic but no hemolytic activity is also an RTX-toxin and is found in *A. pleuropneumoniae* strains associated with HlyII. So far, no reference or field strain has been detected which produces all three RTX toxins. Genetic data indicate that two different hemolysin operons did undergo rearrangements in *A. pleuropneumoniae* by deletion formation. Insertion sequences (IS) or IS-like DNA structures which were found in the vicinity of the hemolysin II genes *cytC* and *cytA* ($\equiv$ *hlyIIC* and *hlyIIA*, Table 2) (Anderson *et al.*, 1992) might be the cause of such deletion formations. Data from *A. pleuropneumoniae* outbreaks, from experimental infections of pigs and mice and vaccination results indicate that HlyI is associated with strains of particular virulence and high mortality. In spite of the detailed genetic data available on *A. pleuropneumoniae* RTX toxins, their exact functions and their target cells remain unknown. Further investigations on the specific lytic and also nonlytic activities of these RTX-toxins (Welch, 1991) will therefore be required to gain a better understanding of the role of these toxins in pathogenicity.

Acknowledgements

This work was supported by grant 31–28401.90 from the Swiss National Science Foundation. I thank Jacques Nicolet and Paola Peveri, University of Berne, and Ruud Segers and Han van den Bosch, Intervet International, Boxmeer NL for reviewing this manuscript, and to Marianne Beck and Urs Stucki University of Berne for their contributions.

References

Anderson, C., Potter, A.A., and Gerlach, G.-F. (1991) Isolation and molecular characterization of spontaneously occurring cytolysin-negative mutants of *Actinobacillus pleuropneumoniae* serotype 7. Infect. Immun. 59: 4110–4116.

Bertram, T.T. (1988) Pathobiology of acute pulmonary lesions in swine infected with *Haemophilus (Actinobacillus) pleuropneumoniae*. Can. Vet. J. 29: 574–577.

Bhatia, B., Mittal, K.R., and Frey, J. (1991) Factors involved in the immunity against *Actinobacillus pleuropneumoniae* in mice. Vet. Microb. 28: 147–158.

Blumentals, I.I., Kelly, R.M., Gorziglia, M., Kaufman, J.B., and Shiloach, J. (1987) Development of a defined medium and two-step culturing method for improved exotoxin A yields from *Pseudomonas aeruginosa*. Appl. Environ. Microbiol. 53: 2013–2020.

Boehm, D.F., Welch, R.A., and Snyder, I.S. (1990) Domains of *Escherichia coli* hemolysin (HlyA) involved in binding of calcium and erythrocyte membranes. Infect. Immun. 58: 1959–1964.

Chang, Y., Young, R., and Struck, K. (1989) Cloning and characterization of a hemolysin gene from *Actinobacillus (Haemophilus) pleuropneumoniae*. DNA 8: 635–647.

Chang, Y., Young, R., and Struck, K. (1991) The *Actinobacillus pleuropneumoniae* hemolysin determinant: unlinked *appCA* and *appBD* loci flanked by pseudogenes. J. Bacteriol. 173: 5151–5158.

Cornelis, G.R., Biot, T., Lambert de Rouvroit, C., Michels, T., Mulder, B., Sluiters, C., Sory, M.-P., Van Bouchaute, M., and Vanooteghem, J.-C. (1989) The *Yersinia yop* regulon. Mol. Microbiol. 3: 1455–1459.

Devenish, J., Rosendal, S., Johnson, R., and Hubler, S. (1989) Immunoserological comparison of 104-kilodalton proteins associated with hemolysis and cytolysis in *Actinobacillus pleuropneumoniae, Actinobacillus suis, Pasteurella haemolytica* and *Escherichia coli*. Infect. Immun. 57: 3210–3213.

Devenish, J., Rosendal, S., Bossé, J.T., Wilkie, B.N., and Johnson, R. (1990a) Prevalence of seroreactors to the 104-kilodalton hemolysin of *Actinobacillus pleuropneumoniae* in swine herds. J. Clin. Microb. 28: 789–791.

Devenish, J., Rosendal, S., and Bossé, J.T. (1990b) Humoral antibody response and protective immunity in swine following immunization with the 104-kilodalton hemolysin of *Actinobacillus pleuropneumoniae*. Infect. Immun. 58: 3829–3832.

Devenish, J., and Rosendal, S. (1991) Calcium binds to and is required for biological activity of the 104-kilodalton hemolysin produced by *Actinobacillus pleuropneumoniae* serotype 1. Can. J. Microbiol. 37: 317–321.

Fales W.H., Morehouse, L.G., Mittal, K.R., Bean-Knudsen, C., Nelson, S.L., Kintner, L.D., Turk, J.R., Turk, M.A., Brown T.P., and Shaw D.P. (1989) Antimicrobial susceptibility and serotypes of *Actinobacillus (Haemophilus) pleuropneumoniae* recovered from Missouri swine. J. Vet. Diagn. Invest. 1: 16–19.

Fedorka-Cray, P.J., Huether, M.J., Stine, D.L., and Anderson, G.A. (1990) Efficacy of a cell extract from *Actinobacillus pleuropneumoniae* serotype 1 against disease in swine. Infect. Immun. 58: 358–365.

338

Felmlee, T., Pellett, S., and Welch, R.A. (1985) Nucleotide sequence of *Escherichia coli* hemolysin. J. Bacteriol. 163: 94–105.

Fenwick, B.W., Osburn, B.I., and Olander, H.J. (1986) Isolation and biological characterization of two lipopolysaccharides and a capsular enriched polysaccharide preparation from *Haemophilus pleuropneumoniae*. Am. J. Vet. Res. 47: 1433–1441.

Frey, J., and Nicolet, J. (1988a) Purification and partial characterization of a hemolysin produced by *Actinobacillus pleuropneumoniae* type strain 4074. FEMS Microbiol. Lett. 55: 41–46.

Frey, J., and Nicolet, J. (1988b) Regulation of hemolysin expression in *Actinobacillus pleuropneumoniae* serotype 1 by Ca^{2+}. Infect. Immun. 56: 2570–2576.

Frey, J., and Nicolet, J. (1990) Hemolysin patterns of *Actinobacillus pleuropneumoniae*. J. Clin. Microbiol. 28: 232–236.

Frey, J., and Nicolet, J. (1991) Immunological properties of *Actinobacillus pleuropneumoniae* hemolysin I. Vet. Microb. 28: 61–73.

Frey, J., Meier, R., Gygi D., and Nicolet, J. (1991a) Nucleotide sequence of the hemolysin I gene from *Actinobacillus pleuropneumoniae*. Infect. Immun. 59: 3026–3032.

Frey, J., Deillon, J.-B., Gygi, D., and Nicolet, J. (1991b) Identification and partial characterization of the hemolysin (HlyII) of *Actinobacillus pleuropneumoniae* serotype 2. Vet. Microbiol. 28: 303–312.

Frey, J., van den Bosch, H., Segers, R., and Nicolet, J. (1992) Identification of a second hemolysin (HlyII) in *Actinobacillus pleuropneumoniae* serotype 1 and expression of the gene in *Escherichia coli*. Infect. Immun. 60: 1671–1676.

Frey, J., Beck, M., Stucki, U., and Nicolet, J. (1993) Analysis of hemolysin operons in *Actinobacillus pleuropneumoniae*. Gene 123: 51–58.

Gygi, D., Nicolet, J., Frey, J., Cross, M., Koronakis, V., and Hughes, C. (1990) Isolation of the *Actinobacillus pleuropneumoniae* haemolysin gene and the activation and secretion of the prohaemolysin by the HlyC, HlyB and HlyD proteins of *Escherichia coli*. Mol. Microbiol. 4: 123–128.

Gygi, D., Nicolet, J., Hughes, C., and Frey, J. (1992) Functional analysis of the Ca^{2+} regulated hemolysin I operon of *Actinobacillus pleuropneumoniae* serotype 1. Infect. Immun. 60: 3059–3064.

Inzana, T. J., Todd, J., Ma, J., and Veit, H. (1991) Characterization af a non-hemolytic mutant of *Actinobacillus pleuropneumoniae* serotype 5: role of the 110 kilodalton hemolysin in virulence and immunoportection. Microb. Pathog. 10: 281–296.

Jacques, M., Foiry, B., Higgins, R., and Mittal, K.R. (1988) Electron microscopic examination of capsular material from various serotypes of *Actinobacillus pleuropneumoniae*. J. Bacteriol. 170: 3314–3318.

Jensen, A.E., and Bertram, T.A. (1986) Morphological and biochemical comparison of virulent and avirulent isolates of *Haemophilus pleuropneumoniae* serotype 5. Infect. Immun. 51: 419–424.

Jansen, R., Briaire, J., Kamp, E., and Smits, M.A. (1991) Cloning and characterization of the *Actinobacillus pleuropneumoniae* cytolysin genes. Conference of Research Workers in Animal Disease. 72nd meeting. Chicago Ill. Abstract nr. 301. p. 53.

Jansen, R., Briaire, J., Kamp, E., and Smits, M.A. (1992) Comparison of the Cytolysin II genetic determinants of *Actinobacillus pleuropneumoniae* serotypes. Infect. Immun. 60: 630–636.

Jansen, R., Briaire, J., Kamp, E., Gielkens, A.L., and Smits, M.A. (1993) Another member of the RTX cytotoxin family produced by *Actinobacillus pleuropneumoniae*: Cloning and characterization of the ApxIII gene. Infect. Immun. 61: in press.

Kamp, E.M., and van Leengoed, L.A.M.G. (1990) Serotype related differences in production and type of heat-labile hemolysin and heat-labile cytotoxin of *Actinobacillus pleuropneumoniae*. J. Clin. Microbiol. 27: 1187–1191.

Kamp, E.M., Popma, J.K., Anakotta, J., and Smits, M.A. (1991) Identification of hemolytic and cytotoxic proteins of *Actinobacillus pleuropneumoniae* by using monoclonal antibodies. Infect. Immun. 59: 3079–3085.

Komal, J.P.S., and Mittal, K.R. (1990) Grouping of *Actinobacillus pleuropneumoniae* strains of serotypes 1 through 12 on the basis of their virulence in mice. Vet. Microbiol. 25: 229–240.

Koronakis, V., Cross, M., Senior, B., Koronakis, E., and Hughes, C. (1987) The secreted hemolysins of *Proteus vulgaris* and *Morganella morganii* are genetically related to each other and to the α-hemolysin of *Escherichia coli*. J. Bacteriol. 169: 1509–1515.

Lo, R.Y.C., Strathdee, C.A., and Shewen, P.E. (1987) Nucleotide sequence of the leukotoxin genes of *Pasteurella haemolytica* A1. Infect. Immun 55: 1987–1996.

Ludwig, A., Jarchau, T., Benz, R., and Goebel, W. (1988) The repeat domain of *Escherichia coli* haemolysin (HlyA) is responsible for its Ca^{2+}-dependent binding to erythrocytes. Mol. Gen. Genet. 214: 553–561.

Macdonald, J., and Rycroft, A.N. (1992) Molecular cloning and expression of *ptxA*, the gene encoding the 120-kilodalton cytotoxin of *Actinobacillus pleuropneumoniae* serotype 2. Infect. Immun. 60: 2726–2732.

Martineau, G.P., Desrosiers, R., Charette, R., and Moore, C. (1984) Control measures and economical aspects of swine pleuropneumonia in Quebec. p. 97–111. *In*: Proceedings of the American Association of Swine Practioners: American Association of Swine Practioners. Kansas City. Mo. U.S.A.

Mittal, K.R. (1990) Cross-reactions between *Actinobacillus* (*Haemophilus*) *pleuropneumoniae* strains of serotypes 1 and 9. J. Clin. Microbiol. 28: 535–539.

Nakai, T., Sawata, A., and Kume, K. (1984) Pathogenicity of *Haemophilus pleuropneumoniae* for laboratory animals and possible role of its hemolysin for production of pleuropneumonia. Jp. J. Vet. Sci. 46: 851–858.

Nicolet, J. (1970) Aspects microbiologiques de la pleuropneumonie contagieuse du porc. Ph.D. thesis. University of Berne, Berne, Switzerland.

Nicolet, J. (1988) Taxonomy and serological identification of *Actinobacillus pleuropneumoniae*. Can. Vet. J. 29: 578–580.

Nicolet, J. (1992) *Actinobacillus pleuropneumoniae*. In: Leman, A.D., Straw, B.E., Mengeling, W.L., D'Allaire, S., Taylor, D.J. (eds.) Diseases of Swine. 7th edition. Iowa State University Press. Ames, Iowa. U.S.A.

Nielsen, R. (1986a) Serology of *Haemophilus* (*Actinobacillus*) *pleuropneumoniae* serotype 5 strains: establishment of subtypes A and B. Acta Vet. Scand. 27: 49–58.

Nielsen, R. (1986b) Serological characterization of *Actinobacillus pleuropneumoniae* strains and proposal of a new serotype: serotype 12. Acta Vet. Scand. 27: 453–455.

Pohl, S., U. Bertschinger, W. Frederiksen, and Mannheim, W. (1983) Transfer of *Haemophilus pleuropneumoniae* and the Pasteurella haemolytica-like organism causing porcine necrotic pleuropneumonia to the genus *Actinobacillus* (*Actinobacillus pleuropneumoniae* com. nov.) on the basis of phenotypic and deoxyribonucleic acid relatedness. Int. J. Syst. Bacteriol. 33: 510–514.

Rapp, V.J., Ross, R.F., and Zimmermann-Erickson, B. (1985) Serotyping of Haemophilus pleuropneumoniae by rapid slide agglutination and indirect fluorescent antibody tests in swine. Am. J. Vet. Res. 46: 1053–1058.

Rosendal, S., Devenish, J., McInnes, J.I., Lumsden, J.H., Watson, S., and Xun, H. (1988) Evaluation of heat sensitive, neutrophil-toxic, and hemolytic activity of *Haemophilus* (*Actinobacillus*) *pleuropneumoniae*. Am. J. Vet. Res. 49: 1053–1058.

Rossi-Campos, A., Anderson, C., Gerlach, G-F., Klashinsky, S., Potter, A.A., and Willson, P.J. (1992) Immunization of pigs against Actinobacillus pleuropneumoniae with two recombinant protein preparations. Vaccine 10: 512–518.

Rycroft, A.N., Williams, D., Cullen, J.H., and Macdonald, J. (1991) The cytotoxin of *Actinobacillus pleuropneumoniae* (pleurotoxin) is distinct from the haemolysin and is associated with at 120-kDa polypeptide. J. Gen. Microbiol. 137: 561–568.

Smits, M.A., Briaire, J., Jansen, R., Smith, H.E., Kamp, E.M., and Gielkens, A.R.J. (1991) Cytolysins of *Actinobacillus pleuropneumoniae* serotype 9. Infect. Immun. 59: 4497–4504.

Shope, R.E. (1964) Porcine contagious pleuropneumonia. Experimental transmission, etiology and pathology. J. Exp. Med. 29: 301–306.

Straley, S.C., and Browmer, W. (1986) Virulence genes regulated at the transcriptional level by Ca^{2+} in *Yersinia pestis* include structural genes for outer membrane proteins. Infect. Immun. 51: 445–454.

Strathdee, C.A., and Lo, R.Y.C. (1989) Cloning, nucleotide sequence and characterization of genes encoding the secretion function of the *Pasteurella haemolytica* leukotoxin determinant. J. Bacteriol. 171: 916–928.

Tomcik, K., Gilchrist, A., Potter, A., Deneer, H., Klashinsky, S., and Willson, P. (1988) Pilus like structures on *Actinobacillus pleuropneumoniae*. Abstract nr. 70. Proceedings of the 69th Conference of Research Workers on Animal Disease. p. 12. Chicago Ill. U.S.A.

Utrera, V., and Pijoan, C. (1991) Fimbriae in *A.pleuropneumoniae* strains isolated from pig respiratory tracts. Vet. Record 128: 357–358.

Van den Bosch, J.F. (1990) *Actinobacillus pleuropneumoniae* subunit vaccine. Patent application EP-0453024.

Van den Bosch, J.F., Pennings, A.M.M.A., Cuijpers, A.N.B., Pubben, A.N.B., Van Vugt, F.G.A., and Van der Linden, M.F.I. (1990) Heterologous protection induced by an *A. pleuropneumoniae* subunit vaccine. Proc. 11th IPVS Congr., p 11, Lausanne, Switzerland.

Van den Bosch, J.F., Jongenelen, I.M.C.A., Pubben, A.N.B., Van Vugt, F.G.A., and Segers, R.P.A.M. (1992) Protection induced by a trivalent *A.pleuropneumoniae* subunit vaccine. Proc. 12th IPVS Congr., p 197, The Hague, Netherlands.

Welch, R.A. (1991) Pore-forming cytolysins of Gram-negative bacteria. Mol. Microbiol. 5: 521–528.

24. Capsular polysaccharide production in *Erwinias*

DAVID L. COPLIN, FRANK BERNARD, DORIS MAJERCZAK and
KLAUS GEIDER

Abstract. *Erwinia stewartii* causes Stewart's bacterial wilt and leaf blight of corn and *Erwinia amylovora* causes fireblight of apple, pear and other rosaceous hosts. The virulence of both pathogens depends on the production of extracellular polysaccharide (EPS) slime and capsules. These erwinias produce acidic heteropolysaccharides composed of galactose, glucuronic acid and glucose. In terms of physical properties and genetics, these polysaccharides are similar to group I capsules of other enteric bacteria. Synthesis of *E. stewartii* EPS is determined by a large *cps* gene cluster, which contains *galE* and five *cps* complementation groups. Two unlinked *cps* loci are also required. Likewise, most of the genes for synthesis of amylovoran by *E. amylovora* are located in a large cluster of *ams* genes, which contains at least five complementation groups. In interspecific complementation tests, similar biochemical functions appear to be encoded by *cpsC* and *amsA* and by *cpsD* and *amsC-E*. Some *cps/ams* merodiploids differ from parental strains in their sensitivity to EPS-specific bacteriophages, suggesting that they produced mixed or altered EPSs. The *E. stewartii cpsA-D* genes are positively regulated by both the RcsA protein and a two-component system consisting of an effector, RcsB, and a sensor, RcsC. The regulation of EPS synthesis in *E. stewartii* is similar to the regulation of colanic acid production in *E. coli*.

Abbreviations: CPS, Capsular Polysaccharide; EPS, Extracellular Polysaccharide; ORF, Open Reading Frame

Introduction

Production of polysaccharides outside of the cell wall is common in many genera of bacteria. These extracellular polysaccharides (EPSs) can either form an organized "capsule" or glycocalyx around the cell, or they can be shed into the cell's milieu as "slime". Although synthesis of EPS may be optional depending on growth conditions, it frequently is a major determinant in the ability of a bacterium to colonize a given niche, host or vector. The protection conferred by capsules and slime is important for the survival of both animal and plant pathogens in the environment. Capsular EPSs are also major virulence factors for human and plant pathogens, in both cases serving to protect the bacterium from host defenses, such as phagocytosis in humans or recognition events in plants. With respect to plant diseases, EPS is responsible for the symptoms of vascular wilts and aids in local lesion formation (reviewed in Coplin & Cook, 1990; Leigh & Coplin, 1992). In leaf spots, bacterial slime formed in the intercellular spaces may hold water and nutrients released from damaged cells and thereby contribute to both water-soaking symptoms and

C.I. Kado and J.H. Crosa (eds.), Molecular Mechanisms of Bacterial Virulence, 341–356.
© 1994 *Kluwer Academic Publishers*.

creation of a favorable environment for bacterial multiplication. EPS has been clearly implicated as a mechanistic factor for vascular occlusion (Van Alfen, 1989). The symptoms caused by wilt-inducing bacteria may be primarily a function of EPS size and viscosity, but very little is known about how the actual structure of EPS influences pathogenicity. EPS⁻ mutants of *Erwinia stewartii*, *Erwinia amylovora*, *Xanthomonas campestris*, and *Pseudomonas solanacearum* have been shown to have reduced virulence (reviewed in Leigh & Coplin, 1992).

E. amylovora and *E. stewartii* cause fireblight on apple, pear and other rosaceous hosts and Stewart's wilt on corn, respectively. These bacteria can be considered both wilt-inducing and necrogenic pathogens. *E. amylovora* grows primarily in cortical tissues, where it produces large cankers, but it also interferes with xylem function. *E. stewartii* grows both in the xylem, causing wilting, and in the intercellular spaces of corn leaves, where it produces water-soaked lesions. Although they infect completely different hosts, these two *Erwinia* species are very closely related and may have similar mechanisms of pathogenicity (Braun, 1990; Coplin *et al.*, 1992; Roberts & Coleman, 1991). Infection by each requires *hrp*-like genes and the production of EPS, but the pathogens do not need to secrete classical toxins or degradative enzymes.

Enteric bacteria synthesize a wide range of capsular polysaccharides (CPS). *E. coli* strains alone synthesize over 70 structurally different types of capsules. A given strain, however, usually makes only one CPS. These CPSs are grouped into two classes based on their acidic component, charge density, substitution with lipid and attachment to LPS, and synthesis at low temperatures. Colanic acid (M-antigen) and K30 CPS, which are complex heteropolysaccharides containing hexuronic sugars, are typical of group I capsules produced by *E. coli*. The CPS and slime produced by *E. stewartii* and *E. amylovora* can be classified in group I with respect to their composition, production at low temperatures, linkage to the *his* operon, and regulation by RcsA (see below).

E. amylovora produces a very large ($50–150 \times 10^6$ Da) EPS, which is the main component of ooze from infected plants. This EPS was originally named amylovorin by Goodman *et al.* (1974), but more recently it has been termed amylovoran to be more consistent with the naming of other polysaccharides (Bellemann & Geider, 1992). A structure for amylovoran has been proposed by Smith *et al.* (1990) that consists of a five-sugar subunit containing galactose and glucuronic acid in a ratio of 4:1, which is decorated with pyruvate and acetate groups. *E. amylovora* may also produce levan (Gross *et al.*, 1992) and a low molecular weight glucan (Rastall *et al.*, 1987). Levan is not synthesized in infected immature pear tissue and very little is known about the glucan. The capsular EPS of *E. stewartii* is also large (45×10^6 Da), viscous, and highly charged. J. Costa and D. Horton (Costa, 1991) conducted a structural analysis of the EPS in collaboration with our laboratory. The polysaccharide has a repeating unit of seven monosaccharides and contains glucose, galactose, and glucuronic acid (3:3:1). This is the only major EPS that we detected in cultures of *E. stewartii* . The EPSs produced by soft rotting erwinias have not been characterized either chemically or genetically.

Genetics of capsular polysaccharide synthesis

The E. stewartii cps gene cluster

The genes required for capsular polysaccharide synthesis (*cps*) in *E. stewartii* reside in a large (10 kb) cluster, which has been cloned in cosmid pES2144 (Dolph *et al.*, 1988) (Fig. 1). The *galE* gene (UDP-galactose-4-epimerase), which provides UDP-sugar precursors for EPS synthesis, is immediately adjacent to this cluster. *Trans* complementation tests between chromosomal mutants and plasmid-borne *cps* mutations have shown that the *cps* genes are arranged in at least five complementation groups (termed *cpsA* though *cpsE*), which are all transcribed in the same direction (Coplin & Majerczak, 1990). All of the *cps* genes are required for EPS synthesis and wilt induction, and they can affect water-soaked lesion formation. The *cps* region is linked to the *his* operon on the *E. stewartii* chromosome and is analogous to the *cps* genes in *E. coli* , which determine the synthesis of colanic acid (Gottesman *et al.*, 1985; Trisler & Gottesman, 1984). The finding that the *galE* gene is not part of a galactose operon in *E. stewartii* is unusual. Its location in the *cps* region implies that this enzyme's role in supplying precursors for polysaccharide biosynthesis is more important than its role in galactose utilization. A similar organization of the *gal* genes has also been reported in *Vibrio cholera* (Huong & Cook, 1986) and *Rhizobium meliloti* (Buendia *et al.*, 1991; Canter-Cremers *et al.*, 1990). Mutational analysis has identified two additional *cps* loci, *cpsF* and *cpsG*, in *E. stewartii*, but they do not appear to be linked to the main *cps* cluster. The *cpsG* region was cloned and localized within a 2.3 kb *Eco*R1 fragment of cosmid pES1429 (Fig. 1); the *cpsF* mutation has not been mapped.

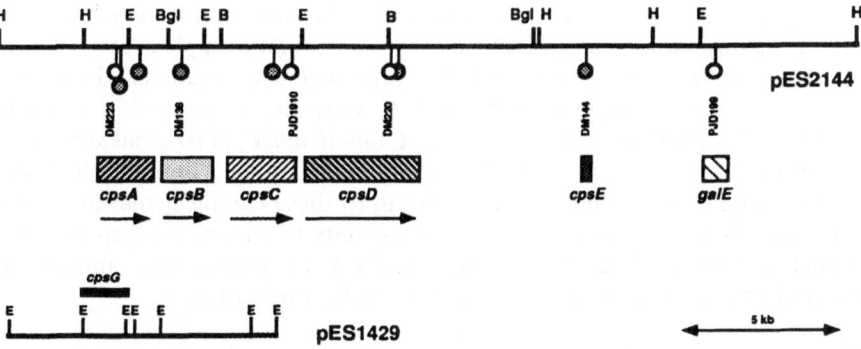

Figure 1. Restriction map of the *cps* regions cloned in plasmids pES2144 (top) and pES1429 (bottom). Arrows indicate the direction of transcription for each complementation group. Tn*3*HoHoI (●) and Tn*5lac* (O) insertion mutations that have been marker exchanged are shown below the pES2144 map.

Very little is known about the products of the *E. stewartii cps* genes. Mucoidy was restored to a strain carrying a mutation at the end of *cpsD* by the cloned *R. meliloti exoA* gene (see below), so this region may encode a similar glucosyl transferase function. Using Tn*phoA* mutagenesis, we were able to create alkaline phosphatase-positive *phoA* fusions in *cpsB* and *cpsC*. This finding is consistent with their products also being glycosyl transferases, polysaccharide polymerases or proteins involved in export of the subunit, since these proteins are believed to be membrane-associated and could generate PhoA$^+$ protein fusions. The CpsC protein and two proteins from the *cpsD* operon have been expressed in *E. coli* minicells and have a molecular masses of ca. 88, 38, and 34 kDa, respectively.

The E. amylovora Ams *gene cluster*
EPS$^-$ mutants of *E. amylovora* are relatively easy to obtain (Belleman & Geider, 1992; Steinberger & Beer, 1988), but they have proven difficult to complement with library clones (Belleman and Geider, 1992). Consequently, we attempted to identify clones containing the *amylovoran synthesis* (*ams*) genes from *E. amylovora* by their ability to restore mucoidy to *E. stewartii cps* mutants. (For simplicity, we will refer to the restoration of EPS synthesis in interspecific merodiploids as "complementation", although it most likely results in production of a different EPS.) A cosmid clone from an *E. amylovora* genomic library was found to complement mutants from four *E. stewartii cps* regions. This clone, pEA109, contained a 14.8 kb insert of chromosomal DNA. pEA109 restored EPS synthesis to *E. stewartii* strains DM138 (*cpsB*), PJD1910 (*cpsC*), DM215 (*cps-192*), DM220 (*cpsD*) and DM144 (*cpsE*), but not to *cpsA* or *galE* mutants. The *cps* mutants complemented by pEA109 produced less EPS than wild-type, parental strain DC283, but they regained full virulence on sweet corn seedlings. *ams* mutants were constructed by Tn*5* mutagenesis of pEA109 followed by marker-exchange of the mutation into the chromosome of a wild-type strain. *ams* transcriptional groups were defined by complementation tests between pEA109 mutants and subclones and chromosomal *E. amylovora ams* mutants (Bernhard *et al.*, 1992) and by additional complementation tests between pEA109 subclones and *E. stewartii cps* mutants. Five *ams* complementation groups, termed *amsA-E*, were delineated within a 6.4 kb region of pEA109 (Fig. 2). Recently, *amsA, amsB, amsC, amsD*, and *amsE* were sequenced (Bernard *et al.*, 1992; P. Bugert and K. Geider, unpublished) and open reading frames (ORFs) corresponding to these complementation groups were identified. The *ams* ORFs have the capacity to encode polypeptides with predicted masses of 80, 34, 42, 40, and 31 kDa, respectively, and are all transcribed from left to right as shown in the bottom of Fig. 2.

Functional correlation between the E. amylovora ams *and E. stewartii* cps *regions*
The 6.4 kb *ams* region from pEA109 was sufficient to complement mutations in the *E. stewartii cps* region from *cpsC* through *cpsE*. Using various subclones and

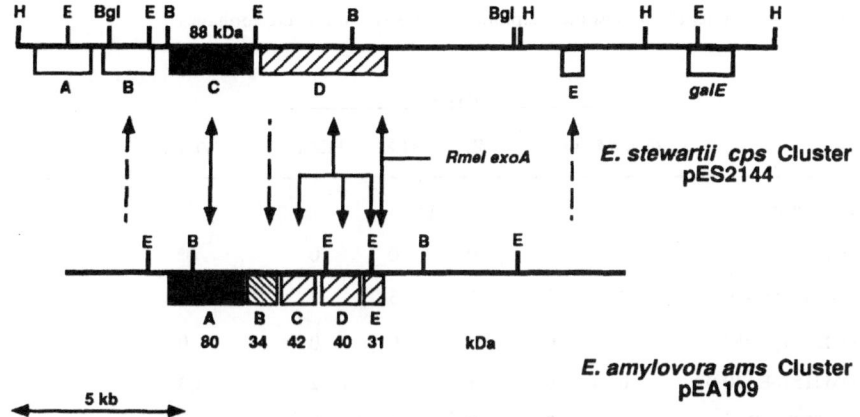

Figure 2. Correspondence between *E. stewartii cps* genes in pES2144 (top), *E. amylovora ams* genes in pEA109 (bottom), and *R. meliloti exoA* as demonstrated by restoration of EPS synthesis in merodiploids. Dashed arrows indicate that cosmid gene responsible for complementation has not been mapped. The positions of *ams* ORFs and the masses of their predicted protein products are shown below the pEA109 map.

Tn5-induced mutants of pEA109, we determined that the *E. stewartii cpsC* region corresponded to *E. amylovora amsA* (Fig. 2). Since DM220 *cpsD* was only complemented by subclones containing *amsC-E*, and all three *ams* genes were required, it appears that the *amsC*, *amsD*, and *amsE* homologues in *E. stewartii* may comprise a single operon. The regions of pEA109 responsible for complementation of *cpsB* and *cpsE* have not been located. Preliminary nucleotide sequencing of the *cps* region has revealed homology between *cpsB* and DNA 450 bp to the left of *amsA*, *cpsC* and *amsA*, and the first gene in *cpsD* and *amsB* (Coplin & Majerczak, unpublished). This suggests that the map order of the *cps* and *ams* genes is very similar. However, there appears to be an additional gene between the *amsA* and *amsB* homologues in *E. stewartii* that is not present in *E. amylovora*. Clones from an *E.amylovora* library that complement *cpsA* and *galE* have been identified, and are linked to the *ams* cluster (K. Geider, unpublished). *E. amylovora ams* mutants were likewise complemented for mucoidy by pES2144, carrying *cpsA-E*+, but no subclones of pES2144 were able to restore EPS synthesis. Moreover, pES2144 failed to restore virulence on immature pear slices to any of the *ams* mutants.

Sensitivity of ams/cps *merodiploids to EPS-specific bacteriophages*
Several EPS-specific bacteriophages were used to determine if complementation of *E. stewartii* and *E. amylovora* mutants by heterologous EPS genes might involve changes in EPS composition or structure. *E. amylovora* phages 4LM, a host range mutant of phage 4L (Bernhard *et al.*, 1990) that lyses *E. stewartii*, and PEa1(h) (Hartung *et al.*, 1988) infected wild-type strains of both *E. amylovora*

Table 1. Sensitivity of *ams/cps* merodiploids to EPS-specific bacteriophages.

Strain	Φ-K9M	Φ-K9	4L	4LM	PEA1	Virulence
			Phage			
E. stewartii						
DM144 *cpsE*	3[a]	0	0	0	0	0[b]
DM144 (pEA109)	3	3	1	3	2	2.1
DM215 *cps-192*[c]	3	0	0	0	0	0
DM215 (pEA109)	0	0	1	3	2	2.3
DM220 *cpsD*	3	0	0	0	0	1.3
DM220 (pEA109)	0	0	1	3	2	2.1
PJD1910 *cpsC*	3	0	0	0	0	0
PJD1910 (pEA109)	0	0	1	3	2	2.5
DC283 (pEA109)	3	3	2	3	2	3.0
E. amylovora						
Ea1/79N (pES2144)	1	1	3	3	3	1.5
Ea1/79N-D4[d]	0	0	0	0	0	0
Ea1/79N-D4 (pES2144)[d]	2	2	2	2	2	0

[a]Phage sensitivity rating: 0 = no reaction; 1 = faint spot; 2 = turbid spot; 3 = clear lysis.

[b]Virulence rating: 0 = no symptoms; 3 = full symptoms.

[c]*cps-192* is a chromosomal Tn*5lac* mutation that has not been physically mapped, but is genetically part of the *cpsD* complementation group.

[d]Similar results were obtained with *E. amylovora* mutants Ea1/79N-D2, D9, D12, D41, D49 and pES2144-containing derivatives.

and *E. stewartii*, but not acapsular mutants of these species (Table 1). Complementation of PJD1910 *cpsC*, DM220 *cpsD*, and DM144 *cpsE* by pEA109 and Ea1/79N-D4 *amsE* by pES2144 restored sensitivity to these phages. These data suggest that phages 4LM and pEa1(h) recognize a common feature of the two EPSs that is not altered by complementation with heterologous EPS genes.

A phage from *E. stewartii*, Φ-K9 (Bradshaw-Rouse *et al.*, 1981), appeared to be specific for *E. stewartii* EPS. Φ-K9 only lysed wild-type strains of *E. stewartii* (Table 1). Complementation by pEA109 restored Φ-K9 sensitivity to DM144 *cpsE* but not to the *cpsC* and *cpsD* mutants. This was confirmed by incubation of purified EPS from the transconjugants with concentrated Φ–K9 suspensions (Bernhard *et al.*, 1992). The phage EPS-depolymerase released dialyzable oligosaccharides from DM144 (pEA109) EPS, but not from DM215 *cps-192* (pEA109) EPS or amylovoran. Chemical analysis of the EPSs further indicated changes in sugar composition. *cps-192*, *cpsC*, and *cpsD* mutants complemented with pEA109 exhibited an increase in galactose and *amsE* mutants complemented with pES2144 contained extra glucose, reflecting the sugar content of the donor species. On the other hand, the sugar composition of EPS from a *cpsE* mutant complemented with pEA109 was unchanged. These findings suggest that *cpsE* and its corresponding gene on pEA109 encode common functions that do not change Φ-K9 recognition or sugar composition, whereas complementation of *cps-192*, *cpsC*, and *cpsD* by the *ams* cluster probably results in altered EPS structure or composition. In pathogenicity tests, both DM144 (pEA109) and DM215 (pEA109) regained water-soaking ability on sweet corn seedlings, indicating that any changes in the EPS structure brought about by expression of the heterologous EPS cluster in these strains did not affect virulence. However, complementation of the *E. amylovora amsE* mutant D4 with pES2144 restored EPS synthesis, but not virulence (Table 1). It is interesting that the Ea1/79N-D4 (pES2144) transconjugants gained sensitivity to Φ-K9. Considering the specificity of Φ-K9, this result suggests that pES2144 can direct synthesis of *E. stewartii*-like EPS in *E. amylovora*.

The E. amylovora amsE, E. stewartii cpsD, *and* R. meliloti exoA *genes encode similar functions*

A large *exo* gene cluster that determines succinoglycan synthesis in *Rhizobium meliloti* has been described by Leigh *et al.* (1985) and cloned in plasmid pRG100 (Zhan *et al.*, 1989 & 1990). Subclones of pRG100 were tested for their ability to restore mucoidy to various *cps* and *ams* mutants. Plasmids pEX20 (*exoNMA*+) and pHZ405 (*exoAL*+), which share only the *exoA* gene, were able to complement mutant DM220, whereas pEX41 (*exoLKH*+), which lacks *exoA*, could not. Mucoid DM220 (pHZ405) transconjugants were fully virulent on corn seedlings. Since *cpsD* corresponds to three *E. amylovora* complementation groups (*amsC*, *amsD*, and *amsE*), the *exo* plasmids were tested for their ability to complement *E. amylovora* mutants D2 (*amsD*), D4 (*amsE*), and D41 (*amsE*). The *amsE* strain exhibited the same complementation pattern as *cpsD*; pRG100, pEX20 and pHZ405 transconjugants were mucoid, whereas pEX41 transconjugants were not. Again, mucoid D2 and D4 transconjugants regained full virulence on pear fruit. These findings suggest that the ExoA, AmsE, and one of the CpsD proteins may have similar biochemical functions. Recently, the *exoA* gene has been shown to encode a glycosyltransferase activity that adds the first glucose residue following galactose to the repeat unit of succinoglycan (T. L. Reuber & G. Walker, personal communication). These experiments suggest that *amsE* and a gene within the *cpsD* operon could encode glucosyltransferases.

348

Regulation of capsular polysaccharide synthesis

The E. coli *model*
Our understanding of EPS regulation in erwinias has been aided by a model
proposed by Gottesman and Stout (1991) for colanic acid synthesis in *E. coli*. In
this model, two effector proteins, RcsA and RcsB (Fig. 3), activate transcription
of the *cps* genes. In culture, EPS synthesis is normally limited by the availability
of RcsA, because it is rapidly degraded by the Lon protease (Torres-Cabassa &
Gottesman, 1987). RcsA is stablized in *lon* mutants and the bacteria form
capsules and slime. RcsA is conserved in the Enterobacteriaceae and has been
shown to regulate other group 1 capsules in clinical isolates of *E. coli*
(Keenleyside *et al.*, 1992) and *Klebsiella* spp. (McCallum & Whitfield, 1991), as
well as in *E. amylovora* (Bernhard *et al.*, 1990; Chatterjee *et al.*, 1990; Coleman
et al., 1990) and *E. stewartii* (Torres-Cabbassa *et al.*, 1987; Poetter & Coplin,
1991). In *E. amylovora*, levan production is also impaired in *rcsA* mutants
(Bernhard *et al.*, 1990). RcsA has homology to the LuxR family of regulators
and is inferred to be a DNA-binding protein (Henikoff *et al.*, 1990; Stout *et al.*,
1991).

Additional control of EPS synthesis is provided by a two-component
regulatory system consisting of an effector, RcsB, and a sensor, RcsC. The

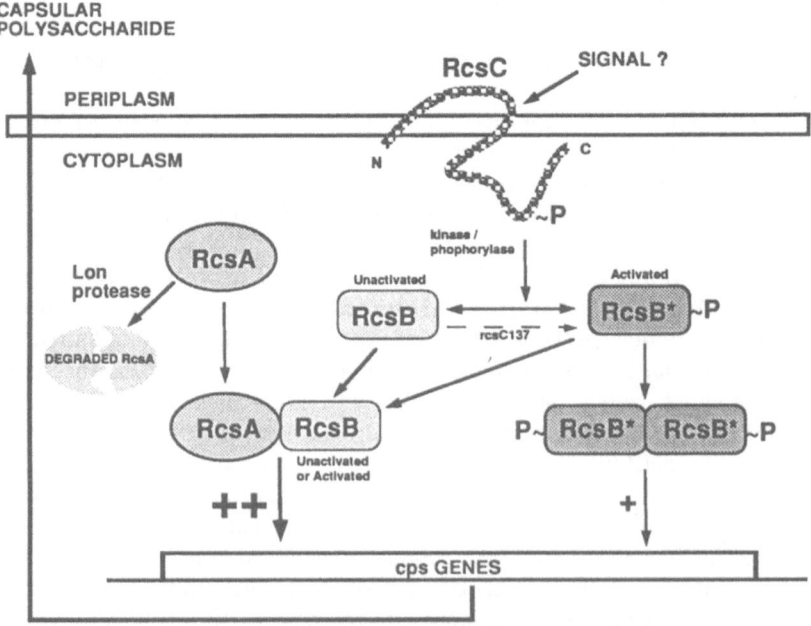

Figure 3. Proposed model for the regulation of capsular polysaccharide in *E. coli* and *Erwinia* sp.
Based on the ideas of Gottesman and Stout (1991) and reproduced, with permission, from the
Annual Review of Microbiology, Vol. 46, © 1992 by Annual Reviews, Inc (Leigh & Coplin, 1992).

sequences of RcsB and RcsC share high identity with a large class of two-component prokaryotic regulators (Stout & Gottesman, 1990), which also includes the plant virulence regulators, LemA, from *Pseudomonas syringae* pv. syringae (Hrabak & Willis, 1992) (see Chapter 35) and VirA from *Agrobacterium tumefaciens* (Leroux *et al.*, 1987). In these regulators, the sensor is a membrane spanning protein with its N-terminal portion in the periplasmic space, and its C-terminal portion in the cytoplasm. The periplasmic domain is believed to be involved either directly or indirectly in recognition and binding of signal molecules, and the cytoplasmic domain is a histidine protein kinase. The sensor autophosphorylates at a specific histidine residue, and the phosphate is then transferred to an aspartate residue in the N-terminal portion of the effector. The phosphorylated effector then binds to its target promoter via a sequence specific DNA-binding domain near its C-terminus. All of the sensors show homology in their C-terminal regions and the effectors in their N-terminal regions. In a few of the sensor proteins, such as VirA, RcsC and LemA, the C-terminal ends are homologous to the N-termini of their corresponding effectors. Although RcsC is assumed to be an environmental sensor in *E. coli*, it is not known what signal is detected, and capsule synthesis in *rcsC* null mutants is not affected in culture.

Complex dominance relationships among the *rcs* genes suggest that transcription of the *cps* genes is most likely initiated by an effector dimer (Fig. 3). This might either be a phosphorylated RcsB dimer or an RcsA-RcsB dimer, but not an RcsA dimer. Since overexpression of RcsB will suppress the phenotype of *rcsA* mutations, it appears that, in the presence of the appropriate environmental stimulus, activated RcsB is capable of stimulating *cps* transcription by itself. Gottesman & Stout (1991) hypothesize that RcsA may therefore be an accessory transcription factor that enhances activation by RcsB. RcsA may interact with RcsB to form a temperature sensitive complex, thereby protecting RcsA from proteolysis by Lon. Moreover, the RcsA:RcsB complex could be an alternative mechanism for activating the *cps* genes in the absence of RcsC-mediated phosphorylation of RcsB. Additional regulation of colanic acid synthesis may lie in the regulation of *rcsB* transcription. The *rcsB* promoter is dependent on the alternate sigma factor, RpoN (Stout & Gottesman, 1990) and may be repressed by LexA (Gervais *et al.*, 1992).

Regulation of EPS synthesis in Erwinia stewartii
E. stewartii produces a bound capsule under all conditions and makes copious slime when presented with a readily fermentable carbohydrate source. *E. coli*, on the other hand, does not normally produce colanic acid and mucoidy is only seen when strains carrying a *lon* mutation are grown below 32°C. Nevertheless, the regulation of EPS synthesis in the two species has a common mechanism. By means of *cps::lacZ* operon fusions, we have shown that the *cpsA-cpsD* genes in *E. stewartii* are under positive regulation by RcsA (Torres-Cabassa *et al.*, 1987) and the RcsB-RcsC two-component regulatory system. We similarly demonstrated regulation of amylovoran and levan synthesis in *E. amylovora* by

RcsA (Bernhard *et al.*, 1991), but *rcsB* and *rcsC* have not been reported in this species.

Regulation by RcsA.

The *rcsA* gene in *E. stewartii* is functionally equivalent to its homologue in *E. coli*. The cloned *E. stewartii* and *E. coli* genes both complement *rcsA* mutants and activate *cps* genes in the heterologous species, and both *rcsA* genes encode proteins of 25 to 27 kDa, which are unstable in an *E. coli lon*[+] strain (Torres-Cabassa *et al.*, 1987). The *rcsA* gene from *E. stewartii* was sequenced (Poetter & Coplin, 1991) and the open reading frame was identified by the presence of a ribosome binding site at -13 and by homology to the *E. coli* (Stout *et al.*, 1991) and *K. pneumoniae rcsA* sequences (Allen *et al.*, 1987). The predicted product of the *E. stewartii rcsA* gene is a 211 amino acid basic protein with a molecular mass of 24.3 kDa, corresponding to our previous estimate from maxicell experiments (Torres-Cabassa *et al.*, 1987). Comparison of our *rcsA* sequences from *E. stewartii* and *E. amylovora* (Bernhard *et al.*, 1991) with those of *E. coli* and *K. pneumoniae* revealed predicted amino acid homologies ranging from 55% between *E. amylovora* and *K. pneumoniae* to 82% between *E. stewartii* and *E. amylovora*. In the 5' non-coding regions, the two erwinias were only homologous from -1 to -70.

Stout *et al.* (1991) conducted homology searches with the *E. coli rcsA* sequence and found that RcsA belongs to the LuxR family of regulatory proteins. These proteins all share a region of homology near their carboxyl terminus, which contains a helix-turn-helix DNA binding motif. We observed the highest degree of homology between the *Erwinia* RcsAs and the other LuxR family activators within this region. Computer analysis of the four RcsA proteins revealed a helix-turn-helix motif centered on amino acid 175. Interestingly, RcsB also belongs to this family and has significant homology with RcsA in this region, suggesting that the two effectors may have similar DNA-binding properties.

Regulation by RcsB and RcsC. A search for new EPS[−] mutants yielded several nonmucoid Tn5-induced mutants that were complemented by cosmid pES2006 from a wild-type genomic library. Intergeneric complementation tests revealed that they contained *rcsB* mutations. pES2006 restored mucoidy to an *E. coli rcsB lon* strain, and the *E. stewartii rcsB*::Tn5 mutants were complemented by pKP4521, which carries the *E. coli* HB101 *rcsB*[+] locus (Table 2). When plasmids carrying *cps::lacZ* reporter gene fusions were introduced into the *rcsB*::Tn5 mutants, and β-galactosidase levels in the transconjugants were compared to wild-type strains carrying the same plasmids, expression of *cpsD::lacZ* and *cpsB::lacZ* was decreased more than two-fold and ten-fold, respectively, in the *rcsB*::Tn5 mutants. "Molecular Koch's postulates" were fulfilled for the *rcsB* gene as follows. An 8.4 kb fragment from the right end of pES2006 was subcloned to produce plasmid pKP2 (Fig. 4). Two Tn5 insertions in pKP2 that inactivated *rcsB* (Fig. 4) were marker-exchanged into the chromosome of a wild-type strain and the resulting mutants were unable to produce EPS or support expression of a plasmid-borne *cpsD::lacZ* fusion.

Table 2. Reciprocal complementation tests between *E. stewartii* and *E. coli rcsB* and *rcsC* mutants and clones.

	EPS Production		
	E. coli	*E. coli*	*E. stewartii*
Plasmid	*rcsB*	*rcsC137*	*rcsB::Tn5*
E. coli			
pKP4521 *rcsBC+*	+	-	+
E. stewartii			
pKP2 *rcsBC+*	+	-	+
pKP2 *rcsB+ rcsC::Tn5*	+	+	+
pKP2 *rcsB::Tn5 rcsC+*	-	-	-
none	-	+	-

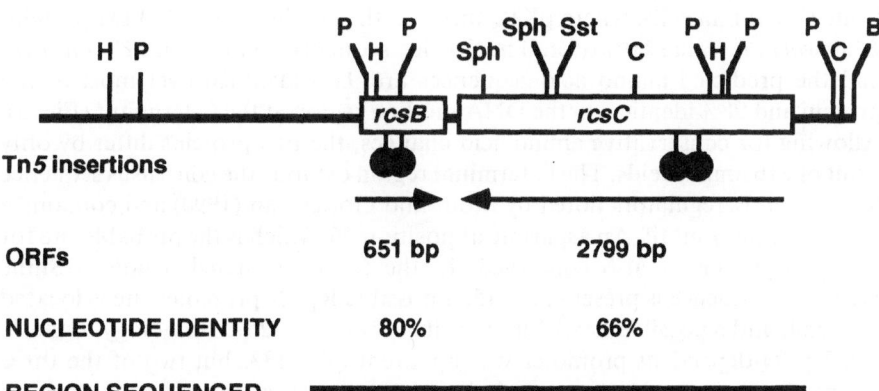

Figure 4. The *E. stewartii rcsB* and *rcsC* region of plasmid pKP2 and its nucleotide homology to *E. coli rcsB* and *rcsC*. Boxes indicate ORFs and arrows designate the direction of transcription. Tn5 insertions (●) are shown below the map.

Finally, mucoidy was restored to the marker-exchanged mutants by pES2006.

In *E. coli*, *rcsB* and *rcsC* are adjacent to one another and convergently transcribed (Stout & Gottesman, 1990), so we expected to find an adjacent *rcsC* locus in *E. stewartii*. A problem in doing complementation tests with *rcsC* is that null mutants in *E. coli* do not have any apparent phenotype, however, complementations can be done with a point mutation, *rcsC137*, that causes over-production of colanic acid. *rcsC137* mutants presumeably cannot dephosphorylate RcsB once it is activated. Consequently, *rcsC137* is recessive to the wild-type allele and *rcsC+/rcsC137* merodiploids are mucoid. pKP2 was able to repress mucoidy of an *E. coli rcsC137* host (Table 2) indicating that *rcsC* is also present in this region, and we were able to isolate two Tn*5* mutations that could not suppress *E. coli rcsC137* strains (Fig. 4, Table 2). We were successful in obtaining a chromosomal *rcsC* insertion mutation using the R6K-based universal suicide plasmid pGP704 (Miller & Mekalanos, 1988). The 1.4 kb *Sst*I-*Hin*dIII fragment from the center of the *rcsC* region was cloned into pGP704 and the resulting plasmid was mobilized into a wild-type strain. Stable maintenance of the ampicillin-resistance marker of pGP704 indicated that a cross-over event occurred between the *rcsC* insert and the chromosome resulting in insertion of pGP704 into the wild-type *rcsC* gene. The *rcsC*::pGP704 mutant had no apparant phenotype; it synthesized normal amounts of EPS and expression of plasmid-borne *cps::lacZ* fusions was not altered. As in *E. coli*, this suggests that *rcsC* can be a redundant part of the EPS regulatory circuit when bacteria are grown in culture.

Nucleotide sequence analysis of the *rcsBC* region has recently been completed in our laboratory. The ORFs for *rcsB* and *rcsC* were identified by homology to the *E. coli* sequence and are shown in Fig. 4. As in *E. coli*, the two genes are convergently transcribed. The *rcsB* ORF can encode a 216 amino acid polypeptide with a predicted mass of 23.7 kDa. This agreed with our results from *E. coli* minicells, where pKP2 directed the synthesis of a 24 kDa protein. *E. stewartii rcsB* has 80% overall nucleotide sequence identity with *E. coli rcsB*, and the predicted amino acid sequences are 93% identical over most of the protein and 98% identical in the DNA binding region at the C-terminus (Fig. 5). Allowing for conservative amino acid changes, the two proteins differ by only 4 out of 216 amino acids. The C-terminal region exhibits the consensus sequence for the LuxR regulators noted by Stout and Gottesman (1990) and contains a helix-turn-helix motif. An aspartate at position 56, which is the probable site for phosphorylation, is also conserved. In the 5′ untranslated region, a Shine Delgarno sequence is present at −15, a possible RpoN promoter site is located at −354, and a possible LexA binding site occurs at −257. Homology to the *E. coli* RpoN-dependent promoter was apparent at −133, but two of the three essential guanine residues were missing, suggesting that this promoter has not been conserved in *E. stewartii*.

Many of the features of *E. coli rcsC* are also conserved in *E. stewartii* (Fig. 5). The *rcsC* ORF can encode a 933 amino acid peptide with a predicted mass of 104.9 kDa. Two hydrophobic regions are present from amino acid positions 4 to 25 and 297 to 320, which may serve to span the cytoplasmic membrane. Both the sensor/kinase and effector domains are highly conserved, with 82% and 76% amino acid identity to *E. coli*, respectively. A conserved histidine residue is located at position 462 that could act as a site for autophosphorylation. Interestingly, the periplasmic domain and the spacer region between the kinase and effector domains were quite different from *E. coli*. Only 43% amino acid identity was present in the periplasmic domain, which suggests that *E. stewartii* and *E. coli* may sense different environmental signals, even though they employ the same effector system. The intergenic region between *rcsB* and *rcsC* is much shorter in *E. stewartii* than in *E. coli* (40 vs. 160 bp) and does not contain the REP sequences and long direct repeats observed by Stout and Gottesman (1990). Whether or not this region somehow controls expression of *rcsB* and *rcsC* in *E. coli* is not known.

Erwinia stewartii Rcs Proteins

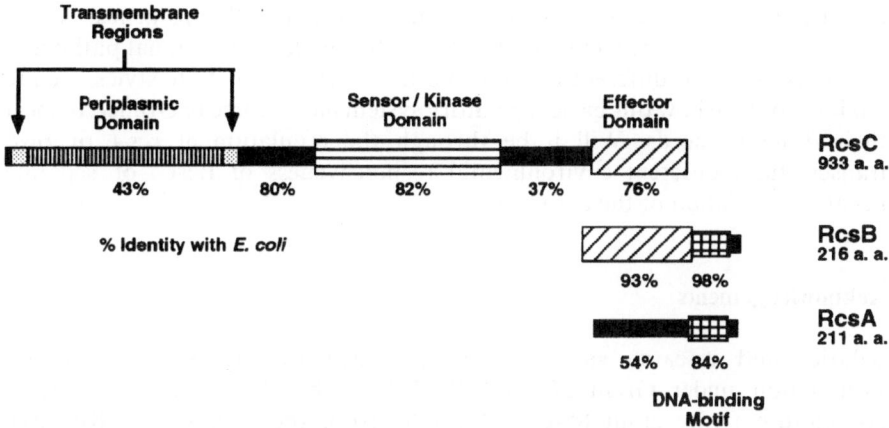

Figure 5. Functional domains of *E. stewartii* RcsA, RcsB, and RcsC proteins determined by nucleotide sequence homologies. The percent amino acid identity of each domain with its homologue in *E. coli* is indicated.

Directions for future research

The studies on EPS synthesis in *Erwinia* spp. discussed in this chapter demonstrate a number of common features of group I capsule synthesis in enteric bacteria. A similar physical arrangement was found for the *ams* and *cps* genes, suggesting that the two gene clusters co-evolved as alternative EPS modules. It will be interesting to compare these clusters with the *cps* genes of *E. coli* when they are completely sequenced. Moreover, a number of the Cps and Ams enzymatic functions appeared to be interchangeable, but the nature of the polysaccharides produced when recipient strains were restored to mucoidy remains to be determined by chemical analysis. Unfortunately, this cannot be done until difficulties in determining the exact structures for amylovoran and *E. stewartii* EPS have been overcome. In the case of *cpsE*, it appears that this mutation can be truly complemented by the *ams* cluster resulting in production of wild-type recipient EPS, but other interspecific crosses seemed to bring about changes in phage sensitivity and the galactose:glucose ratios of the EPS. At this point, we must determine whether these mucoid merodiploids produce mixed or altered EPSs. If EPS structures have been altered, then our results imply that the enzymes involved in subunit assembly and polymerization can act on a range of substrates and suggest that it will be possible to produce "designer" polysaccharides by exchanging EPS genes between species. Such experiments would be able to address how exopolysaccharide structure affects the virulence of bacteria to plants and animals.

Another common feature of group I capsule synthesis is regulation by RcsA and RcsB. The model discussed above offers interesting possibilites for complex control of EPS synthesis by both physiological and environmental factors, because it includes alternate pathways for activation of the *cps* genes, i.e. via RcsA or RcsC-mediated activation of RcsB. Since plant and animal pathogens synthesize EPS for different reasons and lead very different life styles, we are anxious to learn how this basic regulatory system has become fine-tuned to meet their different needs. Will it be through the regulation of *rcsA* or *rcsB* transcription, different environmental responsiveness of RcsC, or separate negative regulation of the *cps* genes?

Acknowledgements

Salaries and research support were provided by the National Science Foundation under Grant No. DMB-8703722, by the U.S.Department of Agriculture under grant 85-CRCR-1-1781 from the Competitive Research Grants Office, by state and federal funds appropriated to the Ohio Agricultural Research and Development Center, The Ohio State University to DLC, and by a grant from the Deutscher Akademischer Austauschdienst to FB. We thank Drs. V. Stout and S. Gottesman for helpful discussions during the course of the research, Dr. M. Nimtz for performing microanalysis of EPS-probes, and Dr. J. Leigh for providing us with cloned *R. meliloti exo* genes.

References

Allen P, Hart CA & Saunders JR (1987) Isolation from *Klebsiella* and characterization of two *rcs* genes that activate colanic acid capsular biosynthesis in *Escherichia coli*. J. Gen. Microbiol. 133: 331–340.

Bellemann P & Geider K (1992) Localization of transposon insertions in pathogenicity mutants of *Erwinia amylovora* and their biochemical characterization. J. Gen. Microbiol. 138: 931–940.

Bernhard F, Bugert P, Coplin DL & Geider K (1992) Characterization of *ams* genes for amylovoran synthesis in *Erwinia amylovora* and their ability to complement *Erwinia stewartii cps* mutants. Fallen Leaf Lake Conference on Bacterial Pathogenesis. S. Lake Tahoe.

Bernhard F, Poetter K, Geider K & Coplin DL (1990) The *rcsA* gene from *Erwinia amylovora*: identification, nucleotide sequence, and regulation of exopolysaccharide biosynthesis. Mol. Pl. Microb. Interact. 3: 429–437.

Bradshaw-Rouse JJ, Whatley MA, Coplin DL, Woods A, Sequeira L & A Kelman A (1981) Agglutination of strains of *Erwinia stewartii* with a corn agglutinin: correlation with extracellular polysaccharide production and pathogenicity. Appl. Environ. Microbiol. 42: 344–350.

Braun EJ (1990) Colonization of resistant and susceptible maize plants by *Erwinia stewartii* strains differing in exopolysaccharide production. Physiol. Mol. Plant. Pathol. 36: 363–379.

Buendia AM, Enekel B, Köplin K, Niehaus K, Arnold W & Pühler A (1991) The *Rhizobium meliloti exoZ/exoB* fragment of megaplasmid 2; ExoB functions as a UDP-glucose 4-epimerase and ExoZ shows homology to NodK of *Rhizobium leguminosarum* biovar viciae strain TOM. Mol. Microbiol. 5: 1519–1530.

Canter-Cremers HCJ, Batley M, Redmond JW, Eydems L, Breedveld MW, *et al.* (1990) *Rhizobium leguminosarum exoB* mutants are deficient in the synthesis of UDP-glucose 4'-epimerase. J. Biol. Chem. 265: 21122–21227.

Chatterjee A, Chun W & Chatterjee AK (1990) Isolation and characterization of an *rcsA*-like gene of *Erwinia amylovora* that activates extracellular polysaccharide production in *Erwinia* species, *Escherichia coli*, and *Salmonella typhimurium*. Mol. Plant-Microbe Interact. 3: 144–148.

Coleman M, Pearce R, Hitchin E, Busfield F, Mansfield JW & Roberts IS (1990) Molecular cloning, expression and nucleotide sequence of the *rcsA* gene of *Erwinia amylovora*, encoding a positive regulator of capsule expression: evidence for a family of related capsule activator proteins. J. Gen. Microbiol. 136: 1799–1806.

Coplin DL & Cook D (1990) Molecular genetics of extracellular polysaccharide biosynthesis in vascular phytopathogenic bacteria. Mol. Plant-Microbe Interact. 3: 271–279.

Coplin DL, Frederick RD & Majerczak DR (1992) New pathogenicity loci in *Erwinia stewartii* identified by random Tn*5* mutagenesis and molecular cloning. Mol. Plant-Microbe Interact. 3: 266–268.

Coplin DL & Majerczak DR (1990) Extracellular polysaccharide genes in *Erwinia stewartii*: directed mutagenesis and complementation analysis. Mol. Plant-Microbe Interact. 3: 286–292.

Costa JB (1991) Structural studies of some viscous, acidic bacterial exopolysaccharides. Ph. D. thesis. The Ohio State University.

Dolph PJ, Majerczak DR & Coplin DL (1988) Characterization of a gene cluster for exopolysaccharide biosynthesis and virulence in *Erwinia stewartii*. J. Bacteriol. 170: 865–871.

Gervais GG, Phoenix P & Drapeau GR (1992) The *rcsB* gene, a positive regulator of colanic acid biosynthesis in *Escherichia coli*, is also an activator of *ftsZ* expression. J. Bacteriol. 174: 3964–3971.

Goodman RN, Huang JS, Huang PY (1974) Host specific phytotoxic polysaccharide from apple tissue infected by *Erwinia amylovora*. Science 183: 1081–1082.

Gottesman S & Stout V (1991) Regulation of capsular polysaccharide synthesis in *Escherichia coli* K12. Mol. Microbiol. 5: 1599–1606.

Gottesman S, Trisler P & Torres-Cabassa AS (1985) Regulation of capsular polysaccharide synthesis in *E. coli* K-12: characterization of three regulatory genes. J. Bacteriol. 162: 1111–1119.

Gross M, Geier G, Rudolf K & Geider K (1992) Levan and levansucrase synthesized by the fireblight pathogen *Erwinia amylovora*. Physiol. Mol. Plant Pathol. 40: 371–381.

Hartung JS, Fulbright DW & Klos EJ (1988) Cloning of a bacteriophage polysaccharide depolymerase gene and its expression in *Erwinia amylovora*. Mol. Plant-Microbe Interact. 1: 87–93.

Henikoff S, Wallace JC & Brown JP (1990) Finding protein similarities with nucleotide sequence databases. Methods Enzymol. 183: 111–132.

Houng H-SH & Cook TM (1986) Cloning of the galactose utilization genes of *Vibrio cholerae*. First Colloquium in Biological Sciences. Ann. N.Y. Acad. Sci. 435: 601–603.

Hrabak EM & Willis DK (1992) The *lemA* gene required for pathogenicity of *Pseudomonas syringae* pv. syringae on bean is a member of a family of two-component regulators. J. Bacteriol. 174: 3011–3020.

Keenleyside WJ, Jayaratne P, MacLachlan PR & Whitfield C (1992) The *rcsA* gene of *Escherichia coli* O9: K30: H12 is involved in the expression of the serotype-specific group I K (capsular) antigen. J. Bacteriol. 174: 8–16.

Leigh JA & Coplin DL (1992) Exopolysaccharides in plant-bacterial interactions. Annu. Rev. Microbiol. 46: 307–346.

Leigh J A, Signer ER & Walker GC (1985) Exopolysaccharide-deficient mutants of *Rhizobium meliloti* that form ineffective nodules. Proc. Natl. Acad. Sci. U.S.A. 82: 6231–6235.

Leroux B, Yanofsky MF, Winans SC, Ward JE, Ziegler SF & Nester EW (1987) Characterization of the *virA* locus of *Agrobacterium tumefaciens*: a transcriptional regulator and host range determinant. Proc. Natl. Acad. Sci. USA 6: 849–856.

McCallum KL & Whitfield C (1991) The *rcsA* gene of *Klebsiella pneumoniae* 01: K20 is involved in expression of the serotype-specific K (capsular) antigen. Infect. Immun. 59: 494–502.

Miller VL & Mekalanos J (1988) A novel suicide vector and its use in construction of insertion mutations: osmoregulation of outer membrane proteins and virulence determinants in *Vibrio cholerae* requires *toxR*. J. Bacteriol. 170: 2575–2583.

Poetter K & Coplin DL (1991) Structural and functional analysis of the *rcsA* gene from *Erwinia stewartii*. Mol. Gen. Genet. 229: 155–160.

Rastall A, Smith ARW, Blake P & Hignett RC (1987) An extracellular glucan from a virulent strain of *Erwinia amylovora*. Proceedings of the Fallen Leaf Lake Conference, South Lake Tahoe, CA.

Roberts IS & Coleman MJ (1991) The virulence of *Erwinia amylovora*: molecular genetic perspectives. J. Gen. Microbiol. 137: 1453–1457.

Smith ARW, Rastall RA, Rees NH & Hignett RC (1990) Structure of the extracellular polysaccharide of *Erwinia amylovora*: a preliminary report. Acta Hortic. 273: 211–219.

Steinberger EM & Beer SV (1988) Creation and complementation of pathogenicity mutants of *Erwinia amylovora*. Mol. Plant-Microbe Interact. 1: 135–144.

Stout V & Gottesmann S (1990) RcsB and RcsC, a two component regulator of capsule synthesis in *Escherichia coli*. J. Bacteriol. 172: 659–669.

Stout V, Torres-Cabassa A, Maurizi MR, Gutnick D & Gottesman S (1991) RcsA, an unstable positive regulator of capsular polysaccharide synthesis. J. Bacteriol. 173: 1738–1747.

Torres-Cabassa AS & Gottesman S (1987) Capsule synthesis in *Escherichia coli* K-12 is regulated by proteolysis. J. Bacteriol. 169: 981–989.

Torres-Cabassa A, Gottesman S, Frederick RD, Dolph PJ & Coplin DL (1987) Control of extracellular polysaccharide biosynthesis in *Erwinia stewartii* and *Escherichia coli* K-12: a common regulatory function. J. Bacteriol. 169: 4525–4531.

Trisler P & Gottesman S (1984) *lon* transcriptional regulation of genes necessary for capsular polysaccharide synthesis in *Escherichia coli* K-12. J. Bacteriol. 160: 184–191.

Van Alfen NK (1989) Reassessment of plant wilt toxins. Annu. Rev. Phytopathol. 27: 551–550.

Zhan H, Gray JX, Levery SB, Rolfe BG & Leigh JA (1990) Functional and evolutionary relatedness of genes for exopolysaccharide synthesis in *Rhizobium meliloti* and *Rhizobium* sp. strain NGR234. J. Bacteriol. 172: 5245–5253.

Zhan H, Gray JX, Levery SB, Battisti L, Rolfe BG & Leigh JA (1991) Heterologous exopolysaccharide production in *Rhizobium* sp. strain NGR234 and consequences for nodule development. J. Bacteriol. 173: 3066–3077.

25. The role of extracellular polysaccharides as virulence factors for phytopathogenic pseudomonads and xanthomonads

KLAUS W.E. RUDOLPH, MICHAEL GROSS, FIROUS EBRAHIM-NESBAT, MATTHIAS NÖLLENBURG, ALIM ZOMORODIAN, KERSTIN WYDRA, MICHAEL NEUGEBAUER, URSULA HETTWER, WAGIH EL-SHOUNY, BERND SONNENBERG and ZOLTAN KLEMENT

Abstract. Bacterial exopolysaccharides (EPS) were investigated for their role as virulence factors of leaf spot diseases caused by pseudomonads and xanthomonads. The capacity of these bacteria to induce persistent water-soaking in leaves plays a crucial role during pathogenesis that seems to be accomplished by a synergistic interaction between bacterial EPS and plant polymers. Under conditions of low EPS production (e.g. in continuously darkened plants) the bacteria were not able to cause typical water-soaked disease symptoms. The main EPS components were alginate and levan (*Pseudomonas*), xanthan (*Xanthomonas*), as well as lipopolysaccharides (LPS) and a small amount of proteins. It is suggested that alginate which is very similar to plant pectate is required for establishing bacterial infections in later disease stages. This concept was confirmed by evaluating transposon mutants with EPS deficiencies. LPS may be involved in specific interactions with plant polymers leading to agglutination and precipitation (incompatibility) or gel-formation (compatibility). Bacteria which are embedded in a gel-like matrix *in planta* are not easily recognized by the plant and are protected against bacteriostatic compounds and desiccation.

Abbreviations: EPS, Extracellular Polysaccharides; LPS, Lipopolysaccharides; *P.s.*pv., *Pseudomonas syringae* Pathovar; X.c.pv., *Xanthomonas campestris* Pathovar; Alg, Alginate; Lev, Levan; HR, Hypersensitive Reaction; WS, Water-Soaking; cv., cultivar: cfu, colony forming units; pv., pathovar; GSPB, Göttinger Sammlung Phytopathogener Bakterien; NTG, Nitroguanidine; GPC-Gel Permeation Chromotography; PAGE, Polyacrylamide Gel Electrophoresis

Introduction

Virulence has been defined as the relative capacity to cause disease. Therefore, virulence factors of pathogens comprise substances which not only participate in symptom expression but also increase disease severity. Although several laboratories have recently concentrated their efforts on the elucidation of the so-called "avirulence genes" which may be involved in the induction of resistance (Keen, 1990), it seems obvious that the mere absence of a resistance reaction does not necessarily result in disease. Phytopathogenic bacteria need specific capabilities to cause disease. Especially in the case of leaf spot causing pseudomonads and xanthomonads, the pathogenic capacity appears to be a very fascinating phenomenon. Although most members of this group do not require specific nutrients and can be cultured on very simple media *in vitro*, they are characterized by their very narrow host specificity. Many species and pathovars of these pseudomonads and xanthomonads can infect only one or very few, related plant species.

C.I. Kado and J.H. Crosa (eds.), Molecular Mechanisms of Bacterial Virulence, 357–378.
© 1994 *Kluwer Academic Publishers.*

The factors responsible for the specific host-parasite interactions involved are not completely understood at present. However, visible symptoms and other characteristics of pathogenesis in all these diseases are rather similar. It is assumed that the decisive mechanisms which result in the compatible interaction (disease) are regulated by the same principles. Four groups of bacterial virulence factors can be differentiated: enzymes, toxins, polysaccharides, and membrane-active substances. For several years we have been studying the role of polysaccharides in pathogenesis (Rudolph *et al.*, 1989). Several lines of evidence suggest that bacterial slimes composed of extracellular polysaccharides (EPS) play a decisive role for the maintenance of compatible interactions. Therefore, we investigated the role of EPS as virulence factor in the case of leaf spot causing pseudomonads and xanthomonads.

Results and discussion

Models

In most of our studies the host-parasite combination bush bean (*Phaseolus vulgaris*)/ *Pseudomonas syringae* pv. *phaseolicola* (halo blight of bean) was selected as a model for the large group of leaf spot diseases caused by pseudomonads. Representatives of the xanthomonads included *Xanthomonas campestris* pvs. malvacearum, glycines and others.

Development of water-soaking symptoms

The bacteria colonize the intercellular spaces of the leaf mesophyll, as shown by light-microscopy in Fig. 1. However, the parenchyma cells are never invaded by the bacteria, although they depend on the nutrients of the host cells to build up high populations. As shown in Fig. 1, plant cells are partly degraded and their walls bend inwardly, so that the intercellular space is filled progressively by masses of bacteria (Rudolph and Mendgen, 1985). When the entire parenchyma between two adjacent veins is completely invaded by bacterial masses, a "water-soaked" angular leaf spot appears. Water-soaking is typical for many bacterial leaf spot diseases and was shown to be necessary for maximum bacterial multiplication in the case of pseudomonads (Rudolph, 1984) and xantho-monads (El-Banoby and Rudolph, 1989). Finally, bacterial multiplication in the intercellular space leads to pressure build-up so that the bacteria are extruded from the tissue. These exudates play a decisive role in epidemic spread of the bacteria and are, thus, obligatory for a fully susceptible (compatible) reaction in many cases.

Previous studies showed that the main hindrance to bacterial multiplication in the intercellular space of leaves is the non-availability of water (Rudolph, 1980). Most bacteria cannot grow on dry surfaces and require a relative humidity between 0.965–0.999% for growth (Young, 1974). The relative

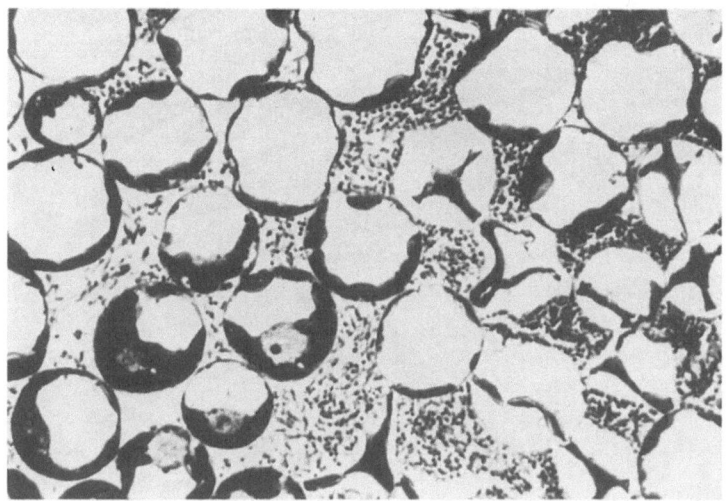

Figure 1. Light microscopic view of a bean leaf infected by *P. syringae* pv. phaseolicola. The intercellular spaces are filled with bacteria. The cytoplasm of most plant cells collapsed, some walls are bended inwardly.

humidity of the intercellular spaces of respiring leaf tissue is greater than 0.96, thus indicating that water may be available for bacterial metabolism. However, the cell walls of leaf parenchyma cells may be partially cutinized or suberized at their outer extremities. Thus, the exposed internal surfaces are hydrophobic to some degree (Häusermann, 1944) and may appear dry rather than moist (Slatyer, 1967). Therefore, it seems obvious that phytopathogenic bacteria need a mechanism by which they can induce and maintain water congestion in the intercellular space. In addition, a very narrow host range is characteristic for most leaf spot causing pseudomonads and xanthomonads indicating the occurrence of specific interactions between plant and bacterium in the compatible combination. It is still an open question whether the mechanisms regulating host specificity are involved in the induction of water congestion (Wydra and Rudolph, 1992a). In any case, the capability of the bacteria to induce persistent water-soaking in susceptible plant tissue seems to play a crucial role during pathogenesis.

Several lines of evidence indicated that only the production of large masses of slime or extracellular polysaccharides (EPS) allow the bacteria to fill the intercellular space of the leaf mesophyll (Rudolph *et al.*, 1989). Thus, electron microscopic studies revealed that the bacterial cells in the intercellular spaces were embedded in masses of slime (Fig. 2). Similar results were obtained for several pseudomonads and xanthomonads (e.g. Jones and Fett, 1985; Mansfield and Brown, 1986; Rudolph, 1993).

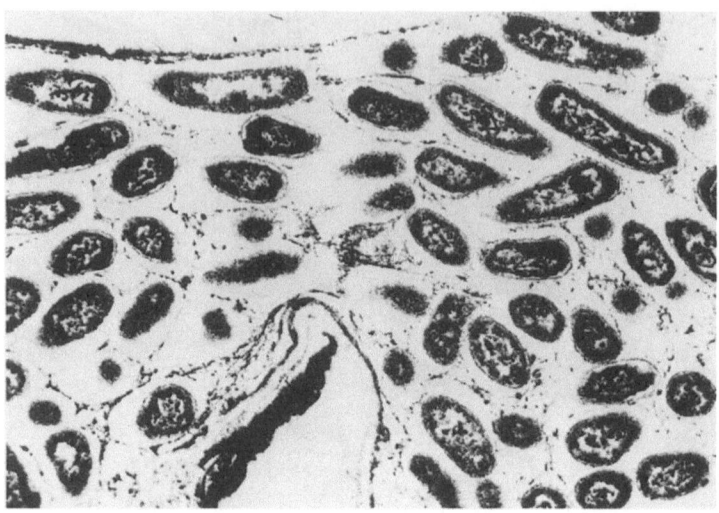

Figure 2. Cells of *P.s.* pv. *phaseolicola* in the intercellular space of a bean leaf are embedded in a network of fibrillar material visualized by staining with ruthenium-red.

Induction of persistent water-soaking by extracellular polysaccharides (EPS)

Under certain experimental conditions persistent water-soaked spots were induced in bean leaves by bacteria-free exudate preparations from infected bean plants (Rudolph, 1978) or by bacteria-free EPS obtained from *in vitro* grown cultures (El-Banoby and Rudolph, 1979a, 1979b, 1980). To which extent these EPS-preparations mimic the narrow host-specificity of intact bacteria has to be elucidated by further studies. Also the earlier finding that bacterial EPS were degraded or inactivated by the intercellular washing fluid from resistant leaves but not from susceptible ones (El-Banoby *et al.*, 1981a) has to be re-evaluated. In recent studies, El-Shouny *et al.* (1992) obtained EPS-preparations from *P. syringae* pv. phaseolicola (N7), *P. syringae* pv. coriandricola and an unidentified *P. syringae* pathovar from carrot (W43). All three EPS-preparations induced longer persisting water-soaking on a halo-blight susceptible bean cultivar compared to a resistant bean line. However, only EPS from *P.s.* pv. phaseolicola induced a few persistent water-soaked spots on the halo-blight susceptible cultivar (Table 1).

Zachowski (1989) obtained crude EPS-preparations from *Xanthomonas campestris* pv. malvacearum by ethanol precipitation which induced persistent water-soaked spots (1 to 4 days) in cotton cotyledons. In the highly resistant cv. "Tamcot Camd-E", EPS of the highly virulent race 18 caused a stronger reaction than EPS of the weakly virulent race 1 (Fig. 3). However, correlations between varietal susceptibility to pathogenic races and sensitivity to the respective EPS preparations were not found in other race/cultivar combinations.

Table 1. Number of persistent water-soaked spots in bush bean (*Phaseolus vulgaris*) trifoliates at different times after inoculation (tpi) with EPS-preparations from *Pseudomonas syringae* pv. phaseolicola (N7), *P. syringae* pv. coriandricola (COR. 21) and an uncharacterized pv. from carrot (W 43)

tpi	Cultivar "Red Kidney"				Breeding line "02"			
	H_2O	N7	Cor.21	W43	H_2O	N7	Cor.21	W43
	s+w	s+w	s+w	s+w	s+w	s+w	s+w	s+w
0 h	20	20	20	20	20	20	20	20
18 h	14	9+5	16	14	5	8+1	4	6
24 h	4+1	5+3	10	9+2	5	7+1	1	3+2
36 h	1	5+3	7	9+1	5	4+1	0	3+2
42 h	0	0+3	0+2	0+3	5	0	0	0+2
48 h	0	0+2	0+2	0+3	2+1	0	0	0+2
3 days	0	2+1	0+2	0	0	0	0	0
4 "	0	5+1	0+2	0	0	0	0	0
5 "	0	4+1	0	0	0	0	0	0
6 "	0	4+1	0	0	0	0	0	0
			Removal of plastic bags					
7 "	0	1+1	0	0	0	0	0	0
8 "	0	1+1	0	0	0	0	0	0

s   : strong spot
w   : weak spot
Red Kidney: susceptible to *P.s.*pv.phaseolicola
02   : resistant against *P.s.*pv.phaseolicola

Likewise, other laboratories induced persistent water-soaked spots by EPS-preparations, for example with so-called lipomucopolysaccharides from *P.s.* pv. *lachrymans* (Keen and Williams, 1971), low-concentrated LPS-solutions from *P.s.* pv. phaseolicola (Epton *et al.*, 1977), or EPS from *X.c.* pv. malvacearum (Borkar and Verma, 1989). The latter authors reported that EPS of a virulent, but not of an avirulent mutant, of *X.c.* pv. malvacearum produced persistent water-soaking in leaves of the sensitive cotton cultivars Acala 44 and 1–1OB but not in leaves of the resistant cv. 1O1–1O2B. Water-soaking was also induced by infiltration in non-host plants, such as tobacco, castor bean and cowpea. Verma *et al.* (1983) described that cells of *X.c.* pv. malvacearum secreted slime through "pores" in the surface layers in form of jet or spray creating a water-soaking of host tissue in advance, thereby helping the rapid spread and multiplication of the bacteria.

362

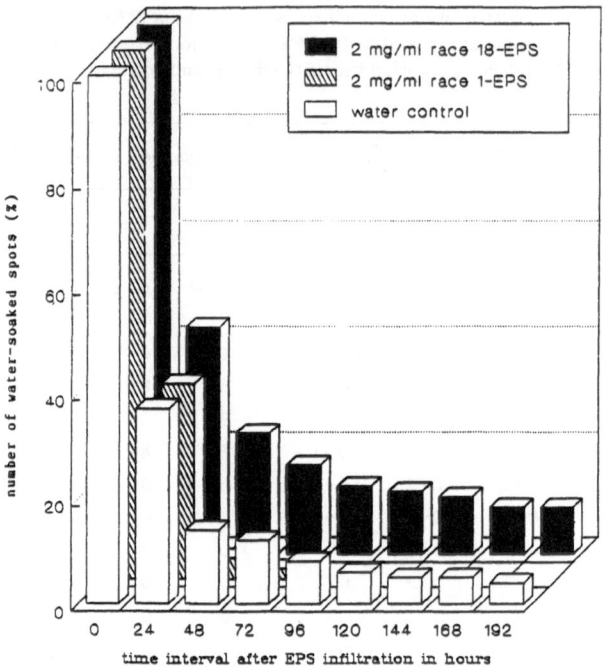

Figure 3. Induction of persistent water-soaked spots (in per cent) by EPS of *Xanthomonas campestris* pv. malvacearum races 1 and 18 in cotton cotyledons of cv. "Tamcot Camd-E".

However, most attempts to induce persistent water-soaking in leaves by EPS infiltration were only partially successful (e.g. Fig. 3), or failed completely (Fett *et al.*, 1986). The plants had always to be covered with plastic bags to e sure high relative humidity, which is not necessary during pathogenesis. Appa ɛ tly, a discrepancy exists between the proposed role of EPS during pathogenesis, *i.e.* induction of water-soaking, and the feasibility to demonstrate this effect experimentally. We explain this discrepancy by the impossibility to infiltrate highly viscoid EPS-solutions into leaves. Thus, we determined that more than 5% of the total dry weight of *P.s.* pv. phaseolicola infected bean leaves were bacterial EPS (Gross and Rudolph, 1987 b). Light microscopy (Fig. 1) revealed that about 25% of the leaf volume was occupied by bacteria embedded in slime. Assuming a relation of 1:1 for bacterial cells/slime one can conclude that 12.5% of the leaf volume was occupied by slime containing at least 5% of the leaf's dry weight. Since the average dry matter content of the leaf is about 12.5% it can be calculated that the bacterial slime *in planta* contained approximately 5% dry matter. A concentration of 5% EPS results in a very viscoid consistency which can not be infiltrated into leaves experimentally. *E.g.*, the experiments of El-Shouny *et al.* (Table 1) were carried out with 1% EPS, and those of Zachowski (Fig. 3) with 0.2% EPS.

Effect of EPS on bacterial multiplication in planta

Additional evidence for the involvement of EPS in pathogenesis resulted from co-infiltration of bacteria and EPS into leaves. Simultaneous application of purified EPS and bacteria revealed the capacity of EPS to enhance bacterial multiplication in bean leaves (Rudolph *et al.*, 1987). However, this effect depended on the composition of different EPS-types involved and the physiological stage of the leaves.

Similar experiments were carried out with EPS from *Xanthomonas campestris* pv. glycines (Hokawat and Rudolph, 1988). Partially purified EPS added to the inoculum suspension at 10 mg ml^{-1} promoted bacterial growth in a susceptible (cv. S.J.5) and a resistant (cv. CLARK 63) soybean cultivar. When 5 mg ml^{-1} were applied, bacterial growth was increased only in the susceptible cultivar but not in the resistant one, in which a decreased bacterial multiplication was observed within 1 day after inoculation.

Also crude EPS from *X.c.* pv. malvacearum promoted bacterial growth and symptom expression in compatible as well as in incompatible combinations (Zachowski, 1989). Addition of 2 mg EPS ml^{-1} of a bacterial suspension (10^4 CFU cm^{-2} leaf) led to a 10-fold higher bacterial population in cotyledons of cotton leaves. The same effect was observed in incompatible combinations, but it was weaker or occurred only during the early phase.

Effect of darkening

Bacteria do not produce copious amounts of slime when grown on a medium with a low content of carbohydrates. To reduce the assimilate level *in planta*, plants were incubated in continuous darkness before and after inoculation. Under these conditions the typical water-soaking symptom did not occur. Instead, within 3 or 4 days a dry necrosis developed which is characteristic for the hypersensitive resistance response. This effect of darkness was observed for *P.s.* pv. phaseolicola and *P.s.* pv. lachrymans on bean and cucumber, respectively, and for *X.c.* pv. malvacearum on cotton (Klement *et al.*, 1987). The atypical reaction of the susceptible cultivar was similar to the incompatible reaction of resistant cultivars, as indicated by depressed multiplication of bacteria in the leaves and occurrence of ion leakage, indicating tissue collapse.

Similar results were obtained by Smith and Kennedy (1970) for bacterial leaf blights of soybean (*Pseudomonas syringae* pv. glycinea) and bean (*Xanthomonas campestris* pv. phaseoli), as well as by Morgham *et al.* (1988) for *X. campestris* pv. malvacearum. Therefore, it is assumed that this effect of darkening occurs generally in interactions of pseudomonads or xanthomonads and green leaves. The conversion of water-soaking to a dry necrotic reaction was even more pronounced when the dark treatment started 2 days before inoculation.

In darkened leaves the concentration of free sugars (fructose, glucose and sucrose) decreased to 4–6% of that in illuminated leaves (Klement *et al.*, 1987). It was therefore concluded that deficiency of soluble sugars in darkened leaves leads to a drastically reduced rate of EPS-production, so that the bacteria are not surrounded by a thick slime layer. In consequence, a resistance-like response is induced by a cell-to-cell contact as known in incompatible plant-bacteria interactions. Indeed, it was shown by electron microscopy that in darkened susceptible cotton cotyledons considerably less EPS were produced by the bacteria, *X.c.* pv. malvacearum, so that the bacteria were more densely packed than in illuminated leaves (Zachowski *et al.*, 1990) (Fig. 4).

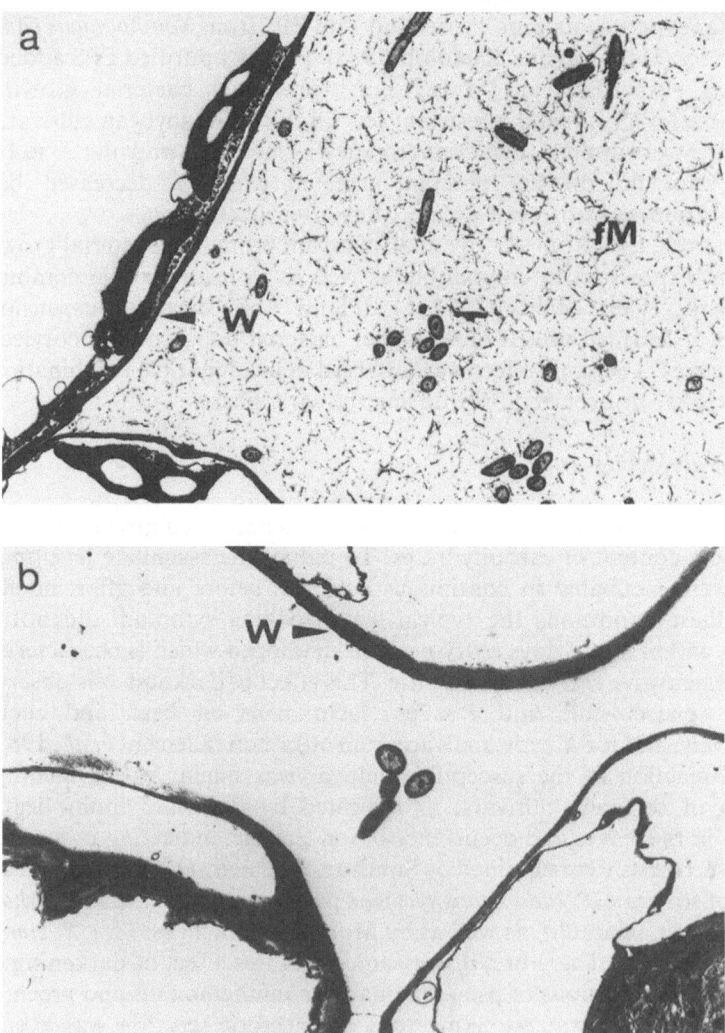

Figure 4. Electron micrographs of cotton cotyledons 48 h after inoculation with *X.c.* pv. malvacearum. Ruthenium Red stained fibrillar material (fM) around the bacteria in the illuminated cotyledons (a); in continuously darkened leaves (b) the bacteria were not surrounded by fibrillar material; W = cell wall.

Components of EPS

Many phytopathogenic pseudomonads produce *levan* (β-2,6-D-fructan). A typical example is the levan of *P.s.* pv. phaseolicola with mainly (β-2,6)-linkages, some (2,1)-branchings, and a molecular weight around $6x10^6$ daltons (El-Banoby *et al.*, 1980; Rudolph and El-Banoby, 1982). In addition to levan, most pathovars of *P. syringae* synthesize *alginate*, an acetylated, β-1,4-linked poly(D-)mannuronic acid (Gross and Rudolph, 1984, 1987a and b; Fett *et al.*, 1986; El-Shouny *et al.*, 1992), a polysaccharide resembling cell wall constituents of brown algae. The major difference between bacterial and algal material is the presence of O-acetyl substituents in bacterial alginates (Sutherland, 1977), whereas guluronic acid may not be a regular component of bacterial alginate. *Pseudomonas marginalis* synthetizes marginalan, a pyruvylated and succinylated glucogalactan (Osman and Fett, 1989). A rather complex exopolysaccharide is synthetized by *P. solanacearum* (Orgambide *et al.*, 1991). Genetic analyses revealed that EPS are virulence determinants in *P. solanacearum* (Cook and Sequeira, 1991) and comprise an important factor in the ability to wilt tomato plants (Denny and Baek, 1991).

The predominant component in the slime of xanthomonads is the so-called xanthan-gum. Xanthan consists essentially of a cellulose backbone, substituted on alternate residues with trisaccharide side-chains (mannose, glucuronic acid, mannose), so that the repeating unit is a pentasaccharide composed of D-glucose, D-mannose and D-glucuronic acid in the ratio of 2:2:1, with varying amounts of pyruvic and acetic acid (Sutherland, 1977). The molecular weight ranges between 1 to 9×10^6, and a double-helix molecular structure is assumed (Miles *et al.*, 1991).

Lipopolysaccharides (LPS) are not only constituents of the outer membrane of Gram-negative bacteria but are also released into the surrounding medium. Therefore, EPS-preparations from pseudomonads and xanthomonads regularly contain a LPS fraction (Gross and Rudolph, 1987 b; Ramm, 1992).

Finally, a small portion of proteins is often detected in the EPS of phytopathogenic bacteria. To which extent the proteins are contaminants or tightly bound constituents of the EPS is largely unknown.

Some components of the EPS were studied in more detail:

Alginate
Many *P. syringae* pathovars synthesize alginate (Fett *et al.*, 1986; Gross and Rudolph, 1987 a). Semiquantitative data (Albers, 1990; El-Shouny *et al.*, 1992) exemplified that additional pathovars of *P.* syringae (pv. aptata, atrofaciens, glycinea, and pisi), and *P. andropogonis*, produce alginate. Especially high amounts of total EPS (3 to 5 g l^{-1}) including alginate were produced by strains of *P. syringae* isolated from *Umbelliferae* (pv. coriandricola and additional unidentified pathovars). On the other hand, the opportunistic pathogens *P. marginalis* and *P. viridiflava*, which do not induce the water-soaking symptom, produced no or very low amounts of alginate *in vitro*.

Interestingly, alginate synthesis in *planta* by *P.s.* pv. phaseolicola was detectable from 1 d after inoculation onwards (Gross and Rudolph, 1987 d). The earlier assumption that the masses of acidic polysaccharides surrounding bacterial cells *in planta* (Fig. 2) are bacterial alginates could be substantiated by the help of alginate-specific antibodies derived from chicken. The biotin-labelled antibodies were used as probes to recognize bacterial alginate in ultra-thin sections of diseased plant tissue. Bound antibodies were detected with streptavidin-gold (Neugebauer, 1992). Electron microscopic examination revealed a specific gold-labelling of the matrix surrounding the bacterial cells with nearly no cross-reactivity of the label towards pectic substances of the plant cell wall (Fig. 5).

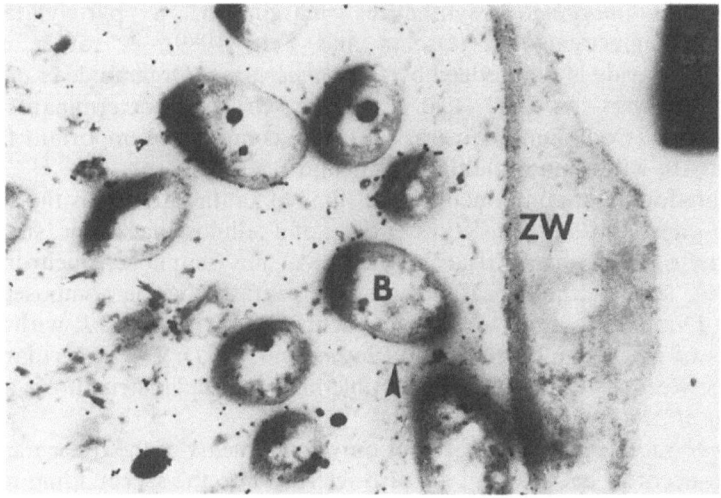

Figure 5. Immuno-gold labelling of bacterial alginate in ultra-thin sections of bean leaf tissue infected with *P.s.* pv. phaseolicola, hcw = host cell wall, B = bacterial cell, ZW = cell wall.

By the use of anti-alginate antibodies it was also possible to determine the alginate content quantitatively in the crude EPS of *in vitro* cultures. The method of rocket-immuno-electrophoresis was adapted for this purpose (Neugebauer *et al.*, 1991) (Fig. 6). Albers (1990) compared EPS production, viscosity of the culture filtrate, induction of water-soaking, and virulence of 33 *P. syringae* strains belonging to 8 different pathovars, and of two strains each of *P. andropogonis* and *P. marginalis*. The studies demonstrated that alginate is the decisive EPS-component which is responsible for the viscosity of the culture fluid. In three pathovars (phaseolicola, tomato, and aptata) the alginate content of the EPS correlated with the water-soaking capacity and virulence of the bacterial strains, whereas other pathovars (lachrymans, pisi) did not reveal this correlation. Similar results were obtained when several strains of two races of *P. syringae* pv. phaseolicola were compared (Neugebauer, 1987; Rudolph *et al.*,

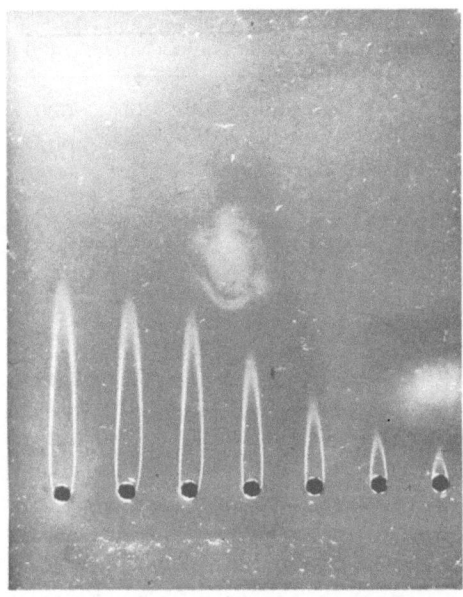

Figure 6. Rocket-immuno-electrophoresis showing a correlation between alginate concentration and the height of rockets; from left to right: 7000, 5600, 4200, 2800, 1400, 700, 350 ng alginate.

1989). The more virulent race 2 isolates of *P. syringae* pv. phaseolicola generally produced more EPS. In addition, all of the race 2 isolates showed a markedly higher proportion of alginate in the EPS than the race 1 isolates. Wydra (1991) also found that quality of bacterial EPS (contents of alginate) and development of water-soaked spots in bacteria-inoculated leaves were positively correlated (Table 2). Determination of the molecular weights of alginate from the two races did not reveal remarkable differences (Sonnenberg *et al.*, 1992). The alginate from race 1 tended to contain more polymers of lower molecular weight than that from race 2.

When crude EPS preparations of different composition were tested on leaves we found that those rich in alginate induced the longest persisting water-soaked spots. Commercially available alginate preparations (obtained from sea weed) did neither cause persistent water-soaking nor enhance bacterial multiplication in the leaves (Rudolph *et al.*, 1989). Most of our results, thus, support the hypothesis that bacterial alginates create a favourable milieu for bacterial multiplication *in planta*.

The *mode of action* of this biological effect apparently involves interactions of macromolecules of host and parasite. In this respect it seems noteworthy that bacterial alginate and plant pectate share common features in that both are modified uronates, and possess similar secondary and tertiary structures (Rees and Welsh, 1977), and characteristic gelling behaviours, particularily in mixed alginate-pectate solutions (Morris and Chilvers, 1984). It has long been known

Table 2. Comparison of symptom expression (water-soaking) by 9 *Pseudomonas phaseolicola* strains with quantity and quality of extracellular polysaccharides (EPS) produced *in vitro*

Bacterial strain	Water-soaking		Virulence		Race	OD	EPS mg/l	Alg %	Lev %	Prot %
	RK	RM	RK	RM						
Kl–S1a	1.9	0	1.8	0	1	0.98*	309	1.5	51	3.4
Kl–S1b	2.3	0	2.2.	0	1	2.65	582	1.5	24	2.0
Ro	1.3	0	1.8	0	1	1.46	670	1	88	1.8
CH	0.8	0	1.0	0	1	1.92	450	10	49	1.5
FV	3.5	2.4	3.3	2.4	2	1.76	519	7	48	1
N7	3.9	3.9	3.7	3.4	2	2.56	1222	76	28	0.7
Ex4a	3.5	1.6	3.5	1.8	2	1.85	508	6.5	59	6.0
Ex4b	3.6	2.6	3.6	2.7	2	1.98	519	10	68	2.0
KW	3.6	2.6	3.7	3.0	2	1.65	531	18.5	61	2.0

*EPS production was determined after 24 h growth in mineral medium with sucrose as carbon source.

OD = optical density (660 nm) of culture solution
Alg = alginate, Lev = levan, Prot = protein
RK = bean cultivar Red Kidney, susceptible
RM = bean cultivar Red Mexican, resistant towards race 1

that a close cell-to-cell contact of host and parasite can trigger plant defence reactions (Stall and Cook, 1979). Also, an agglutination of bacteria during the resistance reaction has been observed in several cases (Duvick and Sequeira, 1984; El-Banoby and Rudolph, 1980). Bacterial cells embedded in a viscoid alginate gel-like matrix seem to be protected against recognition and agglutination by the plant.

We extracted a so-called agglutinin (an arabinogalactan with galacturonic acid) from resistant bean leaves which agglutinated bacterial cells (Wydra and Rudolph, 1989, 1990). Agglutinins from susceptible cultivars were much less active (Fig. 7). Washed bacteria (EPS removed) agglutinated more strongly than unwashed bacteria (Table 3). Agglutination could be weakened or even inhibited by addition of EPS or purified alginate (Wydra 1991). Since purified lipopolysaccharides (LPS) from the bacterial cell membrane precipitated when mixed with bean agglutinin (Fig.8), it was suggested that LPS could be the bacterial receptor for the plant agglutinin. In contrast, in the compatible interaction the less effective plant agglutinin forms a synergistical gelling-system with alginate-rich EPS from highly virulent strains, thus preventing the bacteria

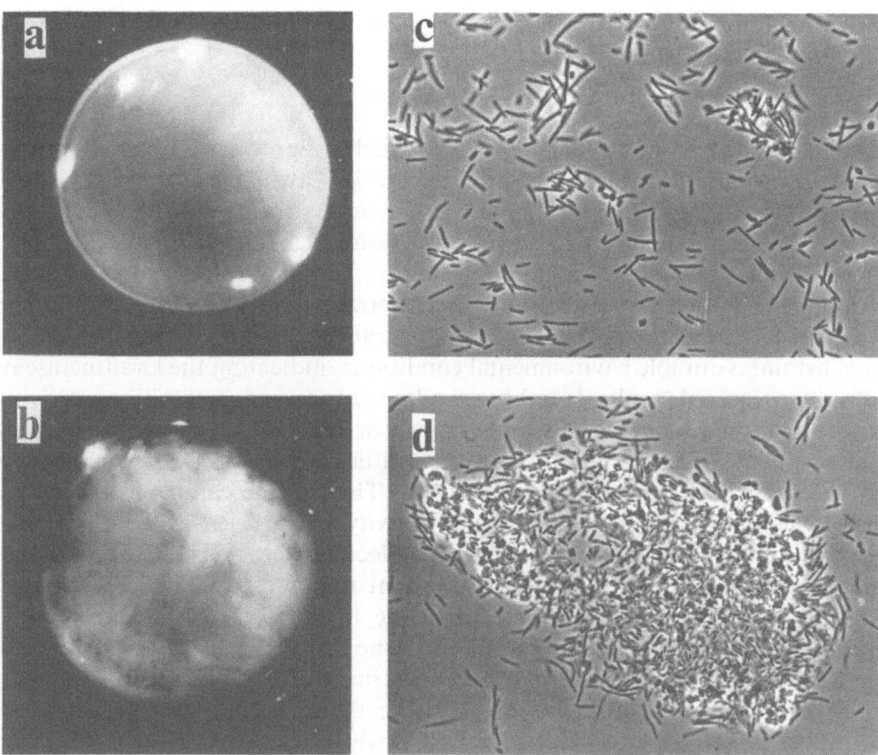

Figure 7. Agglutination of bacterial cells (*P.s.* pv. phaseolicola) by partially purified agglutinins from bean leaves, photographed under a dissecting microscope (a and b) or phase contrast (d and c). Strong agglutination occurred by the resistant cv. (b and d), but not by the susceptible cv. (a and c).

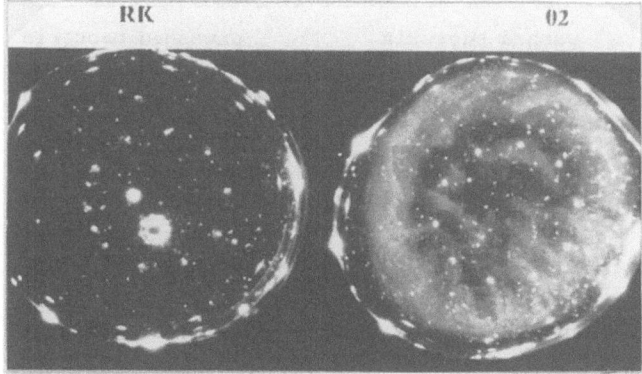

Figure 8. Reaction between lipopolysaccharides (LPS) of *P.s.* pv. phaseolicola 707 and agglutinins extracted from susceptible (RK) and resistant (02) bean leaves. Only the agglutinin from 02 precipitated the LPS.

from agglutination (Wydra and Rudolph, 1992b) and shielding the bacteria against bacteriostatic compounds.

Levan

Levan is the second EPS component which is produced by most phyto-pathogenic pseudomonads when sucrose is available. The exoenzyme *levansucrase* (β-2,6-fructan: D-glucose 6-fructosyltransferase, E.C.2.4.1.10.) splits sucrose and polymerizes the fructose residues to levan without any cost of energy while glucose is released. Levansucrase from *P. syringae* pv. phaseolicola was purified to homogeneity and characterized (Hettwer *et al.*, 1992). The features of the enzyme (Table 4) demonstrate a remarkably high resistance against unfavourable environmental conditions, indicating the levansucrase of *P. syringae* pv. phaseolicola to be well adapted to the adverse milieu in a living host organism. Judged by its kinetics levansucrase may well be active in the intercellular space. Composition of the synthesized levan is not homogenous but depends on environmental conditions. The enzyme can act as a levanase when sucrose is no longer available in heavily infected leaves. Levansucrases from different organisms were similar in molecular weight. These enzymes were produced constitutively, although in different amounts, by all of the *P. syringae* pathovars investigated, but were inducible in *P. fluorescens*. These results suggests that the enzyme plays a role in pathogenesis.

Sucrose is a widely used transport form in sugar assimilation (Giaquinta, 1980) and occurs in considerable amounts in higher plants. Also in the apoplast of leaves (*i. e.* the intercellular space) the predominant soluble sugar is sucrose (Ntsika and Delrot, 1986). Therefore, a highly active, constitutively synthesized levansucrase may have the function to create a protective shield around the

Table 3. Agglutination of *Pseudomonas* spp. by agglutinins of 3 cultivars of *Phaseolus vulgaris* (agglutination indices from 1 to 4)

bacterial strains	washed bacteria			unwashed bacteria		
	RK	RM	O2	RK	RM	O2
P. syringae. pv. phaseolicola						
RO	2	3.5	4	0	0	0.25
FV	2	2.5	3	0	0	0.25
N7a	0.5	1	1.5	0	0	0.5
P. fluorescens	0.5	0.5	1	0.25	0.5	0.25

RK = Red Kidney, highly sucseptible
RM = Red Mexican U1 34, resistant against race 1
O2 = breeding line, resistant against races 1 and 2

Table 4. Properties of levansucrase from *Pseudomonas syringae* pv. phaseolicola

Feature	Specification
optimum pH	6.1 (in McIllvaine's-buffer), from pH 5.5 to 7.5 only slight decrease of activity
optimum temperature	activity increases up to denaturation at 60°C
resistance towards	reducing agents (mercaptoethanol, dithioerythritol, 100 mM) detergents (Tween 20, Elugent, SDS, 1%) proteases
kinetics	Km for cleavage of sucrose: 0.16 M sucrose inhibitors: glucose (1 mM) Hg-salts (irreversible)
molecular weight	denatured: 45 kD native : 80 kD (by PAGE) 68 kD (by GPC)

bacterial cells by synthesizing levan, which retains moisture and thus can protect microcolonies from sudden desiccation. Simultaneously, glucose is provided as carbon source for bacterial growth. This role of levansucrase may be critical during the early phase of pathogenesis, because the synthesis of alginate occurs intracellularly and requires several enzymes (Bannerjee *et al.*, 1985). Indeed, we found that in batch cultures of *P. s.* pv. phaseolicola with sucrose as substrate, the formation of alginate followed the levan synthesis (Gross and Rudolph, 1987c). In later stages of disease alginate may be a more important virulence factor than levan. Thus, in contrast to alginate, all our experiments to induce persistent water-soaking in plant leaves by purified levan or levansucrase failed.

EPS-deficient mutants
Nearly 3000 mutants were obtained from strain GSPB 659 (race 2 of *P. s.* pv. phaseolicola) by transposon (Tn5) and chemical (NTG) mutagenesis (Somlyai *et al.*, 1990). Alginate and levan defective strains were selected by screening for colony morphology, indirect detection of levansucrase activity by adding sucrose to the culture fluid, growth on sucrose as sole carbon source, detection of alginate by polysaccharide blotting and staining with toluidine blue, or detection with specific antisera (Neugebauer *et al.*, 1990). Only prototrophic

Table 5. Prototrophic Tn5 mutants of *Pseudomonas syringae* pv. phaseolicola with deficiencies in EPS production and reduction in virulence

| Strain | Exopolysaccharides (EPS) | | | HR | WS | Growth *in planta* |
	Total (mg l^{-1})	Alg	Lev			(CFU cm^{-2})
678	780	+	+++	−	−	
1887	730	(+)	+++	+/−	−	
353	603	−	+++	+/−	−	6.4×10^1
NTG20	nd**	nd	−	+/−	++	
1715	nd	nd	++	+	++	2.0×10^8
504	760	−	++	++	(+)	6.4×10^1
546	580	(+)	++	++	+	5.2×10^8
1207	648	−	+++	++	++	5.4×10^7
1238	713	−	+++	++	++	
2136	436	−	+++	+++	(+)	
253	855	(+)	+++	+++	++	
336	806	+	+++	+++	++	
441	536	−	+++	+++	++	
558	765	(+)	+++	+++	++	1.0×10^8
429	370	(+)	(+)	+++	++	2.9×10^6
728	823	++	++	+++	−	
673	716	(+)	+++	+++	++	8.3×10^7
659*	844	+++	+++	+++	+++	7.0×10^8

*) wild type
**) not determined

mutants were selected. The virulence of the selected strains was carefully evaluated in comparison with the wild strain. Some of the mutants obtained are listed in Table 5. Mutants with reduced alginate production always showed reduced disease symptoms (especially water-soaking) and reduced growth *in planta*, whereas this correlation was not clear for levan production. In several cases, also the intensity of the hypersensitive reaction (HR) on the non-host tobacco was impaired (Table 5). A few mutants showed a decreased production of alginate as well as of levan. Only the non-pleiotrophic mutants which seem to be deficient specifically in alginate synthesis will be characterized further by complementation experiments.

Xanthan

The acidic heteropolysaccharide xanthan is produced by all phytopathogenic xanthomonads. By staining with ruthenium red an embedding of bacterial cells in the intercellular leaf space in masses of acidic material has been demonstrated in several cases (*e.g.* Fig. 4). The structures appear to be of bacterial origin, by comparison with bacteria-free EPS preparations from *in vitro* cultures of *X.c.* pv. pelargonii (Rudolph *et al.*, 1987). Our experiments also showed that bacterial growth in leaves (*X. camprestris* pv. glycines and *X. campestris* pv. malvacearum) and symptom expression was increased by addition of EPS to the inoculum suspension (Hokawat and Rudolph, 1988; Zachowski, 1989). A synergistic interaction between xanthan and plant polymers (Morris *et al.*, 1977; Tako, 1991) may create a stable gel-like slime which initiates the susceptible reaction. Genetic analyses from other laboratories confirmed the role of xanthan as virulence factor (De Crecy-Lagard *et al.*, 1990; Kingsley and Gabriel, 1991).

Lipopolysaccharides (LPS)
From the three components of LPS, the core and the lipid A represent highly conserved structures which also occur in non-phytopathogenic pseudomonads (Gross *et al.*, 1988). The O-specific side chains, on the other hand, appear to be specific for certain pathovars of *P. syringae* or *X. campestris* (Yakovleva *et al.*, 1992; Kamiunten and Fujita, 1990) and may play a role in host-specificity.

Bacterial mutants of *P. s.* pv. *phaseolicola* which were deficient in the O-specific side chain proved to be non-pathogenic. Electron micrographs revealed the bacteria to be very densely packed *in planta* without any space of EPS between them (Rudolph *et al.*, 1989). Thus, these bacteria were obviously not able to grow into the intercellular space. It is assumed that the rough LPS more easily induces a resistance reaction in the host plant which can be modified by interaction with plant polymers (Wydra and Rudolph, 1990). The intact LPS may play a role in pathogenesis by induction of water-soaking (Epton *et al.*, 1977) and release of nutrients from the plant cytoplasm due to membrane damage (Mazzucchi *et al.*, 1988).

Proteins
All of the crude, lyophilized EPS preparations obtained so far were "contaminated" by 0.5 to 5% (w/w) protein. When the protein portion of the EPS was enriched, a considerable amount of levansucrase was found (Gross and Rudolph, 1987c). It is unknown to which extent other extracellular bacterial proteins are connected to the EPS and participate in the host-parasite interaction. However, surface proteins from phytopathogenic pseudomonads proved to be well suited as pathovar specific antigens (Niepold and Huber, 1988; Baharuddin *et al.*, 1992) and thus seem to play a role in host specificity.

Possible functions of exopolysaccharides

It is suggested that the .extracellular polysaccharides of pseudomonads and xanthomonads form a gel-like matrix with plant polymers into which the bacteria are embedded *in planta*. The matrix promotes bacterial multiplication by several mechanisms:

a) attraction of water and ions so that bacterial microcolonies in the normally air-filled intercellular spaces do not desiccate and can take up nutrients more easily; b) inhibition of a close morphological contact of the bacteria with plant cell walls so that resistance reactions are not triggered; c) inhibition of bacterial agglutination by plant polymers (agglutinins); and d) shielding of bacteria against bacteriostatic compounds (*e.g.* phytoalexins) and stress molecules (*e.g.* oxygen radicals).

Acknowledgements

Financial support was obtained from the Deusche Forschungsgemeinschaft (Ru 207/ 18, 19, 20, 21), the Volkswagen-Stiftung (U. H.), and the Studienmission der Arabischen Republik Ägypten (W. E.). We also thank Linda Rudolph for typing the manuscript.

References

Albers A (1990) Beziehungen zwischen Menge und Zusammensetzung *in vitro* gebildeter extrazellulärer Polysaccharide und der Virulenz phytopathogener Pseudomonaden. Diplomarbeit, Fachbereich Agrarwissenschaften, Universität Göttingen.

Baharuddin B, Niepold F & Rudolph K (1992) Detection of blood disease bacteria in infected banana plants using "monospecific" antibodies. *In*: Proc. 8th Intern. Conf. Plant Pathogenic Bacteria, 9–12 June, Versailles, France.

Banerjee PC, Vanags RI, Chakrabarty AM & Maitra PK (1985) Fructose-1,6-bisphosphate aldolase activity is essential for synthesis of alginate from glucose by *Pseudomonas aeruginosa*. J. Bacteriol. 165: 458–460.

Borkar SG & Verma JP (1989) Exopolysaccharide, a water soaking inducing factor produced by bacterial blight of cotton, *Xanthomonas campestris* pv. *malvacearum*. Cot. Fib. Trop. XLIV: 149–152.

Cook D & Sequeira L (1991) Genetic and biochemical characterization of a *Pseudomonas solanacearum* gene cluster required for extracellular polysaccharide production and for virulence. J. Bacteriol. 173: 1654–1662.

De Crecy-Lagard V, Glaser P, Lejeune P, Sismeiro O, Barber CE, Daniels MJ & Danchin A (1990) A *Xanthomonas campestris* pv. *campestris* protein similar to catabolite activation factor is involved in regulation of phytopathogenicity. J. Bacteriol. 172: 5877–5883.

Denny TP & Baek SR (1991) Genetic evidence that extracellular polysaccharide is a virulence factor of *Pseudomonas solanacearum*. Mol. Plant-Microbe Interact. 4: 198–206.

Duvick JP & Sequeira L (1984) Interaction of *Pseudomonas solanacearum* lipopolysaccharide and extracellular polysaccharide with agglutinin from potato tubers. Appl. Environ. Microbiol. 48: 192–198.

El-Banoy FE & Rudolph K (1979a) A polysaccharide from liquid cultures of *Pseudomonas phaseo icola* which specifically induces water-soaking in bean leaves (*Phaseolus vulgaris* L.). Phytop. th. Z. 95: 38–50.

El-Banoby FE & Rudolph K (1979b) Induction of water-soaking in plant leaves by extracellular polysaccharides from phytopathogenic pseudomonads and xanthomonads. Physiol. Plant Pathol. 15: 341–349.

El-Banoby FE & Rudolph K (1980) Agglutination of *Pseudomonas phaseolicola* by bean leaf extracts (*Phaseolus vulgaris*). Phytopath. Z. 98: 91–95.

El-Banoby FE & Rudolph KWE (1989) Multiplication of *Xanthomonas campestris* pvs. *secalis* and *translucens* in host and non-host plants (rye and barley) and development of water soaking. Bulletin OEPP/EPPO Bulletin 19: 105–111.

El-Banoby FE, Rudolph K & Hüttermann A (1980) Biological and physical properties of an extracellular polysaccharide from *Pseudomonas phaseolicola*. Physiol. Plant Pathol. 17: 291–301.

El-Banoby FE, Rudolph K & Mendgen K (1981a) The fate of extracellular polysaccharide from *Pseudomonas phaseolicola* in leaves and leaf extracts from halo-blight susceptible and resistant bean plants (*Phaseolus vulgaris* L.). Physiol. Plant Pathol. 18: 91–98.

El-Shouny W, Wydra K, El-Shanshoury A, El-Sayed MA & Rudolph K (1992) Exopolysaccharides of phytopathogenic pseudomonads. *In*: Proc. 8th Intern. Conf. Plant Pathogenic Bacteria, 9–12 June, Versailles, France.

Epton HAS, Sigee DC & Passmoor M (1977) The influence on pathogenicity of ultrastructural changes in *Pseudomonas phaseolicola* during lesion development. Acta Phytopathologica Acad. Sci. Hung. 12: 301–303.

Fett WF, Osman SF, Fishman ML & Siebles TS (1986) Alginate production by plant-pathogenic pseudomonads. Appl. Environ. Microbiol. 52: 466–473.

Giaquinta RT (1980) Translocation of sucrose and oligosaccharides. p. 271–320. *In:* Preiss J (ed.) The Biochemistry of Plants. Vol.3. Acad. Press, New York etc.

Gross M & Rudolph K (1984) Partial characterization of slimes produced by *Pseudomonas phaseolicola in vitro* and *in planta* (*Phaseolus vulgaris*). Proc. 2nd Pseudomonas Working Group, Sounion, Greece (pp. 66–67).

Gross M & Rudolph K (1987a) Studies on the extracellular polysaccharides (EPS) produced *in vitro* by *Pseudomonas phaseolicola*. I. Indication for a polysaccharide resembling alginic acid in seven *P. syringae* pathovars. J. Phytopathol. 118: 276–287.

Gross M & Rudolph K (1987b) Studies on the extracellular polysaccharides (EPS) produced *in vitro* by *Pseudomonas phaseolicola*. II. Characterization of levan, alginate, and 'LPS'. J. Phytopathol. 119: 206–215.

Gross M & Rudolph K (1987c) Studies on the extracellular polysaccharides (EPS) produced in vitro by *Pseudomonas phaseolicola* III. Kinetics of levan and alginate formation in batch culture and demonstration of levansucrase activity in crude EPS. J. Phytopathol. 119: 289–297.

Gross M & Rudolph K (1987d) Demonstration of levan and alginate in bean plants (*Phaseolus vulgaris*) infected by *Pseudomonas syringae* pv. *phaseolicola*. J. Phytopathol. 120: 9–19.

Gross M, Meyer H, Widemann C & Rudolph K (1988) Comparative analysis of the lipopolysaccharides of a rough and a smooth strain of *Pseudomonas syringae* pv. *phaseolicola*. Arch. Microbiol. 149: 372–376.

Häusermann E (1944) Über die Benetzungsgröße der Mesophyllinterzellularen. Ber. Schweiz Bot. Ges. 54: 541–578.

Hettwer U, Gross M & Rudolph K (1992) Levansucrase as a possible factor of virulence? Characterization of the enzyme and occurrence in the *Pseudomonas syringae* group. *In*: Proc. 8th Intern. Conf. Plant Pathogenic Bacteria, 9–12 June, Versailles, France.

Hokawat S & Rudolph K (1988) Effect of extracellular polysaccharides of *Xanthomonas campestris* pv. *glycines* on multiplication and survival of the pathogen. Proc. 26th Sci. Conf. Plant Division, 3.–5.2.1988, Kasetsart, Thailand (pp. 23–30).

Jones SB & Fett WF (1985) Fate of *Xanthomonas campestris* infiltrated into soybean leaves: an ultrastructural study. Phytopathology 75: 733–741.

Kamiunten H & Fujita T (1990) Electrophoretic analysis of the lipopolysaccharide from *Xanthomonas campestris* pv. *oryzae* and *Pseudomonas avenae*. Kyushu Byogaichu Kenkyukaiho 36: 14–15.

Keen NT (1990) Gene-for-gene complementarity in plant-pathogen interactions. Annu. Rev. Genet. 24: 447–463.

Keen NT & Williams PH (1971) Chemical and physiological properties of a lipomucopolysaccharide from *Pseudomonas lachrymans*. Physiol. Plant Pathol. 1: 247–264.

Kingsley MT & Gabriel DW (1991) A mutation in *Xanthomonas campestris* pv. *citrumelo* affects both host-specific virulence and exopolysaccharide production. Phytopathology 81: 1195.

Klement Z, Gross M & Rudolph K (1987) Leaf necrosis instead of water-soaking due to light deficiency after inoculation with *Pseudomonas* and *Xanthomonas*. *In*: Civerolo EL, Collmer A, Davis RE, Gillaspie AG (ed.) Plant Pathogenic Bacteria. Proc. 6th Int. Conf. Plant Path. Bact., Maryland, June 2–7, 1985 (pp. 430–436) M. Nijhoff Publishers, Dordrecht, Boston, Lancaster.

Mansfield JW & Brown IR (1986) The biology of interactions between plants and bacteria. p. 71–98. *In*: Bailey J (ed.) Biology and Molecular Biology of Plant-Pathogen Interactions. NATO ASI Series, Vol. H1. Springer-Verlag Berlin Heidelberg.

Mazzucchi U, Gasperini C, Noli E & Medeghini Bonatti P (1988) Increase of free space solutes in tobacco leaves in relation to the localized cellular response following injections of a bacterial protein-lipopolysaccharide complex. J. Phytopathol. 121: 193–208.

Miles MJ, Lee I & Atkins EDT (1991) Molecular resolution of polysaccharides by scanning tunneling. J. Vac. Sci. Technol., B, 9: 1206–1209.

Morgham AT, Richardson PE, Essenberg M & Cover EC (1988) Effects of continuous dark upon ultrastructure, bacterial populations and accumulation of phytoalexins during interactions between *Xanthomonas campestris* pv. *malvacearum* and bacterial blight susceptible and resistant cotton. Physiol. Mol. Plant Pathol. 32: 141–162.

Morris ER, Rees DA, Young G, Walkinshaw MD & Darke A (1977) Order-disorder transition for a bacterial polysaccharide in solution. A role for polysaccharide conformation in recognition between *Xanthomonas* pathogen and its plant host. J. Mol. Biol. 110: 1–16.

Morris VJ & Chilvers GR (1984) Cold setting alginate-pectin mixed gels. J. Sci. Food Agric. 35: 1370–1376.

Neugebauer M (1987) Untersuchungen zur Produktion der Exopolysaccharide von *Pseudomonas syringae* pv. *phaseolicola* unter Berücksichtigung der Morphologie und der Bedeutung für die Virulenz. Diplomarbeit, Fachbereich Agrarwissenschaften, Universität Göttingen.

Neugebauer M (1992) Immunologische Untersuchungen über das saure Polysaccharid (Alginat) von *Pseudomonas syringae* pv. *phaseolicola* mit Antikörpern aus den Eiern immunisierter Hühner. Dissertation, Universität Göttingen.

Neugebauer M, Gross M & Rudolph K (1991) Isolation of antibodies against bacterial exopolysaccharide alginate from the eggs of immunized hens, and their use in studying host-parasite interactions. Proc. 4th Int. Working Group on *Pseudomonas syringae* pathovars. Firenze, June 10–13 (pp. 189–194).

Neugebauer M, Gross M, Nöllenburg M, Somlyai G & Rudolph K (1990) Screening for alg⁻ and lev⁻ strains of *Pseudomonas syringae* pv. *phaseolicola* mutagenized by Tn5 transposon. Proc. 7th Int. Conf. Plant Path. Bact., Budapest (pp. 397–402).

Niepold F & Huber SJ (1988) Surface antigens of *Pseudomonas syringae* pv. *syringae* are associated with pathogenicity. Physiol. Mol. Plant Pathol. 33: 459–471.

Ntsika G & Delrot S (1986) Changes in apoplastic and intracellular leaf sugars induced by the blocking of export in *Vicia faba*. Physiol. Plant 68: 145–153.

Orgambide G, Montrozier H, Servin P, Roussel J, Trigalet-Demery D & Trigalet A (1991) High heterogeneity of the exopolysaccharides of *Pseudomonas solanacearum* strain GMI 1000 and the complete structure of the major polysaccharide. J. Biol. Chem. 266: 8312–8321.

Osman SF & Fett WF (1989) Structure of an acidic exopolysaccharide of *Pseudomonas marginalis* HT041B. J. Bacteriol. 171: 1760–1762.

Ramm M (1992) *Pseudomonas syringae* pv. *phaseolicola* extracellular lipopolysaccharides as possible pathogenicity factors. *In*: Proc. 8th Int. Conf. Plant Pathogenic Bacteria, 9–12 June, Versailles, France.

Rees DA & Welsh EJ (1977) Sekundär- und Tertiärstruktur von Polysacchariden in Lösungen und in Gelen. Angew. Chem. 89: 228–239.

Rudolph K (1978) A host specific principle from *Pseudomonas phaseolicola* (Burkh.) Dowson, inducing water-soaking in bean leaves. Phytopath. Z. 93: 218–226.

Rudolph K (1980) Multiplication of bacteria in leaf tissue. Angew. Botanik 54: 1–9.

Rudolph K (1984) Multiplication of *Pseudomonas syringae* pv. *phaseolicola in planta*. I. Relation between bacterial concentration and water-congestion in different bean cultivars and plant species. Phytopath. Z. 111: 349–362.

Rudolph K (1993) The infection of the plant by *Xanthomonas*. *In*: Swings J & Civerolo EL (ed.) Xanthomonas, Chapman & Hall, London (in press).

Rudolph K & El-Banoby F (1982) Mechanisms of resistance in bean leaves towards bacteria. *In*: Wood RKS (ed.) Active Defense Mechanisms in Plants (p. 352) Plenum Press, New York.

Rudolph K & Mendgen K (1985) Multiplication of *Pseudomonas syringae* pv. *phaseolicola in planta*. II. Characterization of susceptible and resistant reactions by light and electron microscopy compared with bacterial countings. Phytopath. Z. 113: 200–212.

Rudolph K, Ebrahim-Nesbat F, Mendgen K & Thiele C (1987) Cytological observations of leaf spot-causing bacteria in susceptible and resistant hosts. p. 604–612. *In*: Civerolo EL, Collmer A, Davis RE & Gillaspie AG (ed.) Plant Pathogenic Bacteria. Proc. 6th Int. Conf. Plant Path. Bact. Maryland, June 2–7, 1985. M. Nijhoff Publishers, Dordrecht, Boston, Lancaster.

Rudolph KWE, Gross M, Neugebauer M, Hokawat S, Zachowski A, Wydra K & Klement Z (1989) Extracellular polysaccharides as determinants of leaf spot diseases caused by pseudomonads and xanthomonads. p. 177–218. *In*: Graniti A, Durbin RD and Ballio A (ed.) Phytotoxins and Plant Pathogenesis. NATO ASI Series, Vol. H27. Springer Verlag, Berlin, Heidelberg.

Slatyer RO (1967) Plant-water relationships. Experimental Botany 2: 148–149.

Smith MA & Kennedy BW (1970) Effect of light on reactions of soybean to *Pseudomonas glycinea*. Phytopathology 60: 723–725.

Somlyai G, Neugebauer M, Gross M, Nöllenburg M & Rudolph K (1990) Virulence of alginate and levan defective strains of *Pseudomonas syringae* pv. *phaseolicola* mutagenized by Tn5 or MNNG. Abstr. of Papers, 5th Int. Symp. Mol. Gen. of Plant-Microbe Interact., Sept. 9–14, Interlaken, Switzerland.

Sonnenberg B, Neugebauer M & Rudolph K (1992) Investigations on the molecular weight of alginate synthesized by *Pseudomonas syringae* pv. *phaseolicola* races 1 and 2. *In*: Proc. 8th Int. Conf. Plant Path. Bacteria, 9–12 June, Versailles, France.

Stall RE & Cook AA (1979) Evidence that bacterial contact with the plant cell is necessary for the hypersensitve reaction but not the susceptible reaction. Physiol. Plant Pathol. 14: 77–84.

Sutherland I (1977) Surface Carbohydrates of the Prokaryotic Cell. Acad. Press, London, New York, San Francisco.

Tako M (1991) Synergistic interaction between deacylated xanthan and galactomannan. J. Carbohydr. Chem. 10: 619–633.

Verma JP, Formanek H, Singh RP, Trivedi BM & Jindal JK (1983) Chemistry and ultrastructure of surface layers of phytopathogenic bacteria. Indian J. Plant Pathol. 1: 68–74.

Wydra K (1991) Interactions between cell wall polymers from bush beans (*Phaseolus vulgaris*) and cells from *Pseudomonas syringae* pv. phaseolicola and their extracellular polymer products (in German). Diss. Univ. Göttingen.

Wydra K & Rudolph K (1989) An agglutination factor from bush bean leaves against *Pseudomonas syringae* pv. *phaseolicola*. p. 334–340. *In*: Galling G (ed.): Proc. Braunschweig Symp. on Appl. Plant Molecular Biol., Nov. 21–23, 1988, Techn. Universität Braunschweig.

Wydra K & Rudolph K (1990) Interactions between cell wall polymers of *Phaseolus vulgaris* and bacterial cell surface polymers of *Pseudomonas syringae* pv. *phaseolicola*. p. 63–68. *In*: Klement Z (ed.) Plant Pathogenic Bacteria. Proc. 7th Int. Conf. Plant Path. Bact. Budapest, Hungary, 1989.

Wydra K, Rudolph K (1992a) Analysis of toxic extracellular polysaccharides. p. 113–183. *In*: H.F. Linskens and F. Jackson (ed.) Modern Methods of Plant Analysis. New Series, Vol. 13, Plant Toxin Analysis, Springer Verlag Berlin, Heidelberg, New York.

Wydra K & Rudolph K (1992b) A model for mechanisms of resistance and susceptibility on a molecular level in plant-microbe interactions. *In*: Proc. 8th Int. Conf. Plant Path. Bacteria, 9–12 June, Versailles, France.

Yakovleva LM, Gvozdyak RI, Zdorovenko GM, Gubanova NYa & Solyanik LP (1992) Structure and functions of lipopolysaccharides from *Pseudomonas syringae* (serogroup II). *In*: Proc. 8th Int. Conf. Plant Path. Bacteria, 9–12 June, Versailles, France.

Young JM (1974) Effect of water on bacterial multiplication in plant tissue. New Zealand J. Agric. Res. 17: 115–119.

Zachowski MA (1989) Zur Pathophysiologie von *Xanthomonas campestris* pv. *malvacearum* (Smith) Dye, dem Erreger der eckigen Blattfleckenkrankheit, und zur Resistenz verschiedener Baumwollsorten (*Gossypium hirsutum* L.). Dissertation, Universität Göttingen, 113 pp.

Zachowski MA, Ebrahim-Nesbat F, Rudolph K (1990) Effect of continuous darkness on production of exopolysaccharides by *Xanthomonas campestris* pv. *malvacearum* in susceptible cotton cotyledones. p. 173–178. *In*: Klement Z (ed.) Plant Path. Bacteria. Proc. 7th Int. Conf. Plant Path. Bacteria, Budapest, June 11–16, 1989. Akadémiai Kiadó Budapest.

26. Transcriptional regulation of α-hemolysin genetic expression: *hlyM*, a sequence contained in *hlyC*, modulates hemolysin transcription

YOLANDA JUBETE, JUAN CARLOS ZABALA, ANTONIO JUÁREZ and FERNANDO DE LA CRUZ

Abstract. Transcription of the *hly* operon of transmissible plasmids in *Escherichia coli* produces a major *hlyCA* mRNA, and a minor *hlyCABD* mRNA. The upstream sequence *hlyR* enhances α-hemolysin production. Various chromosomal genes, such as the histone-like protein Hha, also affect hemolysin expression. We have analyzed the effects of *hlyR* and *hha* on hemolysin transcription.

hlyCA mRNA levels were 40-times higher in an *hlyR*$^+$ operon that in an *hlyR*$^-$ operon. Increased levels are the result of increased mRNA synthesis, not of slower decay. Therefore, *hlyR* appears to be a transcriptional enhancer. In addition, deletion of an internal *hlyC* fragment increases *hly* transcription in an *hlyR*$^-$ operon to the levels of an *hlyR*$^+$ operon. This region is thus assumed to contain an operator site which was called *hlyM* ("modulator"). *hlyM* does not contain any internal promoter or is capable of acting *in trans*. We think it contains the binding site for a transcriptional repressor.

We also tested the effect of the *hha* mutation on *hly* expression. The *hha* mutation increases *hly* transcription in both an *hlyR*$^-$ and in an *hlyM*$^+$ (but not in an *hlyM*$^-$) operons. It has no effect in *hlyR*$^+$ operons, irrespective of *hlyM*. Thus, both *hlyR* and the *hha* mutation activate transcription and suppress the modulatory effect dictated by *hlyM*.

Abbreviations: RTX, Repeat Toxin

Introduction

Hemolysin is a virulence factor of *E. coli* that causes human extra-intestinal infections. It is the prototype of the **RTX** family of toxins produced by a range of Gram-negative bacteria. These proteins are exported to the culture medium by a *secA*-independent mechanism. Synthesis and secretion of hemolysin is determined by the *hly* operon, which comprises four contiguous genes, *hlyCABD*. *hlyA* encodes the inactive toxin protein, which is activated to the mature toxin by specific fatty acylation directed by HlyC and is dependent upon the cellular acyl carrier protein (Hardie *et al.*, 1991; Issartel *et al.*, 1991). The products of *hlyB* and *hlyD*, together with the product of the chromosomal gene *tolC*, are involved in the secretion of hemolysin (Welch, 1991).

The *hly* operon is located either in transmissible plasmids, or in the bacterial chromosome (de la Cruz *et al.*, 1980; Müller *et al.*, 1983). While the sequences of the *hly* genes and the structure of the operon are highly conserved, the sequences of the upstream region diverge significantly (Hess *et al.*, 1986).

C.I. Kado and J.H. Crosa (eds.), Molecular Mechanisms of Bacterial Virulence, 379–397.
© 1994 *Kluwer Academic Publishers.*

Table 1. Bacterial strains and plasmids.

Strain or plasmid	Phenotype or genotype	Reference
Strains		
HB101	F⁻ *proA2 leuB6 recA13 thi1 ara14 lacY1 galK2 endA rpsL20 lmbd⁻ supE44 hsdRM*	Bolívar et al (1977)
Hha-2	5K *hha-2*::MudI	Godessart et al. (1987)
5K	F⁻ *recA1 strA hsdR hsdD thr leu thi lacZ*	Juarez & Goebel (1984)
MC1061	F⁻ *araD139 (ara-leu) 7696 lacY74 galU galK hsr hsm rpsL*	Casadaban & Cohen (1980)
N100	*galK lac recA pro*	McKenney et al. (1981)
SUY-1	MC1061 *hha-2*::Tn5	This work
Plasmids		
pUC8/9	vectors for cloning; Apʳ	Viera & Messing (1982)
pANN202-312	*hly* operon (*hlyR⁻*) cloned in pACYC184; Cmʳ	Goebel & Hedgpeth (1982)
pANN202-312R	*hly* operon (*hlyR⁺*) cloned in pACYC184; Cmʳ	Godessart et al. (1987)
pHP45	Ω cassette; Spʳ	Prenki & Krisch (1984)
pKL200	*galK* cloned downstream *lac* promoter; Apʳ	Grinsted (unpubl.)
pKO-1	vector for GalK assays; Apʳ	McKenney et al. (1981)
pKO-2	vector for GalK assays; Apʳ	De Boer (1984)
pSP65	vector for riboprobe synthesis; SP6 promoter; Apʳ	Melton et al. (1984)
pSU127	*hly* operon (*hlyR⁺*) cloned in pUC9 vector; Ω cassette; Apʳ Spʳ	This work
pSU129	*hly* operon (*hlyR⁻*) cloned in pUC8 vector; Ω cassette; Apʳ Spʳ	This work
pSU134	1.7-kb *Hind*III-*Bam*HI fragment in pKO-1; Apʳ	This work
pSU137	1.5-kb *Hind*III-*Sma*I fragment in pKO-1; Apʳ	This work
pSU2739	2.7-kb *Eco*RI-*Sma*I fragment in pKO-1; Apʳ	This work
pSU2939	2.9-kb *Eco*RI-*Bam*HI fragment in pKO-2; Apʳ	This work
pSU2946	793-bp *Eco*RI-*Hind*III from *hlyA* in pSP65; Apʳ	This work
pSU2947	525-bp *Hind*III-*Eco*RV from *hlyB* in pSP65; Apʳ	This work
pSU2968	pSU129 carrying a 220-bp deletion (*Sma*I-*Bam*HI) within *hlyC*	This work
pSU2969	2.4-kb *Hind*III fragment in pKO-2	This work
pSU2970	pSU2969 carrying a 220-bp deletion (*Sma*I-*Bam*HI) within *hlyC*	This work

Transcription of the plasmid operon occurs from a single promoter and starts 264-bp upstream of the *hlyC* initiation codon. Two transcripts have been identified: a major 4-kb *hlyCA* transcript, and a minor 8-kb *hlyCABD* transcript (Welch and Pellett, 1988). Between *hlyA* and *hlyB* there is a Rho-independent transcriptional terminator that only a minority of the RNA polymerase molecules can read through, which therefore produces the two transcripts. Furthermore, the *hlyCA* mRNA is more stable (10.2 min) than the *hlyCABD* mRNA (4.4 min) (Welch and Pellett, 1988). Similar patterns of expression have been reported for the leukotoxins of *Pasteurella* (Strathdee and Lo, 1989) and *Actinobacillus* (Spitznagel *et al.*, 1991), and for the *cya* operon of *Bordetella pertussis* (Laoide and Ullmann, 1990).

In the case of the plasmid *hly* operons, an upstream sequence called *hlyR* increases hemolysin synthesis and secretion (Vogel *et al.*, 1988). The primary mechanism by which *hlyR* exerts its action has been reported to be anti-termination at the *hlyA-hlyB* transcription terminator (Koronakis *et al.*, 1988, 1989). Recently two *E. coli* regulatory proteins have been described that affect hemolysin expression. Hha is an 8.6-kDa protein that shuts down transcription of the *hly* operon if *hlyR* is not present (Nieto *et al.*, 1991). SfrB is an 18.3-kDa transcriptional activator required for *hly* expression (Bailey *et al.*, 1992).

In this work we analyze the effects of *hlyR* and *hha* on the transcription of the major *hlyCA* transcript. We provide a different view on *hlyR* mechanism of action and describe *hlyM*, a new regulatory element in the transcriptional control of *hly* expression.

Physical analysis of the *hly* operon transcription. Transcriptional effect of *hlyR*

Plasmids pSU127 and pSU129 are derivatives of pHly152, in which the whole *hly* operon has been cloned in a high copy number vector (pUC8 and pUC9, respectively) (Vieira and Messing, 1982). Their physical maps are shown in Fig. 1. They only differ in that pSU127 contains *hlyR* while pSU129 does not contain it, and both carry an Ω cassette (Prentki and Krisch, 1984) to avoid any upstream transcription from the vector. *E. coli* strains carrying plasmid pSU127 produce more hemolysin (550 HU) than isogenic strains harbouring plasmid pSU129 (10 HU), as expected from previous results (González-Carreró *et al.*, 1985; Vogel *et al.*, 1988). We wanted to know if *hlyR* had a direct effect on transcription from the *hly* promoter (P_{hly}) and therefore if it was responsible for the different level of protein expression.

We looked directly at the mRNA expression in plasmids pSU127 and pSU129 by using Northern blot analysis (Fig. 2). The 4-kb *hlyCA* mRNA was clearly more abundant in the presence of *hlyR*. According to Welch and Pellett (1988) there should be an 8-kb *hlyCABD* mRNA, which we were unable to visualize in Northern blot. The mRNA levels were also quantified by slot blots (Fig. 3a). *hlyR* increased the *hlyCA* mRNA expression by 40-fold (4 experiments). This difference can explain the levels of hemolysin secreted by the

382

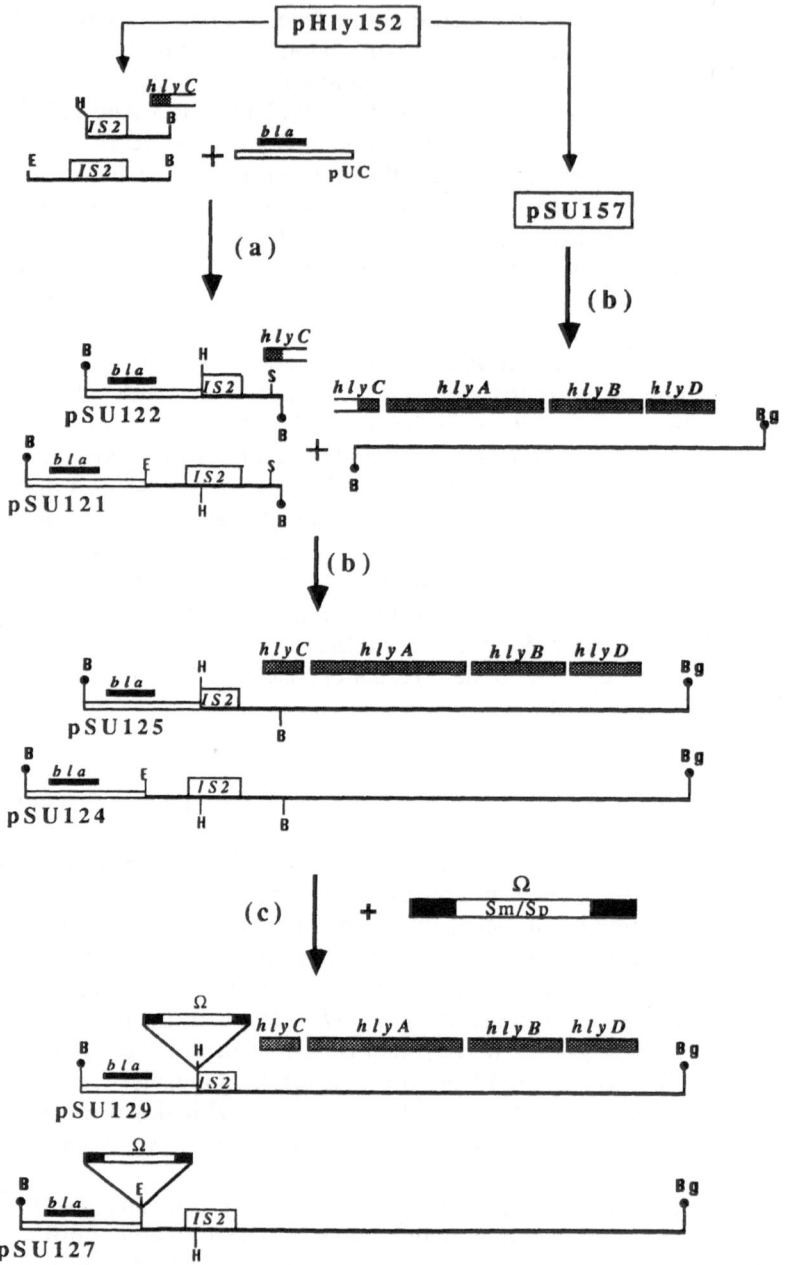

Figure 1. Construction of plasmids pSU127 and pSU129. (a) The *Eco*RI-*Bam*HI fragment [coordinates 9.5-kb to 12.4-kb in the pHly152 map of Zabala *et al.* (1984)] and the *Hind*III-*Bam*HI fragment (coordinates 10.7-kb to 12.4-kb in the same pHly152 map) were cloned in the homologous sites of pUC8 and pUC9 respectively, giving rise to plasmids pSU121 and pSU122; (b) A 7.5-kb

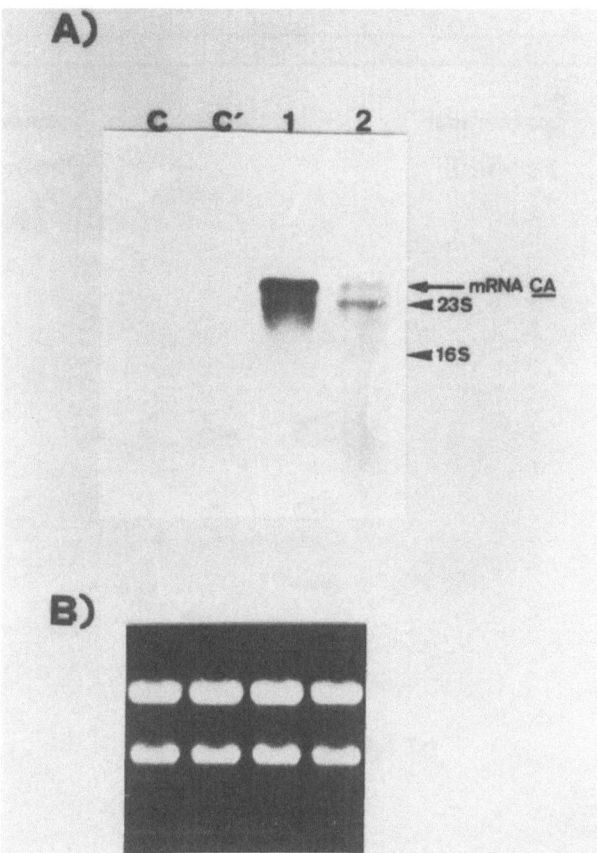

Figure 2. Northern blot analysis of the *hlyR* effect on *hlyCA* mRNA expression. (A) Total bacterial RNA from *E. coli* HB101 derivative strains was isolated as described Mellado *et al.* (1981) and equal amounts (20 μg) were separated under denaturing conditions in a 1.0% agarose-formaldehyde gel, transferred to a nitrocellulose membrane (Schleicher & Schuell), and hybridized with antisense *hlyA* mRNA probe (Probe 1, Fig. 3) synthesized *in vitro* from SP6 promoter (plasmid pSU2946 linearized with *Hind*III). The 4-kb *hlyCA* mRNA is indicated by the arrow. The positions of the 16S and 23S rRNAs served as standards; (B) Ethidium bromide staining of the electrophoresis gel. Tracks: (C) HB101; (C′) HB101 (pUC8); (1) HB101 (pSU127); (2) HB101 (pSU129).

---

*Bam*HI-*Bgl*II fragment (coordinates 12.4-kb to 20.15-kb in the map of pHly152) that was obtained by partial digestion of plasmid pSU157 (González-Carreró *et al.*, 1985) was cloned at the *Bam*HI site of plasmids pSU121 and pSU122 giving rise to plasmids pSU124 and pSU125, respectively; (c) The Ω fragment from plasmid pHP45 (Prentki and Krisch, 1984) was inserted at the *Eco*RI site (coordinate 9.5-kb in the map of pHly152) of plasmid pSU124, and at the *Hind*III site of plasmid pSU125, giving rise to plasmids pSU127 and pSU129, respectively.

Restriction enzyme cleavage sites are *Bam*HI (B), *Bgl*II (Bg), *Eco*RI (E), *Hind*III (H) and *Sma*I (S).

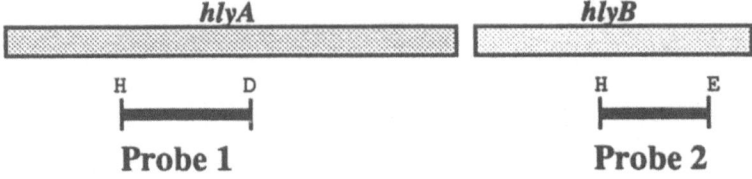

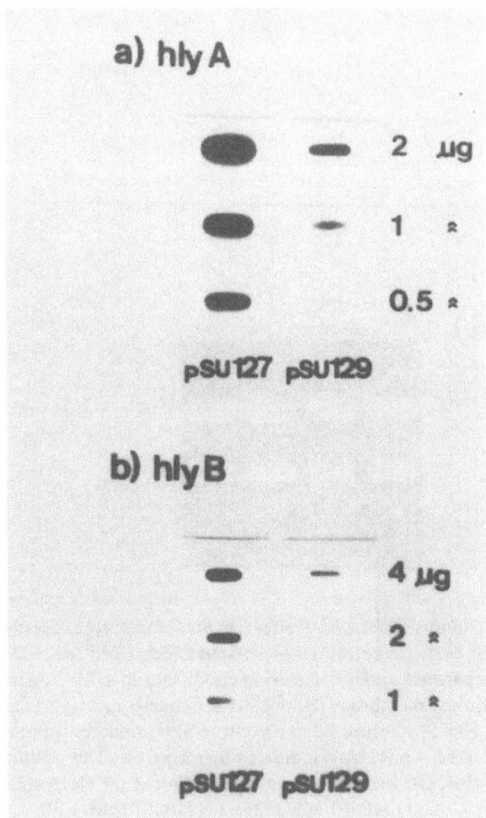

Figure 3. Slot blot analysis. The upper part of the figure shows a schematic diagram of the antisense RNA probes internal to *hlyA* (793-bases long) (Probe 1) and to *hlyB* (525-bases long) (Probe 2). For the slot blot analysis, serial dilutions of total cellular RNA from strains HB101 (pSU127) and HB101 (pSU129) were spotted onto a nitrocellulose filter, which was hybridized with Probe 1 (a) and Probe 2 (b). The hybridization with antisense β-lactamase mRNA probe [*bla*-specific probe (pSU2948 linearized with *Eco*RI)] was used as a control of the amount of RNA. Restriction enzyme cleavage sites are *Eco*RI (E), *Eco*RV (D) and *Hind*III (H).

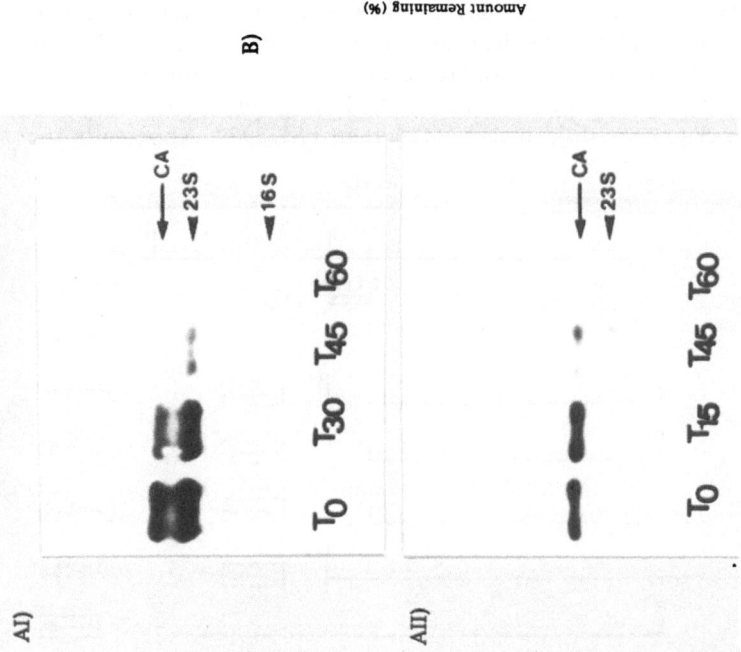

Figure 4. Stability of the *hlyCA* mRNA generated by pSU127 (*hlyR*+) and pSU129 (*hlyR*−). (A) Northern blot analysis. Total cellular RNA from *E.coli* HB101 carrying plasmids pSU127 (AI) and pSU129 (AII) was isolated at 30°C as described Mellado *et al.* (1981) at the indicated time points (in min) after rifampicin was added to a final concentration of 200 μg ml⁻¹. Analysis of the *hlyCA* mRNA decay was performed by Northern blot. The filter-bound RNA was hybridized to a ³²P-labeled antisense *hlyA* mRNA synthetized *in vitro* from SP6 promoter (probe 1, Fig. 3); (B) The radioactivity of the band corresponding to the *hlyCA* mRNA was measured. The amount of remaining cpm (%) were plotted against the time of RNA isolation. The half-lives calculated from these plots were 20.2 ± 2.2 min (3 experiments) for pSU127 and 21.5 ± 2.1 min (2 experiments) for pSU129.

plasmids pSU127 and pSU129. When we used an *hlyB* probe (Fig. 3b), the mRNA levels were 4-times higher in pSU127 than in pSU129.

The higher levels of the *hlyCA* mRNA in the presence of *hlyR* (pSU127) could be the result of higher levels of transcription or, alternatively, of a higher stability of the transcript. In order to distinguish between these two possibilities, we measured the stability of the *hlyCA* mRNA either in the presence or in the absence of *hlyR*. Figure 4 shows that the half-life of the *hlyCA* transcript was very similar in both plasmids (20.2 min in pSU127 and 21.5 min in pSU129). This result indicates that *hlyR* has no effect on the *hlyCA* mRNA stability.

Transcription from P*hly* is attenuated within *hlyC* by *hlyM*

The data reported above indicated that the region of pHly152 containing *hlyR* was responsible for an increase in hemolysin transcription. In order to further confirm this result, we constructed a set of *galK* transcriptional fusions containing different regions around P*hly* (Fig. 5) and assayed their Galacto-kinase (GalK) activity. The results from these assays are shown in Table 2. Plasmid pSU2939 leads to a 9-fold increase in GalK activity versus plasmid pSU134, a situation which reflects the original observation of pANN202-312R vs. pANN202-312 (Vogel *et al.*, 1988). However, a couple of similar plasmids, pSU137 and pSU2739 gave very different results. Both promote the same levels of elevated transcription, 2-fold higher, even than pSU2939. Apparently the

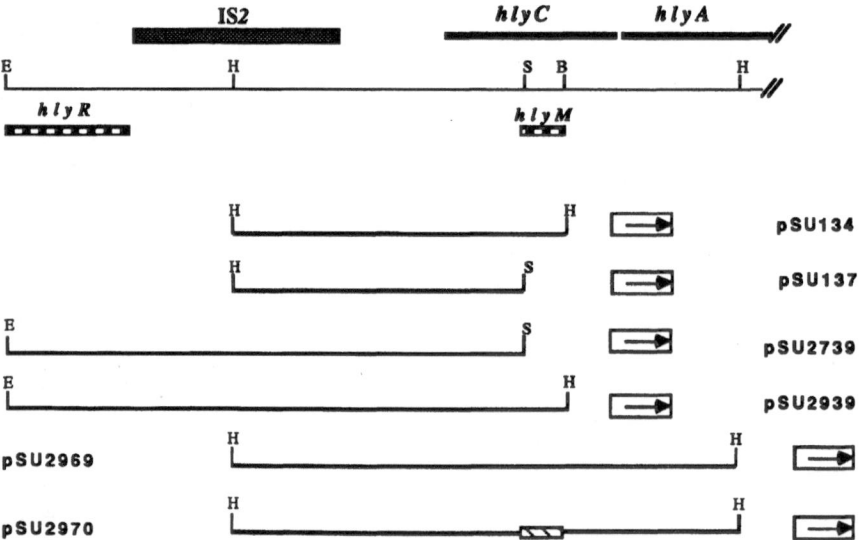

Figure 5. Schematic diagram of *galK* transcriptional fusions made in the *hly* upstream region. The rectangles represent the *galk* gen and the arrow, the transcription direction. The striped rectangle (pSU2970) represents the *Sma*I-*Bam*HI deletion within *hlyC* (see text for more details). Restriction enzyme cleavage sites are *Bam*HI (B), *Eco*RI (E), *Hind*III (H) and *Sma*I (S).

Table 2. Galactokinase activity of transcriptional fusions to the *hly* upstream region.

Plasmid[a]	Copy number[b]	GalK activity[c]	Relative strength (%)[d]
pKL200	13.0	24.5 ± 2.0	100
pSU134	9.0	1.0 ± 0.2	4
pSU137	5.5	21.2 ± 3.0	87
pSU2739	5.0	19.0 ± 2.0	78
pSU2939	5.0	9.0 ± 0.5	37
pSU2969	11.0	1.1 ± 0.1	5
pSU2970	11.0	3.9 ± 0.2	16

[a] Plasmids harboured in strain N-100 (McKenney et al., 1984).

[b] Determined by β-lactamase assay (Lupski et al., 1984). β-lactamase activity of N-100 (pBR322) (from 3 experiments) was considered equivalent to 20 plasmid copies per cell.

[c] Galactokinase assays were performed on toluene-treated cells by the method of McKenney et al. (1984). Units are millimoles of galactose phosphate produced per min, per ml of culture (O.D.$_{600}$, 1) and per plasmid copy (Rosenberg et al., 1983). The column shows the mean galactokinase activity ± standard deviation (from 4 experiments). The GalK activities of the vectors pKO-1 or pKO-2 (0.7 ± 0.1) have been substracted.

[d] Strength relative to pKL200 (Grinsted, unpubl.), which contains the *lac* promoter fused to the *galK* gene.

220-bp DNA segment between the *Sma*I and the *Bam*HI sites within *hlyC* could attenuate transcription from P$_{hly}$. This hypothesis was partially confirmed by measuring the GalK activities of plasmids pSU2969 and pSU2970. Plasmid pSU2969 produces the same low level of transcription as pSU134. However, deletion of the *Sma*I-*Bam*HI fragment in pSU2970 results in 4-fold of increase in transcription. In this case the increase is smaller than in pSU137, but that could be a consequence of the more distant position of the *galK* fusion.

In order to analyze directly the genetic expression *in vivo*, we deleted the *Sma*I-*Bam*HI region from the haemolytic plasmid pSU129 obtaining the plasmid pSU2968. This plasmid produced an intermediate level of HlyA in comparison to pSU127 and pSU129 (data not shown). In addition, we looked at the levels of mRNA expression. The deletion of the *Sma*I-*Bam*HI fragment increased by 4-fold the amount of *hlyCA* mRNA (Fig. 6). Therefore, transcription from P$_{hly}$ is attenuated within *hlyC*, unless *hlyR* is present. The DNA sequence responsible for this attenuation effect was called *hlyM*.

388

The *hha* chromosomal mutation overcomes the attenuation produced by *hlyM*

We knew from previous results (Godessart *et al.*, 1987; Nieto *et al.*, 1991) that the *hha* chromosomal mutation increased the amount of hemolysin both intracellular and extracellular, even in Hly⁺ plasmids lacking *hlyR*. This suggested that Hha protein could interact with specific regions of the *hly* operon and thus regulate its expression. We assayed *galK* fusions to the 5' *hly* region in *hha⁺* and *hha⁻* backgrounds (Table 3). The fusion contained in pSU137, lacking *hlyM*, shows the same GalK activity in *hha⁺* strain as in *hha⁻* strain. However, the same fusion carrying *hlyM* (pSU134) increases the GalK activity

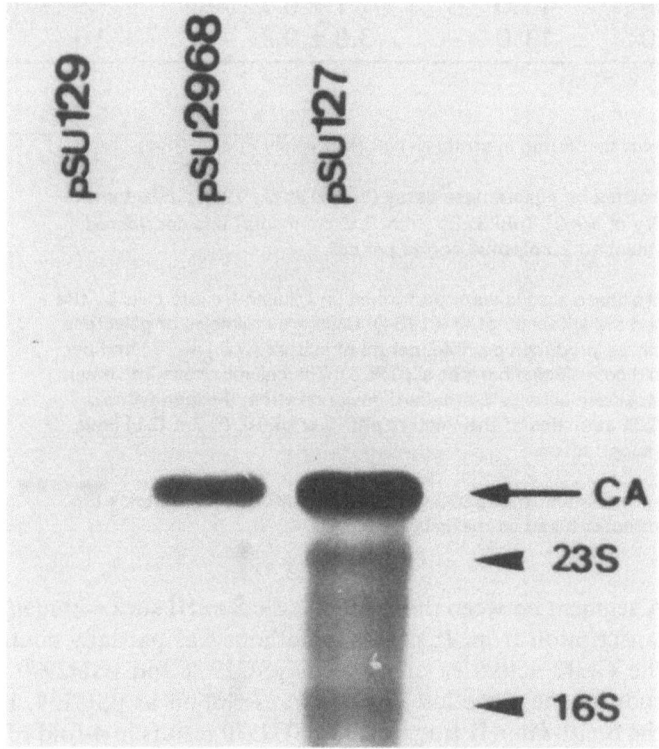

Figure 6. Northern blot analysis of the *hly* transcriptional expresion in an *hlyM⁻* operon. Total bacterial RNA (20 μg) from *E. coli* HB101 carrying pSU127 (*hlyR⁺*), pSU129 (*hlyR⁻*) and pSU2968 (*hlyR⁻*, *hlyM⁻*) was separated on a 1% formaldehyde agarose gel, transferred to a nitrocellulose membrane and hybridized with Probe 1 (Fig. 3). The 4-kb *hlyCA* mRNA is indicated by the arrow.

from 5% in *hha⁺* to 80 % in *hha⁻*. The result was similar when we analyzed the GalK activity in a similar couple of plasmids carrying *hlyR* (pSU2739 and pSU2939). Our results suggest that *hlyM*-mediated attenuation is independent of *hlyR* and therefore both mechanisms act indepently. It is possible that Hha acts on the *hlyM* sequence to attenuate transcription.

Table 3. Galactokinase activity of transcriptional fusions in *hha*[+] and *hha*[−] backgrounds.[a]

	Copy number	GalK activity	Relative strength (%)
Strain *hha*[+]			
pKL200	12.0	23.0 ± 2.0	100
pSU134	12.0	1.2 ± 0.1	5
pSU137	5.5	19.6 ± 1.0	85
pSU2739	5.0	18.2 ± 0.5	79
pSU2939	5.0	9.2 ± 0.4	40
Strain *hha*[−]			
pKL200	20.0	27.5 ± 2.2	100
pSU134	20.0	22.1 ± 1.2	80
pSU137	21.0	22.6 ± 1.0	82
pSU2739	15.0	23.3 ± 1.3	85
pSU2939	14.0	22.5 ± 0.8	82

[a] See legend Table 2 for details

Effect of the *hha* mutation on the *hlyCA* mRNA expression

Since plasmid pSU127 was very unstable in strains carrying the *hha* mutation, we used the lower copy number plasmids pANN202-312 (Goebel and Hedgpeth, 1982) and pANN202-312R (Godessart *et al.*, 1987) in order to analyze the effect of *hha* on *hly* transcription. Additionally, we also determined the effect of *hlyR* on *hly* expression in intermediate copy number plasmids. Northern blot analysis (Fig. 7) showed that the expression of the *hlyCA* mRNA was increased in *hha*[−] strains by 14-fold in the absence of *hlyR*, and by 5-fold in the presence of *hlyR*. We were only able to see the 8-kb *hlyCABD* mRNA in *hha*[−] strains carrying the plasmid pANN202-312R (data not shown). These results were further confirm by slot blot analysis (Fig. 8A). Thus the *hha* mutation increases *hly* transcriptional expression, and its effect is stronger in the absence of *hlyR*.

On the other hand, *hlyR* was responsible for 4-fold of increase in the expression of the *hlyCA* mRNA in the wild-type strain and for 2-fold in the *hha*[−] strain (Figs. 7 and 8A). This result agrees with our previous observations (Nieto *et al.* 1991) and confirms that *hlyR* increases the expression of the major *hlyCA* mRNA, although the level of this increase seems to be dependent on the copy number of the Hly[+] plasmid.

We looked at the effect of the *hha* mutation on *hlyB* expression (Fig. 8B). In the absence of *hlyR* (pANN202-312), *hha* increased its expression by 12-fold, while it was only 4-fold of increase in *hlyR*[+] plasmid (pANN202-312R). In

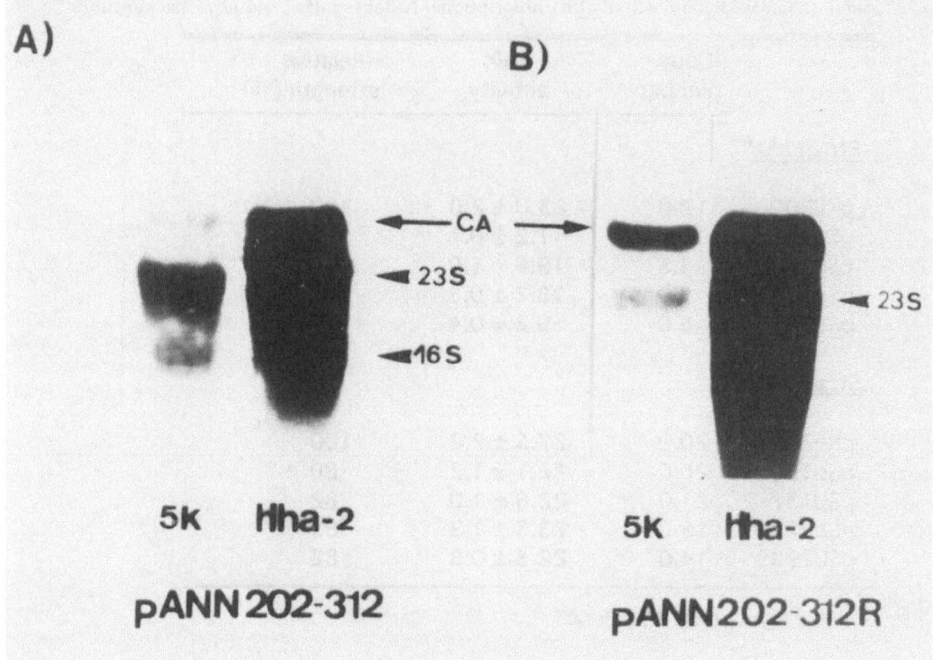

Figure 7. hly transcriptional expression in *hha⁻* cells. Total bacterial RNA (20 µg) from *E. coli* 5K and Hha-2 strains carrying pANN202-312 (a) and pANN202-312R (b) was subjeted to Northern blot analysis and hybridized with antisense RNA probe internal to *hlyA* (Probe 1, Fig. 3). The 4-kb *hlyCA* mRNA is indicated.

addition, *hlyR* was responsible for 4–5–fold of increase in *hha⁺* background and 1–2-fold of increase in *hha⁻* background.

We also demonstrate that the *hha* activator effect is independent of the stability of the *hlyCA* mRNA (Figs. 9 and 10). In fact, the 4-kb *hlyCA* mRNA is less stable in strains carrying the *hha* mutation. The half-life we measured was about 20 min in wild-type strain, while it was about 11 min in *hha⁻* strain, independent of the presence of *hlyR*.

Discussion

Expression of the *hly* operon is governed primarily by the regulatory region upstream of *hlyC* and transcription generates two mRNA species: the majority being *hlyCA*, and the minority being the full operon, *hlyCABD*. The *hlyCA* transcripts terminate at a Rho-independent transcriptional terminator located between *hlyA* and *hlyB*. The plasmid-derived *hly* upstream region shows increasing expression of the hemolysin operon. This effect is largely due to the 600-bp *hlyR* sequence which has no identity to any previously described

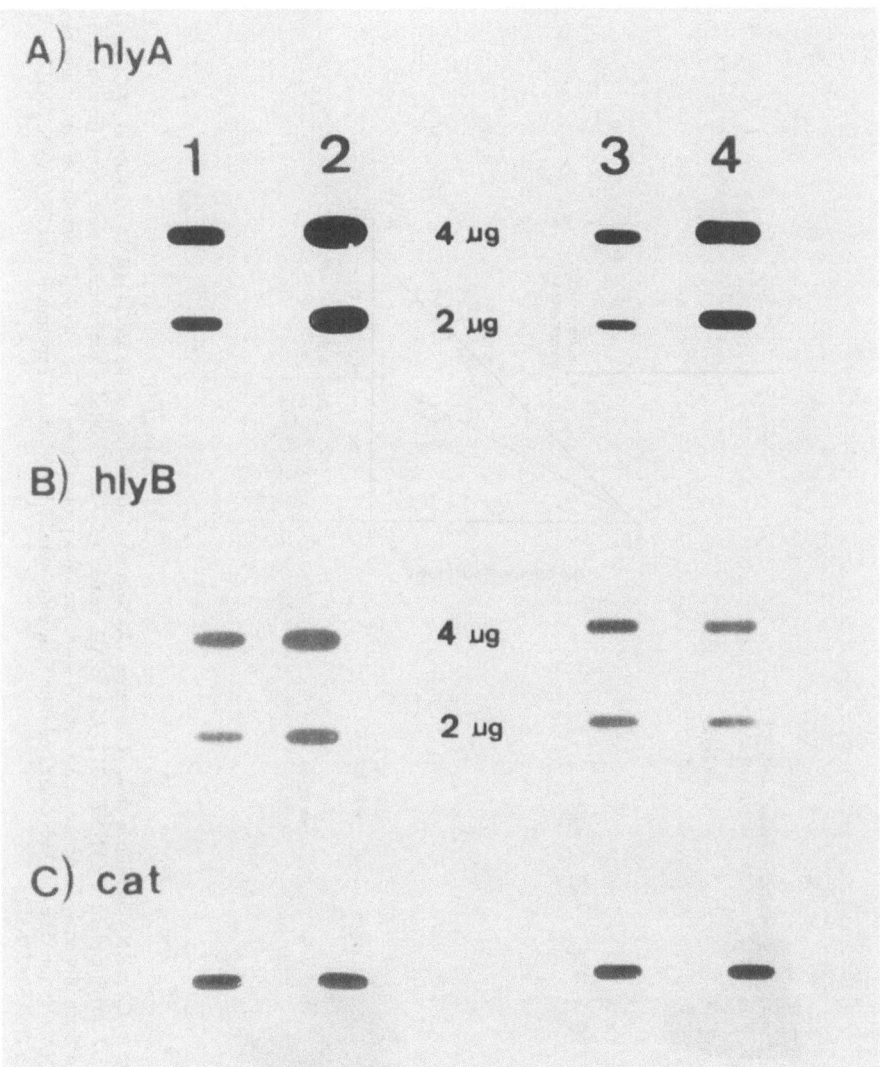

Figure 8. Slot blot analysis of the *hha* effect on *hly* transcriptional expression. Various amounts of total cellular RNA from strains 5K(pANN202-312R) (1), Hha-2(pANN202-312R) (2), 5K(pANN202-312) (3) and Hha-2(pANN202-312) (4) were hybridized with Probe 1 (*hlyA*) (A) and Probe 2 (*hlyB*) (B). The hybridization to antisense *cat* –mRNA probe was used as a control of the amount of RNA.

392

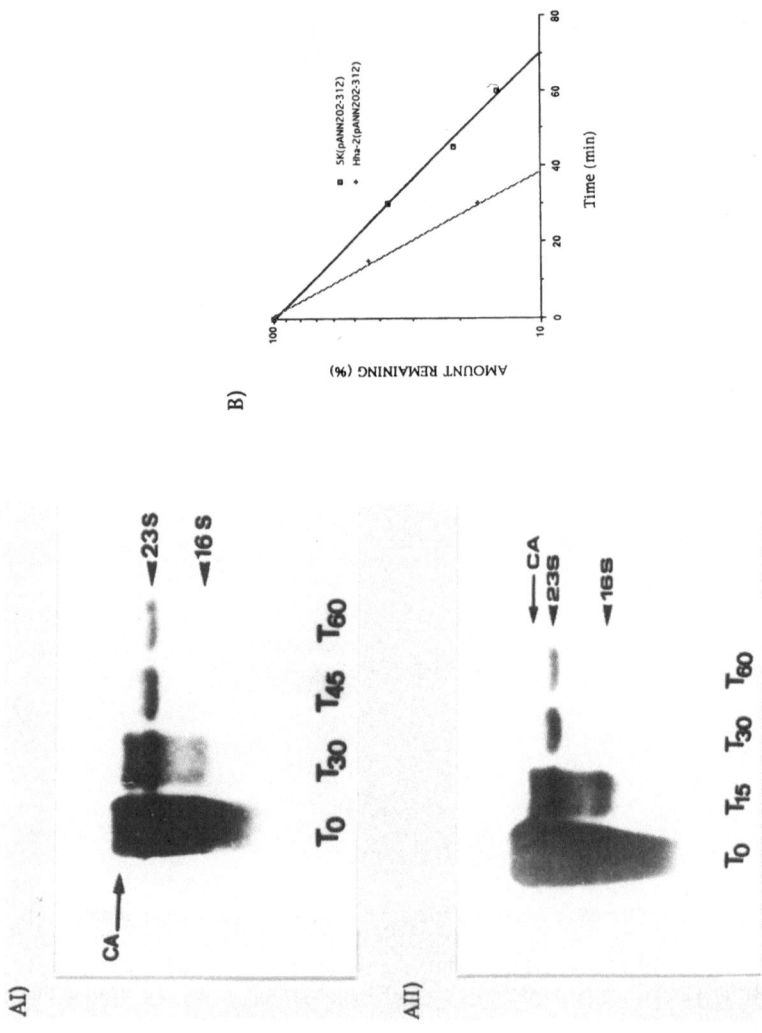

Figure 9. Effect of *hha* mutation on the *hlyCA* mRNA half-life in an *hlyR⁻* operon. (A) Northern blot analysis of the *hlyCA* mRNA stability at 30°C. Total cellular RNA from *E.coli* 5K and Hha-2 carrying plasmid pANN202-312 (AI and AII, respectively) was isolated at the indicated time points (in min) after rifampicin was added, and used for Northern blot analysis. Antisense *hlyA* mRNA was used as a probe (Probe 1, Fig. 3); (B) The amount of remaining cpm (%) were plotted against the time of RNA isolation. The half-lives calculated from these plots were 20.8 ± 1.7 min (3 experiments) for 5K(pANN202-312) and 11.3 ± 1.5 min (3 experiments) for Hha-2(pANN202-312).

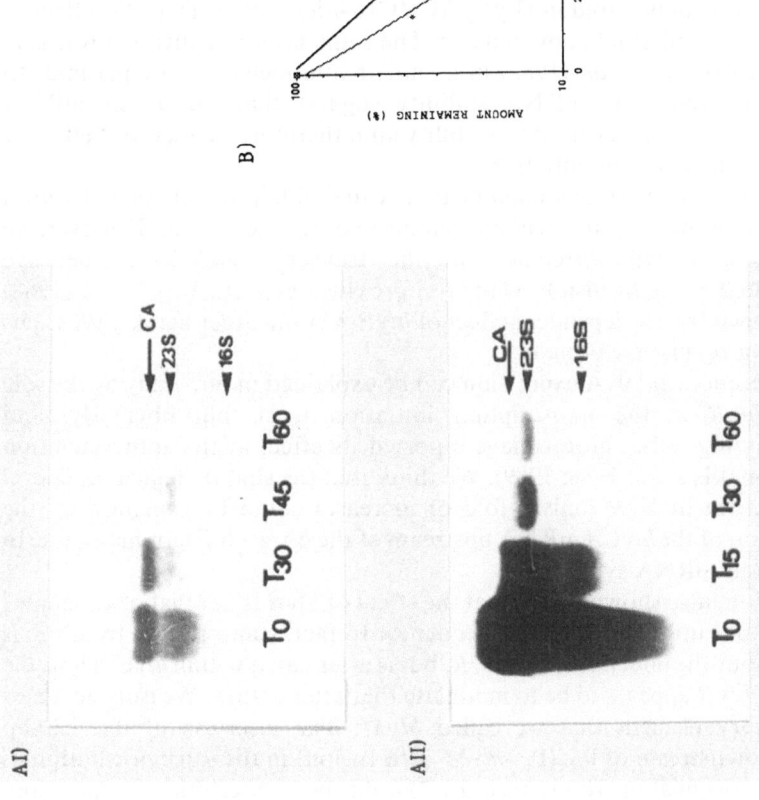

Figure 10. Effect of *hha* mutation on *hlyCA* mRNA half-life in a *hlyR*+ operon. (A) Northern blot analysis of *hlyCA* mRNA stability at 30°C. Strains: 5K(pANN202-312R) (AI) and Hha-2(pANN202-312R) (AII). (B) The half-lives were 20.2 ± 0.8 min (3 experiments) for 5K(pANN202-312R) and 10.7 ± 1.1 min (2 experiments) for Hha-2(pANN202-312R). See legend of Figure 9 for details.

regulatory component (Vogel *et al.*, 1988). Koronakis *et al.* (1988; 1989) proposed that intra-operon transcript elongation through the *hlyA-hlyB* terminator is dependent upon the regulatory region upstream of *hlyC*.

We have studied the effect of *hlyR* on *hly* transcription expression and our results suggest that *hlyR* is a transcriptional upstream activator sequence. The data presented gives a new view about how *hlyR* works. To characterize the transcriptional effect of *hlyR* on the *hly* operon, we constructed *galK* transcriptional fusions with different segments of the *hly* upstream region. The insertion of *hlyR* upstream of the P_{hly}-*galK* fusion increased the expression by 9-fold. We have also looked directly at the *hly* mRNA levels. In our multicopy Hly$^+$ system (pSU127 and pSU129) *hlyR* was responsible for an increase of 40-fold in *hlyCA* mRNA expression. The enhanced expression of the major *hlyCA* mRNA in the presence of *hlyR* allows us to explain the high hemolytic activity found in pSU127. While we found a 40-fold increase in Hly$^+$-pUC derivative, the increase was only 5-fold in Hly$^+$-pACYC184 derivative. Thus, the effect of *hlyR* depends on plasmid copy number. The same level of activation was also found by Koronakis *et al.* (1988) in a similar intermediate copy plasmid. In addition, our results in mRNA stability suggest that the *hlyCA* mRNA accumulation is independent of its stability and, therefore, it is a direct effect of *hlyR* on *hly* transcription initiation.

hlyR is located 1.6-kb upstream of P_{hly}. A difusible product coded by *hlyR* could be related with the molecular mechanism of this activation. However, we did not observe any differences at the transcriptional level when we complemented P_{hly}-*galK* fusions with *hlyR* provided *in trans*. Vogel *et al.* (1988) has also shown the *cis*-dependent effect of *hlyR*. On the other hand, *hlyR* seems not encode a regulatory protein.

The differences in HlyA expression can be explained in our study by the sole effect of *hlyR* at the transcription initiation level, but obviously, and notwithstanding, other groups have reported its effect at the antitermination level (Koronakis *et al.*, 1988; 1989). We think that the slightly higher *hlyB* level in the presence of *hlyR* (only 4-fold of increase) could be explained by the accumulation of the *hlyCA* mRNA upstream of the *hlyA-hlyB* terminator due to an increase in mRNA synthesis.

In addition, also shown here is that the effect of *hlyR* is not that of a "simple" transcriptional upstream activator sequence. In fact, transcription from P_{hly} is strong without the intervention of *hlyR*, but is attenuated within *hlyC*. Thus, the function of *hlyR* appears to be to modulate that attenuation. We isolated a new *hly* regulatory element that we called *hlyM*. The insertion of this 220-bp sequence downstream of P_{hly} (P_{hly}-*hlyM*-*galK* fusion) in the same orientation as it has in the original operon reduces by 21-fold the transcriptional expression. The effect of *hlyM* could be mediated by an antisense RNA, but we did not find it by using primer-extension experiments (de la Cruz, per. comm.), and *hlyM* does not show promoter activity. On the other hand, the attenuation effect seems to be *cis*-dependent.

Recently, two proteins have been identified which modulate the expression of the complete *hly* operon, Hha (Nieto *et al.*, 1991) and HlyT (Bailey *et al.*, 1992). We had previously reported that the *hha* mutation resulted in an increase of the intracellular concentration of the *hlyCA* transcript (Nieto *et al.*, 1991). The *hha* locus (min 10.5) encodes an 8-kDa protein, Hha. It is possible that Hha protein is directly involved in the attenuation process. In fact, the transcriptional fusions show that in the absence of Hha (*hha⁻* strain), the transcriptional level is independent of the presence of *hlyM*. In other words, the *hha* mutation is able to overcome most of the attenuation caused by *hlyM*. We believe that there is some link, still unknown, between Hha, *hlyM* and *hlyR*.

Hypothetically, the binding of Hha to *hlyM* could decrease the *hly* expression and could have some effect on *hlyCA* mRNA stability. In fact, the decay of the *hlyCA* mRNA was significatively faster in the *hha⁻* strain than in the parental strain (from 20 min to 13 min, respectively), independent of the presence of *hlyR*. The results presented here rule out the possibility that the increase of HlyA could be due to a higher stability of the *hlyCA* transcript, and suggest that Hha protein modulates hemolysin expression at the transcriptional level. To our present knowledge the faster decay of *hlyCA* mRNA in *hha⁻* cells is difficult to interpret although it may be a consequence of the pleitropy of the *hha* mutation (Nieto *et al.*, 1987).

Hha has strong homology to the YmoA of *Yersinia enterocilitica* (Cornelis *et al.*, 1991), which plays a role in the thermoregulation of various *Y. enterocolitica* virulence genes. Our study shows that the Hha protein seems to modulate *hly* transcription expression. It is possible that both the Hha and YmoA belong to a new class of proteins which affect prokaryotic gene expression, perhaps by modifying the DNA topology (Carmona *et al.*, submitted). In addition, we know that *hha* mutation has a pleitropic effect on other operons besides *hly*. Therefore, Hha could be a general "regulator or repressor" for different systems.

In summary, our results suggest that *hlyR* enhances the *hlyCA* mRNA expression. *hlyM* is the second regulatory element in the *hly* operon. This 220-bp sequence negatively modulates the transcription initiate in P_{hly}. In *hlyR⁻* operons, *hlyM* determines a low level of α-hemolysin expression. *hly* operons carrying *hlyR* are able to achieve a high level of expression, because *hlyR* overcomes the attenuating effect of *hlyM*. Therefore, *hlyM* could be an "operator" at which a repressor could bind, and gives consequently a lower level of *hly* expression.

Recently, Bailey *et al.* (1992) have identified an *E. coli* cellular locus, *hlyT*, required for the synthesis and secretion of hemolysin encoded *in trans* by intact *hly* operons carrying *hlyR*. Mutation of the *hlyT* locus specifically reduced the level of *hlyA* transcript 20–100-fold and markedly lowered both intracellular and extracellular levels of the HlyA protein. Thus the Hha and HlyT proteins appear to be implicated in *hly* regulation and to affect transcription of different prokaryotic genes, but do not contain any of the structural motifs found in previously characterized transcriptional factors.

References

Bailey MJA, Koronakis V, Schmoll T & Hughes C (1992) *Escherichia coli* HlyT protein, a transcriptional activator of haemolysin synthesis and secretion, is encoded by the *rfaH* (*sfrB*) locus required for expression of sex factor and lipopolysaccharide genes. Mol. Microbiol. 6: 1003–1012.

Bolívar F, Rodriguez RL, Greene PJ, Betlach MC, Heyneker HL, Boyer HW, Crosa JH & Falkow S (1977) Construction and characterization of new cloning vehicles. II. A multipurpose cloning system. Gene 2: 95–113.

Casadaban M & Cohen SN (1980) Analysis of gene control signals by DNA fusion and cloning in *Escherichia coli*. J. Mol. Biol. 138: 179–208.

Cornelis GR, Sluiters C, Delor I, Geib D, Kaniga K, Lambert de Rouvroit C, Sory MP, Vanooteghem JC & Michiels T (1991) *ymoA*, a *Yersinia enterocolitica* chromosomal gene modulating the expression of virulence functions. Mol. Microbiol. 5: 1023–1034.

De Boer HA (1984) A versatile plasmid system for the study of prokaryotic transcription signals in *Escherichia coli*. Gene 30: 251–255.

De la Cruz F, Zabala JC & Ortiz JM (1980) The molecular relatedness among α-hemolytic plasmids from various incompatibility groups. Plasmid 4: 76–81.

Godessart N, Muñoa FJ, Regué M & Juárez A (1987) Chromosomal mutations that increase the production of a plasmid-encoded hemolysin in *Escherichia coli*. J. Gen. Microbiol. 134: 2779–2787.

Goebel W & Hedgpeth J (1982) Cloning and functional characterization of the plasmid-encoded hemolysin determinant of *Escherichia coli*. J. Bacteriol. 151: 1290–1298.

González-Carreró MI, Zabala JC, de la Cruz F & Ortiz JM (1985) Purification of α-hemolysin from an overproducing *E. coli* strain. Mol. Gen. Genet. 199: 106–110.

Hardie KR, Issartel JP, Koronakis E, Hughes C & Koronakis V (1991) *In vitro* activation of *Escherichia coli* prohaemolysin to the mature membrane-targetted toxin requires HlyC and a low molecular-weight cytosolic polypeptide. Mol. Microbiol. 5: 1669–1679.

Hess J, Wels W, Vogel M & Goebel W (1986) Nucleotide sequence of a plasmid-encoded hemolysin determinant and its comparison with a corresponding chromosomal hemolysin sequence. FEMS Microbiol. Lett. 34: 1–11.

Issartel JP, Koronakis V & Hughes C (1991) Activation of E. coli prohemolysin to the mature toxin by acyl carrier protein-dependent fatty acylation. Nature (London) 351: 759–761.

Juárez A & Goebel W (1984) Chromosomal mutation that affects the excretion of hemolysin in *Escherichia coli*. J. Bacteriol. 159: 1083–1085.

Koronakis V, Cross M & Hughes C (1988) Expression of the *E. coli* hemolysin secretion gene *hlyB* involves transcript anti-termination within the *hly* operon. Nucleic Acids Res. 16: 4789–4800.

Koronakis V, Cross M & Hughes C (1989) Transcription antitermination in an *Escherichia coli* haemolysin operon is directed progressively by cis-acting DNA sequences upstream of the promoter region. Mol. Microbiol. 3: 1397–1404.

Laoide BM & Ullmann A (1990) Virulence dependent and independent regulation of the *Bordetella pertussis cya* operon. EMBO J. 9: 999–1005.

Lupski JR, Altaba AR & Godson GN (1984) Promotion, termination and antitermination in the *rps-dnaG-rpoD* macromolecular synthesis operon of *E. coli* K-12. Mol. Gen. Genet. 195: 391–401.

McKenney K, Shimatake H, Court D, Schmeissner U, Brady C & Rosenberg M (1981) A system to study promoter and terminator signals recognized by *Escherichia coli* RNA polymerase. p. 383–415. *In*: Chirikjian JG & Papas T (eds.) Gene amplification and analysis. Vol. 2 Elsevier, North Holland.

Mellado RP, Delius H, Klein B & Murray K (1981) Transcription of sea urchin histone genes in *Escherichia coli*. Nucleic Acids Res. 9: 3889–3906.

Melton DA, Krieg PA, Rebagliati MR, Maniatis T, Zinn K & Green MR (1984) Efficient *in vitro* synthesis of biologically active RNA and RNA hybridization probes from plasmids containing a bacteriophage SP6 promoter. Nucleic Acids Res. 12: 7035–7056.

Müller D, Hughes C & Goebel W (1983) Relationship between plasmid and chromosomal haemolysin determinants of *Escherichia coli*. J. Bacteriol. 153: 846–851.

Nieto JM, Tomás J & Juárez A (1987) Secretion of an *Aeromonas hydrophila* aerolysin by a mutant strain of *Escherichia coli*. FEMS Microbiol. Lett. 48: 413–417.

Nieto JM, Carmona M, Bolland S, Jubete Y, de la Cruz F & Juárez A (1991) The *hha* gene modulates haemolysin expression in *Escherichia coli*. Mol. Microbiol. 5: 1285–1293.

Prentki P & Krisch HM (1984) *In vitro* insertional mutagenesis with a selectable DNA fragment. Gene 29: 303–331.

Rosenberg M, Chepelinsky AB & McKenney K (1983) Studying promoters and terminators by gene fusion. Science 222: 734–739.

Spitznnagel J, Kraig E & Kolodrubetz D (1991) Regulation of leukotoxin in leukotoxic and non-leukotoxic strains of *Actinobacillus actinomycetemcomitans*. Infect. Immun. 59: 1394–1401.

Strathdee CA & Lo RYC (1989) Regulation of expression of the *Pasteurella haemolytica* leukotoxin determinant. J. Bacteriol. 171: 5955–5962.

Vieira J & Messing J (1982) The pUC plasmids, an M13mp7-derived system for insertion mutagenesis and sequencing with synthetic universal primers. Gene 19: 259–268.

Vogel M, Hess J, Then I, Juárez A & Goebel W (1988) Characterization of a sequence (*hlyR*) which enhances synthesis and secretion of hemolysin in *Escherichia coli*. Mol. Gen. Genet. 212: 76–84.

Welch RA & Pellett S (1988) Transcriptional organization of the *Escherichia coli* haemolysin genes. J. Bacteriol. 170: 1622–1630.

Welch RA (1991) Pore-forming cytolysins of Gram-negative bacteria. Mol. Microbiol. 5: 521–528.

Zabala JC, García-Lobo JM, Díaz-Aroca E, de la Cruz F & Ortiz JM (1984) *Escherichia coli* α-hemolysin synthesis and export genes are flanked by direct repetition of IS91-like elements. Mol. Gen. Genet. 197: 90–97.

27. The role of the *syrBCD* gene cluster in the
biosynthesis and secretion of Syringomycin by
Pseudomonas syringae pv. syringae

NEIL B. QUIGLEY and DENNIS C. GROSS

Abstract. Syringomycin is a lipopeptide toxin related structurally to polypeptin antibiotics, and its production is a key determinant of virulence in diseases caused by the plant pathogen *Pseudomonas syringae* pv. syringae. The *syrBCD* gene cluster of strain B301D is required for syringomycin biosynthesis, and two of these genes, *syrB* and *syrC*, are induced synergistically in the presence of specific phenolic glucoside and sugar signals of plant origin. New evidence indicates that the majority of natural strains can be induced to produce syringomycin in a defined medium by addition of arbutin and fructose. Derivatives of strain B301D carrying *syrD* mutations are severely attenuated for virulence, exhibit reduced expression of the *syrB* gene, and are deficient in five large proteins thought to compose the syringomycin biosynthetic complex. DNA homologous to the *syrB* and *syrD* genes was detected in pathogenic natural strains of *P. syringae* pv. syringae but not in other pathovars, indicating that these genes may be used as pathovar-specific gene probes. The *syrD* gene product, SyrD, has strong homology with the ATP-binding cassette transporter proteins, and is proposed to be the central component of a secretion mechanism which is specific for syringomycin and which provides autoresistance to the toxin. This is the first indication that an ATP-binding cassette transporter is involved in secretion of a lipopeptide antibiotic, or of a bacterial phytotoxin.

Abbreviations: ATP-Binding Cassette (ABC)

Introduction

Phytotoxin production has long been known to be an important element of virulence for many plant pathogens. Among phytopathogenic bacteria, the secretion of toxins appears to be a major weapon in the arsenal of virulence factors produced by pathovars of *P. syringae* (Gross, 1991). Two categories of phytotoxins have been identified, namely the chlorosis-inducing toxins produced by several pathovars of *P. syringae* and the necrosis-inducing toxins produced by *P. syringae* pv. syringae. Although the toxins are usually peptide in nature, structural diversity is evident among the toxins that reflects their distinct modes of action, and suggests that they are synthesized by unique biosynthetic pathways. Because nonribosomal multistep pathways appear to be involved in their biosynthesis (Gross, 1991), the underlying genetic organization is likely to be complex. Unfortunately, little is known about the genetic determinants of phytotoxin biosynthesis and the regulatory network linking it to expression of plant pathogenicity. The recent discoveries of gene clusters in *P. syringae* that constitute the core genetic information for toxigenesis offers distinct

C.I. Kado and J.H. Crosa (eds.), Molecular Mechanisms of Bacterial Virulence, 399–414.
© 1994 *Kluwer Academic Publishers.*

opportunities for methodically resolving the specific biosynthetic steps (Kinscherf *et al.*, 1991; Ma *et al.*, 1991; Morgan & Chatterjee, 1988; Peet *et al.*, 1986). Further observations that at least some toxin genes are activated in response to specific plant signals (Ma *et al.*, 1991; Mo & Gross, 1991b) suggests that their is a complex genetic network responsible for the perception and transduction of signals to the transcriptional apparatus of genes involved in virulence.

We are studying the genetic determinants of syringomycin production, the lipodepsipeptide phytotoxin produced by *P. s.* syringae that causes necrotic symptoms in infected plant tissues. The importance of these studies is emphasized by the fact that *P. s.* syringae causes serious diseases of major crop species, as exemplified by bacterial canker of stone fruit trees. Furthermore, *P. s.* syringae is prevalent in nature existing as either a resident epiphyte or as a pathogen that exhibits little restriction in host range. Recent studies on the regulatory and genetic factors that affect toxigenesis in *P. s.* syringae indicated that syringomycin production is a complex process that is highly responsive to environmental and host signals (Mo & Gross, 1991a & 1991b). Three key elements of the host-pathogen interaction that cooperate in syringomycin production are (i) the perception of specific plant signals by *P. s.* syringae, (ii) the induction of *syr* gene expression, and (iii) the secretion of syringomycin. Our present work is directed toward a better understanding of the environmental and genetic factors that affect these three aspects of syringomycin production in the context of a dynamic interaction between virulent *P. s.* syringae strains and their hosts.

Syringomycin is a potent phytotoxin

The plasma membrane of plants is the primary site of action for syringomycin (Takemoto, 1992), reflecting the lipophilic nature of the toxin that permits it to insert within lipid membrane layers. At low toxin concentrations, an assortment of effects on the plasmalemma have been described by Takemoto and associates (Takemoto, 1992; Takemoto *et al.*, 1991), including hyperpolarization, rapid efflux of K^+, net influx of Ca^{2+}, and stimulation of a proton pump ATPase. Consequently, syringomycin is functionally similar to several membrane-active peptide antibiotics, such as the polymyxins and gramicidin S, which also have been found to interfere with energy metabolism (Betina, 1983; Kleinkauf & von Döhren, 1983). Takemoto (1992) recently proposed a model linking the various effects of the toxin on ion transport and electrical potential in plant plasma membranes. The opening of Ca^{2+} channels that leads to higher cytoplasmic levels of the ion was proposed to be the primary effect of syringomycin. This would lead to a cascade of events including the activation of a calcium-dependent protein kinase that in turn activates by phosphorylation the H^+ pumps and K^+ channels. Such a chain of events would lead to a rapid death of cells due to the drastic disruption of the ionic balance of membranes.

Syringomycin is a lipopeptide phytotoxin

The resolution of the syringomycin structure is an important landmark because the structure suggests a biosynthetic mechanism that can be used to guide genetic studies of syringomycin production by *P. s.* syringae. Independent studies by Ballio *et al.* (1990) and Fukuchi *et al.* (1990b & 1992b) confirmed that syringomycin is a cyclic lipodepsinonapeptide composed of a hydroxylated fatty acid linked to a 9-member peptidolactone ring (Fig. 1). Such a structure places syringomycin in a small class of antibiotics called the polypeptins (Sogn, 1976). In *P. s.* syringae, three types of syringomycin-like structures have been identified that are delineated by their amino acid composition (Fig. 1). Although syringomycin itself is the toxin type produced by most strains of the bacterium, syringotoxin is produced by citrus strains (Ballio *et al.*, 1990; Fukuchi *et al.*, 1990a) and syringostatin is produced by a lilac strain from Japan (Fukuchi *et al.*, 1992a). A distinctive structural feature of all three toxin types is the presence of a trio of uncommon amino acids, dehydrothreonine, β-hydroxyaspartic acid, and 4-chlorothreonine (Fig. 1). The N-terminal serine residue is most commonly acylated with either a 12- or 14-member 3-hydroxy fatty acid. The occurrence of two of the nine amino acids in the D-isomeric form (i.e., serine and 2,4-diaminobutyric acid in syringomycin, and 2,4-diaminobutyric acid and homoserine in syringotoxin and syringostatin) is likely to contribute to toxin activity and stability (Kleinkauf & von Döhren, 1987).

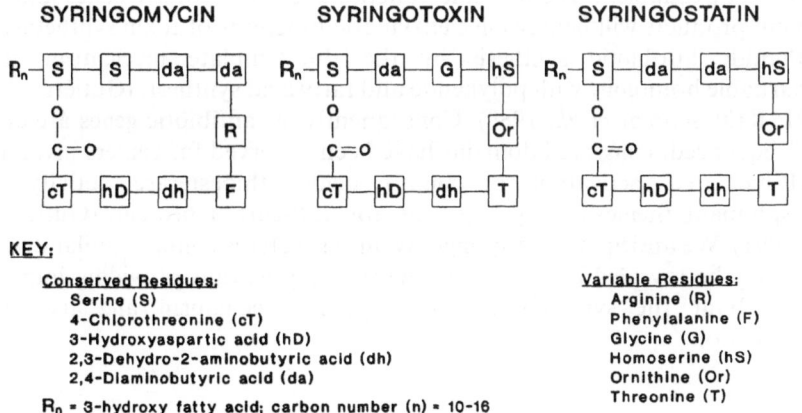

Figure 1. Diagrammatic comparison of the structures of the syringomycin phytotoxin family. Amino acids are depicted by labeled boxes (defined in the key).

Syringomycin synthesis is regulated similarly to many other bacterial antibiotics (Gross, 1985). The biosynthetic route is unknown, but it is probably similar to that of various peptide antibiotics made by multifunctional peptide synthetases (Demain *et al.*, 1983; Kleinkauf & von Döhren, 1982, 1987; Martín & Liras, 1989). The well-characterized peptide synthetases contain multifunctional polypeptide chains between 100 and 500 kDa, and serve as templates for the sequential addition of amino acids to peptides. Amino acid activation, enzyme aminoacylation, amino acid racemization and peptide bond formation are involved in synthesis (Gutiérrez *et al.*, 1991; Krätzschmar *et al.*, 1989). Peptides are transported within the system by 4'-phosphopantetheine, a cofactor that is covalently linked to the multienzyme. For example, the first step in the biosynthesis of the lipopeptide antibiotic, polymyxin, is activation and binding of the N-terminal amino acid, 2,4-diaminobutyric acid, to a 300-kDa enzyme (Komura & Kurahashi, 1980). The activated amino acid is then acylated by acyltransferase and octanoyl coenzyme A before peptide formation and cyclization.

Associated with syringomycin and syringotoxin production is the formation of large proteins up to ~ 470 kDa that are hypothesized to function as toxin synthetases (Xu & Gross, 1988b; Mo & Gross, 1991a). Both physiological and genetic evidence supports such a synthetase model. Indeed, the physiological conditions required for formation of the proteins correspond with those necessary for toxin production. The positive regulation by iron of both toxin production and formation of the large proteins exemplifies the strong association of the putative synthetases with toxigenesis (Xu & Gross, 1988b). Furthermore, deficiencies in one or more protein species or the formation of truncated protein derivatives are observed for most nontoxigenic mutants (Morgan and Chatterjee, 1988; Quigley *et al.*, 1993; Xu & Gross, 1988b). Nevertheless, enzymatic evidence that the large proteins function as synthetases is not yet available. Structural and functional analysis of *syr* genes and their protein products will help resolve enzymatic functions of toxin synthetases.

Peptide antibiotic synthesis by the thioltemplate mechanism shares remarkable homology with polyketide and fatty acid synthesis (Gutiérrez *et al.*, 1991; Krätzschmar *et al.*, 1989). Consequently, as antibiotic genes are cloned and sequenced, conserved domains have been observed for centers involved in ATP-mediated activation of amino acids, thioesterase activity, and phosphopantetheine-binding sequences for acyl-carrier proteins (Gutiérrez *et al.*, 1991). We anticipate that syringomycin synthetases contain similar domains that may be revealed by protein sequence analysis, thus providing important clues about the syringomycin biosynthetic mechanism and its genetic organization.

Characterization of the *syr* gene cluster

The structural genes coding for antibiotic synthetases are commonly organized in clusters located on the chromosome (Gross, 1991; Martín & Liras, 1989). The syringomycin biosynthesis genes also appear to be closely linked on the *P. s.* syringae chromosome. A cluster of three (*syr*) genes that are absolutely required for syringomycin production, but which are not required for prototropy, were identified on a cosmid clone that complemented a specific *syr* mutation (Mo & Gross, 1991a; Quigley *et al.*, 1993). These genes, *syrB*, *syrC*, and *syrD*, were mapped in three steps. First, the complementing cosmid was mutagenized by random Tn3HoHo1 insertions (Stachel *et al.*, 1985a). These mutations were then introduced into the parental strain genome by conjugation and marker exchange. Finally, insertions in the cosmid that disrupted a *syr* gene were identified by evaluating the recombined derivatives for toxigenicity using the sensitive indicator fungus, *Geotrichum candidum*. The approximate sizes and relative positions of these *syr* genes were determined by restriction mapping of *syr*::Tn3HoHo1 insertions (Quigley *et al.*, 1993) (Fig. 2). The relative transcriptional orientation of the *syr* genes was determined using Tn3HoHo1 insertions 132 (*syrB*), 334 (*syrC*), and 105 (*syrD*) that positioned the *lacZ* gene of the transposon in-frame with each *syr* gene (Fig. 2). These experiments established that at least three *syr* genes were clustered on the *P. s.* syringae chromosome, and that *syrD* was transcribed divergently from the *syrBC* gene pair. The proximity of *syrC* to *syrB* suggested that these genes may be co-regulated. Preliminary investigations examining the expression of plasmid-borne *lacZ* reporter gene fusions, 132 and 334, in *syrB* or *syrC* mutant strains indicated that efficient expression of *syrC* required a functional *syrB* gene (Mo & Gross, unpublished).

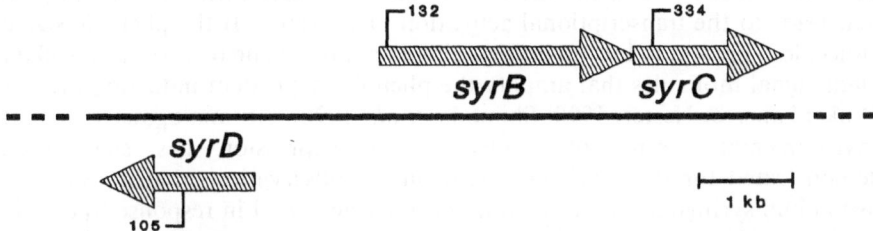

Figure 2. The *syrBCD* cluster. A ~ 7-kb segment of the *P. s.* syringae strain B301D chromosome is represented by a solid line with broken ends. The relative sizes, positions, and transcriptional orientations of the *syrB*, *syrC*, and *syrD* genes are indicated by shaded arrows. The position of a key in-frame *lacZ* fusion (generated by insertion of Tn3HoHo1) in each gene is indicated by a numbered vertical stick.

The Phenotypes of the *syrB*, *syrC*, and *syrD* Mutations

Mutations in the *syrB*, *syrC*, or *syrD* loci all affect syringomycin production. The presence of specific high molecular weight proteins in cell lysates is correlated with production of both syringomycin (Mo & Gross, 1991a; Xu & Gross, 1988b) and syringotoxin (Morgan & Chatterjee, 1988). Up to five large proteins (~ 130 to ~ 470 kDa) are associated with syringomycin production (Xu & Gross, 1988b). Mutations in the *syrB* and *syrC* genes affect the formation of only two of these syringomycin-associated proteins, whereas, *syrD* mutants are deficient in all five proteins (Quigley *et al.*, 1993). Mutations in these three *syr* genes were characterized further by comparing the relative virulence of the corresponding mutant strains and of the wild type parental strain, B301D, using a cherry fruit pathogenicity test (Mo & Gross, 1991a). This comparison revealed significant differences in virulence between the *syrD* mutants and the *syrB* and *syrC* mutants. Although *syrB* and *syrC* mutants were both about 65% as virulent as strain B301D, *syrD* mutants were only about 30% as virulent as the parental strain (Quigley *et al.*, 1993). The protein profile and virulence data suggested that the *syrB* and *syrC* genes have structural roles in toxin biosynthesis and that *syrD* has a regulatory or other nonbiosynthetic role in syringomycin production.

Activation of the *syrB* gene and syringomycin production by plant signals

In studies of host-pathogen interactions, an emerging principle is that specific environmental stimuli co-induce or co-repress virulence operons organized to form a stimulon. This was first exemplified in phytobacteriology by the discovery that specific plant phenolic compounds serve as regulators of virulence (*vir*) genes of *Agrobacterium tumefaciens* (Stachel *et al.*, 1985b). The prototype plant signal was acetosyringone, a phenolic perceived by the bacterium in the wound environment to initiate a signal transduction process that leads to the transcriptional activation of *vir* genes. If the phenolic signal molecule is present in low quantities, simple pyranose sugars serve as ancillary plant signal molecules that amplify the phenolic-dependent induction process (Ankenbauer & Nester, 1990; Shimoda *et al.*, 1990). An analogous system of environmental control of virulence gene expression has since been demonstrated for *P. s.* syringae based on the observations that a *syrB-lacZ* fusion and syringomycin production itself are activated in response to certain phenolic and sugar signals of plant origin (Mo & Gross, 1991b). The discovery that a sensory mechanism is linked to virulence in phytobacteria is not surprising because it ensures a swift response to a dynamic plant environment governing disease development.

The original evidence suggestive that plant signal molecules activated genes required for syringomycin production by *P. s.* syringae resulted from the induction of a *syrB-lacZ* fusion to produce significant β-galactosidase activity both in immature cherry fruits and in a defined medium supplemented with a crude aqueous extract from cherry leaves (Mo & Gross, 1991a). Subsequently, it was shown that certain phenolic glucosides served as the primary signal and that their activity was enhanced by the presence of specific mono- and disaccharides (Mo & Gross, 1991b). Of the phenolics tested, only arbutin, phenyl-β-D-glucoside, and salicin had strong *syrB*-inducing activity ($\sim$ 1,200 U); esculin and helicin had moderate activity (250–400 U). The intact glucosidic linkage was necessary for activity since the aglycone portions of these inducers were inactive. Representative flavones and flavonones that induce *nod* genes in some strains of *Rhizobium* (Peters & Long, 1988), and acetosyringone that induces *vir* genes in *Agrobacterium* (Stachel *et al.*, 1985b) were inactive, suggesting that *P. s.* syringae responded to a different type of phenolic signal molecule. Sucrose and D-fructose were the most active saccharide signal molecules based on a 5-fold amplification of arbutin-mediated *syrB* induction. Many plant species parasitized by *P. s.* syringae contain one or more of the active phenolic β-glucosides in their leaves, bark and flowers (Mo & Gross, 1991b; Paris N, 1963). Indeed, pear (*Pyrus communis*) leaves accumulate arbutin at concentrations equivalent to ~ 150 mM (Miller, 1973a), which is in excess of the amount (i.e., ≥ 10 μM) needed for maximum *syrB*-inducing activity (Mo & Gross, 1991b). Because sucrose and D-fructose are among the principal nonstructural carbohydrates found in plants (Miller, 1973b), the intensity of the induction would be amplified even when levels of the phenolic signal molecule are low.

The strongest evidence for a link between plant signal induction and virulence in *P. s.* syringae is based on induction of the whole syringomycin biosynthetic pathway (Mo & Gross, 1991b). For example, strain B3A-R does not produce the toxin in syringomycin minimal (SRM) medium, but over 250 U of syringomycin are produced if SRM is supplemented with both arbutin and D-fructose. In recent surveys of wild-type strains, over 80% of strains that produced either syringomycin, syringotoxin, or syringostatin were induced by plant signals to produce significantly higher quantities of toxin *in vitro* (Quigley & Gross, 1993). Therefore, the discovery that plant signals activate both the *syrB-lacZ* fusion and syringomycin production documents the concept that processes required for virulence in phytobacteria other than *A. tumefaciens* are responsive to environmental stimuli.

The syrD protein is homologous with the ABC family of secretion proteins

The nucleotide sequence of the *syrD* gene was determined and used to predict the sequence of its protein product, SyrD. A search of the Swiss-Prot protein sequence database (Bairoch & Boeckmann, 1991) showed that SyrD had strong homology with all known ATP-binding cassette (ABC) transporter proteins. These are a family of membrane-associated proteins found in a wide variety of eukaryotes and prokaryotes, and are hypothesized to perform the translocation step in the uptake of essential nutrients or in the secretion of toxins and other compounds (Blight & Holland, 1990). Transport systems for nutrient uptake that involve an ABC protein, are composed of separate proteins for ATP hydrolysis and translocation, whereas these activities both reside in a single bifunctional polypeptide in most ABC systems for the secretion of various target molecules (Higgins *et al.*, 1990). In the secretion group, the characteristic conserved sequences common to all ABC transporter proteins are located in a hydrophilic C-terminal domain. These conserved amino acid residues are thought to form a 'pocket' for ATP binding and hydrolysis; ATP hydrolysis probably creates a conformational change which is transmitted to the hydrophobic transmembrane domain to effect translocation of the target molecule (Higgins *et al.*, 1990). Nutrient uptake systems appear to transmit this message through a close physical association between the two separate proteins (Higgins *et al.*, 1990).

The SyrD protein corresponds in size to, and shows the greatest homology with, the secretion subfamily of ABC transporters (Fig. 3) (Quigley *et al.*, 1993). This specialized subfamily is represented by HlyB, the hemolysin secretion protein of *E. coli*. In bacteria, many of these ABC proteins are involved in the export of cytolytic or proteolytic proteins that enhance pathogenicity. Some of these targets include colicin V of *E. coli* (Fath *et al.*, 1991; Gilson *et al.*, 1990), β-1,2-glucans of *A. tumefaciens* (Cangelosi *et al.*, 1989), and capsule-polysaccharide of *Haemophilus influenzae* (Kroll *et al.*, 1990). The alkaline protease of *P. aeruginosa* is another virulence determinant (Howe & Iglewski, 1984) that is secreted by an ABC transporter protein (Guzzo *et al.*, 1991); this is the only previously reported example of an ABC protein from a pseudomonad. The mammalian Mdr proteins function in multiple drug resistance by re-exporting a wide range of therapeutic drugs which enter the cells by diffusion (Kane *et al.*, 1990). In contrast to genes encoding prokaryote ABC transporters, the *mdr* genes are twice as long as the *hlyB*-like genes (including *syrD*), but appear to be composed of a simple in-frame tandem duplication of an original progenitor gene. The protein products of these genes are covalently linked dimers (Kuchler *et al.*, 1989), an observation which strongly supports the hypothesis that the membrane domains of all ABC transporter proteins function as dimers.

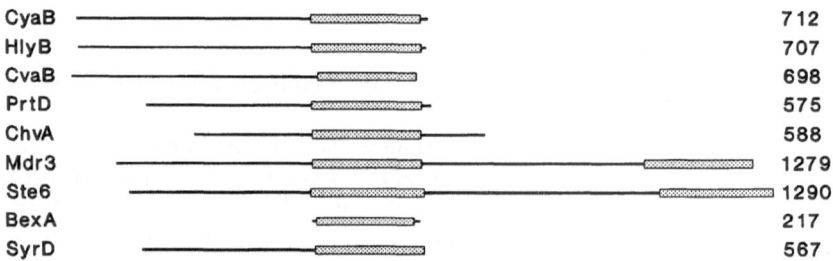

Figure 3. Alignment of SyrD with other ATP-binding cassette transporter proteins. Linear protein sequences (solid lines, N-termini at the left) are aligned via their conserved ATP-binding domains (shaded boxes). The C-terminal duplications of these domains in proteins Mdr3 and Ste6 are less conserved than the corresponding internal domains. The number of residues in each protein is indicated to the right. CyaB (Glaser *et al.*, 1988), HlyB (Felmlee *et al.*, 1985), CvaB (Gilson *et al.*, 1990), PrtD (Létoffé *et al.*, 1990), ChvA (Cangelosi *et al.*, 1989), Mdr3 (Lincke *et al.*, 1991), Ste6 (Kuchler *et al.*, 1989), BexA (Kroll *et al.*, 1990).

In addition to the sequence homologies between SyrD and the ABC proteins, there is other data to support the hypothesis that SyrD functions as an ABC transporter for syringomycin secretion. The hydropathy profiles of the N-terminal domains of bacterial ABC transporters predict that each carries three or four pairs of membrane-spanning hydrophobic α-helices, separated by hydrophilic loops which lie on the cytoplasmic or periplasmic sides of the inner membrane (Blight & Holland, 1990; Delepelaire & Wandersman, 1991). It is thought that target recognition and binding occur at the cytoplasmic loops and that the periplasmic loops participate in the translocation process. SyrD exhibits a similar arrangement and spacing of potential hydrophobic α-helices and hydrophilic domains (Fig. 4). Consistent with the hypothesis that the N-terminal domains of ABC transporter proteins define target specificity, no significant homology exists between the primary sequences of any of these domains, including that of SyrD. Also consistent with the concept that SyrD is an ABC transporter with two functionally different but interacting domains is the following observation. All Tn3HoHo1 insertions in the *syrB* or *syrC* loci abolished syringomycin production, as did an insertion immediately downstream of the 5'-end of the *syrD* gene. In contrast, an insertion in *syrD* immediately downstream of sequences encoding the proposed membrane-spanning translocation domain still resulted in a low but significant amount of syringomycin production. The differential effects on syringomycin production of these two *syrD* mutations could be interpreted by proposing that ATP-binding by the C-terminal domain of SyrD greatly enhances the translocating activity of the N-terminal domain, but that the latter domain can secrete a small amount of syringomycin independent of this effect.

Proteins which are exported by an ABC protein can be distinguished from those exported by the Sec-dependent mechanism by the absence of a leader

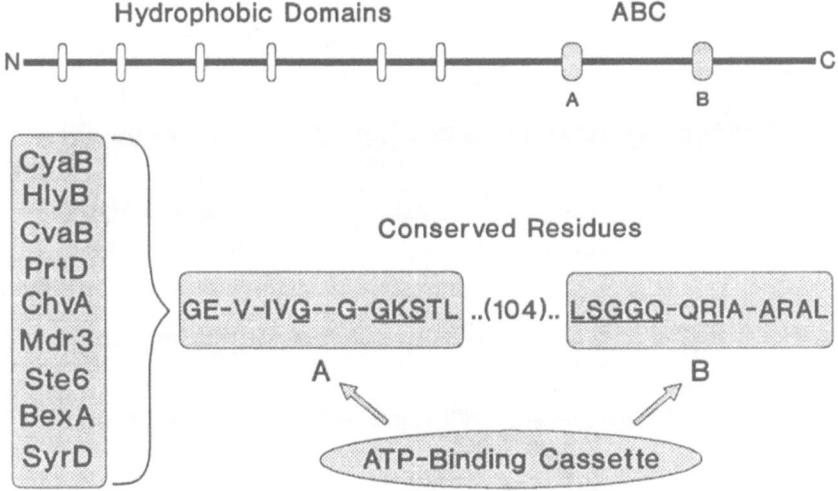

Figure 4. Domain structure and conserved residues of SyrD and other ATP-binding cassette transporter proteins. Top: linear view of SyrD (N- and C-termini indicated) showing the relative positions of the hydrophobic transmembrane domains (open symbols) and the A and B domains of the ATP-binding domain (shaded symbols). Bottom: SyrD residues in the A and B ATP-binding domains that are either absolutely (underlined) or highly conserved among the set of ABC proteins involved in secretion (including those bracketed at the left). The A and B domains are 104 residues apart in SyrD. The positions of variable residues within these domains are indicated by hyphens.

408

peptide (Tai, 1990). In contrast to the Sec-dependent protein export system, the ABC transporter systems are dedicated essentially to the secretion of only one target. This extraordinary specificity suggests that there is tight regulation of intracellular levels of each of these targets. Indeed, syringomycin, which is extremely toxic to producer cells, does not accumulate inside cells of *syrD* mutants (that are hypothesized to be unable to secrete the toxin) (Quigley *et al.*, 1993). Furthermore, we have shown that a *syrD* mutation suppressed the expression of a *syrB-lacZ* reporter fusion by over 50% (Quigley *et al.*, 1993). In view of the autotoxicity of syringomycin, we expect that regulatory mechanisms exist in toxigenic *P. s.* syringae strains to limit the intracellular accumulation of syringomycin following overstimulation of the biosynthetic pathway or when secretion is blocked.

A model for syringomycin secretion by SyrD

The presence of the conserved C-terminal ATP-binding site, and the strong structural similarity between the N-termini of SyrD and the ABC secretion proteins, suggests a model for the possible role of SyrD in syringomycin export (Fig. 5). This model is based primarily on research concerning the mechanism of

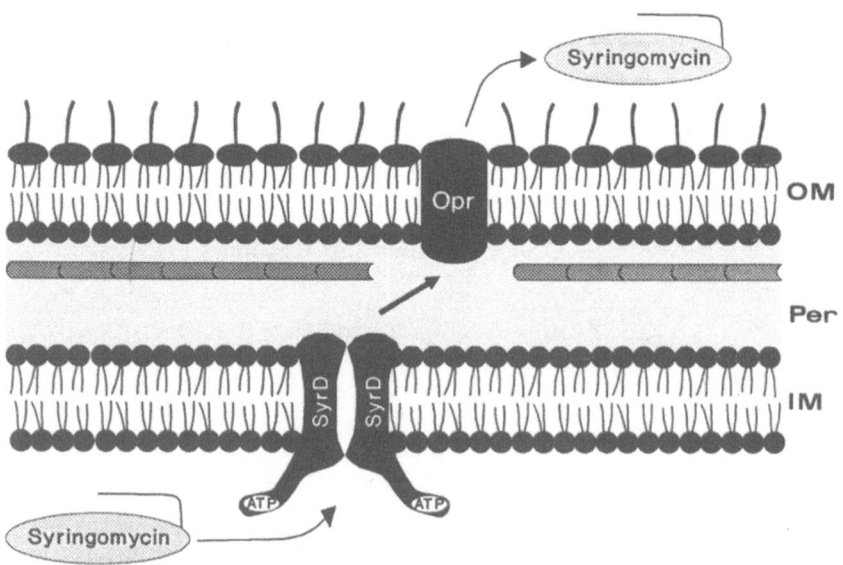

Figure 5. Model for syringomycin secretion. It is proposed that a SyrD dimer is inserted in the inner membrane (IM) by its hydrophobic N-terminal domains, and functions to export syringomycin to the periplasm (Per). Energy for export comes from ATP, which is hydrolyzed by the hydrophilic C-terminal domains of the dimer in the cytoplasm. A TolC-like outer membrane protein (Opr) is proposed to facilitate the passage of syringomycin across the OM and its release from the cell.

hemolysin (HlyA) export in *E. coli* (Blight & Holland, 1990). In this model, the hydrophilic C-terminal half of SyrD is a cytoplasmic domain that binds and hydrolyzes ATP. The largely hydrophobic N-terminal half of SyrD is an inner membrane protein that selectively binds syringomycin and translocates it to the periplasmic space with energy supplied from ATP hydrolysis. By analogy with the Mdr proteins of eukaryotes (Kane *et al.*, 1990; Lincke *et al.*, 1991), it is proposed that syringomycin is translocated to the periplasm by a SyrD dimer. Finally, an outer membrane protein (Opr), analogous to TolC in *E. coli*, facilitates the passage of syringomycin across the outer membrane and its release from the cell. HlyA is secreted by a mechanism involving HlyB and requiring TolC (Wandersman & Delepelaire, 1990). Additionally, HlyA secretion uses a third membrane protein, HlyD, which is thought to lend specificity to the process (Wang *et al.*, 1991). However, a requirement for an HlyD homolog appears to exist only for the export of large protein targets. Multiple drug resistance in eukaryotes (via the Mdr translocators) and secretion of β-1,2-glucan by *R. meliloti* and *A. tumefaciens* (via NdvA and ChvA, respectively) appear not to employ an HlyD homolog (Blight & Holland, 1990); we anticipate that syringomycin secretion does not require an HlyD analog.

The *syrB* and *syrD* genes are conserved among pathogenic strains of *P. s. syringae*

Syringomycin production is a key virulence determinant in *P. s.* syringae but is not required absolutely for pathogenicity (Xu & Gross, 1988a). To determine how closely associated the presence of the *syrB* and *syrD* genes are with pathogenicity, a collection of 47 natural strains that were originally designated as *P. syringae* were examined for pathogenicity, toxigenicity, and the presence of sequences homologous to these genes (Quigley & Gross, 1993). Of this collection, 42 strains were toxigenic and 33 were both toxigenic and pathogenic in cherry fruits (Table 1). All 42 toxigenic strains were found to carry DNA homologous to both probes, although restriction site polymorphisms were evident at the *syrB* and *syrD* loci in these *syr* gene-positive strains. A collection of 13 *P. syringae* strains of other pathovars were included in the probing experiment, including strains producing coronatine, phaseolotoxin, tabtoxin and tagetitoxin. None of these strains carried DNA that was homologous with either probe (Table 1). Together, these results demonstrate that the *syrB* and *syrD* genes are unique to pathovar syringae strains, and that syringomycin production is an important factor in the interaction between these pathogens and their hosts.

Table 1. Identification of *syrB* and *syrD* homologs in *P. syringae* pathovars.

P. syringae pathovar	Number of strains	Pathogenic to cherry[1]	Phytotoxin produced	Homology to *syrB* and *syrD*
syringae	33	Yes	Syringomycin	Yes
morsprunorum	2	Yes	Coronatine	No
tomato	2	No	Coronatine	No
phaseolicola	3	No	Phaseolotoxin	No
tagetis	2	No	Tagetitoxin	No
tabaci	2	No	Tabtoxin	No
pisi	2	No	Unknown	No

[1] Determined using immature cherry (*Prunus avium* L. cv. Bing) fruits.

Summary

Syringomycin production by *P. s.* syringae provides a system with which to study several aspects of the plant-pathogen interaction. Syringomycin is a major virulence determinant in diseases caused by *P. s.* syringae and its expression is stimulated by the presence of specific plant signals in the medium. Although a precise role for syringomycin production in pathogenicity has not been assigned, it is known to be required for the development of full disease symptoms (Xu & Gross, 1988a), and its phytotoxic effects have been well-documented (Takemoto, 1992).

As summarized in Fig. 6, our work has focused on three aspects of syringomycin production. We propose that certain *syr* genes that encode elements of the syringomycin biosynthetic complex are inducible by specific phenolic signals of plant origin, and that this effect is enhanced by specific plant sugars. Prior work demonstrated that both the *syrB* gene and syringomycin production are inducible by certain phenolic glucosides, such as arbutin, and by a few saccharides, such as fructose (Mo & Gross, 1991b; Fig. 6). *P. s.* syringae responds to these signals by a mechanism that is presently unknown, but is possibly of the two-component regulator type controlling *vir* gene expression in *A. tumefaciens* (Winans *et al.*, 1988). In contrast to the *syrB* and *syrC* genes that may have roles in syringomycin biosynthesis, *syrD* encodes a protein proposed to function in syringomycin secretion. The strong homology between SyrD and the ABC transporter proteins, together with its predicted structural similarities to these export proteins, supports this hypothesis.

Mutants of *P. s.* syringae that do not produce syringomycin are still pathogenic, but exhibit significantly reduced virulence. Some pathogenic *P. s.* syringae strains were shown recently to produce, in addition to syringomycin, another secondary metabolite with phytotoxic properties, called syringopeptin (Ballio *et al.*, 1991). Syringomycin and syringopeptin are both lipopeptides and

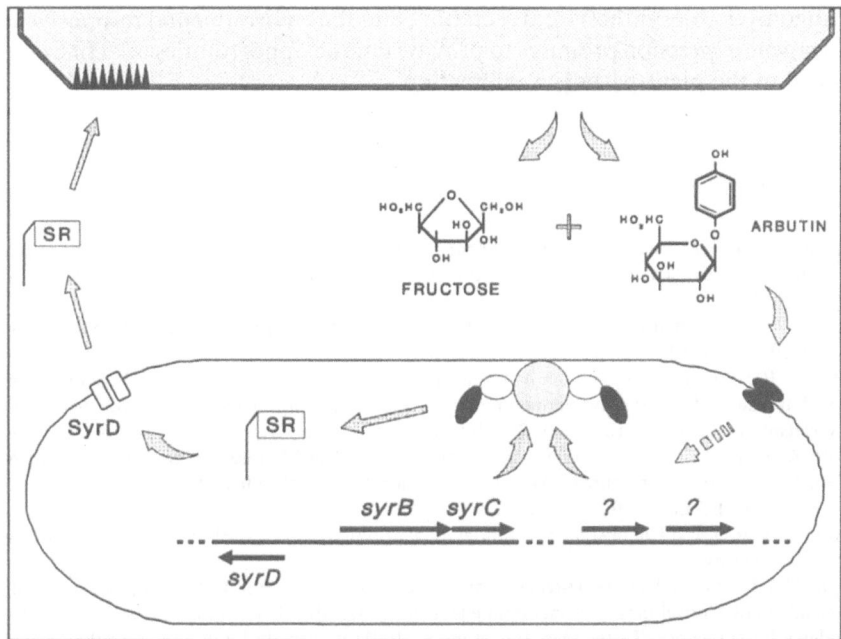

Figure 6. Diagrammatic model for plant signal-mediated induction of syringomycin production by *P. s.* syringae. A stylized bacterial cell (oval) is shown carrying a multienzyme complex for syringomycin biosynthesis (at top). A plant cell (shaded polygon at top) is shown releasing two classes of specific *syr* gene inducer molecules, phenyl glycosides and simple sugars (represented by arbutin and fructose, respectively). The requirement for a membrane-associated signal transduction apparatus (paired darkly-shaded ovals at right) remains to be established. The *syrB* and *syrC* genes are shown along with other hypothetical structural genes for syringomycin biosynthesis on the *P. s.* syringae chromosome (discontinuous solid line). The *syrD* gene product, SyrD, is proposed to function as a dimer in the specific secretion of syringomycin (tailed box labeled 'SR'). Free syringomycin causes plant cell necrosis (connected triangles) via effects on the plasmalemma (see text).

may have cooperative modes of action in affecting the plasmalemma of host cells (Iacobellis *et al.*, 1992). In fact, the residual necrosis caused by the nontoxigenic *syr* mutants may be due to the phytotoxic activity of syringopeptin. Consequently, an important goal is to understand the relative roles of syringomycin and syringopeptin in virulence, to determine if the biosynthesis of both toxins is coordinately regulated, and to determine if they are secreted by the same mechanism.

Future studies on the mechanism and specificity of syringomycin secretion should provide valuable information about the secretion of other lipopeptide antibiotics, including those of medical importance. Furthermore, there is growing evidence from *P. syringae* (Iacobellis *et al.*, 1992) and *P. tolaasii* (Brodey *et al.*, 1991) for an important role of lipopeptide phytotoxins as virulence factors in fluorescent pseudomonads. Further genetic analysis of syringomycin biosynthesis and secretion, and the environmental responsiveness of *syr* gene expression promises to provide unique opportunities to define events critical to the plant-pathogen interaction.

References

Ankenbauer RG & Nester EW (1990) Sugar-mediated induction of *Agrobacterium tumefaciens* virulence genes: structural specificity and activities of monosaccharides. J. Bacteriol. 172: 6442–6446.

Bairoch A & Boeckmann B (1991) The SWISS-PROT protein sequence data bank. Nucleic Acids Res. 19: 2247–2249.

Ballio A, Barra D, Bossa F, Collina A, Grgurina I, Marino G, Moeti G, Paci M, Pucci P, Segre A & Simmaco M (1991) Syringopeptins, new phytotoxic lipodepsipeptides of *Pseudomonas syringae* pv. *syringae*. FEBS Lett. 291: 109–112.

Ballio A, Bossa F, Collina A, Gallo M, Iacobellis NS, Paci M, Pucci P, Scaloni A, Segre A & Simmaco M (1990) Structure of syringotoxin, a bioactive metabolite of *Pseudomonas syringae* pv. *syringae*. FEBS Lett. 269: 377–380.

Betina V. (1983) The Chemistry and Biology of Antibiotics, Vol. 5. Elsevier Sci. Pub. Co., New York (590 pp).

Blight MA & Holland IB (1990) Structure and function of haemolysin B, P-glycoprotein and other members of a novel family of membrane translocators. Mol. Microbiol. 4: 873–880.

Brodey CL, Rainey PB, Tester M & Johnstone K (1991) Bacterial blotch disease of the cultivated mushroom is caused by an ion channel forming lipodepsipeptide toxin. Mol. Plant-Microbe Interact. 4: 407–411.

Cangelosi GA, Martinetti G, Leigh JA, Lee CC, Theines C & Nester EW (1989) Role of *Agrobacterium tumefaciens* ChvA protein in export of β-1,2-glucan. J. Bacteriol. 171: 1609–1615.

Delepelaire P & Wandersman C (1991) Characterization, localization and transmembrane organization of the three proteins PrtD, PrtE and PrtF necessary for protease secretion by the Gram-negative bacterium *Erwinia chrysanthemi*. Mol. Microbiol. 5: 2427–2434.

Demain AL, Aharonowitz Y & Martin JF (1983) Metabolic control of secondary biosynthetic pathways. p. 49–72. *In*: Vining LC (ed.) Biochemistry and Genetic Regulation of Commercially Important Antibiotics. Addison-Wesley Pub. Co., Reading, MA.

Fath MJ, Skvirsky RC & Kolter R (1991) Functional complementation between MDR-like export systems: Colicin V, alpha-hemolysin, and *Erwinia* protease. J. Bacteriol. 173: 7549–7556.

Felmlee T, Pellett S & Welch R (1985) Nucleotide sequence of an *Escherichia coli* chromosomal hemolysin. J. Bacteriol. 163: 94–105.

Fukuchi N, Isogai A, Nakayama J & Suzuki A (1990a) Structure of syringotoxin B, a phytotoxin produced by citrus isolates of *Pseudomonas syringae* pv. *syringae*. Agric. Biol. Chem. 54:3377–3379.

Fukuchi N, Isogai A, Nakayama J, Takayama S, Yamashita S, Suyama K & Suzuki A (1992a) Isolation and structural elucidation of syringostatins, phytotoxins produced by *Pseudomonas syringae* pv. *syringae* lilac isolate. J. Chem. Soc. Perkin Trans. 1 1992: 875–880.

Fukuchi N, Isogai A, Nakayama J, Takayama S, Yamashita S, Suyama K, Takemoto JY & Suzuki A (1992b) Structure and stereochemistry of three phytotoxins, syringomycin, syringotoxin and syringostatin, produced by *Pseudomonas syringae* pv. *syringae*. J. Chem. Soc. Perkin Trans. 1 1992: 1149–1157.

Fukuchi N, Isogai A, Yamashita S, Suyama K, Takemoto JY & Suzuki A (1990b) Structure of phytotoxin syringomycin produced by a sugar cane isolate of *Pseudomonas syringae* pv. *syringae*. Tetrahedron Lett. 31: 1589–1592.

Gilson L, Mahanty HK & Kolter R (1990) Genetic analysis of an MDR-like export system: The secretion of colicin V. EMBO J. 9: 3875–3884.

Glaser P, Sakamoto H, Bellalou J, Ullmann A & Danchin A (1988) Secretion of cyclolysin, the calmodulin-sensitive adenylate cyclase – haemolysin bifunctional protein of *Bordetella pertussis*. EMBO J. 7: 3997–4004.

Gross DC (1985) Regulation of syringomycin synthesis in *Pseudomonas syringae* pv. *syringae* and defined conditions for its production. J. Appl. Bacteriol. 58: 167–174.

Gross DC (1991) Molecular and genetic analysis of toxin production by pathovars of *Pseudomonas syringae*. Annu. Rev. Phytopathol. 29: 247–278.

Gutiérrez S, Díez B, Montenegro E & Martín JF (1991) Characterization of the *Cephalosporium acremonium pcbAB* gene encoding α-aminoadipyl-cysteinyl-valine synthetase, a large multidomain peptide synthetase: Linkage to the *pcbC* gene as a cluster of early cephalosporin biosynthetic genes and evidence of multiple functional domains. J. Bacteriol. 173: 2354–2365.

Guzzo J, Duong F, Wandersman C, Murgier M & Lazdunski A (1991) The secretion genes of *Pseudomonas aeruginosa* alkaline protease are functionally related to those of *Erwinia chrysanthemi* proteases and *Escherichia coli* α-haemolysin. Mol. Microbiol. 5: 447–453.

Higgins F, Hyde SC, Mimmack MM, Gileadi U, Gill DR & Gallagher MP (1990) Binding protein-dependent transport systems. J. Bioenerg. Biomembr. 22: 571–592.

Howe TR & Iglewski BH (1984) Isolation and characterization of alkaline protease-deficient mutants of *Pseudomonas aeruginosa* in vitro and in a mouse eye model. Infect. Immun. 43: 1058–1063.

Iacobellis NS, Lavermicocca P, Grgurina I, Simmaco M & Ballio A (1992) Phytotoxic properties of *Pseudomonas syringae* pv. *syringae* toxins. Physiol. Mol. Plant Pathol. 40: 107–116.

Kane SE, Pastan I & Gottesman MM (1990) Genetic basis of multidrug resistance of tumor cells. J. Bioenerg. Biomembr. 22: 593–618.

Kinscherf TG, Coleman RH, Barta, TM & Willis, DK (1991) Cloning and expression of the tabtoxin biosynthetic region from *Pseudomonas syringae*. J. Bacteriol. 173: 4124–4132.

Kleinkauf H & von Döhren H (1982) A survey of enzymatic peptide formation. p. 3–21. *In*: Kleinkauf H & von Döhren H (eds.) Peptide Antibiotics. Walter de Guyten and Co., New York.

Kleinkauf H & von Döhren H (1983) Peptides. p. 95–145. *In*: Vining LC (ed.) Biochemistry and Genetic Regulation of Commercially Important Antibiotics. Addison-Wesley Pub. Co., Reading, MA

Kleinkauf H & von Döhren H (1987) Biosynthesis of peptide antibiotics. Annu. Rev. Microbiol. 41: 259–289.

Komura S & Kurahashi K (1980) Biosynthesis of polymyxin E. III. Total synthesis of polymyxin E by a cell-free enzyme system. Biochem. Biophys. Res. Commun. 95: 1145–1151.

Krätzschmar J, Krause M & Marahiel MA (1989) Gramicidin S biosynthesis operon containing the structural genes *grsA* and *grsB* has an open reading frame encoding a protein homologous to fatty acid thioesterases. J. Bacteriol. 171: 5422–5429.

Kroll JS, Loynds B, Brophy LN & Moxon ER (1990) The *bex* locus in encapsulated *Haemophilus influenzae*: a chromosomal region involved in capsule polysaccharide export. Mol. Microbiol. 4: 1853–1862.

Kuchler K, Sterne RE & Thorner JW (1989) *Saccharomyces cerevisiae* STE6 gene product: A novel pathway for protein export in eukaryotic cells. EMBO J. 8: 3973–3984.

Létoffé S, Delepelaire P & Wandersman C (1990) Protease secretion by *Erwinia chrysanthemi*: the specific secretion functions are analogous to those of *Escherichia coli* α-haemolysin. EMBO J. 9: 1375–1382.

Lincke CR, Smit JJM, van der Velde-Koerts T & Borst P (1991) Structure of the human *MDR*3 gene and physical mapping of the human *MDR* locus. J. Biol. Chem. 266: 5303–5310.

Ma SW, Morris VL & Cuppels DA. (1991) Characterization of a DNA region required for production of the phytotoxin coronatine by *Pseudomonas syringae* pv. *tomato*. Mol. Plant-Microbe Interact. 4: 69–74.

Martín JF & Liras P (1989) Organization and expression of genes involved in the biosynthesis of antibiotics and other secondary metabolites. Ann. Rev. Microbiol. 43: 173–206.

Miller LP (1973a) Glycosides. p. 297–375. *In*: Miller LP (ed.) Phytochemistry, Vol. 1, the Process and Products of Photosynthesis. Van Nostrand Reinhold Co., New York.

Miller LP (1973b) Mono- and oligosaccharides. p. 145–175. *In*: Miller LP (ed.) Phytochemistry, Vol. 1, the Process and Products of Photosynthesis. Van Nostrand Reinhold Co., New York.

Mo YY & Gross DC (1991a) Expression *in vitro* and during plant pathogenesis of the *syrB* gene required for syringomycin production by *Pseudomonas syringae* pv. *syringae*. Mol. Plant-Microbe Interact. 4: 28–36.

Mo YY & Gross DC (1991b) Plant signal molecules activate the *syrB* gene, which is required for syringomycin production by *Pseudomonas syringae* pv. *syringae*. J. Bacteriol. 173: 5784–5792.

Morgan MK & Chatterjee AK (1988) Genetic organization and regulation of proteins associated with production of syringotoxin by *Pseudomonas syringae* pv. *syringae*. J. Bacteriol. 170: 5689–5697.

Paris R (1963) The distribution of plant glycosides. p. 337–358. *In*: Swain T (ed.) Chemical Plant Taxonomy. Academic Press, Inc., New York.

Peet RC, Lindgren PB, Willis DK & Panopoulos NJ (1986) Identification and cloning of genes involved in phaseolotoxin production by *Pseudomonas syringae* pv. *phaseolicola*. J. Bacteriol. 166: 1096–1105.

Peters NK & Long SR (1988) Alfalfa root exudates and compounds which promote or inhibit induction of *Rhizobium meliloti* nodulation genes. Plant Physiol. 88: 396–400.

Quingley NB & Gross DC (1993) Syringomycin production among strains of *Pseudomonas syringae* pv. syringae: conservation of the *syrB* and *syrD* genes, and activation of phytotoxin production by plant signal molecules. Mol. Plant-Microbe Interact. (accepted Sept. 2, 1993).

Quigley NB, Mo YY & Gross DC (1993) SyrD is required for syringomycin production by *Pseudomonas syringae* pathovar *syringae* and is related to a family of ATP-binding secretion proteins. Mol. Microbiol. 9: 787–801.

Shimoda N, Toyoda-Yamamoto A, Nagamine J, Usami S, Katayama M, Sakagami Y & Machida Y (1990) Control of expression of *Agrobacterium vir* genes by synergistic actions of phenolic signal molecules and monosaccharides. Proc. Natl. Acad. Sci. USA 87: 6684–6688.

Sogn JA (1976) Structure of the peptide antibiotic polypeptin. J. Med. Chem. 19: 1228–1231.

Stachel SE, An G, Flores C & Nester EW (1985a) A Tn*3 lacZ* transposon for the random generation of β-galactosidase gene fusions: application to the analysis of gene expression in *Agrobacterium*. EMBO J. 4: 891–898.

Stachel SE, Messens E, Van Montagu M & Zambryski P (1985b) Identification of the signal molecules produced by wounded plant cells that activate T-DNA transfer in *Agrobacterium tumefaciens*. Nature (London) 318: 624–629.

Tai PC (1990) Protein export in bacteria: An overview. J. Bioenerg. Biomembr. 22: 209–212.

Takemoto JY (1992) Bacterial phytotoxin syringomycin and its interaction with host membranes. p. 247–260. *In*: Verma DPS (ed.) Molecular Signals in Plant-Microbe Communications. CRC Press, Inc., Boca Raton, Florida.

Takemoto JY, Zhang L, Taguchi N, Tachikawa T & Miyakawa T (1991) Mechanism of action of the phytotoxin syringomycin: a resistant mutant of *Saccharomyces cerevisiae* reveals an involvement of Ca^{2+} transport. J. Gen Microbiol. 137:653–659.

Wandersman C & Delepelaire P (1990) TolC, an *Escherichia coli* outer membrane protein required for hemolysin secretion. Proc. Natl. Acad. Sci. USA 87: 4776–4780.

Wang R, Seror SJ, Blight M, Pratt JM, Broome-Smith JK & Holland IB (1991) Analysis of the membrane organization of an *Escherichia coli* protein translocator, HlyB, a member of a large family of prokaryote and eukaryote surface transport proteins. J. Mol. Biol. 217: 441–454.

Winans SC, Kerstetter RA & Nester EW (1988) Transcriptional regulation of *virA* and *virG* genes of *Agrobacterium tumefaciens*. J. Bacteriol. 170: 4047–4057.

Xu GW & Gross DC (1988a) Evaluation of the role of syringomycin in plant pathogenesis by using Tn*5* mutants of *Pseudomonas syringae* pv. *syringae* defective in syringomycin production. Appl. Environ. Microbiol. 54: 1345–1353.

Xu GW & Gross DC (1988b) Physical and functional analyses of the *syrA* and *syrB* genes involved in syringomycin production by *Pseudomonas syringae* pv. *syringae*. J. Bacteriol. 170: 5680–5688.

28. Characterization of genes involved in phaseolotoxin production and its thermal regulation

SURESH S. PATIL, K.B. ROWLEY, Y.X. ZHANG, D.E. CLEMENTS, M. MANDEL and T. HUMPHREYS

Abstract. *Pseudomonas syringae* pv. phaseolicola, the causal agent of halo blight of bean, produces phaseolotoxin, a nonspecific chlorosis-inducing toxin. The wild-type strain of *P.s.* pv. phaseolicola G50-1 produces phaseolotoxin at 18°C but not at 28°C. When a derivative of a previously reported genomic clone isolated from G50-1 which complements only EMS and UV (but not Tn5) mutants is mobilized into these mutants, as well as in the wild type strain, the transconjugants produce toxin at both 18°C and 28°C. Further characterization of the insert in this clone showed that a 485 bp fragment containing motifs characteristic of DNA binding sites, when present in multiple copies in the transconjugant, relieves repression of phaseolotoxin production at 28°C. Results of gel retardation assays showed that a protein(s) present in extracts of cells grown at 28°C binds specifically to the 485 bp fragment and to a 260 bp subfragment that contains these DNA binding motifs. We suggest that thermoregulation of phaseolotoxin production at 28°C by the wild type and perhaps by the Tox⁻ mutant at both temperatures, involves a regulatory protein(s) that represses transcription of one or more phaseolotoxin structural genes, and that the DNA binding site(s) in the fortuitously cloned fragment derepresses toxin synthesis by titrating this protein. We previously showed that a genomic cosmid clone, pHK120, containing a 24 kb fragment of DNA from the wild-type strain of the pathogen restores toxin production to all EMS, UV and Tn5 Tox⁻ mutants. Tn5 mutagenesis of pHK120, marker exchange of pHK120::Tn5 in the wild-type strain and pair-complementation analysis revealed that a minimum of 8 genes (A through H) are clustered in the insert of pHK120.

Abbreviations: EMS, Ethylmethansulfonate; OCT, Ornithine Carbonyltransferase; ORF, Open Reading Frame; TRR, Thermoregulatory Region

Introduction

Phaseolotoxin, [N$^\delta$ (N'-sulfo-diaminophosphinyl)-ornithyl-alanyl-homoarginine], is an extracellular, nonspecific, chlorosis-inducing toxin produced by *Pseudomonas syringae* pv. phaseolicola, the causal agent of halo blight of bean (Moore *et al.*, 1984). In infected plants the effective toxin (Mitchell and Bieleski, 1977) is octicidin, a cleaved product of phaseolotoxin. The mode of action of phaseolotoxin (or octicidin) in bean plants involves specific inhibition of an important ornithine cycle enzyme, ornithine carbamoyltransferase (OCT)(Patil *et al.*, 1970). Inhibition of OCT in planta leads to ornithine accumulation and chlorosis (Patil *et al.*, 1972). Over the past 20 years the *P.s.* pv. phaseolicola bean system has been used as a model system by several laboratories and a substantial amount of work on the chemistry of the toxin and the biochemistry and

C.I. Kado and J.H. Crosa (eds.), Molecular Mechanisms of Bacterial Virulence, 415–428.
© 1994 *Kluwer Academic Publishers.*

physiology of its effects in bean have been described (Patil, 1974; Mitchell, 1984). However, genetic studies on the biosynthesis of phaseolotoxin have begun relatively recently (Peet & Panopoulos, 1986; Kamdar et al., 1991).

Our laboratory has been involved in the characterization of genes involved in the production of phaseolotoxin. We used cosmid cloning and mutagenesis to identify genes involved in the production of this toxin. Eight stable clones were isolated from a genomic cosmid library by en-masse mating to 10 ethylmethanesulfonate (EMS)-induced toxin minus (Tox⁻) mutants. In cross-matings each complemented all 10 mutants as well as an additional 70 EMS-induced (and one UV-induced) Tox⁻ mutants. On the basis of restriction endonuclease analysis and hybridization studies, the clones were grouped into three classes. Clones in a particular class shared common fragments, whereas clones in different classes did not. Clones in class I (but not classes II & III) also complemented Tn5 induced Tox⁻ mutants. Interposon mutagenesis of a representative clone from class III and its marker exchange into the wild-type genome did not alter its Tox⁺ phenotype, indicating that this clone does not harbor structural or regulatory genes involved in phaseolotoxin production. We suggested that the genome of P.s. pv. phaseolicola contains a "hot spot" in one of the functions involved in toxin production which is affected by EMS or UV and that heterologous clones restore toxin production in the Tox⁻ mutants by non-allelic complementation because their inserts encode products that are able to substitute for the product of the mutated gene. We also proposed an alternative explanation for the non-allelic complementation of Tox⁻ mutants by the heterologous clones. It was suggested that the inserts may contain sequences that titrate a repressor protein that is constitutively produced by the Tox⁻ mutants. Furthermore, our data (Kamdar et al., 1991) suggested that the complementation of EMS- and UV-induced mutants occurs because the plasmid containing inserts from Class III (and probably Class I, as well) is present in the mutant transconjugants in multiple copies. Because class I clones complemented not only the EMS- and UV-induced mutants but also several independent Tn5 mutants with insertions in different regions of the genome, we predicted that these clones probably contained a cluster of structural genes involved in phaseolotoxin biosynthesis (Kamdar et al., 1991). In this paper we describe the latest studies on the characterization of a DNA fragment from a class III clone (pDC938) involved in the complementation of the UV-induced Tox⁻ mutant, and the possible mechanism of this suppression. Additionally, we describe the Tn5 mutagenesis of a class I clone (pHK120), the marker exchange of pHK120::Tn5 clones with homologous regions of the chromosome of the wild-type strain, and pair-complementation analysis to define the number of toxin loci present in the pHK120 insert.

Results

Tn3HoHo1 mutagenesis and subcloning of pDC938

Kamdar *et al.* (1991) previously reported the isolation of a clone, pDC938, that contained a 24.4 kb insert from the genomic library of *P.s.* pv. phaseolicola (Fig. 1A). pDC938 complements all phaseolotoxin-deficient EMS- and UV-induced mutants. When the biologically active region was inactivated by insertion of a spectinomycin cassette and marker-exchanged into the chromosome of the wild-type strain, toxin production was unaffected. Therefore, pDC938 does not harbor a structural or regulatory gene involved in phaseolotoxin production.

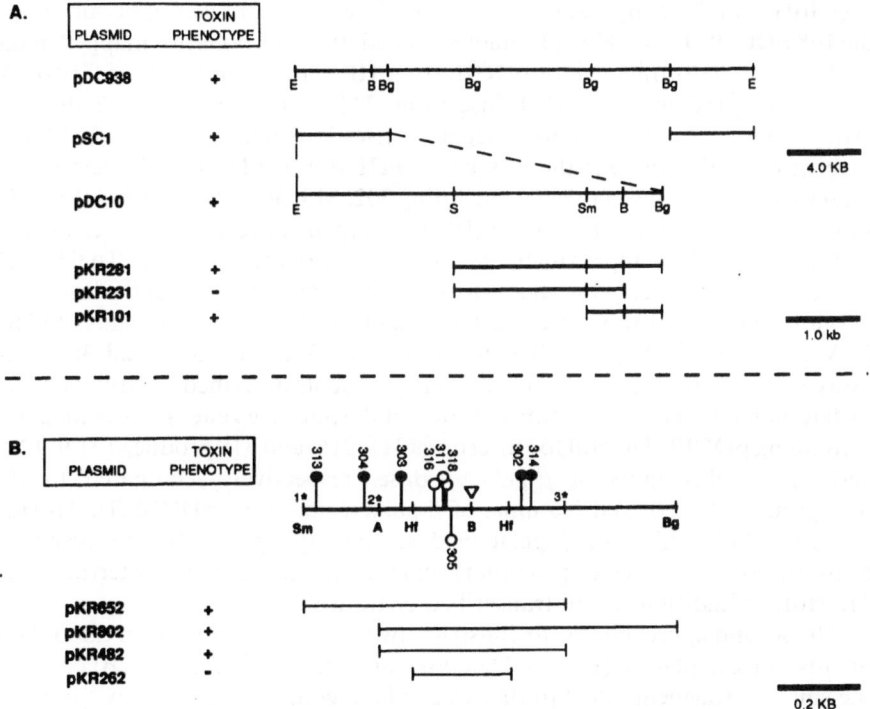

Figure 1. Restriction map of pDC938, Tn*3*HoHo1 mutagenesis of pDC10, and deletion subcloning showing the toxin phenotype (+ or −) of transconjugants in G50-1 UV. A. Schematic drawing of the restriction map and deletion subclones of pDC938. The inserts were cloned into the vector pLAFR3. B. The map of Tn*3*HoHo1 insertions in pDC10 (top line) show Tox⁺ (closed circle) and Tox⁻ insertions (open circle) located between the *Sma*I and *Bgl*II sites in pDC10. The orientation of the *lacZ* gene is indicated by the placement of the symbol above the line (right to left) or below the line (left to right). The lines below represent deletion fragments subcloned into the vector pNP483. *1, *2, and *3 show the positions of the synthetic primers PCR1, PCR2, and PCR3, respectively, used in the PCR to synthesize the 689 bp insert for pKR652 and the 485 bp insert for pKR482. The site of insertion for the spectinomycin cassette (Kamdar *et al.*, 1991) is indicated by the symbol (▽). E, *Eco*RI; B, *Bam*HI; Bg, *Bgl*II; S, *Sma*I; A, *Ava*I; Hf, *Hinf*I.

Fragments from the insert of pDC938 generated by digestion with *Bgl*II were shotgun cloned into *E. coli* and mobilized into the *P.s.* pv. phaseolicola UV-induced mutant (G50-1UV; Kamdar *et al.*, 1991). This mutant was used as a representative of the EMS- and UV-induced mutants because of its superior mating ability. Although most of the subclones contained multiple *Bgl*II fragments, only the subclone containing the 4.4 kb and 4.9 kb *Eco*RI-*Bgl*II fragments (pSCI, Fig. 1A) complemented G50-1UV. Further subcloning of these fragments in pLAFR3 (Staskawicz *et al.*, 1987) resulted in the clone pDC10 (which contained the 4.9 kb fragment) that complemented G50-1UV (Fig. 1A).

To determine the region of DNA in pDC10 involved in the complementation of G50-1UV, the insert from pDC10 was mutagenized with the transposon Tn*3*HoHo1 and the approximate positions of the insertions were determined. Of the 168 pDC10::Tn*3*HoHo1 plasmids screened, the insertions in 8 mapped in the 1.0 kb *Sma*I-*Bgl*II fragment of pDC10 (Fig. 1B). Transconjugants of G50-1UV containing pDC10::Tn*3*HoHo1 insertions 311, 316, 318, and 305 did not produce phaseolotoxin. All four negative insertions mapped within 100 bp of each other and very near the unique *Bam*HI site of pDC10. The rest of the Tn*3*HoHo1 insertions in pDC10, including 302, 314, 303, 304, and 313 (Fig. 1B) showed a Tox$^+$ phenotype in G50-1UV. The map distance between insertion 303 and insertion 302 is approximately 300 bp. The orientation of the Tn*3*HoHo1 insert 305 with respect to the *lacZ* gene was left to right. The rest of the insertions in the 1 kb fragment were oriented from right to left. Transconjugants of G50-1UV containing pDC10::Tn*3*HoHo1 insertions 311, 316, 318, and 305 were assayed for β-galactosidase activity, using protocols described by Miller (1972), to determine the direction of transcription of the putative gene. Transconjugants containing pDC10::Tn*3*HoHo1 insertions 311, 316, and 318 produced 31.9, 12.2, and 11.9 Miller units of β-galactosidase, respectively, compared to the background (1.1 U), and 3.6 units of β-galactosidase for pDC10::Tn*3*HoHo1 insertion 305. This low β-galactosidase activity (probably the result of transcription from a vector promoter) suggests that the sequence interrupted by Tn*3*HoHo1 insertions is not transcribed or translated.

The second approach was to construct two sets of deletions from the ends of the insert from pDC10 (Fig. 1). Deletion subclone pKR482 (Fig. 2B) contains the smallest fragment (485 bp) that was able to complement the Tox$^-$ mutant. When the 500 bp *Bam*HI-*Bgl*II fragment from pDC10 was deleted (pKR231), the transconjugants containing the subclone did not complement the Tox$^-$ mutant (Fig. 1A). This confirmed the previous results (Kamdar *et al.*, 1991), which showed that the insertion of a spectinomycin cassette into the *Bam*HI site of pDC938 (Fig. 1B) abolished the ability of pDC938 to complement G50-1UV. We found that transconjugants containing pDC938 and pDC10 produced wild-type levels of toxin, but transconjugants containing pKR281, pKR101, and pKR482 produced progressively less toxin (data not shown). The reason for this is not known but may be due to a copy number effect (ie. either pDC938 contains more than one copy of the element that suppresses the Tox$^-$

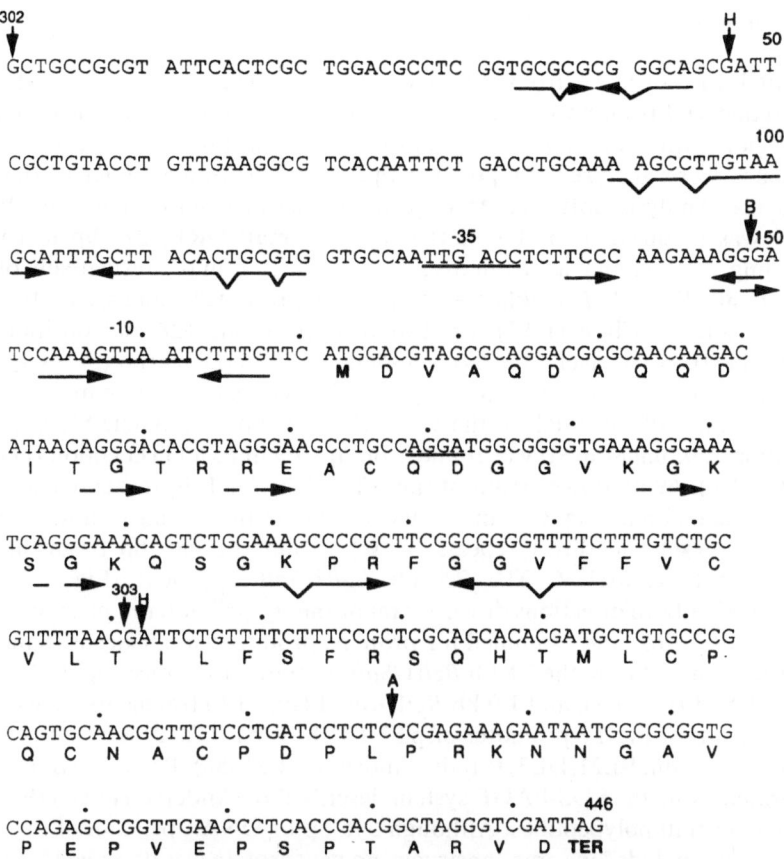

Figure 2. Nucleotide sequence of the TRR. The sequence (446 bp) is shown from Tn*3*HoHo1 insertion site 302, to the end of the putative open reading frame (ORF). The putative −10 and −35 promoter regions and the ribosomal binding sites as determined by comparison with the *E. coli* consensus sequences are underlined. Arrowheads 302 and 303 indicate the location of Tn*3*HoHo1 insertions 302 and 303 (also shown in Fig. 1). Arrows denote inverted repeats (-----▶) and direct repeats (——▶) of AGGGA. A mismatch is indicated by (——∨——▶). B, *Bam*HI; A, *Ava*I; H, *Hinf*I.

phenotype or the plasmid copy number of pKR101 is lower than pDC938). We found (data not shown) that the amount of toxin produced by transconjugants of G50-1UV containing the 1.0 kb *Bgl*II-*Sma*I fragment in pLAFR3 (7–8 copies; Panopoulos personal communication) was less than the amount produced by transconjugants of G50-1UV containing the same fragment in a higher copy number vector (pNP483; ~ 20 copies per cell; Borthakur, personal communication).

Nucleotide sequence analysis

About 1.5 kb of DNA from the region that complemented the Tox⁻ mutant was sequenced from both strands. The sequence (446 bp) which encompassed the region required for complementation is shown in Fig. 2. All 6 frames of the sequence were analyzed for possible open reading frames (ORF) using the Sequence Analysis Software Package for Genetics Computer Group. When ATG was assumed to be the start codon, a small ORF, 273 bp in length, encoding a 91 amino acid putative protein, was found downstream of the *Bam*HI site. Possible *E. coli*-like −35 and −10 promoter regions, as defined by Hawley and McClure (1983), are found at positions 128 and position 156 (underlined), respectively, but no well-defined Shine-Delgarno (SD) sequence (Stormo *et al.*, 1982) is present (Fig. 2). The sequence is presented with the coding strand of this ORF in the 5′-to-3′ direction. A possible SD sequence (position 229; underlined) was found 7 bp upstream of a GTG codon (Darnell *et al.*, 1990), 69 bp downstream of the ATG. The ORF in this case encodes a polypeptide containing 67 amino acids. The sequence also shows several inverted (→) and direct repeats (----▶) within the 485 bp fragment that complements G50-1UV. The 2.8 kb *Bgl*II-*Sal*I fragment (Fig. 1A) was subcloned in both directions downstream of the *trp* promoter in plasmid pAD9 (Das, *et al.*, 1983) and examined for proteins produced in maxicells (Sancar *et al.*, 1977). In addition, the 2.8 kb *Bgl*II-*Sal*I (in both directions; Fig. 1A), 2.3 kb *Bam*HI-*Sal*I (Fig. 1A), and 1.0 kb *Bgl*II-*Sma*I (Fig. 1A) fragments were cloned downstream of the T7 promoter of pT7/T3-18 or pT7/T3-19 for expression in the *E. coli* strain, BL21(DE3) (Studier and Moffatt, 1986). Protein extracts were separated using the SDS-PAGE system described by Anderson *et al.* (1983) for resolving small polypeptides. Although 545 bp of DNA upstream of the ATG codon were included in some constructs, no new protein was detected in extracts prepared from strains containing these plasmids, using either method. That a functional polypeptide is produced can also be ruled out from the Tn*3*HoHo1 mutagenesis data because transconjugants containing pDC10 (Fig. 1A) with a Tn*3*HoHo1 insertion 100 bp from the 3′-end of the putative gene (insertion 303, Fig. 2B) or plasmid inserts in which 62 bp have been deleted from the 3′-end of the ORF (pKR802, pKR482; Fig. 2) were Tox⁺. Therefore, it is unlikely that a functional polypeptide is produced from the 273 bp ORF.

Toxin production at 28°C

Since it appeared that the ability of the functional sequence in pDC10 to complement G50-1UV was not due to the production of a protein that could substitute for the product of the mutated gene in G50-1UV we proceeded to test the alternative hypothesis. According to this hypothesis, the biologically active sequence in pDC10 apparently "titrates" a protein involved in the regulation of toxin production (Kamdar *et al.*, 1991).

To test this hypothesis, transconjugants of G50-1 containing pDC938 were assayed for toxin production at 18°C (Fig. 3A, spot 3) and at 28°C (Fig. 3B, spot 3). The G50-1 (wild type)(Fig. 3, spot 1) and G50-1 containing pLAFR3 (Fig. 3, spot 2) were used as controls. As expected, toxin was produced by all transconjugants when incubated at 18°C (Fig. 3A). However, at 28°C (Fig. 3B), toxin was produced only by the transconjugant containing pDC938. The experiment was repeated using transconjugants containing pDC10 and pKR101 (pLAFR3-derivative plasmids) and pKR102 (high-copy-number pNP483-derivative plasmid; Table 1). Transconjugants of G50-1 which contain either pDC10 or pKR102 produced toxin at 28°C, although much less (roughly 90%) than by transconjugants containing pDC938. No toxin was produced by transconjugants containing pKR101 at 28°C. Finally, transconjugants of G50-1UV containing pDC938 also produced toxin at both temperatures (Table 1). For convenience, the 485 bp region of pDC938 that affects the thermoregulation of phaseolotoxin production in G50-1 is referred to as the thermoregulatory region or TRR.

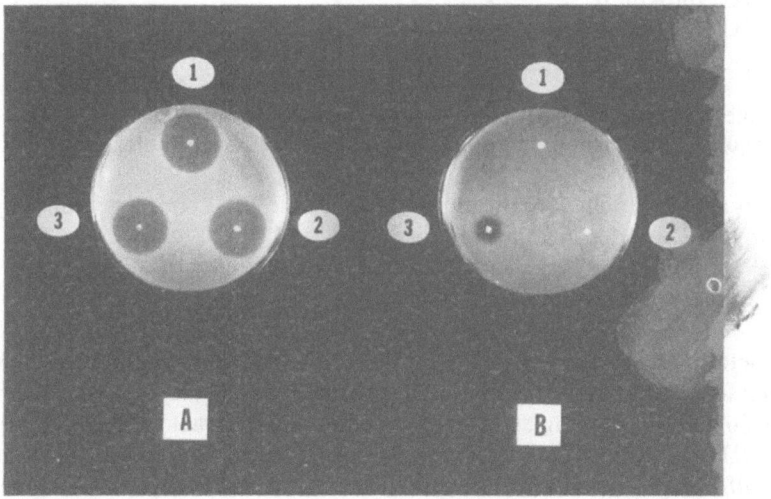

Figure 3. Toxin bioassay of *P. s.* pv. phaseolicola, G50-1(Tox[+]) (spot 1); the transconjugant of G50-1 containing pLAFR3 (spot 2), and the transconjugant of G50-1 containing pDC938 (spot 3) incubated at 18°C (A) and 28°C (B).

Since toxin is produced at 18–20°C but not at 28°C, toxin production maybe inhibited at the higher temperature by a regulatory mechanism possibly involving a repressor. It is known that when multiple copies of the operator region for an operon are present in the cell, regulation is abolished due to titration of the regulatory protein by the DNA binding sites(s) in the operator (Irani, *et al.*, 1983). The presence of the TRR in the wild-type strain (Fig. 2; Table 1) derepresses phaseolotoxin production at 28°C. Also, increasing the copy number of the TRR by subcloning it into a high copy number plasmid, results in increased toxin production in the mutant (data not shown). These

Table 1. Production of phaseolotoxin at 18°C and 28°C by transconjugants of G50-1 (Tox⁺) and G50-1UV (Tox⁻).

Transconjugant	TRR[1]	Temperature[2]	
		18°C	28°C
G50-1(pLAFR3)	-	+	-
G50-1(pDC938)	+	+	+
G50-1(pDC10)	+	+	+
G50-1(pKR101)	+	+	-
G50-1(pNP483)	-	+	-
G50-1(pKR102)	+	+	+
G50-1UV(pLAFR3)	-	-	-
G50-1UV(pDC938)	+	+	+

[1]Denotes presence or absence of thermoregulatory region, TRR.
[2]Symbols denote toxin production as determined by the toxin bioassay.

results are consistent with the hypothesis that the TRR contains a binding site(s) for a protein(s) that regulates toxin production.

DNA-binding site in TRR

To determine whether or not the TRR contains a binding site for a putative repressor, a partially purified extract from G50-1 (Tox⁺) grown at 28°C was used in a DNA-binding assay (Fig. 4). The crude extract was partially purified by $(NH_4)_2SO_4$ fractionation. Initially, the 485 bp EcoRI-HindIII fragment from pKR482 (Fig. 1) was used as the probe because it was the smallest fragment that complemented G50-1UV. This probe formed a complex with a protein(s) in the fraction that precipitated between 5% and 40% $(NH_4)_2SO_4$, and with a protein(s) in the fraction which precipitated between 40% and 60% $(NH_4)_2SO_4$, but not in the fraction that precipitated between 60% and 80% $(NH_4)_2SO_4$. Only the 5% – 40% fraction was used in subsequent experiments (Fig. 4) because most of the DNA-specific protein was found in this fraction. Although the 260 bp fragment from pKR262 (Fig. 1) which included the DNA binding motifs present in the 485 bp fragment, did not suppress the Tox⁻ phenotype of G50-1UV, when this fragment was used as the probe in the binding assay, specific DNA-protein complexes (indicated by the letter B in Fig. 4) were formed with the 5% to 40% $(NH_4)_2SO_4$ fraction (Fig. 4A, Lane 2; Fig. 4B, Lane 2). These complexes were resistant to the addition of 2.1 μg and 6.0 μg of nonspecific (pT7/T3α-19) cold competitor DNA in the reaction mixture (Fig. 4A, Lanes 5

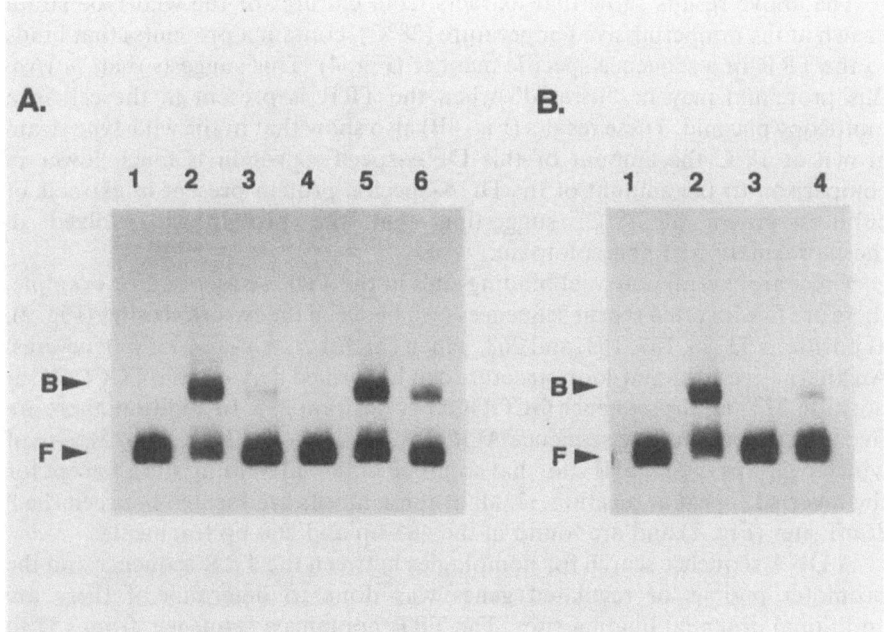

Figure 4. Mobility shift assay using the 260 bp fragment from the TRR as the probe. The end-labeled 260 bp *Eco*RI-*Hin*dIII fragment from pKR260 (~ 3 ng) was incubated with 50 μg of the 5%–40% (NH₄)₂SO₄ protein fraction from cultures of G50-1 or G50-1UV, 20 ng of pT 7/T3α-19, and 1 μg poly (dI-dC). The unbound fragments (F) were separated from DNA-protein complexes (B) by electrophoresis through 4.2% nondenaturing polyacrylamide gels. (A) Labeled probe alone (Lane 1); labeled probe with 50μg of partially purified protein from G50-1 grown at 28°C (Lanes 2–5); no competitor DNA (Lane 2); incubation mixture with specific competitor DNA (pKR260), at a concentration of 2.1 μg (Lane 3) or 6.0 μg (Lane 4);incubation mixture with noncompetitor DNA (pT7/T3α-19) at concentrations of 2.1 μg (Lane 5) or 6.0 μg (Lane 6). B. Partially purified protein extracts from G50-1 grown at 28°C (Lane 2), G50-1 grown at 18°C (Lane 3), and G50-1UV grown at 18°C (Lane 4) were incubated with labeled 260 bp probe (~ 3 ng).

and 6). In contrast, the DNA-protein complexes formed with this probe were almost completely titrated by 2.1 μg of specific (pKR260) competitor DNA (Fig. 4A, Lane 3) and entirely titrated by 6.0 μg (Fig. 4A, Lane 4).

To determine whether the DNA-binding protein is also present in cultures grown at the toxin producing temperature, protein extracts were prepared from cultures of G50-1 (Fig. 4B, Lane 3) and G50-1UV (Fig. 4B, Lane 4) grown at 18°C. Equal amounts of protein (50 μg) from the 5% – 40% (NH₄)₂SO₄ fraction were incubated with the probe. The retarded band corresponding to the DNA-protein complex is barely detectable in Lane 3 and less intense in Lane 4 as compared to Lane 2. This indicates that, whereas almost no DNA-specific protein is found in extracts from the wild-type strain grown at 18°C, there is a significant amount of the protein in extracts from the mutant grown at the same temperature.

The above results show that extracts from cultures of the wild-type strain grown at the nonpermissive temperature (28°C), contain a protein(s) that binds to the TRR in a sequence-specific manner (Fig. 4). This suggests that, *in vivo*, this protein(s) may be "titrated" when the TRR is present in the cell in a multicopy plasmid. These results (Fig. 4B) also show that in the wild-type strain grown at 18°C the amount of this DNA-specific protein is much lower in comparison to the amount of the DNA-specific protein present in extracts of cultures grown at 28°C, suggesting that the protein is involved in thermoregulation of phaseolotoxin.

There are several potential binding sites in the TRR sequence. For example, there are five inverted repeat sequences ($\rightarrow$) between the two *Hin*fI sites (Fig. 2), at positions 33, 90, 137, 153, and 268, which can form stem-and-loop structures. An alternative stem-and-loop structure can be formed between the CCCCGC at position 273 and the sequence GCGGGG at position 235. In addition, there are five direct repeats of the sequence AGGGA (----▶) in the TRR (Fig. 2), one of which overlaps the *Bam*HI site, that could be potential binding sites. Except for the inverted repeat at position 33, all of these motifs are located between the 2 *Hin*fI sites (Fig. 2) and are found in the 485 bp and 260 bp fragments.

A DNA sequence search for homologies between the TRR sequence and the promoter regions of regulated genes was done to determine if there are additional potential binding sites. The TRR contains a sequence, from 137 to 161 of the complementary strand (Fig. 2), that shares 72% homology with a OmpR binding site in the regulatory region of the *ompF* operon of *E. coli* (Slauch and Silhavy, 1991; Fig. 2; Table 2). This sequence in the *ompF* operon is important for negative regulation of the operon at high osmolarity (Slauch and Silhavy, 1991). Furthermore, the homologous sequences contain the same motif CTTTC and GTTTC, separated by 5 bp. Similar sequence motifs are important for the regulation of the *ompC* operon of *E. coli* (Mizuno and Mizushima, 1986). We also found a 16 bp sequence in the TRR, at position 95 (Fig. 2), that is 75% homologous with the integration host factor (IHF) binding site in the *ompF* operon and to the IHF binding site at the *attP* integration site in bacteriophage lambda (Table 2; Craig and Nash, 1984). The critical determinant for specific IHF-DNA interaction, contained in the sequence T.PyAA...PuTTGaT (Craig and Nash, 1984), is present in the TRR sequence (Table 2). IHF is one of several histone-like proteins in bacteria and plays a role in the regulation of transcription of a number of cellular functions (Tsui *et al.*., 1988 and reviewed in Higgins *et al.*., 1990; Drlica and Rouviere-Yaniv, 1987). It has been proposed that IHF is involved in bending DNA to facilitate various protein-protein interactions (Slauch and Silhavy, 1991).

A comparison of the TRR sequence with the 5'-end of the sequence for the *argK* gene revealed a 14 bp sequence at position 253 to 265 that is 79% homologous with a sequence at position 138 to 151, in the *argK* promoter, presumably in the leader region of the transcript (Table 2). Also there is a 13 bp sequence at position 298 to 310 in the TRR that is 85% homologous with a sequence upstream of the proposed −35 region of the *argK* promoter at

Table 2. Sequence comparison of TRR[1] with upstream regions of *ompF* and *argK* and with the IHF sequences of *attP* and *ompF*.

Regulatory Region[2]	Sequence[3]	Reference
TRR	TTAACTTTGGATCCCTTTCTTGGGA[4]	This study
ompF -40 conserved motif	TTAtCTTTGtAgCaCTTTCacGGtA CTTTG.....CTTTC	Slauch & Silhavy, 1991
TRR	TTGTAAGCATTTGCTT	This study
attP	aTcTAAgCATTTGCTT	Craig & Nash, 1984
ompF -68	TcaTAAagATTTGgTT	
Consensus	T..cAA...gTTGaT t a	Craig & Nash, 1984
TRR	CAGGGAAACAGTCT	This study
argK (138-151)	CAcGGAAgCAGTtT	Mosqueda, *et al.*, 1990
TRR	CGTTAAAACGCAG	This study
argK (81-93)	CGTTAAtACaCAG	Mosqueda, *et al.*, 1990

[1]Thermoregulatory region

[2]The sequences for the *ompF* operon are located at the 5′-ends in relation to the transcription initiation site at +1. This site is not known for *argK*. Therefore, the location of the homologous sequence is reported relative to the first nucleotide of the published sequence.

[3]The IHF sequences of the *ompF* promoter and *attP* site and the TRR sequence are aligned with the IHF consensus sequence taken from Craig and Nash (1984). Bases that are homologous to the TRR sequence are capitalized.

[4]Complementary strand of TRR sequence between positions 137 and 161 (Fig. 2).

position 81 to 93 (Table 2). The *argK* gene that encodes the phaseolotoxin-resistant OCT in *P. s.* pv. phaseolicola, (Mosqueda *et al.*, 1990; Hatziloukas and Panopoulos, 1992), is coordinately regulated with phaseolotoxin by temperature (Peet and Panopoulos, 1987). The homologies between the *argK* gene and the TRR region indicate that the two regions may share a recognition site(s) for a common repressor that is regulated by temperature. Recently, we isolated the promoter region of the thermoregulated *argK* gene that contains these sequences and found that specific DNA-protein complexes were formed (umpublished data). Currently, we are doing DNase I footprinting analysis to determine the sequence of the binding site(s) in the *argK* promoter.

Finally, our results show that transconjugants of G50-1UV containing pDC938 also produce toxin at 28°C (Table 1), the nonpermissive temperature for toxin production in the wild-type strain. This suggests that complementation of the Tox⁻ phenotype in this strain is also the result of titration of a repressor by the binding site(s) in the TRR. One possible explanation for this is that the Tox⁻ mutant produces the repressor constitutively. Alternatively, the mutation may occur in the regulatory region of a phaseolotoxin structural gene allowing the repressor to bind with greater affinity to the site, thus inhibiting transcription of the gene at both temperatures. Mobility shift assays (Fig. 4B) showed that there was an increase in the amount of DNA-specific complexes formed from extracts from the Tox⁻ mutants grown at 18°C (Fig. 4, Lane 4) compared with extracts from the wild-type strain grown at the same temperature (Fig. 4, Lane 3). Whether or not this increased level is sufficient to prevent toxin production in the mutant is unknown. Additional studies will be necessary to determine the nature of the mutation in the UV-induced Tox⁻ mutant.

In conclusion, we have subcloned a DNA binding site(s) from the presumed regulatory region of an, as yet, unidentified gene that is thermoregulated in a manner similar to that of phaseolotoxin genes, and this site binds a common regulatory protein involved in thermoregulation of phaseolotoxin production. Studies are underway to purify this protein and to determine the sequence of the binding site in the TRR. We have also begun the isolation of the gene which encodes this protein using the spontaneous mutant of *P.s.* pv. phaseolicola strain PDDCC4612 (unpublished data) that produces toxin at 28°C, as the recipient to screen a genomic library of the wild-type strain (G50-1).

Genetic characterization of pHK120

A genomic cosmid clone, pHK120, containing a 24 kb fragment of DNA from the wild-type strain of *P. syringae* pv. phaseolicola restores toxin production to all EMS, UV and Tn5 Tox⁻ mutants tested. Tn5 mutagenesis of pHK120 and marker exchange of pHK120::Tn5 plasmids resulted in the isolation of 45 chromosomal mutants which harbor Tn5 insertions at known positions. Toxin bioassays of the mutants revealed that 27 mutants distributed throughout the insert of pHK120 had a Tox⁻ phenotype, indicating that a functional locus for toxin production was inactivated in each case. Pair-complementation analysis of these 27 tox⁻ mutants, as well as three random chromosomal Tn5 mutants obtained previously, revealed that there are a minimum of eight toxin loci (A through H) in pHK120. So far, marker-exchange mutants in locus A, C, D and F have been complemented by appropriate subclones of pHK120. However, no appropriate subclones are available to complement mutants in loci B, E, G, and H. Recently, the entire E locus (approximately 6.7 kb) has been sequenced on both strands. Preliminary analysis has shown 5 putative ORFs ranging in size from 480 bp to 1031 bp. A computer search for DNA sequence homologies with other known bacterial DNA sequences in Genbank revealed that the largest putative ORF (1031 bp) shows 53% homology with the acetylornithine aminotransferase gene from *E. coli*, which is not surprising since ornithine is a component of phaseolotoxin.

References

Anderson BL, Berry RW & Telser A (1983) A sodium dodecyl sulfate-polyacrylamide gel electrophoresis system that separates peptides and proteins in the molecular weight range of 2500 to 90,000. Anal. Biochem. **132**: 365–375.

Craig NL & Nash HA (1984) *E. coli* integration host factor binds to specific sites in DNA. Cell **39**: 707–716.

Darnell J, Lodish H & Baltimore D (1990) Molecular Cell Biology. p. 88. W.H. Freeman and Company, New York. Scientific American Books.

Das A, Urbanowski J, Weissbach H, Nestor J & Yanofsky C (1983) *In vitro* synthesis of the tryptophan operon leader peptides of *Escherichia coli, Serratia marcescens*, and *Salmonella typhimurium*. Proc. Natl. Acad. Sci. USA **80**: 2879–2883.

Drlica K & Rouviere-Yaniv J (1987) Histonelike proteins of bacteria. Microbiol. Rev. **51**: 301–319.

Hatziloukas E & Panopoulos NJ (1992) Origin, structure, and regulation of *argK*, encoding the phaseolotoxin-resistant ornithine carbamoyltransferase in *Pseudomonas syringae* pv. phaseolicola, and functional expression of *argK* in transgenic tobacco. J. Bacteriol. p. 5895–5909.

Hawley DK & McClure WR (1983) Compilation and analysis of *Escherichia coli* promoter DNA sequences. Nucleic Acids Res. **11**: 2237–2255.

Higgins CF, Hinton JCD, Hulton CSJ, Owen-Hughes T, Pavitt GD & Seirafi A (1990) Protein H1: a role for chromatin structure in the regulation of bacterial gene expression and virulence? Mol. Microbiol. **4**: 2007–2012.

Irani MH, Orosz L, Busby S, Taniguchi F & Adhya S (1983) Cyclic AMP-dependent constitutive expression of *gal* operon: use of repressor titration to isolate operator mutations. Proc. Natl. Acad. Sci. USA **80**: 4775–4779.

Kamdar HV, Rowley KB, Clements D & Patil SS (1991) *Pseudomonas syringae* pv. phaseolicola genomic clones harboring heterologous DNA sequences suppress the same phaseolotoxin-deficient mutants. J. Bacteriol. **173**: 1073–1079.

Miller JH (1972) Experiments in molecular genetics, p.138. Cold Spring Harbor Laboratory, cold Spring Harbor, N. Y.

Mitchell RE (1984) The relevance of non-host-specific toxins in the expression of virulence by pathogens. Annu. Rev. Phytopathol. **22**: 215–245.

Mizuno T & Mizushima S (1986) Characterization by deletion and localized mutagenesis *in vitro* of the promoter region of the *Escherichia coli ompC* gene and importance of the upstream DNA domain in positive regulation by the OmpR protein. J. Bacteriol. **168**: 86–95.

Moore RE, Niemczura WP, Kwok OCH & Patil SS (1984) Inhibitors of ornithine carbamoyltransferase from *Pseudomonas syringae* pv phaseolicola. Revised structure of phaseolotoxin. Tetrahedron Lett. **25**: 3931–3934.

Mosqueda G, Van den Broeck G, Saucedo O, Bailey AM, Alvarez-Morales A & Herrera-Estrella L (1990) Isolation and characterization of the gene from *Pseudomonas syringae* pv. phaseolicola encoding the phaseolotoxin-insensitive ornithine carbamoyltransferase. Mol. Gen. Genet. **222**: 461–466.

Patil SS (1974) Toxins produced by phytopathogenic bacteria. Ann. Rev. Phytopathol. **12**: 259–279.

Patil SS, Kolattukudy PE, Diamond AE (1970) Inhibition of ornithine carbamoyltransferase from bean plants by the toxin of *Pseudomonas phaseolicola*. Plant Physiol. **46**: 752–753.

Patil SS, Tam LQ & Sakai WS (1972) Mode of action of the toxin from *Pseudomonas phaseolicola* I. Toxin specificity, chlorosis and ornithine accumulation. Plant physiol. **49**: 803–807.

Patil SS, Hayward AC, Emmons R (1974) An ultraviolet-induced nontoxigenic mutant of *Pseudomonas phaseolicola* of altered pathogenicity. Phytopathology **64**: 590–595.

Sancar A, Wharton RP, Seltzer S, Kacinski BM, Clarke ND & Rupp WD (1981) Identification of the *urvA* gene product. J. Mol. Biol. **148**: 45–62.

Slauch JM & Silhavy TJ (1991) *Cis*-Acting *ompF* mutations that result in ompR-dependent constitutive expression. J. Bacteriol. **173**: 4039–4048.

428

Staskawicz B, Dahlbeck D, Keen N & Napoli C (1987) Molecular characterization of cloned avirulence genes from race 0 and race 1 of *Pseudomonas syringae* pv. glycinea. J. Bacteriol. **169**: 5789–5794.

Stormo GD, Schneider TD & Gold KM (1982) Characterization of translational initiation sites in *E. coli*. Nucleic Acids Res. **10**: 2971–2996.

Studier FW & Moffatt BA (1986) Use of bacteriophage T7 RNA polymerase to direct selective high-level expression of cloned genes. J. Mol. Biol. **189**: 113–130.

Tsui P, Helu V & Freundlich M (1988) Altered osmoregulation of *ompF* in integration host factor mutants of *Escherichia coli*. J. Bacteriol. **170**: 4950–4953.

29. Application of capillary liquid chromatography – electrospray mass spectrometry to identify major siderophores of *Erwinia amylovora* as proferrioxamines and their potential role in virulence*

GOTTFRIED J. FEISTNER, ANN H. GABRIK and STEVEN V. BEER

Abstract. A large number of hydroxamate siderophores of *Erwinia amylovora* have been identified using capillary liquid chromatography – electrospray mass spectrometry in combination with ferric ion complexation studies. Specifically, we have identified the trihydroxamate siderophores, proferrioxamines E, D_2, X_1, X_2, and novel X_7, the corresponding open chain proferrioxamines G_1, G_{2a}, G_{2b}, and G_{1t} (a truncated G_1), novel tetrahydroxamate siderophores, $T_1 - T_3$, as well as a partially characterized siderophore T_4. Production of the siderophores was achieved in a minimal medium with iron concentrations equal to or less than 10^{-6} M. No qualitative differences in the siderophore pattern were found between *E. amylovora* strains from different parts of the world or between dihydrophenylalanine-producing and –non-producing strains. Proferrioxamines may be important virulence factors of *E. amylovora* and proferrioxamine biosynthesis may offer targets for new fire blight control agents. Before launching a new program to develop specific inhibitors of proferrioxamine biosynthesis, it is desirable to confirm that *E. amylovora* mutants solely defective in proferrioxamine biosynthesis are indeed avirulent. Sneath *et al.* previously reported on mutants Ea321T140, Ea321T145, and Ea321T160 that were avirulent as well as siderophore-deficient and could partly be complemented by cosmid pCPP610 [*Acta Horticult.* **273**, 255 (1990)]. It was not conclusively shown, however, that avirulence was the result of blocked siderophore biosynthesis and not mutations in regulatory genes with the siderophore deficiencies merely being pleiotropic effects. According to our mass spectrometric analysis, only Ea321T160 has a clear proferrioxamine deficiency that can be reversed by complementation with cosmid pCPP610. Ea321T140 is still capable of expressing large amounts of proferrioxamines, whereas pCPP610 does not restore proferrioxamine production in Ea321T145. Traces of proferrioxamines are also found with Ea321T145 and Ea321T160. These results suggest that, other than previously thought, the mutations have affected regulatory rather than biosynthetic genes and that pCPP610 does not contain all the regulatory and biosynthetic genes that are necessary for proferrioxamine production. The question whether fire blight can be controlled via interference with proferrioxamine biosynthesis thus remains open.

Abbreviations: CAS, Chrome Azurol S; CC-ES-MS, Capillary Chromatography Electrospray Mass Spectrometry; TFA, Trifluoroacetic Acid

Introduction

At least five known factors of *E. amylovora* contribute to the pathogenic mechanisms of fire blight. These are extracellular polysaccharides (Geider *et al.*,

---

* This work was presented, in part, at the 39th and 40th ASMS Conferences on Mass Spectrometry and Allied Topics in Nashville, TN (1991) and Washington, DC (1992). Part 7 in the series 'Secondary Metabolites of *Erwinia*'; part 6 see (Feistner, 1988)

C.I. Kado and J.H. Crosa (eds.), Molecular Mechanisms of Bacterial Virulence, 429–444.
© 1994 *Kluwer Academic Publishers.*

1991; Roberts & Coleman, 1991), a low molecular mass phytotoxin (dihydrophenylalanine) (Feistner, 1988; Schwartz et al., 1991), a protein elicitor of the hypersensitive response (harpin) (Wei et al., 1992); an unknown number of structurally uncharacterized siderophores, and the as yet unknown products of the disease-specific (dsp) genes (Barny et al., 1990). The total number of virulence factors is not known and may be rather large. Also, previous work on dihydrophenylalanine has shown that there is variation in the metabolic profiles between wildtype strains (Schwartz et al., 1991; G.J. Feistner, unpublished), which makes it necessary to screen many strains before general conclusions can be drawn. The need therefore exists to develop an analytical method for metabolic profiling that is fast and allows the detection of a broad range of compound classes. Mass spectrometry lends itself for this purpose because it requires only minute sample amounts, is broadly applicable but at the same time highly specific, and, in combination with chromatography, is capable to analyze many compounds simultaneously.

One of the most promising new mass spectrometric methods for metabolic profiling is electrospray mass spectrometry (ES-MS). It produces mainly molecular ions, covers the wide mass range from small secondary metabolites to large antibodies, and can readily be interfaced to liquid chromatography. Amino acids, peptides and proteins are especially amenable to the electrospray technique. For example, we have had excellent success in detecting neuroendocrine hormones in the mass range of 1 to 22 kDa (Feistner et al., 1991), serum albumin (66 kDa) (Swiderek et al., 1992), as well as pathogenicity-related proteins (7 to 32 kDa) from pear (G.J. Feistner & M.K. Kundlas, unpublished). The performance of ES-MS for potential virulence factors belonging to other compound classes, especially secondary metabolites of bacterial pathogens, has not yet been thoroughly evaluated.

For the research reported here, we have used an on-line capillary chromatographic (CC) ES-MS system that allows the introduction of the entire eluate into the mass spectrometer. The latter feature provides for excellent sensitivity and potentially enables us to detect virulence factors in host tissue. The on-line combination preserves the chromatographic resolution and allows quantification by mass chromatography. With mass chromatography, metabolites that are related to virulence may be identified by differential metabolic profiling between wildtype pathogens and avirulent mutants.

Here we report on our success in using CC-ES-MS and differential metabolic profiling to identify the major siderophores of E. amylovora as proferrioxamines and to study the expression of cosmid pCPP610, which has been reported to encode the genes of E. amylovora for hydroxamate siderophore production (Sneath et al., 1990). In addition to the molecular mass, iron regulation, and iron complexation studies reported here, we obtained structural evidence for the proferrioxamines by tandem mass spectrometry and precursor feeding experiments but these results will be presented elsewhere (Feistner et al., 1993). Two previous studies had independently indicated that a deficiency in siderophore biosynthesis may be linked to a loss of virulence in E. amylovora

(Sneath *et al.*, 1990; Vanneste & Expert, 1990), thus it may be possible to control fire blight via inhibition of proferrioxamine biosynthesis. However, the previous studies had not conclusively shown that the mutants were deficient in siderophore biosynthesis rather than in some regulatory function. Only the existence of an avirulent, truly biosynthetic mutant would allow the prediction that fire blight may be prevented by proferrioxamine biosynthesis inhibitors. With regulatory mutants, siderophore deficiency may be just one of several phenotypic effects of a pleiotropic mutation, and avirulence may be due to something other than a block in proferrioxamine biosynthesis. We have therefore subjected the mutants described by Sneath *et al.* to metabolic profiling by CC-ES-MS. Our results indicate that the mutations have effected regulatory events rather than proferrioxamine biosynthesis. Thus we cannot yet make a prediction of how successful fire blight control via inhibition of proferrioxamine biosynthesis may potentially be.

Materials and methods

Culturing of E. amylovora

All strains were maintained on a minimal medium containing (g per liter): $K_2HPO_4 \cdot 3 H_2O$ (10), $MgSO_4 \cdot 7 H_2O$ (0.2), $(NH_4)_2SO_4$ (2), sucrose (10), nicotinic acid (100 μg), and 4 ml of a 2.5×10^{-3} M ferric citrate stock solution (1×10^{-5} M Fe final). Siderophore production was stimulated in the following minimal medium B (g per liter): $K_2HPO_4 \cdot 3 H_2O$ (2), $MgSO_4 \cdot 7 H_2O$ (0.1), $(NH_4)_2SO_4$ (0.4), sucrose (2), nicotinic acid (100 μg), and 2 ml of a 5×10^{-6} M ferric citrate stock solution (1×10^{-8} M Fe final). For Ea321T140, Ea321T145, and Ea321T160, rifampicin (25 mg) and tetracycline (10 mg) were added to the growth medium, whereas the corresponding pCPP610-bearing strains were maintained on tetracycline (10 mg) and spectinomycin (50 mg). Cultivation was typically achieved in volumes of 50 ml in 250 ml Erlenmeyer flasks under heavy stirring (Bellco nine position magnetic stirrer, stirring speed 8; Bellco Glas Inc., Vineland, N.J.).

Culturing of E. coli DH5 and E. coli DH5(pCPP610)

400 ml Cultures of *E. coli* DH5 with or without pCPP610 were started in minimal medium B fortified with 80 mg thiamine (and in the case of *E. coli* DH5(pCPP610) with 20 mg spectinomycin) by inoculating with a loop-full of bacteria taken from an LB plate. Cultures were grown in a Lab-Line rotary shaker at 37°C and 170 rpm. 50 ml of culture each were withdrawn twice to test whether the addition of Nitsch and Nitsch vitamin solution (50 μl of 1000 ×) or casamino acids (5 mg) would accelerate the poor growth observed. Since no beneficial effect was seen, supernatants of the remaining 300 ml cultures were obtained as described below.

Modified CAS medium (Gross, 1990)
60.5 mg Chrome azurol S were dissolved in 50 ml H_2O, mixed with 10 ml 1 mM $FeCl_3 \cdot 6 H_2O$ in 10 mM HCl, and added to a solution of 72 mg hexadecyltrimethylammonium bromide in 40 ml H_2O to give a solution of a dark blue Fe(III) complex. After autoclaving, this CAS solution was mixed with 1 liter of minimal medium B (with 1.5% agar but without ferric citrate). The medium turns orange upon removal of the iron from the blue complex by a strong Fe(III) chelator such as the proferrioxamines.

Preparation of methanolic extracts
Bacterial cells were removed by centrifugation at 8,600 × g for 10 min, the supernatant dried by rotary evaporation below 40°C, and the residue extracted twice with 50 ml methanol for 15 min (for the *E. coli* supernatants 2 × 100 ml were used). The methanolic extract was again dried by rotary evaporation and the residue dissolved in 1 ml H_2O (0.5 ml for the *E. coli* extracts). Any precipitate of 6-thioguanine, which is poorly soluble in water, was removed by centrifugation. The supernatants were stored at −20°C until analysis by either analytical (2 mm ID) or capillary reversed-phase chromatography.

Identification of proferrioxamines
Extracts were analyzed by reversed-phase chromatography either on a Hewlett-Packard 1090M workstation or on an on-line capillary chromatography – electrospray mass spectrometric system comprised of a Finnigan (San Jose, CA) TSQ700 triple stage quadrupole mass analyzer, a DECstation 2100 workstation, an automatic microgradient chromatography system (T.D. Lee *et al.*, unpublished), and self-constructed capillary columns (Davis & Lee, 1991). In addition to molecular mass measurements, the CC-ES-MS system allowed structural analyses via mass separation in quadrupole Q_1, dissociation in collision cell Q_2, and mass analysis of the product ions in quadrupole Q_3. The individual proferrioxamines were confirmed on the basis of characteristic fragment ions arising from cleavage at the hydroxamate bonds, which in turn resulted in abundant ions for multiples of the ω-*N*-succinylamino-α-*N*-hydroxyaminoalkane monomers. Furthermore, acetylation shifted the molecular masses, and feeding of the biosynthetic precursors putrescine and cadaverine shifted the ratios between the various proferrioxamines in the expected way. Details are given in a subsequent paper (Feistner *et al.*, 1993). Inhibition of proferrioxamine biosynthesis at high iron concentration is demonstrated with Fig. 3, and the iron chelating capability was proven by treating the extracts with ferric citrate, as shown in Fig. 4. Injection volumes generally were 50 μl on the HP1090M and 2–10 μl on the capillary chromatographic system. For *E. coli* DH5(pCPP610), the entire extract was loaded in two batches onto an analytical reversed phase column. After the void volumes had passed through, any absorbed components were eluted with the usual gradient of 0–45% acetonitrile in 0.1% TFA over 40 min, the entire eluate

pooled, dried, and taken up in 20 μl 0.1% TFA for mass spectrometric analysis. To avoid clogging of the electrospray interface transfer capillary by the high concentrations of sugar in the methanolic extracts, the void volumes of the CC-ES-MS runs were diverted into a paper tissue.

Results and discussion

All wildtype strains analyzed (11) were found to produce nine or more different proferrioxamines, which mostly arise from various combinations of succinic acid, 1-amino-5-*N*-hydroxyaminopentane (N-hydroxycadaverine), 1-amino-4-*N*-hydroxyaminobutane (N-hydroxy-putrescine), and 1-amino-3-*N*-hydroxyaminopropane (Fig. 1). Whereas the iron complexes are known as ferrioxamines, the free chelators are variously referred to as desferri-ferrioxamines, desferrioxamines, deferrioxamines, or deferoxamines. Since this inconsistency

$$R_1HN-(CH_2)_m-N(OH)CO(CH_2)_2CONH-(CH_2)_n-N(OH)CO(CH_2)_2CO$$
$$R_2CO(CH_2)_2CO-[N(OH)(CH_2)_5NHCO(CH_2)_2CO]_p-N(OH)-(CH_2)_o-NH$$

PFO	R_1	m	n	o	p	R_2
G_1	H	5	5	5	0	OH
G_{2a}	H	5	5	4	0	OH
G_{2b}	H	4	5	5	0	OH
E	cyclic	5	5	5	0	cyclic
D_2	cyclic	4	5	5	0	cyclic
X_1	cyclic	4	4	5	0	cyclic
X_2	cyclic	4	4	4	0	cyclic
X_7	cyclic	3	5	5	0	cyclic
T_1	cyclic	5	5	5	1	cyclic
T_2	cyclic	4	5	5	1	cyclic
T_3	cyclic	3	5	5	1	cyclic

Figure 1. Structures of proferrioxamines that have been identified in *Erwinia amylovora*. G_{1t} is a truncated G_1 that is missing the C-terminal succinyl moiety. Identification of the positional isomers G_{2a} and G_{2b} is based on differences in the mass spectrometric fragmentation pattern but tentative. T_4 has not been fully characterized but does contain ω-*N*-succinylamino-α-*N*-hydroxyaminopentane and –butane residues.

appears to indicate dissatisfaction with the current nomenclature, the prefix pro- is suggested and used here. It is accepted by the scientific community to indicate precursor molecules in compound classes as different as peptide hormones and the Fe(II) chelator proferrorosamine.

Specifically, we have identified proferrioxamines D_2 and E, the smaller proferrioxamine homologs X_1, X_2, and X_7, the corresponding open chain proferrioxamines G_1, G_{2a}, and G_{2b}, and a truncated form of G_1, which lacks a succinic acid and which we call G_{1t}. In addition, we have identified tetrahydroxamate siderophores T_1-T_3, which to our knowledge have not been described before. The apparent presence of 3-succinylamino-1-N-hydroxy-aminopropane in T_3 and X_7 is worth special mention because previous feeding experiments had indicated that 1,3-diaminopropane does not get incorporated into proferrioxamines by *Streptomyces olivaceus* (Konetschny-Rapp, 1990). Proferrioxamines E (major siderophore) and D_2 (traces) have also been independently identified in France (D. Expert, pers. comm.).

Fig. 2 gives an example of the mass chromatograms that provided the basis for this study and also shows the relative elution order of the proferrioxamines. With so many different siderophores, each of which may be affected by the

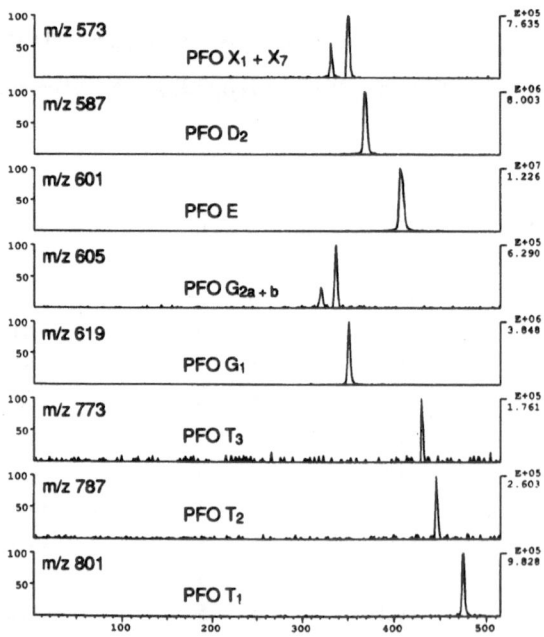

Figure 2. Identification of the major proferrioxamines (PFOs) of *E. amylovora* Ea321 by CC-ES-MS. Shown are relative ion intensities versus scan numbers; absolute ion intensities are indicated on the right. Chromatographic conditions were as follows: C_{18} capillary column (18 cm × 250 μm, 5 μm, 300Å particles), 2 μl extract injected, linear gradient of 1–50% acetonitrile in 0.1% TFA over 50 min, gradient injection at 2 min, flow rate 20 μL/min for first 2.5 min then ramped to 2 ml/min in 0.5 min.

mutations under study, singly or in combination, the speed and comprehensiveness of our CC-ES-MS analysis is a definitive advantage. Speed and minimal handling of the sample also keep any decomposition of the proferrioxamines, which are susceptible to hydrolysis, to a minimum. Detection limits for the proferrioamines were much lower with CC-ES-MS than with standard analytical chromatography and absorbance detection at 215 nm; in fact, the minor proferrioxamines were not detected on our HP1090M workstation. Some mass traces show two peaks and this is due to isobaric but structurally distinct siderophores. For example, proferrioxamine X_7 with two N-hydroxycadaverine and one 1-amino-3-N-hydroxyaminopropane residues is isobaric with proferrioxamine X_1 containing one hydroxycadaverine and two N-hydroxyputrescine residues.

Metabolic profiling by CC-ES-MS was useful in determining the iron concentration at which E. amylovora is stimulated to its highest proferrioxamine production (Fig. 3). E. amylovora did excrete proferrioxamines when the iron concentration of the medium was 10^{-6} M or lower, but not at 10^{-5} M Fe^{3+}. Maximal proferrioxamine production appeared to occur at only 10^{-8} M ferric citrate, which is the concentration we have routinely used in this study. Since a high ammonium content in the growth medium is known to repress the expression of virulence genes in E. amylovora (Wei et al., 1992; Barny & Laurent 1991), we have kept the ammonium concentration of our growth medium as low as possible (3 mM), i.e. as low as compatible with a reasonable growth rate for E. amylovora. The $(NH_4)_2SO_4$ concentration employed in this study is thus even lower than the one in the so-called 'inducing medium' used at Cornell (5 mM) (Wei et al., 1992) or the M9 and Tris media used in studies on E. amylovora in France (Vanneste et al., 1990); this may possibly explain differences in proferrioxamine production between our laboratories.

Fig. 3 also serves to make two additional points. One is that ES-MS is capable of detecting metabolites of E. amylovora other than proferrioxamines, which is important with regard to our interest in using metabolic profiling by CC-ES-MS for the identification of virulence factors other than siderophores. The second point is regarding the mass spectral resolution, which is superior to the chromatographic resolution. When the culture filtrate with 10^{-5} M ferric citrate was run on a standard analytical chromatographic system, a small UV-active peak eluted in the position of proferrioxamine E, thus giving the impression that the siderophore was still produced at this iron concentration. With CC-ES-MS, however, the corresponding, not yet identified compound gave rise to a major ion at m/z 607 that could easily be distinguished from the molecular ion of proferrioxamine E ($[M+H]^+$ 601). Our capillary chromatographic system also resolved the unknown from proferrioxamine E better than the standard analytical column. Proferrioxamine E could have been unambiguously identified by complexation with ferric citrate and absorbance monitoring at 430 nm but the latter methodology is much less sensitive than mass spectrometric detection of the free proferrioxamines.

The capability of the different proferrioxamines to bind ferric ions was

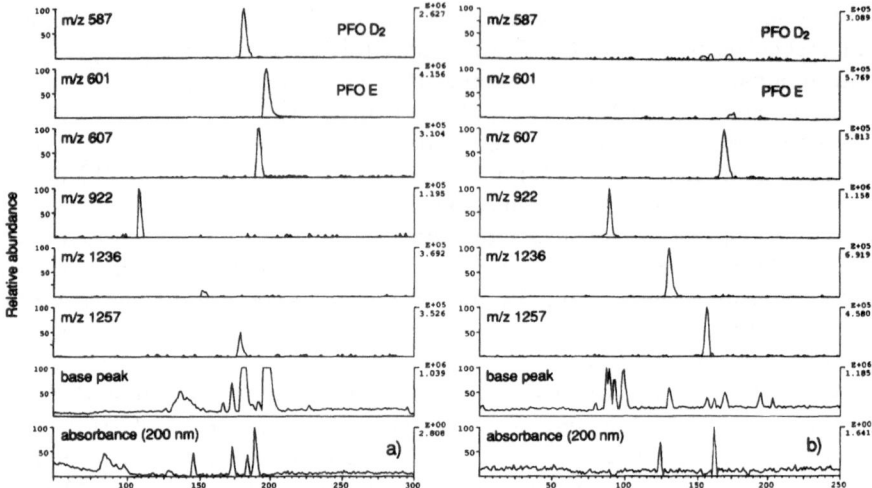

Figure 3. Differential metabolic profiling of *E. amylovora* Ea266 at different iron concentrations; (a) 10^{-8}, and (b) 10^{-5} M ferric citrate; mass chromatograms shown only for selected metabolites; PFO, proferrioxamine. Note the appearance of proferrioxamines D_2 and E only at the lower ferric ion concentration. Other components with ions at *m/z* 922, 1236 and 1257, but of unknown structure, are more abundant at the higher iron concentration, whereas the abundance of *m/z* 607 is more or less constant.

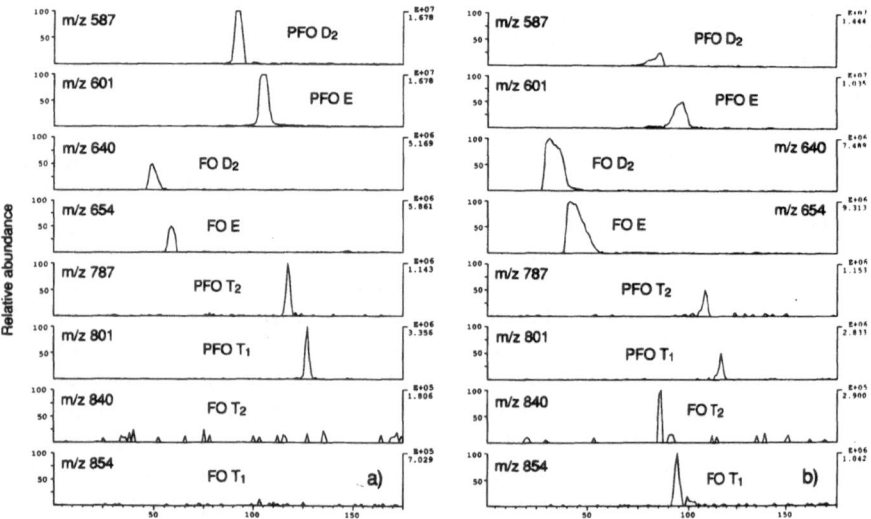

Figure 4. Iron complexation studies for (pro)ferrioxamines D_2, E, T_1 and T_2. Shown are the corresponding mass chromatograms for *E. amylovora* Ea5/84 obtained (a) before and (b) after addition of 775 pmole ferric citrate to 40 μl of the methanolic extract; FO, ferrioxamine; PFO, proferrioxamine.

shown by treating a culture extract with a solution of ferric citrate. Fig. 4 gives an example of the corresponding shifts in retention time and mass (+ Fe − 3H = 53 u) for proferrioxamines D_2, E, T_1 and T_2. From these data it is clear that, despite their considerably larger ring sizes, the tetrahydroxamate siderophores are not only able to bind ferric ions, but must have binding constants similar to those of the trihydroxamate siderophores; T_1 and T_2 would otherwise not be able to successfully compete for the binding of iron in the presence of excess trihydroxamates. This finding could not be predicted because the larger ring sizes may not have fitted the metal ion very well. Apparently the tetrahydroxamates are flexible enough to wrap tightly around the ferric ion. We have also partially characterized three other unusual hydroxamate compounds, T_4-T_6 ([M+H]$^+$ 623, 720, and 734, respectively), but these are apparently only produced by some strains. Furthermore, T_5 and T_6 do not seem to be siderophores or at least do not bind iron as tightly as the trihydroxamates.

As Fig. 5 shows, the proferrioxamine pattern of *E. amylovora* wildtype strains is fairly consistent, regardless of the country where the particular strain was isolated and of whether it is a dihydrophenylalanine producing strain or not. While differences in the absolute amounts of each proferrioxamine were found, the cyclic trihydroxamate proferrioxamines D_2 and E were always the major siderophores. This is an important finding because it allows us to generalize and to suggest that, provided the putative link between proferrioxamine production and virulence can be further substantiated, any new control agents that are based on interference with proferrioxamine biosynthesis are likely to control fire blight worldwide and not just a small subpopulation of *E. amylovora*.

Proferrioxamines are common to a large number of microorganisms. In this context, it is particularly noteworthy that proferrioxamines E and D_2 have previously been identified as the principal siderophores of *Erwinia herbicola* (*Pantoea agglomerans*) and proferrioxamines G_1 and G_2 as the principal siderophores of *Hafnia alvei* (Berner *et al.*, 1988; Reissbrodt *et al.*, 1990). Since *E. herbicola* is an epiphyte, there is no reason to assume that proferrioxamines E and D_2 themselves may be harmful to plants. The consequences of the fact that *E. amylovora* also produces tetrahydroxamate siderophores, are not yet known. However, all proferrioxamines may or may not be virulence-enhancing factors in the sense that they may be necessary for *E. amylovora* to grow rapidly in its hosts. After all, fire blight does only occur under conditions that are optimal for the growth for *E. amylovora*. A previous example where a siderophore has been linked to virulence is chrysobactin, which apparently is required for the pathogenicity of *Erwinia chrysanthemi* towards african violets (see chapter 12). On the other hand, siderophore production does not appear to be important for the virulence of *Pseudomonas syringae* or for tumor formation by *Agrobacterium tumefasciens* (Expert & Gill, 1992).

It is conceivable that *E. amylovora* produces other, not yet identified siderophores that can compensate for any proferrioxamine deficiency. If these would be catechol-type siderophores, then an inhibitor directed only against

438

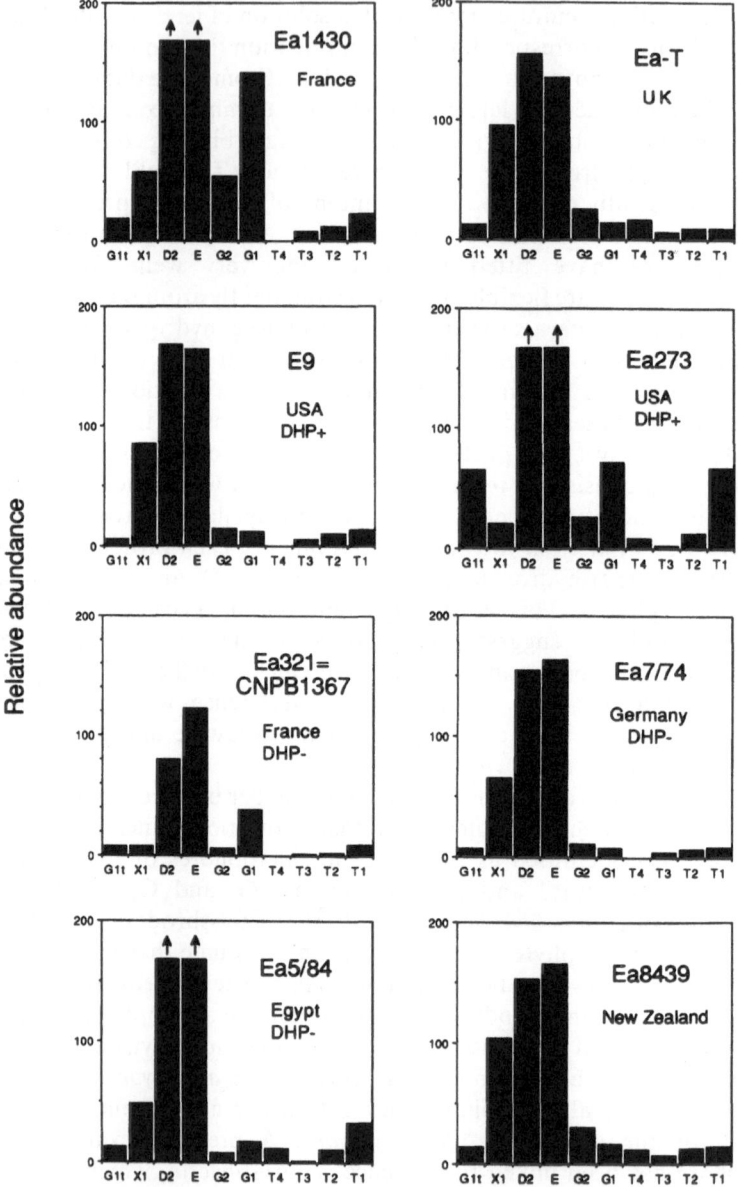

Figure 5. Relative abundance of proferrioxamines for wild-type *E. amylovora* isolated in different parts of the world. Also analyzed were extracts from Ea266, Ea322, Ea1/79, Ea-S (eps⁻), and Ea-9AP (NTG mutant; hrp⁻), and these strains gave similar results. For proferrioxamine structures see Fig. 1. The bar graphs represent the peak heights from the mass chromatograms minus the chemical noise level of approximately 1×10^5 counts. In order to detect the minor siderophores, the molecular ion abundances for some of the major proferrioxamines were allowed to exceed the dynamic range setting of the mass spectrometer (indicated by arrows).

hydroxamate siderophores may not be effective in controlling fire blight. *Pseudomonas aeruginosa*, for example, synthesizes two structurally very different siderophores, namely, pyochelin and pyoverdin. Only a defect in pyoverdin but not in pyochelin production slows the growth of *P. aeruginosa* in serum (Visca *et al.*, 1992). Alternative siderophores of *E. amylovora* could possibly be enterobactin, aerobactin, or chrysobactin, which have been found with other *Erwinia* species. However, a cursory screening of our metabolic profiles of *E. amylovora* has not yet turned up any evidence for the molecular ions of these siderophores.

We therefore sought to confirm the importance of proferrioxamines for fire blight by showing that the previously described siderophore/virulence mutants Ea321T140, Ea321T145, and Ea321T160 are indeed biosynthetic mutants. We also wanted to confirm that cosmid pCPP610 contains the biosynthetic genes for proferrioxamine biosynthesis. Based on their results with the chrome azurol (CAS) siderophore test of Schwyn and Neilands (Gross, 1990) and the modified Czaky hydroxamate test of Gillam (Gillam *et al.*, 1981), Sneath *et al.* had concluded that siderophore production in these mutants was substantially reduced but restored by pCPP610.

Initially, we assumed that the mutations were stable and grew the mutants through five transfers without antibiotics, and analyzed transfers 3–5. Unexpectedly, proferrioxamines E and D_2 were detected in all cases. At that point, all bacteria also produced large halos on chrome azurol S plates. When the cultures from the fifth transfer were placed into media containing the appropriate antibiotics (transfer 6), none of the cosmid-containing mutants grew, whereas the 'siderophore-mutants' gave the results shown in Fig. 6. Ea321T140 still produced large amounts of proferrioxamines D_2 and E, whereas Ea321T145 and Ea321T160 produced substantially reduced amounts and traces of proferrioxamines D_2 and E, respectively. We thus became concerned that, without selective pressure from antibiotics, the cosmid and the transposons may not be entirely stable in the mutants. Thereafter, all mutants were always grown in the presence of the appropriate antibiotics. The antibiotic tetracycline that had to be used gave rise to two very large and broad UV-active peaks that covered the retention times of both the proferrioxamines as well as the ferrioxamines. Thus, the siderophore mutant studies were impossible to perform using absorbance detection, but, since the molecular masses of tetracycline and proferrioxamines are very different, we had no trouble detecting the proferrioxamines in the presence of tetracycline by CC-ES-MS.

Fig. 6 also shows the expression of siderophores in two additional batches (transfers 0 & 1). The results largely agree with those from the sixth transfer described above. Only Ea321T160 behaved as expected in that it showed a clear deficiency in proferrioxamine production and reversal of this deficiency by cosmid pCPP610. Ea321T160 was apparently affected and restored in the synthesis of all proferrioxamines rather than, for example, blocked in the biosynthesis of either N-hydroxycadaverine or N-hydroxyputrescine. The latter would have resulted in the loss/restoration of specific rather than all

440

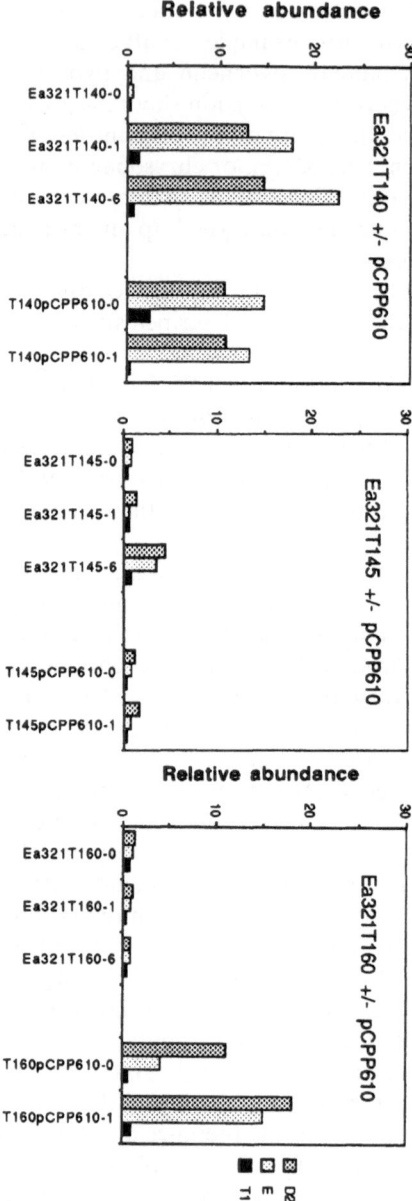

Figure 6. Results from differential metabolic profiling between *E. amylovora* mutants Ea321T140, Ea321T145, and Ea321T160 and the corresponding strains that carry cosmid pCPP610; shown for each mutant are the relative abundances of proferrioxamines D_2, E, and T_1. Extensions refer to the number of consecutive transfers in liquid culture (thus ' – 0' indicates the first liquid culture started with an inoculum from an agar plate). For proferrioxamine structures see Fig. 1.

proferrioxamines. Ea321T140 was found to produce large amounts of proferrioxamines in two out of three batches. When we grew two more batches of Ea321T140 (data not shown), proferrioxamines were detected in both cases. Ea321T140 thus seems capable of proferrioxamine synthesis. Conversely, Ea321T145 was deficient in proferrioxamine production but proferrioxamine synthesis was not restored by cosmid pCPP610. The latter observation is inconsistent with pCPP610 containing all the genes necessary for proferrioxamine production. Also, at least traces of proferrioxamines E and D_2 were seen with all extracts of Ea321T145 and Ea321T160, but not with solvent blanks in between. The mutants therefore only seem to be deficient in, but not entirely lacking, proferrioxamine biosynthesis. It thus seems more likely that pCPP610 restores regulatory rather than biosynthetic genes.

Our results are partly contradictory to the observations reported by Sneath et al. Ea321T140 was supposed to be siderophore-deficient whereas we find that Ea321T140 produces proferrioxamines, and Ea321T145 was supposed to be restored in siderophore production by pCPP610 but we cannot confirm this. The reason for these discrepancies is not known, but since the mutations apparently have affected regulatory genes, and E. amylovora is highly responsive to environmental stimuli, it is conceivable that the differences are due to the use of different growth media.

It is interesting to note that the only mutant whose virulence could partly be restored by pCPP610, namely Ea321T140, is not deficient in siderophore biosynthesis. Again, this indicates that pCPP610 restores a virulence function separate from siderophore biosynthesis. On the other hand, pCPP610 does restore full siderophore-production but not virulence to Ea321T160 (Sneath et al., 1990). The mutations in Ea321T140 and Ea321T160 apparently have caused avirulence for different reasons, and pCPP610 obviously restores more than one gene. Whether the genes that are introduced by pCPP610 are identical to the ones that have been interrupted in the mutants is not yet known. The failure to restore (full) virulence may also be the result of a slower growth rate due to the extra requirement for cosmid replication. Finally, the apparent instability of cosmid pCPP610 in the absence of a selective pressure from antibiotics complicates the interpretation of virulence tests on immature pear slices because these are not performed in the presence of antibiotics.

Based on larger halos around merodiploid colonies in the CAS assay, Sneath et al. also suggested that pCPP610 leads to expression of E. amylovora siderophores in Escherichia coli DH5. We tried to confirm this, but, as would be expected if pCPP610 was harboring regulatory genes and not (all) the genes for proferrioxamine biosynthesis, we could not find any proferrioxamines in the culture filtrate of E. coli DH5(pCPP610). However, E. coli DH5, with or without pCPP610, grows only poorly in our minimal medium, even when supplemented with thiamine and Nitsch and Nitsch vitamin solution or casamino acids. It would be best if pCPP610 were expressed in a better growing host before any final conclusions are made.

Concluding remarks

In pursuing this research we had several goals, namely, to show that ES-MS is useful for analyzing secondary metabolites of bacteria, to demonstrate the feasibility of differential metabolic profiling by CC-ES-MS, to structurally characterize the siderophores of *E. amylovora*, and to prove that the siderophores are essential virulence factors in fire blight. We have been successful in all but the latter aspect. Even where we fall short of our goal, this is not because of a problem with our CC-ES-MS methodology that was to be tested but rather because of the difficult (molecular) biology of fire blight. Metabolic profiling by CC-ES-MS still appears to be the method of choice for determining the importance of proferrioxamines. However, for this approach to work, true proferrioxamine biosynthetic mutants are needed. Perhaps they can be found among the siderophore mutants created by the Paris group.

Due to our mass spectrometric metabolic profiling, we now have precise knowledge of the chemical nature of the siderophores of *E. amylovora*. This specific knowledge will allow us to address the importance of siderophores in fire blight with a more straightforward approach in the future. Gene clusters for the biosynthesis of proferrioxamines have been cloned from other microorganisms and may be taken advantage of for the identification and characterization of the proferrioxamine genes of *E. amylovora*. When subsequently these genes are specifically targeted for deletion by site-directed mutagenesis and marker exchange, pleiotropic mutations will no longer be of any concern.

We would like to emphasize that metabolic profiling by CC-ES-MS is not limited to fire blight, and that our methodology can and should be extended to other pathogens of plants, animals, and humans as well as the hosts themselves. We expect CC-ES-MS to have a major impact on our ability to understand (and eventually treat) plant, animal and human diseases and disorders.

On the other hand, we do not want to give the impression that our metabolic profiling by CC-ES-MS has reached its full potential. While our capillary reversed phase chromatographic system is well suited for the analysis of neutral proferrioxamines and other more lipophilic compounds, it does not yet allow for the analysis of the full spectrum of compounds produced by *E. amylovora*. For example, the chromatographic system will need to be modified for the analysis of very hydrophilic compounds such as 6-thioguanine and dihydrophenylalanine, because the surface capacity of our capillary column is very small and hydrophilic compounds are hardly retained at all. We could give up the sensitivity advantage of our capillary chromatography and work with standard analytical columns and post-column splitting of the effluent. However, this would only partially solve our problem. Better results are expected from employing a combination of various chromatographic principles and/or various precolumn derivatizations. Such a more comprehensive CC-ES-MS analysis is most likely needed for the projected characterization of the previously described *dsp*-mutants (Barny *et al.*, 1990; Bauer & Beer, 1991).

Acknowledgements

We wish to thank Manjit K. Kundlas for preparing and analyzing extracts in the initial phase of the project, Michael T. Davis for assistance with the capillary chromatography, Douglas C. Stahl for implementing the automated capillary gradient system, and Terry D. Lee and John E. Shively for general support of our fire blight research. Wildtype strains of *E. amylovora* were kindly made available by K. Geider (MPI Ladenburg, FRG), J. Laurent (INRA, Paris, France), and J.W. Mansfield (University of London, Wye, UK). This work was supported by a Beckman Grant Award from the City of Hope.

References

Barny M-A & Laurent J (1991) Regulation of pathogenicity genes in *Erwinia amylovora*. J. Exp. Bot. 42 Supplement: 238.

Barny M-A, Guinebretiere MH, Marcais B, Coissac E, Paulin JP & Laurent J (1990) Cloning of a large gene cluster involved in *Erwinia amylovora* CFBP1430 virulence. Mol. Microbiol. 4: 777–786.

Bauer DW & Beer SV (1991) Further characterization of an *hrp* gene cluster of *Erwinia amylovora*. Mol. Plant-Microbe Interact. 4: 493–499.

Berner I, Konetschny-Rapp S, Jung G & Winkelmann G (1988) Characterization of ferrioxamine E as the principal siderophore of *Erwinia herbicola* (*Enterobacter agglomerans*). Biol. Metals 1: 51–56.

Davis MT & Lee TD (1992) Analysis of peptide mixtures by capillary high performance liquid chromatography: A practical guide to small-scale separations. Protein Sci. 1: 935–944.

Expert D & Gill PR (1992) Iron: a modulator in bacterial virulence and symbiotic nitrogen-fixation. p. 229–245. *In*: Verma DPS (ed.) Molecular signals in plant-microbe communication CRC Press Inc., Boca Raton, Florida.

Feistner GJ (1988) (L)-2,5-Dihydrophenylalanine from the fireblight pathogen *Erwinia amylovora*. Phytochemistry 27: 3417–3422.

Feistner GJ, Barofsky DF, Evans CJ, Faull KF & Roepstorff P (1991) Charting of rat pituitary peptides by plasma desorption and electrospray mass spectrometry. p. 45–53. *In*: Standing K & Ens W (eds.) Methods and mechanisms for producing ions from large molecules Plenum, New York.

Feistner GJ, Stahl DC, & Gabrik, AH (1993) Proferrioxamine siderophores from *Erwinia amylovora*. A capillary liquid chromatography – electrospray tandem mass spectrometric study. Org. Mass Spectrom. 28: 163–175.

Geider K, Bellemann P, Bernhard F, Chang J-R, Geier G, Metzger M, Pahl A, Schwartz T & Theiler R (1991) Exopolysaccharides in the interaction of the fire-blight pathogen *Erwinia amylovora* with its host cells. p. 90–93. *In*: Hennecke H and Verma DPS (eds.) Advances in molecular genetics of plant-microbe interactions. Vol. 1. Kluwer Academic Publishers, Dordrecht, The Netherlands.

Gillam AH, Lewis AG & Andersen RJ (1981) Quantitative determination of hydroxamic acids. Anal. Chem. 53: 841–844.

Gross M (1990) Siderophores and fluorescent pigments. In: Klement Z, Rudolph K & Sand DC (eds.) Methods in phytobacteriology. (pp. 434–437) Akademiai Kiado, Budapest.

Konetschny-Rapp S (1990) Neue mikrobielle Eisenkomplexbildner. Screening, Isolierung, Strukturaufklärung und komplexchemische Untersuchungen. Thesis, University of Tübingen, FRG.

Reissbrodt R, Rabsch W, Chapeaurouge A, Jung G & Winkelmann G (1990) Isolation and identification of ferrioxamine G and E in *Hafnei alvei*. Biol. Metals. 3: 54–60.

Roberts IS & Coleman MJ (1991) The virulence of *Erwinia amylovora*: molecular genetic perspectives. J. Gen. Microbiol. 137: 1453–1457.

Schwartz T, Bernhard F, Theiler R & Geider K (1991) Diversity of the fire blight pathogen in production of dihydrophenylalanine, a virulence factor of some *Erwinia amylovora* strains. Phytopathology 81: 873–878.

Sneath BJ, Howson JM & Beer SV (1990) Siderophore genes of *Erwinia amylovora*: cloning and putative role in virulence. Acta Horticult. 273: 255–257.

Swiderek KM, Chen S, Feistner GJ, Shively JE & Lee TD (1992) Applications of liquid chromatography electrospray mass spectrometry (LC-ES/MS). p. 457–465. *In*: R. H. Angeletti (ed.) Techniques in protein chemistry III. Academic Press, New York.

Vanneste JL & Expert D (1990) Detection and characterization of an iron uptake system in *E. amylovora*. Acta Horticult. 273: 249–253.

Vanneste JL, Paulin J-P & Expert D (1990) Bacteriophage *Mu* as a genetic tool to study *Erwinia amylovora* pathogenicity and hypersensitive reaction on tobacco. J. Bacteriol. 172: 932–941.

Visca P, Serino L & Orsi N (1992) Isolation and characterization of *Pseudomonas aeruginosa* mutants blocked in the synthesis of pyoverdin. J. Bacteriol. 174: 5727–5731.

Wei Z-M, Laby RJ, Zumoff CH, Bauer DW, He SY, Collmer A & Beer SV (1992) Harpin, elicitor of the hypersensitive response produced by the plant pathogen *Erwinia amylovora*. Science 257: 85–88.

Wei Z-M, Sneath BJ & Beer SV (1992) Expression of *Erwinia amylovora hrp* genes in response to environmental stimuli. J. Bacteriol. 174: 1875–1882.

Regulation of virulence genes and signal transduction

30. Transcriptional elements in *Pasteurella haemolytica* leukotoxin expression

SARAH K. HIGHLANDER

Abstract. *Pasteurella haemolytica* is the primary bacterial agent in bovine shipping fever pneumonia and produces and secretes a species-specific leukotoxin. The toxin is a member of the RTX family of cytolysins, and is maximally expressed during log-phase growth at 37°C. We are examining the regulatory mechanisms involved in leukotoxin expression using operon fusions and *trans*-complementation in *Escherichia coli*. The leukotoxin promoter region contains a static DNA bend that enhances transcription. A *trans*-activator of this promoter has been isolated and partially characterized. A second, divergent promoter maps immediately upstream of the DNA bend and transcribes a gene encoding a periplasmic binding protein. This permease protein, and its associated nucleotide-binding and membrane spanning components, may be involved in sensory regulation of the leukotoxin in *Pasteurella*.

Abbreviations: FIS, Factor for Inversion Stimulation; RTX, Repeat Toxin; IHF, Integration Host Factor; OD, Optical Density; X-Gal, 5-Bromo-4-Chloro-3-Indolyl-β-D-Galactopyranoside

Introduction

Shipping fever, or pasteurellosis, is a severe respiratory disease of cattle responsible for losses of more than $800 million per year to the American feedlot and dairy industries (Drummond, *et al.*, 1981). Pasteurellosis kills at least five percent of American feedlot cattle and is responsible for morbidity and decreased weight gain, or performance, in at least an additional ten percent of these animals (Martin, *et al.*, 1981). Both viral and bacterial components are involved in development of the disease, though the ultimate cause of death is secondary bacterial infection with *Pasteurella haemolytica* serotype A1. Many healthy calves carry *P. haemolytica* in the nasopharynx (Magwood, *et al.*, 1969). Yet, following stress and exposure to viruses, such as infectious bovine rhinotracheitis virus (IBR) or parainfluenza-3 (PI-3), *P. haemolytica* bacteria multiply rapidly and descend into the lung (Frank *et al.*, 1986, 1987; Jericho, *et al.*, 1986). The standard treatment for this disease has been antibiotic therapy, which, until recently, included prophylactic feeding of broad spectrum antibiotics to all cattle on the lot. As a result, most *P. haemolytica* organisms isolated from feedyard animals are resistant to one or more of the common antibiotics and the disease has become difficult to treat by standard antimicrobial methods.

C.I. Kado and J.H. Crosa (eds.), Molecular Mechanisms of Bacterial Virulence, 447–462.
© 1994 *Kluwer Academic Publishers.*

448

P. haemolytica secretes a 102 kD heat-labile leukotoxin that contributes to its pathogenicity by lysing alveolar macrophages, thereby impairing primary lung immune defense mechanisms (Himmel, *et al.*, 1982; Markham & Wilkie, 1980; Shewen & Wilkie, 1982). Identification of the secreted leukotoxin explained the failure of conventional whole cell, supernatant-free, *P. haemolytica* bacterins to provide immunity against shipping fever (Markham & Wilkie, 1980) and has made the leukotoxin a target for vaccine development (Berget, *et al.*, 1991; Lo, Strathdee, & Shewen, 1987; Yates, *et al.*, 1983).

The *P. haemolytica* leukotoxin gene cluster (*lktCABD*) has been cloned and sequenced (Highlander, *et al.*, 1989; Lo, *et al.*, 1985; Lo *et al.*, 1987; Strathdee & Lo, 1989b). A map of the leukotoxin gene cluster is shown in Fig. 1. The *lktA* gene encodes the 102 kD toxin protein while LktC is responsible for post-translational activation of LktA (Highlander, *et al.*, 1990; Lo *et al.*, 1985). Both the LktB and LktD proteins are required for toxin secretion from the bacterial cell (Highlander, Engler & Weinstock, 1990; Strathdee & Lo, 1989a). In addition to its importance in pasteurellosis, the leukotoxin system is a model for the growing repeat toxin (RTX) family (Strathdee & Lo, 1989a). Members of this toxin family are found in the swine pathogen *Actinobacillus pleuropneumoniae* (Chang, *et al.*, 1989), the urinary tract pathogens *Escherichia coli, Proteus mirabilis, Proteus vulgaris* and *Morganella morganii* (Felmlee, *et al.*, 1985; Strathdee & Lo, 1987; Koronakis, *et al.*, 1987), in *Bordetella pertussis* (Glaser, *et al.*, 1988), in *Neisseria meningitidis* (Thompson, *et al.*, 1991), and in the periodontal pathogen *Actinobacillus actinomycetemcomitans* (Kolodrubetz, *et al.*, 1989). While the diseases caused by these organisms are diverse, the toxins expressed are very similar in their genetic organization and activity. Each protein is a membrane-active cytotoxin and possesses a series of glycine-rich repeats near its carboxy-terminus; the repeats (GGXGDD) form the basis for the RTX nomenclature. Some, such as the *P. haemolytica* leukotoxin, are species-specific with regard to leukotoxic activity (Shewen & Wilkie, 1982), yet all have some non-specific hemolytic activity. Since these gene clusters are so similar, it is anticipated that principles of toxin regulation that are revealed

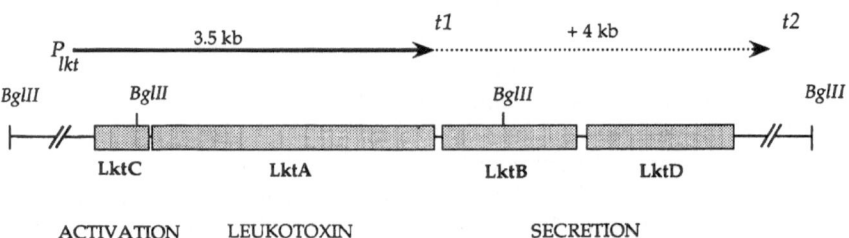

Figure 1. Reading frame map of the *P. haemolytica* leukotoxin gene cluster. The locus is contained within three contiguous *Bgl*II fragments of 3.6, 3.9 and 6.4 kb, respectively. Transcription initiates upstream of the *lktC* gene and continues, via rare antitermination, through terminator *t1*, to terminator *t2*.

from study of the *P. haemolytica* leukotoxin will be useful in studies of other members of the RTX family.

Transcription of the leukotoxin locus is organized into two separate, though perhaps not distinct, parts. A single 3.5 kb transcript encompasses the *lktC* and *lktA* genes and terminates at a putative ϱ-independent terminator within the *lktA-lktB* intergenic region (Highlander, *et al.*, 1990; Strathdee & Lo, 1989a). Full-length *lktCABD* transcripts are believed to arise from rare transcriptional read-through beyond the terminator (Strathdee & Lo, 1989a); this also occurs in the *E. coli* hemolysin system (Koronakis, *et al.*, 1989). Leukotoxin expression in broth cultures of *P. haemolytica* is maximal during early log-phase growth at 37°C (Highlander, *et al.*, 1990; Strathdee & Lo, 1989b), and it has been reported that transcription of the leukotoxin gene cluster is negatively regulated by decreased iron concentration and acidic pH (Strathdee & Lo, 1989b). This report describes studies aimed at characterization of the mechanisms involved in some of these regulatory processes.

Results and Discussion

DNA bending within the leukotoxin promoter region

Primer extension analysis indicated that transcription of the entire leukotoxin gene cluster was initiated 31 bp upstream of the putative start codon for the *lktC* gene (Fig. 2) (Highlander, *et al.*, 1990). A second transcriptional start-site, at bp −290 has also been reported (Strathdee & Lo, 1989b). The primary transcript initiates at a cytosine residue and is preceded by an appropriately spaced σ^{70}-like −10 promoter sequence. No consensus-like −35 sequence is observed, but a series of $CA_6(C/T)A$ repeats, phased at approximately 10 bp intervals are present. As described in other systems (Wu & Crothers, 1984), these repeats can form a static DNA bend. The region also contains consensus-like binding sites for two proteins that are involved in protein-directed DNA bending: integration host factor (IHF) (Yang & Nash, 1989) and factor for inversion stimulation (FIS) (Hubner & Arber, 1989).

Bent DNA fragments, exhibit retarded mobility in polyacrylamide gels (Wu & Crothers, 1984), and this property was used to map the position of the bend near the leukotoxin promoter. A tandem duplication of a 695 bp *Eco*RV + *Hin*dII fragment containing the *lktCA* promoter was constructed, as shown in Fig. 3a, and then digested with restriction endonucleases that cleave the original monomer fragment only once. The plasmid with the duplication (pSH237) was digested with *Ssp*I, *Fok*I, *Pvu*I, *Dde*I and *Eco*RV + *Hin*dII to generate a series of monomer fragments (plus known vector sequences) that were analyzed on polyacrylamide gels (Fig. 3b). The relative mobility of each fragment was compared to that of a 692 bp *Dra*I fragment isolated from the cloning vector pBS+, and these mobilities were plotted versus the position of the restriction site

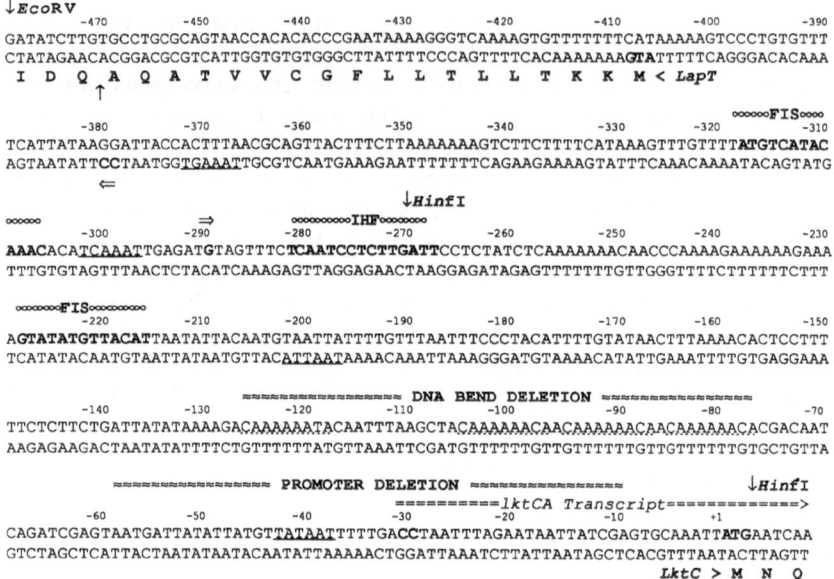

Figure 2. Nucleotide sequence of the leukotoxin promoter region, beginning at the *Eco*RV restriction site (see also Fig. 4). The primary transcript initiates at bp −31. A secondary transcriptional start at bp −290 is marked with a rightward arrow (→). A leftward promoter is also marked at bp −380 (←) (S. K. Highlander, unpublished). Pribnow boxes are underlined. The $CA_6(C/T)A$ repeats that form the DNA bend are marked with a dotted underline. The DNA bend sequence was deleted using a synthetic oligonucleotide, 5'-CTTCTGATTATATAAAAGCGAC AATAAGATCGAG-3', to prime DNA synthesis on uracil-containing single-stranded pSH224 DNA, by the method of Kunkel (Kunkel, 1985). Deletion of the DNA bend sequence was verified by DNA sequencing using custom synthetic oligonucleotide primers and double-stranded DNA templates, as described previously (Highlander *et al.*, 1989). A double deletion, missing the DNA bend sequence and the promoter − 10 sequence, was also detected as product of the mutagenesis. The positions of the bend and promoter deletions are indicated with the ≈≈ symbols. Putative binding sites for IHF and FIS are shown in boldface and are labeled (∞∞). Predicted amino terminal LktC and LapT amino acid sequences appear below the DNA sequence and a potential signal peptide cleavage site within the LapT sequence is indicated by the upright arrow (↑).

within the DNA fragment (Fig. 3c). Fragments with the bending locus near their center run slowest in the gel, while those with the locus near the end have the greatest mobility. The extrapolated minimum on the plot, between the *Ssp*I and *Fok*I restriction sites, corresponds to the position of the bend, following the predictions of Wu and Crothers (Wu & Crothers, 1984). The static bend center lies near nucleotide − 100, or ca. 70 nucleotides upstream of the leukotoxin transcriptional start site and within the $CA_6(C/T)A$ repeat region (Fig. 2).

DNA segments that assume unusual tertiary conformations have been suggested to be involved in regulatory DNA-protein interactions. Bending can be induced by protein binding to the segment, or specific sequences within it can cause static bending, as described here (Wu & Crothers, 1984; Zahn & Blattner,

1985). DNA bending loci have been discovered within the origins of replication (Zahn & Blattner, 1985; Koepsel & Khan, 1986), and they have been found upstream of several strong promoters (Lamond & Travers, 1983; Bauer, *et al.*, 1988). The proximity of the DNA bend to the *lktCA* promoter suggests that it could be involved in transcriptional regulation of the leukotoxin gene cluster.

The DNA bend locus enhances leukotoxin transcription in E. coli

Since a system for study of gene expression in *Pasteurella haemolytica* has not yet been established, we have used gene fusions in *E. coli* as means to examine the interactions of potential regulatory elements involved in leukotoxin expression. Chromosomal fusions of the leukotoxin promoter to β-galactosidase were constructed in *E. coli* using the *P. haemolytica* DNA fragments illustrated in Fig. 4. Strain SH368, which carries the primary promoter within the 274 bp *Hin*fI fragment, produces 15 units of β-galactosidase during log-phase growth in LB broth at 37°C (Miller, 1972). This indicates that the minimal promoter region is poorly expressed in *E. coli* and confirms previous reports of poor expression of leukotoxin from *P. haemolytica* promoters in *E. coli* (Highlander, *et al.*, 1990; Lo *et al.*, 1985). Strain SH370, which additionally carries the upstream promoter described by Strathdee and Lo (Strathdee & Lo, 1989b), produces 23 units of activity under the same growth conditions. The activity of the upstream promoter alone has not been determined.

To examine the role of the DNA bend sequence, the $CA_6(C/T)A$ repeats on the *Hin*fI fragment were removed by oligonucleotide-directed deletion of bases $^-128$ to $^-77$ (Fig. 2), and an operon fusion (SH371) was created using the deleted fragment (Fig. 4). A second deletion, missing the DNA bend and bp $^-60$ to $^-8$, which spans the promoter -10 sequence, was also used to create a chromosomal fusion (SH372). Comparison of the β-galactosidase activities of the bend deletion (SH371) and bend plus promoter deletion (SH372) fusions to strain SH368 suggested that the DNA bend locus is responsible for only a two-fold enhancement of leukotoxin transcription in this system. In other systems, where DNA bends lie in promoter -70 regions, ten-fold increases in transcriptional activity are attributed to the sequences (Lamond & Travers, 1983; Bauer *et al.*, 1988). The *lkt* -70 region may facilitate an interaction with RNA polymerase or allow proper topological looping required for interactions with other transcriptional regulators. Since the leukotoxin promoter is poorly expressed in *E. coli*, it is likely that a *P. haemolytica*-specific gene product is required to fully activate it. In the absence of this factor, the DNA bend alone has only a minor effect on promoter strength.

452

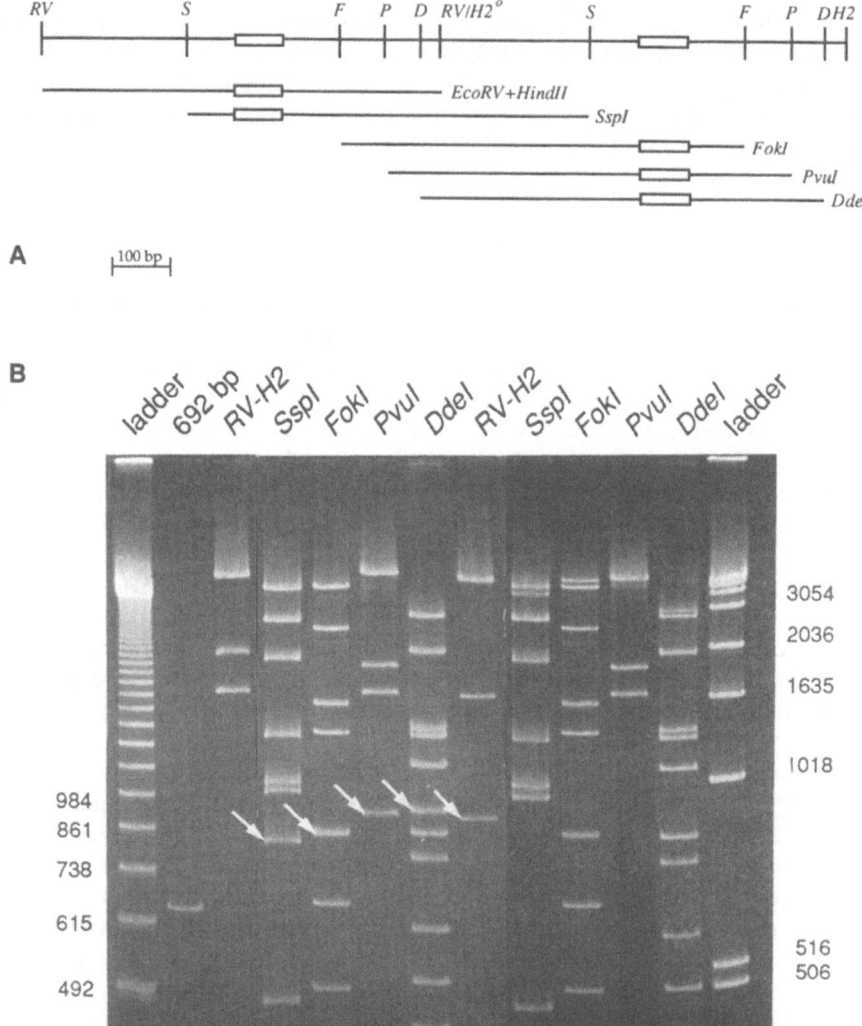

A

B

Permuted Dimer Monomer (control)

Figure 3. Analysis of DNA bending. A head-to-tail dimer of the 695 bp *Eco*RV-*Hind*II fragment carrying the leukotoxin promoter (see Fig. 4) was created by blunt-end ligation of the purified fragment from the promoter-containing plasmid pSH224 (Highlander, *et al.* 1989), into *Eco*RV-linearized pSH224. (a) Circularly permuted DNA fragments containing the *lktCA* promoter. The restriction sites are *Dde*I (D), *Eco*RV (RV), *Fok*I (F), *Hind*II (H2), *Pvu*I (P), and *Ssp*I (S). *RV-H2°* denotes the position of the cloning junction that contains *Eco*RV and *Hind*II half sites. (b) 4% polyacrylamide gel (70:1 acrylamide:bis, 45 mM Tris-borate, pH 8.0, 10 mM EDTA) of digests of the dimer-containing plasmid, pSH237 (left), and the monomer control plasmid, pSH224 (right). A

The leukotoxin genes are flanked by a permease gene cluster

While constructing leukotoxin operon fusions, we discovered that *lkt* promoter-containing fragments could also drive expression of β-galactosidase when present in the reverse orientation of insertion. The transcriptional activities of the leftward (−) promoter-containing fragments were determined for these operon fusions (Fig. 4). As shown, the leftward promoter drives 20 units of β-galactosidase activity (SH460) and is essentially balanced in strength to the leukotoxin promoter. Comparison with the bend deletion (SH461), suggests that the bend also enhances transcription of the leftward promoter. Though the DNA bend was postulated to have a primary influence on leukotoxin expression, it appears that it also can influence the leftward

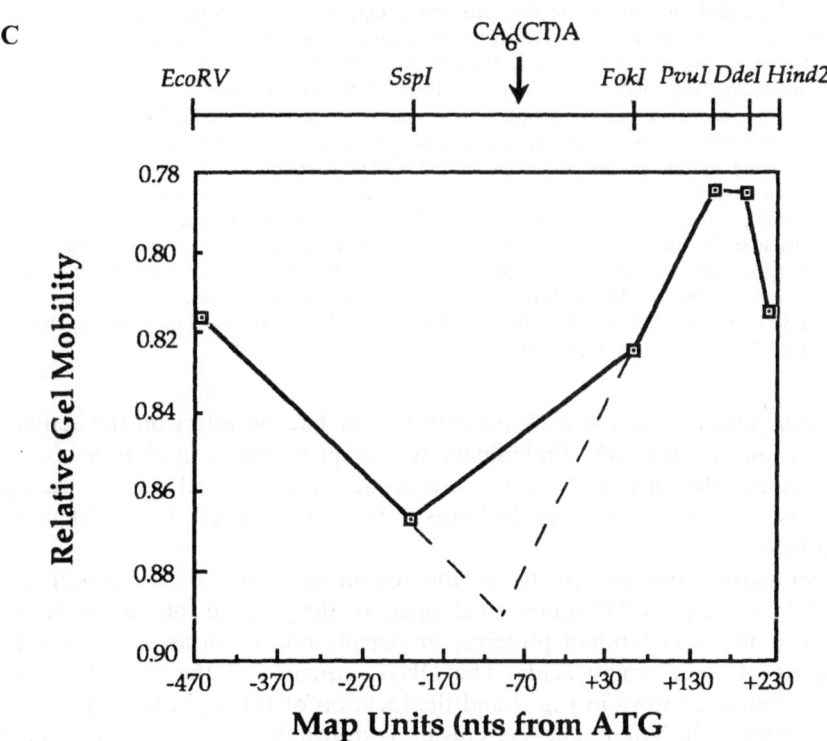

692 bp *Dra*I fragment from plasmid pBS+ is shown as a control, as are the 123 bp (left) and one kilobase (right) molecular weight ladders with their actual lengths in base pairs. Gels were stained with ethidium bromide following electrophoresis and mobility determinations were repeated at least three times for each restriction fragment. (c) Relative mobilities of the permuted fragments. The mobility of each fragment, indicated by the arrows in (b), was compared to that of the 692 bp control fragment and plotted versus the map location (top) of each of the restriction sites. The sequence is numbered such that nucleotide +1 corresponds to the first base of the LktC open reading frame as shown in Fig. 2.

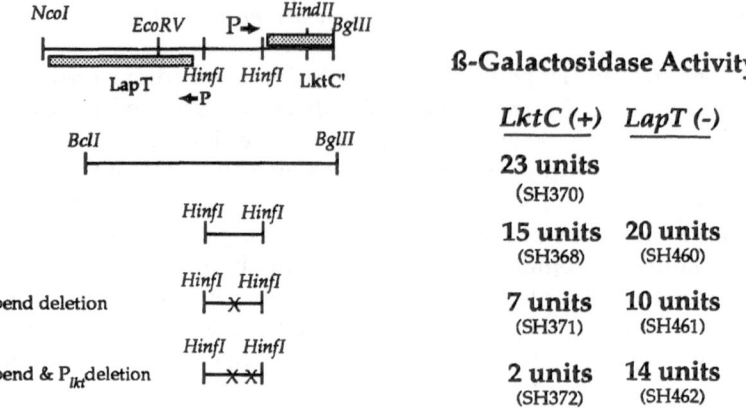

ß-Galactosidase Activity

	LktC (+)	*LapT (-)*
	23 units (SH370)	
	15 units (SH368)	**20 units** (SH460)
	7 units (SH371)	**10 units** (SH461)
	2 units (SH372)	**14 units** (SH462)

Figure 4. Lkt*C* and *lapT* fusions to β-galactosidase. Operon fusions driven by the *Pasteurella* leukotoxin or *lapT* promoter were constructed by cloning various restriction fragments into the operon fusion vector, pGE593 (Eraso & Weinstock, 1992). The vector cloning region contains three translational stop codons, in each reading frame, followed by translation initiation signals, to ensure creation of expressed transcriptional fusions (Eraso & Weinstock, 1992). Plasmids with promoter-containing fragments were transduced, via the lambda phage λRZ-5 (Ostrow, Silhavy, & Garrett, 1986), into the prophage deletion strain GE43 (Weisemann, Funk, & Weinstock, 1984). Homologous recombination between the transduced plasmid and the chromosome causes single copy gene replacement of the truncated β-galactosidase gene on the GE43 chromosome with the operon fusion. β-Galactosidase activities of fusion strains in the presence and absence of the cloned regulatory genes were determined in log-phase growth, in LB broth, with appropriate antibiotics, at 37°C, by the method of Miller (Miller, 1972). Units are averages of values reported for cells grown at 37°C in LB broth, at OD_{600} between 1.0 and 3.0. The corresponding strain number of each operon fusion is shown in parentheses.

promoter. Deletion of the leukotoxin promoter had no effect on the activity of this promoter (SH462). Preliminary transcript mapping in *P. haemolytica* indicates that the activity of the leftward promoter is maximal at 37°C, during log phase growth, similar to leukotoxin transcription (S. K. Highlander, unpublished).

Examination and sequencing of the region upstream of the leukotoxin promoter revealed a 237 amino acid open reading frame that is similar to several amino acid binding proteins, or periplasmic permeases, involved in transport of basic amino acids. The DNA sequence of the 5' end of this reading frame is shown in Fig. 2 and the location of this open reading frame, with respect to the leukotoxin gene cluster, is shown in Figs. 4 and 5. The ATG start codon for the open reading frame is preceded by a potential ribosome binding site, and the 18 amino terminal residues are characteristic of a signal peptide for periplasmic targeting (Fig. 2) (von Heijne, 1983).

The amino acid sequence predicted by this locus is ca. 35% identical to two, related periplasmic amino acid binding proteins: *Salmonella typhimurium* ArgT (Higgins & Ames, 1981) and *Escherichia coli* HisJ (Higgins, *et al.*, 1982). Because of this sequence similarity, I have named the locus *lapT*, for *l*eukotoxin

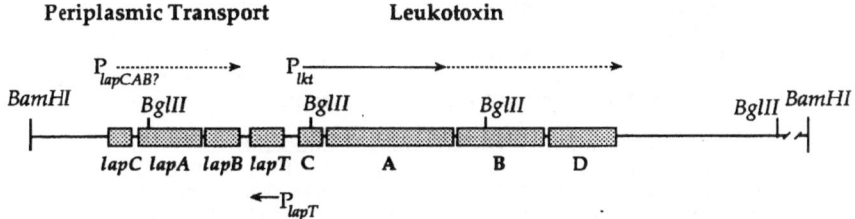

Figure 5. Open reading frame map of the leukotoxin and *l*eukotoxin-*a*ssociated-*p*ermease (*lap*) gene clusters. The *lapT* gene is transcribed in the leftward direction. The location of the *lapCAB* transcript is hypothetical.

*a*ssociated *p*ermease. The predicted LapT sequence is 58% similar to both ArgT and HisJ, when conservative amino acid matches are considered, as shown in the combined pairwise BestFit alignment (Smith & Waterman, 1981) in Fig. 6. Other proteins with sequences similar to LapT are the *Agrobacterium tumefaciens* NocT and OccT octopine periplasmic binding proteins (Zanker, *et al.*, 1992), the *Neisseria gonorrhoeae* HisJ homologue (Lavitola, *et al.*, 1992), and the *E. coli* glutamine periplasmic binding protein, GlnH (Nohno, *et al.*, 1986). In each of these systems, the binding protein plus associated nucleotide-binding and membrane-associated proteins are responsible for substrate-specific uptake into the bacterial cytoplasm. These all are members of the ATP binding cassette (ABC) family of transporters (Hyde, *et al.*, 1990).

In addition to the sequence of *lapT* reported here, three kilobases of DNA sequence beyond the putative binding protein gene have been determined (Highlander, *et al.*, 1990; S. K. Highlander unpublished). This region contains at least three additional significant open reading frames, illustrated in Fig. 5 and presumably encodes the membrane transport apparatus required for activity of the Lap permease. *LapA, lapB,* and *lapC* read left to right and in the same orientation as the leukotoxin gene cluster, while the *lapT* sequence is read from the opposite DNA strand. Characteristics of the peptide sequences predicted from these open reading frames are summarized in Table 1. LapA contains two short sequences that match the consensus "motif A" (GXXGXGKT) and "motif B" (4 hydrophobic amino acids-aspartate) sequences of a family of nucleotide-binding proteins (Mimura, *et al.*, 1991; Higgins, *et al.*, 1988).

Table 1. Features of the leukotoxin-associated permease (Lap) gene cluster.

LapT	237 aa	26 kD	6.1	Signal Peptide 58% similarity to ArgT & HisJ
LapA	442 aa	49 kD	5.3	2 Nucleotide Binding Motifs "Motif A" = GPTGVGKT "Motif B" = GIVFID
LapB	225 aa	26 kD	10.7	4 Membrane-spanning Domains
LapC	173 aa	19 kD	6.3	4 Membrane-spanning Domains

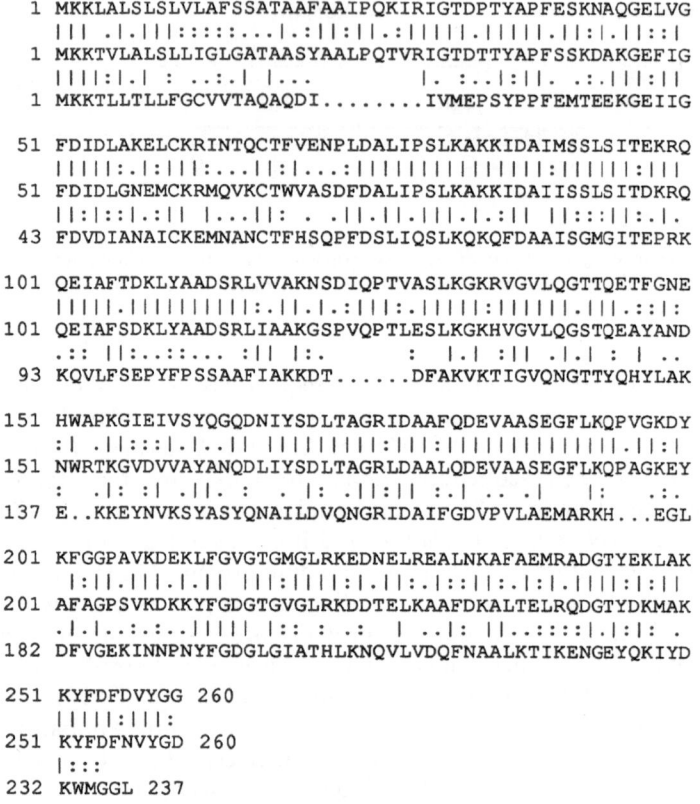

```
  1 MKKLALSLSLVLAFSSATAAFAAIPQKIRIGTDPTYAPFESKNAQGELVG
    |||  .|.|||::::::...|.:||:||.:|||||.|||||.||:|.||::|
  1 MKKTVLALSLLIGLGATAASYAALPQTVRIGTDTTYAPFSSKDAKGEFIG
    ||||:|.|  :  ..:.| |...      |. :..|:||. .:.|||:||
  1 MKKTLLTLLFGCVVTAQAQDI........IVMEPSYPPFEMTEEKGEIIG

 51 FDIDLAKELCKRINTQCTFVENPLDALIPSLKAKKIDAIMSSLSITEKRQ
    |||||:.|:|||...||:|...:||||||||||||||:||||||:||
 51 FDIDLGNEMCKRMQVKCTWVASDFDALIPSLKAKKIDAIISSLSITDKRQ
    ||:|::|.|:|| |...||: . ||.||.|||.|.:|| ||::::||:.|.
 43 FDVDIANAICKEMNANCTFHSQPFDSLIQSLKQKQFDAAISGMGITEPRK

101 QEIAFTDKLYAADSRLVVAKNSDIQPTVASLKGKRVGVLQGTTQETFGNE
    |||||.||||||||:.||.|.:|||:.||||.|||.:::|
101 QEIAFSDKLYAADSRLIAAKGSPVQPTLESLKGKHVGVLQGSTQEAYAND
    .:: ||:..::... :|| |:.       :   |.| :|| .|.| :  |  ..
 93 KQVLFSEPYFPSSAAFIAKKDT......DFAKVKTIGVQNGTTYQHYLAK

151 HWAPKGIEIVSYQGQDNIYSDLTAGRIDAAFQDEVAASEGFLKQPVGKDY
    :|  .||:::|.|.:| ||||||||:|||:|||||||.||:|
151 NWRTKGVDVVAYANQDLIYSDLTAGRLDAALQDEVAASEGFLKQPAGKEY
    :  .|: :| .||. :  .  |: .||:|| :.|  ..  .|   |:  .:.
137 E..KKEYNVKSYASYQNAILDVQNGRIDAIFGDVPVLAEMARKH...EGL

201 KFGGPAVKDEKLFGVGTGMGLRKEDNELREALNKAFAEMRADGTYEKLAK
    |:||.|||.|.|| |||:||||:|.||:.|::||:.|:|.||||:|:||
201 AFAGPSVKDKKYFGDGTGVGLRKDDTELKAAFDKALTELRQDGTYDKMAK
    .|.|..:.:...|||||  |::  :  .:   |  ..|: ||..:::|.|:|: .
182 DFVGEKINNPNYFGDGLGIATHLKNQVLVDQFNAALKTIKENGEYQKIYD

251 KYFDFDVYGG 260
    |||||:|||:
251 KYFDFNVYGD 260
    |:::
232 KWMGGL 237
```

Figure 6. Alignment of HisJ (top), ArgT (middle), and LapT (bottom) amino acid sequences. The figure was generated by joining pairwise alignments between LapT and ArgT and LapT and HisJ using the BestFit algorithm (Smith & Waterman, 1981). Perfect amino acid matches are shown with vertical bars and chemically similar matches with the double and single dots. Sequences were analyzed using the University of Wisconsin Genetics Computer Group package (GCG) through the University of Texas Medical School, Department of Microbiology and Baylor College of Medicine.

Similar nucleotide-binding motifs are also found in the leukotoxin LktB and *E. coli* hemolysin HlyB proteins involved in secretion of these toxins from Gram-negative bacteria (Felmlee *et al.*, 1985). Database searches using the predicted LapB and LapC sequences did not reveal similarities to known sequences, but hydrophobicity predictions for both peptides revealed domains that could be membrane-associated (Kyte & Doolittle, 1982). Both the histidine and lysine-arginine-ornithine permease systems of *S. typhimurium* include a nucleotide-binding and two membrane-spanning proteins (Ames, 1986). It is likely that the reading frames near *lapT* are analogous to these components.

The function and substrate for the putative LapT protein are unknown. We tried to complement auxotrophic *hisJ⁻ S. typhimurium* strains with plasmids that express the LapT protein and attempted to identify LapT using anti-HisJ antibody with Western blots. Neither of these approaches was successful, suggesting that LapT is not a histidine permease. Periplasmic binding proteins have been described that transport amino acids, sugars, peptides, and ions (Mimura *et al.*, 1991) and molecules in any of these classes could be the Lap T substrate. Arginine is a probable substrate for the leukotoxin-associated permease since *P. haemolytica* requires it for maximal growth (Wessman, 1966) and its growth can be inhibited by the arginine analogue, canavanine (S. K. Highlander, unpublished). Arginine could also be involved in *P. haemolytica* pathogenesis because macrophages require L-arginine for production of nitrite and nitrate during anti-bacterial immunostimulation and respiratory burst (Iyengar, *et al.*, 1987). The permease might cause the inhibition of phagocytic killing by sequestering arginine from the macrophage, or, arginine may simply supplement a leaky auxotrophy in the bacterium and promote increased growth rate and enhanced leukotoxin synthesis. In either case, the proximity of the permease genes to the leukotoxin gene cluster and its transcriptional regulatory region, suggest that either its function, or its expression, is involved in leukotoxin regulation.

Identification of a P. haemolytica *activator of leukotoxin expression in* E. coli

Since the leukotoxin promoter is poorly expressed in *E. coli*, I sought to identify new *P. haemolytica* genes that might influence, and perhaps activate expression. The operon fusion strain SH370 (Fig. 4), was used as a target to screen for cosmid clones that affected leukotoxin transcription. A *P. haemolytica BamHI* cosmid library was constructed using the pLAFRX vector (G.M. Weinstock, personal communication), derived from the pLAFR2 (Friedman, *et al.*, 1982), and amplified using λcI857S7am. The cosmid library was then transduced into SH370 and transductants were isolated that varied in color from the original light blue colonies on LB agar containing the β-galactosidase substrate, X-Gal. One of the cosmid clones, plasmid pSH2001, carried a ca. 20 kb *BamHI* fragment and formed deep blue, erose colonies on X-Gal. Subcloning limited the activator phenotype to a 5.0 kb *EcoRI* fragment (Fig. 7.). The ability of the cloned 5.0 kb *EcoRI* fragment, on plasmid pSH2007, to enhance transcription of the SH368 operon fusion is shown in Fig. 7a. Plasmid pSH2007 causes a three to four-fold activation of the leukotoxin promoter in this strain. Unfortunately, the activity of the locus on pSH2007 in other strains has been difficult to quantitate since *E. coli* cells carrying multiple copies of the activator locus exhibit severe growth-inhibition. Subclones of the region that don't exhibit the growth inhibition have rearrangements within the central 0.4 kb *DraIII* restriction fragment of the *P. haemolytica* insert (Fig. 7b), suggesting that the activator locus maps within this region.

Determination of DNA sequences within the 5.0 kb *EcoRI* activator fragment is in progress. We have discovered an unusual DNA repeat sequence and two potential open reading frames that span the 0.4 kb *DraIII* fragment

A

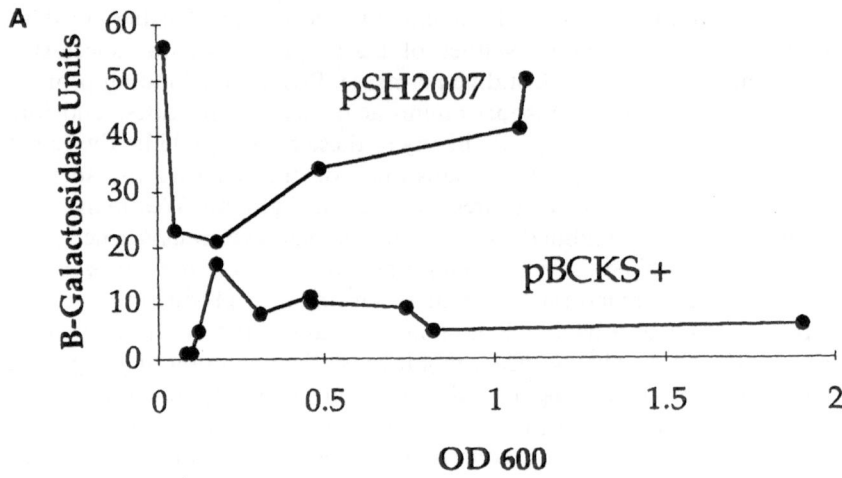

B

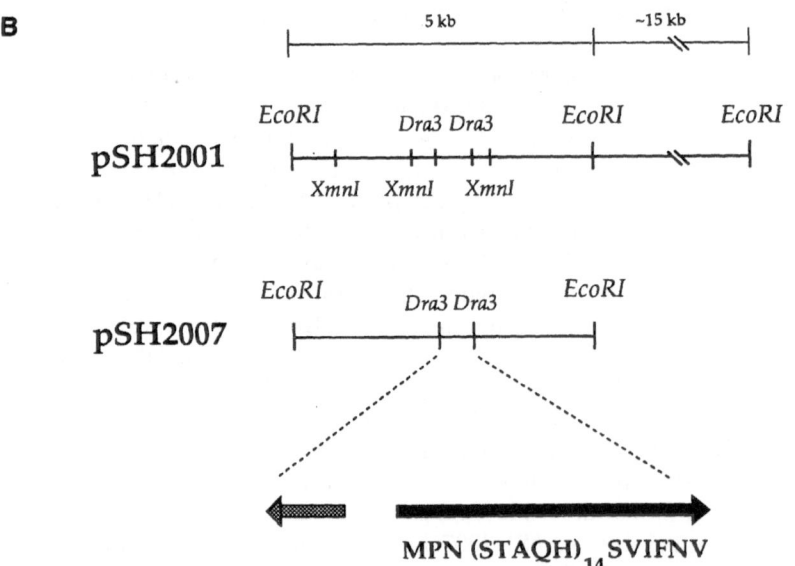

Figure 7. (a) Leukotoxin promoter activity of operon fusion strain SH368 in the presence of the activator clone, pSH2007. β-galactosidase units in the presence of the vector plasmid, pBCKS + (Stratagene, La Jolla, CA), are also illustrated. (b) Restriction map of the cosmid clone pSH2001 and activator subclone, pSH2007. Only selected restriction sites are shown. pSH2001 carries a 20 kb *P. haemolytica* fragment in the vector pLAFRX, and pSH2007 carries a 5 kb *Eco*RI fragment of the pSH2001 insert in the multicopy vector pBCKS+. Two open reading frames observed within the hyper-recombinogenic *Dra*III fragment are illustrated. Duplication of this region results in loss of the activation phenotype displayed in (a).

(Fig. 7b). The wild-type fragment contains 39 repeats of the sequence GCACA; rearranged subclones of the fragment contain more than 50 repeats of this sequence within the *Dra*III fragment. The repeat region is immediately preceded by a potential ATG start codon and consensus-like ribosome binding site that could encode a repeat peptide with the sequence MPN(STAQH)$_{14}$SVIFNV... Database searches for similarities to this peptide failed to provide any linear correlation with known protein sequences. A complete characterization of this region is in progress.

Conclusions

P. haemolytica is a commensal organism of cattle, yet under appropriate conditions of stress to the animal, it becomes virulent. Transcriptional studies have shown that expression of the major virulence factor of *P. haemolytica*, the macrophage-specific leukotoxin, is regulated by temperature, growth-phase, and medium composition (Highlander, *et al.*, 1990; Strathdee & Lo, 1989b). Here, I have described several *cis* and *trans* elements that may be involved in this switch from non-pathogenic to pathogenic states. While a single, global regulator of leukotoxin expression has not been identified, DNA topology and *trans*- activation are general mechanisms that are expected to influence leukotoxin transcription in *P. haemolytica*. The role of a periplasmic binding protein as an environmental sensor that could be involved in escape from macrophage killing is also intriguing. It is clear that continued meaningful studies of leukotoxin regulation will require a manipulable gene transfer system be established for *P. haemolytica*. A new vector for this organism has just been described (Frey, 1992) and it is hoped that this and other similar vectors will allow continued analysis of the leukotoxin in its native host.

Acknowledgements

I thank Dr. George Weinstock, in whose laboratory the DNA bending studies were performed, for his continued interest in, and support of, this work. I also thank Elizabeth Wickersham and Orlando Garza for their expert technical help with DNA sequencing and β-galactosidase assays. This research was supported by Cactus Feeders, Inc., Dumas, Texas, by USPHS grant RR-05425, and by USDA grant 91–0236.

Note added in proof

We have recently shown that the LapT protein is an arginine-specific periplasmic binding protein (L. Caskey and S. K. Highlander, unpublished).

References

Ames, G. F.-L. (1986) Bacterial periplasmic transport systems: structure, mechanism, and evolution. Annu. Rev. Biochem. 55: 397–425.

Bauer, B.F., Kar, E.G., Elford, R.M., & Holmes, W.M. (1988) Sequence determinants for promoter strength in the leuV operon of Escherichia coli. Gene 63: 123–134.

Berget, P., Engler, M., Highlander, S., & Weinstock, G. (1991) Pharmaceutical compositions of a 105 kD P. haemolytica derived antigen useful for treatment of shipping fever. U.S. Patent 4, 957, 739.

Chang, Y.F., Young, R., & Struck, D.K. (1989) Cloning and characterization of a hemolysin gene from Actinobacillus (Haemophilus) pleuropneumoniae. DNA 8: 635–647.

Drummond, R.O., Lambert, G., Smalley, H.E., & Terrill, C.E. (1981) p. 111–127. In: D. Pimentel (ed.) p. 111–127. CRC handbook of pest management in agriculture. CRC Press, Inc., Boca Raton.

Eraso, J.M., & Weinstock, G.M. (1992) Anaerobic control of colicin E1 production. J. Bacteriol. 174: 5101–5109.

Felmlee, T., Pellett, S., & Welch, R.A. (1985) Nucleotide sequence of an Escherichia coli chromosomal hemolysin. J. Bacteriol. 163: 94–105.

Frank, G.H., Briggs, R.E., & Gillette, K.G. (1986) Colonization of the nasal passages of calves with Pasteurella haemolytica serotype 1 and regeneration of colonization after experimentally induced viral infection of the respiratory tract. Am. J. Vet. Res. 47: 1704–1707.

Frank, G.H., Briggs, R.E., & Gillette, K.G. (1987) Pasteurella haemolytica serotype 1 and regeneration of colonization after experimentally induced viral infection of the respiratory tract. Am. J. Vet. Res. 48: 1674–1677.

Frey, J. (1992) Construction of a broad host range shuttle vector for gene cloning and expression in Actinobacillus pleuropneumoniae and other Pasteurellaceae. Res. Microbiol. 143: 263–269.

Friedman, A.M., Long, S.R., Brown, S.E., Buikema, W.J., & Ausubel, F.M. (1982) Construction of a broad host range cosmid cloning vector and its use in the genetic analysis of Rhizobium mutants. Gene 18: 289–296.

Glaser, P., Ladant, D., Sezer, O., Pichot, F., Ullmann, A., & Danchin, A. (1988) The calmodulin-sensitive adenylate cyclase of Bordetella pertussis: cloning and expression in Escherichia coli. Mol. Microbiol. 2: 19–30.

Higgins, C.F., & Ames, G.F. (1981) Two periplasmic transport proteins which interact with a common membrane receptor show extensive homology: complete nucleotide sequences. Proc. Natl. Acad. Sci. USA 78: 6038–6042.

Higgins, C.F., Gallagher, M.P., Mimmack, M.L., & Pearce, S.R. (1988) A family of closely related ATP-binding subunits from prokaryotic and eukaryotic cells. BioEssays 8: 111–116.

Higgins, C.F., Haag, P.D., Nikaido, K., Ardeshir, F., Garcia, G., & Ames, G. F.-L. (1982) Complete nucleotide sequence and identification of membrane components of the histidine transport operon of S. typhimurium. Nature (London) 298: 723–727.

Highlander, S.K., Chidambaram, M., Engler, M.J., & Weinstock, G.M. (1989) DNA sequence of the Pasteurella haemolytica leukotoxin gene cluster. DNA 8: 15–28.

Highlander, S.K., Engler, M.J., & Weinstock, G.M. (1990) Secretion and expression of the Pasteurella haemolytica leukotoxin. J. Bacteriol., 172: 2343–2350.

Highlander, S.K., Wickersham, E.A., Garza, O. & Wemstock, G.M. (1993) Expression of the Pasteurella Laemolytica leukotoxin is inhibited by a locus that encodes an ATP-binding cassette homolog. Infect. Immun. 61: 3942–3951.

Himmel, M.E., Yates, M.D., Lauerman, L.H., & Squire, P.G. (1982) Purification and partial characterization of a macrophage cytotoxin from Pasteurella haemolytica. Am. J. Vet. Res. 43: 764–767.

Hubner, P., & Arber, W. (1989) Mutational analysis of a prokaryotic recombinational enhancer element with two functions. EMBO J. 8: 577–585.

Hyde, S.C., Emsley, P., Hartshorn, M.J., Mimmack, M.M., Gileadi, U., Pearce, S.R., Gallagher, M.P., Gill, D.R., Hubbard, R.E., & Higgins, C.F. (1990) Structural model of ATP-binding proteins associated with cystic fibrosis, multidrug resistance and bacterial transport. Nature (London) 346: 362–365.

Iyengar, R., Stuehr, D.J., & Marletta, M.A. (1987) Macrophage synthesis of nitrite, nitrate, and N-nitrosamines: precursors and the role of the respiratory burst. Proc. Natl. Acad. Sci. USA 84: 6369–6373.

Jericho, K.W.F., Lejeune, A., & Tiffin, G.B. (1986) Bovine herpesvirus-1 and *Pasteurella haemolytica* aerobiology in experimentally infected calves. Am. J. Vet. Res. 47: 205–209.

Koepsel, R.R., & Khan, S.A. (1986) Static and initiator protein-enhanced bending of DNA at a replication origin. Science 233: 1316–1318.

Kolodrubetz, D., Dailey, T., Ebersole, J., & Kraig, E. (1989) Cloning and expression of the leukotoxin gene from *Actinobacillus actinomycetemcomitans*. Infect. Immun. 57: 1465–1469.

Koronakis, V., Cross, M., & Hughes, C. (1989) Transcription antitermination in an *Escherichia coli* haemolysin operon is directed progressively by *cis*-acting DNA sequences upstream of the promoter region. Mol. Microbiol. 3: 1397–1404.

Koronakis, V., Cross, M., Senior, B., Koronakis, E., & Hughes, C. (1987) The secreted hemolysins of *Proteus mirabilis, Proteus vulgaris*, and *Morganella morganii* are genetically related to each other and to the alpha-hemolysin of *Escherichia coli*. J. Bacteriol. 169: 1509–1515.

Kunkel, T.A. (1985) Rapid and efficient site-specific mutagenesis without phenotypic selection. Proc. Natl. Acad. Sci. USA 82: 488–492.

Kyte, J., & Doolittle, R. (1982) A simple method for displaying the hydrophobic character of a protein. J. Mol. Biol. 157: 105–132.

Lamond, A.I., & Travers, A.A. (1983) Requirement for an upstream element for optimal transcription of a bacterial tRNA gene. Nature (London) 305: 248–250.

Lavitola, A., Vanni, M., Martin, P.M.V., & Bruni, C.B. (1992) Cloning and characterization of a *Neisseria* gene homologous to *hisJ* and *argT* of *Escherichia coli* and *Salmonella typhimurium*. Res. Microbiol. 143: 295–305.

Lo, R.Y.C., Shewen, P.E., Strathdee, C.A., & Greer, C.N. (1985) Cloning and expression of the leukotoxin gene of *Pasteurella haemolytica* A1 in *Escherichia coli* K-12. Infect. Immun. 50: 667–671.

Lo, R.Y.C., Strathdee, C.A., & Shewen, P.E. (1987) Nucleotide sequence of the leukotoxin genes of *Pasteurella haemolytica* A1. Infect. Immun. 55: 1987–1996.

Magwood, S.E., Barnum, D.A., & Thomson, R.G. (1969) Nasal bacterial flora of calves in healthy and in pneumonia-prone herds. Can. J. Comp. Med. 33: 237–243.

Markham, R.J.F., & Wilkie, B.N. (1980) Interaction between *Pasteurella haemolytica* and bovine alveolar macrophages: cytotoxic effect on macrophages and impaired phagocytosis. Am. J. Vet. Res. 41: 18–22.

Martin, S.W., Meek, A.H., Davis, D.G., Johnson, J.A., & Curtis, R.A. (1981) Factors associated with morbidity and mortality in feedlot calves: The Bruce County beef project, year two. Can. J. Comp. Med. 45: 102–112.

Miller, J.H. (1972) Experiments in molecular genetics. Cold Spring Harbor Laboratory, Cold Spring Harbor, NY.

Mimura, C.S., Holbrook, S.R., & Ames, G.F.-L. (1991) Structural model of the nucleotide-binding conserved component of periplasmic permeases. Proc. Natl. Acad. Sci. USA 88: 84–88.

Nohno, T., Saito, T., & Hong, J. (1986) Cloning and complete nucleotide sequence of the *Escherichia coli* glutamine permease operon (*glnHPQ*). Mol. Gen. Genet. 205: 260–269.

Ostrow, K.S., Silhavy, T.J., & Garrett, S. (1986) *cis*-acting sites required for osmoregulation of *ompF* expression in *Escherichia coli* K-12. J. Bacteriol. 168: 1165–1171.

Shewen, P.E., & Wilkie, B.N. (1982) Cytotoxin of *Pasteurella haemolytica* acting on bovine leukocytes. Infect. Immun. 35: 91–94.

Smith, T.F., & Waterman, M.S. (1981) Comparison of bio-sequences. Adv. Appl. Math. 2: 482–489.

Strathdee, C.A., & Lo, R.Y.C. (1987) Extensive homology between the leukotoxin of *Pasteurella haemolytica* A1 and the alpha-hemolysin of *Escherichia coli*. Infect. Immun. 55: 3233–3236.

Strathdee, C.A., & Lo, R.Y.C. (1989a) Cloning, nucleotide sequence, and characterization of genes encoding the secretion function of the *Pasteurella haemolytica* leukotoxin determinant. J. Bacteriol. 171: 916–928.

Strathdee, C.A., & Lo, R.Y.C. (1989b) Regulation of expression of the *Pasteurella haemolytica* leukotoxin determinant. J. Bacteriol. 171: 5955–5962.

Thompson, S.A., Elkins, C., West, A., & Sparling, P.F. (1991) Cloning of a gene from *Neisseria meningitidis* with sequence specificity to the RTX family of cytotoxins. In: Annual Meeting American Society for Microbiology, (poster B37). Dallas, TX.

von Heijne, G. (1983) Patterns of amino acids near signal-sequence cleavage sites. Eur. J. Biochem. 133: 17–21.

Weisemann, J.M., Funk, C., & Weinstock, G.M. (1984) Measurement of *in vivo* expression of the *recA* gene of *Escherichia coli* by using *lacZ* gene fusions. J. Bacteriol, 160: 112–121.

Wessman, G.E. (1966) Cultivation of *Pasteurella haemolytica* in a chemically defined medium. App. Microbiol. 14: 597–602.

Wu, H.M., & Crothers, D.M. (1984) The locus of sequence-directed and protein-induced DNA bending. Nature, 308: 509–513.

Yang, C.C., & Nash, H.A. (1989) The interaction of *E. coli* IHF protein with its specific binding sites. Cell 57: 869–880.

Yates, W.D.G., Stockdale, P.H.G., Babiuk, L.A., & Smith, R.J. (1983) Prevention of bovine pneumonic pasteurellosis with an extract of *Pasteurella haemolytica*. Can. J. Comp. Med. 47: 250–256.

Zahn, K., & Blattner, F.R. (1985) Sequence-induced DNA curvature at the bacteriophage lambda origin of replication. Nature (London) 317: 451–453.

Zanker, H., von Lintig, J., & Schroder, J. (1992) Opine transport genes in the octopine (*occ*) and nopaline (*noc*) catabolic regions in Ti plasmids of *Agrobacterium tumefaciens*. J. Bacteriol. 174: 841–849.

31. Signal transduction in *Yersinia pseudotuberculosis*

THOMAS BERGMAN, ELENA DUBININA, ÅKE FORSBERG,
EDOUARD GALYOV, SEBASTIAN HÅKANSSON, ROLAND
NORDFELTH, CATHRINE PERSSON, MARJA RIMPILÄINEN,
ROLAND ROSQVIST and HANS WOLF-WATZ

Abstract. Virulent *Yersinia* possess a common 70kb virulence plasmid which encodes a number of indispensable Yops virulence determinants. Yops proteins are regulated by external stimuli temperature and calcium concentration. At 37° *yop* transcription is induced and the rate of transcription is regulated by the Ca^{2+} concentration of the growth medium. In parallel to this transcriptional regulation, Yops are secreted into the growth medium by a specific Ca^{2+} regulated plasmid-encoded secretion system. One mutant (*yop*N) has been isolated and studies using this mutant suggest that the surface-located YopN protein senses the Ca^{2+} concentration and transmits this signal accordingly. The final step of the regulatory hierarchy involves a *yop* transcriptional repressor. LcrH is suggested to be this repressor since overproduction of LcrH leads to repression of transcription of *yop* genes. Several Yop proteins have been shown to be essential virulence determinants. Two of these, YopH and YopE, act in concert to block phagocytosis by macrophages. YopH exhibits a protein tyrosine phosphatase activity suggesting that YopH dephosphorylates host proteins. YopE is a cytotoxin. Recent studies have shown that interaction between the target cell surfaces and the pathogen triggers YopE expression and its polarized transfer through the plasma membrane of the target cell. The previous findings with respect to Yop regulation and Yop secretion is in agreement with the polarized transfer of Yop proteins into the target cell.

Abbreviations: TS, Temperature Sensitive

1. The genus *Yersinia*

The genus *Yersinia* encompasses three pathogenic species *Y. pseudotuberculosis*, *Y. pestis* and *Y. enterocolitica* (Bercovier *et al.*, 1980a, and 1980b). Hybridization studies have revealed that *Y. pestis* and *Y. pseudotuberculosis* are almost identical (Bercovier, 1980b) and it has been suggested that *Y. pestis* should be classified as a subspecies of *Y. pseudotuberculosis*. *Y. pestis* is the causative agent of plague and *Y. pestis* was responsible for the pandemic known as Black Death which led to the death of 25% of the population of Europe (Pollitzer, 1954). Plague is a systemic disease in humans and depending on the mode of transmission it occurs in two forms; the bubonic form and the pneumonic form. Bubonic plague is transmitted to man by the bite of a flea that has previously been infected by a rodent. This mode of transmission has attracted lots of interest since it involves a temperature shift from 26°C (flea) to 37°C (man or rodent). As will be shown later this is an important step in the pathogenesis of

C.I. Kado and J.H. Crosa (eds.), Molecular Mechanisms of Bacterial Virulence, 463–475.
© 1994 *Kluwer Academic Publishers.*

Yersinia. Symptoms include painful lympho-adenopathy and the bacteria spread quickly throughout the lymphatic system and later enter the blood stream, leading to septicemia. Thus it can be concluded that the pathogen is able to grow in the hostile environment of the lymphatic system.

Yersinosis includes infections caused by both *Y. pseudotuberculosis* and *Y. enterocolitica.* Both species are spread in nature and have been recovered from many different animal as well as environmental sources. A post infection complication is the development of reactive arthritis (Aho *et al.*, 1981).

The three pathogenic species all harbour a related plasmid of about 70 kb (Portnoy *et al.*, 1984). This plasmid is essential for virulence as well as for the characteristic Ca^{2+} requirement for growth at 37°C exhibited by virulent *Yersinia.* (Zink *et al.*, 1980; Gemski *et al.*, 1980 and Portnoy *et al.*, 1981). In addition to the common virulence plasmid, *Y. pestis* harbours two other plasmids having molecular weights of about 110 kb and 12 kb respectively (Ferber and Brubaker, 1981). The larger plasmid has been associated with the production of the murine toxin and fraction 1 antigen (Portsenko *et al.*, 1983). The small plasmid has been shown to encode pesticin and a prolease (Ferber and Brubaker, 1981; Sodeinde *et al.*, 1988).

2. Calcium dependence

Higuchi *et al.* (1959) were the first to describe that virulent *Y. pestis* has an unusual temperature dependent requirement for Ca^{2+}. The same Ca^{2+} dependency (CD) has also been shown for virulent strains of *Y. pseudotuberculosis* (Brubaker, 1967) and *Y. enterocolitica* (Gemski *et al.*, 1980). When virulent *Y. pestis* is grown at 37°C in the absence of Ca^{2+}, the bacteria cease to grow within 2 hours (growth restriction). However, growth is normal at 37°C in the presence of at least 2.5 mM Ca^{2+} and is independent of Ca^{2+} at temperatures below 30°C. This phenomenon is called Ca^{2+} dependence (CD) (Higuchi *et al.*, 1959) or low calcium response (lcr)(Goguen *et al.*, 1984).

Calcium independent (CI) mutants can be obtained by using a selective plating medium which lacks Ca^{2+} and contains magnesium-oxalate (Higuchi and Smith, 1961). In most mutants the virulence plasmid is lost (Ben-Gurion and Schafferman, 1981; Ferber and Brubaker, 1981; Gemski *et al.*, 1980; Portnoy *et al.*, 1981). However, in case of *Y. pestis*, the virulence plasmid may contain insertions or deletions due to the insertion element, IS100 (Portnoy and Falkow, 1981). These insertions and deletions are all located within a 20 kb region on the plasmid called the low-calcium (*lcr*) region (Fig. 1).

The calcium region is conserved among the virulence plasmids of the three *Yersinia* species (Portnoy *et al.*, 1984) and insertion mutants in the region result in avirulent strains (Cornelis *et al.*, 1989; Goguen *et al.*, 1984, Portnoy *et al.*, 1983; Bölin and Wolf-Watz, 1984).

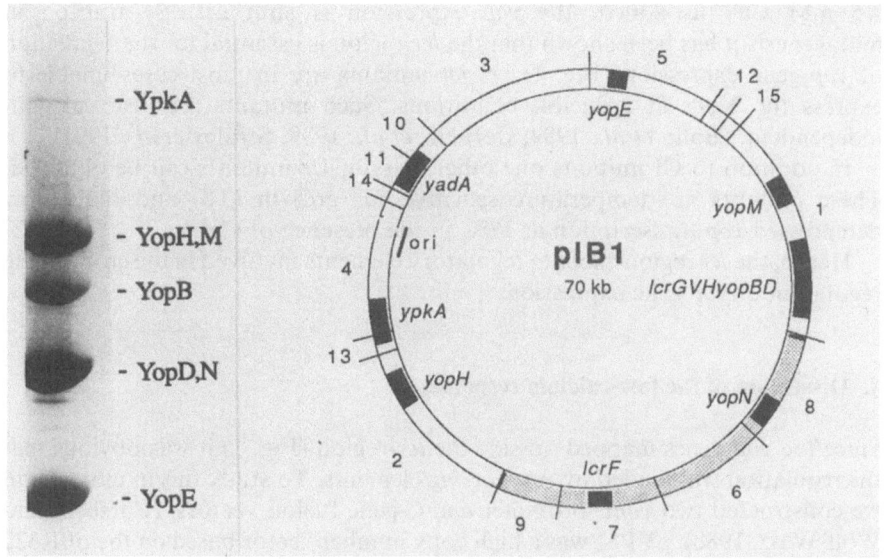

Figure 1. Protein profile of secreted Yop proteins and restriction endonuclease map of the virulence plasmid pIB1 of *Y. pseudotuberculosis*. The numbers refer to the commonly used *Bam*HI resistriction map. The genes that have been mapped are indicated by name. The stippled area indicate the *lcr* region. The specific Yop secretion operon is located between *lcr*F and *yop*H (Michels *et al.* 1991).

3. Expression and secretion of Yops

In response to the Ca^{2+} and temperature stimuli a number of plasmid encoded Yops (*Yersinia outer proteins*)(Bölin *et al.*, 1985) are expressed and secreted to the culture supernatant (Fig. 1). These proteins were first described in *Y. enterocolitica* by Portnoy *et al.*, (1981) as being outer membrane proteins. It was, however, later shown by Heesemann *et al.*, (1984) that the Yop proteins of *Y. enterocolitica* were secreted to the culture supernatant and subsequently it was shown that the Yops were secreted in all three species (Wolf-Watz el al, 1985).

This suggested that the plasmids exhibited identical functions in the three species. This was further supported when it could be shown that the plasmids of *Y. pestis* and *Y. pseudotuberculosis* were interchangeable (Wolf-Watz, *et al.*, 1985). When the *yop*-genes were mapped on the virulence plasmid pIBI of *Y. pseutotuberculosis*, they were found scattered around the plasmid and usually contained within monocistronic operons (Fig. 1) (Forsberg *et al.*, 1987). This was also found to be the case in *Y. pestis* and *Y. enterocolitica* (Bölin *et al.*, 1988).

Although the *yop*-genes are contained within different transcriptional units, they are coordinately regulated by the extracellular calcium concentration and temperature. At 26°C the *yop*-genes are expressed at a very low level. At 37°C the expression is derepressed if the extracellular Ca^{2+} concentration is low. If

2.5 mM Ca²⁺ is added, the *yop* expression is shut off. By transposon mutagenesis it has been shown that the *lcr* region is essential for the regulation of *yop*-gene expression (Fig. 1) i.e. *lcr*-mutants are in most cases unable to express the Yops at inducible conditions. Such mutants are also calcium-independent. (Bölin *et al.*, 1984; Cornelis *et al.*, 1989, Straley *et al.*, 1991).

In addition to CI mutants one other class of *lcr*-mutants can be obtained. These mutants are temperature sensitive for growth (TS) and they show derepressed *yop* transcription at 37°C in the presence of Ca²⁺.

Hence, the *lcr* region encodes regulatory elements involved in the co-ordinate regulation of *yop* gene expression.

4. Dissection of the low-calcium response

Since the *yop*-genes mapped outside the *lcr*-region (Fig. 1) it was obvious that the regulation was guided by *trans acting* elements. To study this in more detail we constructed two *yop*E promoter *ampC*-gene fusion vectors. (Forsberg and Wolf-Watz, 1988). pYP51 was a high copy number vector based on the pBR322 replicon which had a copy number of about 60 copies per cell. The other vector pYP52 was based on the replicon of pACYC184 and exhibited a considerable lower copy number (about 7 copies per cell). When either of these vectors were put into the cured strain of *Y. pseudotuberculosis* YPIII no significant β-lactamase expression could be monitored suggesting that the virulence plasmid pIB1 encoded an activator (Forsberg and Wolf-Watz, 1988). This was confirmed when the plasmids were put into the wild-type strain YPIII(pIB1) (Fig. 2). In this case a clear temperature controlled activation of β-lactamase could be seen, strongly indicating that the plasmid pIB1 encoded a trans-activator (Fig. 2).

The existence of this activator was confirmed by Cornelis coworkers (1989b) who also demonstrated that this transcriptional activator (VirF) showed a high homology with the AraC regulator of the arabinose operon of *E.coli*. Moreover, it was also shown that the expression of VirF was temperature regulated. The corresponding gene in *Y. pseudotuberculosis* is called *lcr*F. (Fig. 1). When the *lcr*F mutant YPIII(pIB73) was used in similar experiments as described above, no expression could be detected from the reporter gene (Fig. 2). As expected this strain did not express the Yops.

When the low copy number plasmid pYP52 was used *in trans* in the wild-type strain it was evident that the reporter gene was expressed at 37°C in a Ca²⁺ controlled manner (Fig. 2) identical to that of the *yop* genes. In contrast when the high copy number vector pYP51 was used no effect of Ca²⁺ was observed (Fig. 2). In this case the expression of the *yop*E promoter fusion reporter gene was derepressed in absence as well as in presence of Ca²⁺ (Fig. 2). Moreover, this strain YPIII(pIB1, pYP51) was temperature sensitive for growth at 37°C and showed derepressed transcription of the virulence plasmid encoded *yop* genes at 37°C in presence of Ca²⁺ (Forsberg and Wolf-Watz, 1988).

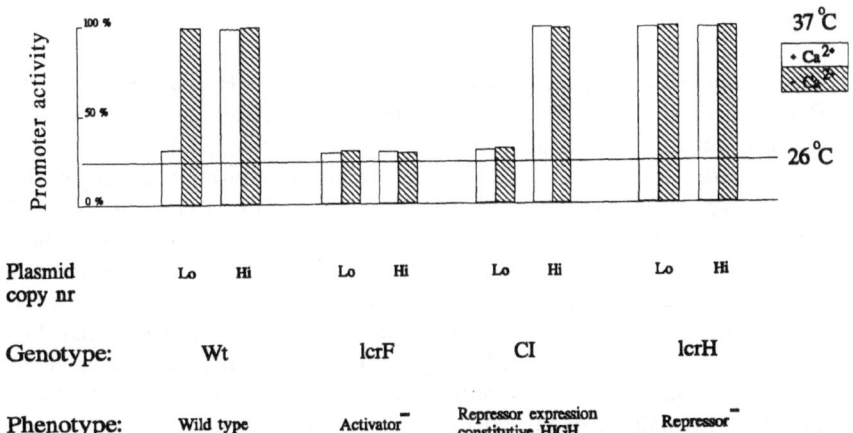

Figure 2. Transregulation of the *yop*E promoter by the virulence plasmid pIB1. Two different vectors pYP51 and pYP52 carrying a *yop*E promoter fusion (*amp*C-gene reporter) were put into different strains of *Y. pseudotuberculosis* YPIII carrying either the wild-type plasmid pIB1 or mutants of the virulence plasmid as indicated. The expression of the reporter gene during different growth conditions was monitored in the different constructs. The reporter-plasmids pYP51(Hi) and pYP52(Lo) differed in copy number; pYP51 was present in about 60 copies per cell while pYP52 was present in about 7 copies per cell. When these plasmids were present in the cured strain YPIII a very low level of expression was monitored.

As a ten fold difference in copy number between pYP51 and pYP52 was established it is likely that a negative element regulating *yop* gene expression was titered out by the high copy number plasmid pYP51. Thus, we concluded that the *lcr* region encodes a repressor and that the level of repressor is affected by the concentration of calcium in the growth medium.

Two different classes of CI mutants could be identified by this analysis (Fig. 2). One class of mutants was in the positive control loop discussed above. In the other class of mutants (Fig. 2 CI), transcription from the *yop*E promoter could only be induced when the high copy number plasmid pYP51 was introduced. In this case transcription was as high as in the wild-type. Thus, these *lcr*-mutants express the repressor at a high constitutive level (as Ca^{2+} always was present). demonstrating the presence of calcium regulated control functions determining repressor concentration.

From the results presented above it could be predicted that mutants unable to express a functional repressor would be TS and show derepressd Yop expression at 37°C. Such mutants could be isolated and three different loci have so far been implicated in this phenotype. Here we will only dicuss two of those: *yop*N mutants and *lcr*H mutants.

4.1 Is the LcrH the repressor?

The *lcr*H gene is a constituent of the large 4,5 kb *lcr*GVH *yop*BD operon (Fig. 1) (Bergman *et al.*, 1991; Price and Straley, 1989). Insertion mutants of this operon are TS for growth at 37°C and show derepressed *yop* transcription (Bergman *et al.*, 1991). This phenotype has been coupled to the inability of these mutants to express the LcrH protein in sufficient amounts (Bergman *et al.*, 1991; Price and Straley, 1989). Overexpression of LcrH *in trans*, induced from the *tac* promoter, in *lcr*H mutants renders the strain CI. Thus, overexpression of LcrH in this construct results in a change in the phenotype from TS to CI which is the expected result if LcrH is the functional repressor. Hence, we suggest that LcrH is the repressor which interacts with *yop*-operator sequences and consequently LcrH is at the end of the Ca^{2+} controlled negative pathway (Fig. 3).(Note, is has not been formally proven that LcrH is the repressor)

4.2 YopN is a surface located protein involved in Yop regulation

The Yops do not undergo post-translational processing upon secretion (Forsberg and Wolf-Watz, 1988) and it has been shown that the N-terminal end of the Yops contains a specific secretion signal (Michiels *et al.*, 1990). In line with these observations it has been shown that the Yops are exported by a specific transport system encoded by the plasmid (Rosqvist *et al.*, 1990; Michiels *et al.*, 1991).

The expression of this export system is controlled by Ca^{2+} and temperature as the Yops are. (Rosqvist *et al.*, 1990, Michiels *et al.*, 1991). Thus, the TS *lcr*H mutant YPIII(pIB13) is only able to export the Yops in a Ca^{2+} depleted medium although Yop expression is derepressed in the presence of Ca^{2+}. In this latter case the Yops can be recovered from the whole cell fraction (Bergman *et al.*, 1991). Thus, LcrH is most likely not involved in the regulation of the Yop-export apparatus. Yother and Goguen (1985) were first to describe a mutant obtained by chemical mutagenesis with a repressor-negative phenotype (TS). They designated this mutant calcium-blind since it was growth-restricted and expressed high amounts of the V-antigen irrespective of the Ca^{2+} concentration of the growth medium. We isolated a specific mutant YPIII(pIB82) in the same locus (*yop*N) (Fig. 1) and it was confirmed that this mutant showed a TS phenotype (Forsberg *et al.*, 1991). We could show that this mutant was unable to express the YopN protein which also was shown to be surface located. As expected from its TS phenotype the *yop*N mutant was derepressed for *yop* expression in presence of Ca^{2+}. In addition, the mutant was able to export the Yops in presence of Ca^{2+} demonstrating that the export function also was derepressed in the *yop*N mutant (Forsberg *et al.*, 1991). Hence, the YopN protein is a surface located protein involved in regulation of Yop expression as well as Yop export. We suggest from these findings that YopN is a surface located protein which senses the appropriate environment for Yop expression and transmits this signal accordingly (Fig. 3).

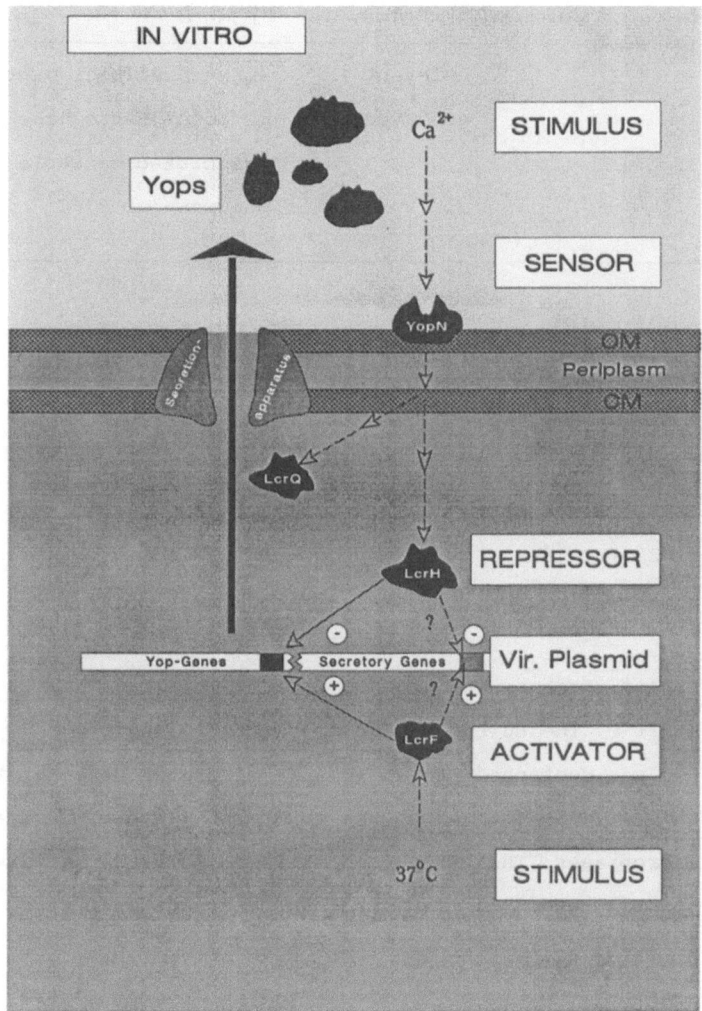

Figure 3. Model of the regulation of Yop expression and Yop secretion.

5. YopE and YopH are essential virulence determinants

Several of the Yops have been shown by site-directed mutagenesis to be virulence determinants (Straley and Cibull, 1989; Cornelis *et al.*, 1989; Rosqvist *et al.*, 1990). Two of these, YopH and YopE are important in the process of *Y. pseudotuberculosis* to block phagocytosis by macrophages (Table 1) (Rosqvist *et al.*, 1988; Rosqvist *et al.*, 1990). It was first shown that a plasmid containing wild-type strain YPIII(pIB102) was not ingested by macrophages in contrast to its cured derivative strain YPIII which was phagocytosed by the macrophages.

Table 1. Resistance to phagocytosis by different strains of *Y. pseudotuberculosis.*

Strain	Relevant genotype	% Extracellular bacteria of total Macrophage-associated bacteria pregrown at 37°C
YPIII	plasmid-	6 ± 2 (5)
YPIII (pIB29)	*yop*H	23 ± 3 (5)
YPIII (pIB29, pYOP21)	*yop*H/*yop*H+	63 ± 4 (5)
YPIII (pB102)	wt	58 ± 7 (3)
YPIII (pIB522)	*yop*E	54 ± 5 (3)
YOPIII (pIB251)	*yop*E/*yop*H	7 ± 3 (3)

The different strains were grown overnight in a Ca^{2+}-depleted medium at 26°C. The cultures were diluted 1/20 and the incubation was continued. After 1 h, the cultures were shifted to growth at 37°C and 2h after the shift the bacteria were added to the macrophages. The extra- and intra-cellular bacteria were determined by the double immuno-fluorescence antibody test.

± = Standard deviation of the mean: numbers in parentheses are replicates.

(Table 1), indicating that plasmid encoded proteins were essential in this process. When the *yop*H mutant YPIII(pIB29), which has the whole *yop*H deleted, was tested for its ability to block phagocytosis it was found that this strain showed a lowered ability to do so (Table 1). The level of phagocytosis was in between the wild type strain and the cured strain indicating that other Yop-proteins also were involved in the process. Therefore a *yop*H/*yop*E double mutant YPIII(pIB251) was constructed and tested. This mutant was as unable to block phagocytosis as the cured strain YPIII. (Table 1). Thus, YopE and YopH act in concert in the ability of the pathogen to block phagocytosis by macrophages. Interestingly, YopH was recently shown to possess a protein

tyrosine phosphatase activity (Guan and Dixon, 1990) which causes dephosphorylation of target cell proteins (Bliska *et al.*, 1991). It would come as no
surprise if it will be shown that YopH interacts with the second message signalling
pathway of the macrophage to generate a false signal preventing phagocytosis.

During these studies it was observed that YopE generated a cytotoxic effect
on cultured macrophages (Rosqvist *et al.*, 1990). Earlier studies had also shown
that *Y. pestis* and *Y. enterocolitica* could induce a cytotoxic response on cultured
cells (Portnoy *et al.*, 1981, Goguen *et al.*, 1986). The cytotoxic effect could also
be demonstrated on cultured HeLa cells (Rosqvist *et al.*, 1991). To investigate
this process in more detail we first asked the question if the pathogen must
adhere to the cells to generate a cytotoxic effect on the target cell. In this case we
took advantage of the fact that *Y. pestis* strain EV76 is unable to adhere to the
surface of the HeLa cell (Rosqvist *et al.*, 1990). In addition this strain is not
cytotoxic for HeLa cells. However, when a hybrid-plasmid encoding the lectin
YadA of *Y. pseutotuberculosis* was put into EV76 it could be demonstrated that
this construct was not only able to adhere to the HeLa cells but it was also
cytotoxic. Hence, adhesion to the target cells is a first step in the ability of
Yersinia to induce cytotoxicity.

Since the Yops are released in vast amounts to the culture supernatant we
asked next if purified Yops could induce this cytotoxic effect *in vitro*. When
purified Yops were added to a monolayer of HeLa cells no effect could be
recorded. However, if the proteins were delivered inside the HeLa cell by
microinjection a cytotoxic effect was seen (Table 2) (Rosqvist *et al.*, 1990).

Table 2. Abilities of different strains of *Y. pseudotuberculosis* to mediate cytotoxicity on monolayers
of growing HeLa cells.

Bacterium Y.pseudotuberculosis strain	Description	Cytotoxic activity	
		In vivo[a]	In vitro[b]
YPIII	Plasmid free	−	−
YPIII (pIB102)	Wild type	+	+
YPIII (pIB522)	yopE	−	−
YPIII (pIB522, pAF55)	yopE/yopE+	+	+
YPIII (pIB15)	yopD	−	+

[a] Ability of viable bacteria to induce a cytotoxic response on
cultured HeLa cells within 6 h of incubation at 37°C.

[b] Ability of microinjected Yop proteins to induce a cytotoxic
response on cultured HeLa cells within 3 h of incubation at
37°C.

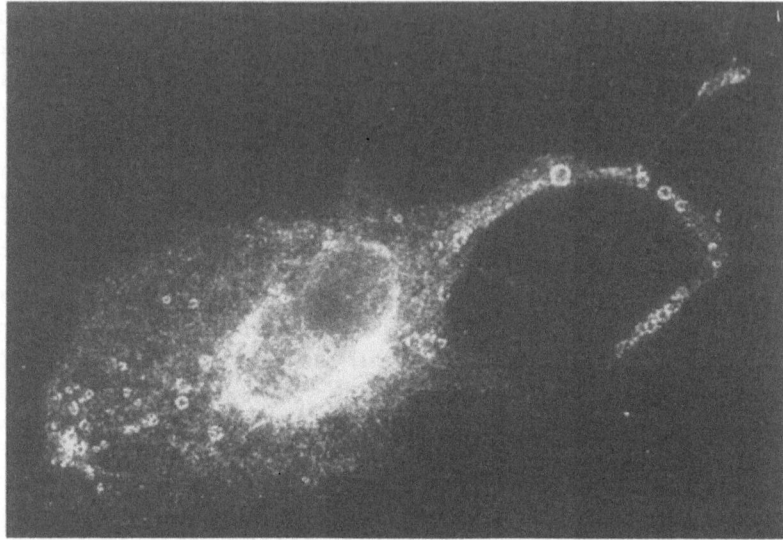

Figure 4. Visualization of the YopE protein after infection of HeLa cells with *Y. pseudotuberculosis* strain YPIII(pIB1). The bright areas show the presence of the YopE protein detected by immunofluorescence microscopy using a BioRad confocal microscope. The YopE protein was visualized using anti YopE antibodies in combination with FITC conjugated antibodies. Note: No bacteria are seen. Only the YopE protein is detected within the cell. Using light microscopy bacteria can be detected at the surface the HeLa cell (not shown).

By the use of different specific mutants it was demonstrated that the YopE protein was responsible for this effect (Table 2). In addition it was also observed that the YopD protein exhibited a function in this process, since the *yop*D mutant YPIII(pIB15) was not cytotoxic *in vivo* but Yops isolated from this strain induced a cytotoxic response after microinjection (Table 2). This result may indicate that YopD is involved in the translocation process of YopE through the plasma membrane of the target cell.

6. Polarized transfer of YopE

From the results presented above it was obvious that YopE (and possibly other Yops) must be transferred from the pathogen through the plasma membrane of the target cell to reach its target. To study this process in more detail we used immunofluorescence microscopy to follow the fate of YopE after infection of HeLa cells. To our surprise we were unable to detect YopE on the surface of the bacteria. In fact no bacteria were seen that reacted with the YopE specific antibody (Fig. 4).

All YopE antigen that could be detected was found within the HeLa cell (Fig. 4) indicating that the majority of the YopE protein had been transferred through the plasma membrane of the cell. When we repeated this experiment using the *yop*N mutant YPIII(pIB82) described above two observations were made: i) the strain showed a delayed cytotoxic response and ii) the YopE antigen was deposited on the whole surface of the pathogen and only small amounts of YopE was found within the target cell. Since the *yop*N mutant shows derepressed expression as well as derepressed Yop secretion we argue that the surface located YopN protein interacts with a specific cell receptor of the target cell. This interaction induces Yop expression and polarized transfer of the Yops into the target cells, as seen here for YopE. This idea is consistent with both the observations made on the regulation and secretion of Yops *in vitro* and the *in vivo* interaction between *Y. pseudotuberculosis* and HeLa cells. More work is needed to show if this idea is valid. Such work is in progress in our laboratory.

Acknowledgements

This works was supported by grants form the Swedish Medical Research Council grant 07490–5A. Swedish Natural Science Reserach Councial grant No 4426–301.

References

Aho K, Ahonen P,Laitinen O & Lerisalo M (1981) Arthritis associated with *Yersinia enterocolitica* infection. p. 113–124. *In*: Bottone EJ (ed.) (113–124) CRS Press, Boca Raton, Florida.

Ben-Gurion R & Shafferman A (1981) Essential virulence determinants of different *Yersinia* species are carried on a common plasmid. Plasmid 5: 183–187.

Bercovier H, Brenner DJ, Ursing J Stiegerwalt AG, Fanning GR, Alonso JM, Carter G & Mollaret HH (1980a) Characterization of *Yersinia enterocolitica* sensu stricto. Curr. Microbiol.4: 201–206.

Bercovier H, Mollaret HH, Alonso JM, Brault J, Fanning GR, Stiegerwalt AG & Brenner DJ (1980b) Intra- and interspecies relatedness of *Yersinia pestis* by DNA hybridization and its relationship to *Yersinia pseudotuberculosis*. Curr. Microbiol. 4: 225–229.

Bergman T, Håkansson S, Forsberg Å, Norlander L, Machellaro A, Bäckman A & Wolf-Watz H (1991) Analysis of the V-antigen *lcrGVHyopBD* operon of *Yersinia pseudotuberculosis*: evidence for a regulatory role of LcrH and LcrV. J. Bacteriol. 173: 1607–1616.

Bliska JB, Guan K, Dixon JE & Falkow S (1991) Tyrosine phosphate hydrolysis of host proteins by an essential *Yersinia* virulence determinant. Proc. Natl. Acad. Sci. USA 88: 1187–1191.

Brubaker RR (1967) Growth of *Pasturella pseudotuberculosis* in simulated intracellular and extracellular environments. J. Infect. Dis. 117: 403–417.

Bölin I, Forsberg Å, Norlander L, Skurnik M & Wolf-Watz H (1988) Identification and mapping of the temperature-inducible, plasmid-encoded proteins of *Yersinia* spp. Infect. Immun. 56: 343–348.

Bölin I, Portnoy D & Wolf-Watz H (1985) Expression of the temperature-inducible outer membrane proteins of yersiniae. Infect. Immun. 48: 234–240.

Bölin I & Wolf-Watz H (1984) Molecular cloning of the temperature inducible protein 1 of *Yersinia pseudotuberculosis*. Infect. Immun. 43: 72–78.

Cornelis GR, Biot T, Lambert de Rouvroit C, Michiels T, Mulder B, Sluiters C, Sory MP, VanBouchaute M & Vanooteghem JC (1989) The *Yersinia* yop-regulon. Mol. Microbiol. 3: 1455–1459.

Cornelis G, Sluiters C, Lambert de Rouvriot C & Michiels T (1989b) Homolgy between *virF*, the transcriptional activator of the *Yersinia* virulence regulon and *araC*, the *Escherichia coli* arabinose regulator. J. Bacteriol. 171: 254–262.

Ferber DM & Brubaker RR (1981) Plasmids in *Yersinia pestis*. Infect. Immun. 31: 839–841.

Forsberg Å, Bölin I, Norlander L & Wolf-Watz H (1987) Molecular cloning and expression of calcium-regulated plasmid encoded proteins of *Y. pseudotuberculosis* Microbial. Pathogen. 2: 123–137.

Forsberg Å, Vitanen A, Skurnik M & Wolf-Watz H (1991) The surface located YopN protein is involved in calcium signal transduction in *Yersinia pseudotuberculosis*. Mol. Microbiol. 5: 977–986.

Forsberg Å & Wolf-Watz H (1988) The virulence protein Yop5 of *Yersinia pseudotuberculosis* is regulated at transcriptional level by plasmid pIB1 encoded trans acting elements controlled by temperature and calcium. Mol. Microbiol. 2: 121–133.

Gemski P, Lazere J, Casey T & Wholmieter JA (1980) Presence of a virulence associated plasmid in *Yersinia pseudotuberculosis*. Infect. Immun. 28: 1044–1047.

Goguen J, Yother J & Straley S (1984) Genetic analysis of the low calcium response in *Yersinia pestis* Mud1 (Ap lac) insertion mutants. J. Bacteriol. 160: 842–848.

Goguen JD, Walker WS, Hatch TP & Yother J (1986) Plasmid determined cytotoxicity in *Yersinia pestis* and *Yersinia pseudotuberculosis*. Infect. Immun. 51: 788–794.

Guan K & Dixon JE (1990) Protein tyrosine phosphatase activity of an essential virulence determinant in *Yersinia*. Science 249: 553–556.

Heesemann J, Algermissen B & Laufs R (1984) Genetically manipulated virulence of *Yersinia enterocolitica*. Infect. Immun. 46: 105–110.

Higuchi K & Smith JL (1961) Studies on the nutrition and physiology of *Pasturella pestis* VI. A different plating medium for estimation of mutation rate to avirulence. J. Bacteriol. 81: 605–608.

Higuchi K, Kupferberg L & Smith L (1959) Studies on the nutrition and physiology of *Pasturella pestis* III. Effects of calcium ions on the growth of virulent and avirulent strains of *Pasturella pestis*. J. Bacteriol. 77: 317– 321.

Michiels T, Vanooteghem J-C, Lambert de Rouvroit C, China B, Gustin A, Boudry P & Cornelis GR (1991) Analysis of *virC* an operon involved in secretion of Yop proteins by *Yersinia enterocolitica*. J. Bacteriol. 173: 4994–5009.

Michiels T, Wattiau P, Brasseur R, Ruysschaert JM & Cornelis GR (1990) Secretion of Yop proteins by yersiniae. Infect. Immun. 58: 2840–2849.

Pollizer R (1954) Plague. World Health Organization series no 22. WHO Geneva.

Portnoy DA, Blank HF, Kingsbury DT & Falkow S (1983) Genetic analysis of essential plasmid determinants of pathogenicity of *Yersinia pestis*. J. Infect. Dis. 148: 297–304.

Portnoy DA, Mosely S & Falkow S (1981) Characterization of plasmids and plasmid-associated determinants of *Yersinia enterocolitica* pathogenesis. Infect. Immun. 31. 775–782.

Portnoy DA & Falkow S (1981) Virulence associated plasmids from *Yersinia enterocolitica* and *Yersina pestis*. J. Bacteriol. 148: 877–893.

Portnoy DA, Wolf-Watz H, Bölin I, Beeder AB & Falkow S (1984) Characterization of common virulence plasmids in *Yersinia* species and their role in the expression of outer membrane proteins. Infect. Immun. 43: 108–114.

Price S & Straley SC (1989) *lcrH*, a gene necessary for virulence of *Yersinia pestis* and for the normal response of *Y. pestis* to ATP and calcium. Infect. Immun. 57: 1491–1498.

Protsenko OA, Anisimov PI, Mozharov OT, Konnov NP, Popov YA & Kokushikin AM (1983) Detection and characterization of *Yersinia pestis* plasmids determining pestecin, fraction 1 antigen and "mouse" toxin synthesis. Genetica 19: 1081–1090.

Rosqvist R, Bölin I & Wolf-Watz H (1988) Inhibition of phagocytosis in *Yersinia pseudotuberculosis*: a virulence plasmid encoded ability involving the Yop2b protein. Infect. Immun. 56: 2139–2143.

Rosqvist R, Forsberg Å, Rimpiläinen M, Bergman T & Wolf-Wtaz H (1990) The cytotoxic protein YopE of *Yersinia* obstructs the primary host defence. Mol. Microbiol. 4: 657–667.

Rosqvist R, Forsberg Å & Wolf-Watz (1991) Intracelluar targeting of the *Yersinia* YopE cytotoxin in mamalian cells induces actin microfilament disruption. Infect. Immun. 59 (4562–4569).

Sodeinde OA, Sample AK, Brubaker RR & Goguen JD (1988) Plasminogen activator/coagulase gene of *Yersinia pestis* is responsible for degradation of plasmid-encoded outer membrane proteins. Infect. Immun. 56: 2749–2752.

Straley S & Cibull ML (1989) Differential clearance and host-pathogen interactions of YopE- and YopK-/YopL- *Yersinia pestis* in BALB/c mice. Infect. Immun. 57: 1200–1210.

Straley SC (1988) The plasmid-encoded outer membrane proteins of *Yersinia pestis*. Rev. Infect. Dis. 10: 323–326.

Wolf-Watz H, Bölin I, Forsberg Å & Norlander L (1986) Possible determinants of virulence in Yersiniae. p. 329–334. *In:* Protein-carbohydrate interactions. Lark D (ed) Academic Press London.

Wolf-Watz H, Portnoy D, Bölin I & Falkow S (1985) Transfer of the virulence plasmid of *Yersinia pestis* to *Yersinia pseudotuberculosis*. Infect Immun. 48: 241–243.

Yother J & Goguen JD (1985) Isolation of calcium blind mutants of *Yersinia pestis*. J. Bacteriol. 164: 704–711.

Zink DL, Feeley JC, Wells JG, Vanderzant C, Vickery JC, Roolf WD & O'Donovan GH (1980) Plasmid mediated tissue invasiveness in *Yersinia enterocolitica*. Nature (London) 283: 224–226.

32. Regulation of *Agrobacterium tumefaciens* virulence gene expression

ANATH DAS

Abstract. Expression of *Agrobacterium tumefaciens* virulence (*vir*) genes is positively regulated by *virA*, *virG* and a plant phenolic inducer(s). *VirA* and *virG* are members of the bacterial two-component regulatory systems. VirA functions as the sensor and VirG is the transcriptional activator. A cis-acting regulatory element, the *vir* box, serves as the VirG binding site and is essential for *vir* gene expression. The *vir* box is 14 residues in length, has a dyad symmetry and has the consensus sequence dPuPyTNCAATTGNAAPy. Expression of different *vir* genes requires the participation of one or more *vir* boxes. Inducer independent mutations in *virA* that support *virB* expression mapped to four regions of the VirA protein: the first transmembrane domain, near the active site histidine, a glycine-rich region that probably constitutes the ATP-binding domain and the carboxy-terminal domain that is homologous to the amino-terminal receiver domain of *virG*. Inducer and *virA* independent mutations in *virG* mapped to the receiver domain. The *virA* and *virG* mutations led to an 11 to 1350 fold increase in the basal level of *virB* expression. All constitutive mutations require low pH for maximal *vir* gene expression. The phenotype of *virG* and some of its derivatives are suppressed by *virA* in the absence of an inducer.

Abbreviations: AS, Acetosyringone; T-DNA, Transferred DNA

Introduction

Agrobacterium tumefaciens causes crown gall tumor disease on many dicotyledoneous plants. Tumor formation results from the transfer and stable integration of a segment of bacterial plasmid DNA, the T-DNA, into the plant nuclear genome. The T-DNA (10–25 kb in size) is carried on a large (200–300 kb) tumor-inducing (Ti-) plasmid present in all virulent *A. tumefaciens* strains. Expression of the T-DNA encoded biosynthetic genes for the phytohormones auxin and cytokinin in a transformed plant cell leads to uncontrolled cell growth, the phenotypic characteristic of the crown gall tumor disease (reviewed in Zambryski, 1988; Ream, 1989; Kado, 1991).

The T-DNA is defined by a conserved 24 bp repeat sequence, the border sequence. This element is required in *cis* and is the only essential *cis*- or *trans*-acting element encoded within the T-DNA. Some Ti-plasmids also contain a second 24 bp conserved sequence, the overdrive, that enhances the tumor-forming ability of the bacterium. The overdrive sequence is located outside the T-DNA right border and functions like an enhancer element (Peralta and Ream, 1986).

C.I. Kado and J.H. Crosa (eds.), Molecular Mechanisms of Bacterial Virulence, 477–489.
© 1994 *Kluwer Academic Publishers.*

Functions essential for T-DNA transfer to plant cells and for its integration into the host nuclear genome are encoded within the virulence (*vir*) region of the Ti-plasmid. The octopine Ti-plasmid *vir* region resides within about a 35 kb segment and is composed of at least eight transcription units, *virA-H* (Garfinkel and Nester, 1980; Ooms *et al.*, 1980; Stachel and Nester, 1986; Kanemoto *et al.*, 1989; Melchers *et al.*, 1990). Of these, *virA* and *virG* are two regulatory loci that, in conjunction with a plant cell factor(s), positively control expression of all *vir* genes (Stachel *et al.*, 1986; Stachel and Zambryski, 1986; Winans *et al.*, 1988). Two plant factors that were isolated from *Nicotiana tabacum* culture cells are phenolics, acetosyringone (3,5-dimethoxy-4-hydroxy acetophenone) and α-hydroxy-acetosyringone (Stachel *et al.*, 1985). Several other phenolic inducers have since been isolated from various plants (reviewed in Kado, 1991; Winans, 1992). In free-living bacteria, only *virA* and *virG* are expressed, whereas expression of the other *vir* genes requires the interaction of bacteria with plant cells. Other factors that control *vir* gene expression are pH, temperature and certain monosaccharides (Stachel *et al.*, 1985; Rogowsky *et al.*, 1987; Alt-Moerbe *et al.*, 1988; Huang *et al.*, 1990). *Vir* gene induction is inefficient at 37^0C and pH above 6.0. Both of these effects are largely mediated by *virA* (Melchers *et al.*, 1989; Jin *et al.*, 1992). The effect of the monosaccharides are mediated by a chromosomal locus, *chvE*, that encodes a sugar-binding protein. Interaction of *chvE* and *virA* is essential for the positive regulation of *vir* gene expression by monosaccharides (Cangelosi *et al.*, 1991).

DNA sequence analysis indicated that *virA* and *virG* are members of the bacterial two-component regulatory system (Leroux *et al.*, 1987; Winans *et al.*, 1986). In two-component regulatory systems, one component, the sensor, functions to sense an external stimulus (stimuli). In response to the signal the sensor protein is phosphorylated at a conserved histidine. The phosphate moiety is then transferred to a conserved aspartic acid of the regulator (reviewed in Ronson *et al.*, 1987; Stock *et al.*, 1989). The phospho-regulator is believed to be the active form that controls cellular processes, usually transcription activation.

Analysis of the deduced polypeptide sequence of VirA indicate that the N-terminal region contains two transmembrane domains (TM1 and TM2) that are believed to span the inner membrane (Leroux *et al.*, 1987). The cytoplasmic C-terminal region of VirA shares homology with the sensor components of the two-component regulatory systems. In the extreme C-terminus of VirA exists an additional domain that is homologous to the N-terminal domain of VirG and other regulators of the two-component regulatory system. This region is not found in most members of the sensor family.

Deletion analysis indicate that removal of 178 residues of the 221 residue periplasmic domain of VirA leads to reduced pH dependence (Melchers *et al.*, 1989; Pazour *et al.*, 1991) and to the loss of monosaccharide-induced *vir* gene expression (Cangelosi *et al.*, 1991). Studies using hybrid proteins composed of segments of VirA and that of the *E. coli* Tar receptor indicate that the acetosyringone responsive element may lie within the second transmembrane domain and the adjacent cytoplasmic region (Melchers *et al.*, 1989). Analysis of deletions in the extreme C-terminal domain gave interesting results. Removal of part of the domain renders the protein nonfunctional (Melchers *et al.*, 1989), but complete removal of the domain leads to an increase in the basal level of expression (Chang and Winans, 1992; A. D., unpublished results). These results suggest that the extreme C-terminal domain of VirA may negatively control the basal level of *vir* gene expression (see below).

The VirG protein is predicted to be a cytosolic protein composed of at least two domains (Winans *et al.*, 1986). The N-terminal domain is conserved in all members of the regulator family and is the signal receiving domain. A conserved aspartic acid residue (at position 52) within this domain is phosphorylated by the VirA kinase (Huang *et al.*, 1990; Jin *et al.*, 1990a). The C-terminal domain is conserved only in a subset of these proteins, viz. OmpR, PhoB, ArcA, PhoM-2, PhoP, TctD and ToxR, and most likely constitutes the transcriptional activator domain (Stock *et al.*, 1989). VirG is a DNA-binding protein that binds specifically to sequences within the *vir* gene promoter-regulatory regions (Jin *et al.*, 1990b; Pazour and Das, 1990a). Analysis of the DNA-binding ability of *E. coli* TrpE′-′VirG fusions and truncated VirG indicate that the C-terminal domain encodes the sequence-specific DNA-binding motif (Powell and Kado, 1990; Roitsch *et al.*, 1990).

Cis-acting sequences necessary for *vir* gene expression

Analysis of the non-transcribed region sequences of the Ti-plasmid pTiA6 *vir* genes indicated that two hexameric sequences, dCGATGA and dGCAATT, are conserved in the promoter-regulatory region of several *vir* genes (Das *et al.*, 1986). It was postulated that one or both of these sequences may function as a *cis*-acting regulatory element for *vir* gene expression. Further analysis of these sequences by Winans *et al.* (1987) indicated that one of these, dGCAATT, is a subset of a larger dodecameric conserved sequence, dTNCAATTGAAAPy, termed the *vir* box. The cis-acting sequences essential for *vir* gene expression were identified by deletion analysis of the *virB* promoter-regulatory region (Das and Pazour, 1989). For acetosyringone-induced expression of *virB* 68 base pairs upstream of the transcription start site were sufficient . Within this region there are two *vir* box sequences at positions −68 to −57 and −48 to −37. A deletion that had lost up to residue −63 showed no inducible expression indicating that removal of half of the first *vir* box sequence was sufficient for the loss of function. Subsequent studies from several laboratories including ours demonstrated the requirement of the *vir* box in *virC*, *virD*, *virE* and *virG* expression (Jin *et al.*, 1990b; Pazour and Das, 1990b; Winans, 1990).

Analysis of the *vir* gene non-transcribed region sequences showed the presence of multiple *vir* box sequences in most *vir* genes (Winans, 1987). While *virA* and *virE* contain a single *vir* box the divergent *virC* and *virD* operons share five potential *vir* boxes. To identify the functional *vir* box(es), site-specific mutagenesis was used to alter the most conserved central six residues (dCAATTG) of the *vir* box. Only one of the five *vir* boxes of *virD* was necessary for *virD* expression. In contrast, maximal expression of both *virB* and *virC* requires the participation of two *vir* boxes. In both cases the promoter-distal *vir* box is essential for *vir* gene induction while the second one is necessary but not sufficient for maximal induction (Pazour and Das, 1990b). Similar studies with *virG* showed two of the three *vir* boxes are essential for *virG* expression (Winans, 1990). A summary of the functional analysis of the *vir* box is presented in Fig. 1.

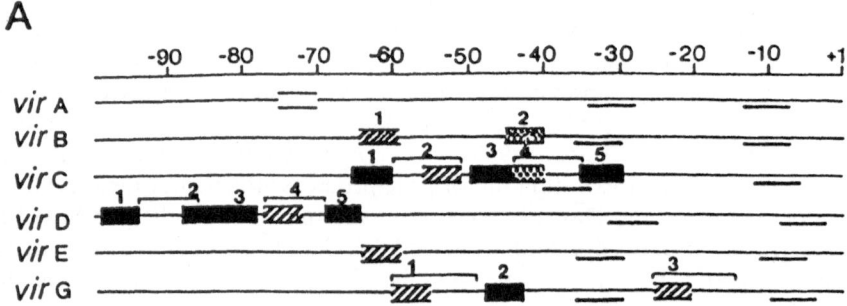

Figure 1. Relative position of the *vir* boxes of the *vir* genes: Numbers at the top indicate the location of the *vir* boxes relative to the transcription initiation site (+1, according to Das *et al.*, 1986). Stripped box, required for function; dotted box, necessary but not sufficient for function; dark box, not required; open box, function not known. Number on top of each box identifies individual *vir* box. Brackets indicate regions protected by VirG in an *in vitro* DNaseI footprinting assay (Pazour and Das, 1990a). Reprinted from Pazour and Das, 1990b.

To identify the *vir* box(es) the transcriptional activator VirG binds to DNase I footprinting assays were performed. VirG bound specifically to the subset of *vir* boxes that are necessary for *vir* gene expression (Jin *et al.*, 1990b; Pazour and Das, 1990a). A closer inspection of the *vir* box sequences and functional reconstruction of the *virE vir* box by molecular cloning indicated that the *vir* box is a tetradecameric sequence, dPuPyTNCAATTGNAAPy, and contains a dyad symmetry (Das and Pazour, 1989; Pazour and Das, 1990b). The presence of a dyad symmetry suggests that VirG probably functions as a dimer or a multimer.

Mutational analysis of trans-acting factors

A mutational approach was used to identify derivatives of *virA* and *virG* that function independent of the inducing agent (termed constitutive mutations). *A. tumefaciens* A136 (which lacks a Ti-plasmid) harboring plasmid pGP159 (contains *virA*, *virG* and a *virB-lacZ* reporter gene) was mutagenized with nitrosoguanidine or ultraviolet radiation (Pazour *et al.*, 1991). A mutation in *virA*, *virG* or the *virB* promoter-regulatory region of the *virB-lacZ* reporter gene that activates *virB* transcription will lead to the formation of a blue colony on plates containing the chromogenic substrate 5-bromo-4-chloro-3-indolyl β-D-galactoside (X-gal). Seventy-one blue colonies were isolated from a screen of ~600,000 survivors of the mutagenesis experiments. Further analysis of the plasmid DNAs indicated that all mutations were in the *virA* locus.

Since this approach was unsuccessful in isolating mutations in *virG*, a different strategy that utilizes both a genetic selection and a genetic screen was used (Pazour *et al.*, 1992). *A. tumefaciens* A136 harboring plasmid pGP358R that contains *virG*, *virB-bla* and *virB-lacZ* gene fusions was mutagenized with nitrous acid or nitrosoguanidine. A mutation in *virG* that is independent of both *virA* and the phenolic inducer will allow the formation of a carbenicillin resistant (from *virB-bla*), blue (from *virB-lacZ*) colony on solid medium containing X-gal. Mutants with the desired phenotype were isolated at a frequency of 1 in 10^7 to 10^8.

Analysis of *virA* mutants

DNA sequence analysis of the mutant genes showed that mutations in *virA* mapped to five different positions. Four strains (NG1, NG8, NG34, NG37) had a mutation that led to the substitution of a leucine to phenylalanine at codon 24. This residue lies within the first transmembrane domain (TM-1, residues 18–39). This domain can be substituted with an analogous but non-homologous region of the *E. coli* Tar receptor indicating that this domain is important for proper membrane topology of VirA (Melchers *et al.*, 1989). The mapping of a constitutive mutation to this domain would imply its involvement in the signaling process as well. Genetic studies with *tar* indicate that the TM-1 region interacts with the TM-2 and a cytoplasmic region (Oosawa and Simon, 1986). Such interactions are presumed to be involved in the signaling process. It is likely that a similar interaction(s) plays a role in VirA signaling.

Two strains (NG53 and UV3) had mutations that mapped near the active site histidine (position 474). NG53 had an alanine 469 to valine substitution and UV3 had a glycine 471 to arginine substitution. A substitution of glycine 471 to glutamic acid also gave a similar phenotype (Aukenbauer *et al.*, 1991). The proximity of these residues to the conserved histidine is likely to affect the structure of the domain resulting in a locked active conformation of the mutant proteins.

Mutant NG31 had a leucine to phenylalanine substitution at codon 658. This residue lies within a glycine-rich region that is conserved among the sensor proteins. This region is homologous to the $GXGX_2GX_nK$ (X, any amino acid) sequence, a conserved region found within a 40 residue sequence in the catalytic domains of all protein kinases and is thought to constitute the nucleotide-binding region (Soderling, 1990). The mutation in NG31 is in a nonconserved residue and alters the sequence $GGT\underline{GLG}X_{35}K$ to $GGT\mathbf{GFG}X_{35}K$ (mutation in bold and invariant glycines [Stock *et al.*, 1989] are underlined).

Two other mutants, NG9 and NG10, had a substitution of alanine 735 to valine. This residue lies within a region that is homologous to the N-terminal domain of the response regulators. This mutation alters a nonconserved residue within an α-helical region, αA of Stock *et al.* (1989). This domain is not present in most members of the sensor family and its role in signal transduction is not fully understood. The aspartic acid residue that is phosphorylated in VirG and its other homologs is found conserved in this domain of VirA. Alteration of this residue to a non-phosphorylable residue, asparagine, did not abolish inducible *virB* expression indicating that intramolecular phosphotransfer is not likely to be involved in *vir* gene induction.

Analysis of *virG* mutants

In 43 of the 44 *virG* mutants obtained in our study, a single point mutation that led to the substitution of asparagine 54 to aspartic acid (N54D) was found (Pazour *et al.*, 1992). This residue lies close to the aspartic acid 52 that is phosphorylated by the VirA kinase. The other mutant had a substitution of isoleucine 106 to leucine (I106L). This residue is in close proximity to a conserved lysine residue at position 102.

The crystal structure of CheY, a VirG homologue, has been determined (Stock *et al.*, 1989; Volz and Mutsumura, 1991). CheY and the N-terminal half of VirG share 24 percent residue identity and 42 percent conservative substitutions within a 120 amino acid segment. Analysis of the CheY structure suggests that the active site aspartic acid (asp 57) interacts with an asparagine residue at position 59 and a conserved lysine residue at position 109. These two residues interact with another conserved aspartic acid residue (asp 12) through solvent molecules. Phosphorylation of the active site aspartic acid that accompanies activation will force rearrangement of this bonding network possibly leading to a large scale conformational change at the backbone of the conserved lysine or elsewhere in the polypeptide chain (Volz and Mutsumura, 1991).

Since VirG is homologous to CheY it can be assumed that these two proteins are structurally similar. On this basis, both asparagine 54 and lysine 102 of VirG lie close to the active site aspartic acid at position 52. The *virG* N54D mutation will clearly disrupt various interactions at the active site. The other substitution, I106L, may affect the orientation of the conserved lysine 102 (equivalent to lysine 109 in CheY) leading to its constitutive phenotype. These mutations can function by mimicking the active conformation that results from phosphorylation, by stabilizing the aspartyl phosphate, or by changing a site(s) distal to the site of phosphorylation, e.g., a site necessary for interaction with RNA polymerase. The last two possibilities are unlikely because the half-life of the VirG-phosphate *in vitro* is rather long (> 1 hr [Jin *et al.*, 1990a]) and mutational analysis indicate that the constitutive phenotype of both mutants is dependent on the active site aspartic acid (Pazour *et al.*, 1992). Therefore, it appears most likely that the *virG* constitutive mutants function by locking the protein in an active conformation. This view is supported by the recent observations of Jin *et al.* (1993b) that VirG N54D is not phosphorylated by the VirA kinase.

Characterization of *virA* and *virG* mutants

The effect of the mutations in *virA* and *virG* on *virB* transcription was assessed by measuring β-galactosidase activity of a *virB-lacZ* fusion (Pazour *et al.*, 1991;1992). The mutations led to an 11 to 1350 fold increase in the basal level of *virB* expression (Table 1). The level of expression in strains bearing the *virG* N54D mutation was about 3 fold higher than that observed in fully induced *A. tumefaciens* indicating that under normal conditions *virB*, and most likely other *vir* genes as well, are not maximally expressed. Whether this increased expression in *virG* N54D translates to an increased virulence remains to be seen.

Induction of *vir* gene expression is dependent on the low pH ($\leq$ 6.0) of the growth medium. Low pH is thought to be required for the interaction of VirA with the inducing molecule(s). VirA functions as the pH sensor because this requirement can be partially relieved in a VirA mutant that has lost the periplasmic domain (Melchers *et al.*, 1989). If this is the only explanation, the

Table 1. Characterization of *virA* and *virG* mutations.

Mutant	Site of Mutation	*virB-lacZ* expression[1], ß-gal units
wild type	---	9
NG1	*virA* L24F	1,560
NG53	*virA* A469V	1,270
UV3	*virA* G471R	103
NG31	*virA* L658F	419
NG9	*virA* A735V	5,060
N54D	*virG* N54D	12,126
I106L	*virG* I106L	6,493
wild type + inducer		4,170

[1] Effect of mutations in *virA* and *virG* on *virB-lacZ* expression in the absence of an inducing agent is shown. Assays of *virG* mutants were performed in the absence of both inducer and *virA*. The level of *virB-lacZ* expression in the presence of inducer (AS) in a strain containing the wild type *virA* and *virG* is shown on the last line as a reference. Data from Pazour et al., 1991; 1992.

inducer independent *virA* mutants as well as the inducer and *virA* independent *virG* mutants would behave in a pH independent manner. Studies on the effect of pH on the phenotype of the *virA* and *virG* mutants showed that the mutants retained the low pH requirement for full induction, although significant induction of *vir* gene expression was observed in the *virG* mutants at a higher (6.8) pH (Pazour *et al.*, 1991;1992). These results suggest that pH has an effect on other step(s) of the signaling process. The most surprising of these studies is the pH dependence of the *virG* mutants. VirG is a cytosolic protein and is expected to be unaffected by pH since alteration of external (growth medium) pH does not considerably alter the intracellular pH (Ingraham, 1987). Yet the *virG* mutants are not fully active in growth medium of pH 6.8 indicating that VirG activity is dependent on alteration of intracellular conditions that result from low extracellular pH. Extracellular pH is known to alter intracellular concentration of ions such as K^+ which might directly affect VirG activity.

Effect of other substitutions at codon 54 and 106 of *virG*

To study the amino acid preference at codon 54 and 106 of *virG*, site-specific mutagenesis was used to alter these residues to other amino acids (Pazour *et al.*, 1992). At position 54 no substitution other than an asparagine to aspartic acid led to a constitutive phenotype. Several substitutions (N → E, F, I, L, P, R, S, W or Y) abolished *virG* activity while others (N → G, H, K, M or T) attenuated *virG* activity. Substitutions of N54 to A, C, Q or V had no detectable effect on *virG* activity. In contrast, no substitution at position 106 abolished *virG* activity and several substitutions (I → F, L, N, P or Y) led to a constitutive phenotype. In most I106 substitutions the level of *virB* expression in the presence of the inducer AS was considerably higher than that in a wild type *virG* containing strain again indicating that the *vir* genes are not normally activated to the maximal level.

Effect of *virA* on *virG* activity

The *virG* constitutive mutations were isolated by mutagenesis of an *A. tumefaciens* strain lacking *virA*. To study if *virA* has an effect on the constitutive phenotype of *virG* N54D and *virG* I106L substitutions, the *virA* gene was reintroduced into the respective plasmids and *virB* expression was monitored in the absence of acetosyringone. Analysis of these strains indicate that *virA* negatively affects *virG* activity in the absence of the inducer (Fig. 2).

In strains containing wild type *virG*, *virB* expression was 10 fold lower in the presence of *virA*. Similar results were also obtained in strains harboring *virG* I106L and *virG* I106P. In contrast, no significant effect of *virA* was observed in strains harboring *virG* I106F, *virG* I106N, *virG* I106Y or *virG* N54D. How *virA* affects *virG* activity is not known. The C-terminal domain of VirA (residues 712

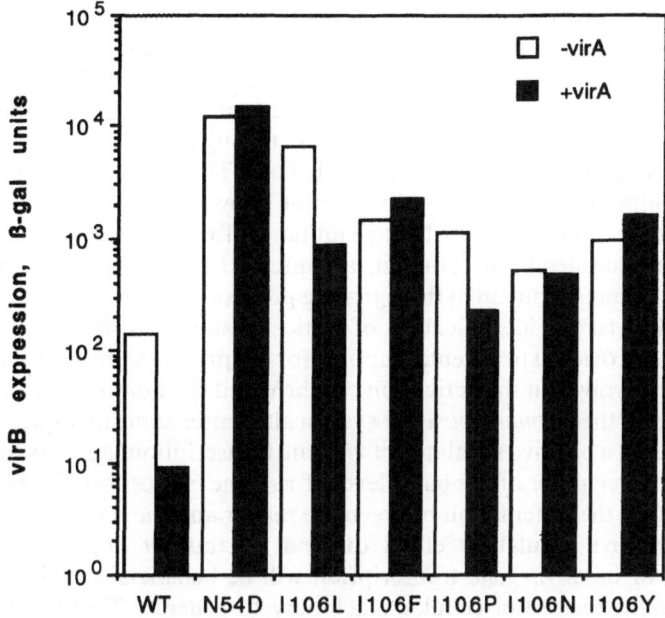

Figure 2. Effect of VirA on VirG activity: The effect of VirA on VirG activity was monitored by measuring *virB-lacZ* expression. VirG and its mutant derivatives were cloned into plasmid pAD1092 (*-virA*, *+virB-lacZ*) or pGP119 (*+virA*, *+virB-lacZ*), the resultants plasmids were introduced into *A. tumefaciens* A136 and the strains were assayed for ß-gal activity. Mutants are indicated by the single letter amino acid code. Numbers identify the VirG residue altered. WT, wild type. Reprinted from Pazour *et al.*, 1992.

to 829), homologous to the response regulator receiver domain, has a negative effect on the basal level of *virB* expression (Chang and Winans, 1992; A. Das, unpub. results). Whether this domain is responsible for the negative effect of *virA* observed in the *virG* I106 substitutions is currently under investigation.

Perspective and summary

In two-component regulatory systems a signal(s) is recognized by a sensor protein leading to its activation by phosphorylation of a conserved histidine residue. The phospho-sensor in turn transmits the signal to the regulator by phsophorylation of one of its conserved aspartic acid. This theme is conserved in the signal transduction process leading to the activation of *A. tumefaciens vir* gene transcription.

Activation of *A. tumefaciens vir* gene transcription requires the presence of one or more copies of a conserved 14 residue *vir* box sequence. This region serves as the binding site for the regulator VirG. The palindromic nature of this sequence suggests that VirG binds as a dimer or a multimer. Phosphorylation of

VirG is not essential for DNA-binding since VirG isolated from an overproducing *E. coli* strain, and VirGD52N, a non-phosphorylable mutant, both bind specifically to the *vir* box (Jin *et al.*, 1990b, 1993; Pazour and Das, 1990a; Roitsch *et al.*, 1990). VirG overproduced in *E. coli* is unlikely to be phosphorylated although phosphorylation through cross-talk (Ninfa *et al.*, 1988) by a VirA homologue cannot be ruled out. The effect of phosphorylation on the affinity of VirG for the *vir* box is not known.

Characterization of the constitutive mutants of *virA* indicate the importance of the first transmembrane domain, a conserved glycine-rich region and the extreme C-terminal domain in the signaling process. Similar analyses of the *virG* mutants allowed the identification of critical residues in the signal receiving domain and provide experimental support for the proposed interactions that are likely to be involved in the activation of CheY and the homologous regulators.

Analysis of the *A. tumefaciens vir* system also raises some new questions. For example, *virA*, a positive regulator of *vir* gene transcription, appears to function as a negative regulator of the basal level of *vir* gene expression. Current studies would suggest that interaction between the sensor and the regulator is required for the negative regulatory effect of *virA* (Pazour *et al.*, 1992). A down regulation of basal *vir* gene transcription will be beneficial to the cell since it would prevent the use of cellular machinery to generate T-strand DNA that cannot be transferred to a host. In addition to the negative effect of *virA*, *virG* down regulates its expression by using an inefficient codon, UUG, for translation initiation (Pazour and Das, 1990a; Parsons *et al.*, 1988; Reddy *et al.*, 1985). In free living cell *A. tumefaciens* produces only enough VirG to serve a monitoring function. When the bacterium encounters a host plant transcription of *virG*, and consequently that of all *vir* genes, is activated. A second regulatory gene, *ros*, may function directly in the prevention of abortive T-strand synthesis (Close *et al.*, 1985; Cooley *et al.*, 1991). Ros specifically turns off the basal level of *virC* and *virD* expression. The proteins of the *virD* operon catalyzes the T-DNA processing events (Yanofsky *et al.*, 1986).

Another unresolved question is the role of pH in *vir* gene transcription. *Vir* gene transcription is usually not observed at a pH above 6.0. VirA is likely to be the major pH sensor since the removal of its periplasmic domain results in partial pH-independence. However, VirA is not entirely responsible for the pH effect. VirG mutants that function in a VirA and inducer independent manner still require an acidic growth medium for full activity. Since VirG is a soluble cytoplasmic protein and extracellular pH does not significantly affect intracellular pH some other factor(s) is likely to be involved. One probable factor is intracellular ionic concentration which is known to be affected by extracellular pH.

Acknowledgements

I am indebted to Gregory Pazour for his contribution to this study. I thank Shouguang Jin and Eugene Nester for sharing unpublished results, Bonnie Allen for typing, and Ann Vogel for a critical reading of the manuscript. Works in the author's laboratory is supported by a grant from the National Institutes of Health (GM 37555) and an American Cancer Society Faculty Research Award (FRA 386).

References

Alt-Moerbe J, Nedderman P, Von Lintig J, Weiler EW & Schroder J (1988) Temperature-sensitive step in the Ti-plasmid vir-region induction and correlation with cytokinin secretion by Agrobacteria . Mol. Gen. Genet. 213: 1–8.

Ankenbauer R, Best E, Palanca C & Nester E (1991) Mutants of the Agrobacterium tumefaciens virA gene exhibiting acetosyringone-independent expression of the vir regulon. Mol. Plant-Micro. Inter. 4: 400–406.

Cangelosi GA, Ankenbauer RG & Nester EW (1990) Sugars induce the Agrobacterium virulence genes through a periplasmic binding protein and a transmembrane signal protein. Proc. Natl. Acad. Sci. USA 87: 6708–6712.

Chang C-H & Winans S (1992) Functional roles assigned to the periplasmic, linker, and receiver domains of the Agrobacterium tumefaciens VirA protein. J. Bacteriol. 174: 7033–7039.

Close T, Tait R & Kado C (1985) Regulation of Ti-plasmid virulence genes by a chromosomal locus of Agrobacterium tumefaciens. J. Bacteriol. 164: 774–781.

Cooley M, D'Souza M & Kado C (1991) The virC and virD operons of the Agrobacterium Ti plasmid are regulated by the ros chromosomal gene: analysis of the cloned ros gene. J. Bacteriol. 173: 2608–2616.

Das A, Stachel S, Ebert P, Allenza P, Montoya A & Nester E (1986) Promoters of the Agrobacterium tumefaciens virulence genes. Nucl. Acids Res. 14: 1355–1364.

Das A & Pazour G (1989) Delineation of the regulatory region sequences of Agrobacterium tumefaciens virB operon. Nucl. Acids Res. 17: 4541–4550.

Garfinkel DJ & Nester EW (1980) Agrobacterium tumefaciens mutants affected in crown gall tumorigenesis and octopine catabolism. J. Bacteriol. 144: 732–743.

Huang MLW, Cangelosi GA, Halperin W & Nester EW (1990) A chromosomal Agrobacterium tumefaciens gene required for effective plant signal transduction. J. Bacteriol. 172: 1814–1822.

Huang Y, Morel P, Powell B & Kado C (1990) VirA, a coregulator of Ti-specified virulence genes, is phosphorylated in vitro. J. Bacteriol. 172: 1142–1144.

Ingraham J (1987) Effect of temperature, pH, water activity, and pressure on growth. In: Ingraham J. et al. (eds.) Escherichia coli and Salmonella typhimurium : Cellular and Molecular Biology, Amer. Soc. for Microbiol., Washington, DC, Vol 2: 1543–1554.

Jin S, Prusti R, Roitsch T, Ankenbauer R & Nester EW (1990a) Phosphorylation of the VirG protein of Agrobacterium tumefaciens by the autophosphorylated VirA protein: Essential role in the biological activity of VirG. J. Bacteriol. 172: 4945–4950.

Jin S, Roitsch T, Christie P & Nester EW (1990b) The regulatory VirG protein specifically binds to a cis-acting regulatory sequence involved in transcriptional activation of Agrobacterium tumefaciens virulence genes. J. Bacteriol. 172: 531–537.

Jin S, Song Y & Nester EW (1992) Function of the regulatory VirA protein of Agrobacterium tumefaciens is sensitive to elevated temperatures. Sixth Intl. Symp. on Mol. Plant-Microbe Interactions. Abs. #18.

Jin S, Song Y, Pan SQ & Nester EW (1993) Characterization of a virG mutation that confers constitutive virulence gene expression in Agrobacterium . Mol. Micro. biol. 7: 555–562.

Kado C (1991) Molecular mechanisms of crown gall tumorigenesis. Crit. Rev. Plant Sci. 10: 1–32.

Kanemoto RH, Powell AT, Akiyoshi DE, Regier DA, Kerstetter RA, Nester EW, Hawes MC & Gordon MP (1989) Nucleotide sequence and analysis of the plant-inducible locus *pin*F from *Agrobacterium tumefaciens*. J. Bacteriol. 171: 2506–2512.

Leroux B, Yanofsky M, Winans S, Ward J, Zeigler S & Nester E (1987) Characterization of the *vir*A locus of *Agrobacterium tumefaciens*: a transcriptional regulatory and host range determinant. EMBO J. 6: 849–856.

Melchers LS, Maroney MJ, den Dulk-Ras A, Thompson D, van Vuuren H, Schilperoort RA & Hooykaas PJJ (1989) Octopine and nopaline strains of *Agrobacterium tumefaciens* differ in virulence: molecular characterization of the *vir*F locus. Plant Mol. Biol. 14: 249–259.

Melchers LS, Regensburg-Tuink TJG, Bourret RB, Sedee NJA, Schilperoort RA & Hooykaas PJJ (1989) Membrane topology and functional analysis of the sensory protein VirA of *Agrobacterium tumefaciens*. EMBO J. 8: 1919–1925.

Ninfa AJ, Ninfa EG, Lupas A, Stock A, Magasanik B & Stock J (1988) Crosstalk between bacterial chemotaxis signal transduction proteins and the regulators of transcription of the Ntr regulon: evidence that nitrogen assimilation and chemotaxis are controlled by a common phosphotransfer mechanism. Proc. Natl. Acad. Sci. USA 85: 5492–5496.

Ooms G, Klapwijk PM, Poulis JA & Schilperoort RA (1980) Characterization of Tn904 insertions in octopine Ti plasmid mutants of *Agrobacterium tumefaciens*. J. Bacteriol. 144: 82–91.

Oosawa K & Simon M (1986) Analysis of mutations in the transmembrane region of the aspartate chemoreceptor in *Escherichia coli*. Proc. Natl. Acad. Sci. USA 83: 6930–6934.

Parsons GD, Donly BC & Mackie GA (1988) Mutations in the leader sequence and initiation codon of the gene for ribosomal protein S20 (*rpsT*) affect both translational efficiency and autoregulation. J. Bacteriol. 170: 2485–2492.

Pazour GJ & Das A (1990a) VirG, an *Agrobacterium tumefaciens* transcriptional activator, initiates translation at a UUG codon and is a sequence-specific DNA-binding protein. J. Bacteriol. 172: 1241–1249.

Pazour GJ & Das A (1990b) Characterization of the VirG binding site of *Agrobacterium tumefaciens*. Nucl. Acids Res. 18: 6909–6913.

Pazour GJ, Ta C & Das A (1991) Mutants of *Agrobacterium tumefaciens* with elevated *vir* gene expression. Proc. Natl. Acad. Sci. USA 88: 6941–6945.

Pazour GJ, Ta C & Das A (1992) Constitutive mutations of *Agrobacterium tumefaciens* transcriptional activator *virG*. J. Bacteriol. 174: 4169–4174.

Peralta EG, Hellmiss R & Ream W (1986) *Overdrive*, T-DNA transmission enhancer on the *A. tumefaciens* tumour-inducing plasmid. EMBO J. 5: 1137–1142.

Powell BS & Kado CI (1990) Specific binding of VirG to the *vir* box requires a C-terminal domain and exhibits a minimum concentration threshold. Mol. Micro. 4: 2159–2166.

Ream W (1989) *Agrobacterium tumefaciens* and interkingdom genetic exchange. Annu. Rev. Phytopathol. 27: 583–618.

Reddy P, Peterkofsky A & McKenney K (1985) Translational efficiency of the *E. coli* adenylate cyclase gene: mutating the UUG initiation codon to GUG and AUG results in increased gene expression. Proc. Natl. Acad. Sci. USA 82: 5656–5660.

Rogowsky P, Close T, Chimera J, Shaw J & Kado C (1987) Regulation of the *vir* genes of *Agrobacterium tumefaciens* plasmid pTiC58. J. Bacteriol. 169: 5101–5112.

Roitsch T, Wang H, Jin S & Nester E (1990) Mutational analysis of the VirG protein, a transcriptional activator of *Agrobacterium tumefaciens* virulence genes. J. Bacteriol. 172: 6054–6060.

Ronson CW, Nixon BT & Ausubel FM (1987) Conserved domains in bacterial regulatory proteins that respond to environmental stimuli. Cell 49: 579–581.

Soderling TR (1990) Protein kinases: Regulation by autoinhibitory domains. J. Biol. Chem. 265: 1823–1826.

Stachel SE, Messens E, Van Montagu M & Zambryski P (1985) Identification of the signal molecules produced by wounded plant cells that activate T-DNA transfer in *Agrobacterium tumefaciens*. Nature (London) 318: 624–629.

Stachel S & Nester E (1986) The genetic and transcriptional organization of the *vir* region of the A6 Ti plasmid of *Agrobacterium tumefaciens* . EMBO J. 5: 1445–1454.

Stachel S, Nester E & Zambryski P (1986) A plant cell factor induces *Agrobacterium tumefaciens vir* gene expression. Proc. Natl. Acad. Sci. USA 83: 379–383.

Stachel S & Zambryski P (1986) *vir*A and *vir*G control the plant-induced activation of the T-DNA transfer process of *A. tumefaciens* . Cell 46: 325–333.

Stock A, Mottonen J, Stock J & Schutt C (1989) Three-dimensional structure of CheY, the response regulator of bacterial chemotaxis. Nature (London) 337: 745–749.

Stock J, Ninfa A & Stock A (1989) Protein phosphorylation and regulation of adaptive responses in bacteria. Microbiol. Rev. 53: 450–490.

Volz K & Matsumura P (1991) Crystal structure of *Escherichia coli* CheY refined at 1.7-Å resolution. J. Biol. Chem. 266: 15511–15519.

Winans SC (1990) Transcriptional induction of an *Agrobacterium* regulatory gene at tandem promoters by plant-released phenolic compounds, phosphate starvation, and acidic growth media. J. Bacteriol. 172: 2433–2438.

Winans SC (1992) Two-way chemical signaling in *Agrobacterium*-plant interactions. Microbiol. Rev. 56: 12–31.

Winans SC, Ebert, PR, Stachel SE, Gordon MP, & Nester EW (1986) A gene essential for *Agrobacterium* virulence is homologous to a family of positive regulatory loci. Proc. Natl. Acad. Sci. USA 83: 8278–8282.

Winans SC, Jin S, Komari T, Johnson KM & Nester EW (1987) The role of virulence regulatory loci in determining *Agrobacterium* host range. p. 573–582. *In:* von Wettstein D & Chua N-H (eds.) Plant Molecular Biology. Plenum Press, New York.

Winans S, Kerstetter R & Nester E (1988) Transcriptional regulation of the *vir*A and *vir*G genes of *Agrobacterium tumefaciens* . J. Bacteriol. 170: 4047–4054.

Yanofsky M, Porter S, Young C, Albright L, Gordon M & Nester E (1986) The virD operon of *Agrobacterium tumefaciens* encodes a site-specific endonuclease. Cell 47: 471–477.

Zambryski P (1988) Basic processes underlying *Agrobacterium*-mediated DNA transfer to plant cells. Annu. Rev. Genet. 22: 1–30.

33. Regulation of the iron transport genes encoded by the pJM1 virulence plasmid in *Vibrio anguillarum*

JORGE H. CROSA, LUIS A. ACTIS, PATRICIA SALINAS, MARCELO E. TOLMASKY and LILLIAN S. WALDBESER

Abstract. An important component of the virulence of *Vibrio anguillarum* 775 is the pJM1 plasmid-mediated iron uptake system which consists of the siderophore anguibactin and the iron transport proteins FatA-D. Expression of this system is repressed when the iron concentration of the cell is high. We have now shown that under iron-rich conditions two elements may a play a role in the control of the expression of the *fatA* gene: a Fur-like product and a 600 nucleotides antisense RNA, RNAα, which is encoded in the *fatA-fatB* intergenic region and is induced at high iron concentrations. Under iron-limiting conditions two positive regulators, Taf and AngR, are necessary for full expression of the siderophore biosynthetic genes. AngR, a 110 kDa protein, possesses two helix-turn-helix domains with adjacent leucine zippers that may play a role in the regulatory activity of this protein as assessed by site-directed mutagenesis. We have also identified a novel catalytic activity of AngR suggesting that, in addition to its regulatory role, it acts as an enzyme in the biosynthesis of anguibactin.

Abbreviations: AMP, Adenosme Monophoshate; EDDA, Ethylenediamine-di Co-hydroxyphenyla-cetic Acid; MIC, Minimal Inhibitory Concentration; SDS-page, Sodium Dodecyl Salphate-Poly Acrylamide Gel Electrophorhesis

Introduction

Vibriosis caused by infection with *Vibrio anguillarum* is one of the most devastating diseases affecting salmonid fish. It is characterized as a hemorrhagic septicemia as well as severe hypoxia and disfunction of various organs, that leads to a high rate of fish mortality.

Our laboratory has reported that some pathogenic strains of *V. anguillarum* possess the 65 kb pJM1 plasmid that encodes an iron uptake system that is an important virulence factor of this bacterium (Crosa, 1980; 1989; Crosa *et al.*, 1980). The pJM1 plasmid-mediated iron uptake system encodes anguibactin, a 348 Da catechol-type siderophore (Actis *et al.*, 1986; Jalal *et al.*, 1989), and four iron transport proteins: the 86 kDa outer membrane protein FatA, the 40 kDa periplasmic protein FatB, and the ca. 40 kDa cytoplasmic membrane proteins FatC and FatD (Kostler *et al.*, 1991). Fig. 1 shows a model of this iron uptake system. We have further characterized the pJM1 plasmid and have identified genetic units involved in the biosynthesis of anguibactin (Tolmasky *et al.*, 1990), as well as two positive transcriptional activators, AngR and Taf (Tolmasky & Crosa, 1984; Tomalsky *et al.*, 1985; 1988), which intervene in the enhancement

C.I. Kado and J.H. Crosa (eds.), Molecular Mechanisms of Bacterial Virulence, 491–504.
© 1994 *Kluwer Academic Publishers.*

492

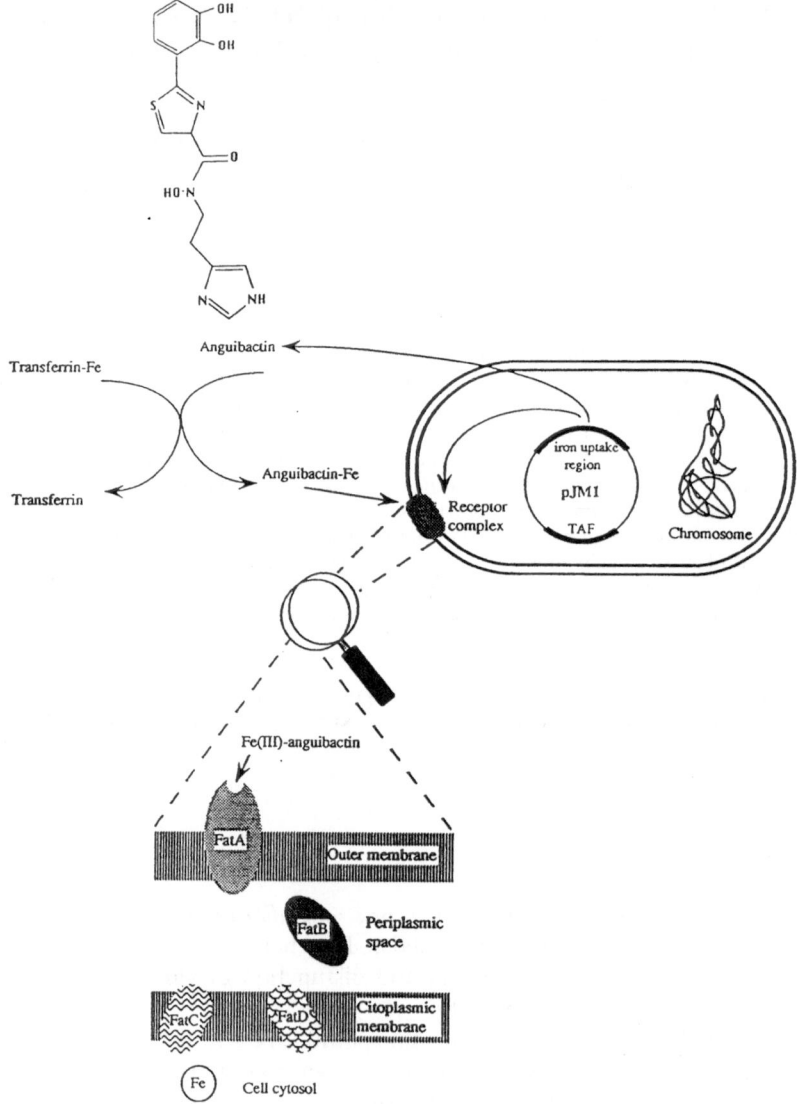

Figure 1. Model of the iron uptake system encoded by the pJM1 plasmid in *V. anguillarum* 77.

at the level of transcription of the expression of siderophore biosynthetic genes (Fig. 2A). The expression of the whole iron uptake system is negatively regulated by the iron concentration of the cell. In this report we will discuss our recent findings concerning this mechanism of regulation by iron in *V. anguillarum* 775 as well as strong evidence suggesting that, in addition to its regulatory activity, AngR may play an important role in the biosynthesis of the siderophore anguibactin.

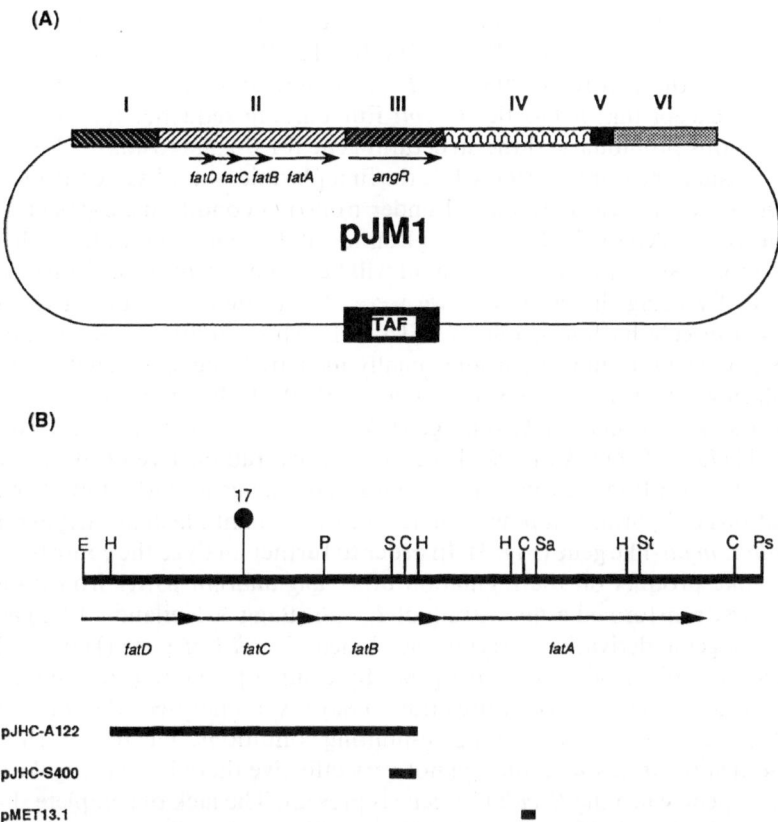

Figure 2. A, genetic and functional map of the virulence plasmid pJM1. The roman numbers indicate the genetic units, defined by insertion mutagenesis, involved in the expression of the iron uptake system. B, physical and genetic map of the pJM1 iron transport region, and recombinant clones pJHC-A122, pJHC-S400, and pMET13.1 harboring sequences located within this region. The solid bars represent pJM1 DNA. The symbol ♇ represents the location of Tn*3*-HoHo1 insertion in mutant #17. The orientation of the arrows indicates the direction of the transcription. Restriction sites are indicated as: E, *Eco*RI; H, *Hind*III; S, *Sal*I; C, *Cla*I; Sa, *Sac*I; St, *Stu*I; P, *Pvu*I; Ps, *Pst*I.

Iron regulation of the expression of pJM1 iron transport genes

In *Escherichia coli* and other enteric bacteria iron represses the expression of iron-regulated genes via the product of the *fur* gene (Hantke, 1984). Complexes of the Fur protein with Fe^{+2} bind to a specific operator sequence, located adjacent to the promoter affecting the initiation of transcription of those genes (deLorenzo *et al.*, 1988). In an attempt to understand the mechanism of iron regulation in *V. anguillarum* we first intended to assess whether a Fur-like product was part of the iron regulatory network in this bacterium. To achieve this aim we first used two plasmids, constructed by Calderwood and Mekalanos (Calderwood & Mekalanos 1988). One (pRT240) carries the β-galactosidase gene under the control of the *ompF* promoter, and the other (pSC27.1) is identical except that it has the *E. coli* Fur binding sequence incorporated in between this promoter and the structural gene for β-galactosidase. We call the latter plasmid the Fur reporter, while the first plasmid is used as a control. If Fur is present and the cells are placed under iron-rich conditions, β-galactosidase synthesis encoded by the Fur probe plasmid will be repressed while synthesis of β-galactosidase by the control plasmid will be independent of the iron status of the cell. By using this approach we were able to detect a Fur activity in *V. anguillarum* cells harboring pSC27.1 (Table 1). This activity was not encoded by the plasmid pJM1 but was chromosomally-mediated since it was detected using the plasmidless strain H775-3 (Crosa *et al.*, 1980). Following, we proceeded to clone the *V. anguillarum fur*-like gene from a gene library constructed from strain H775-3 (Tolmasky *et al.*, 1992). By using a radioactive probe consisting of a restriction fragment internal to the *E. coli fur* gene under low stringency conditions of hybridization, we isolated a recombinant plasmid carrying the *V. anguillarum fur*-like gene (Fig. 3). In order to further analyze the contribution of a Fur-like product on the regulation of *V. anguillarum* pJM1 iron transport genes, we transformed a *fur⁻* strain of *E. coli* (Bagg & Neilands, 1985) as well as the isogenic derivative carrying the cloned *E. coli* Fur gene (Hantke, 1984) with a plasmid encoding the *fatA* gene. By using a probe that recognizes *fatA* mRNA and ribonuclease protection assays we analyzed the *fatA* RNA synthesized under iron-rich and iron-limiting conditions. The results, shown in Fig. 4, demonstrate that although not very effective there is some regulation of the *fatA* gene when the *E. coli* Fur gene is present. The lack of complete shut off may be due to the fact that the *V. anguillarum* Fur-like gene shows low homology with the *E. coli* counterpart or due to the foot that the expression of *fatA* in *E.coli* is not the same as in *V. anguillarum*. An unexpected result, however, was obtained when we used these very same strains and constructs to

Table 1. β-Galactosidase activities of *V. anguillarum*.

Strain	β-galactosidase activity (U)[1]		Inhibition
	+ 30 μM FeCl3	+ 1 μM EDDA	Fe/EDDA
H775-3a (pRT240)	245	177	1.38
H775-3a (pSC27.1)	109	452	0.24

[1]Miller units (Miller, 1972)

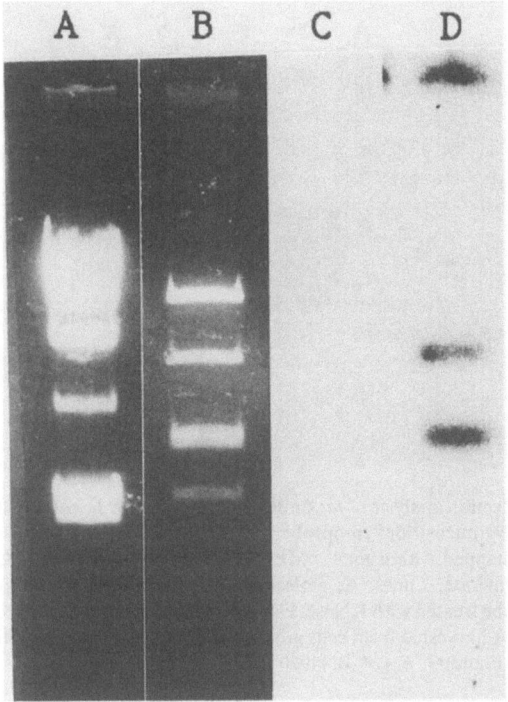

Figure 3. Identification of an *E. coli* recombinant clone harboring the *V. anguillarum fur* gene. Agarose gel electrophoresis of the products of *Hind*III digestion of λ DNA (lane A) and plasmid DNA from a recombinant clone from the *V. anguillarum* H775-3 gene library isolated by colony blot (lane B) using as probe the *Hind*III-*Bgl*I fragment from pMH15 which contained the *E. coli fur* gene (10). Lanes C and D, autoradiography of Southern blot hybridization of lanes A and B, using as probe the *Hind*III-*Bgl*I fragment from pMH15.

detect FatA protein with a specific anti-FatA serum. Table 2 shows that FatA is synthesized only under iron-limiting conditions independent of whether the *E. coli* Fur protein is present. Since a good amount of *fatA* mRNA is synthesized under iron rich conditions (Fig. 4, lanes D and F), especially in the *fur⁻* strain of *E. coli* (Fig 4, lane D), there must be another mechanism that operates under iron-rich conditions to shut-off the synthesis of the FatA protein.

A series of complementation experiments that we had performed earlier to identify the iron transport components, between subclones of the iron transport region and specific transposition mutants in this region, provided us now with a clue to the nature of this puzzle. Plasmids that carried fragments containing the intergenic region between *fatB* and *fatA* although complementing the mutations resulted in a decreased synthesis of the Fat A protein (Fig. 5). One of these plasmids, pJHC-A122, was examined in more detail. It contained a 2.4 kb *Hind*III fragment and, as seen in Fig. 5, its presence leads to a dramatic reduction in FatA synthesis in both the wild type strain of *V. anguillarum*

496

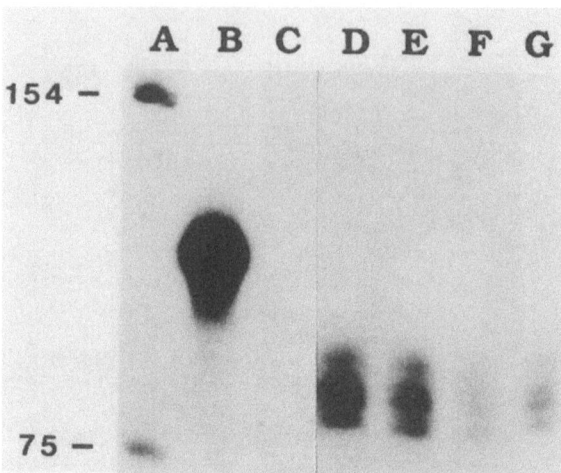

Figure 4. RNase protection analysis of *fatA* mRNA expression in *E. coli* BN4020 in the presence or absence of Fur. The 94 nucleotides riboprobe was made from pMET13.1 (see Fig. 2B) carrying a *Cla*I-*Sal*I fragment mapped within *fatA*. pMET13.1 was linearized with *Cla*I and transcribed by using T3 RNA polymerase. Lanes: A, molecular weight marker; B, riboprobe without RNase treatment; C, riboprobe treated with RNase; D and F, RNA harvested from cells grown in iron-rich media; E and G, RNA harvested from cells grown under iron-limiting conditions; D and E, *E. coli* BN4020 (pJHC-T7); F and G, *E. coli* BN4020 (pJHC-T7, pMH15).

carrying the pJM1 plasmid as well as in a recombinant strain carrying the cloned pJM1 iron uptake region (plasmid pJHC-T7) (Tolmasky & Crosa, 1984). To investigate the nature of the inhibiting phenomenon we inserted the Ω fragment (Prentki & Krish, 1984), which carries transcription and translation termination signals, at several positions in pJHC-A122 (Fig. 6A) and analyzed their influence on the expression of *fatA* (Fig. 6, B and C). It is clear from Fig. 6B that only insertion of Ω at the *Hind*III right site (pJHC-LW96) abolished the inhibitory phenomenon, while no effect was detected with Ω inserted in the left *Hind*III site (pJHC-LW95). Since the direction of transcription of the FatA mRNA is towards the right (Actis *et al.*, 1988) it was possible that the inhibitory substance was an antisense RNA transcribed opposite to the direction of transcription of the *fatA* gene. To assess whether this was the case, we prepared an RNA probe that could detect antisense RNA and performed ribonuclease protection experiments with RNA obtained from *V. anguillarum* cells harboring various derivatives (Fig. 6C). Our results indicated that there was indeed an antisense RNA that we designated RNAα. It was only synthesized under iron rich conditions in the case of wild type strains of *V. anguillarum,* while it was constitutively synthesized in the case of strains containing pJHC-A122 (Salinas *et al.*, 1992). We have now determined that the expression of RNAα in pJHC-A122 is under the control of the Tc^r promoter, while in the natural surroundings this antisense RNA is synthesized under the control of a promoter that responds positively to an increase in the iron concentration of the cell.

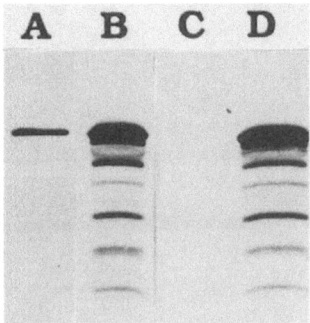

Figure 5. Immunoblot analysis of the effect of pJHC-A122 on the synthesis of FatA in *V. anguillarum.* Total membrane proteins were prepared from bacteria grown under iron-limiting conditions. Lanes: A, strain H775-3a (pJHC-T7, pJHC-A122); B, strain H775-3a (pJHC-T7); C, strain 775 (pJM1, pJHC-A122); D, strain 775 (pJM1).

Primer extension experiments of RNA extracted from *V. anguillarum* carrying pJM1 identified three possible locations of RNAα initiation sites. Since Northern blot experiments led to a size of 600 nucleotides, RNAα must span the intergenic region between *fat*B and *fat*A and must be complementary to the 5' end of *fat*A mRNA. Therefore, the presence of RNAα must account for our results with the *fur*⁻ mutant of *E. coli* (Table 2) and suggest that RNAα must act mostly posttranscriptionally to inhibit synthesis of FatA, although some inhibition of *fat*A transcription is also detected.

Our present results fit a model in which under iron-rich conditions a Fur-like product might act by inhibiting transcription initiation of the *fat*A gene since under iron-rich conditions there is no detectable *fat*A mRNA in *V. anguillarum.* It is possible that RNAα that is synthesized only under iron rich conditions may also contribute to this repression. However, the bulk of RNAα action appears

Table 2. Effect of Fur on the synthesis of FatA in *E. coli* BN4020.

Strain (Plasmids)	Expression of FatA [1]	
	+ FeCl$_3$	+ EDDA
BN4020 (pJHC-T7)	-	+
BN4020 (pJHC-T7, pMH15)	-	+

[1]The expression of FatA was determined by immunoblot assays using anti-FatA serum.

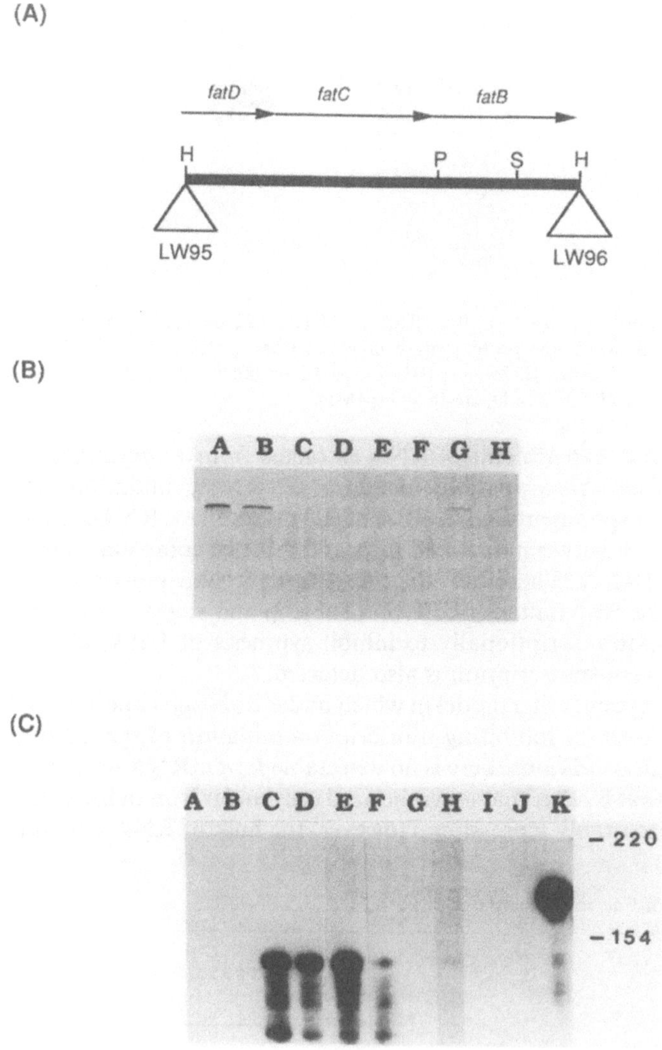

(A)

(B)

(C)

Figure 6. Effect of the insertion of the Ω fragment in pJHC-A122 on the regulation of the expression of *fatA* in *V. anguillarum* cells harboring mutant #17. A, restriction map of cloned pJM1 regions in pJHC-A122. The solid bar represents pJM1 DNA. The open triangles represent the site of the insertion of the Ω fragment in derivatives pJHC-LW95 and pJHC-LW96. Restriction sites are indicated as: H, *Hind*III; P, *Pvu*I; S, *Sal*I. B, immunoblot analysis of FatA protein. Lanes: A, C, E, and G, cells grown under iron-limiting conditions; B, D, F, and H, cells grown in iron-rich media. All lanes are loaded with equal amount of total membrane protein from *V. anguillarum* H775-3a harboring mutant #17 together with either pJHC-LW96 (lanes A and B), pJHC-LW95 (lanes C and D), pJHC-A122 (lanes E and F), or no additional plasmid (lanes G and H). C, RNase protection analysis of RNAα. The 133 nt riboprobe was made from pJHC-S400 (see Fig. 2B) linearized with

to be directed to prevent translation of *fatA* transcripts that had already been initiated, acting thus as a fine tuning control of this mechanism of iron regulation.

The AngR protein

It was recently reported that many *V. anguillarum* strains from various geographical locations harbor pJM1-like plasmids (Tolmasky *et al.*, 1985; 1990). Several of these strains showed higher anguibactin activity when compared to the prototype *V. anguillarum* 775 strain (Tolmasky *et al.*, 1988). It was found that the high siderophore activity of one of these strains, *V. anguillarum* 531A, was determined by *angR*, a gene present in pJM1 as well as in pJM1-like plasmids (Tolmasky *et al.*, 1988). This gene encodes the 110 kDa protein that acts as a *trans*-activator of other(s) gene(s) of this iron-uptake system (Salinas *et al.*, 1989; Tolmasky *et al.*, 1988). Sequence analysis showed that *angR* encodes a protein of 1048 amino acids that has a helix-turn-helix motif typical of prokaryotic DNA binding proteins, with homology to the DNA binding domain of the P22 phage protein Cro (Farrell *et al.*, 1990). Further inspection of this amino acid sequence demonstrated that it also has two leucine zippers each followed by a helix-turn-helix motif.

To investigate the role of these AngR sequences we generated, by site-directed mutagenesis, mutations in the leucine zipper and helix-turn-helix domains of AngR. We also used two mutants previously generated by Tn*3*-HoHo1 transposition (Tolmasky *et al.*, 1988). The nature and/or location of these mutations is shown in Fig. 7. The mutations in plasmids pMET16.2-pMET16.5 consist on substitutions of the original amino acid for Pro, which is known to disrupt helix structures (Branden *et al.*, 1991). The mutations harbored by plasmids pMET16.6-pMET16.8 resulted in substitutions of one or more Leu or Val residues, producing a disruption of the leucine zipper. All these derivatives as well as the wild type *angR* gene were transferred by conjugation to *V. anguillarum* 775::Tn*1*-6B#4. This strain harbors pJHC-T2612#4 which is a clone containing all the pJM1 iron uptake region with a Tn*3*-HoHo1 insertion in *angR*, and the plasmid pJHC9-8 which carries the Taf factor, necessary for full production of anguibactin. The iron uptake proficiency, determined as the minimal inhibitory concentration (MIC) for the iron chelator ethylene-diamine-di(*o*-hydroxyphenyl) acetic acid (EDDA), and the proteins encoded by each one of these plasmids derivatives, analyzed by *in vitro* transcription-translation and

---

*Hind*III and transcribed by T7 RNA polymerase. Lanes: A, C, E, G, and I, cells grown under iron-limiting conditions; B, D, F, H, and J, cells grown in iron-rich media. RNA was obtained from *V. anguillarum* H775-3a harboring mutant #17 together with either pJHC-LW96 (lanes A and B), pJHC-LW95 (lanes C and D), pJHC-A122 (lanes E and F), or no additional plasmid (lanes G and H). RNA from *V. anguillarum* H775-3a (plasmidless) was used as negative control (lanes I and J). Lane K, riboprobe without RNase treatment.

Table 3. Properties of *angR* mutants.

Recombinant plasmid[1]	mutation	aa#	Relevant information	MIC for EDDA (μM)[2]
None				2.5
pJHC-S2571			*angR* from pJHC1	20.0
pJHC-S2771			*angR* from pJM1	10.0
pMET16.2	T to P ACT to CCT	888	2nd. helix-turn-helix helix 3	2.5
pMET16.3	T to P ACT to ACCT (change of phase)	888	2nd. helix-turn-helix helix 3	2.5
pMET16.4	L to P CTA to CCA	889	2nd. helix-turn-helix helix 3	2.5
pMET16.5	Q to P CAG to CCG	304	1st. helix-turn-helix helix 3	2.5
pMET16.6	V to D CAG to CCG L to R CTT to CGT	151 158	1st. leucine zipper	2.5
pMET16.7	V to K GTC to AAG	852	2nd. leucine zipper	2.5
pMET16.8	I to P ATT to CCT	859	2nd. leucine zipper	2.5
pJHC-T5.12#16[3]		965		2.5
pJHC-T7#55[4]		1047		10.0

[1] *V. anguillarum* strains harbored pJHC9–8, pJHC-T2612#4 in addition to the indicated recombinant clones. In the cases of pJHC-T5.12#16 and pJHC–T7#55 the strains carried pJHC9–8 in addition to either derivative.

[2] MIC for EDDA was determined by bioassays as previously described (25).

[3] In these mutants insertion of Tn*3*–HoHo1 took place after the aa indicated.

sodium dodecylsulfate- polyacrylamide gel electrophoresis (SDS-PAGE), was then investigated. Table 3 shows that the mutations harbored in plasmids pMET16.2-pMET16.8, as well as transposition mutant #16 exhibited reduced MICs for EDDA, similar to those displayed by the iron-uptake deficient control strain. Conversely, transposition mutant #55 showed an MIC for EDDA similar to the wild type strain. Plasmids pMET16.2-pMET16.8 were able to express the AngR protein *in vitro* (data not shown). In the case of pMET16.3 the change of phase resulted in the synthesis of a smaller protein, as predicted by the DNA sequence. The transposition mutant #16 encoded an AngR protein truncated at amino acid 965, as predicted from the nucleotide sequence of these mutant. Although not detected by SDS-PAGE, transposition mutant #55 also resulted in a truncated AngR protein in which only the last amino acid is missing. Therefore, at least a portion of the carboxy terminus of AngR downstream of the second helix-turn-helix motif is needed for its function.

It was also recently shown that AngR shares homology in a specific domain with several proteins of three groups of ATP-utilizing enzymes: the acid-thiol ligases, the activating enzymes for the biosynthesis of enterobactin, and the synthetases for tyrocidine, gramicidine S, and penicillin, also known as the firefly luciferase family. Among these enzymes are: the 57 kDa polypeptide of the 4-chlorobenzoate dehalogenase (4-CBA) from *Pseudomonas sp.*, the gramicidine S synthetase and the tyrocidine synthetase from *Bacillus brevis*, the coumarate CoA:ligase from *Petroselinum crispum*, the luciferase from *Photinus pyralis*, the D-alanine-activating enzyme from *Lactobacillus casei*, and the 2,3-dihydroxy-benzoate-AMP ligase (EntE) from *E. coli* (Heaton & Neuhaus, 1992; Scholten *et al.*, 1991; Toh, 1991). The latter case is of particular interest for our studies since EntE catalyzes the activation of 2,3 dihydroxybenzoic acid, an essential step in the biosynthesis pathway of the catechol siderophore enterobactin.

Therefore, it was of interest to determine whether AngR could complement the *entE* mutation in *E. coli* AN93 (Schmitt & Payne, 1988). Fig. 8 shows that transformation of this enterobactin-deficient mutant with plasmid pJHC-S2572 carrying the *angR* gene, led to restoration of the iron-uptake proficient phenotype measured by the ability of this complemented mutant strain to grow in the presence of high concentrations of EDDA. Conversely, neither pJHC-S2570, which carries a truncated *angR* (Salinas *et al.*, 1992) nor the vector pKK223-3 were able to complement this mutation. These results suggest that besides its regulatory role, AngR has an enzymatic function related to that of 2,3-dihydroxybenzoate-adenosine monophosphate (AMP) ligase.

Concluding remarks

We have shown in this work that the iron regulation of the expression of the pJM1-mediated iron uptake system of *V. anguillarum* presents novel components: a Fur-like activity acting mainly at transcription initiation and a mostly posttranscriptional player, antisense RNAα, only expressed under iron-rich conditions. In addition, the transcriptional activator AngR is essential for anguibactin production under iron limitation. This 110 kDa protein has two helix-turn-helix motifs preceded by leucine zippers. These motifs proved to be essential since site-directed mutations in these domains lead to a loss of the transcription enhancement properties. However, AngR also shows an enzymatic activity that plays an important role in the biosynthesis of anguibactin. The complexities of the control mechanisms of expression of this system underscores its importance for the virulence phenotype of *V. anguillarum*.

502

1   MNQNEHPFAFPETKLPLTSNQNWQLSTQRQRTEKKSITNFTYQEFDYENISRDTLERCLTTIIKHHPI

69 FGAKLSDDFYLHFPSKTHIETFAVNDLSNALKQDIDKQLADTRSAVTKSRSQAIISIMFSI

130 L P K N I I R L H V R F N S V V V D N P S V T
                                  GTT
                               16.6 D
                                  GAT

153 L F F E Q L
              CTT
          16.6 R
              CGT

159 TQLLSGSPLSFLNQEQTISAYNHKVNNELLSVDLESARWNEYILTLPSSANLPTICEPEKLDETDITR

227 RCITLSQRKWQQLVTVSKKHNVTPEITLASIFSTVLSLWGHQKYLMMRFDITKINDYTGIIGQ

290 F T E P L L V G M S G F E Q S F L S L V K N
                          CAG
                      16.5 P
                          CCG

312 NQKKFEEAYHYDVKVPVFQCVNKLSNISDSHRYPANITFSSELLNTNHSKKAVWGCRQSANTWLSLHA

380 VIEQEQLVLQWDSQDAIFPKDMIKDMLHSYTDLLDLLSQKDVNWAQPLPTLLPKHQESIRNKINQQGD

448 LELTKELLHQRFFKNVESTPNALAIIHGQESLDYITLASYAKSCAGALTEAGVKSGDRVAVTMNKGIG

516 QIVAVLGILYAGAIYVPVSLDQPQERRESIYQGAGINVILINESDSKNSPSNDLFFFLDWQTAIKSEP

584 MRSPQDVAPSQPAYIIYTSGSTGTPKGVVISHQGALNTCIAINRRYQIGKNDRVLALSALHFDLSVYD

652 IFGLLSAGGTIVLVSELERRDPIAWCQAIEEHNVTMWNSVPALFDMLLTYATCFNSIAPSKLRLTMLS

720 GDWIGLDLPQRYRNYRVDGQFIAMGGATEASIWSNVFDVEKVPMEWRSIPYGYPLPRQQYRVVDDLGR

788 DCPDWVAGELWIGGDGIALGYFDDELKTQAQFLHIDGHAWYRTGDMGCYWPDTLEF

845 L G R R D K Q V K V G G Y R I E L G E I E V A
              GTC   ATT
          16.7 K   16.8 P
              AAG   CCT

868 L N N I P

873 G V Q R A V A I A V G N K D K T   L A A F I V
                              ACT CTA
                          16.2 P   P 16.4
                              CCT CCA
                          16.3 phase change
                              ACCT

896 MDSEQAPIVTAPLDAEEVQLLLNKQLPNYMVPKRIIFLETFPLTANGKVDHKALTRMTNREKKTSQSI

963 NKP IITASEDRVAKIWNDVLGPTELYKSSDFFLSGGDAYNAIEVVKRCHKAGYLIKLSMLYRYSTIE
        ●16

1031 AFAIIMDRCRLAPQEEAE L
                      ●55

Figure 7. Deduced amino acid sequence of AngR. Below mutated amino acids, the nucleotide codon together with the mutant number and the new nucleotide codon are shown. Leu, Ile and Val

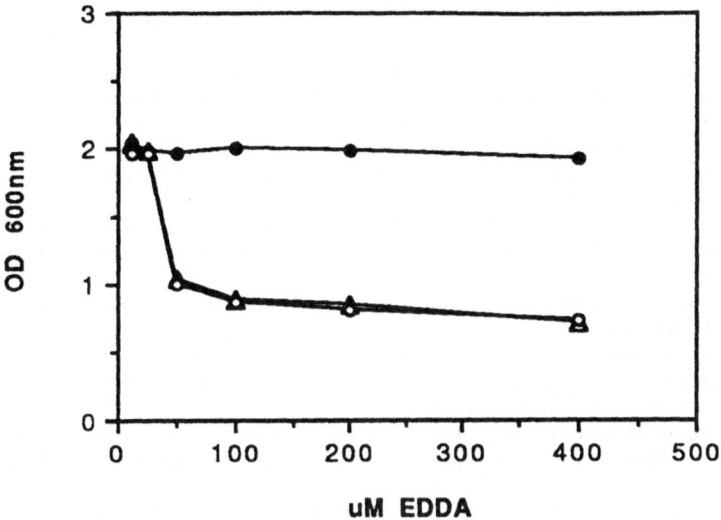

Figure 8. Complementation of the iron uptake activity of *E. coli* AN93 (*entE*) by the *V. anguillarum* *angR*$_{775}$ gene. *E. coli* strains harboring either pJHC-S2572 (—●—), pJHC-S2570 (—▲—) or the vector pKK223-3 (–○–) were cultured overnight in L-Broth containing different EDDA concentrations. Cell growth was recorded spectrophotometrically at 600nm.

Acknowledgements

This work was supported by NIH grant AI19018 to JHC.

References

1. Actis, L.A., W. Fish, J.H. Crosa, K. Kellerman, S. Ellenberger, F. Hauser, and J. Sanders-Loher. 1986. Characterization of anguibactin, a novel siderophore from *Vibrio anguillarum* 775 (pJM1). J. Bacteriol. 167: 57–65.
2. Actis, L.A., M.E. Tolmasky, D. Farrell, and J.H. Crosa. 1988. Genetic and molecular characterization of essential components of the *Vibrio anguillarum* plasmid-mediated iron-transport system. J. Biol. Chem. 263: 2853–2860.
3. Bagg, A. and Neilands, J.B. 1985. Mapping of a mutation affecting regulation of iron uptake systems in *Escherichia coli* K-12. J. Bacteriol. 161: 450–453.
3a. Branden, C., and Tooze, J. 1991. p. 11–31. *In:* Introduction to protein structure. Garland Publishing, Inc., New York and London.
4. Calderwood, S., and Mekalanos, J.J. 1988. Confirmation of the Fur operator site by insertion of a synthetic oligonucleotide into an operon fusion plasmid. J. Bacteriol. 170: 1015–1017.
5. Crosa, J.H. 1980. A plasmid associated with virulence in the marine fish pathogen *Vibrio anguillarum* specifies an iron-sequestering system. Nature (London) 284: 566–568.

residues in the leucine zippers as well as helixes in the helix-turn-helix motifs are underlined. Amino acids 598 to 610 (in shadow font) conform the region that shares homology with fragments of enzymes belonging to the firefly luciferase family. The symbol ● represent the location of Tn*3*-HoHo1 insertions, with the promoterless *lacZ* gene (Tolmasky *et al.*, 1988).

504

6. Crosa, J.H. 1989. Genetics and molecular biology of siderophore-mediated iron transport in bacteria. Microbiol. Rev. 53: 517–530.

7. Crosa, J.H., L. Hodges, and M. Schiewe. 1980. Curing of a plasmid is correlated with an attenuation of virulence in the marine fish pathogen *Vibrio anguillarum*. Infect. Immun. 27: 897–902.

8. deLorenzo, V., F. Giovannini, M. Herrero, and J.B. Neilands. 1988. Metal iron regulation of gene expression: Fur repressor-operator interaction at the promoter region of the aerobactin system of pColV-K30. J. Mol. Biol. 203: 875–884.

9. Farrell, D., P. Mikesell, L. Actis, and J.H. Crosa. 1990. A regulatory gene, *angR*, of the iron uptake system of *Vibrio anguillarum*: similarity with phage P22 *cro* and regulation by iron. Gene 86: 45–51.

10. Hantke, K. 1984. Cloning of the repressor protein gene of iron-regulated systems in *Escherichia coli* K12. Mol. Gen. Genet. 197: 337–341.

11. Heaton, M., and F. Neuhaus. 1992. Biosynthesis of D-alanyl-lipoteichoic acid: cloning, nucleotide sequence, and expression of the *Lactobacilli casei* gene for the D-alanine-acitvating enzyme. J. Bacteriol. 174: 4707–4717.

12. Jalal, M., D. Hossain, J. van der Helm, J. Sanders-Loehr, L.A. Actis, and J.H. Crosa. 1989. Structure of anguibactin, a unique plasmid-related bacterial siderophore from the fish pathogen *Vibrio anguillarum*. J. Amer. Chem. Soc. 111: 292–296.

13. Koster, W.L., L.A. Actis, L. Waldbeser, L., M.E. Tolmasky, M.E., and J.H. Crosa. 1991. Molecular characterization of the iron transport system mediated by the pJM1 plasmid in *Vibrio anguillarum* 775. J. Biol. Chem. 266: 23829–23833.

14. Miller, J.H. 1972. In *Experiments in Molecular Genetics*. Cold Spring Harbor Lab., Cold Spring Harbor, N.Y.

15. Prentki, P., and M. Krish. 1984. *In vitro* insertional mutagenesis with a selectable DNA fragment. Gene 29: 303–313.

16. Salinas, P., M.E. Tolmasky, and J.H. Crosa. 1989. Regulation of the iron uptake system in *Vibrio anguillarum*: evidence for a cooperative affect between two transcriptional activators. Proc. Natl. Acad. Sci. U.S.A. 86: 3529–3533.

17. Salinas, P., L.S. Waldbeser, and J.H. Crosa. 1992. Regulation of the expression of bacterial iron transport genes: possible role of an antisense RNA as a repressor. Gene, in press.

18. Schmitt, M., and S. Payne. 1988. Genetics and regulation of enterobactin genes in *Shigella flexneri*. J. Bacteriol.170: 5579–5587.

19. Scholten, J.D., K. Chang, P. Babbit, H. Charest, M. Sylvestre, and D. Dunaway-Mariano. 1991. Novel enzymatic hydrolytic dehalogenation of a chlorinated aromatic. Science 253: 182–185.

20. Toh, H. 1991. Sequence analysis of firefly luciferase family reveals a conservative sequence motif. Protein Seq. Data Anal. 4: 111–117.

21. Tolmasky, M.E., and J.H. Crosa. 1984. Molecular cloning and expression of genetic determinants for the iron uptake system mediated by the *Vibrio anguillarum* plasmid pJM1. J. Bacteriol. 160: 860–866.

22. Tolmasky, M.E., and J.H. Crosa. 1990. Regulation of plasmid-mediated iron transport and virulence in *Vibrio anguillarum*. Biol. Metals 4: 33–35.

23. Tolmasky, M.E., L.A. Actis, A. Toranzo, J. Barja, and J.H. Crosa. 1985. Plasmids mediating iron uptake in *Vibrio anguillarum* strains isolated from turbot in Spain. J. Gen. Microbiol. 131: 1989–1997.

24. Tolmasky, M.E., L.A. Actis, and J.H. Crosa. 1988. Genetic analysis of the iron uptake region of the *Vibrio anguillarum* plasmid pJM1: molecular cloning of genetic determinants encoding a novel *trans* activator of siderophore biosynthesis. J. Bacteriol. 170: 1913–1919.

25. Tolmasky, P. Salinas, L.A. Actis, and J.H. Crosa. 1988. Increased production of the siderophore anguibactin mediated by pJM1-like plasmids in *Vibrio anguillarum*. Infect. Immun. 56: 1608–1614.

26. Tolmasky, M.E., A.E. Gammie, and J.H. Crosa. 1992. Characterization of the *recA* gene of *Vibrio anguillarum*. Gene 110: 41–48.

34. Conservation of the *lema* gene, a virulence regulator from the plant pathogen *Pseudomonas syringae*, within a human pathogenic bacterium

D. KYLE WILLIS, THOMAS G. KINSCHERF and JESSICA J. RICH

Abstract. The *lemA* gene of plant pathogenic strains of *Pseudomonas syringae* is required for disease lesion formation as well as for production of several phytotoxins and extracellular protease activity. The sequence of this gene places the predicted LemA protein within a family of transmembrane bacterial sensors. The *lemA* gene is ubiquitous within *P. syringae* and we have cloned alleles of this locus from *P. syringae* strains that are pathogenic on bean, tobacco, tomato, oats, coffee, and *Arabidopsis thaliana*. By Southern blot analysis, we have detected physical homology to the *lemA* gene within all tested members of the genus *Pseudomonas* including *Pseudomonas aeruginosa*. Using an internal restriction fragment of the lemA gene from *P. syringae* pv. syringae as a probe, we have cloned the *P. aeruginosa lemA* gene. Surprisingly, the *lemA*$_{PAO}$ gene complemented the *P. syringae* pv. syringae *lemA* mutant strain NUVS1. Introduction of the *lemA*$_{PAO}$ gene into NUVS1 restored lesion formation on bean, and the production of protease and syringomycin. We are in the process of inactivating this locus in *P. aeruginosa* strain PAO in order to determine the effect of a *lemA* mutation on the known virulence factors of this human pathogen. We have also detected homology to the *lemA* gene in other genera that cause animal disease.

Abbreviations: HPK, Histidine Protein Kinase; HR, Hypersensitive Reaction; RR, Response Regulator

The *lemA* gene, a ubiquitous regulator of lesion formation and toxin production in *Pseudomonas syringae*

The *lemA* locus was identified through Tn*5* mutagenesis of *P. syringae* pv. syringae strain B728a, a causal agent of brown spot disease of bean. A B728a *lemA* mutant derivative, such as NPS3136, no longer produces lesions on bean but still induces a strong hypersensitive reaction (HR) on several non-host plants, including tobacco (Willis *et al.*, 1990). Under growth chamber conditions, NPS3136 is still able to grow both epiphytically and within the leaf tissue of bean and attains population levels that are not significantly different from the parental strain, indicating that genes involved in the growth of NPS3136 in bean plants are not affected (Willis *et al.*, 1990). Thus, the analysis of this mutant demonstrates that disease symptom formation can be genetically separated from other plant interaction phenotypes.

In addition to being deficient in lesion formation on bean, *lemA* mutants no longer produce protease (Prt⁻) or syringomycin (Syr⁻), two compounds secreted by the parental strain B728a (Hrabak and Willis, 1993). We have not been able to assign a role for either syringomycin or protease in lesion formation

C.I. Kado and J.H. Crosa (eds.), Molecular Mechanisms of Bacterial Virulence, 505–509.
© 1994 *Kluwer Academic Publishers*.

by *P. syringae* pv. syringae strain B728a on bean. Mutants have been isolated that are either Prt⁻ or Syr⁻ but are still able to form lesions on bean. In addition, we have isolated B728a mutants that produce both syringomycin and protease but do not produce lesions on bean indicating that the production of protease and syringomycin is not sufficient for the manifestation of disease on bean (J. J. Rich and D. K. Willis, unpublished data).

DNA sequence analysis predicts that the *lemA* gene is 3.1 kb in length and contains a single open reading frame (ORF) with a predicted protein product of 113 kDa (Hrabak and Willis, 1992). The comparison of both the DNA and predicted protein sequence to the UWGCG computer database reveals a high degree of sequence identity between *lemA* and members of the "two component" family of bacterial positive transcriptional regulators. The closest relatives of *lemA* within this group are: *bvgS*, a positive regulator of virulence within the whooping cough bacterium *Bordetella pertussis* (Arico *et al.*, 1989; Roy and Falkow, 1991; Weiss *et al.*, 1983); *rcsC*, a regulator of capsular synthesis in *Escherichia coli* (Stout and Gottesman, 1990); and *phoR*, a regulator of phosphate assimilation in *Bacillus subtilus* and *E. coli* (Lee *et al.*, 1989). Members of this family of transcriptional regulators act as sensors of the bacterial environment and have common structural features (Stock *et al.*, 1989; Stock *et al.*, 1990). They consist of two enzymatic activities that are usually contained in two separate polypeptides. The first protein has an external portion that extends into the periplasm and is bounded by two transmembrane (hydrophobic) domains. This "sensor" interacts with molecules (signals) in the extracellular environment. Once a signal molecule is detected by the external portion of the protein, the sensor phosphorylates itself at a conserved histidine residue via a histidine protein kinase (HPK) activity. Next, this phosphoryl group is transferred to an aspartic acid residue on the second polypeptide called the response regulator (RR). This "activated" RR is then thought to bind DNA directly and act as a positive factor causing the transcription of a number of individual genes in a coordinate manner. The predicted *lemA* protein gene product has all of the characteristic structural features of this family of regulators (i.e. two transmembrane domains, a HPK domain with a positionally conserved His residue, and a RR domain with conserved Asp residues). The lemA protein may represent a novel class of bacterial regulators. In all of the "two component" systems described to date, a cytoplasmic RR has been described. This includes regulators that, like *lemA*, contain both activities in a single polypeptide. Even in those proteins that contain both the HPK and RR components, a second cytoplasmic RR protein has been found. We have not as yet detected a second cytoplasmic RR component of the *lemA* system. The possibility remains that the lemA protein is unique and fulfills both functions.

Conservation of the *lemA* gene within *P. syringae* pv. syringae

Using *lemA* DNA as a probe, we have detected restriction fragment length polymorphism (RFLP) among strains of bean pathogenic *P. syringae* pv. syringae isolated in Wisconsin (Rich *et al.*, 1992). In a screen of 80 isolates, seven RFLP patterns were detected using two restriction enzymes. This indicates that at least seven molecularly distinct populations are present in Wisconsin. In addition to homology at the DNA level, we have established that all seven of the RFLP classes identified require the *lemA* locus for lesion formation. Exchange mutagenesis of the *lemA*$_{Pss}$*1*::Tn*5* mutation into representative strains from each class resulted in the loss of lesion forming ability, syringomycin production and protease production in all cases examined. Exchange mutants from three classes were restored to full function by a cosmid clone containing the *lemA* locus from B728a. Significantly, exchange of the *lemA*$_{Pss}$*1*::Tn*5* mutation into the non-pathogenic *P. syringae* strain Cit7 led to the loss of syringomycin and protease production by this strain, demonstrating a general role for the *lemA* gene within *P. syringae*.

P. syringae pv. phaseolicola contains a functional *lemA* allele

We have isolated a cosmid clone with sequence homology to the *lemA* locus from a genomic library of *P. syringae* pv. phaseolicola (Rich *et al.* 1992). This clone (designated as pKW12) restores pathogenicity, syringomycin production and protease production to the *lemA*$_{Pss}$*1*::Tn*5* mutant NPS3136 and, therefore, contains a functional analog of the *lemA* gene. We have isolated Tn*5* insertions in pKW12 that eliminate the ability of this plasmid to complement a *lemA*$_{Pss}$*1*::Tn*5* mutation. The recombinational exchange of one of these pKW12 insertion mutations, designated as *lemA*$_{Psp}$*272*::Tn*5*, into B728a resulted in a mutant strain that was indistinguishable from the original *lemA* mutant NPS3136 (i.e. Lem$^-$ Syr$^-$ Prt$^-$ HR$^+$) indicating that *P. syringae* pv. phaseolicola contains a true *lemA* homolog. However, exchange of the *lemA*$_{Psp}$*272*::Tn*5* mutation into *P. syringae* pv. phaseolicola gave a different result. A *P. syringae* pv. phaseolicola (*lemA*$_{Psp}$*272*::Tn*5*) mutant was found to be unaffected in its ability to produce phaseolotoxin, both in culture and in bean leaves, and remained pathogenic on bean. The isolation of a functional homolog of *lemA* from a pathovar that produces neither syringomycin or protease further supports the hypothesis that *lemA* is a regulatory rather than an enzymatic or structural locus.

Tabtoxin production is regulated by the *lemA* gene

Mutational analysis of *P. syringae* pv. coronafaciens strain Pc27R (Barta *et al.*, 1992) identified a locus required for tabtoxin production that is not linked to the known tabtoxin biosynthetic region (Kinscherf *et al.*, 1991). All seven Tn*5*-generated Tox$^-$ Pc27R mutants retained the ability to form lesions on oats. This phenotype mimics that of *P. syringae* pv. striafaciens, a pathovar that we have found to be deleted for the tabtoxin biosynthetic region. The Tn5 insertions within the 7 Tox⁻ mutants are linked on a 17 kb *Kpn*I fragment. Surprisingly, hybridization analysis revealed DNA homology between clones containing the 17 kb fragment and the *P. syringae* pv. syringae *lemA* gene (*lemA$_{Pss}$*). Furthermore, a clone containing the *lemA$_{Pss}$* gene restores tabtoxin production to all of the *P. syringae* pv. coronafaciens Tox⁻ mutants. Pc27R genomic clones that restore tabtoxin production to the Tn5-generated Tox⁻ mutants show homology to the *lemA$_{Pss}$* gene and restore lesion forming ability to the *lemA$_{Pss}$1*::Tn5 mutant NPS3136. Thus, *P. syringae* pv. coronafaciens contains a functional *lemA* homolog (designated as *lemA$_{Psc}$*) that is required for tabtoxin production but not lesion formation in this pathovar. This conclusion is supported by our finding that a functional *lemA* gene is required for tabtoxin production in *P. syringae* BR2 and that pRTBL823, the cosmid containing the functional tabtoxin biosynthetic region, does not express TBL in *P. syringae* strain Cit7 containing a *lemA$_{Pss}$1*::Tn5 mutation.

We have used the tabtoxin biosynthetic region as a probe in the northern blot analysis of transcription within *P. syringae* pv. coronafaciens strain Pc27R and two of the Tn5-generated Tox⁻ mutants (Barta *et al.*, 1992). A transcript, co-migrating with a 1 kb dsDNA molecular weight marker, is present in RNA extracted from Pc27R but absent from RNA extracted from either of the Tox⁻ mutants. This transcript is restored by plasmids containing the *lemA* homolog from either *P. syringae* pv. syringae or *P. syringae* pv. coronafaciens. The gene that encodes this transcript, *tblA*, has been isolated and sequenced (Barta *et al.*, 1993) The predicted *tblA* ORF encodes a protein of 26 kDa. A search of the GenBank database did not reveal significant similarity between the predicted TblA protein and genes products with known function. Whether the role of *lemA* as a positive regulator of transcription is direct (binding DNA and functioning as a bacterial transcriptional activator protein), or indirect (activating an as yet unknown second component such as *tblA*) remains to be discovered.

Analysis of *lemA* function in *Pseudomonas aeruginosa* and *Aeromonas hydrophila*

The *lemA* gene is apparently conserved throughout the genus *Pseudomonas*. We have cloned active *lemA* alleles from *P. syringae* pathovars syringae, tomato, phaseolicola, and coronafaciens. Through Tn5 mutagenesis, we have identified active *lemA* alleles in *P. syringae* strain BR2 and the non-pathogenic *P. syringae* strain Cit7. The ubiquitous nature of the *lemA* gene within *P. syringae* has prompted us to investigate the physical and functional conservation of this locus within other bacteria. We have found hybridization to an internal fragment of the *lemA* gene within the DNA of *P. solanacearum*, *P. diminuta*, *P. facilis*, *P. acidovorans*, *P. aeruginosa*, *Xanthomonas campestris*, *Aeromonas hydrophila*, and *A. salmonicida* We have recently cloned the *P. aeruginosa lemA* homolog, and we have established that the *lemA$_{PAO}$* gene restores lesion formation, syringomycin production, and protease production to the *P.*

syringae pv. syringae *lemA1*::Tn5 mutant strain NUVS1 (T. G. Kinscherf, J. J. Rich, and D. K. Willis, unpublished data). The effect of a *lemA* mutation on the production of various *P. aeruginosa* factors such as alginate, elastase, protease, and exotoxin A will be determined by the recombinational exchange mutagenesis of the *P. aeruginosa lemA* gene. If we are successful in identifying virulence regulatory genes within *P. aeruginosa* using genes from *P. syringae*, we may be able to use plant pathogenic pseudomonads as a model system for the study of regulators of gene expression within bacterial pathogens of humans, animals, and fish.

References

Arico, B., Miller, J.F., Roy, C., Stibitz, S., Monack, D., Falkow, S., Gross, R. and Rappuoli, R. 1989. Sequences required for expression of *Bordetella pertussis* virulence factors share homology with prokaryotic signal transduction proteins. Proc. Natl. Acad. Sci. USA 86: 6671–6675.

Barta, T.M., Kinscherf, T.G. and Willis, D.K. 1992. Regulation of tabtoxin production by the *lemA* gene in *Pseudomonas syringae*. J. Bacteriol. 174: 3021–3029.

Barta, T.M., Kinscherf, T.G., Uchytil, T.F. and Willis, D.K. 1993. DNA sequence and transcriptional analysis of the *tblA* gene required for tabtoxin biosynthesis by *Pseudomonas syringae*. Appl. Env. Microbiol. 59: 458–466.

Hrabak, E.M. and Willis, D.K. 1992. The *lemA* gene required for pathogenicity of *Pseudomonas syringae* pv. syringae on bean is a member of a family of two-component regulators. J. Bacteriol. 174: 3011–3020.

Hrabak, E.M. and Willis, D.K. 1993. Involvement of the *lemA* gene in production of syringomycin and protease by *Pseudomonas syringae* pv. *syringae* Molec. Plant-Path. Interact. 6: 368–375.

Kinscherf, T.G., Coleman, R.H., Barta, T.M. and Willis, D.K. 1991. Cloning and expression of tabtoxin biosynthesis and resistance from *Pseudomonas syringae*. J. Bacteriol. 173: 4124–4132.

Lee, T.-Y., Makino, K., Shinagawa, H., Amemura, M. and Nakata, A. 1989. Phosphate regulon in members of the family *Enterobacteriaceae*: comparison of the *phoB-phoR* operons of *Escherichia coli*, *Shigella dysenteriae*, and *Klebsiella pneumoniae*. J. Bacteriol. 171: 6593–6599.

Rich, J.J., Hirano, S.S. and Willis, D.K. 1992. Pathovar specific requirement for the *Pseudomonas syringae lemA* gene in disease lesion formation. Appl. Env. Microbiol. 58: 1440–1446.

Roy, C.R. and Falkow, S. 1991. Identification of *Bordetella pertussis* regulatory sequences required for transcriptional activation of the *fhaB* gene and autoregulation of the *bvgAS* operon. J. Bacteriol. 173: 2385–2392.

Stock, J.B., Ninfa, A.J. and Stock, A.M. 1989. Protein phosphorylation and regulation of adaptive responses in bacteria. Microbiol. Rev. 53: 450–490.

Stock, J.B., Stock, A.M. and Mottonen, J.M. 1990. Signal transduction in bacteria. Nature (London) 344: 395–400.

Stout, V. and Gottesman, S. 1990. RcsB and RcsC: a two-component regulator of capsule synthesis in *Escherichia coli*. J. Bacteriol. 172: 659–669.

Weiss, A.A., Hewlett, E.L., Myers, G.A. and Falkow, S. 1983. Tn5-induced mutations affecting virulence factors of *Bordetella pertussis*. Infect. Immun. 42: 33–41.

Willis, D.K., Hrabak, E.M., Rich, J.J., Barta, T.M., Lindow, S.E. and Panopoulos, N.J. 1990. Isolation and characterization of a *Pseudomonas syringae* pv. *syringae* mutant deficient in lesion forming ability on bean. Molec. Plant-Microbe Interact. 3: 149–156.

510

Note added in proof

We have recently identified the *P. syringae* pv. syringae *gacA* analog as the putative *lemA* response regulator (Willis, D.K., Kinscherf, T.G., Rich, J.J. and Kitten, T. 1993. Functional relationships of regulators within the genus *Pseudomonas*. Page 35 in: Proceedings of the Forth International Symposium on *Pseudomonas*. Vancouver, British Columbia, Canada; Laville, J., Voisard, C., Keel, C., Maurhofer, M., Défago, G. and Haas, D. 1992. Global control in *Pseudomonas fluorescens* mediating antibiotic synthesis and suppression of black root rot of tobacco. Proc. Natl. Acad. Sci. USA 89: 1562–1566.). A mutation in the *P. syringae* pv. syringae *gacA* results in the same phenotype as a *lemA* mutation (lem⁻, Syr⁻, Prt⁻). In addition, we have successfully mutated the *Pseudomonas aeruginosa lemA* locus by recombinational exchange (T.G. Kinscherf and D.K. Willis, unpublished data).

35. Virulence regulation in *Bordetella pertussis*

ROY GROSS, THILO M. FUCHS, HEIKE DEPPISCH and NICHOLAS
H. CARBONETTI

Abstract. For several years, *Bordetella pertussis*, the etiological agent of whooping cough, has been
the subject of intensive investigations with the aim of developing new vaccines. These studies have
revealed much about the complex interactions between the invading pathogen and its eukaryotic
host. The different stages of infection with *B. pertussis* include the adherence to host cells, the
colonization of the host tissue and possibly invasion of epithelial cells. Frequently, pathogenic
bacteria have to encounter variable environmental conditions, each of them requiring the
expression of specific factors at the right time. The coordination of the expression of the relevant
factors according to changes in the environment is guaranteed by the unification of most of the
virulence related genes into regulons. In the case of *B. pertussis*, the control of the virulence regulon
is mediated by the *bvg* locus, which codes for a two-component regulatory system. Recent results
demonstrate that regulatory factor(s) in addition to the the two-component system are required for
expression of some of the virulence factors such as pertussis toxin and adenylate cyclase toxin. This
suggests that the virulence associated factors may be expressed differentially during infection,
although they are regulated coordinately on the top level of control.

Abbreviations: *B. Pertussis, Bordetella pertussis*; CYA, Adenylate Cyclase; *E. Coli, Escherichia Coli*;
HLT, Heat-Labile Toxin; FHA, Filamentous Hemagglutinin; FIM, Fimbriae; PRN, Pertactin;
PTX, Pertussis Toxin; TCT, Tracheal Cytotoxin

The *Bordetella* Species

Bordetella pertussis belongs to the group of "classical" bacterial pathogens such
as *Corynebacterium diphteriae* or *Clostridium tetani*, as its first scientific
description dates back to the "golden era" of bacteriology. Nearly ninety years
ago Bordet and Gengou (1906) identified the etiological agent of whooping
cough and called it *Haemophilus pertussis*. Nowadays it is classified together
with three other species in the genus *Bordetella* (Pittman, 1984). The *Bordetellae*
are minute, aerobic Gram-negative coccobacilli colonizing the upper
respiratory tract and thereby causing disease with similar symptoms but
different host specificities.

B. pertussis and *B. parapertussis* are nonmotile, obligate human pathogens,
the latter responsible for mild pertussis-like disease. *B. bronchiseptica* is
flagellated and is able to survive in the environment (Porter *et al.*, 1991).
Predominately it infects various mammalian species such as dogs, where it causes
the kennel cough. *B. avium* is a pathogen for birds causing e.g. turkey coryza.

C.I. Kado and J.H. Crosa (eds.), Molecular Mechanisms of Bacterial Virulence, 511–524.
© 1994 *Kluwer Academic Publishers.*

The human and mammalian pathogens with a remarkable high DNA GC ratio of 67–70 % are very closely related. Indeed, some authors question their classification in different species. However, there is evidence that *B. bronchiseptica* is the most closely related organism to the common ancestor of the three species (Gross *et al.*, 1989a). A major difference between the various species is the production of virulence factors. *B. pertussis* is the only species which produces the pertussis toxin with an ADP-ribosylating activity. Interestingly, also *B. bronchiseptica* and *B. parapertussis* contain the genes coding for this toxin, but they can not express them due to mutations in their promoter regions (Gross and Rappuoli, 1988).

As *B. parapertussis* can be frequently coisolated with *B. pertussis* from patients suffering from whooping cough, it was suggested that it might be a less virulent form of *B. pertussis*. However, comparison of the nucleotide sequences of several virulence genes of the various species (Arico *et al.*, 1987; Gross *et al.*, 1989a) and multilocus enzyme gel electrophoresis (Musser *et al.*, 1986) demonstrated that *B. parapertussis* is not a variant of *B. pertussis* (Rappuoli *et al.*, 1987). *B. avium* neither contains sequences similar to the pertussis toxin operon nor produces the adhesin FHA or the toxin CYA, and with a GC content of about 61 % it is the most distantly related species (Gentry-Weeks *et al.*, 1991) (Table 1).

Table 1. Expression of virulence factors by *Bordetella* species.

B. pertussis	B. parapertussis	B. bronchiseptica	B. avium
PTX	–	–	–
CYA	CYA	CYA	–
FHA	FHA	FHA	–
PRN	PRN	PRN	–
FIM	FIM	FIM	?
TCT	TCT	TCT	TCT
HLT	HLT	HLT	HLT

PTX: pertussis toxin; CYA: adenylate cyclase toxin; FHA: filamentous hemagglutinin; PRN: pertactin; FIM: fimbriae; TCT: tracheal cytotoxin; HLT: heat-labile toxin

The disease whooping cough

Mainly young children are affected by this severe illness which still poses an important public health problem especially in the less developed countries. Since recently pertussis has also gained new attention in the industrialized countries as mass vaccination was not carried out anymore in the last two decades due to little confidence in the safety of the whole cell vaccine which was in use for many decades. Pertussis epidemics are therefore no longer history in the respective countries. Worldwide, more than 50 million cases and 600,000 deaths are reported annually (Hewlett, 1989).

Vaccination against pertussis seems to be responsible for a shift in the peak age of disease since the 1940s. Whereas in the prevaccine era pertussis mainly affected children from 1 year of age on, newborns were protected due to maternal antibodies and adults due to immunity after repeated exposition to the pathogen. In contrast, in the 1980s, more than fifty percent of the pertussis cases in the United States concerned children less than one year of age, the group of infants, which is at the greatest risk for a severe course of the disease. Since immunity after vaccination is frequently protective only for a limited period, adults are increasingly affected, and those with atypical, undiagnosed pertussis are thought to be a major source for passive transfer to children, as the pathogen is unable to survive in the environment.

The progress of the disease can often be divided into several phases. During the incubation period of up to more than three weeks the bacteria adhere to the cilia of epithelial cells in the upper respiratory tract, where they multiply and elaborate a variety of highly toxic compounds. After the catarrhal phase with unspecific clinical manifestations such as rhinorrhea, lacrimation and fever, the paroxysmal phase begins. It is initiated by a dry cough, often followed by the typical whoop, which gave the disease its name. During this stage several pathological effects can be observed such as peripheral lymphocytosis and hyperinsulinemia. Severe cases can be affected by pulmonary and neurological complications. The disease comes to an end with a long convalescent period.

During the onset of the disease the patients do not show specific symptoms which would allow an immediate diagnosis of whooping cough which is rather made in later phases of infection when the more serious symptoms such as cough paroxysm occur. At this point, however, the bacteria have already damaged the lung epithelium and intoxicated the patient with its various toxins. *B. pertussis* is sensitive to many antibiotics, but antibiotic therapy is of little use because the disease syndrome appears in late stages of infection in which the toxins have already been released by the bacteria. This demonstrates how important vaccination is in the case of pertussis, as the prevention of the disease is much more efficient than a therapy of the ongoing disease. Therefore, several groups have focused on the development of new and safer vaccines against pertussis with less side effects than the whole cell vaccines. Several acellular vaccines with a defined composition are currently undergoing clinical trials. The most promising candidates contain as the main component purified pertussis toxin, detoxified either by chemical means or after genetic engineering of the structural genes of pertussis toxin (Ad hoc group for the study of pertussis vaccine, 1988; Rappuoli *et al.*, 1990).

The virulence factors of *Bordetella pertussis*

One of several components involved in virulence is the filamentous hemagglutinin (FHA), a cylindrical protein with subunits of 220 kD, which can be found in the supernatant or cell wall associated. Previous reports suggest that FHA plays an important role in adherence of *B. pertussis* to ciliated respiratory cells and macrophages. The pathogen-host interaction may occur by binding of glycoconjugates through a lectin-like activity of FHA (Relman *et al.*, 1989). Interestingly, this protein also shows similarities with fibronectin, a prototype protein for interactions with integrin receptors, because of two Arg-Gly-Asp-tripeptides in the deduced amino acid sequence of FHA, the so-called RGD-motif (Relman *et al.*, 1990; Saukkonen *et al.*, 1992). Furthermore, FHA stimulates an immune response in humans after clinical disease and vaccination and appears to be an immunoprotective antigen in mouse model systems (Ad hoc group for the study of pertussis vaccine, 1988; Sato and Sato, 1984).

A 69 kDa outer membrane protein, named pertactin (PRN), also seems to be involved in adhesion and can induce protective immunity (Charles *et al.*, 1989; Leininger *et al.*, 1990). Similar to FHA, PRN has an RGD-motif, and a functional role of this motif in adhesion and invasion was recently investigated (Leininger *et al.*, 1991; Leininger *et al.*, 1992). In addition, *B. pertussis* produces several types of fimbriae (FIM). Although their role in pathogenesis is poorly understood, the fimbriae may be involved in adhesion and colonization of the host (Mooi *et al.*, 1992; Willems *et al.*, 1990).

The pertussis toxin (PTX) is an additional factor which seems to be involved in adhesion of the bacteria to ciliated epithelial cells and macrophages (van't Wout *et al.*, 1992; Saukkonen *et al.*, 1992). It is composed of two subunits, an enzymatically active A subunit and a B subunit composed of several proteins, which is involved in the binding of the toxin to eukaryotic cells. This typical A/B toxin is believed to be responsible for most of the systemic effects of the disease, such as lymphocytosis, histamine sensitization and enhancement of insulin secretion. PTX belongs to the large group of ADP-ribosylating toxins. Its toxic activity resides in the A-subunit and has as its target regulatory G-proteins in the eukaryotic cell membrane, which after modification by pertussis toxin deregulate the second messenger equilibrium (Nicosia *et al.*, 1986; Locht and Keith, 1986; Katada *et al.*, 1986). Genetically inactivated PTX will be the main component of new generation whooping cough vaccines (Rappuoli *et al.*, 1990).

Another factor which contributes to the virulence of *Bordetella* is the calmodulin-dependent adenylate cyclase toxin (CYA). It is a bifunctional protein possessing cAMP synthesizing and haemolytic activities (Glaser *et al.*, 1988). The adenylate cyclase part is able to enter mammalian cells where it disturbs the cAMP level. Intracellularly expressed CYA was recently shown to be lethal for certain cell types (Wels *et al.*, 1992). CYA is believed to be involved in the defense against local immune reactions and therefore may serve as a protective antigen (Guiso *et al.*, 1990). Similar to PTX mutants, mutants in the CYA operon are not able to colonize the lung epithelium (Khelef *et al.*, 1992).

Several other toxic compounds such as the heat labile toxin and the tracheal cytotoxin are produced by *B. pertussis* (Wardlaw and Parton, 1988). The latter appears to be a tetrapeptide-disaccharide derivative of the bacterial cell wall which causes deciliation of the epithelial cells and is therefore considered to be the main reason for cough paroxysm (Cookson *et al.*, 1989). Purified heat labile toxin, also called dermonecrotic toxin, has a vasoconstrictive activity and causes skin lesions. The role of this cytoplasmic protein in the pathogenesis of pertussis is currently under investigation (Endoh *et al.*, 1990).

The *bvg* locus is responsible for the coordinate regulation of virulence

With the exception of the tracheal cytotoxin all of the above mentioned virulence associated factors of *B. pertussis* are regulated coordinately. The first description of regulatory events regarding *Bordetella* virulence was made by Leslie and Gardner (1931). These authors observed that upon cultivation of *B. pertussis in vitro*, the bacteria have the tendency to loose spontaneously their virulence properties, a phenomenon which was then called phase variation. The process of phase variation is nearly irreversible. Lacey (1960) observed that such avirulent bacteria can also be generated in a reversible manner by changing the growth conditions. In the presence of certain ions or chemical compounds (e.g. sulphate or nicotinic acid) or at low temperature the bacteria are phenotypically very similar to the avirulent phase variants. This phenomenon was termed phenotypic modulation.

Recently, a single genetic locus was identified, which is responsible for these regulatory phenomena. After insertional inactivation of this *bvg* locus (also termed *vir* locus) by Tn5-induced mutagenesis the bacteria lost their capacity to express the virulence factors and had a similar phenotype as the avirulent phase variants (Weiss and Falkow, 1984).

Several lines of evidence indicate that the *bvg* locus is responsible for both the phase variation and the phenotypic modulation: i) all so far characterized spontaneous phase variants revealed mutations, often small deletions, in the *bvg* locus and could be complemented by the intact operon (Monack *et al.*, 1989); ii) phenotypic modulation is abolished in *bvg* mutants (Gross and Rappuoli, 1989); and iii) mutants which express the virulence factors constitutively and thus are insensitive to external signals contain mutations in the *bvgS* gene (Miller *et al.*, 1992).

The cloning and sequencing of the *bvg* locus revealed the presence of two genes, *bvgA* and *bvgS* (Aricò *et al.*, 1989; Stibitz and Yang, 1991), the products are similar to the large family of regulatory proteins belonging to the two-component systems (Gross *et al.*, 1989b). These systems are wide spread in the prokaryotes and are often used when bacteria have to adapt to changing growth conditions, such as during chemotaxis and during the regulation of sporulation or nitrogen fixation (Stock *et al.*, 1990).

Two-component systems generally consist of two proteins: i) A sensor protein located in the cell membrane with a periplasmic and a cytoplasmic domain, called "transmitter", which is conserved in all sensor proteins. The sensor protein is able to perceive environmental stimuli and to transform them into a cellular signal. ii) A regulator protein, frequently a transcriptional activator, which interacts with the sensor protein. All regulators contain a conserved domain called "receiver" at the N-terminus.

For several two-component systems it was shown that the transmitter domain of the sensor contains a histidine kinase which is regulated in its activity by the external conditions. When the right stimulus is detected the kinase becomes active resulting in an autophosphorylation at conserved histidine residues of the sensor. Subsequently, the phosphate is transfered to aspartic acid residues in the receiver domain of the regulatory element. This covalent modification modulates the activity of the regulatory component (Stock *et al.*, 1990).

These signal transduction domains are also present in the *Bordetella* proteins (Gross *et al.*, 1989b). The BvgA protein is a typical member of the regulatory factors containing the conserved N-terminal receiver domain. Also the topology of the BvgS transmembrane protein is very similar to that of other sensor proteins (Stibitz and Young, 1991). However, in addition to its transmitter domain the BvgS protein exceptionally contains the receiver domain of the response regulators and a helix-turn-helix motif at its very C-terminus (Arico *et al.*, 1989). Some other sensor proteins have been identified recently containing both of the signal transduction domains (Hrabak and Willis, 1992), but the function of the additional receiver domain is not yet clear.

Although for the *Bordetella bvg* system a phosphotransfer has not been demonstrated so far, it is very likely that the *bvg* system uses the phosphotransfer mechanism for signal transduction. Indirect evidence for this assumption comes from the observation that the *Bordetella bvg* system can communicate with the *Escherichia coli* OmpR system (Gross and Carbonetti, 1993). The two-component system EnvZ/OmpR regulates the expression of the porin genes *ompC* and *ompF* whereby the composition of the outer membrane changes according to osmolarity conditions of the growth medium (Igo *et al.*, 1989). The expression of the OmpC porin can be influenced by the *bvg* locus in certain *E. coli ompR* mutants, which are defective in osmoregulation. After introduction of the *Bordetella* operon, OmpC expression depends on external stimuli which are known to be perceived by the *bvg* locus. It is likely that the putative, strongly conserved phosphotransfer domains of the *bvg* system are involved in this "cross-talk" phenomenon (Wanner, 1992; Gross and Carbonetti, 1993).

The expression of the *bvg* locus itself is regulated in response to changing environmental stimuli via a complex positive autoregulation mechanism. It was shown that under non permissive growth conditions a weak *bvg* independent promoter ("p" in Fig. 1) leads to low level production of *bvg* gene products. Under permissive conditions, however, additional *bvg* promoters are induced ("P" in Fig. 1) which are dependent from activation by the BvgA regulatory protein (Roy *et al.*, 1991; Scarlato *et al.*, 1990). Therefore, if the environmental situation becomes permissive, the basal amount of BvgS protein probably activates the few present BvgA molecules by phosphorylation. The activated form of BvgA is then able to promote transcription from the dependent and more efficient promoters, which leads to a strong increase in the amount of the regulator itself. Due to the permanent activation by the BvgS sensor protein, the amount of activated BvgA protein increases rapidly. According to a model proposed by Scarlato and coworkers (1990), this increase then allows transcription from the other *bvg* dependent virulence promoters (Fig. 1).

Differential regulation of *Bordetella* virulence factors

While an intact *bvg* locus is essential for the expression of the virulence factors of *B. pertussis*, the fine analysis of the various promoters indicates that the regulation model described above is not sufficient to explain how the regulation network actually functions.

Attempts to express virulence associated loci solely in the presence of their own promoters in *Escherichia coli* have been unsuccessful. However, the *fha* and

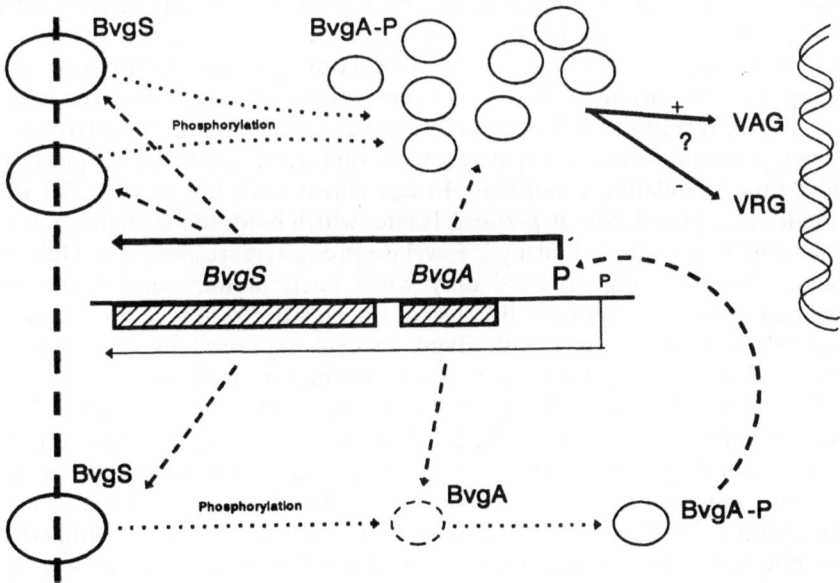

Figure 1. Model for the autoregulation of the *bvg* locus of *B. pertussis* according to Scarlato and coworkers (1990), the details of which are described in the text. BvgS is the transmembrane sensor protein, BvgA the transcriptional activator. Abbreviations: VAG: *bvg* activated virulence genes such as PTX, CYA, FIM, PRN, FHA; VRG: *bvg* repressed genes; p: constitutive promoter of the *bvg* operon, which leads to low level transcription (thin arrow); P: BvgA dependent promoters of the *bvg* operon, which leads to strong transcription (thick arrow).

bvg promoters, fused to a single copy *lacZ* gene, were strongly activated when the *bvg* locus was supplied *in trans* (Miller *et al.*, 1989). In contrast, similar *lacZ* fusions to *ptxA* and *cyaA*, the promoter proximal genes of the operons for PTX and CYA, were not activated in *E.coli* with the *bvg* genes *in trans* (Miller *et al.*, 1989; Goyard and Ullmann, 1991). DNA binding studies revealed that an inverted repeat upstream of *fhaB* and a similar sequence upstream of *bvgA* take part in binding of the activator protein BvgA. Homologous sequences are not present in the *ptx* or *cya* promoters, and consequently, no binding of BvgA protein to these promoters could be observed (Huh and Weiss, 1991; Gross and Carbonetti, 1993; Roy and Falkow, 1991).

Instead, it has been shown that efficient transcription of the *ptx* genes requires two tandem repeats which are located in the regulatory region of the *ptx* promoter far upstream from the transcriptional start point at sequence position -117 to -157 (Gross and Rappuoli, 1988). A similar sequence is present at the same site in the upstream region of the adenylate cyclase toxin, and it was shown to be involved in the regulation of the *cya* operon (Huh and Weiss, 1991; Goyard and Ullmann, 1992). In an attempt to investigate binding of regulatory factors *in vivo* to the *ptx* promoter, the tandem repeats or part of them were cloned behind a *lac* promoter, and interference of transcription from this promoter was measured in *Bordetella* in response to changes in growth conditions. Interestingly, under permissive conditions a strong negative effect on the expression of the *lac* promoter could be observed in the case of the presence of the tandem repeat. Under modulating growth conditions or when the repeats contained mutations no effect on transcription was observed (Gross *et al.*, 1992). It seems therefore that a *bvg* dependent protein binds *in vivo* to the direct repeats and interferes negatively with transcription from the *lac* promoter under non modulating conditions. In agreement with this result, a 23 kDa protein was retained from *B. pertussis* lysates which binds to the tandem repeats of the pertussis toxin and adenylate cyclase promoters, respectively (Huh and Weiss, 1991). Binding occurred only when bacteria were cultivated under permissive growth conditions. Because of the negative results of the various *in vitro* DNA binding assays with BvgA protein expressed in *E. coli* it was concluded that the 23-kDa protein may be distinct from BvgA.

These data suggest that the regulation of some virulence factors such as FHA involves a direct activation by the BvgA protein, whereas the transcriptional activation of the *ptx* and *cya* promoters requires factor(s) in addition to the Bvg proteins. Alternatively, it is possible that BvgA plays a direct role in transcriptional activation of these promoters, but in a modified form, which may not be present under the experimental conditions of the various *in vitro* DNA binding assays carried out with BvgA expressed in *E. coli*.

Interestingly, the same two groups of virulence factors, FHA and BVG on the one hand and PTX and CYA on the other hand, were found by the analysis of the kinetics of transcription during modulation. When the growth parameters were changed from non- permissive to permissive conditions, several hours were required until a *ptx* or a *cya* message could be detected, whereas the *fha* and *bvg*

Table 2. Comparison of some properties of regulatory mutants of *Bordetella pertussis*. Tohama I is the virulent wild type strain, BP359 is an avirulent Tohama I derivative due to a Tn5 insertion in the *bvgA* gene, BC75 and TIII are spontaneous avirulent variants of Tohama I. BC75 was obtained from B. Cookson (Cookson *et al.*, 1990). Specific mRNA was detected after hybridization of RNA slot blots to radioactively labeled restriction fragments containing the respective genes. Proteins were detected using specific antisera on Western blots.

Virulence Factor	Tohama I	TIII/BC75	BP359
Pertussis toxin (106 kDa)			
protein	+++	-	-
mRNA	+++	(+)	-
Adenylate cyclase (190 kDa)			
protein	+++	(+)	-
mRNA	+++	(+)	-
Filamentous hemagglutinin (220 kDa)			
protein	+++	++	-
mRNA	+++	+++	-
Pertactin (69 kDa)			
protein	+++	+++	-
mRNA	+++	+++	-
Fimbriae (21 kDa)			
protein	+++	+++	-
mRNA	+++	+++	-
BvgA (23 kDa)			
protein	+++	+++	-
mRNA	+++	+++	-
BvgS (134 kDa)			
protein	+	+	-

transcripts appeared within minutes (Gross and Rappuoli, 1989; Scarlato *et al.*, 1991). Obviously, the structural differences of the various promoters indeed reflect differences in their regulation.

Several spontaneous avirulent mutants of *B. pertussis* were recently identified, which could not be complemented by the *bvg* locus *in trans* (Gross and Carbonetti, 1993; Cookson *et al.*, 1990). The phenotype of two of the mutants, BC75 and TIII, was analysed in detail looking for expression of the virulence factors, both at the transcriptional and the translational level. As shown in Table 2, the virulence factors could be classified into two groups: The adhesins such as FHA, pili or pertactin and the Bvg system itself are expressed and regulated normally, whereas the toxins PTX and CYA are strongly reduced in their expression. In consequence, this type of mutant is strongly impaired in its capacity to colonize the respiratory tract of mice (Khelef, Guiso and Gross, unpublished results).

Mutation(s) responsible for the phenotype of these mutants are located outside of the *bvg* locus, because the exchange of the *bvg* system of the mutants with the wild type operon had no influence on the mutant phenotypes

520

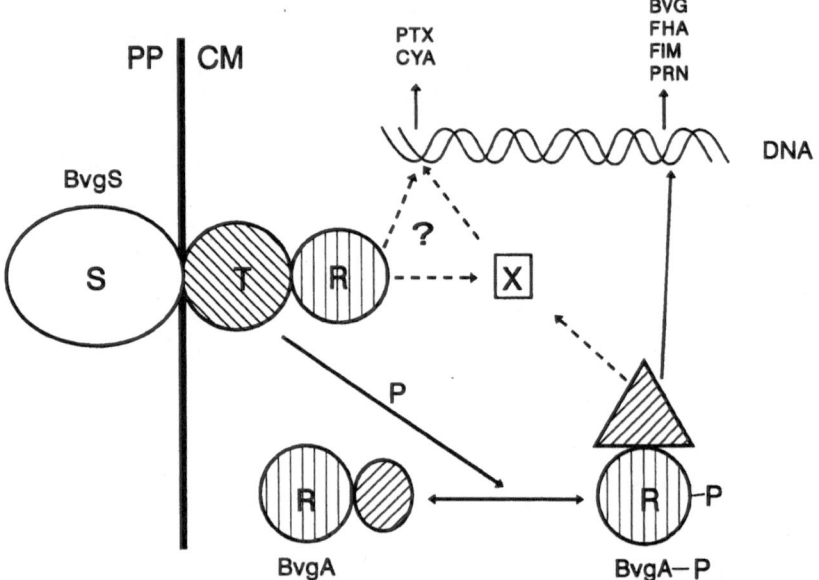

Figure 2. Model for the regulation of the virulence factors of *B. pertussis*. The sensor protein BvgS perceives external signals such as temperature and certain chemicals such as nicotinic acid and $MgSO_4$ with its periplasmic domain (S). The signal is transduced to its C-terminal domain containing the two-component "transmitter" domain (T). The putative kinase activity present in this domain is active in the absence of the above mentioned signals and leads to a phosphorylation of the "receiver" (R) domain of the activator protein BvgA. BvgA in the activated form (BvgA-P) is able to promote transcription from the *bvg*, *fha* and maybe also from the *fim* and *prn* promoters. Additional factor(s) "X" are necessary for the transcription of the toxin operons *ptx* and *cya*. However, so far it is not known, on which level the additional regulatory factor(s) are linked to the bvg system. Furthermore, it should be mentioned that the BvgS sensor protein is one of the very few examples of the two-component sensory proteins which contain in addition to the transmitter domain (T) also a receiver domain (R) and a hypothetical DNA binding domain with a helix-turn-helix motif. Therefore, it is possible that the BvgS protein is more directly involved in gene regulation than thought so far.

(Carbonetti and Gross, unpublished results). Thus additional factor(s) are likely involved in the regulation of PTX and CYA.

As shown in Fig. 2, the *bvg* locus is the master regulator of the virulence regulon and seems to be sufficient to express several virulence factors directly, such as the *bvg* locus itself, the filamentous hemagglutinin, and maybe also the fimbriae and the pertactin. The expression of the pertussis toxin and adenylate cyclase toxin requires additional factor(s), which are not encoded by the *bvg* locus. It is not clear yet, how these factors interact in the activation of the toxin promoters. In addition, a more direct role of the BvgS regulatory domain containing a helix-turn-helix motif can not be ruled out. Experiments are under way to characterize the role of the BvgS receiver domain and to identify the additional regulatory locus required for toxin expression.

Finally, the expression of several genes in *B. pertussis* appears to be regulated negatively by the *bvg* locus. In contrast to all above described factors, these loci are expressed in phase variants and under modulative conditions. At least one of the products of these negatively regulated genes seems to play a role in virulence (Beattie *et al.*, 1992). How the *bvg* system is involved in this regulation is not yet clear. However, the transcription interference assay with the tandem repeats of the *ptx* promoter demonstrates that the protein binding at the *ptx* promoter, which is probably involved in its activation, may also function as a repressor of transcription (Gross *et al.*, 1992).

Conclusion

The combination of all data regarding the differences in the regulation of the various virulence factors of *B. pertussis* reveals that there exist at least three different groups of these factors. Two groups consist of *bvg* activated virulence genes (*vag* genes): i) factors which are directly activated by the *bvg* locus and which are believed to be involved in bacterial adhesion to eukaryotic cells; and ii) factors with mainly toxic properties which are thought to be involved in the evasion from the host defense mechanisms, which need regulatory element(s) in addition to *bvg* gene products for transcriptional activation. The third group consists of the *bvg* repressed genes (*vrg* genes), the functions of which are under investigation. All of them appear to be part of a single regulatory network, the *bvg* regulon. The control of this regulon is complex, including autoregulation of the *bvg* two-component system and additional regulatory factor(s), which are not yet characterized. Although so far there is no experimental evidence that the *in vitro* data on the differential regulation of *Bordetella* virulence factors have *in vivo* relevance, it is an attractive idea that virulence factors such as adhesins are expressed in a different manner than toxins which might play a role in other phases of infection.

Acknowledgements

This work was supported by grants from Deutsche Forschungsgemeinschaft (GR1243/21) and Human Frontier Science Program.

References

Ad hoc group for the study of pertussis vaccine (1988) Placebo-controlled trial of two acellular pertussis vaccines in Sweden – protective efficacy and adverse events. The Lancet 1: 955–960.
Aricò B, Gross R, Smida J and Rappuoli R (1987) Evolutionary relationships in the genus *Bordetella*. Mol Microbiol 1: 301–308.
Aricò B, Miller JF, Roy C, Stibitz S, Monack D, Falkow S, Gross R and Rappuoli R (1989)

Sequences required for *Bordetella pertussis* virulence factors share homology with prokaryotic signal transduction proteins. Proc Natl Acad Sci USA 86: 6671–6675.

Beattie DT, Shakin R and Mekalanos JJ (1992) A *vir*-repressed gene of *Bordetella pertussis* is required for virulence. Infect Immun 60: 571–577.

Bordet J and Gengou O (1906) Le microbe de la coqueluche. Ann Inst Pasteur (Paris) 20: 731–740.

Charles IG, Dougan G, Pickard D, Chatfield S, Smith M, Novotny P, Morrissey P and Fairweather NF (1989) Molecular cloning and characterization of protective outer membrane protein P.69 from *Bordetella pertussis*. Proc Natl Acad Sci USA 86: 3554–3558.

Cookson BT, Berg DE and Goldman WE (1990) Mutagenesis of *Bordetella pertussis* with transposon Tn*5tac*1: conditional expression of virulence-associated genes. J Bacteriol 172: 1681–1687.

Cookson BT, Cho HL, Herwaldt LA and Goldman WE (1989) Biological activities and chemical composition of the purified tracheal cytotoxin of *Bordetella pertussis*. Infect Immun 57: 2223–2229.

Endoh M, Nagai M, Burns DL, Nakasa Y and Manclark CR (1990) Effects of exogenous agents on the action of *Bordetella* heat-labile toxin. p. 30–36. *In*: Manclark C (ed) Proceedings of the sixth international symposium on pertussis. Food and Drug Administration, Washington.

Gentry-Weeks CR, Provence DL, Keith JM and Curtiss R III (1991) Isolation and characterization of *Bordetella avium* phase variants. Infect Immun 59: 4026–4033.

Glaser P, Ladant D, Sezer O, Pichow F, Ullmann A and Danchin A (1988) The calmodulin-sensitive adenylate cyclase of *Bordetella pertussis*: cloning and expression in *Escherichia coli*. Mol Microbiol 2: 19–30.

Goyard S and Ullmann A (1991) Analysis of *Bordetella pertussis cya* operon regulation by use of *cya-lac* fusions. FEMS Microbiol Lett 52: 251–256.

Goyard S and Ullmann A (1992) Functional characterization of the *Bordetella pertussis cya* promoter. Abstract at the "Workshop on Molecular Pathogenesis of Bacteria – Basic and Applied Aspects", Schierke, Germany.

Gross R, Aricò B and Rappuoli R (1989a) Genetics of pertussis toxin. Mol Microbiol 3: 119–124.

Gross R, Aricò B and Rappuoli R (1989b) Families of bacterial signal transducing proteins. Mol Microbiol 3: 1661–1667.

Gross R and Carbonetti NH (1993) Differential regulation of *Bordetella* virulence factors. Zbl Bakt S3, in press.

Gross R, Carbonetti NH, Rossi R and Rappuoli, R (1992) Functional analysis of the pertussis toxin promoter. Res Microbiol 143, 671–681.

Gross R and Rappuoli R (1988) Positive regulation of pertussis toxin expression. Proc Natl Acad Sci USA 85: 3913–3917.

Gross R and Rappuoli R (1989) Pertussis toxin promoter sequences involved in modulation. J Bacteriol 171: 4026–4030.

Guiso N, Szatanik M and Rocancourt M (1990) A protective antigen against lethality and bacterial colonization in murine respiratory and intracerebral models. p. 207–211. *In*: Manclark C (ed) Proceedings of the sixth international symposium on pertussis. Food and Drug Administration, Washington.

Hewlett EL (1989) *Bordetella* species. p. 1756–1762. *In*: Mandell GL, Douglas RG and Bennet JE (eds.) Principles and practice of infectious diseases. Churchill Livingstone, New York.

Hrabak EM and Willis DK (1992) The *lemA* gene required for pathogenicity of *Pseudomonas syringae* pv. syringae on bean is a member of a family of two-component regulators. J Bacteriol 174: 3011–3020.

Huh YJ and Weiss AA (1991) A 23 Kd protein, distinct from BvgA, expressed by virulent *Bordetella pertussis* binds to the promoter region of *vir*-regulated toxin genes. Infect Immun 59: 2389–2395.

Igo MM, Ninfa AJ, Stock JB and Silhavy TJ (1989) Phosphorylation and dephosphorylation of a bacterial activator by a transmembrane receptor. Genes Dev 3: 1725–1734.

Katada T, Oinuma M and Ui M (1986) Mechanism for inhibition of the catalytic activity of adenylate cyclase by the guanine nucleotide-binding proteins serving as the substrate of islet-activating protein, pertussis toxin. J Biol Chem 261: 5215–5221.

Khelef N, Sakamoto H and Guiso N (1992) Both adenylate cyclase and hemolytic activities are required by *Bordetella pertussis* to initiate infection. Microb Pathogen 12: 227–235.

Lacey BW (1960) Antigenic modulation of *Bordetella pertussis*. J Hyg 58: 57–93.

Leininger E, Kenimer JG and Bremann MJ (1990) Surface proteins of *Bordetella pertussis*. p. 100–105. *In*: Manclark C (ed) Proceedings of the sixth international symposium on pertussis. Food and Drug Administration, Washington.

Leininger E, Roberts M, Kenimer JG, Fairweather CIG, Novotny N and Brennan MJ (1991) Pertactin, an Arg-Gly-Asp-containing *Bordetella pertussis* surface protein that promotes adherence of mammalian cells. Proc Natl Acad Sci USA 88 (2): 345–349.

Leininger E, Ewanowich CA, Bhargava A, Peppler MS, Kenimer JG and Brennan, MJ (1992) Comparative role of the Arg-Gly-Asp sequence present in the *Bordetella pertussis* adhesins pertactin and filamentous hemagglutinin. Infect Immun 60: 2380–2385.

Leslie PH and Gardner AD (1931) The phases of *Haemophilus pertussis*. J Hyg 31: 423–455.

Locht C and Keith JM (1986) Pertussis toxin gene: nucleotide sequence and genetic organization. Science 232: 560–563.

Miller JF, Roy CR and Falkow S (1989) Analysis of *Bordetella pertussis* virulence gene regulation by use of transcriptional fusions in *Escherichia coli*. J Bacteriol 171: 6345–6348.

Miller JF, Johnson SA, Black WJ, Beattie DT, Mekalanos JJ and Falkow S (1992) Constitutive sensory transduction mutations in the *Bordetella pertussis bvgS* gene. J Bacteriol 174: 970–979.

Monack D, Arico B, Rappuoli R and Falkow S (1989) Phase variants of *Bordetella bronchiseptica* arise by spontaneous deletions in the *vir* locus. Mol Microbiol 3: 1719–1728.

Mooi FR, Jansen WH, Brunnings H, Gielen H, van der Heide HGJ, Walvoort HC and Guinee PAM (1992) Construction and analysis of *Bordetella pertussis* mutants defective in the production of fimbriae. Microb Pathogen 12: 127–135.

Musser JM, Hewlett EL, Peppler MS and Selander RK (1986) Genetic diversity and relationships in populations of *Bordetella* species. J Bacteriol 166: 230–237.

Nicosia A, Perugini M, Franzini C, Casagli MC, Borri MG, Antoni G, Neri P, Ratti G and Rappuoli R (1986) Cloning and sequencing of the pertussis toxin genes: Operon structure and gene duplication. Proc Natl Acad Sci USA 83: 4631–4635.

Pittman M (1984) Genus *Bordetella*. p. 388–393. *In*: Krieg NR and Holt JG (eds.) Bergey's Manual of systematic bacteriology. Williams and Wilkins, v.l. Baltimore.

Porter JF, Parton R and Wardlaw AC (1991) Growth and survival of *Bordetella bronchiseptica* in natural waters and in buffered saline without added nutrients. Appl Env Microbiol 57: 1202–1206.

Rappuoli R, Arico B and Gross R (1987) Conversion from *Bordetella parapertussis* to *B pertussis*. The Lancet 3, 511.

Rappuoli R, Pizza M, Podda A, De Magistris MT and Nencioni L (1991) Towards third-generation whooping cough vaccines. TIBTECH 9: 232–237.

Relman DA, Domenighini M, Tuomanen E, Rappuoli R and Falkow S (1989) Filamentous hemagglutinin of *Bordetella pertussis*: Nucleotide sequence and crucial role in adherence. Proc Natl Acad Sci USA 86: 2637–2641.

Relman D, Tuomanen E, Falkow S, Golenbock DT, Saukkonen K and Wright SD (1990) Recognition of a bacterial adhesion by an integrin: macrophage CR3 (aMb2, CD11b/CD18) binds filamentous hemagglutinin of *Bordetella pertussis*. Cell 61: 1375–1382.

Roy CR, Miller JF and Falkow S (1989) The *bvgA* gene of *Bordetella pertussis* encodes a transcriptional activator required for coordinate regulation of several virulence genes. J Bacteriol 171: 6338–6344.

Roy CR and Falkow S (1991) Identification of *Bordetella pertussis* regulatory sequences required for transcriptional activation of the *fhaB* gene and autoregulation of the *bvgAS* operon. J Bacteriol 173: 2385–2392.

Sato H and Sato Y (1984) *Bordetella pertussis* infection in mice: correlation of specific antibody against two antigens, pertussis toxin, and filamentous hemagglutinin with mouse protectivity in an intracerebral or aerosol challenge system. Infect Immun 46: 415–421.

Saukkonen K, Burnette WN, Mar VL, Masure HR and Tuomanen EI (1992) Pertussis toxin has eukaryotic-like carbohydrate recognition domains. Proc Natl Acad Sci USA 89: 118–122.

Scarlato V, Prugnola A, Aricò B and Rappuoli R (1990) Positive transcriptional feedback at the *bvg* locus controls expression of virulence factors in *Bordetella pertussis*. Proc Natl Acad Sci USA 87: 6753–6757.

Scarlato V, Aricò B, Prugnola A and Rappuoli R (1991) Sequential activation and environmental regulation of virulence genes in *Bordetella pertussis*. EMBO J 10: 3971–3975.

Stibitz S and Yang MS (1991) Subcellular localization and immunological detection of proteins encoded by the *vir* locus of *Bordetella pertussis*. J Bacteriol 174: 4288–4296.

Stock JB, Stock AM and Mottonen JM (1990) Signal transduction in bacteria. Nature (London) 344: 395–400.

Wanner BL (1992) Is cross regulation by phosphorylation of two-component response regulator protein important in bacteria? J Bacteriol 174: 2053–2058.

Wardlaw AC and Parton R (1988) Pathogenesis and immunity in pertussis. John Wiley & Sons, Chichester.

Weiss AA and Falkow S (1984) Genetic analysis of phase change in *Bordetella pertussis*. Infect Immun 43: 263–269.

Wels W, Baldrich M, Chakraborty T, Gross R and Goebel W (1992) Expression of bacterial cytotoxin genes in mammalian target cells. Mol Microbiol 6, 2651–2659.

Willems R, Paul A, van der Heide HGJ, ter Avest AR and Mooi FR (1990) Fimbrial phase variation in *Bordetella pertussis*: a novel mechanism for transcriptional regulation. EMBO J 9: 2803–2809.

van't Wout J, Burnette WN, Mar VL, Rozdzinski E, Wright SD and Tuomanen EI (1992) Role of carbohydrate recognition domains of pertussis toxin in adherence of *Bordetella pertussis* to human macrophages. Infect Immun 60: 3303–3308.

36. Regulation of *Shigella* virulence expression

M. YOSHIKAWA, C. SASAKAWA, T. TOBE, N. NAKATA, N. OKADA,
Y. HOMMA, I. FUKUDA and K. KOMATSU

Abstract. Extensive investigation on the molecular pathogenesis of bacillary dysentery has been made through the introduction of recombinant DNA technology and various cellular and molecular biological technologies for *in vitro* characterization of the virulence. These studies have indicated that the pathogenicity of shigellae is multifactorial, involving numerous components of the bacteria and requires various genes dispersed around the chromosome as well as on the large plasmid. Expression of each of these genes is under the control of a composite regulatory network. Although our knowledge to understand the pathogenesis of bacillary dysentery at the molecular level has substantially accumulated, a number of questions concerning the molecular basis of certain steps in the infection process still remain.

Abbreviations: CAT, chloramphenicol acetyltransferase; Chl, contact hemolysis as a virulence phenotype; EIEC, enteroinvasive *Escherichia coli*; Fp-test, focus-plaque test; Inv, invasion as a virulence phenotype; IPTG, isopropyl-1-thio-β-D-galactoside; Pcr, Congo red binding as a virulence phenotype; Ser, Sereny test as a virulence phenotype; Vir, virulence as a phenotype

I. Introduction

The invasion of shigellae into colonic epithelial cells is one of the critical steps in the pathogenesis of bacillary dysentery (LaBrec *et al.*, 1964). Virulent translucent strains of *Shigella flexneri* invade intestinal epithelia, whereas the avirulent opaque variants do not, although both have the same parenteral LD_{50} values. Virulent shigellae may invade colonic epithelia by pathogen-induced internalization and multiplication in the cytoplasm after lysis of the phagocytic vacuole. Successive spread of the bacteria to adjacent cells, accompanying cell degeneration, results in severe inflammation, spread of the bacteria to deeper tissue, accentuating the inflammation and, ultimately, in ulceration and bloody diarrhea.

In the 1960s, several chromosomal regions of *S. flexneri* were implicated in virulence, based on intergeneric conjugation between *Escherichia coli* K-12 Hfr and *S. flexneri* (Falkow *et al.*, 1963; Formal *et al.*, 1965, 1971; Gemski *et al.*, 1972; Sansonetti *et al.*, 1983). In addition, Sansonnetti and colleagues have established in the early 1980s that the possession of a large plasmid is essential for Inv+ phenotypes (Kopecko *et al.*, 1980; Sansonetti *et al.*, 1981, 1982, 1983). This discovery has provoked extensive investigation into the molecular pathogenesis of bacillary dysentery, using newly developed recombinant DNA

C.I. Kado and J.H. Crosa (eds.), Molecular Mechanisms of Bacterial Virulence, 525–541.
© 1994 *Kluwer Academic Publishers.*

technology and various cellular and molecular biological technologies for *in vitro* characterization of the virulence. Thus, rapid progress in the molecular elucidation of *vir* genes carried on the large plasmid and the chromosome has been achieved and our knowledge of the *vir* genes has substantially accumulated, thereby increasing our understanding of the pathogenesis of bacillary dysentery at the molecular level.

II. Overview of the virulence genes

On the large 230 kb plasmid of *S. flexneri* 2a seven *vir* regions were initially identified by random Tn*5* insertion mutagenesis (Sasakawa *et al.*, 1986a, 1986b) (Fig. 1 and Fig. 3). The 31 kb DNA sequence was divided into five contiguous segments designated Region-1 (*virB*), Region-2 (*ipaBCD or ipa*), Region-3, –4 and –5 (Sasakawa *et al.*, 1988). The *virB* gene in Region-1 is a

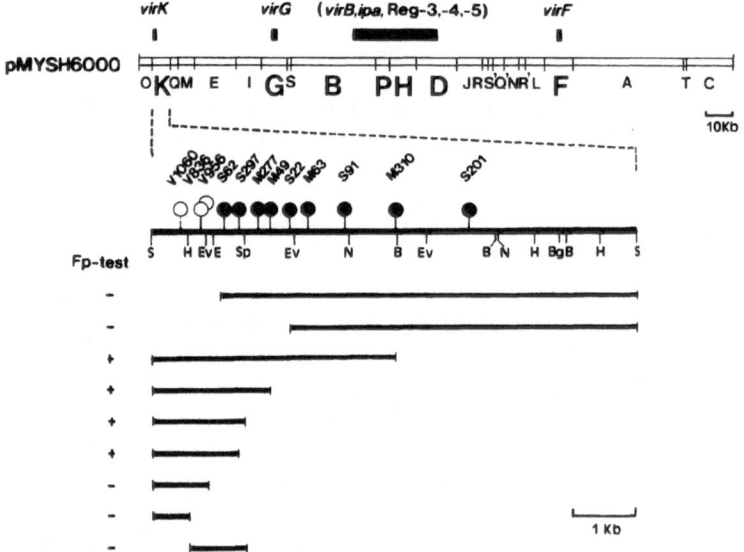

Figure 1. The *Sal*I restriction map of the large invasion plasmid of pMYSH6000 of *Shigella flexneri* 2a YSH6000, localization of virulence-associated genetic loci and mapping of *virK* gene. The virulence-associated loci on the plasmid pMYSH6000 are shown by black boxes over the open bar representing the *Sal*I map of the plasmid (Sasakawa *et al.*, 1986b). The open and closed circles over the physical map of the enlarged *Sal*I fragment K indicate the sites of Tn*10* and Tn*5* respectively. The bottom lines represent the cloned DNA segments derived from the *Sal*I fragment K. + and – shown on the left side indicate positive and negative results of the complementation test of the three Tn*10*-inserted *virK⁻* mutants with each of the cloned fragments determined by the Fp test. Symbols: B, *Bam*HI; Bg, *Bgl*II; E, *Eco*RI; Ev, *Eco*Rv; H, *Hind*III; N, *Nco*I; S, *Sal*I; Sp, *Sph*I. Reproduced from Molecular Microbiology (Nakata *et al.*, 1992) with permission of the publisher.

positive regulator for the expression of the other four *vir* regions, *ipaBCD* (*ipa*), Region-3, −4 and −5 (Adler *et al.*, 1989; Tobe *et al.*, 1991). The *ipa* operon in Region-2 encodes three Inv-associated antigens, IpaB, IpaC and IpaD (Baudry *et al.*, 1987; Buysse *et al.*, 1987; Venkatesan *et al.*, 1988; Sasakawa *et al.*, 1989) which are believed to be directly involved in Inv phenotypes (Hale *et al.*, 1985; Mills *et al.*, 1988; Hromockyj and Maurelli, 1989). The Region-3.4 and Region-5 operons code collectively for specialized transport systems, such as *mxiA* (Membrane Expression of Invasion plasmid antigens) (Andrews *et al.*, 1991), *spa* (Surface Presentation of Antigens) (Venkatesan *et al.*, 1992) and *virH* (C. Sasakawa, unpubl.), all required for correct localization of the Ipa proteins. The *virF* gene (Sakai *et al.*, 1986a, 1986b) regulates the expression of the *virG* gene and the *virB* gene positively at the transcriptional level (Sakai *et al.*, 1988; Adler *et al.*, 1989; Kato *el al.*, 1989). The *virG* (*icsA*) gene encodes a 116 kD immunogenic, surface-exposed outer membrane protein, and is essential for invading bacteria to spread intra- as well as intercellularly (Makino *et al.*, 1988; Lett *et al.*, 1989). The bacteria that express the VirG (IcsA) protein elicit a polar deposition of F-actin around them in the cytoplasm of infected epithelial cells (Bernardini *et al.*, 1989). A plasmid gene, *icsB*, is involved in the lysis of the protrusions, a step necessary for intercellular spread (Allaoui *et al.*, 1992).

Various classes of *vir* loci have also been identified on the chromosome by classical as well as molecular genetic technology (Table 1, Fig. 2 and Fig. 3). Some of them appear to be involved in the expression of the *vir* genes on the large plasmid. For example, a locus near the *trp* gene, designated *virR* (*osmZ*), controls the temperature-dependent production of the Inv-associated antigens

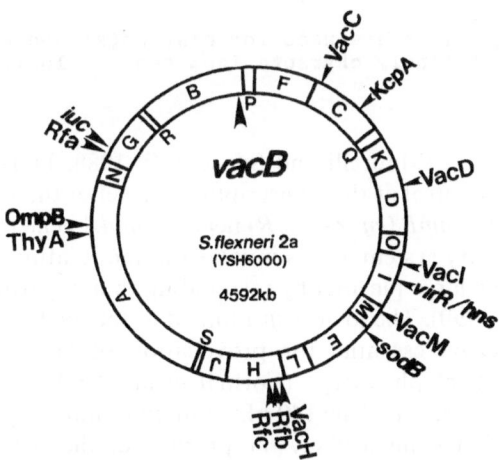

Figure 2. The circular physical map of the chromosome of *Shigella flexneri* 2a YSH6000 with *Not*I restriction fragments and direct assignment of chromosomal virulence-associated genes so far reported indicated with external arrowheads. The location of *vacB*::Tn5 is indicated with an internal arrowhead. Modified from Journal of Bacteriology (Tobe *et al.*, 1992) with permission of the publisher.

Table 1. Virulence-associated chromosomal genetic loci and the phenotypes encoded[a]

Genes or loci[b]	Phenotypes or functions
vacJ	Intercellular spreading
rfa, rfb, rfc	LPS synthesis / Spreading
thyA	Thymine synthesis / Spreading
iuc	Iron acquisition
sodB	Superoxide dismutase
VacC, VacM	*ipa* expression at transcriptional level / Invasion indirectly
KcpA	*virG* expression at post-transcriptional level / Spreading indirectly
vacB	Expression of all plasmid virulence genes at post-transcriptional level / Both invasion and spreading indirectly
hns (*virR/osmZ*)	Thermoregulation of invasion phenotypes / Transcription of *virB* affected
ompB (*envZ-ompR*)	Osmoregulation / Two components signal transduction system
VacD, VacH, VacI	Invasion delayed

a
   For reference see the text.
b
   Italic designations are used for cistron(s) and Roman designations for not yet finely characterized genetic loci.

IpaB, IpaC and IpaD (Maurelli and Sansonetti, 1988; Dorman *et al.*, 1990) indirectly through control at the transcriptional level of the *virB* gene (Tobe *et al.*, 1991). The locus *ompB* (*envZ-ompR*) near the *malA* gene belongs to the two component regulatory system and is involved in modulating the expression of the *vir* genes on the large plasmid by responding to a hypertonic environment (Bernardini *et al.*, 1990). The locus *kcpA* linked to the *purE* gene was originally identified as a region essential for provocation of keratoconjunctivitis or expression of the Ser⁺ phenotype (Formal *et al.*, 1971). The *kcpA* locus is required for the expression of the 140 kD immunodominant protein (Pal *et al.*, 1989), a protein that is most likely the product of the *virG* (*icsA*) gene (M. Yoshikawa, unpublished).

Recently, avirulent chromosomal mutants have been isolated by random Tn*5* insertion mutagenesis in *S. flexneri* 2a strain YSH6000, characterized and classified into various classes (Okada *et al.*, 1991a) (Table 1, Fig. 2 and Fig. 3). A *Not*I restriction map of the *S. flexneri* 2a YSH6000 chromosome has been

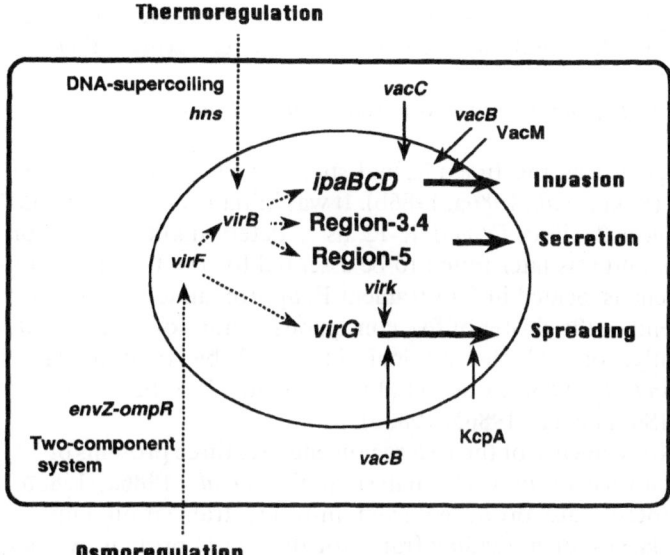

Figure 3. Schematic model to correlates interaction between plasmid and chromosomal genes. The chromosomal regulatory genes or loci are shown outside the ellipse and the plasmid regulatory and target genes or loci inside the ellipse. Italic designations are used for cistrons and Roman designations for not yet finely characterized genetic loci. Thick solid arrows indicate the phenotypes or functions encoded by the genes or genetic loci. Dotted arrows are used when the transcription of the target plasmid genes or loci is affected. Thin solid arrows are used when the expression of the target plasmid genes or loci is affected at the step after transcription.

made (Okada *et al.*, 1991b). The sites of Tn*5* insertions in avirulent mutants isolated were localized on this map (Fig. 2). Two Vac (*v*irulence-*a*ssociated *c*hromosomal) loci, VacC on *Not*I fragment C and VacM on *Not*I fragment M thus identified (Okada *et al.*, 1991b) are required for the full expression of the plasmid-coded Ipa proteins at transcriptional level (Okada *et al.*, 1991a). A regulatory gene, *vacB*, located on the *Not*I fragment B is required for full expression of the *vir* loci on the large plasmid at a post-transcriptional level (Tobe *et al.*, 1992).

III. Regulation of the expression of virulence genes

During the process of infection shigellae may encounter severe environmental changes. The most inhospitable environment for the bacteria, which may expose them to the risk of death, are the host defense mechanisms. Bacterial virulence factors represent the mechanisms by which the bacteria cope with host defense mechanisms and escape from the risk of death. To achieve this, the bacteria must adapt to the environment, and the expression of bacterial virulence factors must be regulated so as to adapt to the environmental changes. Thus, shigellae possess a complex regulatory network to selectively express their *vir* factors.

A. *The plasmid gene virF as a key transcriptional activator*

The virF gene was the first plasmid-specified *vir* gene to be identified and sequenced (Sakai *et al.*, 1986a, 1986b). It was initially cloned from pMYSH6000 of *S. flexneri* 2a into *E. coli* K-12 as a determinant coding for the Pcr⁺ phenotype, and was later found to be essential for the Inv⁺ phenotype as well. The *virF* gene is located in *Sal*I frament F, approximately 40 kb away from the Region-5 end of the cluster of five contiguous regions on pMYSH6000 (Fig. 1). Subsequently, one of the *vir* loci, identified by random Tn*5* insertions (Sasakawa *et al.*, 1986b), was found to be identical to the *virF* gene previously identified (Sakai *et al.*, 1986a, 1986b).

The DNA sequence of the *virF* region encodes three proteins of 30, 27 and 21 kD, as identified by minicell analysis (Sakai *et al.*, 1986a, 1986b). The two smaller proteins are produced from in-frame translation initiation codons within the largest open reading frame for the 30 kD protein. A small deletion, covering only the first translation initiation codon, resulted in the loss of *virF* function accompanied by the disappearance of the 30 kD protein, and by the expression of only the smaller two proteins, indicating that only the 30 kD *virF* product, consisting of 262 amino acid residues, is responsible for the *virF* function (Sakai *et al.*, 1988). The DNA sequence of the *virF* gene is strikingly rich in A and T (63%) residues, resulting in a codon usage markedly different from that of the chromosome of shigellae (Sakai *et al.*, 1986b). This is a property later found to be shared with other plasmid-encoded *vir* genes of shigellae (Baudry *et al.*, 1988; Lett *et al.*, 1989; C. Sasakawa, unpublished). There is a likelyhood that the difference in GC content of the plasmid and the chromosome reflects different evolutionary origins.

Four immunodominant proteins, one of 130 kD encoded by *virG* and directly involved in intra- and intercellular spread (Makino *et al.*, 1986; Sakai *et al.*, 1988; Lett *et al.*, 1989; Bernardini *et al.*, 1989), and three of 62, 41 and 37 kD encoded by Region-2 (*ipa* region) implicated in Inv⁺ capacity (Hale *et al.*, 1985; Mills *et al.*, 1988; Hromockyj and Maurelli, 1989) were either not fully expressed or not expressed at all without the *virF* function (Sakai *et al.*, 1988). A cloned *virF* gene restored *virF⁻* mutants to producing these proteins and to a fully virulent phenotype (Sakai *et al.*, 1988). Thus, the expression of these proteins seems to be positively regulated by the *virF* product. This was confirmed quantitatively by means of a *virG-lacZ* fusion. A 3.8 kb RNA transcript specific for the *virG* gene was produced only in the presence of the *virF* gene as shown by Northern blotting analysis (Sakai *et al.*, 1988). Similar results were obtained for Region-1 to –5. These results indicate that the *virF* product acts directly or indirectly as a transcriptional activator for the expression of Region-1 to –5, and also for the *virG* gene (Sakai *et al.*, 1988) (See Fig. 3).

A *virF* gene probe from pMYSH6000 hybridized with the large plasmids of all virulent shigellae and EIEC strains tested, suggesting that the *virF* gene is one of the *vir* genes common amongst all of them (Sakai *et al.*, 1988). Indeed, an analogous gene has been found on pSS120 in *S. sonnei* (Kato *et al.*, 1989).

Apparently inconsistent with this data, the 44 kb cosmid clone of pWR100 (pHS4108) was Inv+ in spite of the lack of the positive regulatory gene *virF* (Maurelli *et al.*, 1985). This is presumably due to high copy number of the cosmid vector used to obtain pHS4108. Consistent with this interpretation is the fact that a 37 kb corresponding sequence of pSS120 in *S. sonnei* cloned into a low copy cosmid vector, pJK292, confers the Inv+ phenotype on plasmid-cured strains of *S. sonnei* only in the presence of the positive regulatory gene *virF* (Kato *et al.*, 1989).

The amino acid sequence of the 30 kD VirF protein deduced from the DNA sequence has recently been shown to be homologous with the sequence of positive regulators such as CfaD for colonization factor antigen I (CFA/I) in *E. coli* (Savelkoul *et al.*, 1990), Rns for colonization factor antigen II (CFA/II) in *E. coli* (Caron *et al.*, 1989), and a distinct VirF for the *yop* genes in *Yersinia enterocolitica* (Cornelis *et al.*, 1989), all belonging to the *E. coli* AraC superfamily of DNA binding proteins (Ramos *et al.*, 1990).

B. A regulatory element encoded by virB on Region-1

The *virB* gene was recognized by random Tn*5* insertions as a *vir* gene on Region-1 within the 31 kb DNA segment of pMYSH6000 (Adler *et al.*, 1989; Sasakawa *et al.*, 1986b), and is located downstream of the *ipa* operon (Fig. 1). From the nucleotide sequence of Region-1, an open reading frame has been identified that encodes a 35.4 kD protein comprising 309 amino acid residues, close to the molecular mass of 33 kD seen in minicell analysis (Adler *et al.*, 1989).

By immunoblotting analysis and S1 nuclease mapping it was found that *virB* is a transcriptional activator for the expression of the *ipaBCD* genes located on Region-2, but not for the *virG* gene (Adler *et al.*, 1989) (Fig. 3). Similar results were subsequently obtained for Regions-3, −4 and −5. Northern blotting analysis performed to clarify the functional relationship between *virF* and *virB* as the transcriptional activators for Region-2 to −5, has indicated that *virF* is a transcriptional activator for *virG* and *virB* and that the VirB protein produced by this activation is a transcriptional activator for the *ipaBCD* genes, and also for Region-3 to −5 (Adler *et al.*, 1989) (Fig. 3). This seems to imply that all the genetic determinants on the plasmid involved directly or indirectly in Inv ability are dually regulated by *virF* and *virB*. To further confirm this, a *virF*− mutant of YSH6000 containing a *tac-virB* fusion to control the expression of the *virB* gene was constructed. The Inv test and the Chl test (Sansonetti *et al.*, 1986) were positive with this strain when IPTG was added (Tobe *et al.*, 1991). Thus, the *virB* gene was transcribed in the absence of the *virF* function under the control of the *tac* promotor only if the *tac* promotor was activated by IPTG. However, the same strain did not show a positive Fp-test or a Ser+ phenotype even at 37°C (Tobe *et al.*, 1991) because of the lack of *virG* function, for which *virF* function is essential (Sakai *et al.*, 1988) (Fig. 3), further confirming the regulatory circuit proposed by Adler *et al.* (1989). Thus, the *virB* gene is now known to be a positive regulator whose expression is governed by the *virF* gene (Adler *et al.*, 1989) and by temperature (Tobe *et al.*, 1991) (Fig. 3 and Section IVa).

A *virB* gene probe from pMYSH6000 hybridized with the large plasmid of all virulent shigellae and EIEC strains, suggesting that the *virB* gene is one of the *vir* genes common to all of them (Sasakawa *et al.*, 1986b). A similar regulator gene was subsequently identified in *S. flexneri* 5 (*ipaR;* Buysse *et al.*, 1990) and in *S. sonnei* (*invE;* Watanabe *et al.*, 1990). The *invE* (*virB*) gene product is significantly homologous with the *parB* product of phage P1 (Watanabe *et al.*, 1990).

C. A plasmid-coded element, virK, *on* SalI *fragment K required for the expression of the* virG *gene*

The plasmid-coded gene *virK* was identified by Tn*10* insertion mutagenesis of *S. flexneri* 2a YSH6000T, an antibiotic sensitive derivative of the wild type YSH6000 (Nakata *et al.*, 1992). Three avirulent mutants were found to have a single copy insertion of Tn*10* within *Sal*I fragment K, which is located far away from the previously identified *vir* loci on the plasmid (Nakata *et al.*, 1992). The locus was designated *virK* (See Fig. 1). No *vir* loci were previously found using Tn*5* within *Sal*I fragment K (Sasakawa *et al.*, 1986b). The mutants formed foci but not plaques in the Fp-test, i.e. they were Inv+ but incapable of intra- and intercellular spread (Okada *et al.*, 1991a).

The sites of transposon insertion in three *virK*⁻ Tn*10* insertion mutants and in nine previously isolated virulent Tn*5* insertions (Sasakawa *et al.*, 1986b) within *Sal*I fragment K were determined precisely (Nakata *et al.*, 1992) (See Fig. 1). Tn*10* insertions were located within a 1.0 kb segment near the end of *Sal*I-K proximal to *Sal*I-O, while Tn*5* insertions were scattered separately from Tn*10* insertions and mapped outside the 1.0 kb segment defined by Tn*10* insertions (Nakata *et al.*, 1992). The mutant phenotypes of three *virK*⁻ Tn*10* insertion mutants were complemented with the cloned *Sal*I fragment K. Examination of the ability of various restriction fragments derived from *Sal*I fragment K, and cloned into pBR322 to complement these *virK*⁻ Tn*10* insertion mutations defined a 1.4 kb *virK* segment from the left end of *Sal*I-K proximal to *Sal*I-O. Thus, the location of the *virK* locus was approximately 50 kb apart from the 5' end of the *virG* gene (Nakata *et al.*, 1992) (Fig. 1).

Three *virK*⁻ Tn*10* insertion mutants exhibited a phenotype similar to the *virG*⁻ mutant; both formed foci but not plaques in the Fp-test (Makino *et al.*, 1986; Nakata *et al.*, 1992) (Fig. 3). In an infection experiment with tissue-cultured MK2 cells in the presence of gentamicin, the invading bacteria with either mutation initially exhibited normal multiplication within the cytoplasm. Subsequently, the wild type moved freely and actively within the cytoplasm and reinfected adjacent cells, whereas both mutants were localized within the cytoplasm and were finally converted to a spherical shape (Makino *et al.*, 1986;

Nakata *et al.*, 1992). Although intracellular spreading of the *virK⁻* mutants was significantly greater than that of the *virG⁻* mutant and most of the cyloplasmic space of MK2 cells became filled with bacteria in the case of the *virK⁻* mutants but not in the *virG⁻* mutant, the invading bacteria with either mutation were nevertheless incapable of spreading intercellularly. In accordance with these observations, the size of foci formed by *virK⁻* mutants in the Fp-test were consistently larger than those of the *virG⁻* mutant (Nakata *et al.*, 1992). Depositions of F-actin in the *virG⁻* mutant, an event that is mediated by the VirG protein and is essential for intra- and intercellular spread of the bacteria (Bernardini *et al.*, 1989; C. Sasakawa, unpublished), were not seen at all, whereas, in *virK⁻* mutants, some deposition of F-actin surrounding the bacteria was seen, and occasionally even a polar deposition was observed, although not so marked as in the wild type. Thus, the invading bacteria with *virK⁻* mutation were capable of at least some intra- but not intercellular spread (Nakata *et al.*, 1992).

The levels of the VirG protein expressed by *virK⁻* mutants, as measured by the use of whole cell lysates in an immunoblot assay with a VirG specific antiserum, were lower than that of the wild type (Fig. 3) but, unlike the *virG⁻* mutant, not completely abolished (Nakata *et al.*, 1992). However, no decrease in *virG*-specific mRNA was observed in *virK⁻* mutants as compared to the wild type, indicating that the *virK* gene does not affect transcription of the *virG* gene (Nakata *et al.*, 1992) (Fig. 3).

The 1642 bp segment containing the putative *virK* gene was sequenced (Nakata *et al.*, 1992). Two ORFs, ORF-1 and ORF-2 were found. ORF-1 was considered not to be directly involved in virulence because one of the previously isolated Tn5 insertions was virulent in spite of the finding that Tn5 is inserted within ORF-1. The 316 amino acids deduced from the nucleotide sequence of ORF-2 indicated a protein of 36.7 kD, very close to the molecular mass of 36 kD as estimated by SDS-PAGE of the proteins expressed from a *virK* gene fragment placed downstream of the T7 RNA polymerase-dependent promoter. Furthermore, the sites of Tn10 insertion in three *virK⁻* Tn10 insertion mutants were located within ORF-2, and none of three linker-insertion mutations in ORF-2 complemented the mutation in three *virK⁻* Tn10 insertion mutants. The hydropathy profile of the amino acid sequence of the VirK protein deduced from the nucleotide sequence indicated a protein with a slightly hydrophilic nature, with no typical signal sequence at the N-terminus. Thus, it is most likely that the VirK protein is a cytoplasmic polypeptide involved in increasing the level of the VirG protein (Fig. 3). A search of the GenBank and EMBL sequence data bases revealed no significant homology (Nakata *et al.*, 1992).

A *virK* gene probe from pMYSH6000 hybridized to all of the plasmid DNAs from 10 *S. dysenteriae*, 14 *S. flexneri*, 9 *S. boydii*, 4 *S. sonnei* and 9 EIEC strains. Hybridization with a *virK* gene probe of the *Sal*I-digested plasmid DNAs of two representative strains from each serogroup of shigellae and EIEC showed that, although the size of the *Sal*I fragments was variable, *virK* homologues were conserved amongst all of them (Nakata *et al.*, 1992). These results suggest that *virK* is essential for both shigellae and EIEC, and responsible for bacterial ability to spread into adjacent cells, by involvement in the full expression of the VirG protein at a post-transcriptional level (Nakata *et al.*, 1992) (Fig. 3).

D. A chromosomal gene, vacB, involved in the expression of the plasmid-coded vir genes at a post-transcriptional level

By random Tn*5* insertion mutagenesis of *S. flexneri* 2a YSH6000, a Pcr⁻ chromosomal single copy Tn*5* insertion mutant without any detectable molecular alteration in any of the *Sal*I fragments of the large plasmid was found (Tobe *et al.*, 1992). By the method previously employed in constructing a *Not*I-restriction map of the *S. flexneri* chromosome and in assigning each of nine *vir* loci tagged by Tn*5* insertions to one of the *Not*I-fragments by using pulsed-field gel electrophoresis (Okada *et al.*, 1991a, 1991b), the mutation was localized to *Not*I fragment B, at 45 kb away from the end adjacent to *Not*I-fragment P containing the *thr* gene at 100 min on the chromosome map of *E. coli* K-12. This locus was named *vacB* (Tobe *et al.*, 1992) (Fig. 2).

The level of the Inv capacity of the *vacB*⁻ mutant was less than one tenth that of the wild type as determined by Inv and Chl tests. However, the mutant exhibited a significantly higher Inv capacity than a plasmid-cured derivative of the wild type YSH6000. The mutant had poor spreading ability into adjacent cells as judged by the Fp-test when compared with the wild type; even after 3 days incubation the plaques were tiny, suggesting that the mutation affected the ability to spread intercellularly (Tobe *el al.*, 1992) (Fig. 3).

Since the ability of shigellae to invade epithelial cells and to spread into adjacent cells is dependent on the *ipa* genes and the *virG* gene, respectively (See Fig. 3), the effect of the *vacB* mutation on the expression of these genes was examined by immunoblotting. The production of IpaB, IpaC, IpaD and VirG in the *vacB*⁻ mutant was markedly decreased as compared with the wild type. Furthermore, the amount of IpaB protein in the *vacB*⁻ mutant was one fourth of that in the wild type strain, as measured by immunoprecipitation with anti-IpaB antiserum. However, there was no substantial difference in the levels of mRNA between the wild type and the *vacB*⁻ mutant for any of the six *vir* operons, *virF*, *virB*, *ipa*, Region-3.4, Region-5 and *virG*. When examined by making an operon fusion with the CAT structural gene and measuring the CAT activity, no marked difference was found between the wild type and the *vacB*⁻ mutant in promoter activities from *ipa*, Region-3.4 and Region-5 operons. Moreover, the amount of Ipa proteins expressed by the plasmid harboring the *tac-ipa* operon fusion in the *vacB*⁻ mutant was lower than that in the wild type, even after full induction of transcription of the *ipa* operon by IPTG. Thus, these results most likely indicate that the *vacB* gene is not involved in the expression of the transcriptional activators *virF* and *virB*, but affects the expression of *ipa*, Region-3.4, Region-5, and *virG* operons on the large plasmid at a post-transcriptional level (Tobe *el al.*, 1992) (Table 1 and Fig. 3).

Cloning and nucleotide sequencing of the *vacB* region showed that it contained an ORF consisting of a 2280 bp DNA sequence, located 669 bp downstream from the 3' end of the *purA* gene. The molecular weight of the VacB protein deduced from the nucleotide sequence was 86.9 kD, very close to the molecular mass of 90 kD as measured by using the T7 RNA polymerase-dependent expression system. The hydropathy profile of the amino acid sequence of the VacB protein deduced from the nucleotide sequence revealed it to be slightly hydrophilic with no typical signal sequence at the N-terminus. Thus, it is likely that the VacB protein is a cytoplasmic polypeptide. A search of the GenBank and EMBL sequence data bases using the DNASIS program (Hitachi Soft. Engin. Co, Tokyo, Japan) revealed a difference of only 2 bp from a reported DNA sequence of the region containing *purA* at 95 min on the *E. coli* K-12 map (Tobe *et al.*, 1992) (Fig. 2). Southern hybridization using a *vacB* gene probe from strain YSH6000 revealed that not only various serotypes of shigellae and EIEC strains but even *E. coli* K-12 have a DNA sequence homologous to the *vacB* gene. Disruption of the *vacB* gene homologues of these shigellae and EIEC strains resulted in a diminished Chl activity and in a reduced production of the Ipa proteins, indicating that the *VacB* gene plays an important role for the full expression of the *vir* phenotypes of both shigellae and EIEC (Tobe *et al.*, 1992) (See Fig. 3).

The molecular mechanisms underlying regulation of the *vir* genes on the plasmid by *vacB* are still obscure. Since the half lives of mRNA for *ipa*, Region-3.4 and *virG* operons were the same in the *vacB* mutant as in the wild type (Tobe *et al.*, 1992), the stability of mRNA transcribed from these operons is apparently not affected by the *vacB* mutation. The *vacB* gene product may be required for the efficient translation of the plasmid genes or the stability of their protein products. A low GC content and different codon usage of the genes on the plasmid as compared with those on the chromosome have repeatedly been pointed out (Sakai *et al.*, 1986b; Baudry *et al.*, 1988; Lett *et al.*, 1989). This suggests that the translation of the plasmid genes may be inefficient, or that their products are unstable. An alternative explanation may be that the *vacB* product is required for the assembly or translocation across the cytoplasmic membrane of other Vir proteins. This latter explanation is plausible because the VirG protein is located in the outer membrane and the IpaB, IpaC and IpaD proteins are excreted outside the bacteria from the periplasmic space (C. Sasakawa, unpublished).

E. A chromosomal locus, kcpA *required for the expression of the* virG *gene at a step after transcription*

By intergeneric conjugation between *E. coli* HfrH and *S. flexneri*, a locus designated *kcpA*, essential for keratoconjunctivitis provocation in guinea pigs or in rabbits was identified near *purE* (Formal *et al.*, 1971) (Fig. 2). The expression of an immunodominant antigen of 140 kD encoded by *S. flexneri* 5 is under the regulatory control of the chromosomal *kcpA* locus (Pal *et al.*, 1989). A *kcpA⁻* mutant constructed by intergeneric conjugation produced a reduced amount of the VirG protein (M. Yoshikawa, unpubl.). Like the $virG^-$ mutant, the $kcpA^-$ mutant is Inv$^+$ and Pcr$^+$, but is incapable of intra- and intercellular spread, resulting in Ser$^-$ phenotype (Makino et al., 1986; Bernardini et al., 1989; Pal et al., 1989; Yamada et al., 1989). The localized deposition of F-actin trailing from one pole of the bacteria was also greatly reduced (M. Yoshikawa, unpubl.). By selecting for the ability of an S. flexneri chromosomal gene library to complement the KcpA$^-$ phenotype, the putative KcpA$^-$ gene was cloned and sequenced (Yamada et al., 1989). The VirG protein was produced when the putative kcpA clone was intoduced into the $kcpA^-$ mutant. Since the virG-specific mRNA was normally produced by the wild type, the $kcpA^-$ mutant and the transformant with the putative kcpA clone, it was concluded that the kcpA gene is responsible for the full expression of virG at a step after transcription (M. Yoshikawa, unpubl.) (Table 1 and Fig. 3). The sequence of the cloned kcpA gene has subsequently been shown to be identical to the C-terminal sequence of osmZ (virR or hns) located at 27.5 min on the E. coli K-12 chromosome (Hulton et al., 1990) (Fig. 2). Although cloning of the functional osmZ gene into a multicopy vector was difficult (Hulton et al., 1990), the putative kcpA gene could be cloned presumably because of its mutation leading to a different translation initiation codon and to a product smaller than the osmZ product. It has been suggested by Hulton et al. (1990) that a truncated osmZ gene was cloned and the product complemented the $kcpA^-$ lesion. Recently, the kcpA locus, genetically mapped at approximately 12 min near purE, was localized on the NotI-C fragment of the chromosomal NotI map of S. flexneri 2a YSH6000, and the osmZ gene linked to trp was found to be on the NotI-I fragment, located more than 10 min away from the NotI-C fragment (Okada et al., 1991b) (Fig. 2).

IV. Environmental effects on regulation

A. Regulation by temperature

The Vir phenotype of shigellae is regulated by temperature; they are Inv$^+$ and fully virulent when grown at 37°C but not when grown at 30°C (Maurelli et al., 1984). Maurelli and Sansonetti (1988) proposed that the expression of the plasmid-coded vir genes was repressed at 30°C by a trans-acting repressor-like regulator. To confirm this, they constructed lacZ operon fusions with either ipaBCD or some other uncharacterized Inv-associated operons on the large plasmid of S. flexneri 2a strain 2457T, to place the expression of lacZ under the control of temperature. Using these strains carrying a lacZ operon fusion on the plasmid they isolated chromosomal Tn10 insertion mutants which expressed a Lac$^+$ phenotype at both 37°C and 30°C constitutively (Maurelli and Sansonetti, 1988; Hromockyj and Maurelli, 1989). As expected, upon transduction of such a chromosomal Tn10 insertion mutation into the wild type strain carrying an intact plasmid, the transductant became Inv$^+$ even when grown at 30°C. The mutated gene, designated *virR*, was located near the *galU* gene at 28 min on the chromosome map (Fig. 2). Dorman *et al.* (1990) have shown that the *virR* gene is identical to the *osmZ* (*hns*) gene of *E. coli* (Fig. 2), which mediates its regulatory effect through changes in DNA supercoiling (Higgins *et al.*, 1988). According to their proposal, a thermal change causes conformational alteration of DNA sequences involved in the expression of the plasmid-coded *vir* genes.

To find the target gene(s) on the large plasmid which mediate this thermoregulation, Tobe *et al.* (1991) investigated the effects of temperature on transcription, and found that the activation of *virB* depended upon temperature much more than that of *virF*. By increasing *virF* transcription, the transcription of *virB* was activated at 37°C much more efficiently than at 30°C, whereas levels of the transcription of the *ipa* and two other Inv-associated genes under varying levels of *virB* transcription were not affected by temperature at all. As described above, the activation by IPTG of *virB* gene transcription in a *virF⁻* mutant containing a *tac-virB* construct resulted in deregulation of the temperature-dependent Inv phenotypes, indicating that the temperature-regulated Inv phenotypes are mediated through the transcriptional activation of the *virB* gene on the large plasmid (Tobe *et al.*, 1991) (Table 1 and Fig. 3). In the 44 kb cosmid clone containing the *virB* gene but not the *virF* gene, the expression of the Ipa proteins was temperature-regulated (Maurelli *et al.*, 1985). The LacZ⁺ phenotype expressed by *invE* (*virB*)::Tn3-*lac* fusion was dependent not only on *virF* function but also on temperature (Watanabe *el al.*, 1990). These observations are consistent with the interpretation of Tobe *et al.* (1991) that regulation by temperature is mediated directly through the transcriptional activation of the *virB* gene (Fig. 3). It is still not understood how the *virB* gene is activated by temperature and the *virF* gene.

B. Regulation by osmolarity

The *malA*-linked *ompB* (*envZ-ompR*) locus restored the Ser⁺ phenotype to a spontaneous avirulent colonial variant M90X of *S. flexneri* 5 (Bernardini *et al.*, 1990) (Table 1, Fig. 2 and Fig. 3). The *envZ* gene product is a transmembrane osmolarity sensor that transmits the signal to the *ompR* gene product, resulting in modulation of the transcription of the porin protein genes, *ompF* and *ompC* (Mizuno and Mizushima, 1990). The expression of the plasmid-coded *vir* genes as judged by the β-galactosidase activity expressed from a *lac* fusion construct was markedly enhanced under conditions of high osmolarity, such as are encountered in the colon. By using the *lac* fusion construct, the *envZ* mutation was shown to decrease the expression of the *vir* genes in both low and high osmolarity conditions, but did not affect the expression in response to osmotic changes (Bernardini *et al.*, 1990). Deletions in *ompR* severely impaired virulence and abolished the expression of the *vir* genes in both low and high osmolarity conditions (Bernardini *et al.*, 1990). Thus, it can be speculated that the products of the two component *envZ-ompR* regulatory loci modulate the expression of the plasmid-encoded Inv phenotype when exposed to the hypertonic contents of the colon (Bernardini *et al.*, 1990).

V. Concluding remarks

The pathogenesis of shigellosis is a complex process which requires numerous genes on the large plasmid as well as on the chromosome. The product of each of these genes is not simultaneously required during all stages of infection and hence are produced in a defined amount at a critical time in the infection process, responding to stimuli from the surrounding environment. Although investigation of the molecular pathogenesis of shigellosis is relatively advanced, the full characterization of the genes involved is still in progress and a number of questions concerning the molecular basis of certain steps in the infection process still remain.

Acknowledgements

The authors acknowledge Mr. Ruari Mac Siomoin for the critical reading of this manuscript. The studies from the authors' laboratory have been supported by grants provided by the Ministry of Education, Science and Culture, the Japanese Government (59480159, 61440035, 62304036, 01440031, 03304030 and 03557022).

References

Adler B, Sasakawa C, Tobe T, Makino S, Komatsu K, & Yoshikawa M (1989) A dual transcriptional activation system for the 230 kb plasmid genes coding for virulence-associated antigens of *Shigella flexneri*. Mol Microbiol 3: 627-635.

Allaoui A, Mounier J, Prevost MC, Sansonetti PJ, & Parsot C (1992) *icsB*: a *Shigella flexneri* virulence gene necessary for the lysis of protrusions during intercellular spread. Mol Microbiol 6: 1605-1616.

Andrews GP, Hromockyj AE, Coker C, & Maurelli AT (1991) Two novel virulence loci, *mxiA* and *mxiB*, in *Shigella flexneri* 2a facilitate excretion of invasion plasmid antigen. Infect Immun 59: 1997-2005.

Baudry B, Maurelli AT, Clerc P, Sadoff JC, & Sansonetti PJ (1987) Localization of plasmid loci necessary for the entry of *Shigella flexneri* into HeLa cells, and characterization of one locus encoding four immunogenic polypeptides. J Gen Microbiol 133: 3403-3413.

Baudry B, Kaczorek AT, Clerc P, Sadoff JG, & Sansonetti PJ (1988) Nucleotide sequence of the invasion plasmid antigen B and C genes (*ipaB* and *ipaC*) of *Shigella flexneri*. Microb Pathog 4: 345-357.

Bernardini ML, Mounier J, d'Hauteville H, Coquis-Rondon M, & Sansonetti PJ (1989) Identification of *icsA*, a plasmid locus of *Shigella flexneri* that governs intra- and intercellular spread through interaction with F-actin. Proc Natl Acad Sci USA 88: 3867-3871.

Bernardini ML, Fontane A, & Sansonetti PJ (1990) The two component regulatory system OmpR-EnvZ controls the virulence of *Shigella flexneri*. J. Bacteriol 172: 6274-6281.

Buysse JM, Stover CK, Oaks EV, Venkatesan M, & Kopecko DJ (1987) Molecular cloning of invasion plasmid antigen (*ipa*) genes from *Shigella flexneri*: analysis of *ipa* gene products and genetic mapping. J Bacteriol 169: 2561-2569.

Buysse JM, Venkatesan MM, Mills JA, & Oaks EV (1990) Molecular characterization of a *trans*-acting, positive effector (*ipaR*) of invasion plasmid antigen synthesis in *Shigella flexneri* serotype 5. Microb Pathog 8: 197-211.

Caron J, Coffield LM, & Scott JR (1989) A plasmid-encoded regulatory gene, *rns*, required for expression of CS1 and CS2 adhesins of enterotoxigenic *Escherichia coli*. Proc Natl Acad Sci USA 86: 963-967.

Cornelis GR, Biot T, Lambert de Rouvroit C, Michiels T, Mulder B, Sluiters C, Sory MP, van Bouchaute M, & Vanooteghem JC (1989) The *Yersinia yop* regulon. Mol Microbiol 3: 1455-1459.

Dorman JC, Bhriain, NN, & Higgins CF (1990) DNA supercoiling and environmental regulation of virulence gene expression in *Shigella flexneri*. Nature (London) 344: 789-792.

Falkow S, Schneider H, Baron LS, & Formal SB (1963) Virulence of *Escherichia-Shigella* genetic hybrids in guinea pig. J Bacteriol 86: 1251-1258.

Formal SB, Gemski PJr, Baron LS, & LaBrec EH (1971) A chromosomal locus which controls the ability of *Shigella flexneri* to evoke keratoconjunctivitis. Infect Immun 3: 73-79.

Formal SB, LaBrec EH, Palmer A, & Falkow S (1965) Restoration of virulence to a strain of *Shigella flexneri* by mating with *Escherichia coli*. J Bacteriol 80: 835-838.

Gemski PJr, Sheahan DG, Washington O, & Formal SB (1972) Virulence of *Shigella flexneri* expressing *Escherichia coli* somatic antigens. Infect Immun 6: 104-111.

Hale TL, Oaks EV, & Formal SB (1985) Identification and antigenic characterization of virulence-associated, plasmid-coded proteins of *Shigella* spp. and enteroinvasive *Escherichia coli*. Infect Immun 50: 620-629.

Higgins CF, Dorman CJ, Stirling DA, Waddel L, Booth IR, May G, & Bremer E (1988) A physiological role for DNA supercoiling in the osmotic regulation of gene expression in *S. typhimurium* and *E. coli*. Cell 52: 569-584.

Hromockyj AE, & Maurelli AT (1989) Identification of *Shigella* invasion genes by isolation of temperature-regulated *inv::lacZ* operon fusions. Infect Immun 57: 2963-2970.

Hulton CS, Seirafi A, Hinton JC, Sidebotham JM, Waddell L, Pavitt GD, Owen-Hughes T, Spassky A, Bue H, & Higgins CF (1990) Histon-like protein H1 (H-NS): DNA supercoiling and gene expression in bacteria. Cell 63: 631-642.

Kato JI, Ito K, Nakamura A, & Watanabe H (1989) Cloning of regions required for contact hemolysis and entry into LLC-MK2 cells from *Shigella sonnei* form I plasmid: *virF* is a positive regulator for these phenotypes. Infect Immun 57: 1391-1398.

Kopecko DJ, Washington O, & Formal SB (1980) Genetic and physical evidence for plasmid control of *Shigella sonnei* form I cell surface antigen. Infect Immun 29: 207-214.

LaBrec EH, Schneider H, Magnani TJ, & Formal SB (1964) Epithelial cell penetration as an essential step in the pathogenesis of bacillary dysentery. J Bacteriol 88: 1503-1518.

Lett MC, Sasakawa C, Okada N, Sakai T, Makino S, Yamada M, Komatsu K, & Yoshikawa M. (1989) *virG*, a plasmid-coded virulence gene of *Shigella flexneri*: identification of the VirG protein and determination of the complete coding sequence. J Bacteriol 171: 353-359.

Makino S, Sasakawa C, Kamata K, Kurata T, & Yoshikawa M (1986) A genetic determinant required for continuous reinfection of adjacent cells on large plasmid in *S. flexneri* 2a. Cell 46: 551-555.

Maurelli AT, Baudry B, d'Hauteville H, Hale TL, & Sansonetti PJ (1985) Cloning of plasmid DNA sequences involved in invasion of HeLa cells by *Shigella flexneri*. Infect Immun 49: 164-171.

Maurelli AT, Blackmon B, & Curtiss RIII (1984) Temperature-dependent expression of virulence genes in *Shigella* species. Infect Immun 43: 195-201.

Maurelli AT, & Sansonetti PJ (1988) Identification of a chromosomal gene controlling temperature-regulated expression of *Shigella* virulence. Proc Natl Acad Sci USA 85: 2820-2824.

Mills JA, Buysse JM, & Oaks EV (1988) *Shigella flexneri* invasion plasmid antigens B and C: epitope location and characterization by monoclonal antibodies. Infect Immun 56: 2933-2941.

Mizuno T, & Mizushima S (1990) Signal transduction and gene regulation through the phosphorylation of two regulatory components; the molecular basis for the osmotic regulation of the porin genes. Mol Microbiol 4: 1077-1082.

Nakata N, Sasakawa C, Okada N, Tobe T, Fukuda I, Suzuki T, Komatsu K, & Yoshikawa M (1992) Identification and characterization of virK, a virulence-associated large plasmid gene essential for intercellular spreading of Shigella flexneri. Mol Microbiol 6: 2387-2395.

Okada N, Sasakawa C, Tobe T, Yamada M, Nagai S, Talukder KA, Komatsu K, Kanegasaki S, & Yoshikawa M (1991a) Virulence-associated chromosomal loci of Shigella flexneri identified by random Tn5 insertion mutagenesis. Mol Microbiol 5: 187-195.

Okada N, Sasakawa C, Tobe T, Talukder KA, Komatsu K, & Yoshikawa M (1991b) Construction of a physical map of the chromosome of Shigella flexneri 2a and the direct assignment of nine virulence-associated loci identified by Tn5 insertions. Mol Microbiol 5: 2171-2180.

Pal T, Newland JM, Tall BD, Formal SB, & Hale TL (1989) Intracellular spread of Shigella flexneri associated with the kcpA locus and a 140-kilodalton protein. Infect Immun 57: 477-486.

Ramos JL, Rojo F, Zhou L, & Timmis KN (1990) A family of positive regulators related to the Pseudomonas putida TOL plasmid xylS and the Escherichia coli araC activators. Nucl Acids Res 18: 2149-2152.

Sakai T, Sasakawa, C, Makino S, Kamata K, & Yoshikawa M (1986a) Molecular cloning of a genetic determinant for Congo red binding ability which is essential for the virulence of Shigella flexneri. Infect Immun 51: 476-482.

Sakai T, Sasakawa C, Makino S, & Yoshikawa M (1986b) DNA sequence and product analysis of the virF locus responsible for Congo red binding and cell invasion in Shigella flexneri 2a. Infect Immun 54: 395-402.

Sakai T, Sasakawa C, & Yoshikawa M (1988) Expression of four virulence antigens of Shigella flexneri is positively regulated at the transcriptional level by the 30 kiloDalton virF protein. Mol Microbiol 2: 589-597.

Sansonetti PJ, Kopecko DJ, & Formal SB (1981) Shigella sonnei plasmids: evidence that a large plasmid is necessary for virulence. Infect Immun 34: 75-83.

Sansonetti PJ, Kopecko DJ, & Formal SB (1982) Involvement of a plasmid in the invasive ability of Shigella flexneri. Infect Immun 35: 852-860.

Sansonetti PJ, Hale TL, Dammin GJ, Kapfer C, Collins HHJr, & Formal SB (1983) Alterations in the pathogenicity of Escherichia coli K-12 after transfer of plasmid and chromosomal genes from Shigella flexneri. Infect Immun 39: 1392-1402.

Sansonetti PJ, Ryter A, Clerc P, Maurelli AT, & Mounier J (1986) Multiplication of Shigella flexneri within HeLa cells: lysis of the phagocytic vacuole and plasmid-mediated contact hemolysis. Infect Immun 51: 461-469.

Sasakawa C, Kamata K, Sakai T, Murayama SY, Makino S, & Yoshikawa M (1986a) Molecular alteration of the 140-megadalton plasmid associated with the loss of virulence and Congo red binding activity in Shigella flexneri. Infect Immun 51: 470-475.

Sasakawa C, Makino S, Kamata K, & Yoshikawa M (1986b) Isolation, characterization, and mapping of Tn5 insertions into the 140-megadalton invasion plasmid defective in the mouse Sereny test in Shigella flexneri 2a. Infect Immun 54: 32-36.

Sasakawa C, Kamata K, Sakai T, Makino S, Yamada M, Okada N, & Yoshikawa M (1988) Virulence-associated genetic regions comprising 31 kilobase of the 230-kilobase plasmid in Shigella flexneri 2a. J Bacteriol 170: 2480-2484.

Sasakawa C, Adler B, Tobe T, Okada N, Nagai S, Komatsu K, & Yoshikawa M (1989) Functional organization and nucleotide sequence of virulence region-2 on the large virulence plasmid in Shigella flexneri 2a. Mol Microbiol 3: 1191-1201.

Savelkoul PH, Willshaw GA, McConell MM, Smith HR, Hamer AM, & van der Zeist BAM (1990) Expression of CFA/I fimbriae is positively regulated. Microb Pathog 8: 91-99.

Tobe T, Nagai S, Okada N, Yoshikawa M, & Sasakawa C. (1991) Temperature-regulated expression of invasion genes in Shigella flexneri is controlled through the transcriptional activation of the virB gene on the large plasmid. Mol Microbiol 5: 887-893.

Tobe T, Sasakawa C, Okada N, Honma Y, and Yoshikawa M (1992) *vacB*, a novel chromosomal gene regulating expression of the virulence genes on the large plasmid of *Shigella flexneri*. J Bacteriol 174: 6359-6367.

Venkatesan MM, Buysse JM, & Kopecko DJ (1988) Characterization of invasion plasmid antigen genes (*ipaBCD*) from *Shigella flexneri*. Proc Natl Acad Sci USA 85: 9317-9321.

Venkatesan MM, Buysse JM, & Oaks EV (1992) Surface presentation of *Shigella flexneri* invasion plasmid antigens requires the products of the *spa* locus. J Bacteriol 174: 1990-2001.

Watanabe H, Arakawa K, Ito K, Kato JI, & Nakamura A (1990) Genetic analysis of an invasion region by use of a Tn*3-lac* transposon and identifcation of a second positive regulator gene, *invE*, for cell invasion of *Shigella sonnei*: significant homology of InvE with ParB of plasmid P1. J Bacteriol 172: 619-629.

Yamada M, Sasakawa C, Okada N, Makino S, & Yoshikawa M (1989) Molecular cloning and characterization of chromosomal virulence region *kcpA* of *Shigella flexneri*. Mol Microbiol 3: 207-213.

37. Molecular genetic analysis of global regulation of extracellular enzyme synthesis in *Erwinia carotovora* subspecies *carotovora*

S.E. JONES, P. GOLBY, S.K. STEPHENS, V. MULHOLLAND,
A.R.T. COX, N. BUNCE, P.J. REEVES, M. GIBSON and G.P.C. SALMOND

Abstract. *Erwinia carotovora* subspecies *carotovora* (*Ecc*) is a phytopathogenic enterobacterium. Members of the genus *Erwinia* cause tissue maceration or soft rot diseases in a wide spectrum of plant hosts. Extracellular enzymes produced by the soft rot erwiniae include pectinases, cellulases, and proteases. Tissue maceration has been attributed largely to the pectinases. Because the exoenzymes appear to be the major pathogenicity factors, the co-ordinate regulation of such enzymes is of particular interest. Mutants pleiotropically defective in the production of extracellular enzymes have been isolated. Two cosmid clones able to complement this phenotype were obtained from an *Ecc* gene library, and the complementing DNA has been sequenced. For one mutant it has been demonstrated that cloned DNA functions as an extragenic suppressor of the mutant phenotype. The region of DNA responsible for the suppression effect has been localised and does not contain a coding region.

Abbreviations: ECC, *Erwinia corotovora* subspecies carotovora; ORF, Open Reading Frame; Rex, Regulation of Exoenzymes

Introduction

Erwinia carotovora subsp. *carotovora* (*Ecc*) is a member of the "soft rot" *Erwinia* group, which cause several plant diseases including soft rot, blackleg and leaf wilt in a variety of plant species (Perombelon, 1982; Perombelon and Kelman, 1980; Perombelon and Kelman, 1987). Soft rot erwiniae produce an array of extracellular enzymes including pectinases, cellulases, hemicellulases and proteases (Kotoujansky, 1987). Pathogenicity has been correlated to the ability specifically to produce pectinases e.g. pectate lyase (Pel) and endopolygalacturonase (Peh) which have been shown to macerate plant tissue directly (Collmer and Keen, 1986; Lei *et al.*, 1985; Mount *et al.*, 1970; Roberts *et al.*, 1986; Willis *et al.*, 1987). Although not directly responsible for tissue maceration the remaining extracellular enzymes may augment the action of the pectinases by inflicting further stress on macerated plant tissue or may provide the bacterium with carbon sources by degrading polymers.

Global regulation of extracellular enzymes has been demonstrated in other bacterial systems e.g. saprophytic *Bacillus subtilis* (Wang and Doi, 1990), animal pathogen *Staphylococcus aureus* (Peng *et al.*, 1988), and plant pathogen *Xanthomonas campestris* pathovar *campestris* (De Crecy-Lagard *et al.*, 1990). Since extracellular enzymes are major pathogenicity determinants in soft rot

C.I. Kado and J.H. Crosa (eds.), Molecular Mechanisms of Bacterial Virulence, 543–547.
© 1994 *Kluwer Academic Publishers.*

Erwinia species, it seems likely that several levels of regulation may occur. Several regulatory loci have already been identified in *Ecc* and *Erwinia chrysanthemi* (*Echr*) which are specifically involved in the regulation of pectinases, for example *pehR, pecS, pecI, pecL,* and *kdgR*. (Hugouvieux-Cotte-Pattat and Robert-Badouy, 1989; Reverchon *et al.*, 1990; Reverchon *et al.*, 1991; Saarilahti *et al.*, 1992). Global regulatory loci have also been identified which regulate production of all the extracellular enzymes, for example *aepA* (Murata *et al.*, 1991) and *expB* (Pirhonen *et al.*, 1991). Such levels of regulation may be necessary to ensure that a measured and rapid response to various environmental stimuli occurs and may be of considerable importance to the success of pathogenic invasion.

Erwinia is a useful model for the study of global regulation of pathogenicity factors since it is amenable to many *E. coli* genetic techniques.

Results

Random chemical mutagenesis using ethyl methane sulphonate (EMS) and transposon mutagenesis via Tn*phoA* (Hinton and Salmond, 1987) generated several mutants defective in the production of all the extracellular enzymes, identified by screening suspected mutants on enzyme indicator plates. Unlike classical secretion mutants (Out⁻) these mutants were defective in synthesis of the extracellular enzymes and were designated Rex⁻ (regulation of *exo*enzymes). These mutants appear phenotypically similar to the Aep⁻ (Murata *et al.*, 1991) and Exp⁻ (Pirhonen *et al.*, 1991) mutants isolated in other *Ecc* species, and are possibly similar to some of the *Erwinia carotovora* subsp. *atroseptica* mutants which show reduced virulence (Rvi⁻) on potato plants (V. Mulholland, unpub. obs.).

Analysis of mRNA isolated from wild-type and two Rex⁻ mutants, one transposon-induced and one chemically-induced, showed that exoenzyme production in these mutants is down-regulated at the level of transcription. Hence it appeared that Rex⁻ mutants may be defective in a gene which encodes a transcriptional activator.

By allelic complementation of the Rex⁻ phenotype, utilizing a wild-type *Ecc* gene library constructed in cosmid vector pHC79, two clones (clones A and B) were identified which could restore the wild-type phenotype to Rex⁻ mutants. Insert DNA from the clones did not cross-hybridize. Subcloning and sequencing of the complementing DNA revealed two open reading frames (ORF's) in insert DNA from clone A, and one ORF in insert DNA from clone B.

However, Southern blot analysis using clone A DNA as a probe showed that, for one transposon-induced mutant phenotypically complemented by clone A, the transposon was inserted in a gene which did not correspond to the cloned DNA. Further subcloning revealed that, for clone A, DNA upstream of the ORF's could mediate suppression of the Rex⁻ phenotype (Fig. 1). The presence of a repetitive DNA element has been noted in the suppressor region.

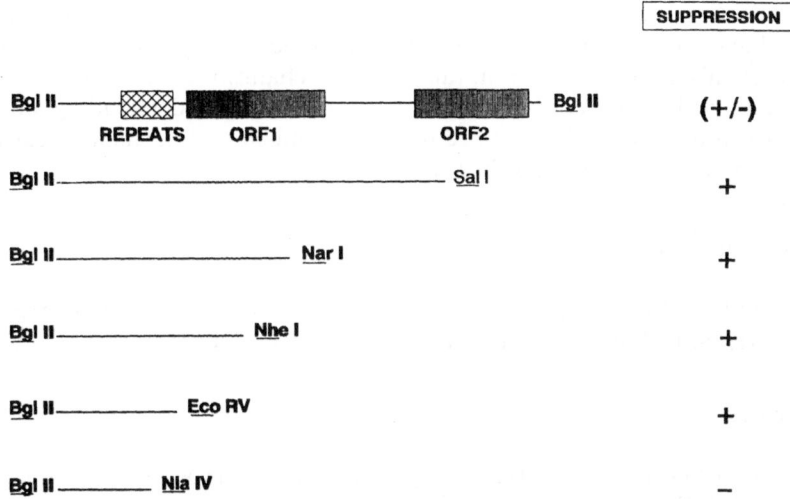

Figure 1. Subcloning of insert DNA from clone A. Suppression of the Rex-phenotype was determined by spectrophotometric enzyme assay data.

Suppression may be mediated by sequestration of a protein which normally represses transcription of the extracellular enzymes, or by sequestration of a protein which functions as a repressor of an activator of enzyme expression. Current work aims to resolve how the suppressor DNA functions (P. Golby *et al.*, in preparation).

The predicted protein product of the ORF from clone B showed no significant homologies when it was screened against the EMBL GenBank protein data-base. For one transposon-induced mutant phenotypically complemented by clone B, Southern blot analysis using clone B DNA as a probe revealed that the cloned DNA did not correspond to the transposon insertion. Hence this clone represents another suppressor of the Rex⁻ phenotype. As yet it is unknown whether the suppressor effect of this clone is due to the action of the encoded protein product or is mediated through the DNA sequence for clone B.

Recently it has been shown that exogenous addition of a small molecule, *N*-(3-oxohexanoyl)-L-homoserine lactone, can restore exoenzyme production in a subset of Rex⁻ mutants (S.E. Jones *et al.*, in press). Hence Rex⁻ mutants can be classified into distinct subsets based on this response.

The global regulatory system also responds to thermoregulation by a mechanism unknown. The synthesis of all extracellular enzymes of *Ecc* is down-regulated at several degrees below the maximal growth temperature. However preliminary experiments suggest that this effect cannot be overcome by the exogenous addition of *N*-(3-oxohexanoyl)-L-homoserine lactone. Hence the mechanism of exoenzyme thermoregulation is not directly mediated by this small molecule.

In summary, insert DNA from two clones which can suppress the Rex⁻

phenotype has been sequenced. The effect of addition of a small molecule on extracellular enzyme expression has also been noted. Current work to elucidate the mechanism of suppression, and the mechanism of action of N-(3-oxohexanoyl)-L-homoserine lactone, may enable an understanding of the molecular interactions which co-ordinate the expression of the major virulence factors (extracellular enzymes) in *Ecc*.

Acknowledgements

We thank the AFRC for generous financial support (PG88/503, PG88/513 and PG88/501). S. E. Jones was supported by an AFRC studentship.

References

Collmer, A. and Keen, N.T. 1986. The role of pectic enzymes in plant pathogenesis. Annu. Rev. Phytopathol. 24: 383–409.

De Crecy-Lagard, V., Glaser, P., Lejeune, P., Sismeiro, O., Barber, C.E., Daniels, M.J., and Danchin, A., 1990. A *Xanthomonas campestris* pv. *campestris* protein similar to catabolite activation factor is involved in regulation of phytopathogenicity. J. Bacteriol. 172: 5877–5883.

Hinton, J.C.D. and Salmond, G.P.C., 1987. Use of Tn*phoA* to enrich for extracellular enzyme mutants of *Erwinia carotovora* subsp. *carotovora* Mol. Microbiol. 1: 381–386.

Jones, S.E., Yu, B., Bainton, N.J., Birdsall, M., Bycroft, B.W., Chhabra, S.R., Cox, A.J.R., Golby, P., Reeves, P.J., Stephens, S.K.S., Winson, M.K., Salmond, G.P.C., Stewart, G.S.A.B., and Williams, P. 1993. The *lux* autoinducer regulates the production of exoenzyme virulence determinants in *Erwinia carotovora* and *Pseudomonas aeruginosa*. EMBO J. (in press).

Hugouvieux-Cotte-Pattat, N. and Robert-Badouy, J. 1989. Isolation of *Erwinia chrysanthemi* mutants altered in pectinolytic enzyme production. Mol. Microbiol. 3: 1587–1597.

Kotoujansky, A. 1987. Molecular genetics of pathogenesis by soft rot Erwinias. Annu. Rev. Phytopathol. 25: 405–430.

Lei, S.-P., Lin, H-C., Heffernan, L., and Wilcox, G. 1985. Cloning of the pectate lyase genes from *Erwinia carotovora* and their expression in *Escherichia coli*. Gene 35: 63–70.

Mount, M.S., Bateman, D.F., and Basham, H.G. 1970. Induction of electrolyte loss, tissue maceration and cellular death of potato tissue by an endopolygalcturonate *trans*-eliminase. Phytopathology 69: 117–120.

Murata, H., McEvoy, J.L., Chatterjee, A., Collmer, A., and Chatterjee, A.K. 1991. Molecular cloning of an *aepA* gene that activates production of extracellular pectolytic, cellulolytic and proteolytic enzymes in *Erwinia carotovora* subsp. *carotovora*. Mol. Plant-Microbe. Interact. 4: 239–246.

Peng, H-L., Novick, R.P., Kreiswirth, B., Kornblum, J., and Schlievert, P. 1988. Cloning characterisation and sequencing of an accessory gene regulator (*agr*) in *Staphylococcus aureus*. J. Bacteriol. 170: 4365–4372.

Perombelon, M.C.M. 1982 The impaired host and soft rot bacteria. p. 55–68. *In*: G. Lacy and M. Mount (ed.) Phytopathogenic Prokaryotes, Vol 2. Academic Press, New York.

Perombelon, M.C.M, and Kelman, A. 1980. Ecology of the soft rot Erwinias. Annu. Rev. Phytopathol., 18: 361–387.

Perombelon, M.C.M, and Kelman, A. 1987. Blackleg and other potato diseases caused by soft rot Erwinias: proposal for revision of terminology. Plant Disease, 71: 283–285.

Pirhonen, M., Saarilahti, H., Karlsson, M-B, and Palva, E.T. 1991. Identification of pathogenicity determinants of *Erwinia carotovora* subsp. *carotovora* by transposon mutagenesis. Mol. Plant-Microbe. Interact. 4: 276–283.

Reverchon, S., Hugouvieux-Cotte-Pattat, N., Condemine, G., Bourson, C., Arpin, C., and Robert-Badouy, J. 1990. Pectinolysis regulation in *Erwinia chrysanthemi*. p. 739–744. *In*: Plant Pathogenic Bacteria, Z. Klement (ed.) Budapest: Akademiai Kiado.

Reverchon, S., Nasser, W., and Robert-Badouy, J. 1991. Characterisation of *kdgR*, a gene of *Erwinia chrysanthemi* that regulates pectin degradation. Mol. Microbiol. 5: 2203–2216.

Roberts, D.P., Berman, P.M., Allen, C. Stromberg, V.K., Lacy, G.H., and Mount, M.S. 1986. Requirement of two or more *Erwinia carotovora* subsp. *carotovora* pectolytic gene products for maceration of potato tuber tissue by *Escherichia coli*. J. Bacteriol 167: 279–284.

Saarilahti, H.T., Pirhonen, M., Karlsson, M-B., Flego, D., and Palve, E.T. 1992. Expression of *pehA-bla* fusions in *Erwinia carotovora* subsp. *carotovora* and isolation of regulatory mutants affecting polygalacturonase production. Mol. Gen. Genetics. 234: 81–88.

Wang, L-F., and Doi, R-H. 1990. Complex character of *senS*, a novel gene regulating expression of extracellular protein genes of *Bacillus subtilis*. J. Bacteriol. 172: 1939–1947.

Willis, J.W., Engwall, J.K., and Chatterjee, A.K. 1987. Cloning of genes for *Erwinia carotovora* subsp. *carotovora* pectolytic enzymes and further characterisation of the polygalacturonases. Phytopathol. 77: 1199–1205.

Mechanisms against host defenses

38. Physiology of resistant interactions between *Xanthomonas oryzae* pv. oryzae and rice

JAN E. LEACH, AILAN GUO, PETER REIMERS, SEONG HO CHOI, CHRISTOPHER M. HOPKINS and FRANK F. WHITE

Abstract. Resistant interactions between *Xanthomonas oryzae* pv. *oryzae* and rice are characterized by increases in the activities of three extracellular peroxidases (two anionic and one cationic), lignin deposition, host cell death, and a decrease in the rate of bacterial multiplication. The timing and dynamics of these events is dependent on the specific avirulence gene-resistance gene interaction. In susceptible interactions, increases in peroxidase activity, lignin deposition, and host cell death are delayed, and bacterial multiplication is not inhibited. In the absence of light, the events associated with resistance do not occur, and a response similar to the susceptible response is observed.

Abbreviations: HR, Hypersensitive Reaction; LS, Mild Water Soaking; NV., Pathovar; WS, Water Soaking

Introduction

Rice, like many plant species, employs a diverse array of defenses to minimize losses during pathogen attack. For example, deposition of melanin, increases in peroxidase and lipoxygenase activities, and the accumulation of phenolic compounds, oxygenated polyunsaturated fatty acid derivatives, and the phytoalexins oryzalexin and momilactone are associated with the colonization of rice leaves by *Magnaporthe grisea*, the blast pathogen (Sridhar and Ou, 1974; Toyoda and Suzuki, 1960; Li *et al.*, 1991; Ohta *et al.*, 1991). Phenolic compounds toxic to *X. oryzae* pv. *oryzae* (Ishiyama) Dye (Swings *et al.*, 1990), the bacterial blight pathogen of rice, also are found in greater amounts in healthy leaves of resistant cultivars compared to healthy leaves of susceptible cultivars (Horino and Kaku, 1989). Phytoalexins are thought to be produced in response to infection with some strains of *X. oryzae* pv. *oryzae* (Nakanishi and Watanabe, 1977). Beyond these observations, however, little was known about the physiology of defense responses in this important monocot.

For the past several years, we have been investigating the induction of resistance in rice to avirulent strains of *Xanthomonas oryzae* pv. *oryzae*. This work has been facilitated by the availability of near-isogenic rice cultivars that contain single bacterial blight resistance genes (Ogawa *et al.*, 1988), and the recent cloning and characterization of two avirulence genes (*avrXa7* and *avrXa10*) from *X. oryzae* pv. *oryzae* (Hopkins *et al.*, 1992). *X. oryzae* pv. *oryzae* strains containing the cloned avirulence genes acquire the ability to elicit

C.I. Kado and J.H. Crosa (eds.), Molecular Mechanisms of Bacterial Virulence, 551–560.
© 1994 *Kluwer Academic Publishers.*

resistance when inoculated to rice cultivars with the corresponding resistance genes (*Xa-7* and *Xa-10*). The avirulence genes from *X. oryzae* pv. *oryzae* are members of a family of avirulence genes from *Xanthomonas* that are typified by the *avrBs3* gene cloned from *X. campestris* pv. *vesicatoria* (Bonas *et al.*, 1989). In this report, we summarize our current knowledge of the physiological events that occur during race-specific interactions between *X. oryzae* pv. *oryzae* and rice.

Defense responses in interactions between *X. oryzae* pv. oryzae and rice

Deposition of lignin in incompatible interactions

Antibacterial compounds have been isolated from healthy leaves of susceptible- and resistant-rice cultivars (Horino & Kaku, 1989) and from leaves after exposure to avirulent strains of *X. oryzae* pv. oryzae (Nakanishi & Watanabe, 1977). Some of the antibacterial compounds were oxidized lignin components with aldehyde and phenol groups (Horino & Kaku, 1989). We observed that bacteriostasis was correlated with the early (24 h post-infiltration) accumulation of bright yellow-green fluorescent compounds and host cell death (24–48 h post-infiltration) in the incompatible interactions (Reimers & Leach, 1991). In the compatible interactions, bacterial multiplication was not inhibited and host cell death and the accumulation of the fluorescent compounds were not observed until late in the interaction (48–72 h after infiltration). Furthermore, we demonstrated that lignin-like polymers accumulated in inoculated leaves during the incompatible interaction between rice cultivars carrying the *Xa-5*, *Xa-7*, and *Xa-10* genes for bacterial blight resistance and strains of *X. oryzae* pv. oryzae carrying the corresponding avirulence genes (Reimers and Leach 1991; Guo and Leach, unpublished). [We have since demonstrated that the polymers were indeed lignin (J.E. Leach & R. Hammerschmidt, unpublished] The spatial and temporal patterns of phenolic polymer deposition were correlated with the accumulation of fluorescent compounds, host cell death, a decrease in bacterial multiplication rates and the onset of bacteriostasis in the incompatible interaction (Reimers & Leach, 1991). In susceptible or compatible interactions, the deposition of lignin polymers did not occur, the bacteria continued to multiply, and the infiltrated leaf wilted and died.

Lignin and other phenolic polymers serve as physical barriers, and, as such, are thought to prevent fungal penetration of host cells (Ride, 1983; Tiburzy & Reisener, 1990). Bacterial plant pathogens such as *X. oryzae* pv. *oryzae* are found primarily in the vascular tissues or extracellular spaces; they do not penetrate host cells. Consequently, cell wall lignification probably would not be an effective defense against these organisms unless lignified materials prevented bacterial spread by blocking vessels or filling extracellular spaces. However, the lignin biosynthetic process itself could be an important component of the defense response. That is, we suggest that bacterial multiplication and movement may be inhibited by toxic phenolic compounds (Horino & Kaku,

1989; Venere, 1980), phenolic free radicals, or activated oxygen (Elstner, 1982), all of which are associated with lignification (Gross, 1980). Hence, we initiated an investigation of enzymes involved in lignin biosynthesis, such as the extracellular peroxidases.

Increase of peroxidase activity in incompatible interactions
Peroxidases (EC 1.11.1.7, donor: H_2O_2 oxidoreductase) are a family of isoenzymes found in all higher plants. Plant peroxidases have been implicated in regulation of cell elongation (Goldberg *et al.*, 1986), phenol oxidation (Scmid & Feucht, 1980), polysaccharide cross-linking (Fry, 1986), indole-3-acetic acid oxidation (Hinnman & Lang, 1985), cross-linking of extensin monomers (Everdeen *et al.*, 1988), and wound-healing (Espelie *et al.*, 1986). In addition, peroxidases are involved in the last enzymatic step of lignin biosynthesis, that is, the oxidation of hydroxy cinnamyl alcohols into free radical intermediates, which subsequently are coupled into the lignin polymer (Gross, 1980; Grisebach, 1981). Although a role for peroxidases in defense responses has not been clearly demonstrated, increases in peroxidase activity have been correlated with infection in many species, including cotton (Mellon & Lee, 1985; Venere, 1980), tomato (Mohan & Kolattukudy, 1990; Robb *et al.*, 1991), cucurbits (Hammerschmidt *et al.*, 1982; Smith & Hammerschmidt, 1988), rice (Sridhar & Ou, 1974; Toyoda & Suzuki, 1960), barley (Kerby & Sommerville, 1989), and wheat (Flott *et al.*, 1989; Moerschbacher *et al.*, 1988; Schweizer *et al.*, 1989; Seevers *et al.*, 1971).

After infiltration of rice cultivars containing the *Xa-10* gene for bacterial blight resistance with race 2 strains of *X. oryzae* pv. oryzae, an incompatible combination, there is an increase in total peroxidase activity in tissues exposed to the bacteria (Reimers *et al.*, 1992). This increase was correlated with several changes in the peroxidase isoenzyme profile, including the appearance of a cationic peroxidase with a pI of 8.6 and increased activities of two anionic peroxidases. Later during the interactions, total peroxidase activities increased in both compatible and control (infiltrated with water) treatments, but the final activities were less than those observed in the incompatible combination. Similarly, the activities of the three peroxidases were detected in all three treatments by 48 h after infiltration, but at reduced levels in compatible and water-infiltrated control treatments relative to the incompatible combination. Fig. 1 demonstrates the timing of induction of the extracellular cationic peroxidase activity. The increase in specific peroxidase isoenzyme activities was accompanied by a decline in the number of viable bacteria recovered from the inoculated tissue and an accumulation of lignin in the inoculation site. Thus, the increase in three specific peroxidase isoenzyme activities is correlated with the resistance response of rice to *X. oryzae* pv. oryzae.

Effect of light on resistant interactions
To determine if the resistant response in interactions between rice and *X. oryzae* pv. oryzae is dependent on light, we exposed infiltrated plants to different

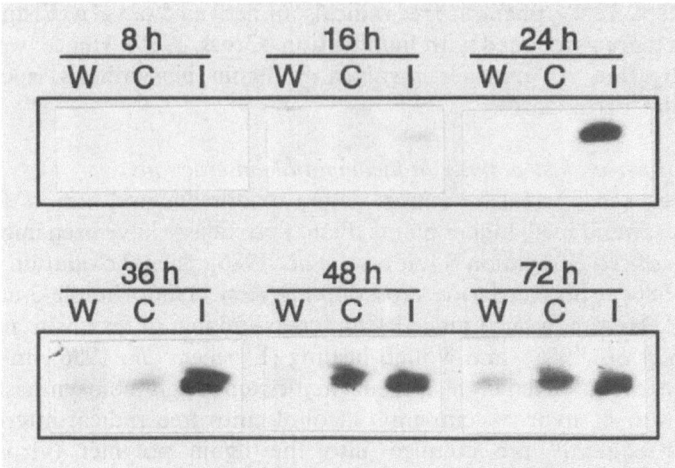

Figure 1. Accumulation of the pI 8.6 cationic peroxidase activity in extracellular extracts of Cas 209 (*Xa-10*) rice leaves infiltrated with water (W), and *X. oryzae* pv. *oryzae* strains PXO61 (C = compatible) and PXO86 (I = incompatible). Infiltrated leaves were extracted at 8, 16, 24, 36, 48, and 72 h for activity of the extracellular peroxidase as described (Reimers *et al.*, 1992). Proteins (0.2 μg/lane) were separated in nondenaturing, cathodic polyacrylamide gel with previously described buffering conditions (Thomas & Hodes, 1981). Peroxidase activity was detected with a mixture of the substrates guaiacol and 3-amino-9-ethylcarbazole (Graham *et al.*, 1965).

durations of light and darkness. In the absence of light, or if tissues are exposed to 4 h or less of light after inoculation, interactions that would normally result in resistance are altered, that is, the bacterial populations reach high levels, and tissues become watersoaked, similar to susceptible interactions (Table 1). Furthermore, the accumulation of peroxidase and lignin that are correlated with resistance are not observed in the absence of light. The presence of the plasmid containing the avirulence gene *avrXa10* (pXO5–15) in a normally virulent strain is sufficient to reduce the growth rate in the *Xa-10* cultivar, but not in a susceptible rice cultivar, if adequate light is provided (Table 1). However, in the absence of light, the effects of the cloned avirulence gene are not observed, and the tissues are susceptible. Absence of light early in the compatible interactions (PXO99) has little or no apparent effect on the phenotype of the response. These results suggest that light is essential for defence responses in rice plants.

The timing of defensive responses is dependent on specific avirulence gene/resistance gene combinations

To determine if the defensive responses induced in rice are the same with different avirulence/resistance gene combinations, we introduced the avirulence genes *avrXa7* and *avrXa10* individually into a normally virulent strain of *X. oryzae* pv. *oryzae* and investigated their effects in near-isogenic rice cultivars carrying *Xa-7* and *Xa-10*, respectively. The phenotypes of the interactions

Table 1. Effects of light on symptoms, accumulation of extracellular cationic peroxidase activity, deposition of lignin, and final bacterial numbers in interactions between strains of *X. oryzae* pv. oryzae and rice cultivars with the *Xa-10* bacterial blight resistance gene.

Treatment[a]	Symptoms[b]	Cationic peroxidase[c]	Lignification[d]	Bacteria[e] (log cfu/leaf)
PXO86 (incompatible)				
4 h light/20 h dark	LS	+	+	8.7
24 h light	HR	+ + + +	+ + + +	7.9
4 h dark/20 h light	HR	+ + + +	+ + + +	7.3
24 h dark	LS	+	+	8.9
PXO99 (compatible)				
24 h light	WS	-	+	9.1
24 h dark	WS	-	+	9.4
PXO99(pXO5-15)[f] (incompatible)				
4 h light/20 h dark	WS	+	+	8.6
24 h light	HR	+ + + +	+ + + +	7.3
4 h dark/20 h light	HR	+ + +	+ + +	7.4
24 h dark	WS	-	+	8.7

[a]Infiltrated plants (Reimers & Leach, 1991) were placed in growth chambers and exposed to light or darkness for the duration indicated. After the first 24 h treatment, the plants were returned to their normal 12 h light/12 h dark cycle for the remainder of the experiment.
[b]Symptoms (LS = mild watersoaking; WS = watersoaking; HR = hypersensitive response) were assessed at 24 h after infiltration. Tissues with LS symptoms were watersoaked at 48 h after infiltration.
[c]The activity of the cationic peroxidase in extracellular extracts from infiltrated rice leaves was assessed at 24 h after infiltration as described (Reimers et al., 1992). Estimates of activity are based on intensity of staining in native cathodic gels; - = no activity, + + + + = maximum activity detected.
[d]Lignification was determined at 48 h after infiltration by the Weisner test as described by Beardmore et al. (1983). Estimates of activity are based on visual assessment of color intensity in stained leaves; + = little color, + + + + = maximum color detected.
[e]Bacterial numbers were measured at 96 h after infiltration as described (Reimers & Leach, 1991) and are reported as log cfu/leaf.
[f]Plasmid pXO5-15 contains the *avrXa10* gene, which confers on *X. oryzae* pv. *oryzae* the ability to elicit a resistance response in cultivars with *Xa-10* (Hopkins et al., 1992).

between rice and *X. oryzae* pv. oryzae varied with the resistance gene involved in the interaction. Interactions involving *avrXa10* and *Xa-10* resulted in a dark brown color throughout the infiltrated site within 24–48 h, whereas in combinations with *avrXa7* and *Xa-7*, a dark ring formed around the perimeter of the watersoaked site at 48 h, and the tissue within the site became tan to brown after 72 h. Compatible interactions in both combinations remained watersoaked in the infiltration site through 5 days, after which time the watersoaked lesion had spread, and the leaf had wilted.

Incompatible interactions involving cultivars carrying *Xa-10* were characterized by an increase in the extracellular cationic peroxidase and the deposition of lignin within 16–24 h, which coincided with a decrease in the rate of bacterial multiplication (Reimers *et al.*, 1992; Hopkins *et al.*, 1992; Guo & Leach, unpublished). In incompatible interactions with *avrXa7* and *Xa-7*, the increase in cationic peroxidase activity, lignin deposition, and the decrease in the rate of bacterial multiplication were delayed (48 h) when compared to *Xa-10* (Hopkins *et al.*, 1992; Guo & Leach, unpublished). Since the avirulence genes were contained in the same strain of *X. oryzae* pv. oryzae, i.e., the same genetic background, and the host resistance genes were near-isogenic in IR24, we conclude that the timing of the physiological and phenotypical reactions is dependent on the specific avirulence-resistance gene combination and that the increases in peroxidase activities, deposition of lignin, and browning of tissues in incompatible interactions are likely controlled by the specific host resistance genes.

Role of peroxidases in resistance

Although extensively studied, the role of plant peroxidases in growth, development, or response to environmental or pathogen stress is still not understood (Gaspar *et al.*, 1982). This is primarily because studies are complicated by the large number of isoenzyme forms, the high catalytic activity of the isoenzymes, and the enzyme's lack of substrate specificity. Studies are further complicated because many of the reaction products are toxic, and the activity of the enzyme is usually tightly controlled at the level of gene expression or through cellular localization and substrate availability. One approach to understand the role of a specific peroxidase and to avoid some of the problems inherent in investigations of isoenzyme families is to generate plants with modifications in a single peroxidase. Molecular analyses of several peroxidases have been initiated in recent years. Peroxidase genes have been cloned and sequenced (Dudler *et al.*, 1991; Intapruk *et al*, 1991; Ito *et al.*, 1991; Hertig *et al.*, 1991; Lagrimini *et al.*, 1987; Mazza & Welinder, 1980; Roberts *et al.*, 1988; Schweizer *et al.*, 1989). Lagrimini *et al.* (1990, 1991) demonstrated an anionic peroxidase from tobacco under the control of the 35S CaMV promoter exhibited wound-inducible accumulation of the peroxidase in transgenic tobacco. The transgenic tobacco plants that overexpressed peroxidase were characterized by excessive polymerization of polyphenols and rapid lignification of pith tissues after wounding; resistance of these tissues to disease was not discussed. The expression of a tomato anionic peroxidase in tobacco has been reported to confer resistance to *Phytophthora parasitica*, although the details of this work have not yet been published (Kolattukudy *et al.*, 1992; Kolattukudy, 1992).

To determine if peroxidases play a role in resistance in rice, we have begun a molecular characterization of the cationic, extracellular peroxidase induced during the incompatible interactions. Although the isoenzyme has a very high activity, the concentrations of the protein in plant tissues or in extracellular fluids are very low, which complicated attempts to purify significant quantities of the protein. Sufficient quantities were purified for N-terminal sequencing; however, the N-terminus was blocked. Sequence data has been obtained for three fragments of the cationic peroxidase derived by cyanogen bromide digestion (Guo & Leach, unpubl). A combination of oligonucleotide primers with sequences unique to the cationic peroxidase and primers derived from conserved sequences (based on comparison with other peroxidases) have been synthesized. PCR-amplified portions of the peroxidase gene have been isolated and are being sequenced. PCR products that match the amino acid sequence of the cationic peroxidase will be used as probes to identify candidate clones containing the cationic peroxidase sequences from a cDNA library constructed from rice induced for extracellular cationic peroxidase production (Anuratha *et al.*, 1992), and to monitor the induction of the peroxidase gene by northern hybridization analyses.

Summary

During the past several years, we have been investigating the induction of resistance in rice to avirulent strains of *X. oryzae* pv. oryzae. The cloning of avirulence genes and the development of cultivars near-isogenic for resistance to *X. oryzae* pv. oryzae has added a new dimension to our understanding of resistant interactions. We have shown that resistance is light dependent and is associated with the increased activities of three peroxidases, the deposition of lignin, host cell death, and the a decrease in the rate of bacterial multiplication. The responses induced in such "near-isogenic" interactions vary in their timing and dynamics, and are clearly the result of the interplay between avirulence genes and corresponding resistance genes.

Our current efforts in molecular characterization of the cationic, extracellular peroxidase may provide insight into the significance of this peroxidase to the defense response in rice. This, in turn, may clarify our understanding of the physiological and molecular bases of resistance in bacterial diseases of monocots such as rice.

References

Anuratha CS, Huang JK, Pingali A & Muthukrishnan S (1992) Isolation and characterization of a chitinase and its cDNA clone from rice. J Plant Biochem Biotech 1: 5–10.
Beardmore J, Ride JP & Granger JW (1983) Cellular lignification as a factor in the hypersensitive resistance of wheat to stem rust. Physiol Plant Pathol 22: 209–220.
Bonas U, Stall RE & Staskawicz BJ (1989) Genetic and structural characterization of the avirulence gene *avrBs3* from *Xanthomonas campestris* pv. *vesicatoria*. Mol Gen Genet 218: 127–136.
Dudler R, Hertig C, Rebmann G, Bull J & Mauch F (1991) Nucleotide sequence of a peroxidase-encoding wheat gene. Plant Mol Biol 16: 329–331.
Elstner EF (1982) Oxygen activation and oxygen toxicity. Annu Rev Plant Physiol 33: 73–96.
Espelie KE, Franceshci VR & Kolattukudy PE (1986) Immunocytochemical localization and time

558

course of appearance of an anionic peroxidase associated with suberization in wound-healing potato tuber tissue. Plant Physiol 81: 487–492.

Everdeen DS, Keifer S, Willard, JJ, Muldoon EP, Dey PM, Li XB & Lamport DTA (1988) Enzymic cross-linkage of monomeric extensin precursors in vitro. Plant Physiol 87: 616–621.

Flott BE, Moerschbacher BM & Reisener HJ (1989) Peroxidase isoenzyme patterns of resistant and susceptible wheat leaves following stem rust infection. New Phytol 111: 413–421.

Fry SC (1986) Cross-linking of matrix polymers in the growing cell walls of angiosperms. Annu Rev Plant Physiol 37 :165–186.

Gaspar T, Penel C, Thorpe T & Greppin H (1982) Peroxidases: A Survey of Their Biochemical and Physiological Roles in Higher Plants. University of Geneva Press, Geneva.

Goldberg R, Imberty A, Liberman M & Prat R (1986) Relationships between peroxidatic activities and cell wall plasticity. p. 208–220. In: Greppin H, Penel C & Gaspar T (eds.) Molecular and Physiological Aspects of Plant Peroxidases. University of Geneva, Geneva, Switzerland.

Graham RC Jr, Lundholm U & Karnovsky MJ (1965) Cytochemical demonstration of peroxidase activity with 3-amino-9-ethylcarbazole. J Histochem Cytochem 13: 150–152.

Grisebach H (1981) Lignins. p. 457–478. In: Conn EE (ed.) The Biochemistry of Plants. Academic Press, New York.

Hammerschmidt R, Nuckles E & Kuc J (1982) Association of enhanced peroxidase activity with induced systemic resistance of cucumber to Colletotrichum lagenarium. Physiol Plant Pathol 20: 73–82.

Gross GG (1980) The biochemistry of lignification. Adv Bot Res 8: 25–63.

Herbers K, Conrads-Strauch J & Bonas U (1992) Race-specificity of plant resistance to bacterial spot disease determined by repetitive motifs in a bacterial avirulence protein. Nature, London 356: 172–174.

Hertig C, Regmann G, Bull J, Mauch F & Dudler R (1991) Sequence and tissue-specific expression of a putative peroxidase gene from wheat (Triticum aestivum L.). Plant Mol Biol 16: 171–174.

Hinnman RL & Lang J (1965) Peroxidase catalyzed oxidation of indole-3-acetic acid. Biochemistry 4: 144–158.

Hopkins CM, White FW, Choi SH, Guo A & Leach JE (1992) A family of avirulence genes from Xanthomonas oryzae pv. oryzae. Mol Plant-Microbe Interact 5: 451–459.

Horino O & Kaku H (1989) Defense mechanisms of rice against bacterial blight caused by Xanthomonas campestris pv. oryzae. p. 135–152. In: Bacterial Blight in Rice. International Rice Research Institute, Los Baños, Philippines.

Intapruk C, Higashimura N, Yamamoto K, Okada N, Shinmyo A & Takano M (1991) Nulceotide sequences of two genomic DNAs encoding peroxidase from Arabidopsis thaliana. Gene 98: 237–241.

Ito H, Hiraoka N, Ohbayashi A & Ohashi Y (1991) Purification and characterization of rice peroxidases. Agric Biol Chem 55: 2445–2454.

Kerby K & Somerville SC (1989) Enhancement of specific intercellular peroxidases following inoculation of barley with Erysiphe graminis f. sp. hordei. Physiol Mol Plant Pathol 35: 323–337.

Kolattukudy P (1992) Plant-fungal communication that triggers genes for breakdown and reinforcement of host defensive barriers. p. 65–83. In: Verma DP (ed.) Molecular Signals in Plant-Microbe Communication. CRC Press, Boca Raton, FL.

Kolattukudy PE, Mohan R, Bajar MA & Sherf BA (1992) Plant oxygenases, peroxidases and oxidases. Biochem. Soc. Transact. 20: 333–337.

Lagrimini LM (1991) Wound-induced deposition of polyphenols in transgenic plants overexpressing peroxidase. Plant Physiol 96: 577–583.

Lagrimini LM, Burkhart W, Moyer M & Rothstein S (1987) Molecular cloning of complementary DNA encoding the lignin-forming peroxidase from tobacco: Molecular analysis and tissue-specific expression. Proc Natl Acad Sci USA 84: 7542–7546.

Li WX, Kodma O & Akatsuka T (1991) Role of oxygenated fatty acids in rice phytoalexin production. Agri Biol Chem 55: 1041–1047.

Mazza G, Welinder KG (1980) Covalent structure of turnip peroxidase 7. Cyanogen bromide fragments, complete structure and comparison to horseradish peroxidase C. Eur J Biochem 108: 481–489.

Mellon JE & Lee LS (1985) Elicitation of cotton isoperoxidases by *Aspergillus flavus* and other fungi pathogenic to cotton. Physiol Plant Pathol 27: 281–288.

Moerschbacher BM, Noll UM, Flott BE & Reisener HJ (1988) Lignin biosynthetic enzymes in stem rust infected, resistant and susceptible near-isogenic wheat lines. Physiol Mol Plant Pathol 33: 33–46.

Mohan R & Kolattukudy PE (1990) Differential activation of expression of a suberization-associated anionic peroxidase gene in near-isogenic resistant and susceptible tomato lines by elicitors of *Verticillium albo-atrum*. Plant Physiol 92: 276–280.

Nakanishi K & Watanabe M (1977) Studies on the mechanisms of resistance of rice plants against *Xanthomonas oryzae*. IV. Extraction and partial purification of antibacterial substances from infected leaves. Ann Phytopathol Soc Japan 43: 449–454.

Ogawa T, Yamamoto T, Khush GS, Mew TW & Kaku H (1988) Near-isogenic lines as international differentials for resistance to bacterial blight of rice. Rice Genetics Newsletter 5: 106–107.

Ohta H, Shida K, Peng YL, Furusawa I, Shishiyama J, Aibara S & Morita Y (1991) A lipoxygenase pathway is activated in rice after infection with the rice blast fungus, *Magnaporthe grisea*. Plant Physiol 97: 94–98.

Reimers PJ, Guo A & Leach JE (1992) Increased activity of a cationic peroxidase associated with incompatible interactions between *Xanthomonas oryzae* pv. *oryzae* and rice (*Oryza sativa*). Plant Physiol 99: 1044–1050.

Reimers PJ & Leach JE (1991) Race-specific resistance to *Xanthomonas oryzae* pv. *oryzae* conferred by bacterial blight resistance gene *Xa-10* in rice (*Oryza sativa*) involves accumulation of a lignin-like substance in host tissues. Physiol Mol Plant Pathol 38: 39–55.

Ride JP (1983) Cell walls and other structural barriers in defence. p. 215–236. *In*: Callow JA (ed.) Biochemical Plant Pathology. Wiley-Interscience, New York.

Robb J, Lee SW, Mohan R & Kolattukudy PE (1991) Chemical characterization of stress-induced vascular coating in tomato. Plant Physiol 97: 528–536.

Roberts E, Kutchan T & Kolattukudy PE (1988) Cloning and characterization of cDNA for a highly anionic peroxidase from potato and the induction of its mRNA in suberizing potato tubers and tomato fruits. Plant Mol Biol 11: 15–26.

Schmid PS & Feucht W (1980) Tissue-specific oxidation browning of polyphenols by peroxidase in cherry shoots. Gartenbauwissenschaft 45: 68–73.

Schweizer P, Hunziker W & Mösinger E (1989) cDNA cloning, in vitro transcription and partial sequence analysis of mRNAs from winter wheat (*Triticum aestivum* L.) with induced resistance to *Erysiphe graminis* f. sp. *tritici*. Plant Mol Biol 12: 643–654.

Seevers PM, Daly JM & Catedral FF (1971) The role of peroxidase isozymes in resistance to wheat stem rust disease. Plant Physiol 48: 353–360.

Smith JA & Hammerschmidt R (1988) Comparative study of acidic peroxidases associated with induced resistance in cucumber, muskmelon, and watermelon. Physiol Mol Plant Pathol 33: 255–261.

Sridhar R & Ou SH (1974) Biochemical changes associated with the development of resistant and susceptible types of rice blast lesions. Phytopathol Z 79: 222–230.

Swings J, Van den Mooter M, Vauterin L, Hoste B, Gillis M, Mew TW & Kersters K. (1990) Reclassification of the causal agents of bacterial blight (*Xanthomonas campestris* pv. *oryzae*) and bacterial leaf streak (*Xanthomonas campestris* pv. *oryzicola*) of rice as pathovars of *Xanthomonas oryzae* (ex Ishiyama 1922) sp. nov., nom. rev. Int J Syst Bacteriol 40: 309–311.

Thomas JM & Hodes ME (1981) A new discontinuous buffer system for the electrophoresis of cationic proteins at near-neutral pH. Anal Biochem 118: 194–196.

Tiburzy R & Reisener HJ (1990) Resistance of wheat to *Puccinia graminis* f. sp. *tritici*: Association of the hypersensitive reaction with the cellular accumulation of lignin-like material and callose. Physiol Mol Plant Pathol 36: 109–120.

Toyoda S & Suzuki N (1960) Histochemical studies on rice blast lesions. IV. Changes in the activity of oxidases in infected tissue. Ann Phytopathol Soc Japan 25: 172–177.

Urs NVR & Dunleavy JM (1975) Enhancement of the bactericidal activity of a peroxidase system by phenolic compounds. Phytopathology 65: 686–690.

Van Huystee RB & Cairns WL (1990) Appraisal of studies on induction of peroxidase and associated porphyrin metabolism. Bot Rev 46: 429–446.

Venere RJ (1980) Role of peroxidase in cotton resistant to bacterial blight. Plant Sci Lett 20: 47–56.

Whalen MC, Stall RE & Staskawicz BJ (1988) Characterization of a gene from a tomato pathogen determining hypersensitive resistance in non-host species and genetic analysis of this resistance in bean. Proc Natl Acad Sci USA 85: 6743–6747.

39. Mechanisms for the virulence of *Nocardia*

BLAINE L. BEAMAN

Abstract. *Nocardia* is a Gram positive, strictly aerobic actinomycete found in soil, water and on vegetable matter. Pathogenic species cause a variety of acute and chronic diseases in humans and animals with the lungs and skin being primary sites for these infections. Virulent strains of *Nocardia* are facultative intracellular pathogens that are not killed by normal macrophages, monocytes or polymorphonuclear neutrophils. The mechanisms whereby these organisms avoid being killed are multiple and complex, and include: 1) the ability to inhibit phagosome-lysosome fusion; 2) neutralize phagosomal acidification; 3) modulate lysosomal enzymes; and, 4) resist toxic oxidative metabolites. The virulence of most nocardiae is growth stage dependent, and the relative degrees of pathogenicity correlate with specific changes in the structure of components of the cell wall during the growth cycle. The most important of these is the glycolipid, trehalose dimycolate. In addition, virulent strains of *N. asteroides* secrete a unique superoxide dismutase and contain increased amounts of catalase that protect these organisms from the oxidative killing mechanisms of phagocytes. Furthermore, *Nocardia* should be considered a primary pathogen of the brain since blood borne organisms have a specific predilection for this region of the body. Studies have shown that there is specific binding or adherence to subpopulations of capillary endothelial cells in regions of the brain. Following attachment, the endothelial cells phagocytize these nocardiae and they pass through the basement membrane to enter the brain parenchyma. The nocardiae invade neurons and grow within most types of brain cells without inducing an inflammatory response. The mechanisms for adherence, invasion and growth within the brain are not known.

Abbreviations: SOD, Superoxide Dismutase

Introduction

In 1888, E. Nocard recovered a strictly aerobic, branching filamentous organism from bovine farcy on Guadeloupe Island (Nocard, 1888), and one year later, Trevisan named this organism *Nocardia farcinica* (Lechevalier, 1989). However, the first human infection was not recognized until 1890 when Eppinger isolated a strictly aerobic, branching filamentous organism from the brain of a fatal disease that was called "pseudotuberculosis" (Eppinger, 1891). Later, this organism was named *N. asteroides*, and it was ultimately assigned the status of "type species" for the genus *Nocardia* (Lechevalier, 1989). Furthermore in 1891, Eppinger reported that the organism he isolated from the fatal human infection caused the same kind of disease following injection into rabbits and guinea pigs. These observations fulfilled Koch's postulates for establishing *N. asteroides* as the etiology of the disease (Eppinger, 1891). Since

C.I. Kado and J.H. Crosa (eds.), Molecular Mechanisms of Bacterial Virulence, 561–572.
© 1994 *Kluwer Academic Publishers.*

this time, there have been thousands of cases of disease caused by the nocardiae described in humans and a large variety of animals (Beaman et al., 1976; Beaman and Sugar, 1983). Also, it has been shown that nocardiae represent a portion of the normal flora of soil, vegetable matter, fresh water, and marine environments (ocean floor) (Lechevalier, 1989). Therefore, it is generally believed that infection results from contact of a susceptible host with these organisms from the environment since nocardiae have not been found as a normal inhabitant of animal or human surfaces (Beaman, 1976; Beaman, 1983; Beaman et al., 1976).

The following species of Nocardia are considered to be pathogens for humans and animals: N. asteroides, N. farcinica, N. nova, N. brasiliensis, N. otididisca-viarum (N. caviae), and N. transvalensis (Lechevalier, 1989). Nocardia asteroides is the most frequently recognized pathogen in the United States, and N. brasiliensis is most frequently recognized as causing disease in tropical and subtropical regions of the world (Beaman et al., 1976).

Diseases caused by pathogenic Nocardia

Pathogenic species of Nocardia cause a variety of acute and chronic infections in humans and animals with the lungs, brain, and skin being the primary sites for these infections. Nocardial disease presents a wide spectrum of clinical manifestations that vary from self-limited subclinical or inapparent infection to a relentlessly chronic, destructive disseminated form of disease that defies therapeutic approaches (Beaman et al., 1976; Beaman and Sugar, 1983; Frazier et al., 1975; Hessen and Santoro, 1988; Hodges and Rossett, 1978).

In general, the diseases caused by Nocardia can be grouped into categories that are based upon the tissues involved. The most commonly recognized form of disease in the U.S. is pulmonary nocardiosis which probably results from inhalation or aspiration of nocardiae into the lungs. The clinical and pathologic presentation is variable with no distinguishable characteristics. The pulmonary disease may present as a mild or subclinical infection that is self-limited or it may progress to pneumonia. It may cause a flu-like illness or it may mimic other pulmonary diseases such as tuberculosis. Thus, there may be pneumonia, abscess formation, or both, and occasionally a granulomatous response may be induced. As a consequence, pulmonary nocardiosis is often misdiagnosed as a viral infection, mycoplasmal infection, other pyogenic infections, tuberculosis, various forms of mycoses, actinomycosis, and various forms of cancer. Often pulmonary nocardiosis is recognized as an opportunistic disease in immunocompromised patients; however, it also occurs as a fatal disease in otherwise healthy individuals (Beaman et al., 1976; Brechot et al., 1987; Frazier et al., 1975; Hodges and Rosett, 1978).

Approximately 40 percent of the described cases of nocardiosis involving the lung also caused disease in at least one additional body location (Beaman et al., 1976). Therefore, spread of Nocardia from the lungs by way of the bloodstream to other locations occurs with high frequency, and this form of disease is called systemic nocardiosis (Beaman *et al.*, 1976). Any body site can become infected during dissemination, but most often, the brain represents the primary target for blood borne dissemination (Beaman, 1992a; Beaman *et al.*, 1976). Thus, 22 percent of all cases of nocardial disease (excluding mycetomas) reported in the world literature represent central nervous system involvement. This central nervous system (CNS) nocardiosis usually represents lesions within the brain with infection of the cerebral hemispheres being most common. CNS nocardiosis may occur with no evidence of disease in any other region of the body, and it can represent a primary disease in an otherwise healthy person. The clinical and pathological presentation of CNS nocardiosis is extremely variable (Beaman, 1992b).

Cutaneous and lymphocutaneous nocardiosis involves primarily abscess formation (occasionally granulomatous inflammation) in the dermal and epidermal layers, usually as the result of trauma (e.g., thorn prick) or occasionally during systemic nocardiosis (Beaman, 1983; Wlodaver *et al.*, 1988). Cutaneous nocardiosis may present as a localized pustule or abscess that may be self limited, but occasionally enlarges and becomes progressive. If the lymphatics become involved, then the lymph nodes in the region become enlarged presenting as lymphocutaneous nocardiosis. This form of disease may mimic sporotrichosis caused by the fungus *Sporothrix*, and therefore it is often called sporotrichoid nocardiosis (Hessen and Santoro, 1988; Wlodaver *et al.*, 1988).

If the cutaneous lesions enlarge and expand by direct extension into the surrounding muscle and bone, a mycetoma is produced. Nocardial mycetoma (actinomycetoma) can be defined as a localized and progressive disease usually characterized by enlargement with inflammatory infiltration which becomes a grotesque and disfiguring purulogranulomatous lesion that contains multiple sinus tracts that drain pus. This purulent discharge usually possesses characteristic granules that are colonies of the etiologic agent (i.e. most frequently *N. brasiliensis*). Nocardial induced mycetomas may develop at any site on the body, usually as a result of trauma; however, they most frequently occur on the feet or hands. They progress over a period of years, and they often do not respond well to chemotherapy (Barneston and Milne, 1978; Serrano *et al.*, 1988; Tight and Bartlett, 1981).

Virulence factors of *N. asteroides*: interactions with phagocytes

The virulence of *Nocardia asteroides* is growth stage dependent, as a consequence, cells in log phase of growth are more virulent than cells from the same culture which are in stationary phase (Beaman and Maslan, 1978). This change in virulence corresponds to the ability of the nocardiae to grow within phagocytes (Beaman, 1977; Beaman, 1979). Thus, virulent strains of *Nocardia* are facultatively intracellular pathogens that are not killed by normal macrophages, monocytes or polymorphonuclear neutrophils (Beaman, 1977;

Beaman, 1979; Beaman *et al.*, 1985; Beaman and Smathers, 1976; Davis-Scibienski and Beaman, 1980a; Filice *et al.*, 1980). The mechanisms whereby these organisms avoid being killed are multiple and complex.

Once inside the phagocyte, there are several mechanisms for killing microorganisms, and in order to grow within these host cells, the nocardiae must have the ability to circumvent or neutralize each of these components. Thus, virulent strains of nocardia, such as *N. asteroides* GUH-2 which are not killed by macrophages, inhibit the fusion between the phagosome containing the bacterial cell and lysosomes (Davis-Scibienski and Beaman, 1980[a,b,c]). This inhibition of phagosome-lysosome fusion corresponds with virulence since the less virulent strains of *Nocardia* such as *N. asteroides* 10905 (an organism that differs significantly in virulence from *N. asteroides* GUH-2), do not inhibit phagosome-lysosome fusion (Davis-Scibienski and Beaman, 1980a). These latter strains are killed by phagocytes. Of course, there may be numerous properties related to virulence that differ in individual clinical isolates of *N. asteroides*; however, the cellular component that appears to be most likely involved in the inhibition of phagosome-lysosome inhibition is the glycolipid, trehalose-6-6'-dimycolate (cord factor). When this compound is incorporated into liposomes, it inhibits calcium dependent fusion between phospholipid vesicles (Spargo *et al.*, 1993). In addition, liposomes that contain cord factor block phagosome-lysosome fusion within macrophages that phagocytized dead cells of the yeast, *Saccharomyces* (Spargo *et al.*, 1993). Extraction of the cord factor from the cells of the virulent strain of *N. asteroides* GUH-2 by ether also removes its ability to inhibit phagosome-lysome fusion (Spargo *et al.*, 1993). Furthermore, the cell envelope of the virulent *N. asteroides* is rich in cord factor whereas no cord factor can be detected in the cells of the less virulent *N. asteroides* 10905 (Ioneda *et al.*, 1993).

Another potential microbicidal mechanism involves the acidification of the phagosome (Black *et al.*, 1986a). In professional phagocytes the phagosome becomes acidified by at least two pathways. The first involves the fusion of the phagosome with acidic lysosomes that have a pH less than pH 5. This acidification is phagosome-lysosome fusion dependent (Black *et al.*, 1986a). The second process is a phagosomal membrane associated acidification that is independent of phagosome-lysosome fusion (Black *et al.*, 1986a). Once a particle is ingested, the pH in the phagosome decreases to pH 5 or less, and this low pH kills many microorganisms. Indeed, pH 5 inhibits the growth of most strains of *Nocardia*, and some strains, such as *N. asteroides* GUH-2, are gradually killed at this low pH (Black *et al.*, 1986a). However, by mechanisms that are not yet established, virulent strains (i.e. *N. asteroides* GUH-2) prevent completely the acidification of the phagosome. In contrast, less virulent strains, such as *N. asteroides* 19247, do not block phagosomal acidification to the same extent (Black *et al.*, 1986a).

Phagocytic cells such as polymorphonuclear neutrophils (PMN), monocytes, and macrophages kill microorganisms by a combination of both oxygen dependent (oxidative killing) and oxygen independent mechanisms (Beaman and Beaman, 1984). Phagocytosis of microorganisms by these cells is accompanied by a marked increase in oxygen consumption which is not cytochrome-linked (Beaman and Beaman, 1984). Toxic oxygen metabolites such as superoxide anion (O_2^-), singlet oxygen, hydroxyl radicals, and hydrogen peroxide (H_2O_2) are released in abundance following this respiratory burst of the phagocyte (Beaman and Beaman, 1984). This process represents the basis for oxygen dependent microbicidal activities of phagocytes. In PMNs, azurophilic lysosomal granules contain myeloperoxidase, and following phagocytosis of a particle both a respiratory burst and a fusion of these azurophilic granules with the phagosomes occur. The myeloperoxidase then catalyzes the oxidation of halides (specifically chlorine) in the presence of H_2O_2 to produce the very toxic hypochlorite ion (ClO^-). This myeloperoxidase-halide-hydrogen peroxide system is a major microbicidal mechanism for PMNs (Beaman and Beaman, 1984).

Virulent strains of *N. asteroides* are resistant to the toxic end products of oxygen metabolism (Filice, 1983). Thus, even though *N. asteroides* stimulates an oxidative respiratory burst during phagocytosis by PMNs, monocytes and macrophages, virulent strains such as *N. asteroides* GUH-2 are not killed (Filice *et al.*, 1980). In contrast, less virulent strains such as *N. asteroides* 10905 are sensitive to these oxygen metabolites and are subsequently killed by interactions with PMNs (Beaman *et al.*, 1985). The major differences between those strains not killed by PMNs and those that are killed lie in the relative amounts of intracytoplasmic catalase combined with the secretion of a unique superoxide dismutase (SOD) (Beaman *et al.*, 1985). Thus, *N. asteroides* GUH-2 not only inhibits phagosome-lysosome fusion, but its cells contain large amounts of catalase (relative to most other bacteria), and they secrete a unique SOD that becomes associated with the bacterial surface (Beaman *et al.*, 1983). In contrast, cells of the less virulent *N. asteroides* 10905 possess significantly less catalase (at least 5 fold less) and do not secrete SOD (Beaman *et al.*, 1983; Beaman *et al.*, 1985). By utilizing mutants of *N. asteroides* that contain more or less catalase than the parental strain, combined with antibody that neutralizes the activity of SOD it was shown that these two enzymes are critical for the resistance of *Nocardia* to the microbicidal products of the respiratory burst of phagocytes both *in vitro* and *in vivo* (Beaman *et al.*, 1985; Beaman and Beaman, 1990). Therefore, high levels of intracellular catalase combined with the specific secretion of SOD represent two important virulence factors for *N. asteroides* (Beaman *et al.*, 1985; Beaman and Beaman, 1990).

Virulence factors of *N. asteroides*: interactions in the brain

Nocardia should be regarded as a primary pathogen of the brain in humans and other vertebrates (Beaman, 1992 a;c). Cells of *Nocardia* become localized in the central nervous system (mostly in the brain) following entry into the blood stream, and there appears to be a predilection for specific regions of the brain

(Ogata and Beaman, 1992a). Recently, murine models for CNS nocardiosis have been established, and nocardial interactions in the murine brain have become the focus of several investigations (Beaman, 1992a;Beaman, 1993; Beaman and Ogata, 1993; Beaman and Beaman, 1993; Kohbata and Beaman, 1991; Ogata and Beaman, 1992 a,b).

Most of these studies on interactions within the murine brain have utilized *N. asteroides* GUH-2 and mutants derived from this strain as a model (Ogata and Beaman, 1992 a,b). The injection of a suspension of single cells of *N. asteroides* GUH-2 into the tail vein of mice results in a dose dependent adherence of bacteria in the brain. Furthermore, the specific number of organisms deposited in the brain is growth stage dependent. Thus, IV injection of 10^6 CFU of log phase cells of GUH-2 results in 800 to 1000 CFU in the brain at 1 to 3 h, whereas 10^6 CFU of stationary phase cells from the same culture injected IV yields only 10 to 20 CFU per brain after 1 to 3 h (Ogata and Beaman, 1992 a,b). In addition, the log phase cells of GUH-2 deposited in the brain invade and grow rapidly, while in contrast, stationary phase cells grow much less readily (Beaman, 1992a).

If 500 CFU or more (lethal dose) of log phase cells of *N. asteroides* GUH-2 become deposited in the brain at 3 h, then the nocardiae continue to increase in numbers until the animal dies 2 to 5 days later (Ogata and Beaman, 1992b). On the other hand, if fewer than 500 CFU (non-lethal dose) of log phase cells become localized within the murine brain, then the numbers of CFU increase for 24 to 48 h, and then remain constant or decrease slightly for 4 to 5 days followed by a gradual decrease so that 12 to 14 days after infection no bacteria can be recovered from the brain (Kohbata and Beaman, 1991; Ogata and Beaman, 1992b). At 10 to 14 days after infection, when the nocardiae are being cleared from the brain, the mice develop a variety of neurological signs that indicate damage to the CNS (Beaman, 1992a). Many of these signs represent specific movement disorders that develop when the brain appears to be sterile (Beaman, 1992a). These disorders are usually permanent and may be progressive for the remainder of the life of the animal (Beaman, 1992a). Twenty-four hours after infection with *N. asteroides* GUH-2, there is extensive nocardial invasion and growth within the murine brain tissue without the induction of an apparent inflammatory response (Kohbata and Beaman, 1991; Ogata and Beaman, 1992b). During a non-lethal infection, inflammation or cellular infiltration with PMNs, monocytes and lymphocytes are not observed; however, at 14 days about 10 to 15 percent of these mice develop an L-dopa responsive headshake, stooped posture, hypoactivity and tremulous movement. It was found that these mice share many features with Parkinson's Disease in humans. These include: the L-dopa responsive movement disorder, neurodegeneration and loss of Nissal staining in the substantia nigra, decrease of tyrosine hydroxylase activity in the substantia nigra and "Lewy-like" inclusion bodies in neurons (Kobhata and Beaman, 1991).

These data suggest that there is a specific adherence mechanism in the brain for cells of *N. asteroides* (Ogata and Beaman, 1992a). This was investigated further by injecting BALB/c mice with a non-lethal dose of either *N. asteroides*

GUH-2 or two mutants (NG-49; I-38 syn) that differed in their interactions in the brain. The brains were then removed and microdissected into the following 8 different regions: substantia nigra, hippocampus, striatum, hypothalamus, cerebellum, cerebral cortex, pons-medulla and the midbrain. The determination of viable bacterial counts within each of these regions at 3 h after injection of nocardiae showed that the parental GUH-2 bound throughout the brain, but in contrast, the mutants had a different distribution pattern within the brain. Thus, mutant I-38 syn, which induced a greater frequency of an L-dopa responsive movement disorder in mice than the parental strain of GUH-2, had increased binding in the substantia nigra and decreased binding in the cerebellum (Ogata and Beaman, 1992a). Furthermore, mutant NG-49, that did not induce neurological signs in mice had decreased binding in the substantia nigra and cerebral cortex while there was increased binding in the striatum and cerebellum when compared to the parental strain of GUH-2 (Ogata and Beaman, 1992a). These results suggest that there are specific receptors on the surface of the capillary endothelium that facilitates attachment of the nocardiae (Ogata and Beaman, 1992a).

The distribution and rate of growth of *N. asteroides* GUH-2 and the mutants NG-49 and I-38 syn in regions of the brain was determined following microdissection (Ogata and Beaman, 1992b). These organisms grow at different rates that depended upon the specific region of the brain. Growth within a region appears to be a necessary precursor for cellular damage that results in the variety of neurological disorders observed in the mice. Furthermore, there is a relationship between growth in the substantia nigra and the induction of the Parkinsonian features that developed in some of the mice after a non-lethal injection with *N. asteroides* GUH-2 (Ogata and Beaman, 1992b).

An ultrastructural study was done in order to visualize whether or not there is a specific adherence of the nocardial cell surface to the endothelial cell membrane and to elucidate the manner in which the nocardiae penetrate the capillary wall (Beaman and Ogata, 1993). Following a perfusion of a suspension of log phase cells of GUH-2 into the brain, it was shown that within 10 to 25 min the nocardiae attach to the cytoplasmic membrane of endothelial cells in specific regions of the brain (not a uniform attachment). This attachment appears to occur at the growing tip of the nocardial filament. The outermost layer of the cell wall of the nocardia has electron dense regions that bind to the membrane of the endothial cell. The attached nocardiae then induce a "cup-like" deformation which is followed by a phagocytic response. Within 25 min the bacteria are phagocytized by the capillary endothelial cell and become internalized within phagocytic vesicles. During this process of uptake of the nocardiae by the endothial cell, there appears to be no ultrastructural evidence of damage to either the bacteria or the host cell (Beaman and Ogata, 1993).

Twenty-four hours after IV injection of BALB/c mice with a lethal dose of *N. asteroides* GUH-2, the brains were perfused and examined by electron microscopy. The purpose was to visualize nocardial growth within brain tissue to ascertain the types of brain cells involved, the extent and types of damage, and the host response to nocardial invasion and growth (Beaman, 1993). As suggested by light microscopy, electron microscopic analysis revealed that the nocardiae grow perivascularly without an apparent inflammatory response. There is no evidence of infiltration of PMNs, monocytes or lymphocytes at the sites of nocardial growth. The nocardiae grow within most cell types within the brain, but there is propensity for growth within the soma of neurons and along axons. The nocardial cells are frequently surrounded by layers of membrane with the innermost membrane tightly adherent to the nocardial surface. The nocardiae appear not to induce much damage to host cells since the integrity of the cells containing these bacteria remain intact and exhibit little or no cytopathic effect (CPE). The nocardiae growing within the brain cells are not completely inert because there are some areas wherein a neurodegeneration can be seen. Myelin sheaths along axons that are associated with the nocardial cells become damaged and axonal degeneration is evident. There are a few compact, phagocytic cells within the brain tissue that probably represent microglia (macrophage equivalent cells). These "microglia" phagocytize nocardiae within the brain as well as areas of the brain tissue that contain nocardia. The nocardiae appear to grow within these phagocytic cells initially; however, it is likely that microglia are ultimately responsible for the elimination of the nocardiae from the brain during a non-lethal infection (Beaman, 1993).

In order to better understand the interactions of *N. asteroides* with cellular components of the brain, tissue cultures derived from newborn murine brains were established (Beaman and Beaman, 1993). After 14 days in culture, astroglia and microglia were characterized and their interactions with nocardiae were studied by scanning electron and light microscopy. It was shown that *N. asteroides* GUH-2 adheres to Type 2 astroglia but not to Type 1 astroglia. Furthermore, the nocardiae actively penetrates into the astroglia wherein these bacteria grow. In contrast, these bacteria do not adhere to the microglial surface the same way that they do with astroglia, but instead the microglia phagocytizes them. Once internalized within the microglia, nocardial growth is inhibited. It is likely that after 14 days in tissue culture, these microglia are non-specifically activated. These data suggest that it is the microglia and not astroglia that are important in stopping nocardial growth in the brain following a non-lethal injection (Beaman and Beaman, 1993).

Conclusions

1. *Nocardiae* are environmental bacteria that cause a variety of diseases in humans and other animals (Schaal and Beaman, 1983).
2. *Nocardia* have a predilection for causing disease in the lungs, brain and skin even though any body site may be affected (Beaman, 1992c; Schall and Beaman, 1983).
3. Virulent strains of *N. asteroides* are facultative intracellular pathogens that grow in macrophages and monocytes from humans, mice, rabbits, and

guinea pigs (Beaman, 1976; Beaman, 1977; Beaman, 1979; Beaman, 1983; Black *et al.*, 1983; Black *et al.*, 1985).

4. The mechanisms for the virulence of *Nocardia* are multiple and complex (Beaman, 1976; Beaman, 1983; Beaman, 1992a; Schaal and Beaman, 1983).

5. Murine models of nocardial disease have been established in order to investigate both host-parasite interactions and the mechanisms for nocardial virulence (Beaman, 1983; Beaman, 1992a,b,; Beaman and Maslan, 1978; Beaman and Moring, 1988; Kohbata and Beaman, 1991; Ogata and Beaman, 1992a,b; Silva *et al.*, 1988).

6. Virulent strains of *N. asteroides* inhibit phagosome-lysosome fusion in macrophages whereas non-virulent strains do not (Davis-Scibienski and Beaman, 1980a,b,c).

7. This inhibition of phagosome-lysosome fusion, as well as the alteration of membrane rigidity, fusability and degree of hydration is the result of both the presence of and the structure of the surface glycolipid, trehalose dimycolate (cord factor) (Beaman and Moring, 1988; Beaman, Moring *et al.*, 1988; Ioneda *et al.*, 1993; Spargo *et al.*, 1991; Spargo *et al.*, 1993; Vistica and Beaman, 1983).

8. Cord factor appears to be a virulence factor for *Nocardia* (Beaman and Moring, 1988; Beaman, Moring *et al.*, 1988; Ioneda *et al.*, 1993; Vistica and Beaman, 1983).

9. Virulent strains of *N. asteroides* inhibit phagosomal acidification which is important for its survival within phagocytes (Black *et al.*, 1986a).

10. Virulent *N. asteroides* that grow within macrophages modulate lysosomal enzymes, reduce lysosomal acid phosphatase content, and utilize acid phosphatase as a carbon source whereas less virulent strains do not (Beaman *et al.*, 1988; Black *et al.*, 1983; Black *et al.*, 1985; Black *et al.*, 1986b).

11. Virulent strains of *N. asteroides* induce an oxidative metabolic burst in neutrophils yet they are not killed by these professional phagocytes (Beaman *et al.*, 1985; Filice, 1983; Filice *et al.*, 1980).

12. Virulent *N. asteroides* secrete a unique superoxide dismutase (SOD) that becomes associated with the bacterial cell surface whereas non-virulent strains do not (Beaman *et al.*, 1983).

13. This secreted and surface associated SOD combined with high levels of intracellular catalase is important for the ability of *N. asteroides* to resist the microbicidal effects of the oxidative metabolic burst of phagocytes, thus SOD and catalase are virulence factors (Beaman *et al.*, 1985; Beaman *et al.*, 1983; Beaman and Beaman, 1984; Beaman and Beaman, 1990).

14. *Nocardia* is a primary pathogen of the brain of vertebrates, and it has a specific mechanism for adherence and invasion of the brain (Beaman, 1983; Beaman, 1992a,b; Beaman *et al.*, 1976; Beaman and Ogata, 1993; Beaman and Sugar, 1983; Kohbata and Beaman, 1991; Ogata and Beaman, 1992a,b; Schaal and Beaman, 1983).

15. Certain strains of *N. asteroides* have the ability to induce a variety of

neurodegenerative responses in an experimental murine model (Beaman, 1992a,b; Kohbata and Beaman, 1991).

16. *N. asteroides* strain GUH-2 induces a permanent and progressive L-dopa responsive movement disorder that share many features with idiopathic Parkinson's Disease in humans (Kohbata and Beaman, 1991).

17. Nocardial cells that invade the brain possess a specific surface component that binds to the cell membrane of subpopulations of capillary endothelial cells in regions of the brain (Beaman and Ogata, 1993).

18. Following attachment to the brain capillary, the endothelial cells phagocytize the nocardiae wherein these bacteria then enter the brain parenchyma (Beaman and Ogata, 1993).

19. Once inside the brain, virulent strains of *N. asteroides* grow within most types of brain cells without inducing an inflammatory response in the host; however, there is an induction of axonal degeneration (Beaman, 1993).

20. The nocardiae adhere to, and grow within, astroglia *in vitro*, whereas they are phagocytized and their growth inhibited by Microglia *in vitro* (Beaman and Beaman, 1993).

21. The specific mechanisms for nocardial invasion in the brain with subsequent induction of neurodegeneration are not known (Beaman, 1992a).

Acknowledgements

Much of the work discussed in this review was supported by Public Health Service Research Grant RO1-AI20900 from the National Institute of Allergy and Infectious Diseases. I thank Marilyn Kiene for typing this manuscript.

References

Barneston RS & Milne LHR (1978) Mycetoma: A review. Brit. J. Dermatol. 99: 227–231.

Beaman BL (1976) Possible mechanisms of nocardial pathogenesis. p. 386–417. *In*: Goodfellow M, Brownell GH & Serrano JA (eds.) Biology of the Nocardiae. Academic Press, London, Great Britain.

Beaman BL (1977) The *in vitro* response of rabbit alveolar macrophages to infection with *Nocardia asteroides*. Infect. Immun. 15: 925–937.

Beaman BL (1979) Interaction of *Nocardia asteroides* at different phases of growth with *in vitro*-maintained macrophages obtained from the lungs of normal and immunized rabbits. Infect. Immun. 26: 355–361.

Beaman BL (1983) Actinomycete pathogenesis. p. 457–479. *In*: Goodfellow M, Mordarski M & Williams ST (eds.) The biology of Actinomycetes. Academic Press, Inc. London.

Beaman BL (1992a) *Nocardia* as a pathogen of the brain: mechanisms of interactions in the murine brain – a review. Gene 115: 213–217.

Beaman BL (1992b) *Nocardia*: An environmental bacterium possibly associated with neurodegenerative diseases in humans. p. 147–166. *In*: Isaacson, RL & Jensen K (eds.) The vulnerable brain and environmental risks. Vol. 2. Plenum Press, New York.

Beaman BL (1993) Ultrastructural analysis of the growth of *Nocardia asteroides* during invasion of the murine brain. Infect. Immun. 61: 274–283.

Beaman BL, Black CM, Doughty F & Beaman L (1985) Role of superoxide dismutase and catalase as determinants of pathogenicity of *Nocardia asteroides*: importance in resistance to microbicidal activities of human polymorphonuclear neutrophils. Infect. Immun. 47: 135–141.

Beaman BL, Burnside J, Edwards B & Causey W (1976) Nocardial infections in the United States, 1972–1974. J. Infect. Dis. 134: 286–289.

Beaman BL & Maslan S (1978) Virulence of *Nocardia asteroides* during its growth cycle. Infect. Immun. 20: 290–295.

Beaman BL & Moring SE (1988) Relationship among cell wall composition, stage of growth, and virulence of *Nocardia asteroides* GUH-2. Infect. Immun. 56: 557–563.

Beaman BL, Moring SE & Ioneda T (1988) Effect of growth stage on mycolic acid structure in cell walls of *Nocardia asteroides* GUH-2. J. Bacteriol. 170: 1137–1142.

Beaman BL & Ogata SA (1993) An ultrastructural analysis of attachment and penetration of capillaries in the *pons* and *substantia nigra* regions of the murine brain by *Nocardia asteroides*. Infect. Immun. 61: 955–965.

Beaman BL, Scates SM, Moring, SE, Deem R & Mishra HP (1983) Purification and properties of a unique superoxide dismutase from *Nocardia asteroides*. J. Biol. Chem. 258: 91–96.

Beaman BL & Sugar AM (1983) *Nocardia* in naturally acquired and experimental infections in animals. J. Hyg. Camb. 91: 393–419.

Beaman BL & Smathers M (1976) Interaction of *Nocardia asteroides* with cultured rabbit alveolar macrophages. Infect. Immun. 13: 1126–1131.

Beaman L & Beaman BL (1984) The role of oxygen and its derivatives in microbial pathogenesis and host resistance. Annu. Rev. Microbiol. 38: 27–48.

Beaman L & Beaman BL (1990) Monoclonal antibodies demonstrate that superoxide dismutase contributes to protection of *Nocardia asteroides* within the intact host. Infect. Immun. 58: 3122–3128.

Beaman L & Beaman BL (1993) Interaction of *Nocardia asteroides* with murine glia cells in culture. Infect. Immun. 61: 343–347.

Beaman L, Paliescheskey M & Beaman BL (1988) Acid phosphatase stimulation of the growth of *Nocardia asteroides* and its possible relationship to the modulation of lysosomal enzymes in macrophages. Infect. Immun. 56: 1652–1654.

Black CM, Beaman BL, Donovan RM & Goldstein E (1983) Effect of virulent and less virulent strains of *Nocardia asteroides* on acid phosphatase activity in alveolar and peritoneal macrophages maintained *in vitro*. J. Infect. Dis. 148: 117–124.

Black CM, Beaman BL, Donovan RM & Goldstein E (1985) Intracellular acid phosphatase content and ability of different macrophage populations to kill *Nocardia asteroides*. Infect. Immun. 47: 375–383.

Black CM, Paliescheskey M, Beaman BL, Donovan RM & Goldstein E (1986a) Acidification of phagosomes in murine macrophages: blockage by *Nocardia asteroides*. J. Infect. Dis. 154: 952–958.

Black C, Paliescheskey M, Beaman BL, Donovan RM & Goldstein E (1986b) Modulation of lysosomal protease-esterase and lysozyme in Kupffer cells and peritoneal macrophages infected with *Nocardia asteroides*. Infect. Immun. 54: 917–919.

Brechot JM, Capron F, Prudent J & Rochemaure J (1987) Unexpected pulmonary nocardiosis in a non-immunocompromised patient. Thorax 42: 479–480.

Davis-Scibienski C & Beaman BL (1980a) Interaction of *Nocardia asteroides* with rabbit alveolar macrophages: association of virulence, viability, ultrastructural damage, and phagosome-lysosome fusion. Infect. Immun. 28: 610–619.

Davis-Scibienski C & Beaman BL (1980b) Interaction of *Nocardia asteroides* with rabbit alveolar macrophages: effect of growth phase and viability on phagosome-lysosome fusion. Infect. Immun. 29: 2–29.

Davis-Scibienski C & Beaman BL (1980c) Interaction of alveolar macrophages with *Nocardia asteroides*: immunological enhancement of phagocytosis, phagosome-lysosome fusion, and microbicidal activity. Infect. Immun. 30: 578–587.

Eppinger H (1891) Uber eine neue pathogene *Cladothrix* und eine durch sie hervorgerufene pseudotuberculosis (cladothrichia), Beitr. Pathol. Anat. Allg. Pathol. 9: 287–328.

Filice GA (1983) Resistance of *Nocardia asteroides* to oxygen-dependent killing by neutrophils. J. Infect. Dis. 148: 861–867.

Filice GA, Beaman BL, Krick JA & Remington JS (1980) Effects of human neutrophils and monocytes on *Nocardia asteroides*: Failure of killing despite occurrence of the oxidative metabolic burst. J. Infect. Dis. 142: 432–438.

Frazier AR, Rosenow EC & Roberts GD (1975) Nocardiosis: A review of 25 cases occurring during 24 months. Mayo Clin. Proc. 50: 657–663.

Hessen MT & Santoro J (1988) Lymphocutaneous nocardiosis in Pennsylvania. Pa. Med. 91: 54–58.

Hodges GR & Rosett W (1978) Recent experiences with nocardial infections. Amer. J. Med. Sci. 276: 279–285.

Ioneda T, Beaman BL, Viscaya L & de Almeida ET (1993) Composition and toxicity of diethyl ether soluble lipids from *Nocardia asteroides* GUH-2 and *N. asteroides* 10905. Biochem. Biophys. Acta (Accepted for publication).

Kohbata S and Beaman BL (1991) L-dopa responsive movement disorder caused by *Nocardia asteroides* localized in the brains of mice. Infect. Immun. 59: 181–191.

Lechevalier HA (1989) Nocardioform actinomycetes, p. 2348–2404. *In*: Williams ST, Sharpe ME & Hold JG (eds.) Bergey's manual of systematic bacteriology. Vol. 4. Williams & Wilkins, Baltimore, MD.

Nocard E (1888) Note sur la maladie des boeufs de la Gouadeloupe connue sous le nom de farcin. Ann. Inst. Pasteur (Paris) 2: 293–302.

Ogata SA & Beaman BL (1992a) Adherence of *Nocardia asteroides* within the murine brain. Infect. Immun. 60: 1800–1805.

Ogata SA & Beaman BL (1992b) Site-specific growth of *Nocardia asteroides* in the murine brain. Infect. Immun. 60: 3262–3267.

Serrano JA, Beaman B, Mejia MA, Viloria JE & Zamora R (1988) Histological and microbiological aspects of actinomycetoma cases in Venezuela. Rev. Inst. Med. Trop., São Paulo. 30: 297–304.

Schaal KP & Beaman, BL (1983) Clinical significance of Actinomycetes. p. 389–424. *In*: Goodfellow, M, Mordarski, M, & Williams ST (ed.) Biology of the Actinomycetes. Academic Press, London.

Silva CL, Tincani I, Brandao Filho SL & Faccioli LH (1988) Mouse cachexia induced by trehalose dimycolate from *Nocardia asteroides*. J. Gen. Microbiol. 134: 1629–1633.

Spargo BJ, Crowe LM, Ioneda T, Beaman BL & Crowe JH (1991) Cord factor (α,α-trehalose 6,6'-dimycolate) inhibits fusion between phospholipid vesicles. Proc. Natl. Acad. Sci. USA 88: 737–740.

Spargo BJ, Stillwell GE, Rudolph AS, Crowe LM, Crowe JH, Ioneda, T & Beaman BL (1993) The role of cord factor (α,α-trehalose-6,6'-dimycolate) in the inhibition of phagosome-lysosome fusion in macrophages. J. Gen. Microbiol. (Accepted for publication).

Tight RR & Bartlett MS (1981) Actinomycetoma in the United States. Rev. Infect. Dis. 3: 1139–1150.

Wlodaver CG, Tolomeo T & Benear JB (1988) Primary cutaneous nocardiosis mimicking sporotrichosis. Arch. Dermatol. 124: 659–660.

Vistica CA & Beaman BL (1983) Pathogenic and virulence characterization of colonial mutants of *Nocardia asteroides* GUH-2. Canadian J. Microbiol. 29: 1126–1135.

40. Developments in the interaction of bacterial avirulence genes and plant disease resistance genes

N.T. KEEN, H. SHEN, J. LORANG and D.Y. KOBAYASHI

Abstract. We attempted to clone all of the avirulence genes from *Pseudomonas syringae* pv. *tomato* that cause it to give a non-host hypersensitive response (HR) in soybean plants. Three different classes of clones were isolated from a *P.s. tomato* cosmid library which functioned in *P. syringae* pv. *glycinea* to elicit the HR in soybean. One clone, *avrA*, was identical to a gene previously cloned from race 6 of *P.s. glycinea*. The second locus, *avrE*, occurs adjacent to the *hrpS* end of the *hrp* cluster of *P.s. tomato* and is complex, requiring two divergent transcriptional units for activity. The third gene, *avrD*, functions by causing bacterial hosts to secrete a low molecular weight lactone elicitor that is specifically perceived by soybean plants carrying the matching *Rpg4* disease resistance gene. Ronald *et al.* [J. Bacteriol. 174, 1604 (1992)] showed that *P.s. tomato* also contains an avirulence gene, *avrPto*, that functions in tomato cultivars carrying the *Pto* disease resistance gene and in all tested soybean cultivars except Peking. A *P.s. tomato* strain mutated in *avrA*, *avrD*, *avrE* and *avrPto* yielded a null reaction on the soybean cultivar Peking, but gave HR on all other soybean cultivars tested. This mutant also exhibited reduced virulence on the normal host plant, tomato. The results suggest that *P.s. tomato* carries several avirulence genes which may contribute to virulence on the normal host plant and are also responsible for elicitation of the non-host HR on soybean plants.

Abbreviations: *avr*, Avirulence Gene; HR, Hypersensitive Response; Psg, *Pseudomonas syringae* Pv. Glycinea; Pst, *Pseudomonas syringae* Pv. Tomato; TMV, Tobacco Mosaic Virus

Introduction

Disease resistance genes permit the specific detection by plants of certain but not all biotypes of a pathogen species and the consequent initiation of active plant defense responses, collectively called the *hypersensitive response* (HR). Pathogen biotypes which are detected by a certain disease resistance gene also carry a complementary Mendelian gene, called an avirulence gene (Flor, 1942; Keen, 1990). Thus, the pathogen avirulence gene and its matching disease resistance gene in the plant comprise a 'gene-for-gene' interaction determining plant recognition of pathogens leading to active defense. As such, plant disease resistance genes have superficial features in common with certain genes of vertebrate immune systems, particular those of the major histocompatibility complex (Dangl, 1992). Unfortunately, the biochemical functions of plant disease resistance genes are unknown since they have not yet been cloned and characterized. Information is available, however, on pathogen-produced molecules, called elicitors, that are specifically perceived by plant cells

C.I. Kado and J.H. Crosa (eds.), Molecular Mechanisms of Bacterial Virulence, 573–579.
© *1994 Kluwer Academic Publishers.*

expressing particular disease resistance genes. The production of these cultivar-specific elicitors is determined by avirulence gene activity. Since they elicit active defense responses only in certain plant cultivars, such elicitors appear to play the role of antigens in counterpart vertebrate systems.

Avirulence genes and cultivar-specific elicitors

Characterized specific elicitors include the capsid protein of tobacco mosaic virus (TMV)(Culver and Dawson, 1992), a small peptide produced by isolates of the fungus *Cladosporium fulvum* carrying *avr9* (van den Ackerveken *et al.*, 1992) and a low molecular weight elicitor produced by *Pseudomonas syringae* isolates expressing avirulence gene D (Keen *et al.*, 1990; Smith *et al.*, 1993). These elicitors are formed in very different ways. The TMV capsid protein is the primary translation product of the coat protein gene and remains within the infected plant cell cytoplasm. The elicitor produced by *C. fulvum* isolates carrying *avr9* is a peptide of 28 amino acids that is cleaved from a 63 amino acid primary translation product of the *avr9* gene. This elicitor is extracellularly secreted by the fungus such that it occurs in the intercellular spaces of infected plant leaves. The *avrD* elicitor is a structurally unique secondary metabolite which appears to be synthesized in bacterial cells as a consequence of a catalytic activity by the *avrD* protein product. Similar to the *avr9* peptide, the *avrD* elicitor is efficiently secreted extracellularly.

One of the paradoxes regarding avirulence genes in pathogens is that they behave negatively – that is, their expression *reduces* the number of plant cultivars and/or species which are hosts for the pathogen. Teleologically, avirulence genes should therefore be rapidly lost from pathogen populations, but they have in fact been evolutionarily retained. This indicates that avirulence genes must have functions in the pathogen which are naturally selected. Only in the case of the tobacco mosaic virus coat protein gene, however, do we know the nature of that function. The coat protein is important for dissemination and survival of the virus and therefore the gene encoding it is strongly selected for. However, TMV RNA devoid of a functional coat protein gene is experimentally infectious, indicating that the coat protein gene is not essential for basic pathogenicity and viral multiplication. With microbial avirulence genes, *avrBs2* (Kearney and Staskawicz, 1990) and *pthA* (Swarup *et al.*, 1992) appear to be required for high virulence, but certain other avirulence genes do not seem to play a similar role (Keen, 1990).

Two observations are germane to the probable function of avirulence genes in microbial pathogens. First, most investigated *Pseudomonas syringae* and *Xanthomonas campestris* pathovars harbor several different avirulence genes (Bonas *et al.*, 1989; Debener *et al.*, 1991; DeFeyter and Gabriel, 1991; Minsavage *et al.*, 1990; Staskawicz *et al.*, 1987). Certain of these avirulence genes also function in heterologous *P. syringae* or *X. campestris* pathovars and cause them to elicit HR on certain cultivars of their normal host plant species

(Carney and Denny, 1990; Fillingham *et al.*, 1992; Kobayashi *et al.*, 1989; Ronald *et al.*, 1992; Whalen *et al.*, 1988, 1991). These observations suggest that different plant species contain functionally and perhaps structurally conserved disease resistance genes. They also raise the possibility that avirulence genes may define host ranges in plant-pathogen systems above the race-cultivar level.

The second important observation concerning avirulence gene function is that at least nine different *P. syringae* avirulence genes have highly conserved promoter elements (Salmeron and Staskawicz, 1993; H. Shen and N. Keen, manuscript in preparation; R. Innes, personal communication). Since these promoters are also expressed at high level only when the bacteria are growing in plants, the avirulence genes may have functions related to bacterial development in or on plants.

The observations noted above suggest a resolution to the conundrum that mutation of single avirulence genes frequently does not reduce pathogen virulence or viability. The collection of *X. campestris* and *P. syringae* avirulence genes present in a particular bacterium may, even though they are structurally dissimilar, provide complementary biologic functions. Thus, elimination of this function would require mutagenizing all or most of the avirulence genes in a particular bacterial strain. We set out several years ago to test this hypothesis by cloning all the avirulence genes present in *P. syringae* pv. tomato and sequentially mutating them. The experiment also addressed another important biological question – whether the Pst avirulence gene mutations affected elicitation of the HR in a non-host plant species, soybean.

Most *P. syringae* pathovars are pathogenic on only one or a few plant species. Thus, pathovar *phaseolicola* attacks beans, pathovar *tabaci* attacks tobacco, etc. Pathovars usually elicit the HR on heterologous plant species, raising the possibility that active defense responses define the host range of the bacteria. Alternatively, there is evidence that some pathogens require positive-acting virulence functions in order to attack their normal host plant species (Swarup *et al.*, 1992; Waney *et al.*, 1991). By sequentially mutating the collection of avirulence genes in *P. syringe* pv. tomato that function against soybean plants when introduced into the heterologus pathogen, *P. syringae* pv. glycinea, it may be possible to determine whether they are involved in restricting the host range of the wildtype bacterium to exclude soybean plants.

Avirulence genes of *P. syringae* pv. *tomato* which function in *P. syringae* pv. glycinea

Kobayashi *et al.* (1989) constructed a cosmid library of *P. syringae* pv. tomato (Pst) strain PT23 genomic DNA and isolated three classes of clones which, when introduced into *P. syringae* pv. glycinea (Psg), caused the HR on several soybean cultivars. Subsequent screening of the same library led to the recovery of additional clones of the same genes, but no new avirulence genes (J. Lorang and H. Shen, unpublished data). The PT23 isolate of Pst also should harbor the *avrPto* avirulence gene which complements a resistance gene in tomato plants called *Pto*, but we did not recover clones carrying this gene. Ronald *et al.* (1992), however, cloned the *avrPto* gene from a related strain of Pst and it has recently been characterized (Salmeron and Staskawicz, 1993).

One of the avirulence genes from Pst which functioned in Psg to elicit the HR in soybean turned out to be virtually identical to *avrA* (Table 1), initially cloned from Psg race 6 (Staskawicz *et al.*, 1984). The *avrA* gene from Pst was therefore the first avirulence gene shown to function in two different bacterial pathogens (Kobayashi *et al.*, 1989), although several additional examples have been subsequently discovered. In soybean, *avrA* complements the *Rpg2* disease resistance gene (Keen and Buzzell, 1990).

The second avirulence gene recovered from the Pst cosmid library was called *avrD* (Table 1). This gene complements the soybean disease resistance gene *Rpg4* (Keen and Buzzell, 1990). The *avrD* gene has been extensively studied in our laboratory and is the only bacterial avirulence gene thus far shown to generate a discrete elicitor which initiates the HR only in soybean cultivars carrying the cognate *Rpg4* disease resistance gene (Kobayashi *et al.*, 1990; Keen *et al.*, 1990). This elicitor has recently been isolated and its structure was elucidated as an unusual glycosyl lipid (Smith *et al.*, 1993).

The third avirulence gene locus cloned from the Pst library was called *avrE* and is of interest for several reasons. First, this gene is located immediately

Table 1. Reactions of soybean cultivars inoculated with *Pseudomonas syringae* pv. *glycinea* R4 only or containing various cloned avirulence genes from *P. syringae* pv. *tomato*.

Cultivar	Psg Only	avrA	avrD	avrE	avrPto
Acme	C	HR	C	HR	HR
Chippewa	C	HR	HR	HR	HR
Flambeau	C	C	HR	HR	HR
Hardee	C	C	C	HR	HR
Harosoy	C	HR	HR	HR	HR
Lindarin	C	HR	HR	HR	HR
Merit	C	HR	C	HR	HR
Norchief	C	C	HR	HR	HR
Peking	C	HR	C	HR	C
Centennial	C	HR	HR	HR	HR

adjacent to the *hrp* cluster of genes (J. Lorang, unpublished). The *avrE* locus occurs next to *hrpS* (Grimm and Panopoulos, 1989; Rahme *et al.*, 1991), a regulatory gene related to the *ntrA* family of bacterial two component regulators required for the expression of all *hrp* regulons as well as certain avirulence genes such as *avrD* (Shen and Keen, manuscript in preparation), *avrRpt2* (R. Innes, personal communication) and *avrPto* (Salmeron and Staskawicz, 1993). Secondly, *avrE* is complex, comprising approximately 8 kb and containing two divergent transcriptional units which also require the *hrpS* and *hrpL* regulatory genes for expression (J. Lorang, unpublished observations). Finally, *avrE* is not cultivar specific as in the case of *avrA* and *avrD*, but produces the HR on all soybean cultivars tested (Table 1). It is therefore not possible to determine by crossing experiments if soybean harbors a single Mendelian disease resistance gene complementing the bacterial *avrE* locus.

The *avrPto* gene was cloned from a related strain of Pst by Ronald *et al.* (1992) and interacts in a race-cultivar manner with tomato plants carrying the resistance gene, *Pto*. Avirulence gene *avrPto* also causes HR on several soybean cultivars when introduced into Psg, but not on cultivar Peking (Table 1; R Ransom and J. Lorang, unpublished results). It has not been determined if a single Mendelian disease resistance gene occurs in soybean that complements *avrPto*.

Mutants deficient in avirulence genes

We prepared Pst strain PT23 marker exchange mutants deficient in various combinations of *avrA*, *avrD*, *avrPto* and *avrE* and tested them for responses on both soybean and tomato plants. These results indicate that Pst mutants deficient in either *avrA* or *avrE*, but not *avrD* or *avrPto* (Ronald *et al.*, 1992), exhibit decreased virulence on tomato plants.

The mutational studies with Pst strain PT23 also disclosed that the triple mutant deficient in *avrA*, *avrD* and *avrE* does not cause the HR when inoculated onto soybean cv. Peking but causes HR in the other cultivars listed in Table 1. The mutant strain causes a null reaction on Peking – that is, these inoculated plants are devoid of any symptoms. This result was of interest because Peking is the only cultivar which does not react hypersensitively to Psg carrying the cloned *avrPto* gene (Table 1). We subsequently marker exchange mutated *avrPto* in the Pst triple mutant, creating a quadruple mutant strain devoid of all four avirulence genes. However, the *avrPto* mutation did not further alter the phenotype of the triple mutant strain. In addition, like the *avrPto* mutant strain made by Ronald *et al.* (1992), our quadruple mutant PT23 strain retained the ability to elicit the HR on tomato cultivar 76R, which carries the *Pto* resistance gene.

The mutational experiments discussed above indicate that certain avirulence genes in Pst (*avrA* and *avrE*) may be required for high virulence on its normal host plant, tomato. On the other hand, mutation of *avrD* or *avrPto* by themselves do not appear to affect the virulence of Pst on tomato. Sequential mutant strains mutated in *avrA*, *avrE*, *avrD* and *avrPto* also produced a null reaction on the soybean cultivar, Peking. This is unlike the wild type Pst strain or mutant strains devoid in only *avrA* or *avrE*, which produce the HR on all tested soybean cultivars. Furthermore, the HR on Peking could be restored to the triple mutant by complementation with either the cloned *avrA* or *avrE* genes alone. The null reaction observed on cv. Peking may therefore indicate that the ability of wild type Pst to produce the HR on all soybean cultivars is due to a collection of avirulence genes which are active against various soybean cultivars (Table 1). The *avrA*, *avrD*, *avrE* and *avrPto* quadruple mutant strain of Pst retained the ability to produce HR on tomato cultivar 76R, carrying the *Pto* resistance gene and on all soybean cultivars except Peking. This indicates the presence of other avirulence genes in Pst that function in soybean and tomato.

Our experiments have resulted in two important observations. The first is that certain but not all avirulence genes in Pst are important for bacterial virulence on tomato plants. It will be of interest to determine whether the remaining avirulence genes, including *avrD*, are important in other life cycle stages. The second finding is that sequential mutation of avirulence genes in Pst altered the non-host HR of soybean plants. Thus, mutation of *avrA*, *D* and *E* in strain PT23 eliminated the HR on cultivar Peking, but the HR still occurred on other cultivars. As shown in Table 1, this phenotype is precisely that of the Pst *avrPto* gene. However, further mutation of *avrPto* in the triple mutant PT23 strain to form a quadruple mutant did not alter the soybean reactions. Similar to the results of Ronald *et al.* (1992), this suggests that Pst probably contains another avirulence gene giving the *avrPto* phenotype.

References

Bonas U, Stall RE & Staskawicz B (1989) Genetic and structural characterization of the avirulence gene *avrBs3* from *Xanthomonas campestris* pv. *vesicatoria*. Mol. Gen. Genet. 318: 127–136.

Carney BF & Denny TP (1990) A cloned avirulence gene from *Pseudomonas solanacearum* determines incompatibility on *Nicotiana tabacum* at the host species level. J. Bacteriol. 172: 4836–4843.

Culver JN & Dawson WO (1992) Tobacco mosaic virus elicitor coat protein genes produce a hypersensitive phenotype in transgenic *Nicotiana sylvestris* plants. Mol. Plant-Microbe Interact. 4: 458–463.

Dangl JL (1992) The major histocompatibility complex a la carte: are there analogies to plant disease resistance genes on the menu? Plant J. 2: 3–11.

Debener T, Lehnackers H, Arnold M & Dangl, J. (1991) Identification and molecular mapping of a single *Arabidopsis thaliana* locus determining resistance to a phytopathogenic *Pseudomonas syringae* isolate. Plant Cell 1: 289–302.

De Feyter R. & Gabriel DW (1991) At least six avirulence genes are clustered on a 90-kilobase plasmid in *Xanthomonas campestris* pv. *malvacearum*. Mol. Plant-Microbe Interact. 4: 423–432.

Fillingham AJ, Wood J, Bevan JR, Crute IR, Mansfield JW, Taylor JD & Vivian A (1992) Avirulence genes from *Pseudomonas syringae* pathovars *phaseolicola* and *pisi* confer specificity towards both host and non-host species. Physiol. Mol. Plant Pathol. 40: 1–15.

Flor HH (1942) Inheritance of pathogenicity in *Melampsora lini*. Phytopathology 32: 653–669.

Grimm C, & Panopoulos N (1989) The predicted protein product of a pathogenicity locus from *Pseudomonas syringae* pv. *phaseolicola* is homologous to a highly conserved domain of several prokaryotic regulatory proteins. J. Bacteriol. 171: 5031–5038.

Kearney B, & Staskawicz BJ (1990) Widespread distribution and fitness contribution of *Xanthomonas campestris* avirulence gene *avrBs2*. Nature (London) 346: 385–386.

Keen, NT (1990) Gene-for-gene complementarity in plant-pathogen interactions. Annu. Rev. Genet. 24: 447–463.

Keen NT & Buzzell RI (1990) New disease resistance genes in soybean against *Pseudomonas syringae* pv. *glycinea*: evidence that one of them interacts with a bacterial elicitor. Theor. Appl Genet. 81: 133–138.

Keen NT, Tamaki S, Kobayashi D, Gerhold D, Stayton M, Shen H, Gold S, Lorang J, Thordal-Christensen H, Dahlbeck D, Staskawicz B (1990) Bacteria expressing avirulence gene D produce a specific elicitor of the soybean hypersensitive reaction. Mol. Plant-Microbe Interact. 3: 112–121.

Kobayashi DY, Tamaki SJ & Keen NT (1989) Cloned avirulence genes from the tomato pathogen *Pseudomonas syringae* pv. *tomato* confer cultivar specificity on soybean. Proc. Natl. Acad. Sci., USA 86: 157–161.

Kobayashi DY, Tamaki SJ, & Keen NT (1990) Molecular characterization of avirulence gene D frmm *Pseudomonas syringae* pv. *tomato*. Mol. Plant-Microbe Interact. 3: 97–102.

Minsavage GV, Dahlbeck D, Whalen MC, Kearney, B, Bonas U, Staskawicz BJ & Stall RE. (1990) Gene-for-gene relationships specifying disease resistance in *Xanthomonas campestris* pv. *vesicatoria*-pepper interactions. Mol. Plant-Microbe Interact. 3: 41–47.

Rahme LG, Mindrinos MN & Panopoulos NJ (1991) Genetic and transcriptional organization of the *hrp* cluster of *Pseudomonas syringae* pv. phaseolicola. J. Bacteriol. 173: 575–586.

Ronald PC, Salmeron J, Carland, FC & Staskawicz B (1992) Cloned avirulence gene *avrPto* induces disease resistance in tomato cultivars containing the *Pto* resistance gene. J. Bacteriol. 174: 1604–1611.

Salmeron JM & Staskawicz BJ (1993) Molecular characterization and *hrp*-dependence of the avirulence gene *avrPto* from *Pseudomonas syringae* pv. tomato. Mol. Gen. Genet. 239: 6–161.

Smith MJ, Mazzola EP, Sims JJ, Midland SL, Keen NT, Burton V & Stayton MM (1993) The syringolides: bacterial C-glycosyl lipids that trigger plant disease resistance. Tetrahedron Lett. 34: 223–226.

Staskawicz BJ, Dahlbeck D & Keen NT (1984) Cloned avirulence gene of *Pseudomonas syringae* pv. *glycinea* determines race-specific incompatibility on *Glycine max* (L.) Merr. Proc. Nat. Acad. Sci., USA 81: 6024–6028.

Staskawicz BJ, Dahlbeck D, Keen, NT & Napoli S (1987) Molecular characterization of cloned avirulence genes from race 0 and race 1 of *Pseudomonas syringae* pv. *glycinea*. J. Bacteriol. 169: 5789–5794.

Swarup S, Yang Y, Kingsley MT & Gabriel DW (1992) An *Xanthomonas citri* pathogenicity gene, *pthA*, pleiotropically encodes gratuitous avirulence on nonhosts. Mol. Plant-Microbe Inter. 5: 204–213.

Van den Ackerveken GFJM, van Kan JAL & De Wit PJGM (1992) Molecular analysis of the avirulence gene *avr9* of the tomato pathogen *Cladosporium fulvum* fully supports the gene-for-gene hypothesis. Plant J. 2: 359–366.

Waney VR, Kingley MT & Gabriel DW (1991) *Xanthomonas campestris* pv. *translucens* genes determining host-specific virulence and general virulence on cereals identified by Tn5-*gusA* insertion mutagenesis. Mol. Plant-Microbe Interact. 4: 623–627.

Whalen MC, Stall RE, & Staskawicz BJ (1988) Characterization of a gene from a tomato pathogen determining hypersensitive resistance in a non-host species and genetic analysis of this resistance in bean. Proc. Nat. Acad. Sci., USA 85: 6743–6747.

Whalen MC, Innes RW, Bent AG & Staskawicz BJ (1991) Identification of *Pseudomonas syringae* pathogens of *Arabidopsis* and a bacterial locus determining avirulence on both *Arabidopsis* and soybean. Plant Cell 3: 49–59.

41. Anti-host-defense systems are elaborated by plant pathogenic bacteria

CLARENCE I. KADO

Abstract. Plant pathogenic bacteria possess sophisticated mechanisms to help evade host defenses. These evasive mechanisms are highly evolved and conserved. Distinctive cell surface components are elaborated to prevent recognition by the host defense system that would result in the hypersensitivity reaction (HR) if the bacterial cell is recognized. It has been long viewed that the HR is a host-defense response. Since there are many different compounds that cause HR, it is plausible that the detection system of the host is non-specific and equipped with an "universal receptor" or "universal detector". It is also plausible that the compatible plant pathogenic bacteria elaborate a compound that suppresses HR. The *hrpX* gene of *Xanthomonas campestris* pathovars and *X. oryzae* confer an anti-defense system since that loss of this gene results in HR in the compatible host. Likewise, *Agrobacterium tumefaciens* and *A. rhizogenes* produce substances that suppress HR in a wide range of plants. Tn5 insertional mutants have revealed the presence of a chromosomal locus involved in this suppression. These findings strongly suggest that there are mechanisms operating that suppress defense responses in the plant and that evade triggering majors defense responses.

Abbreviations: HR, Hypersensitive Reaction

Introduction

Necrogenic plant pathogenic bacteria cause diseases whose symptoms gradually appear and culminate by the collapse and necrosis of the invaded tissues. In contrast, water soaking followed by rapid collapse and death of tissues is observed when the same pathogen is mechanically infiltrated into interdermal spaces of non-host plants. This curious reaction is instigated by artificially introducing large doses of the bacteria into tissue regions that have never encounter such bacteria. This rapid host response coincides with the rapid necrotic response caused by a number of fungus-host systems, of which the reaction has been thought to be a natural defense response of the plant. Although this rapid non-host response with bacteria-plant interactions termed hypersensitivity reaction (HR) (Klement and Goodman, 1967) is not an universal mechanism for resistance because there are systems where HR is not the primary determinant of resistance (Ralton *et al.*, 1988), HR is believed to be a component of resistance in compatible cultivated varieties of given species of plant based on genetic gene-for-gene analyses (Staskawicz *et al.*, 1984; Keen and Staskawicz, 1988; Keen, 1990).

C.I. Kado and J.H. Crosa (eds.), Molecular Mechanisms of Bacterial Virulence, 581–591.
© 1994 *Kluwer Academic Publishers.*

In this chapter, we have used the recent studies on the necrogenic *Xanthomonas campestris* pathovars and on non-necrogenic *Agrobacterium tumefaciens* to provide some insight on potential mechanisms by which these organisms evade triggering the hypersensitive reaction (HR).

Is the hypersensitivity response a gauge for plant resistance against bacterial pathogens?

HR is thought to be paradigm to the hypersensitivity defense response in mammals and humans, where exposure to an antigen is recognized and results in a rapid inflammatory reaction. In the mammalian system, the bacterial antigen is recognized by specific IgE antibody molecules which trigger the basophilic leucocytes in the blood to secrete a variety of biologically active amines, particularly histamine and serotonin. These amines enhance permeability of blood vessels to promote defense systems in the form of phagocytes, antibodies and complement to enter the sites of inflammation. In plants, HR is viewed by many investigators as caused by the recognition of a set of signal molecules [termed "elicitins" (*sensu* Keen and Holiday, 1982)] elaborated by the pathogen which is recognized by specific receptors in the incompatible cultivar or non-host plant, which in turn results in a cascade of reactions resulting in the rapid necrosis of tissues surrounding the infection site. In the compatible host, the pathogen or its set of signal molecules remain unrecognized by the receptors in the plant. Such receptors must be always present to recognize incompatible necrogenic pathogens.

Although receptors in mammalian and human systems have been clearly identified, receptors in plants remain to be demonstrated (Ralton *et al.*, 1988). Part of the reason why such receptors have not been identified is due to the absence of characterized elicitors in bacterial systems. Non-specific elicitors such as 1,3-β-D-glucans, glycoproteins, chitosan, arachidonic acids, oligogalacturonides, gluthathione and even *Escherichia coli* DH5α induce HR in both susceptible and resistant plants (Lamb *et al.*, 1989; Darvill and Albersheim, 1984; Dixon, 1986; Jakobek *et al.*, 1993; Kamdar *et al.*, 1993). Also, the mechanism by which the HR is initiated is poorly understood.

Since HR defective mutants of various plant pathogenic bacteria have been isolated, including mutations in avirulence (*avr*) genes, a general hypothesis has been brought forth that these genes encode products or elicitins that trigger the HR. This hypothesis is predicated on the existence of a receptor made by the plant that specifically recognizes the elicitin produced by the pathogen (Staskawicz *et al.*, 1984; Keen and Staskawicz, 1988; Keen, 1990). That is, for every gene for resistance (R gene) in the host there is a corresponding gene *avr* gene in the pathogen. Further support of this hypothesis requires identification of a specific receptor in the cultivar containing the R gene or in the non-host containing the universal R gene. Because no direct gene product has been isolated that has been shown to bind specifically to a receptor in a resistant cultivar or non-host, the plant HR system pictured to be a paradigm of antigen-antibody interactions in mamalian systems needs further scrutiny.

Arguments against the elicitor-receptor hypothesis

If the elicitor-receptor hypothesis is to be accepted, then the curious phenomenon of the HR response in incompatible cultivars or non-hosts plants must be explained from a long term evolutionary interaction between the bacterial cell and the plant cell. As argued as a paradigm of the antigen-antibody mechanism, the pathogen elaborates an elicitor that binds to a specific receptor in the resistant host or non-host and not in the susceptible host plant unless a counterpart resistance gene (or better resistance genes) is present. This means that a specific receptor must always be present in the non-host to recognize fortuitous encounter with the elicitor of a bacterium that is not a pathogen of that plant. This is a reasonable concept since the plant needs a broad-spectrum recognition system as part of its defense repertoire. Thus, selection pressure to maintain a receptor of this type is omni present owing to the endless attacks by various pathogens and pests. On the other hand, the effectiveness or efficiency of pathogenesis by a pathogen would be lowered appreciably if they continued to possess signals (elicitors) recognized by the plant. This dilemma remains to be explained satisfactorily.

With respect to time and space, there are no examples where necrogenic pathogens forcibly infiltrate into mesophillic intercellular spaces during invasion. Necrogenic pathogens normally invade through wounds and invade vascular systems locally and block interveinal tissues from getting water and nutrients. Forced experimental infiltration does not reflect a natural system and responses generated by such procedures should be viewed as man-made responses. Thus, the question is raised here on whether or not research is being directed toward an artificially devised system.

Search for a direct product of hrp genes

The first bacterial gene product causing an HR was isolated from *Erwinia rubrifaciens* in 1976 (Gardner and Kado, 1976). It was identified as a calcium dependent pectin trans-eliminase (now called pectin lyase) of 41 kDa. The tyrosyl residue in the protein is essential for lyase activity. The enzyme is found in the periplasmic space of *E. rubrifaciens* and is induced above basal levels in the presence of sucrose and released in the presence of polygalacturonic acid or pectin. Only 100 to 1000 ng of this enzyme was needed to initiate HR in tobacco leaves. Electrolyte efflux from tobacco leaf cells exposed to the enzyme can be detected within 60 min. Interestingly, the enzyme has no killing effect on tobacco protoplasts. Thus, it may be that the exocellular pectin present on intact tobacco cells is required for HR by serving as the pectin lyase substrate for generating signal molecules leading to HR.

The 44 kDa protein complexed with cell membranous material called "harpin" was recently isolated from *Erwinia amylovora* as a product of the *hrpN* gene (Wei *et al.*, 1992). This gene product has been shown to cause HR when it is expressed in *Escherichia coli* DH5α. Interestingly, *E. coli* harboring the *hrpN* gene does not cause HR in the homologous host (apple and pear), perhaps because of its inability to secrete harpin. The gene is required for pathogenicity and yet its direct product does not cause HR in pear. Is it therefore the *hrpN* product that causes HR in non-hosts such as tobacco and remains unrecognized by apple and pear?

Xanthomonas campestris *pathovars harbor hrpX, whose product is required for pathogenicity and prevents VHR in the compatible host*

The plant inducible *hrpX* gene was isolated from *X. campestris* pv. campestris (Kamoun and Kado, 1990). The counterpart *hrpXo* gene was recently isolated and characterized from *X. oryzae* pv. oryzae (Kamdar *et al.*, 1993). As shown in Table 1, *hrpX* is highly conserved in *X. campestris* pathovars as well as in *X. oryzae* pv. oryzae. The *hrpX* product is functionally conserved since *hrpXo* will complement a *hrpXc* mutant from *X. campestris* pv. campestris and a *hrpXa* mutant from *X. campestris* pv. amoriaceae.

Table 1. Conservation of hrpX among xanthomonads.

Strain/pathovar	Disease	Host	Hybridization
X. c. campestris	black rot	Cruciferae	strong
pellargonii	leaf spot	Geranaceae	strong
phaseoli	leaf spot	Bean	strong
pisi	leaf spot	Pea	strong
pruni	leaf spot	Prunus spp.	strong
translucens	leaf steak	Graminae	strong
vesicatoria	leaf spot	Tomato/Pepper	strong
vitians	leaf spot	Lettuce	strong
X. o. oryzae	leaf blight	Rice	strong
A. tumefaciens	crown gall	many	none
Rhizobium	nodules	Leguminosae	none
P. syringae	canker	many	none
Erwinia amylovora	fire blight	Rosaceae	weak
E. coli	none	none	none

[a] The hybridization probe was an internal fragment of the *hrpXo* gene. Data adapted from Kamdar *et al.*, 1993.

In each complementation, the disease phenotype is typically induced by the recipient wild-type species (Kamdar *et al.*, 1993). This complementation is reciprocal, i.e., the *hrpXc* gene will complement the *hrpXo* mutant of *X. oryzae* pv. oryzae.

HrpXo encodes a polypeptide of 52.3 kDa whose amino sequence bears loose resemblance to proteins that contain targets for covalent linkage to fatty acid.

Although the function of the HrpXo protein has not been determined, some insight is obtained from genetic experiments which show it is required for pathogenicity. Mutations in *hrpX* causes the loss of pathogenicity, the loss of *X. campestris* to grow *in planta*, and the inability to cause HR in non-hosts. But unlike other *hrp* genes *hrpX* mutants now *causes* HR in the vascular system of the compatible host (Kamoun *et al.*, 1992). Bacteria with the ability to elicit HR in the compatible and natural host (crucifers in the case of *X. campestris* pv. campestris) due to the loss of a gene suggests that one or more naturally occurring surface components are now exposed and the bacteria is consequently recognized by the plant defense system. It is also plausible that the bacterial cell surface material is no longer modified due to the absence of the *hrpX* gene product. The loss of *hrpX* is apparently tissue specific. For example, with *X. campestris* pv. campestris, the HR is localized to the vascular system of the host, tissue regions where this pathogen normally infects and invades. It can be hypothesized that like the rest of the plant, the vascular system also contains the universal receptor that will detect foreign bacteria unless the bacteria can mask itself from being recognized by the host defense system.

The hrpX genes are functionally interchangeable

Nucleotide sequence studies between *hrpXo* and *hrpXc* have revealed about 80% of their sequences are identical. (T. Oku & T. Coutinho, unpublished). The homology also extends between *hrpXo* and *hrpX* genes in a number of *X. campestris* pathovars (Table 1). *hrpXo* is therefore highly conserved. As indicated this gene has functional equivalence between *Xanthomonas* species. (Table 2). Thus, it is likely that functionality conferred by this gene is also conserved among these species.

Table 2. Reciprocal complementation of HR and pathogenicity by *hrpXc* and *hrpXo*.

Avirulent strain	Complementation by		
	none	*hrpXc*	*hrpXo*
X.c. pv. campestris SW11	-	+	+
X.c. pv. campestris JS111	-	+	+
X.c. pv. armoraciae XL9	-	+	+
X.c. pv. armoraciae HW9	-	+	+
X.o. pv. oryzae KP8	-	+	+

HR was scored on attached leaves of *Datura stramonium* (Jimson weed). Pathogenicity was assayed for typical black rot symptoms on cabbage EJW and for blight on rice CAS209 plants.

Although *hrpX* genes show strong homologies, there are regions within the gene and its promoter region that are distinct. The presence major dissimilarities upstream of the coding region (Fig. 1) suggests that the individual *hrpX* genes may be controlled differently among xanthomonads. This is reasonable since it has been shown that the *hrpXc* gene is plant inducible.

```
BamHI
GGATCCGCTGCATACAATCGTGTGCGCC ACGGAGCTCGGCGATTGTTGTC  TTTTGCTC
|||||||||||||||||||||||| |||||||*|| *|    |||||   | |||| **|||||||||
GGATCCGCTGCATACAATCGTTTGCGCCCACC ACAACGGCGCCTTTTGTTGTTTTTGCTC


61
CGCCCCCCAAGAGAGAGACCGGC ATG ATC CTT TCG ACC TAC TTT GCA GCG
|** |||||    |||||| ||||| ||| ||| ||| ||| ||| ||| |1   ||| |||
C   TCCCCTTCAGAGAGCCCGGC ATG ATC CTT TCG ACC TAC TTC GCA GCG


109
ATC TCT GCG TTG TCT TAC GCA GAA CGT CTT CCT ACC TAT ACG AGC
||| ||| ||| ||| ||| ||| ||| ||   ||| ||| ||  | | ||| ||| |||
ATC TCT GCG TTG TCT TAC GCA GAC CGT CTT CCG ATC TAT ACG AGC


154
AGG ATG CTG GTT GGT GCT TGG CCG CAG GGA CTG CAA
||| ||| ||  ||| ||| ||| ||| ||  ||| ||  ||| ||
AGG ATG CTT GTT GGT GCT TGG CCA CAG GGT CTG CAG
                                        PstI
```

Figure 1. Sequence upstream of the coding region of *hrpXo* and *hrpXc* (lower sequence) showing the potential translational start site, ribosome binding and promoter region (underlined). Non-matched nucleotides are indicated by an asterisk.

The hrpX *genes are different from the common hrp genes*

DNA hybridizations between *hrpX* and the *hrp* genes of *Pseudomonas syringae* pv. syringae, *P. syringae* pv. phaseolicola, *P. solanacearum*, *Erwinia amylovora* and *X. campestris* pv. vesicatoria showed no homologies indicating that *hrpX* is distinct from the clustered *hrp* genes (Kamoun *et al.*, 1992; Kamdar *et al.*, 1993). These latter genes are thought to be involved in protein secretory/exporting functions (Gough *et al.*, 1992; Fenselau *et al.*, 1992; Van Gijsegem *et al.*, this volume) since the clustered *hrp* sequences show similarities to peptide secretory genes in animal bacterial pathogens (Pugsley, 1993; Salmond and Reeves, 1993). The potential function of the clustered *hrp* genes is therefore to provide the means for elicitors, toxins or unidentified pathogenic factors to be secreted out of the cell during the course of infection. Equally likely would be that the *hrp* gene cluster provides the secretory mechanism for anti-host defense compounds such as HrpX.

Agrobacterium tumefaciens and A. rhizogenes are not recognized by the host defense system

A. tumefaciens and *A. rhizogenes* cause crown gall tumors and hairy root galls respectively on a wide range of different plant species. Infection by these organisms does not result in HR. The question is therefore raised on why these bacteria are able to infect such diverse host systems without triggering HR? There are few examples where a specific strain of *A. tumefaciens* can cause necrosis like HR. Inoculation of young grapevines of certain cultivars causes a necrotic reaction (Yanofsky *et al.*, 1985). In most cases, however, *A. tumefaciens* does not cause HR when infiltrated into leaf panels of various types of plants. Instead, a slight yellowing typical of chlorosis occurs at the site of infiltration. Interestingly, when *A. tumefaciens* cells are infiltrated with a HR producing *Pseudomonas syringae* pv. phaseolicola strain, no HR results in the test host (tobacco) (Robinette and Matthysse, 1990). This phenomenon is very interesting and suggests that *A. tumefaciens* somehow prevents the host defense system from recognizing *P. syringae* pv. syringae. Only live *A. tumefaciens* cells bearing the Ti plasmid seem to prevent HR. The precise mechanism for this suppression of HR in the plant has not been determined. However, transposon mutants of *A. tumefaciens* that causes loss of their ability to suppress *P. syringae* from eliciting HR have been analyzed (Robinette and Matthysse, 1990). Mutations in the *tms* gene located in the T-DNA of the Ti plasmid in *A. tumefaciens* caused loss of this suppressive activity. The *tms* locus is comprised of two genes, one encoding tryptophan monooxygenase and the other encoding indoleacetamide hydrolase, both of which catalyze the production of indole-3-acetic acid (auxin). The possibility that auxin might be the inhibitor of HR was explored (Robinette and Mathhysse, 1990). Injection of the site infiltrated with *P. syringae* with auxin, however, did not inhibit HR. Thus, other factors in addition to those encoded by the *tms* region of the T-DNA is required for inhibiting HR. Idetification of these factors should prove interesting.

Agrobacterium tumefaciens contains anti-host defense genes

Since *A. tumefaciens* does not induce HR, we have asked the fundamental question on why this bacterium does not elicit HR at least in some hosts? It appears that *A. tumefaciens* is equipped with genes that either encode surface components which are not recognized by the plant or produce components that suppress HR generating genes. This means of avoiding detection by the host defense system is apparently very sophisticated because *A. tumefaciens* infects a wide range of plants without causing HR (De Cleene and Deley, 1977). The common *hrp* cluster found in *P. syringae*, *P. solanacearum* and *E. amylovora* is not present in *A. tumefaciens* (Laby and Beer, 1992). Since the common *hrp* cluster is involved in protein secretion, *A. tumefaciens* either cannot secrete the HR inducing components because it lacks the specific secretory system or is using a totally distinct secretory system to elaborate anti-defense compounds, or does not process HR inducing factors on its cell surface. Approaches to answer which

588

of these possbilities is correct can be derived from the results of genetic experiments. Knowledge of the *A. tumefaciens* genes involved in HR circumvention comes from the analysis of a large number Tn5 insertional mutants that were screened for HR inducing ability. Five mutants able to elicit HR were isolated. Of these, mutant RT102 appeared to be the strongest inducer (Fig. 2). The Tn5 insertion is located in the chromosome and therefore suggests that *A. tumefaciens* harbors chromosomal genes involved in the prevention of HR in plants. Although the nature of the gene products causing or leading to HR suppression have not been identified, the above studies are the first to show that there are genes in *Agrobacterium* that prevent or suppress elicitation of HR in plants and that the HR elicitor may be the "naked" bacterial cell stripped of surface components that aid in evading the host defense system. Surface components may be in the form of appendages like flagella or pili that are recognized by the plant defense system. The characterization of this locus should prove valuable in elucidating the nature of host-defense circumvention by certain plant pathogenic bacteria like *Agrobacterium* species.

Figure 2. Tn5 mutants of *Agrobacterium tumefaciens* infiltrated into leaf panels of a Jimson weed leaf. One mutant eliciting the HR is shown on the left side of the leaf while other sites, infiltrated with different Tn5 mutants, were negative for HR.

Conclusions

The mechanism by which plant pathogenic bacteria cause the hypersensitivity reaction in plants may not be due to a specific compound (elicitor) but rather HR may be due to the loss of either surface components on the bacterial cell that mimic host components HR suppressive and therefore are no longer able to avoid detection by the host defense system or prevent HR. This hypothesis is supported in bacterial pathogens interacting with mammalian species and is in contrast to the concept that the hypersensitivity response is due to a specific gene product such as an elicitor interacting with a specific receptor. Instead, the loss of these protective components on the bacterial cell surface leads to the recognition of the entire bacterial cell by the detection system which may be composed of an "universal receptor" or "universal detector". Arguments used in support of the concept of host-defense evasion come from the features of the *hrpX* gene of *X. campestris* pathovars and the HR preventative features of the *A. tumefaciens* invasion system. Earlier studies on *P. pisi* showed that it suppresses the expression of phenylalanine ammonia lyase genes in pea (T. Yamada, personal commun.). Recently, Jakobek et al (1993) reported that *P. syringae* pv. phaseolicola suppresses the expression of genes as measured for the presence of transcripts encoding phenylalanine ammonia lyase, chalcone synthase, chalchone isomerase and chitinase in common bean.

The study of the hypersensitivity response by injecting large doses of bacteria into tissues which do not normally encounter these organisms is an experimentally induced phenomenon. It is however interesting that such a dramatic necrotizing reaction takes place in plants. Whether this HR response is actually a component of defense against plant pathogenic bacteria or whether the response is secondary to invasion remains to be directly demonstrated. Clues to this possibility have been recently found when a *hrp* defective mutant of *P. syringae* pv. tabaci was still able to cause the expression of genes required for HR production in bean (Jakobek and Lindgren, 1993). Thus, strong consideration should be given to existing paradigms in mammalian pathogen systems which suppress effectively host defenses. Our studies together with those recently reported support the concept that the bacterial plant pathogen possess sophisticated means of evading host defense responses.

Acknowledgements

I thank the members of the Davis Crown Gall Group for helpful discussions and Robert Trelford for unpublished results. The research on xanthomonads was supported by a grant from the Aisin Seiki Company. The research on *Agrobacterium tumefaciens* was supported by NIH grant GM 45550 from the National Institute of General Medicine, Department of Health and Human Services.

References

Azad, H R, Kado C I (1984) Relation of tobacco hypersensitivity to pathogenicity of *Erwinia rubrifaciens*. Phytopathology 74: 61–64.

Darvill, A G, Albersheim, P (1984) Phytoalexins and their elicitors: a defense against microbial infection in plants. Annu. Rev. Plant Physiol. 35: 243–293.

DeCleene, M, De Ley, J (1977) The host range of crown gall. Botan. Rev. 42: 389–466.

Dixon, R A (1986) The phytoalexin response: elicitation, signalling and the control of host gene expression. Biol. Rev. 61: 239–291.

Fenselau, S, Balbo, I, Bonas, U (1992) Determinants of pathogenicity in *Xanthomonas campestris* pv. vesicatoria are related to proteins involved in secretion in bacterial pathogens of animals. Mol. Plant-Microbe Interact. 5: 390–396.

Gardner, J M, Kado, C I (1976) Polygalacturonic acid *trans* –eliminase in the osmotic shock fluid of *Erwinia rubrifaciens*: characterization of the purified enzyme and its effect on plant cells. J. Bacteriol 127: 451–460.

Gough, C L, Genin, S, Zischek, C, Boucher, C A (1992) *hrp* genes of *Pseudomonas solanacearum* are homologous to pathogenicity determinants of animal pathogenic bacteria and are conserved among plant pathogenic bacteria. Mol. Plant-Microbe Interact. 5: 384–389.

Jakobek, J L, Lindgren, P B (1993) Generalized induction of defense responses in bean is not correlated with the induction of the hypersensitive reaction. Plant Cell 5: 49–56.

Jakobek, J L, Smith, J A, Lindgren, P B (1993) Suppression of bean defense responses by *Pseudomonas syringae*. Plant Cell 5: 57–63.

Kamdar, H V, Kamoun, S, Kado, C I (1993) Restoration of pathogenicity of avirulent *Xanthomonas oryzae* pv. oryzae and *X. campestris* pathovars by reciprocal complementation with the *hrpXo* and *hrpXc* genes and identification of HrpX function by sequence analysis. J. Bacteriol. 175: 2017–2025.

Kamoun, S, Kado, C I (1990) A plant-inducible gene of *Xanthomonas campestris* pv. campestris encodes an exocellular component required for growth in the host and hypersensitivity on nonhosts. J. Bacteriol. 172: 5165–5172.

Kamoun, S, Kamdar, H V, Tola, E, Kado, C I (1992) Incompatible interactions between crucifers and *Xanthomonas campestris* involve a vascular hypersensitive response: role of the *hrpX* locus. Mol. Plant-Microbe Interact. 5: 22–33.

Keen, N T (1990) Gene-for-gene complementarity in plant-pathogen interactions. Annu. Rev. Genet. 24: 447–463.

Keen, N T, Holiday, M J (1982) Recognition of bacterial pathogens by plants. *In*: Mount, M S, Lacy G C (ed) Phytopathogenic Prokaryotes, Vol II, p. 179–221, Academic Press, New York.

Keen, N T, Staskawicz, B (1988) Host range determinants in plant pathogens and symbionts. Annu. Rev. Microbiol. 42: 421–440.

Klement, Z, Goodman, R N (1967) The hypersensitive reaction to infection by bacterial plant pathogens. Annu Rev Phytopathol 5: 17–44.

Laby, R J, Beer, S V (1992) Hybridization and functional complementation of the *hrp* gene cluster from *Erwinia amylovora* strain Ea321 with DNA of other bacteria. Mol. Plant-Microbe Interact. 5: 412–419.

Lamb, C J, Lawton, M A, Dron, M, Dixon, R A (1989) Signals and transduction mechnisms for activation of plant defenses against microbial attack. Cell 56: 215–224.

Pugsley, A P (1993) The complete general secretory pathway in Gram-negative bacteria. Microbiol. Revs. 57: 50–108.

Ralton, J E, Howlett, B J, Clarke, A E (1988) Receptors in host-pathogen interactions, *In*: Carbohydrate-Protein Interaction, Clarke, A E, Wilson, I A (eds.), Springer-Verlag, Berlin/New York.

Robinette, D, Matthysse, A G (1990) Inhibition by *Agrobacterium tumefaciens* and *Pseudomonas savastanoi* of development of the hypersensitivity response elicited by *Pseudomonas syringae* pv. phaseolicola. J. Bacteriol. 172: 5742–5749.

Salmond, G P C, Reeves, P J (1993) Membrane traffic wardens and proteins secretion in Gram-negative bacteria. TIBS 18: 7–12.

Staskawicz, B J, Dahlbeck, D, Keen, N T (1984) Cloned avirulence gene of *Pseudomonas syringae* pv. *glycinea* determines race-specific incompatibility of *Glycine max* (L.) Merr. Proc. Natl. Acad. Sci. USA 81: 6024–6028.

Wei, Z-M, Laby, R J, Zumoff, C H, Bauer, D W, He, S-Y, Collmer, A, Beer, S V (1992) Harpin, elicitor of the hypersensitive response produced by the plant pathogen *Erwinia amylovora*. Science 257: 85–88.

Yamada, T, Palm, C J, Brooks, B, Kosuge, T (1985) Nucleotide sequences of the *Pseudomonas savastanoi* indoleacetic acid genes show homology with *Agrobacterium tumefaciens* T-DNA. Proc. Natl. Acad. Sci. USA 82: 6522–6526.

Yanofsky, M, Lowe, B, Montoya, A, Ribin, R, Krul, W, Gordon, M, Nester, E (1985) Molecular and genetic factors controlling host range in *Agrobacterium tumefaciens*. Mol Gen Genet 201: 237–246.

42. Organization, regulation and function of *Pseudomonas syringae* pv. syringae *hrp* genes

STEVEN W. HUTCHESON, SUNGGI HEU, HSIOU-CHEN HUANG, MICHAEL C. LIDELL and YINGXIAN XIAO

Abstract. The pathogenicity and host range of *Pseudomonas syringae* is controlled in part by *hrp* genes. Inactivation of these genes produces nonpathogenic mutants that are unable to elicit the hypersensitive response (HR) in resistant plants or nonhost plant species, a response indicative of incompatibility. In *P.s.* syringae 61, the *hrp* genes cluster in a 25 kb region of the genome. Transposon mutagenesis and complementation analyses in merodiploids have revealed 16 apparent translational units organized as seven apparent transcriptional units. Associated with the *hrp* cluster is the *hrm*A locus. Phenotypic expression of the *hrp/hrm* gene cluster in non-phytopathogenic bacteria, such as *E. coli*, enables these bacteria to elicit the HR in a wide variety of plant species. Tn*pho*A mutagenesis has identified two *hrp* genes that produce membrane-associated proteins. Eight of the *hrp* operons are regulated by nutritional conditions. At least two *hrp* loci have regulatory activity. Nucleotide sequence analysis of the region has revealed several loci sharing substantial homology with *Yersinia* virulence genes. The role of these genes in determining *P. syringae* pathogenicity and elicitation of the HR are discussed.

Abbreviations: HR, Hypersensitive Response; *hrp*, Hypersensitive Response-Pathogencity; *hrm*, Hypersensitive Response Modulation

Introduction

Plants, like many multicellular organisms, have the capacity to respond to invading bacterial pathogens to prevent further invasion of the tissue. Plant defense responses to invading pathogens include: 1) production of hydrogen peroxide and superoxide (active oxygen species; Baker *et al.* 1993); 2) accumulation of sesquiterpanoid, flavanoid and polyketide phytoalexins with antimicrobial activity (Dixon and Lamb, 1990); 3) secretion of hydrolytic enzymes (e.g., glucanases and chitinases) which may degrade bacterial cell walls (Bowles, 1990); and 4) modifications to their cell walls, such as incorporation of hydroxyproline-rich glycoproteins and increased ligninification (Bowles, 1990). The combined activity of these responses is thought to cause bacteriostasis and limit further spread of the pathogen. Disease occurs when these apparent defense mechanisms fail to activate early in the interaction or the pathogen has the capacity to defeat the plant's defense through enzymatic modification, dampening, inactivation or other protective mechanisms.

An interesting model system to investigate the phenomenon of disease resistance in plants is the interaction of *Pseudomonas syringae* strains with

C.I. Kado and J.H. Crosa (eds.), Molecular Mechanisms of Bacterial Virulence, 593–603.

susceptible and resistant hosts. *P. syringae* is a ubiquitous fluorescent pseudomonad that grows epiphytically on a wide variety of plant species (Hirano and Upper, 1991). Most strains of *P. syringae* lack the ability to actively penetrate plant tissue or plant cells, but instead use natural portals, such as stomata or wounds, to enter plant tissue (Huang, 1986). Once internal to a susceptible plant, the bacteria colonize the intercellular spaces of the tissue but remain external to living plant cells. "Water soaking" symptoms typically develop as populations increase and the tissue eventually become necrotic. In resistant plants or non-host plant species, only transient multiplication of the bacteria is observed which appears to be arrested when defenses of the surrounding cells become active. In many plant species, the responding plant cells rapidly become necrotic. In such interactions, introduction of inocula sufficient to cause 50% or more of the cells to respond produces a visible necrosis, known as the hypersensitive response (HR; Klement, 1982), which is used as a bioassay for signal factors. At least forty host range variants (pathovars) of *P. syringae* have been identified which are indistinguishable by biochemical or serological criteria, but cause disease in characteristic host plants (Dye *et al.*, 1980).

Activation of plant defenses and the associated HR by *P. syringae* strains appears to occur via a poorly understood recognition process. Recognition of *P. syringae* by plants differs from mammalian paradigms because no known surface feature of the bacteria is correlated with recognition. The bacteria, however, are thought to carry a subtle signal that is perceived by a defense surveillance system of the plant cell to initiate a response in resistant plants. The activity of *avr* and/or *hrp* gene products is thought to produce this postulated recognition signal. *Avr* genes are well established host-range determinants whose presence restricts the host range of a pathogen to those host plants which lack a corresponding resistance gene specific for the *avr* gene (Keen, 1990). *Avr* genes appear to explain the races of a pathogen and may also control its host range at the plant species level. *Hrp* genes control the ability of a pathogen to elicit an HR in any plant host, phenotypic expression of *avr* genes, pathogenicity in susceptible plant species, and multiplication in both susceptible and resistant hosts (Lindgren *et al.*, 1986; 1988).

Although *avr* and *hrp* genes have known for some time, the biochemical components that mediate this hypothesized recognition process are just now being elucidated. The genetic complexity, environmental regulation of contributing genes, and instability of the postulated recognition factor have hindered progress toward understanding the recognition event. Significant progress has resulted from the isolation of a large gene cluster from a strain of *P. syringae* pv. syringae 61 (P.s. syringae 61) by Huang *et al.* (1988) that enables *Escherichia coli* to elicit the HR in tobacco. The genetic dissection of this cluster and its properties have illuminated several aspects of the recognition process and are described below.

Properties of the *hrp/hrm* gene cluster

The phenotype of *P.s.* syringae 61 Tn*5* mutants that map within the cloned DNA fragment suggested the presence of *hrp* genes within the cluster (Baker *et al.*, 1987; Huang *et al.*, 1988). Fourteen apparent translational units organized as eight apparent transcriptional units have been defined within the fragment by transposon mutagenesis and complementation analysis in merodiploids (Huang *et al.*, 1991; Xiao *et al.*, 1992) (Fig. 1). Thirteen of these apparent translational units have been shown to be *hrp* genes by phenotypic criteria and a subset showed sequence homology to the *P.s.* phaseolicola *hrp*L-R cluster (Rahme *et al.*, 1991) by hybridization analysis (Huang *et al.*, 1991). This *hrp* cluster is highly conserved and is present in all *P. syringae* strains examined thus far (150 strains representing more than 27 host range variants; T. Denny, unpublished results; Lindgren *et al.*, 1988). The remaining locus exhibits a unique phenotype in which *P.s.* syringae 61 mutants retained pathogenicity in the susceptible host, but produced a delayed necrotic reaction in resistant plants. We designated this locus, *hrm*A (for hypersensitive response modulation), and hybridization analysis indicates that this locus is only present in a subset of *P. syringae* strains (S. Heu and S.W. Hutcheson, submitted).

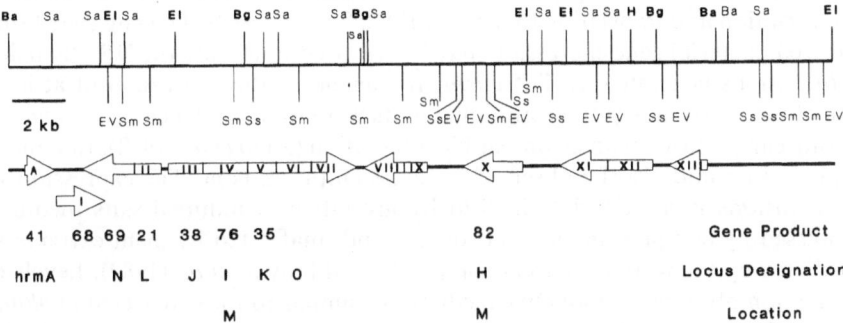

Figure 1. Organization of the *hrp/hrm* gene cluster of *Pseudomonas syringae* pv. syringae 61. (A) Restriction map of the region (modification of Xiao *et al.*, 1992). Abbreviations: Ba, *Bam*H1; Bg, *Bgl*II; EI, *Eco*R1; EV, *Eco*RV; H, *Hind*III; Sa, *Sal*I; Sm, *Sma*I; Ss, *Sst*I. (B) Apparent transcriptional, translational and regulatory organization of the cluster. Transcriptional units and their orientation are indicated by the unfilled arrows (from Huang *et al.*, 1991; Xiao *et al.*, 1992; this report). Translational units labeled with roman numerals exhibit a *hrp* phenotype when mutated. The *hrm*A locus is designated by the letter A. Deduced masses of the gene products (kDa) based on translation of ORF's together with newly assigned genetic designations are indicated below. The ORF's assigned to *hrp*JIKO are separated by intergenic regions of less than 50 nucleotides that *lack* obvious transcriptional termination sequences, and thus, are assumed to represent a single large operon. Compartmentation of gene products was determined by Tn*pho*A mutagenesis and sequence analysis. M indicates membrane-spanning or exported proteins.

Most genetic models describing incompatible interactions of *P. syringae* strains with resistant hosts invoke the *avr* genes. Although the entire cluster and two subcloned regions of the cluster caused several *P. syringae* strains to be incompatible with their normally susceptible hosts (Huang *et al.*, 1988), no subset of genes (contiguous or dispersed) within the cluster, including *hrm*A, exhibited the qualities of an *avr* gene when tested for an effect on *P.s.* glycinea race 4 pathogenicity, a well characterized system for identifying *avr* genes (Heu *et al.*, 1991ab; S. Heu and S.W. Hutcheson, submitted; c.f., Kobayashi *et al.*, 1990; Keen and Buzzell, 1991). Insertional inactivation of any *hrp* or *hrm*A gene within the cluster, however, abolishes the HR-eliciting activity, irrespective of the bacterial host. The HR-eliciting activity generated by the *hrp/hrm* gene cluster is produced in a diverse collection of pathogenic and nonpathogenic bacteria and elicits a response in a wide variety of monocotyledonous and dicotyledonous plant species (Table 1). Either all bacteria tested carry a cryptic *avr* determinant whose phenotypic expression is dependent on the entire *hrp/hrm* gene cluster or the *hrp/hrm* gene cluster forms a minimum genetic unit sufficient to generate a factor or activity that elicits the HR.

Environmental regulation of *hrp* genes

Early work on the development of the HR noted the presence of an induction stage during which the plant response is sensitive to inhibitors of bacterial transcription and translation (Klement, 1982; Sasser, 1982). This suggested that induced bacterial gene expression may be required to initiate the HR. By using *uid*A fusions generated by Tn5-*gus*A1 mutagenesis, the expression of at least five of the apparent *hrp* transcriptional units were found to be induced *in planta* during early phases of an incompatible interaction (Xiao *et al.*, 1992). Enhanced expression could also be observed in some culture media. The expression of most fusions is increased > 20-fold by growth in a minimal salts medium, repressed by complex amino acid sources and unaffected by plant extracts as predicted by the work of Yucel *et al.* (1989) and Huynh *et al.* (1989). Levels of expression observed in inductive media were similar to those detected *in planta* during the incompatible interaction with tobacco and growth in minimal salts media abolishes or substantially reduces the length of the induction stage required to initiate the HR (Hutcheson *et al.*, 1989; Li *et al.*, 1992). In parallel experiments, we have observed that several plasmid-borne *hrp-uid*A fusions are expressed > 50-fold higher than their chromosomal equivalents. This enhanced expression may explain the rapid plant responses elicited by strains carrying the cloned cluster (Xiao *et al.*, 1992). It, therefore, appears that enhanced production of the HR-eliciting signal occurs in culture from strains carrying the cloned *hrp/hrm* gene cluster.

Several *hrp* genes within the cluster appear to have regulatory activity. Heu *et al.* (1992) has observed that *hrm*A expression is dependent upon the expression of *hrp*"II" and *hrp*"XIII" (sensu Xiao *et al.*, 1992). Inactivation of these genes blocked *hrp*-directed expression of a *hrm*A'-*lacZ* fusion in *Escherichia coli*. *Hrp*-directed expression could be reconstituted in *E. coli* by the addition of both loci cloned individually on separate plasmids. Hybridization

Table 1. Plant species exhibiting hypersensitivity to bacteria carrying the *P. syringae* pv. syringae *hrp/hrm* gene cluster.

Bacterium[a]	Incompatible Plant Hosts[b]
P. **fluorescens** 55[c]	Solanaceae: _Nicotiana tabacum_ L.[d,e] _N. tomentosporium_ Ruiz & Paron _N. glauca_ Graham _N. sylvestris_ Speg. & Comes _Lycopersicon esculentum_ P. Mill. _Capsicum annuum_ L. Brassaceae: _Arabidopsis thaliana_ (L.) Heynh. _Raphanus sativus_ L. Geraniaceae: _Pelargonium_ spp. Fabaceae: _Glycine max_ L. _Pisum sativa_ L. _Phaseolus vulgaris_ L. Cucurbitaceae: _Cucumis sativa_ L. Poaceae: _Zea mays_ L. _Triticum aestivum_ L.
P. **solanacearum** 441	_N. tabacum_ L.
X. **campestris** pv. **vesicatoria** 336	_L. esculentum_ P. Mill. _Capsicum annuum_ L.
P. **syringae** pv. **tabaci** pv. **lachrymans** pv. **syringae** 226 pv. **syringae** 61 pv. **glycinea** race 4	_N. tabacum_ L.[e] _Cucumis sativa_ L. _L. esculentum_ P. Mill.[e] _P. vulgaris_ L.[e] _G. max_ L.

[a] Strain carrying pHIR11

[b] Leaves were infiltrated with a bacterial suspension at 1×10^9 bacteria/ml and scored for tissue collapse and necrosis typical of the HR at 24 h.

[c] Similar results were obtained with _E. coli_ MC4100 and P.s. syringae 61

[d] Cultivars Hicks and Samsun

[e] as reported by Huang et al., 1988

analysis indicates that *hrp*"XIII" is a homolog of *hrp*S (henceforth designated *hrp*; Heu and Hutcheson, unpublished results). *Hrp*S is an unusual member of the NtrB/C family of two component regulatory proteins and shares homology with the central domain of the effector component (Grimm and Panopoulos, 1988; Felley *et al.*, 1991). Positional criteria suggests *hrp*"II" is a homolog to the *hrp*L locus of the *P.s.* phaseolicola *hrp* cluster which has been reported to have regulatory activity (henceforth designated *hrp*L; c.f., Huynh *et al.*, 1989; Felley *et al.*, 1991; Rahme *et al.*, 1992)

Identification of exported *hrp* gene products

In the absence of plant cell penetration, recognition of *P. syringae* is likely to be an extracellular event and to involve secreted or membraneous proteins. Li *et al.* (1992) examined the role of the outer membrane in the communication of HR-eliciting signal produced by the *hrp/hrm* cluster. She was unable to detect alterations in the outer membrane of *E. coli* K-12 strains carrying the *hrp/hrm* gene cluster by immunological techniques or through comparative protein analyses. Her results suggested instead that porins may function in the deployment of the HR-eliciting signal, but did not exclude secreted or membrane proteins that may be produced below the sensitivity of the assay. Huang *et al.* (1991) has reported that two *hrp* genes produce membrane-spanning or exported products. Several *hrp*H::Tn*phoA* and *hrp*I::Tn*phoA* mutants were identified that exhibited phenotypes indicative of insertion into membrane-spanning or periplasmic domains.

Nucleotide sequence of *hrp* genes and their homology with *Yersinia* virulence determinants

To further characterize selected genes within the *P.s.* syringae 61 *hrp/hrm* cluster, the nucleotide sequence for regions carrying *hrm*A, *hrp*L, *hrp*JIK and *hrp*H have been obtained. Sequence data revealed that *hrp*L is part of a multigene operon and is oriented opposite to the direction predicted by the results of Xiao *et al.*, 1992. The *hrp*JIK genes appear to be part of a larger operon. A summary of these analyses is presented in Fig. 1.

Sequence analysis of *hrp*H has identified an ORF encoding a 82 kDa polypeptide with properties similar to envelope proteins (Huang *et al.*, 1992). The *hrp*H product exhibits 33% identity over 334 amino acid residues with YscC of the VirC region of *Yersinia enterocolitica*, a bacterial pathogen of humans and other mammals, that may function in protein secretion (Michiels *et al.*, 1991; Huang *et al.*, 1992). It has not been established whether *hrp*H has a similar function. The nucleotide sequence of the *hrp*I locus revealed an ORF encoding a 67 kD polypeptide with strong homology (58% identity/91% similarity over 320 aa residues) with LcrD, a proposed regulatory protein controlling temperature and Ca^{++}-dependent expression of several virulence proteins of *Yersinia pestis* (Plano *et al.*, 1991; H.C. Huang, Y. Xiao, R.-H. Lin, Y. Lu, S. Hutcheson, and A. Collmer, submitted). Lower homology was observed with

FlbF of *Caulobacter crescentus*, a regulatory protein controlling cell cycle-dependent flagellar development (Ramakrishman *et al.*, 1991). The observation that two of the *hrp* genes share homology with yersiniae virulence genes is highly provocative and suggests a common ancestry, and possibly, similar functions for the gene products. Consistent with a common ancestry is the observation that a region of *hrm*A exhibits 63% identity over 133 nt with the *yop*E locus of *Y. enterocolitica* and *hrp*"I" shares 59% identity over 152 nt with *yop*51 (S. Heu, Y. Xiao and S.W. Hutcheson, unpublished results). The *hrp*N locus shares partial homology (39% similarity/17% identity over 280 amino acid residues) with *pro*V, an ATP-binding periplasmic protein that functions in proline and glycine-betaine transport (Gowrishankar, 1989; Y. Xiao, Y. Lu and S.W. Hutcheson, unpublished results). There are, however, significant differences between the functional domains of *pro*V and *hrp*N which suggests *hrp*N may have a different function. The deduced amino acid sequences of *hrm*A, *hrp*L, *hrp*J and *hrp*K have not shown significant homology with any proteins of known function stored in GenBank, EMBL, or Swissprot databases thus far.

Concluding remarks

Several observations, when taken together, are consistent with the conclusion that *hrp* genes play an important role in recognition process that determines the pathogenicity and host range of *P. syringae* strains: 1) all *P. syringae* strains examined thus far carry a homolog to the *hrp*L-R cluster (Lindgren *et al.*, 1988; Huang *et al.*, 1991; T. Denny, unpublished results); 2) mutation of any *hrp* gene within the cluster abolishes pathogenicity and the ability of *P. syringae* strains to elicit the HR which can be restored by genetic complementation (Panopoulos *et al.*, 1985; Lindgren *et al.*, 1986; 1988; Neipold and Mills, 1986; Huynh *et al.*, 1989; Huang *et al.*, 1991; Xiao *et al.*, 1992; Heu *et al.*, 1991); 3) expression of these genes is repressed during culture in rich media and increases > 10-fold during the initial phase of the HR that is associated with recognition (Xiao *et al.*, 1992); 4) amplified expression from a plasmid-borne *hrp*/*hrm* gene cluster appears sufficient to shorten or abolish the time required for recognition and hasten development of the HR (Xiao *et al.*, 1992); 5) environmental signals that affect development of the HR also affect the expression of *P.s.* syringae 61 *hrp* genes (Yucel *et al.*, 1989; Xiao *et al.*, 1992); 6) two *hrp* genes produce exported or membrane-spanning proteins which may be involved in protein secretion and suggests the cluster has the capacity to interface with the external environment (Huang *et al.*, 1991); and 7) the cluster enables heterologous nonpathogenic bacteria to elicit an HR in tobacco whose early physiological features (K^+/H^+ and active oxygen responses) are similar to those produced by the parental *P.s.* syringae 61 strain (Huang *et al.*, 1988; Glazener *et al.*, 1991).

The *hrp*/*hrm* gene cluster isolated from *P.s.* syringae 61 thus appears to form a minimum genetic unit sufficient to generate a factor or activity that elicits the HR. This opens the possibility that, in addition to their role in pathogenicity, the *hrp* genes control the host range of *P. syringae* strains at the host species level. Consistent with this conclusion is the observation that at least two regions of the *P.s.* syringae 61 *hrp/hrm* gene cluster affect the host range of certain *P. syringae* strains (Huang *et al.*, 1988; Heu *et al.*, 1991;1992).

Several hypotheses can be presented to explain the role of *hrp* genes in determining the host range of *P. syringae* strains. Although all *P. syringae* strains examined thus far carry a set of *hrp* genes homologous to the *hrp*L-R cluster originally characterized in *P.s.* phaseolicola (Rahme *et al.*, 1991), sequence variation between the *hrp* clusters is apparent. Differences in hybridization efficiencies and restriction fragment length polymorphisms have been reported (c.f.: Panopoulos *et al.*, 1985; Lindgren *et al.*, 1988; Huang *et al.*, 1988;1991; Huynh *et al.* 1989). Marker exchange mutagenesis can only be completed between closely related strains (Lindgren *et al.* 1988; Huang *et al.*, 1988;1991; Hutcheson, unpublished results). Whether this apparent sequence variation is sufficient to produce products with unique activities, however, has not been established. A subclone carrying *hrp* genes isolated from P.s phaseolicola has been reported to complement a single marker exchange mutant of *P.s.* glycinea and *P.s.* tabaci (Lindgren *et al.* 1988). Other work has shown that some strains with distinct host ranges share apparently identical *hrp* clusters. For example, the EcoR1 fragments of the *hrp* cluster isolated from *P.s.* syringae 226, a tomato pathogen, were similar to those of *P.s.* syringae 61, a weak bean pathogen (Huang *et al.*, 1988). Portions of the *P.s.* syringae 61 *hrp* cluster, however, still affect the host range of *P.s.* syringae 226. Direct comparison of the nucleotide sequence between host range variants has not been reported as yet for any *hrp* genes.

A second possibility could be that *hrp* genes are differentially expressed in tissue of susceptible and resistant plants. Environmental signals reported to affect the expression of the *P.s.* phaseolicola *hrp* genes include osmotic conditions and a putative plant factor (Rahme *et al.*, 1991;1992) whereas the *hrp*-directed expression of the *avr*B locus in *P.s.* glycinea race 0 (and unspecified *hrp* genes; data not shown) was affected by an apparent carbon source-dependent catabolite repression and peptone (Huynh *et al.*, 1989). Expression of the *P.s.* syringae 61 *hrp* genes was unaffected by plant factors and carbon source, but was inhibited by low levels of complete amino acid sources (Xiao *et al.*, 1992). A direct comparison of the expression of *hrp* genes in susceptible and resistant hosts has been reported for *E. amylovora*. Wei *et al.* (1992) reported that the expression of several *E. amylovora hrp* genes is induced slowly in susceptible plant species whereas rapid induction was observed in tobacco, a non-host plant species in which an HR is induced. In contrast, Rahme *et al.* (1992) did not observe significant differences in the expression of several *P.s.* phaseolicola *hrp* genes in susceptible and resistant hosts.

A third possibility could be that other genes, such as *avr* genes or *hrm*A analogs, affect the expression of one or more *hrp* genes or modify the activity of the *hrp* products. This hypothesis is largely unexplored. It is not known whether the HR elicited by the activity of *avr* or *hrp* genes is formed by the same mechanism. Orlandi *et al.* (1992) has reported that early physiological responses (K$^+$/H$^+$ response; active oxygen burst) induced by a *P.s.* glycinea race 4 derivative carrying *avr*A are similar to that reported for other incompatible interactions. The converse in which *hrp* genes affect *avr* expression, however, has been reported. Huynh *et al.* (1989) showed that the expression of a plasmid-borne *avr*B-*lac*Z fusion in *P.s.* glycinea race 0 was dependent upon two *hrp* genes which acted as positive regulatory elements. As described above, we have observed that *hrm*A expression is dependent upon *hrp*L and *hrp*R.

The homology of *hrp*H and *hrp*I with known yersiniae virulence factors involved in protein secretion is consistent with the involvement of one or more secreted proteins in determining the pathogenicity of *P. syringae* strains. Secreted proteins have recently been identified which appear to explain *Erwinia amylovora* and *P. solanacearum* virulence (Wei *et al.*, 1992; see Van Gijsegem *et al.*, this volume). Both bacteria carry *hrp* clusters that are partially homologous to the *P. syringae hrp* cluster and *P. solanacearum* has been shown to carry LcrD and YscC homologs (Genin *et al.*, 1992). To date, no analogous protein has been identified in *P. syringae* strains. Either low level production of a protein with high specific activity or instability of the postulated factor have hindered its identification if this hypothesis is correct. It is clear that further work is necessary to elucidate the function of specific *hrp* genes and the mechanism by which a plant defense response is initiated.

Acknowledgements

This work has been supported by funds provided by the Center for Agricultural Biotechnology of the Maryland Biotechnology Institute, the Maryland Agricultural Experiment Station of the Maryland Institute for Agriculture and Natural Resources, National Science Foundation Grants DCB8716967 and DMB9121226, USDA Cooperative Agreement 58–1275–8–033 and USDA NRICGP Grant 91–37303–6425.

References

Baker CJ, Atkinson MM and Collmer A (1987) Concurrent loss in Tn5 mutants of *Pseudomonas syringae* pv. syringae of the ability to induce the hypersensitive response and host plasma membrane K$^+$/H$^+$ exchange in tobacco. Phytopathol. 77: 1268–1272

Baker CJ (1993) Active oxygen metabolism during plant/bacterial recognition. *In*: Bills D and Kung SD (eds.) Biotechnology and Plant Protection. Bacterial pathogenesis and disease resistance (in press). Singapore Press, Singapore

Bowles DJ (1990) Defense-related proteins in higher plants. Annu. Rev. Biochem. 59: 873–907

Dixon RA and Lamb C (1990) Molecular communication in interactions between plant and microbial pathogens. Annu. Rev. Plant Physiol. 41: 339–367

Dye DW, Bradbury JF, Goto M, Hayward AC, Lelliott RA and Schroth MN (1980) International standards for naming pathovars of phytopathogenic bacteria and a list of pathovar names and pathotype strains. Rev. Plant Pathol. 59: 153–168

Fellay R, Rahme LG, Mindrinos MN, Frederick RD, Pisi A and Panopoulos NJ (1991) Genes and signals controlling the *Pseudomonas syringae* pv. phaseolicola-plant interaction. p. 45–52. *In:* Hennecke H and Verma DPS (eds) Advances in Molecular Genetics of Plant-Microbe Interactions. Vol. 1. Kluwer Academic Publishers Dordrecht

Gough CL, Genin S, Zischek C and Boucher CA (1992) *hrp* genes of *Pseudomonas solanacearum* are homologous to pathogenicity determinants of animal pathogenic bacteria and are conserved among plant pathogenic bacteria. Molec. Plant Microbe Interactions 5: 384–389

Glazener JA, Huang HC and Baker CJ (1991) Active oxygen induction in tobacco cell suspensions treated with *Pseudomonas fluorescens* containing the cosmid pHIR11 and with strains containing Tn*pho*A mutations in the *hrp* cluster. Phytopathology 81: 1196

Grimm C and Panopoulos NJ (1989) The predicted protein product of a pathogenicity locus from *Pseudomonas syringae* pv. phaseolicola is homologous to a highly conserved domain of several prokaryotic regulatory proteins. J. Bacteriol. 171: 5031–5038

Heu S and Hutcheson SW (1991a) *Pseudomonas syringae* pv. syringae *hrp/hrm* genes encode avirulence functions in *P. syringae* pv. glycinea race 4. Phytopathol. 81: 702–703

Heu S and Hutcheson SW (1991b) Molecular characterization of the *Pseudomonas syringae* pv. syringae 61 *hrm*A locus. Phytopathology 81: 1245

Hirano SS and Upper CD (1990) Population biology and epidemiology of *Pseudomonas syringae*. Annu. Rev. Phytopathology 28: 155–177

Huang H-C, Schuurink R, Denny TP, Atkinson MM, Baker CJ, Yucel I, Hutcheson SW and Collmer A (1988). Molecular cloning of a *Pseudomonas syringae* pv. syringae gene cluster that enables *Pseudomonas fluorescens* to elicit the hypersensitive response in tobacco plants. J. Bacteriol. 170: 4748–4756

Huang H-C, Hutcheson SW and Collmer A (1991) Characterization of the *hrp* cluster from *Pseudomonas syringae* pv. syringae 61 and Tn*pho*A tagging of exported *hrp* proteins. Molec. Plant-Microbe Interact. 4: 469–476

Huang, H-C, He SY, Bauer DW, and Collmer A (1992) The *Pseudomonas syringae* pv. syringae 61 *hrp*H product: an envelope protein required for elicitation of the hypersensitive response in plants. J. Bacteriol. 174: 6878–6885

Huang JS (1986) Ultrastructure of bacterial penetration in plants. Annu. Rev. Phytopathol. 24: 141–157

Hutcheson SW, Collmer A and Baker CJ (1989) Elicitation of the hypersensitive response by *Pseudomonas syringae*. Physiol. Plantarum 76: 155–166.

Huynh T, Dahlbeck D and Staskawicz BJ (1989) Bacterial blight of soybean: Regulation of a pathogen gene determining host cultivar specificity. Science 245: 1374–1377

Keen NT (1990) Gene for gene complementarity in plant-pathogen interactions. Annu. Rev. Genet. 24: 447–463

Keen NT and Buzzell RI (1991) New resistance genes in soybean against *Pseudomonas syringae* pv. glycinea: evidence that one of them interacts with a bacterial elicitor. Theor. Appl. Genet. 81: 133–138

Klement Z (1982) Hypersensitivity. p. 149–177. *In:* Mount MS and *lacy* GH (eds.) Phytopathogenic Prokaryotes Vol. 2. Academic Press, N.Y.

Kobayashi DY, Tamaki SJ and Keen NT (1989) Cloned avirulence genes from the tomato pathogen *Pseudomonas syringae* pv. tomato confer cultivar specificity on soybean. Proc. Natl. Acad. Sci. USA 86: 157–161

Li TH, Benson SA and Hutcheson SW (1991) Phenotypic expression of the *Pseudomonas syringae* pv. syringae 61 *hrp/hrm* gene cluster in *Escherichia coli* requires a functional porin. J. Bacteriol. 174: 1742–1749

Lindgren PB, Peet RC and Panopoulos NJ (1986) Gene cluster of *Pseudomonas syringae* pv. phaseolicola controls pathogenicity of bean plants and hypersensitivity on nonhost plants. J. Bacteriol. 168: 512–522

Lindgren PB, Panopoulos NJ, Staskawicz BJ and Dahlbeck D (1988) Genes required for pathogenicity and hypersensitivity are conserved and interchangeable among pathovars of *Pseudomonas syringae*. Mol. Gen. Genet. 211: 499–506

Michiels T, Vanooteghem J-C, Lampert de Rouvroit C, China B, Gustin A, Boudry P and Cornelis GR (1991) Analysis of *virC*, an operon involved in the secretion of Yop proteins by *Yersinia enterocolitica*. J. Bacteriol 173: 4994–5009

Orlandi EW, Hutcheson SW and Baker CJ (1992) Early physiological responses associated with race specific recognition in soybean treated with *Pseudomonas syringae* pv. glycinea. Physiol. Molec. Plant Pathol., (in press)

Panopoulos NJ, Lindgren PB, Willis DK and Peet RC (1985) Clustering and conservation of genes controlling the interactions of *Pseudomonas syringae* pathovars with plants. p. 69–85. *In*: Sussex I, Ellingboe AH, Crouch M, Mulmberg R (eds) Plant Cell/Cell Interactions. Curr. Commun. Mol. Biol. Cold Spring Harbor Laboratory Press, NY

Plano GV, Barve SS and Straley SC (1991) LcrD, a membrane-bound regulator of the *Yersinia pestis* low calcium response. J. Bacteriol. 173: 7923–7303

Rahme LG, Mindrinos MN and Panopoulos NJ (1991) Genetic and transcriptional organization of the *hrp* cluster of *Pseudomonas syringae* pv. phaseolicola. J. Bacteriol. 173: 575–586

Rahme LG, Mindrinos MN and Panopoulos NJ (1992) Plant and environmental sensory signals control the expression of *hrp* genes in *Pseudomonas syringae* pv. phaseolicola. J. Bacteriol. 174: 3499–3507

Ramakrishnan G, Zhao J-L and Newton A (1991) The cell cycle-regulated flagellar gene *flbF* of *Caulobacter crescentus* is homologous to a virulence locus (*lcrD*) of *Yersinia pestis*. J. Bacteriol. 173: 7283–7292

Sasser M (1982) Inhibition by antibacterial compounds of the hypersensitive reaction induced by *Pseudomonas pisi* in tobacco. Phytopathology 72: 1513–1517

Straley SC (1991) The low-Ca2+ response virulence regulon of human-pathogenic yersiniae. Microbial Pathogenesis 10: 87–91

Wei Z-M, Sneath BJ and Beer SV (1992) Expression of *Erwinia amylovora hrp* genes in response to environmental stimuli. J. Bacteriol. 174: 1875–1882

Wei ZM, Laby RJ, Zumoff CH, Bauer DW, He SY, Collmer A and Beer SV (1992) Harpin, elicitor of the hypersensitive response produced by the plant pathogen, *Erwinia amylovora*. Science 257: 85–88

Xiao Y, Lu Y, Heu S and Hutcheson SW (1992) Organization and environmental regulation of the *Pseudomonas syringae* pv. syringae 61 *hrp* cluster. J. Bacteriol. 174: 1734–1741

Yucel I, Xiao Y and Hutcheson SW (1989) Influence of *Pseudomonas syringae* culture conditions on the initiation of the hypersensitive response of cultured tobacco cells. Appl. Environ. Microbiol. 55: 1724–1729

43. The role of indoleacetic acid production by *Pseudomonas syringae* pathovars on their pathogenicity on host plants

FRANK F. WHITE and MARK MAZZOLA

Abstract. The role of indoleacetic acid (IAA) has been documented in a variety of systems involving tissue hyperplasia. However, its role, if any, in diseases that do not involve hyperplasia, is unknown. IAA biosynthesis by *Pseudomonas syringae* pv. syringae, the causal agent of brown leaf spot disease of bean, is produced via the indoleacetamide pathway, similar to the pathway in *P.s.* pv. savastanoi. The *iaaM* and *iaaH* genes are present on a 2.8 kb *Eco*RI fragment in strain Y30 and possess approximately 90% sequence identity with the respective homologues from *P.s.* pv. savastanoi. An *iaaM* mutant of *P.s.* pv. syringae showing greatly reduced IAA levels was generated by marker-exchange. Bean leaves were then spray-inoculated with either the mutant or parent strain. Both strains reached the same density under various plant growth regimens by six days. However, more rapid growth was observed for the *iaa⁻* strain in the first two days days after inoculation when compared to the wild type strain. The data indicate that the abililty to produce IAA does not have a pronounced effect on pathogenicity symptoms under the experimental condition. However, the growth dynamics of the strains appear to differ.

Abbreviations: IAA, Iodoacetic Acid; ORF, Open Reading Frame; NV., Pathovar.

Introduction

The ability to synthesize indole-3-acetic acid (IAA) is common to a variety of plant pathogenic bacteria (Gross and Cody, 1985). The involvement of IAA biosynthesis in plant disease symptomatology has been demonstrated in tissue hyperplasia and root induction by *Agrobacterium* species and *P.s.* pv. savastanoi. IAA is also produced in appreciable quantities by phytopathogenic bacteria that produce disease with no apparent tissue hyperplasia (Fett *et al.*, 1987; White and Ziegler, 1991). The role of IAA in nonhyperplastic diseases is unknown, although some evidence exists for involvement of IAA in other aspects of plant-microbe interactions. Strains of *P.s.* pv. savastanoi with mutations in *iaaL*, the gene controlling the biosynthesis of indole-3-acetyl-lysine production, were reported to have reduced growth within host tissues (Glass and Kosuge, 1988), and the induction of a hypersensitive response on tobacco after inoculation with *P.s.* pv. phaseolicola was inhibited in the presence of strains of *A. tumefaciens* or *P.s.* pv. savastanoi that harbored genes for IAA biosynthesis (Robinette and Matthysse, 1990). In an initial survey, strains of *P.s.* pv. syringae and *P.s.* pv. pisi were positive for hybridization with the *iaa* genes from *P.s.* pv. savastanoi (Ziegler *et al.*, 1987), and the putative genes from

C.I. Kado and J.H. Crosa (eds.), Molecular Mechanisms of Bacterial Virulence, 605–613.
© 1994 *Kluwer Academic Publishers.*

606

P.s. pv. syringae have been shown to direct the biosynthesis of IAA (White and Ziegler, 1991). In order to investigate further the role of IAA biosynthesis in disease, we have examined additional pathovars of *P. syringae* for IAA biosynthesis and the effect that mutations in the the biosynthetic pathway have on the host/pathogen interaction.

Results and discussion

Total genomic DNA was isolated from three pathovars of *P. syringae*, digested with *Eco*RI, and analyzed by Southern hybridization using the 2.8 *Eco*RI fragment containing the *iaa* operon from *P.s.* pv. savastanoi. The probe contains entirely coding sequence from the *iaaM* and *iaaH* genes with the exception of the small intergenic region (Yamada *et al.*, 1985). Since the strain of *P.s.* pv. syringae in the initial study was isolated from bean, additional isolates of *P.s.* pv. syringae from bean and from plants other than bean, were examined (Table 1). Strains of *P.s.* pv. syringae which were isolated from bean synthesized IAA and contained an *Eco*RI fragment of similar sequence and size as the 2.8 kb *Eco*RI fragment from the *iaa* operon of *P.s.* pv. savastanoi (Fig. 1). Also positive were strains of *P.s.* pv. syringae from pear (Fig. 1), peach, and lilac

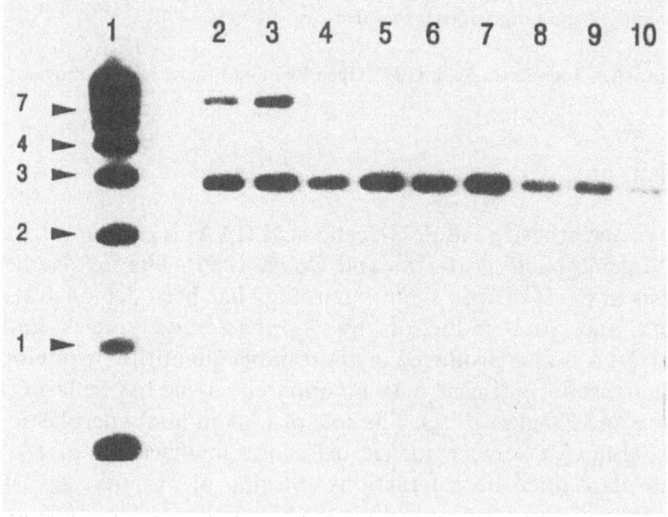

Figure 1. The presence of *iaaM* and *iaaH* in DNA from strains of *P.s.* pv. syringae from bean and pear. DNA from strains were as follows: Lane 1, 1 kb size standard; lane 2, *P.s.* pv. savastanoi TK1050; lane 3, *P.s.* pv. savastanoi EW2009; lane 4, *P.s.* pv. syringae Y30 (from M. Schroth); lane 5, *P.s.* pv. syringae Y30 (from D. Legard); lane 6, *P.s.* pv. syringae B86–7; lane 7, *P.s.* pv. syringae B86–13; lane 8, *P.s.* pv. syringae BBS6–3; lane 9, *P.s.* pv. syringae BBS102–6; lane 10, *P.s.* pv. syringae PS955. DNA samples were digested with *Eco*RI and probed with ^{32}P-labelled 2.8-kb *Eco*RI fragment from the *iaa* operon of *P.s.* pv. savastanoi. Arrows at left indicate selected size standards in kilobases.

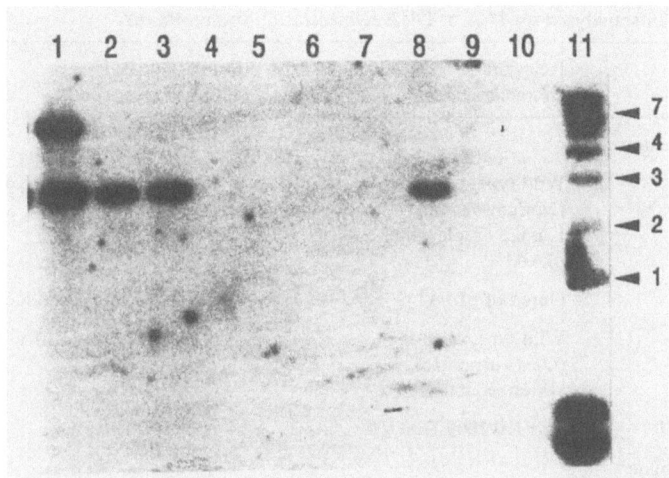

Figure 2. Presence of *iaaM* and *iaaH* in DNA from strains of *P. syringae* from diverse plant hosts. DNA from strains were as follows: Lane 1, *P.s.* pv. savastanoi EW2009; lane 2, *P.s.* pv. syringae Y30 (bean; from M. Schroth); lane 3, *P.s.* pv. syringae Y30 (bean; from D. LeGard); lane 4, *P.s.* pv. syringae B76 (tomato); lane 5, *P.s.* pv. syringae B64 (wheat); lane 6, *P.s.* pv. syringae B61 (wheat); lane 7, *P.s.* pv. syringae 176 (corn); lane 8, *P.s.* pv. syringae PS955 (pear); lane 9, *P.s.* pv. pisi G28–6 (pea); lane 10, *P.s.* pv. pisi pisi-4 (pea); lane 11, 1 kb standard. DNA samples from strains were digested with *Eco*RI. Arrows at the right indicate selected size standards in kilobases.

(blot data not shown). Strains from wheat, corn, and tomato were negative (Fig. 2). Since one strain of *P.s.* pv. pisi was positive in the initial screen, two additional strains were tested, and both were negative for hybridization to the *iaa* probe (Fig. 2).

Clones from a genomic library of *P.s.* pv. syringae that hybridized with the 2.8 kb *Eco*RI fragment from *P.s.* pv. savastanoi were isolated and transferred to an *iaa*⁻ strain of *P.s.* pv. savastanoi (EW2009–3). Two clones with an intact 2.8 kb fragment that hybridized to the probe were obtained, and both were found to direct IAA biosynthesis (Table 2). The bacterial transposon Tn5 was used to interrupt the 2.8 kb *Eco*RI fragment of pY305–4. Insertions within this fragment eliminated IAA biosynthesis in transconjugants of EW2009–3 (Table 2).

The region of the 2.8 kb *Eco*RI fragment from *P.s.* pv. syringae (strain Y30) was subcloned and sequenced. As expected from the hybridization results, the fragment contains open reading frames similar to the *iaaM* and *iaaH* of *P.s.* pv. savastanoi. The ORF for *iaaM* potentially encodes a protein of 556 amino acids, one shorter than the corresponding protein of *iaaM* from *P.s.* pv. savastanoi (Yamada *et al.*, 1985). The open reading frame of *iaaH* is 10 amino acids shorter than the *iaaH* of *P.s.* pv. savastanoi (446 vs. 456 aa). Overall similarity in amino acid sequences for *iaaM* and *iaaH* open reading frames of *P.s.* pv. syringae and *P.s.* pv. savastanoi is 94% and 90%, respectively. As with the IaaM protein of *P.s.* pv. savastanoi, the putative protein of *P.s.* pv. syringae also lacks the amino

Table 1. Strains analyzed for IAA or DNA hybridization to *iaaM/iaaH*.

Designation[a]	Relevant characteristics	IAA[b]	EcoRi[c] 2.8 kb	Reference or source
P.s. pv. savastanoi				
EW2009	Wild type, oleander (Nerium oleander L.), iaa on plasmid pIAA1	+	+	Comai and Kosuge 1980
EW2009–3	Cured of pIAA1	−	−	Comai and Kosuge 1980
TK1050	Wild type, olive (Olea europa L.), chromosomal iaa	+	+	Glass and Kosuge 1986
EW2009–3rif	Rifr EW2009–3	−	−	This study
P.s. pv. syringae				
Y30	Wild type, bean (Wisconsin)	+	+	D. Legard, NYSAES, Cornell Univ., Geneva, NY and M. Schroth, Univ. of California, Berkeley, CA
B86–7	Wild type, snapbean (New York)	+	+	D. Legard, NYSAES, Cornell Univ., Geneva, NY
B86–1	Wild type, snapbean (New York)	+	+	D. Legard, NYSAES, Cornell Univ., Geneva, NY
BBS 6–3	Wild type, snapbean (New York)	+	+	D. Legard, NYSAES, Cornell Univ., Geneva, NY
BBS 102–6	Wild type, snapbean (New York)	+	+	D. Legard, NYSAES, Cornell Univ., Geneva, NY
PS955	Wild type, pear	+	+	Currier and Morgan 1983
W4N27	Wild type, pear	ND	+	D. Gross, WSU, Pullman, WA
W4N65	Wild type, pear	ND	+	D. Gross, WSU, Pullman, WA
W4N72	Wild type, pear	ND	+	D. Gross, WSU, Pullman, WA
B301-D	Wild type, pear	ND	+	D. Gross, WSU, Pullman, WA
B-3A	Wild type, peach	ND	+	D. Gross, WSU, Pullman, WA

Table 1. (continued).

Designation[a]	Relevant characteristics	IAA[b]	EcoRi[c] 2.8 kb	Reference or source
SY12	Wild type, lilac	ND	+	D. Gross, WSU, Pullman, WA
B76 (132)	Wild type, tomato	−	−	T. Denny, Univ. of Georgia, Athens
B64 (138)	Wild type, wheat	−	−	T. Denny, Univ. of Georgia, Athens
B61 (144)	Wild type, wheat	−	−	T. Denny, Univ. of Georgia, Athens
176	Wild type, corn	−	−	T. Denny, Univ. of Georgia, Athens
P. syringae pv. pisi G28–6 (Pisi-1)	Wild type, pea in New York	−	−	D. Legard, NYSAES, Cornell Univ., Geneva, NY
Pisi-4	Wild type, pea in Wisconsin	−	−	D. Legard, NYSAES, Cornell Univ., Geneva, NY

[a] Original designation if known, otherwise the designation of source was used. Designation in parentheses is strain collection number from source.
[b] Presence of IAA determined by thin layer chromotography.
[c] Presence of 2.8-kb EcoRI fragment determined by Southern hybridization analysis with pLUC1 probe.

Table 2. IAA production directed by iaa region of *P.s.* pv. syringae.

Strain	μM IAA[a]
EW2009-3rif (pLAFR3)	0
EW2009-3rif (pY305-4)	33 ± 6
EW2009-3rif (pY305-10)	61 ± 4
EW2009-3rif (pY305-4::Tn5-1)	2 ± 4
EW2009-3rif (pY305-4::Tn5-2)	7 ± 3
EW2009	65 ± 4

[a]IAA concentration in culture supernatants of the transconjugants grown on liquid King's medium was determined by a colorimetric assay (Gordon and Weber, 1951). A dilution series of authenic IAA was used as standards. Absorbance at 530 nm of culture supernatant of EW2009-3rif was assumed to be due to other reactive indoles or IAA via other pathways. The background absorbance of EW2009-3rif (pLAFR3) was subtracted from all values for IAA concentration which were also normalized to cell culture density as determined by absorbance at 600 nm. Values in table were derived by linear regression analysis of absorbance values from three culture supernatants of each strain.

610

terminal sequence of approximately 250 aa found in the IaaM protein of *A. tumefaciens* (Klee *et al.*, 1984). A third ORF (663 bp) was tentatively identified at 78 bp downstream from the termination codon of *iaaH* that is not found in *P.s.* pv. savastanoi. A map of the *iaa* operon of *P.s.* pv. syringae is shown in Fig. 3.

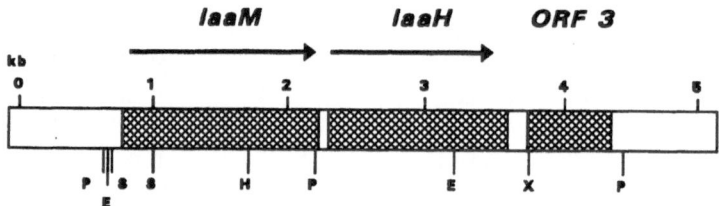

Figure 3. Map of the *iaa* operon of *P.s.* pv. syringae. Symbols: E, *Eco*RI; H, *Hin*DIII; P, *Pst*I; S, *Sal*I; X, *Xho*I.

The promoter region of *P.s.* pv. savastanoi has been reported to lie more than 400 bp upstream of the *iaaM* initiation codon and encompasses a 228 bp ORF (Gaffney *et al.*, 1990). The upstream sequence of *iaaM* of *P.s.* pv. syringae does not contain the 228 bp ORF. Sequence relatedness between the promoter region of the two pathovar operons drops off dramatically within 200 bp of the initiation codon. (The 5' *Eco*RI sites of the fragments of the two are not conserved. The sites appear to be fortuitously close and, therefore, create a similar 2.8 kb fragment in the two genomes.) The location of the promoter for the operon of *P.s.* pv. syringae has not yet been mapped.

The method of marker-exchange mutagenesis was used to incorporate a Tn*5* insertion in *iaaM* of *P.s.* pv. syringae from the clone pY305–10 into the genome of strain Y30. The strain (53–29) was analyzed by Southern hybridization for the presence of the Tn*5* insertion in the proper region of the genome. Digestion of DNA from the mutant with *Hin*DIII revealed the presence of two approximately 2-kb fragments in place of one 1.8-kb fragment of the wild type strain (Fig. 4A). The two fragments are the proper size if the transposon had inserted within the middle of the 1.8 kb fragment which contains the *iaaM* gene. The level of IAA in culture filtrates of the mutant was below the level of detection as determined by thin layer chromotography (Fig. 4B).

Strain Y30, the *iaa* deficient strain (53–29), and a restored strain [RL2 = 53–29(PY305–10)] were spray inoculated on to bean leaves at 10^7 cfu/ml, and total bacterial populations were assayed every 2 days for 10 days. In this study the plant leaves were senescing by the 10-day point. For the first two days, the mutant 53–29 had reached a tenfold higher concentration than either the wild type or restored strain (Fig. 5). However, by day 6 strain Y30 had attained nearly the same maximum population as the mutant. The restored strain exhibited a growth pattern that was similar to strain Y30 with the exception that maximum density of growth was not as high as either the wild type or the mutant.

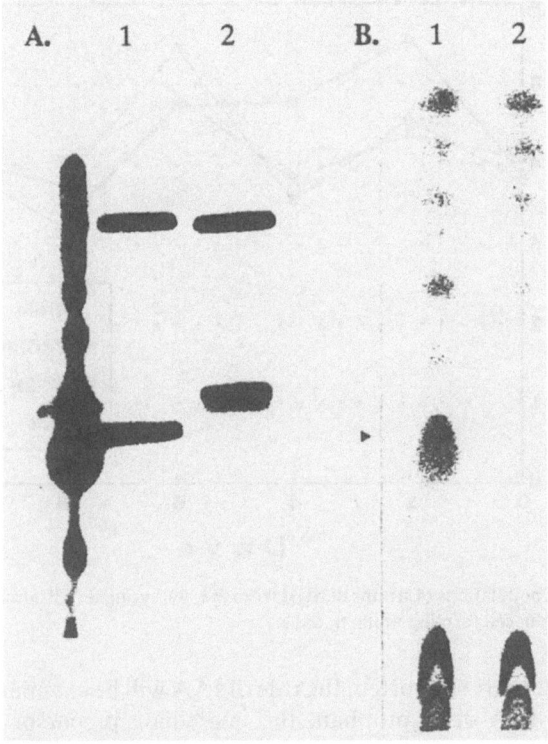

Figure 4. Southern hybridization analysis of DNA from strain 53–29 (lane 1) and Y30 (lane2) using the 2.8-kb *Eco*RI probe (containing *iaaM* and *iaaH*) from *P.s.* pv. syringae Y30. (A 1-kb ladder standard is at the left of lane 1.) **B**. Thin layer chromatography of culture filtrates from Y30 (lane 1) and 53–29 (Iaa⁻, lane 2). Position of IAA is indicated by the arrow.

The results indicate that the *iaa* region of *P.s.* pv. syringae is closely related to the genes from *P.s.* pv. savastanoi, and the genes are present in all strains of closely related isolates of *P.s.* pv. syringae. The consistency of presence of the gene leads us to suspect that the *iaa* genes of *P.s.* pv. syringae has some function in the physiology of the bacterium. The results of the growth studies are too preliminary to be conclusive. However, the initial experiments provide evidence that the growth dynamics *in planta* are altered in the mutant strain. The effect does not appear to alter the basic disease symptoms of *P.s.* pv. syringae on bean. In some experiments larger lesions are detected at a given time point after plant inoculation with the mutant when compared to the wild type strain, and larger populations of mutant bacteria were detected in the first two days after inoculation. In addition, bacterial growth upon isolation from the plant appears to be slower for the wild type and restored strain when compared to the mutant.

Further growth studies will be performed in order to determine whether IAA has a growth regulatory role in the *P.s.* pv. syringae/plant host interaction. If the

612

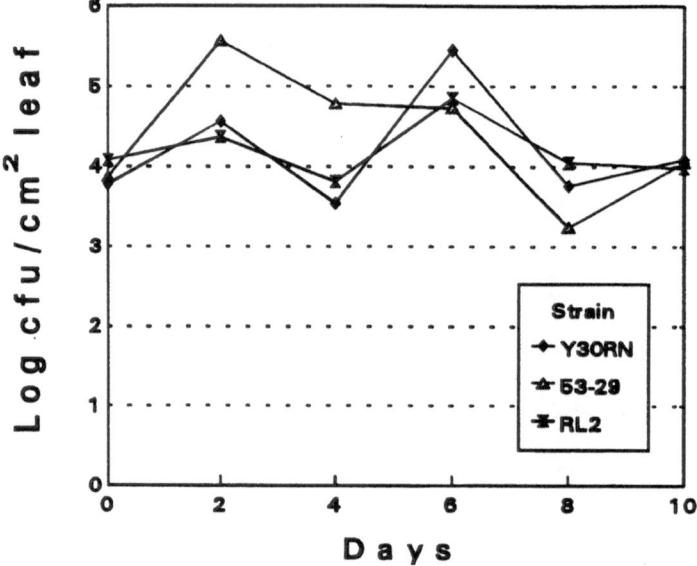

Figure 5. Total leaf populations of strains derived from *P.s.* pv. syringae Y30 after spray-inoculation of bean. Symbols: cfu, cell forming units; d, days.

preliminary results are supported, the role of IAA will be examined to determine if the levels of IAA or tryptophan, the immediate precursor to IAA in the biosynthetic pathway, are the important factors in bacterial cell growth. The generality of the system will also be investigated using additional strains of *P.s.* pv. syringae and pathovars of *P. syringae*. The *iaa* genes have also been tentatively identified in all strains of *P.s.* pv. glycinea, and mutants in the IAA pathway of this pathovar will be generated and tested (Mazzola and White, unpublished results).

Acknowledgements

Contribution no. 93–252-B from the Kansas Agricultural Experiment Station, Kansas State University, Manhattan.

References

Comai, L., and Kosuge, T. 1980. Involvement of plasmid deoxyribonucleic acid in indoleacetic acid synthesis in *Pseudomonas savastanoi*. J Bacteriol 143:950–957.

Currier, T.C., and Morgan, M.K. 1983. Plasmids of Pseudomonas syringae: no evidence of a role in toxin production or pathogenicity. Can J Microbiol 29:84–89.

Fett, W.F., Osman, S.F., and Dunn, M.F. 1987. Auxin production by plant-pathogenic pseudomonads and xanthomonads. Appl Environ Microbiol 53:1839–1845.

Gaffney, T.D., daCosta e Silva, O., Yamada, T., and Kosuge, T. 1990. Indoleacetic acid operon of *Pseudomonas syringae* subsp. *savastanoi*: transcriptional analysis and promoter identification. J Bacteriol 172:5593–5601.

Glass, N.L., and Kosuge, T. 1988. Role of indoleacetic acid-lysine synthetase in regulation of indoleacetic acid pool size and virulence of *Pseudomonas syringae* subsp. *savastanoi*. J Bacteriol 170:2367–2373.

Gordon, S.A., and Weber, R.P. 1951. Colorimetric estimation of indoleacetic acid. Plant Physiol 26:191–195.

Gross, D.C., and Cody, Y.S. 1985. Mechanisms of plant pathogenesis by *Pseudomonas* species. Can J Microbiol 31:403–410.

Klee, H., Montoya, A., Hordyski, F., Lichtenstein, C., Garfinkel, D., Fuller, S., Flores, C., Peschon, J., Nester, E.W., and Gordon, M.P. 1984. Nucleotide sequence of the *tms* genes of the pTiA6NC octopine Ti-plasmid: two gene products involved in plant tumorigenesis. Proc Natl Acad Sci USA 81: 1728–1732.

Robinette, D., and Matthysse, A.G. 1990. Inhibition by *Agrobacterium tumefaciens* and *Pseudomonas savastanoi* of development of the hypersensitive response elicited by *Pseudomonas syringae* pv. *phaseolicola*. J Bacteriol 172:5742–5749.

Yamada, T., Palm, C.J., Brooks, B., and Kosuge, T. 1985. Nucleotide sequences of the *Pseudomonas savastanoi* indoleacetic acid genes show homology with *Agrobacterium tumefaciens* T-DNA. Proc Natl Acad Sci USA 82:6522–6526.

White, F.F., and Ziegler, S.F. 1991. Cloning of the genes for indoleacetic acid synthesis from *Pseudomonas syringae* pv. *syringae*. Mol Plant-Microbe Interact 4:207–210.

44. Molecular analysis of virulence associated gene regions from the ovine footrot pathogen, *Dichelobacter nodosus*

JULIAN I. ROOD, CATHERINE L. WRIGHT, VOLKER HARING and MARGARET E. KATZ

Abstract. *Dichelobacter nodosus*, a strictly anaerobic Gram negative rod, is the causative agent of ovine footrot. The *D. nodosus*-encoded factors believed to be involved in virulence include N-MePhe type fimbriae and extracellular proteases. Labelled genomic DNA from the reference virulent strain A198 and a benign isolate C305 was used to screen an A198 gene bank. Three recombinant plasmids which differentiate virulent and benign isolates of *D. nodosus* were isolated. One plasmid, pJIR318, hybridized with virtually all of the virulent and intermediate isolates of *D. nodosus* that were tested. Therefore, the genomic region carried on pJIR318 was designated as the virulence associated protein, or *vap*, region. The *vapD* gene was shown to encode a protein with similarity to a protein predicted from an ORF encoded by the *Neisseria gonorrhoeae* cryptic plasmid. There are three copies of the *vap* region in the *D. nodosus* strain A198 genome. The other plasmids, pJIR313 and pJIR314B, hybridized with nearly all virulent *D. nodosus* strains but not with benign isolates. To determine the extent of these virulence-related regions a λ gene bank of total *D. nodosus* A198 DNA was screened using pJIR313 as a probe. The resultant recombinant hybridized with both pJIR313 and pJIR314B. Southern hybridization showed that these regions were separated by 9 kb of virulence specific DNA. Chromosome walking experiments led to the cloning and delineation of the complete virulence related locus, *vrl*, which encompassed ca. 25 kb. Using probes which contained the ends of the *vrl* region, and the adjacent non-virulence specific DNA, a region that was equivalent to the site of *vrl* insertion or deletion from a benign *D. nodosus* isolate was cloned. Sequence analysis has not yet revealed the functional significance of the *vrl* region.

Abbreviations: ORF, Open Reading Frame

Introduction

Footrot is a very significant bacterial disease of sheep. The most severe, or virulent, form of the disease is typified by separation of the hoof from the underlying epidermal tissues in the feet of a large percentage of sheep in a flock. The resultant loss of body condition and severe lameness leads to significant financial losses due to decreased wool quality and production and the cost of the labour intensive treatment. Virulent footrot is highly contagious, especially under warm, moist climatic conditions. Intermediate footrot is a milder form of the disease with a smaller percentage of sheep affected, less pronounced lameness and reduced economic effects. Benign footrot is the least contagious type of ovine footrot syndrome and presents as an interdigital dermatitis that is usually restricted to one foot of the sheep and causes minimal production losses (Stewart, 1989).

C.I. Kado and J.H. Crosa (eds.), Molecular Mechanisms of Bacterial Virulence, 615–624.
© 1994 *Kluwer Academic Publishers*.

The microorganism regarded as the primary pathogen in ovine footrot is the Gram negative rod, *Dichelobacter nodosus*, which was formerly known as *Bacteroides nodosus* (Dewhirst *et al.*, 1990). Organisms such as *Fusobacterium necrophorum* and other enteric bacteria have also been implicated in the disease but footrot does not occur in the absence of *D. nodosus* (Stewart, 1989). *D. nodosus* is a slow growing, strict anaerobe which requires specialised medium for growth (Skerman, 1989). No plasmids have been reported from *D. nodosus* and there are no known methods for the genetic manipulation of this organism. Molecular studies must therefore rely upon the cloning and analysis of genes in *Escherichia coli* and other Gram negative bacteria.

Under optimal climatic conditions the form of ovine footrot which is observed in the field is generally regarded as a reflection of the virulence of the causative bacterium. Therefore, isolates of *D. nodosus* often are designated as virulent, intermediate, or benign, depending on the lesion from which they were isolated. Little is known about the molecular basis for the pathogenesis of ovine footrot. Molecular research on *D. nodosus* has primarily involved the development of recombinant fimbrial vaccines (Egerton *et al.*, 1987; Stewart & Elleman, 1987) and gene probe methods for differential diagnosis (Katz *et al.*, 1991). Virulent and benign strains produce extracellular proteases of different thermostability (Depiazzi & Richards, 1979; Depiazzi & Rood, 1984) and with distinct isoenzyme profiles (Every, 1982; Kortt *et al.*, 1983). They also differ in their ability to produce elastase (Stewart, 1979; Stewart *et al.*, 1986). The extracellular proteases therefore can be regarded as potential virulence factors of *D. nodosus*.

Isolates of *D. nodosus* can be divided into a number of different serogroups depending on the antigenicity of the fimbriae present on the cell surface (Claxton, 1989). These fimbriae are also considered to be virulence factors as they are believed to be involved in the translocation of *D. nodosus* cells through the infected lesion. The genes encoding these N-MePhe, or Type 4, fimbriae have been cloned and extensively studied and belong to a class of fimbriae that are commonly found in other pathogenic bacteria such as *Pseudomonas aeruginosa, Moraxella bovis,* and *Neisseria gonorrhoeae* (Elleman, 1988; Hobbs *et al.*, 1991; Mattick *et al.*, 1991).

Isolation of recombinant plasmids containing virulence associated regions of the *D. nodosus* genome

Research in this laboratory has involved the development of gene probes suitable for the differential diagnosis of ovine footrot. A gene bank was constructed from the virulent reference strain of *D. nodosus*, A198, and was probed separately with labelled genomic DNA from strain A198 and from the benign strain C305. Comparison of the respective colony hybridization blots led to the identification of a number of clones which hybridized with DNA from the virulent strain but not with the labelled benign DNA (Katz *et al.*, 1991).

Three of these recombinant plasmids (pJIR318, pJIR313 and pJIR314B) were isolated and studied in detail. Restriction analysis showed that these plasmids were distinct from each other and from previously cloned regions encoding either fimbrial subunit genes (Anderson *et al.*, 1984; Elleman & Hoyne, 1984) or proteases (Lilley *et al.*, 1992). Approximately 100 *D. nodosus* isolates, with known virulence characteristics, were tested to see if they contained DNA which hybridized with the three plasmids. All 29 of the virulent isolates and 94% (34/36) of the intermediate strains hybridized with pJIR318. However, 33% (12/36) of the benign isolates also hybridized with this probe. Virtually all of the virulent strains reacted with pJIR313 (94%) and pJIR314B (100%) in dot blots. In contrast, only two benign isolates hybridized with these plasmids. Based on these data a gene probe-based scheme for the analysis of field isolates was proposed and is currently being tested (Katz *et al.*, 1991).

Genetic organization of the *vap* Region

The plasmid pJIR318 contained 2.3 kb of *D. nodosus*-derived DNA inserted into pUC18. Since the pJIR318-related DNA was present in virtually all virulent and intermediate isolates of *D. nodosus* it was designated as a virulence associated region and analysed further. Southern blots showed that there were three copies of this genomic region in the A198 genome. These copies were found on 6.2 kb, 4.6 kb and 3.5 kb *Hin*dIII fragments in strain A198. Other *D. nodosus* isolates also had multiple copies of this region although strains with only one copy were also detected (Katz *et al.*, 1991). Pulsed field gel electrophoresis experiments have shown that all three copies were located on the A198 chromosome, as were the genes encoded by pJIR313 and pJIR314B (S. La Fontaine and J. Rood, unpublished results).

The complete sequence of the *D. nodosus*-derived DNA present in pJIR318 has been determined (Katz *et al.*, 1992). Four genes which have appropriate ribosome binding sites were identified. These genes were designated as *vapA-vapD* for virulence associated proteins. A complete genetic map of pJIR318 is shown in Fig. 1. An additional open reading frame, ORF118, was found upstream of *vapA* but it was not preceded by a ribosome binding site.

Based on the sequence data *vap* gene-specific probes were constructed from pJIR318 and used to probe *Hin*dIII-digested DNA from strain A198. The results (Fig. 1) showed that the 6.2 kb and 4.5 kb *Hin*dIII fragments had copies of each of the *vapA-vapD* genes. However, the 3.5 kb fragment only had the *vapD* gene. The three copies of the *vap* region present in A198 were therefore not identical.

Searches of various sequence databases were carried out for each of these open reading frames. Neither the *vapA* gene nor the amino acid sequence of the putative VapA protein (11.3 kDa) had any similarity with sequences in the databases. In contrast, 46% amino acid sequence identity was detected between the VapD protein and the protein deduced from the sequence of ORF5 from the

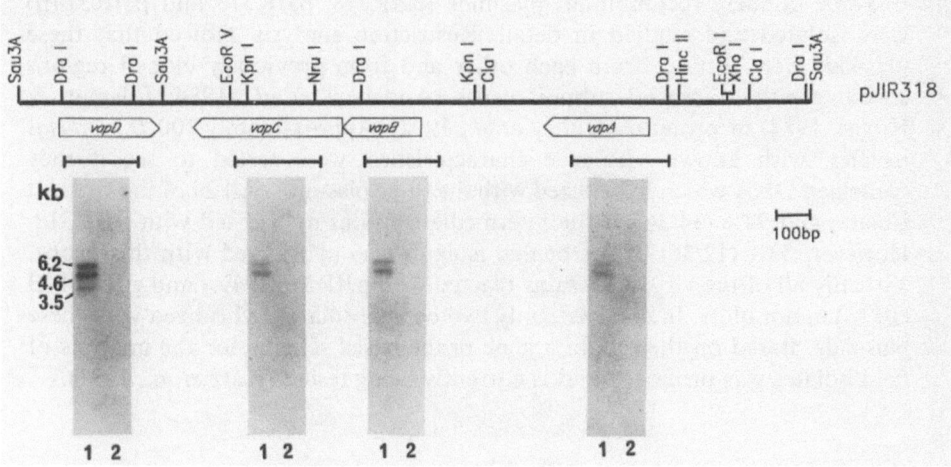

Figure 1. Genetic map and Southern hybridization analysis of pJIR318. The four open reading frames (*vapA-vapD*) with good consensus ribosome binding sites are indicated below the restriction map with open arrows. Identical Southern blots of *Hin*dIII digested *D. nodosus* genomic DNA prepared from the virulent strain, A198 (lane 1) and the benign strain, C305 (lane 2) were hybridized with probes from each *vap* gene (shown by bars below the map). The approximate size of the hybridizing bands is indicated in kb.

Neisseria gonorrhoeae cryptic plasmid. The function of both ORF5 and the cryptic plasmid is unknown but they are present in virtually all clinical isolates of *N. gonorrhoeae* (Roberts *et al.*, 1979). SDS-PAGE experiments showed that the VapD protein was expressed at significant levels in the *E. coli* clones carrying pJIR318. The protein was purified from these gels and subjected to amino terminal sequence analysis. The results confirmed the designation of the start codon of the *vapD* gene (Katz *et al.*, 1992). Using specific antiserum raised against gel purified VapD, it was shown that all of the *D. nodosus* strains tested which hybridized with pJIR318 produced the VapD protein. Isolates that did not hybridize with this plasmid did not produce VapD. In addition, no VapD cross-reacting proteins were detected in cell extracts of *N. gonorrhoeae* (Katz *et al.*, 1992). The finding that there is significant similarity between VapD and the ORF5-encoded protein therefore has not led to insights into the functional significance of either protein, although it is tempting to speculate that both gene products may play a role in pathogenesis.

The *vapB* and *vapC* genes appeared to be arranged in an operon on pJIR318. The initiation codon of *vapC* overlapped the stop codon of *vapB*, suggesting that the expression of these genes also may be translationally coupled (Katz *et al.*, 1992). Analysis of the nucleic acid sequence databases revealed that the *vapBC* region had significant similarity (53% identity) with the *trbH* region of the *E. coli* F plasmid. Examination of the F-derived sequence indicated that on the opposite strand to the *trbH* gene there were two ORFs which had an analogous genetic arrangement to that of *vapBC* and which potentially encoded proteins of similar size. These proteins had 34% and 41% amino acid sequence identity to the putative VapB and VapC proteins, respectively (Katz *et al.*, 1992). Note that there was no equivalent *trbH*-like gene in pJIR318. The role of the *trbH* region on F is unknown but it is not believed to be involved in either conjugation or plasmid maintenance (Yoshioka *et al.*, 1990).

The finding of sequence similarity between the *D. nodosus vap* genomic region and its encoded polypeptides, and plasmids from other bacterial species, suggests that the *vap* region may have arisen from the integration into the *D. nodosus* chromosome of a plasmid(s), or part of a plasmid(s), which originated in other bacteria. Subsequent gene duplication events and genomic rearrangements would account for the presence of multiple copies of the *vap* genes.

Delineation of the *vrl* region

Southern hybridization experiments showed that, unlike the *vap* region, the genomic regions carried on pJIR313 and pJIR314B were only present in one copy on the *D. nodosus* genome (Katz *et al.*, 1991). The 1.3 kb and 0.6 kb inserts present in pJIR313 and pJIR314B, respectively, were also sequenced and shown to contain incomplete ORFs. Coupled with the observation that almost identical results were obtained when *D. nodosus* isolates from all virulence classes were hybridized with each of these plasmids (Katz *et al.*, 1991), these results suggested that these plasmids may have represented different parts of a single gene.

To test this hypothesis, lambda gene banks (kindly provided by E.K. Moses) constructed with *D. nodosus* A198 DNA were screened using labelled pJIR313 DNA as a probe in plaque hybridization assays. Two identical positive clones which hybridized with both pJIR313 and pJIR314B (a subclone of pJIR314) were obtained. A detailed restriction map was prepared from one of these phage clones, λR29 (Fig. 2). Comparative Southern hybridization analysis showed that the gene regions represented by pJIR313 and pJIR314B were separated by 9 kb of DNA (Fig. 2). It therefore appeared to be highly unlikely that these regions were part of a single gene.

To determine whether the region between the pJIR314B and pJIR313 inserts on λR29 had the same virulence specificity as these plasmids, three subclones spanning this region were constructed. These plasmids (pJIR590, pJIR589 and pJIR592) were used in dot blot experiments to probe test filters containing DNA from nine virulent *D. nodosus* isolates and nine benign *D. nodosus* strains. Only the DNA from the virulent strains hybridized with these plasmids. Further experiments showed that pJIR571 (Fig. 2) also only hybridized with DNA from virulent isolates of *D. nodosus*. The virulence specific region therefore had been shown to extend to the right end of the λR29 clone. The genomic region encompassed by these plasmids was subsequently designated as the virulence related locus, or *vrl* region, to distinguish it from the *vap* region.

620

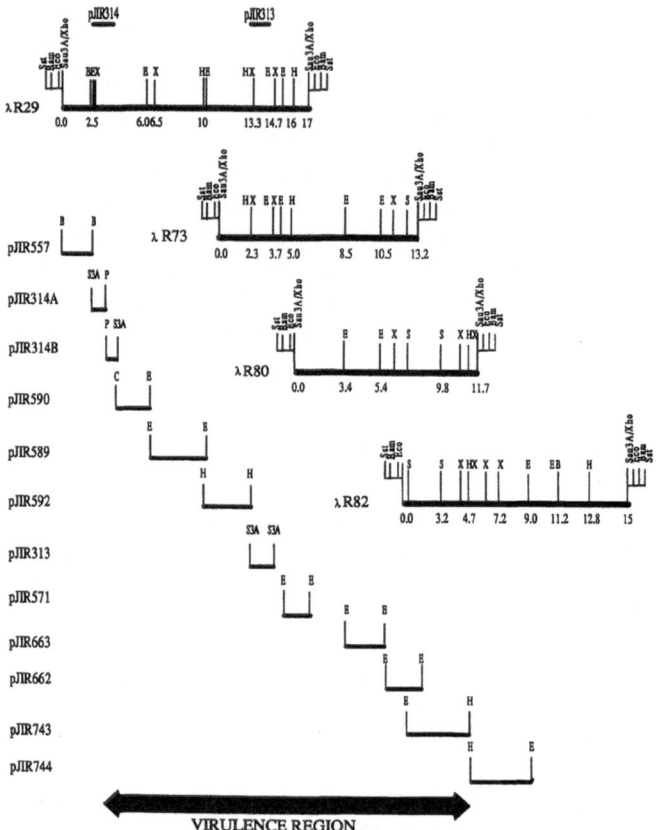

Figure 2. Delineation of the extent of the *vrl* region. Restriction maps of four overlapping λ clones are presented. Only relevant restriction sites are shown. Map locations are indicated in kb. The relative locations of the various subclones are indicated together with the terminal restriction sites used for cloning. The regions of genomic DNA contained within pJIR313 and pJIR314 are indicated by the bars above the λR29 map. The extent of the virulence region is indicated by the thick arrow. Restriction site abbreviations are as follows: E (*Eco*RI); B (*Bam*HI); X (*Xho*I); S3A (*Sau*3AI); C (*Cla*I); S (*Sst*I); P (*Pst*I).

The full extent of the *vrl* region was delineated by a series of chromosome walking experiments. Using pJIR568 (a λR29 subclone containing the right hand 1 kb *Hind*III-*Bam*HI fragment) as a probe, the overlapping recombinants λR73 and λR80 were isolated. These clones were mapped and appropriate subclones (Fig. 2) probed against the same test filter panel of virulent and benign *D. nodosus* isolates. In this way the *vrl* region was shown to extend to the end of the region present on λR80. Finally, pJIR674 (a λR80 subclone containing the right hand 0.4 kb *Hind*III-*Bam*HI fragment) was used to isolate λR82 from the gene bank (Fig. 2). Hybridization analysis, using the subclones pJIR743 and pJIR744 as probes, showed that although pJIR743 only hybridized with DNA from the nine virulent isolates, pJIR744 hybridized with both virulent and benign DNA preparations. The *vrl* region had therefore been shown to encompass approximately 25 kb of A198-derived DNA which extended from the plasmid pJIR314B to pJIR743, inclusive (Fig. 2). The junction plasmids which must contain the regions where the *vrl* locus commences therefore were pJIR314A (Katz *et al.*, 1991) and pJIR744.

Detailed sequence analysis of the *vrl* region has commenced but is only at the preliminary stages. Although several incomplete ORFs have been identified none of these genes have sequence similarity with sequences deposited in the databases. In the absence of methods for genetic manipulation of *D. nodosus* cells, functional analysis of the *vrl* region must await the completion of these sequence studies. Other experiments which involve the use of expression vectors to detect protein products encoded within the *vrl* region are also under way.

Cloning of the *vrl* junction region from a benign *D. nodosus* isolate

It is possible that the 25 kb *vrl* region represents a virulence locus which has been inserted into the genome of a benign *D. nodosus* isolate, perhaps from a plasmid originating in another species. Alternatively, benign strains may be derived from virulent isolates by deletion of the *vrl* region. To obtain further insights into the relationship between the virulent and benign isolates of *D. nodosus* it was decided to clone a benign-derived DNA fragment which hybridizes to the junction of the *vrl* region, to determine the DNA sequence of that region, and to compare it to sequences obtained from the *vrl* border regions from the virulent isolate.

From previous experiments it was known that pJIR314A contained the left hand junction of the *vrl* region. A subclone of this plasmid, pJIR689, which contained the 750 bp *Eco*RI-*Pst*I fragment (Fig. 3), was used to probe Southern blots of DNA from the benign *D. nodosus* strain C305. The smallest fragment which hybridized to the probe was a 2 kb *Eco*RI fragment. To isolate this junction fragment, *D. nodosus* C305 DNA was digested with *Eco*RI, ligated to pUC18 and used to transform *E. coli* strain DH5α. Four clones which hybridized to pJIR689, and which contained 2 kb *Eco*RI fragments with identical restriction patterns, were detected, one such clone was designated as pJIR787.

Examination of the deduced restriction map of pJIR787 revealed significant restriction site identity to pJIR314A. Both plasmids had a close succession of *Cla*I, *Ssp*I, and *Xho*I sites located about 100 bp from an *Eco*RI site (Fig. 3). However, there was no obvious similarity between the restriction maps of pJIR787 and pJIR744, the plasmid marking the right hand junction of the *vrl* region. The *Asp*718 site, and the second *Ssp*I site, were unique to pJIR787 and were not found either in pJIR314A or in pJIR744. In addition, the *Eco*RV site in pJIR744 (Fig. 3) was absent in pJIR787. When the 2 kb *Eco*RI fragment from pJIR787 was used to probe a series of plasmids spanning the *vrl* region and the

622

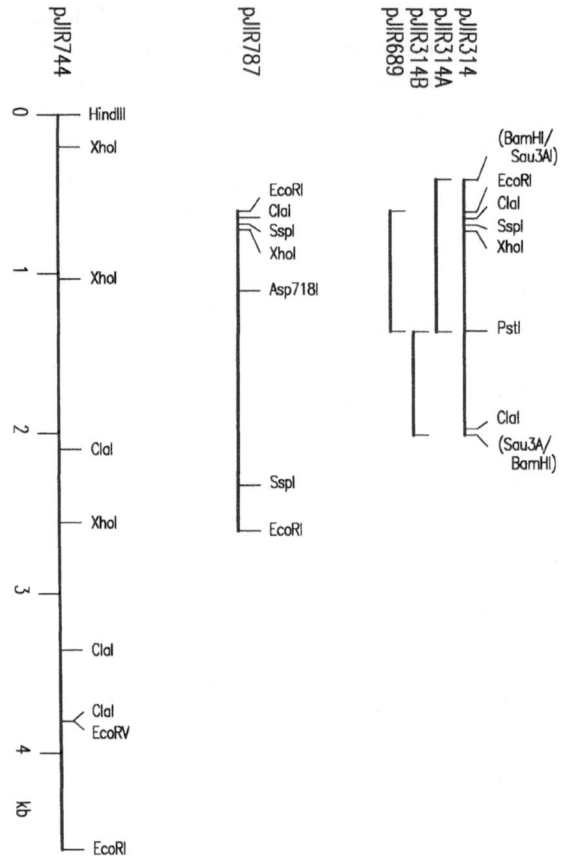

Figure 3. Restriction maps of junction plasmids. Linear maps of the A198-derived recombinant plasmids pJIR314, pJIR314A, pJIR314B (Katz *et al.*, 1991), pJIR689 and pJIR744 are shown. The C305-derived recombinant, pJIR787, is aligned with pJIR314. The plasmids pJIR314 and pJIR314A contain the left hand junction of the *vrl* region whereas pJIR744 contains the right hand junction region. Note that pJIR314A and pJIR314B are subclones of pJIR314. All plasmids are drawn in the same orientation as Fig. 2. The scale in kb is indicated on the pJIR744 map.

adjacent sequences from A198, hybridization was only detected with pJIR314A, the left hand junction plasmid. It was concluded that pJIR787 only contained sequences corresponding to the left junction of the *vrl* region.

These results suggest that the difference observed at the *vrl* locus between benign and virulent *D. nodosus* strains is not the result of a simple insertion or deletion event. It could be due to the displacement of one segment of DNA by another or to an insertion or deletion event and an intrachromosomal rearrangement. If the displacement hypothesis is valid then it is possible that some regions of pJIR787 are benign specific because they would only be present in benign isolates derived from the displacement process. In contrast, with the rearrangement theory all of the pJIR787 sequences should be found in both benign and virulent strains, although at different locations in the genome. Current studies are aimed at testing these hypotheses and cloning sequences of *D. nodosus* strain C305 which correspond to the right hand junction of the *vrl* region. Hopefully, these experiments will enable the elucidation of the precise relationship between strains A198 and C305 at the *vrl* locus and the construction of a model which explains their evolution.

Acknowledgements

We thank Eric Moses for provision of the lambda gene bank and Pauline Howarth and Tracy Warner for their expert technical assistance. This research was generously supported by grants from the Australian Wool Research and Development Corporation.

References

Anderson BJ, Bills MM, Egerton JR & Mattick JS (1984) Cloning and expression in *Escherichia coli* of the gene encoding the structural subunit of *Bacteroides nodosus* fimbriae. J. Bacteriol. 160: 748–754.

Claxton PD (1989) Antigenic classification of *Bacteroides nodosus*. p. 155–166. *In*: Egerton JR, Yong WK & Riffkin GG (eds.) Footrot and foot abscess of ruminants. CRC Press, Inc., Boca Raton.

Depiazzi LJ & Richards RB (1979) A degrading proteinase test to distinguish benign and virulent ovine isolates of *Bacteroides nodosus*. Aust. Vet. J. 55: 25–28.

Depiazzi LJ & Rood JI (1984) The thermostability of proteases from virulent and benign strains of *Bacteroides nodosus*. Vet. Microbiol. 9: 227–236.

Dewhirst FE, Paster BJ, La Fontaine S & Rood JI (1990) Transfer of *Kingella indologenes* (Snell and Lapage 1976) to the genus *Suttonella* gen. nov. as *Suttonella indologenes* comb. nov.; Transfer of *Bacteroides nodosus* (Beveridge 1941) to the genus *Dichelobacter* gen. nov. as *Dichelobacter nodosus* comb. nov.; and assignment of the genera *Cardiobacterium*, *Dichelobacter*, and *Suttonella* to *Cardiobacteriaceae* fam. nov. in the gamma division of *Proteobacteria* based on 16S ribosomal ribonucleic acid sequence comparisons. Int. J. Syst. Bacteriol. 40: 426–433.

Egerton JR, Cox PT, Anderson BJ, Kristo C, Norman M & Mattick JS (1987) Protection of sheep against footrot with a recombinant DNA-based fimbrial vaccine. Vet. Microbiol. 14: 393–409.

Elleman TC (1988) Pilins of *Bacteroides nodosus*: Molecular basis of serotypic variation and relationships to other bacterial pilins. Microbiol. Rev. 52: 233–247.

Elleman TC & Hoyne PA (1984) Nucleotide sequence of the gene encoding pilin of *Bacteroides nodosus*, the causal organism of ovine footrot. J. Bacteriol. 160: 1184–1187.

Every D (1982) Proteinase isoenzyme patterns of *Bacteroides nodosus*: distinction between ovine virulent isolates, ovine benign isolates and bovine isolates. J. Gen. Microbiol. 128: 809–812.

Hobbs M, Dalrymple BP, Cox PT, Livingstone SP, Delaney SF & Mattick JS (1991) Organization of the fimbrial gene region of *Bacteroides nodosus*: class I and class II strains. Mol. Microbiol. 5: 543–560.

Katz ME, Howarth PM, Yong WK, Riffkin GG, Depiazzi LJ & Rood JI (1991) Identification of three gene regions associated with virulence in *Dichelobacter nodosus*, the causative agent of ovine footrot. J. Gen. Microbiol. 137: 2117–2124.

Katz ME, Strugnell RA & Rood JI (1992) Molecular characterization of a genomic region associated with virulence in *Dichelobacter nodosus*. Infect. Immun. 60: 4586–4592.

624

Kortt AA, Burns JE & Stewart DJ (1983) Detection of the extracellular proteases of *Bacteroides nodosus* in polyacrylamide gels: a rapid method of distinguishing virulent and benign ovine isolates. Res. Vet. Sci. 35: 171–174.

Lilley GG, Stewart, DJ & Kortt, AA (1992) Amino acid and DNA sequences of an extracellular basic protease of *Dichelobacter nodosus* show that it is a member of the subtilisin family of proteases. Eur. J. Biochem. 210: 13–21.

Mattick JS, Anderson BJ, Cox PT, Dalrymple BP, Bills MM, Hobbs M & Egerton JR (1991) Gene sequences and comparison of the fimbrial subunits representative of *Bacteroides nodosus* serotypes A to I: class I and class II strains. Mol. Microbiol. 5: 561–573.

Roberts M, Piot P & Falkow S (1979) The ecology of gonococcal plasmids. J. Gen. Microbiol. 114: 491–494.

Skerman TM (1989) Isolation and identification of *Bacteroides nodosus*. p. 85–104. *In*: Egerton JR, Yong WK & Riffkin GG (eds.) Footrot and foot abscess of ruminants. CRC Press, Inc., Boca Raton.

Stewart DJ (1979) The role of elastase in the differentiation of *Bacteroides nodosus* infections in sheep and cattle. Res. Vet. Sci. 27: 99–105.

Stewart DJ (1989) Footrot of sheep. p. 5–45. *In*: Egerton JR, Yong WK & Riffkin GG (eds.) Footrot and foot abscess of ruminants. CRC Press Inc, Boca Raton.

Stewart DJ & Elleman TC (1987) A *Bacteroides nodosus* pili vaccine produced by recombinant DNA for the prevention and treatment of foot-rot in sheep. Aust. Vet. J. 64: 79–81.

Stewart DJ, Peterson JE, Vaughan JA, Clark BL, Emery DL, Caldwell JB & Kortt AA (1986) The pathogenicity and cultural characteristics of virulent, intermediate and benign strains of *Bacteroides nodosus* causing ovine foot-rot. Aust. Vet. J. 63: 317–326.

Yoshioka Y, Fujita Y & Ohtsubo E (1990) Nucleotide sequence of the promotor-distal region of the *tra* operon of plasmid R100, including *traI* (DNA helicase I) and *traD* genes. J. Mol. Biol. 214: 39–53.

45. Genes governing the secretion of factors involved in host-bacteria interactions are conserved among animal and plant pathogenic bacteria

FRÉDÉRIQUE VAN GIJSEGEM, MATTHIEU ARLAT, STÉPHANE GENIN, CLARE L. GOUGH, CLAUDINE ZISCHEK, PATRICK A. BARBERIS and CHRISTAN BOUCHER

Abstract. The *hrp* gene cluster of several phytopathogenic bacteria is needed for the expression of virulence on host plants and for the elicitation of a hypersensitive response, associated with resistance on non-host plants. In *Pseudomonas solanacearum*, the *hrp* gene cluster has been sequenced and was shown to contain 19 putative ORFs. Seven of the proteins predicted from these ORFs have characteristics of membrane proteins. For eight of the Hrp proteins, homologies with proteins involved in the secretion of virulence determinants in the mammalian pathogens *Yersinia* and *Shigella* have been found. These proteins include five of the putative membrane proteins, a protein sharing homologies with several ATPases and the HrpB protein which was proven to be a positive regulator of the *hrp* gene cluster expression. These results prompted us to analyze whether the *hrp* gene cluster was involved in the secretion of factors able to induce a hypersensitive-like reaction in non-host plants. Such a factor has been found in the supernatant of *P. solanacearum* grown in conditions which allow the expression of the *hrp* genes. This factor is heat-resistant and proteinase K sensitive indicating that it might be a protein. Analysis of several *hrp* mutants indicates that the *hrp* gene cluster could indeed be involved in the secretion of this active factor and that the synthesis of this factor is regulated by the *hrpB* gene.

Abbreviations: HR, Hypersensitive Response; *hrp*, Hypersensitive Response and Pathogenicity; Ipas, Invasion Plasmid Antigens; Yops, Yersinia Outer Membrane Proteins

Introduction

The interaction between a microbial pathogen and its plant host is a complex multistep process. To successfully infect a plant, the pathogen first has to be able to penetrate the plant and overcome active plant defense responses. It must then maintain and multiply within the plant tissue and finally induce the set of events which leads to disease symptoms. Plants protect themselves from pathogen entry by an array of passive physical and chemical barriers, essentially cuticles, waxes and reinforced cell walls. If such barriers are overcome by a pathogen, two types of interactions may take place with the plant : either the pathogen is able to multiply thoroughly in the plant and to produce disease symptoms, the interaction is then called compatible ; or there is not extensive pathogen multiplication and there are no observable disease symptoms, the interaction is then called incompatible. The incompatible interaction itself may be observed at two levels ; plant resistance to a pathogen might be expressed at the species level (all genotypes of a given plant species are resistant to a given pathogen species)

C.I. Kado and J.H. Crosa (eds.), Molecular Mechanisms of Bacterial Virulence, 625–642.

or at the cultivar level (only certain genotypes of the same plant species are resistant to certain races of the pathogen). This aspect of plant-bacteria interactions is addressed in more detail in other chapters of this volume.

In the case of an incompatible interaction between a bacterial pathogen and a plant, when a relatively high concentration of bacteria is inoculated into the plant, there is a very rapid necrosis of the plant cells in contact with or in close vicinity of the pathogen, followed by plant cell death. This resistance response, which is thought to contain the pathogen and to avoid further spreading, is called the hypersensitive reaction or HR (Klement, 1982). The occurrence of the HR has been reported to require metabolically active bacteria, as inoculation with killed cells or coinoculation of the pathogen with drugs which inhibit bacterial transcription or *de novo* protein synthesis failed to induce this reaction (Meadows & Stall, 1981). Furthermore, the inoculation of a non-pathogenic bacterium like *Escherichia coli* or *Pseudomonas fluorescens* does not induce any necrosis whatever the concentration of the inoculum used. To tackle the bacterial factors needed for the elicitation of the HR, mutagenesis has been performed in several phytopathogenic bacterial genera in order to isolate mutants which are no longer able to induce the HR in incompatible plants. Whatever the type of bacterial genus mutagenized, the same type of mutants were isolated ; they were impaired in their ability to produce the HR on the incompatible host but were also unable to provoke the disease on a compatible host. These mutants were therefore called *hrp* for *h*ypersensitive *r*esponse and *p*athogenicity. Such *hrp* mutants were isolated in most Gram-negative phyto-pathogenic bacterial genera including *Pseudomonas*, *Xanthomonas* and *Erwinia* species (reviewed in Willis *et al.*, 1991 and more recently in this volume). The function of the genes impaired in these *hrp* mutants have been poorly understood.

For more than a decade, the interaction between plants and the phyto-pathogenic bacterium *P. solanacearum* has been studied in our laboratory. It is the causal agent of bacterial wilt of more than 200 solanaceous plant species (including potato, tomato and tobacco) in Musaceae plants like banana (Hayward, 1991) and plants of 30 different families. The host range of *P. solanacearum* varies from strain to strain, some of them being able to cause disease in a large number of plant species while other isolates being restricted to a narrow spectrum of susceptible plants. Several *P. solanacearum* pathogenicity factors have been identified and characterized (reviewed in Boucher *et al.*, 1992; Denny and Schell, 1992).

In this paper we will focus on the data concerning the induction of the HR caused by *P. solanacearum* strain GMI1000. This strain is pathogenic on most solanaceous plants but induces a typical HR following infection into leaves of the non-host resistant tobacco plant. When avirulent mutants of strain GMI1000 were isolated by direct screening on axenic tomato seedlings after transposon mutagenesis (Boucher *et al.*, 1985), most of them appeared to be *hrp* mutants as they were found to be impaired in the ability both to provoke disease on the compatible plant tomato and to elicit the HR on the incompatible host tobacco. These *hrp* mutants were also shown to be altered in their ability to colonize and to multiply in compatible tomato plants (Trigalet and Démery, 1986). The genes impaired in most of these *hrp* mutants were shown to be clustered in a region of the megaplasmid (more than 1000 kb) present in strain GMI1000 (Boucher *et al.*, 1986 ; 1987). The scope of this paper deals with the characterization of the *P. solanacearum hrp* gene cluster and the paper gives some insights into the possible function(s) of the *hrp* genes.

Genetic organization of the *hrp* gene cluster

Cloning and saturation mutagenesis with the Tn5-B20*lac* fusion transposon allowed us to propose that the *P. solanacearum hrp* cluster is organized in at least six transcription units covering a total of about 23 kb of DNA (Arlat *et al.*, 1992 ; Fig. 1 and 2). Mutations in the four leftmost transcription units completely abolish the ability of the corresponding mutants to elicit an HR on tobacco and to provoke disease on tomato while mutations in units 5 and 6 led to a leaky phenotype (Arlat *et al.*, 1992). Sequencing of the 20 kb DNA fragment which encompasses the transcription units 1 to 4 has led to the identification of 19 open reading frames (ORFs) with high coding probabilities as deduced from codon usage bias. Seven of these ORFs exhibit characteristics of membrane proteins, two of which (HrpA and HrpU) have at their amino terminal end a typical signal sequence which might be recognized by the general *sec*-dependent protein export machinery (d'Enfert *et al.*, 1989). A third putative membrane polypeptide, HrpI, possesses a suitable amino-terminal acylation site sequence for being cleaved by the signal peptidase II (Wu & Tokunaga, 1986).

The genetic organization of the *hrp* gene cluster as deduced from the DNA sequence data fits exactly with the one which was deduced from the *lac* fusion genetic analysis. The regions preceeding units 1, 2 and 4 have been cloned in front of a reporter gene and, in these constructs, act as active promoters

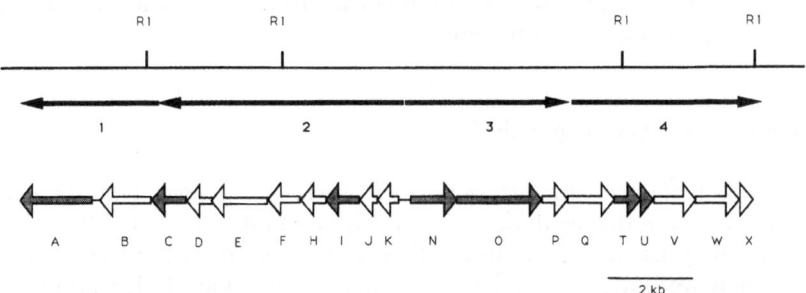

Figure 1. Genetic organization of the *hrp* gene cluster: The top line is an *Eco*R1 (R1) restriction map of the left end of the cluster. Solid arrows show the individual transcription units. Open arrows represent the individual open reading frames coding for the putative Hrp proteins. When a protein is predicted to have transmembrane domain(s) the corresponding arrow is shaded.

628

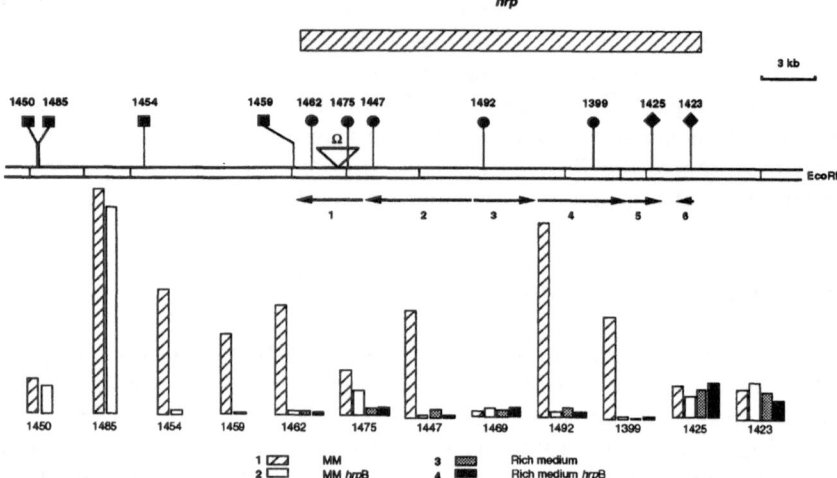

Figure 2. Regulation of *hrp* gene expression and flanking regions by *hrpB* : The top of the figure represent the *EcoRI* restriction map of the *hrp* gene cluster and of the left flanking region together with the localization of the *lacZ* reporter gene fusions which have been generated using transposon Tn5-B20 (Arlat *et al.* 1992). The triangle above the map shows the location of the insertion of the omega interposon generating a *hrpB* mutation. Histograms represent the relative level of expression of each gene fusion under inducing condition in the wild-type (1) versus a *hrpB* mutant background (2) and in rich medium in the wild-type (3) versus a *hrpB* mutant background (4) respectively. Mutant 1469 carries a non fusion insertion in transcription unit 2 and was used as a negative control.

regulated like the corresponding *hrp* genes (see below). There is a non-coding region of 85 or 170 bp between the putative *hrp*A and *hrp*B genes. However, neither complementation analysis nor promoter cloning analysis allowed us to show a promoter activity in this DNA fragment. No sequences related to consensus *E. coli* or *Pseudomonas* promoter sequences were detected in any of the promoter regions and there is also no evidence for a sequence which might be conserved among these different regions.

Regulation of *hrp* gene expression

Expression analysis using *hrp* gene *lacZ* fusions revealed that there is a difference between the regulation of transcription units 1 to 4 and that of units 5 and 6. The genes belonging to the first four units are not expressed in rich medium and are induced in minimal medium to different levels depending of the carbon source provided for growth. In contrast, fusions in units 5 or 6 are expressed to the same extent in rich and in minimal medium (Arlat *et al.*, 1992; Genin *et al.*, 1992 ; Fig. 2). There is no evidence for the existence of an inducing factor of plant origin since the level of expression of *hrp* genes *in planta* is comparable to that found in minimal medium supplemented with the best inducing carbon source (Arlat *et al.*, 1992).

Computer-assisted analysis showed that the predicted HrpB polypeptide possesses a helix-turn-helix motif typical of DNA-binding proteins. Protein homology searches revealed that the predicted HrpB polypeptide bears in its carboxyterminal part similarities with a family of regulatory proteins, for example the *E. coli* AraC arabinose operon transcriptional activator protein (Wallace *et al.*, 1980). The best homologies were found with the TOL plasmid meta-cleavage pathway operon regulatory XylS protein of *P. putida* (43% similarity over 25% of the protein ; Mermod *et al.*, 1987) and the *Yersinia enterocolitica* VirF protein which regulates the expression of virulence determinants (Cornelis *et al.*, 1989) (Genin *et al.*, 1992). The actual positive regulatory function of the HrpB protein was demonstrated by showing that in *hrp*B *hrp-lac*Z double mutants, there is no longer *hrp* gene activation in minimal medium for genes belonging to units 2, 3 and 4. Expression of *hrp* units 5 and 6 is not affected by the inactivation of the *hrp*B gene (Fig. 2). When the *hrp*B gene is introduced into *P. solanacearum* on a multicopy plasmid however, a slight stimulation of expression of unit 5 is observed during growth in inducing conditions. The activator activity of HrpB is not restricted to the *hrp* gene cluster: *lac* fusions located to the left of the *hrp* gene cluster and exhibiting a wild-type Hrp phenotype are also HrpB-regulated. Finally, the transcription of the *hrp*B gene itself is also induced in minimal medium and is partly autoregulated (Genin *et al.*, 1992). This implies that other regulatory factors are needed for *hrp*B gene expression.

Search for homologies to Hrp proteins

In an attempt to discover the possible functions of the other *hrp* genes, data base protein homology searches were accomplished with proteins predicted from all the putative ORFs deduced from the DNA sequence. Besides the HrpB polypeptide, similarities were found for seven additional putative Hrp polypeptides. They are summarized in Table 1 and Table 2.

The more exciting homologies are certainly those which were found with genes involved in the virulence of the enterobacteria *Yersinia* and *Shigella* responsible for disease in mammals (Table 1).

In *Yersinia*, besides the homology already described between the activator protein VirF and HrpB, homologies were found with 5 other putative Hrp proteins. Three of them belong to the *ysc* operon (YscC, YscJ and YscL), which was shown to be involved in the secretion into the extracellular medium of a set of virulence determinants called the Yop proteins (Cornelis *et al.*, 1989a) (Michiels *et al.*, 1991). The YscJ protein, which was shown to be a membrane located lipoprotein (Michiels *et al.*, 1991), is homologous to the putative HrpI lipoprotein which also has characteristics of membrane proteins. The YscC protein has no obvious transmembrane domains as seen in the HrpA

Table 1. Homologies between *P. solanacearum* Hrp proteins and proteins involved in virulence of mammalian pathogens.

P.solanacearum	Yersinia	Shigella	function
HrpA 568 aa	YscC [a] 34% over 683 aa		secretion
HrpB 477 aa	VirF[b] 20% over 483 aa		positive regulator
HrpE 439 aa	partial LcrB1[c] 38% over 150 aa	Spa47[d] 44% over 421 aa	ATPase
HrpF 277AA	YscL[a] 23% over 170AA		
HrpI 269 aa	YscJ[a] 35% over 237 aa		lipoprotein
HrpO 690 aa	LcrD[e] 43% over 720 aa	VirH[f] 34% over690AA	regulator?
HrpQ 355 aa		partial Spa33[d] 26% over 69 aa	
HrpT 218 aa		Spa24[d] 40% over 209 aa	secretion

[a] Michiels *et al*,1992
[b] Cornelis *et al*, 1989
[c] Viitanen *et al*, 1990
[d] Venkatesen *et al*, 1992
[e] Plano *et al*, 1991
[f] Sasakawa unpublished, GenBank Accession.D10999

polypeptide but, like HrpA, it shares significant similarity with two described outer membrane proteins,PulD and pIV (see below). No special features were detected in the putative YscL peptide but complementation experiments demonstrated that this protein is required for Yop protein synthesis (Michiels *et al.*, 1991). The two other *Yersinia* proteins showing homology to presumed Hrp polypeptides are coded by genes being part of the *lcr* region which is involved in the regulation of Yop production by low Ca^{2+} and temperature (Plano *et al.*, 1991)). In mutants affected in this region , both the regulation of the synthesis and of the secretion of the Yop proteins were altered.

Shigella polypeptides homologous to Hrp proteins are all clustered in an operon, the *spa* operon, which is needed for the cell surface presentation of

Table 2. Homologies between *P. solanacearum* and additional proteins.

Hrp protein	Homologous protein	Organism	Homology	Reference
HrpA 568 aa	PulD pIV	*Klebsiella pneumoniae* phage I2-2	22% over 691 aa 24% over 582 aa	d'Enfert *et al*, 1989 EMBL accession number X14336
HrpE 439 aa	FliI FlaA-ORF4	*Salmonella typhimurium* *Bacillus subtilis*	45% over 445 aa 48% over 445 aa	Vogler *et al*, 1991 Galizzi *et al*, 1991
HrpO 690 aa	FlbF FlhA	*Caulobacter crescentus* *Bacillus subtilis*	34% over 710 aa 34 % over 696 aa	Ramakrishnan *et al*, 1991, Sanders *et al*, 1992 EMBL accession X63698
HrpQ 355aa	FliN (138 aa) MotD (137 aa)	*Salmonella typhimurium* *Escherishia coli*	25 % over 138 aa 26% over 137 aa	Kihara *et al*, 1989 Malakooti *et al*, 1989

invasion plasma antigens (Ipas). These Ipas are pathogenicity factors which have been shown to be responsible for the *Shigella flexneri* invasion phenotype in epithelial cells (Halle, 1991). DNA sequencing of the *spa* operon revealed the presence of 6 putative ORFs (Venkatesan *et al.*, 1992; Sasakawa, unpublished). Four of the proteins predicted from these ORFs show similarities with predicted Hrp proteins (Table 1). For 3 of them the similarity extends to the entire length of the respective proteins, the carboxyterminal region of the Spa33 predicted product is homologous to the corresponding region of the HrpQ polypeptide. Both the Spa47 polypeptide and its HrpE homolog show significant similarity to several ATPases from various sources. The best homologies were found with flagellar ATPases such as the FliI protein of *S. typhimurium* which is thought to play a role in the energization of the construction of the flagellum, and the FlaA protein of *Bacillus subtilis* (Galizzi *et al.*, 1991) (Table 2). All these proteins exhibit similarities on the whole length of the proteins.

Additional homologies found between Hrp proteins and other proteins are summarized in Table 2. It should be noted that most of these similarities are with proteins which are somehow involved in the transport of proteinaceous complexes across membranes. For example, the outer membrane located PulD protein is involved in the secretion of pullulanase in *Klebsiella pneumoniae* (d'Enfert *et al.*, 1989), the protein pIV of the filamentous bacteriophoge I2-2 is thought to be implicated in extrusion through the bacterial cell envelope of the phage particles (Russel,1991) and the *E. coli* MotD (Malakooti *et al.*, 1989) and the *S. typhimurium* FliN (Kihara *et al*, 1989) proteins are part of the flagellar motor and are also thought to participate in the transport of the flagellum components through membranes.

Involvement of the *hrp* gene cluster in the secretion of an extracellular HR-inducing factor

All the homologies described in the previous section point to a role of the *hrp* gene cluster in the transport of molecules to the extracellular medium or to the bacterial surface. To substantiate this hypothesis, we attempted to address two questions. First, does *P. solanacearum* produce an extracellular factor able to elicit an HR on the incompatible plant tobacco? Secondly, what is the role of the *hrp* gene cluster in the production and/or transport of this factor?

A. Production of an HR-inducing extracellular factor

When the wild-type strain GMI1000 was grown in conditions which are known to allow *hrp* gene expression (minimal medium with glutamate as sole carbon source) (Arlat *et al.*, 1992), it was possible to detect a HR-like inducing activity in the non-dialysable fraction (MW > 10,000) of a 20-fold concentrate of the supernatant. This activity was revealed following infiltration into tobacco leaf parenchyma and subsequent incubation for 24 to 48 hours at 25°C. It resulted in the development of a necrotic lesion restricted to the infiltrated tissues which was highly reminiscent of the reaction induced by live bacteria both with respect to the kinetics of the response and the aspect of the necrotic lesions. Such activity was not found in the supernatant of any of the *hrp* mutants which mapped in transcription units 1 to 4, thus indicating that this activity was *hrp*-dependent and not the result of a non specific toxic effect of the supernatant on plant tissues. This activity was still present in the supernatant of mutants mapping in transcription units 5 and 6 which had previously been shown to induce only a partial HR. Kinetic studies showed that this HR-inducing activity began to be detectable at the end of the logarithmic growth phase.

The activity present in the supernatant of the wild-type strain GMI1000 is assumed to be due to the presence of (a) heat stable protein(s) since it was resistant to heating for 8 min at 100°C but was lost after treatment with proteinase K.

B. Possible involvement of hrp genes in controlling secretion of the HR-like eliciting factor

Although no HR-like inducing activity could be detected in the supernatant of any of the *hrp* mutants which mapped in transcription units 1 to 4, an activity was found in filter-sterilized, sonicated cell lysates prepared from most *hrp* mutants tested as well as from the wild-type strain (Fig. 3). The only mutants which had no intracellular activity were the *hrpB* mutant and a mutant deleted of the entire *hrp* gene cluster (Δ *hrp*). These mutants fail to produce a functional positive *hrp* regulator which indicates again that the activity does not result from a non-specific toxic effect of the cell lysate, but rather that it is *hrp* gene specific. Similar to the extracellular activity, the intracellular activity was found to be heat-resistant and sensitive to proteinase K indicating that both intra- and extra-cellular activities could be due to the same protein(s) present both within and outside the bacterial cells. If this is true, our data establish that in a similar way to their *Yersinia* or *Shigella* counterparts, several *hrp* genes are involved in the transport through the bacterial cell envelopes of extracellular protein(s)

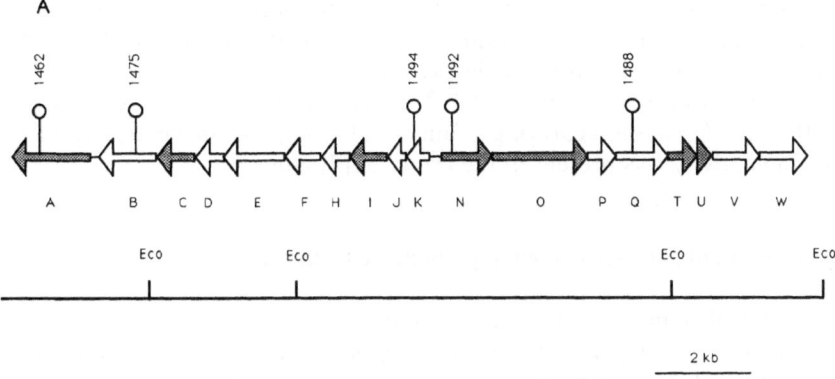

Figure 3. Cellular localization of the HR-inducing activity in wild-type strain and various *hrp* mutants: A) Mapping of the mutations corresponding to the different *hrp* mutants used, B) Presence (+) or absence (−) of activity in the supernatant and the crude cell lysate of the corresponding strains grown under *hrp* genes inducing conditions.

which, in the case of *P. solanacearum*, have a HR-like inducing activity. This secretion mechanism seems to be quite specific as the secretion of other proteins by *P. solanacearum*, the polygalacturonase and endoglucanase depolymerizing enzymes, is not affected in any *hrp* mutants.

The requirement of a functional *hrpB* gene for the production of the intracellular activity suggests that the (or at least some of the) gene(s) involved in the synthesis of the active compound(s) is (are) part of the *hrp* gene regulon. However, the fact that all the other mutants which map within the *hrp* gene cluster had retained an intracellular activity, suggests that this (these) gene(s) map(s) outside of the *hrp* gene cluster. This latter conclusion is reinforced by the fact that in a *hrp* mutant carrying a functional *hrpB* gene on a plasmid, the HR-inducing activity was present in the crude cell lysate.

Conservation of *hrp* genes in phytopathogenic bacteria

As we recalled in the introduction, most Gram-negative phytopathogenic bacteria possess *hrp* genes. In all cases reported, most of these *hrp* genes are clustered in a DNA region ranging from 22 to about 40 kb (reviewed in Willis *et al.*,1991, see chapters of this volume). Hybridization using parts of the *hrp* gene cluster as probes revealed that the *hrp* gene clusters are conserved among various pathovars of *Pseudomonas syringae* (Lindgren *et al.*, 1988) and among strains of *P. solanacearum* (Boucher *et al.*, 1988 ; Cook el al, 1989). Furthermore, the *hrp* gene cluster of the enterobacterium *Erwinia amylovora* was shown to be conserved between both phytopathogenic and non-pathogenic *Erwiniae* species (Laby and Beer, 1990). More recently, the conservation and the colinearity of the *hrp* gene clusters were demonstrated between *P. solanacearum* and *Xanthomonas campestris* (Boucher *et al.*, 1988 ; Arlat *et al.*, 1991). This conservation is also reflected at the protein level as it was recently reported that, as with *P. solanacearum*, genes of the *X. campestris* pv. *vesicatoria hrp* gene cluster code for proteins sharing homologies with the YscC, YscJ and LcrD proteins of *Yersinia* (Fenseleau *et al.*, 1992). Homologies between the corresponding *hrpA*, *hrpI* and *hrpO* genes of *P. solanacearum* genes and the *X. campestris* genome were also detected at the DNA level (Gough *et al.*, 1992). Hybridization studies using as a probe a DNA fragment corresponding to the part of HrpO of *P. solanacearum* which is the most conserved to the LcrD protein, showed that the *hrpO* gene is also conserved in *P. syringae* pv. *phaseolicola* and *E. amylovora* (Gough *et al.*, 1992). These results clearly suggest that the function of the *hrp* genes might be closely related in the different bacteria able to induce an HR on a non-host plant. The fact that Wei *et al.* (1992) reported the isolation of an HR-inducing protein, the harpin, produced by *E. amylovora* also points to that conclusion even if the harpin does not seem to be secreted, as is the case for the *P. solanacearum* HR-inducing factor, but rather is associated with the bacterial cell envelope as reported for *Shigella* invasion plasmid antigens.

Discussion and perspectives

Analysis of the DNA sequence data from *P. solanacearum hrp* gene cluster has revealed several similarities between putative Hrp proteins and proteins involved in the transport through membranes of various macromolecules such as *Yersinia* Yop proteins, *Shigella* Ipas or *E. coli* and *S. typhimurium* flagellar components. Preliminary data also indicate that, when grown in inducing conditions (i. e., in minimal media), *P. solanacearum* produces and most probably secretes a heat stable proteinase K sensitive HR-inducing factor. The synthesis of this presumed proteinaceous factor may be regulated by the *hrpB* gene, the positive regulator of expression of most of the *hrp* genes, the other *hrp* genes are probably involved in the secretion but not in the synthesis of this factor.

Two protein secretory machineries are well characterized in Gram-negative bacteria including several phytopathogenic species (reviewed in Pugsley *et al.*, 1990). They are exemplified by the *K. pneumoniae* pullulanase secretion system (Pul system) and by the *E. coli* hemolysin export apparatus. No homologies were found between the Hrp, Ysc or Spa proteins and the Hly proteins involved in the hemolysin export system. Even though homology was found between one of the Pul proteins and the HrpA/YscC homologs, the fact that neither the Yop proteins nor the *Shigella* Ipas harbor typical aminoterminal signal sequences indicates that the secretion of these proteins does not occur via a sec-dependent pathway as in the pullulanase system. Of course, we have to wait for the sequence of the *P. solanacearum* HR-inducing factor to ascertain that it does not exhibit a signal sequence, but it is very likely that we have here a third type of secretion mechanism shared both by plant and mammalian Gram-negative pathogens.

Another exciting concept arising from this work is the wide conservation of *hrp* genes between most Gram-negative phytopathogenic bacteria irrespective of the plants they infect or of the kind of symptoms they can induce. The *hrp* genes could in fact be a core of common essential pathogenicity genes. If the principal role of these genes is the formation of a secretory complex, the important key in the various interactions would then be the secreted products. Here again we can draw a parallell between plant and mammalian pathogens. Indeed, as extracellular molecules, the Yop proteins mediate the interaction between *Yersinia* and animal cells, probably partly by disrupting the normal signal transduction machinery of the host cells as one of the Yops, YopH, was found to have tyrosine phosphatase activity (Guan & Dixon, 1990). In the same way, the *P. solanacearum* HR-inducing factor interferes with the plant cell programme in a way which ultimately leads to cell death in the HR. So, most phytopathogenic bacteria, and at least some animal pathogens, seem to have developed a common strategy of host infection. This raises the question as to how the corresponding genes have evolved in order to be present in such taxonomically distant organisms as enterobacteriae, pseudomonads or xanthomonads. It should be noted that genes encoding for the secretion machinery in *P. solanacearum*, *Yersinia* and *Shigella* are all clustered on a plasmid suggesting that horizontal transfer might have occurred. However, this does not seem to be true since the GC content of these different clusters is totally in accordance with their bacterium of origin. For example, the *hrpA* gene is 68%

636

GC rich in accordance with the 66–69% GC content of *P. solanacearum* as deduced by buoyant density experiments (Palleroni & Doudoroff, 1971) while the *yscC Yersinia* counterpart gene is only 45% GC rich. This is also reflected by the wide differences between the codon usage found in these genes.

To complete the picture which begins to emerge from our data, of the way in which plant and bacterial pathogens interact, many questions still remain open. The first one is to determine the number and the nature of the products which might be secreted by the *hrp* secretory complex. In this respect, it is striking that the products of several avirulence genes, which are involved in plant-bacteria race-specific resistance, are predicted to be hydrophilic proteins without a signal sequence and that the phenotypic expression of race-specific avirulence is usually *hrp*-dependent. These Avr products might therefore be good candidates for being secreted by the *hrp*-encoded secretion machinery. Other questions are also presently under investigation in our laboratory concerning the function of the *hrp*-secreted products. Is it the same product which is needed for the elicitation of the HR on non-host plants and for expression of the disease in host plants? What is the role of this product(s) in establishment of the disease? And finally, is it possible that these products might be used in the future in disease plant protection programmes?

Acknowledgements

This work was supported by grants from European Economic Community (BIOT-CT90–0168) and North Atlantic Treatry Organization ([30]880310/88).

References

Arlat, M., Gough, C.L., Barber, C.E., Boucher C. & Daniels, M.J. (1991) *Xanthomonas campestris* contains a cluster of *hrp* genes related to the larger *hrp* cluster of *Pseudomonas solanacearum*. Mol. Plant-Microbe Interact.. 4: 593–601.

Arlat, M., Gough, C.L., Zischek, C., Barberis, P.A., Trigalet, A., & Boucher, C.A. (1992) Transcriptional organisation and expression of the large *hrp* gene cluster of *Pseudomonas solanacearum*. Mol. Plant-Microbe Interact. 5: 187–193.

Boucher, C.A., Barberis, P.A. & Arlat, M. (1988) Acridine orange selects for deletion of *hrp* genes in all races of *Pseudomonas solanacearum*. Mol. Plant-Microbe Interact. 1: 282–288.

Boucher, C.A., Barberis, P.A., Trigalet, A.P. & Démery, D.A. (1985) Transposon mutagenesis of *Pseudomonas solanacearum*: Isolation of Tn5-induced avirulent mutants. J. Gen. Microbiol. 131: 2449–2457.

Boucher, C.A., Martinel A., Barberis, P., Alloing, G. & Zischek C. (1986) Virulence genes are carried by a megaplasmid of the plant pathogen *Pseudomonas solanacearum*. Molec. Gen. Genet. 205: 270–275.

Boucher, C.A., Gough, C.L. & Arlat, M. (1992) Molecular genetics of pathogenicity determinants of *Pseudomonas solanacearum* with special emphasis on *hrp* genes. Annu. Rev. Phythopath. 30: 443–461.

Boucher, C.A., Van Gijsegem, F., Barberis, P.A., Arlat, M. & Zischek, C. (1987) *Pseudomonas solanacearum* genes controlling both pathogenicity and hypersensitivity on tobacco are clustered. J. Bacteriol. 169: 5626–5632.

Cook, D., Barlow, E. & Sequeira L. (1989) Genetic diversity of *Pseudomonas solanacearum*: detection of restriction fragment lenght polymorphism with DNA probes that specify virulence and the hypersensitive response. Mol. Plant-Microbe Interact. 2: 113–121.

Cornelis, G.R., Biot, T., Lambert de Rouvroit, C., Michiels, T., Mulder, B., Sluiters, C., Sory, M.P., Van Bouchaute, M. & Vanooteghem, J.C. (1989a) The *Yersinia* Yop regulon. Mol. Microbiol. 3: 1455–1459.

Cornelis, G., Sluiters, C., Lambert de Rouvroit, C. & Michiels, T. (1989b) Homology between VirF, the transcriptional activator of the *Yersinia* virulence regulon, and AraC, the *Escherichia coli* arabinose operon regulator. J. Bacteriol. 171: 254–262.

d'Enfert, C., Reyss, I., Wandersman, C., & Pugsley, A.P. (1989) Protein secretion by Gram-negative bacteria. J. Biol. Chem. 264: 17462–17468.

Denny, T.P. & Schell, M.A. (1992) Virulence and pathogenicity of *Pseudomonas solanacearum*: genetic and biochemical perspectives. Fourth International Symposium on Biotechnology and Plant Protection, 21–23 october 1991, University of Maryland, College Park, USA, in press.

Fenselau, S., Balbo, I. & Bonas, U. (1992) Determinants of pathogenicity in *Xanthomonas campestris* pv. *vesicatoria* are related to proteins involved in secretion in bacterial pathogens of animals. Mol. Plant-Microbe Interact. 5: 390–396.

Galizzi, A., Scoffone, F., Crabb, W.D., Caramori, T., & Albertini A.M. (1991) The *flaA* locus of *Bacillus subtilis* is part of a large operon coding for the flagellar structures, motility functions, and an ATPase-like polypeptide. J. Bacteriol. 173: 3573–3579.

Genin, S. ,Gough, C.L., Zischek, C. & Boucher C.A. (1992) Evidence that the *hrpB* encodes a positive regulator of *hrp* genes from *Pseudomonas solanacearum*. Mol. Microbiol. 6: 3065–3076.

Gough, C.L., Genin, S., Zischek, C. & Boucher, C.A. (1992) *hrp* genes of *Pseudomonas solanacearum* are homologous to pathogenicity determinants of animal pathogenic bacteria and are conserved among plant pathogenic bacteria. Mol. Plant-Microbe Interact. 5: 384–389.

Guan, K. and Dixon, J.E. (1990) Protein tyrosine phosphatase activity of an essential virulence determinant in *Yersinia*. Science 249: 553–556.

Halle, T.L. (1991) Genetic basis of virulence in *Shigella* species. Microbiol. Revs. 55: 206–224.

Hayward, A.C. (1991) Biology and epidemiology of bacterial wilt caused by *Pseudomonas solanacearum*. Annu. Rev. Phytopath. 29: 65–67.

Kihara, M., Homma, M., Kutsukake, K. & Macnab, R. (1989) Flagellar switch of *Salmonella typhimurium*: gene sequences and deduced protein sequences. J. Bacteriol. 171: 3247–3257.

Klement, Z. (1982) Hypersensitivity. p. 150–178. *In*: M. S. Mount & G. S. Lacy (eds.) Phytopathogenic Pokaryotes. Academic Press. New York.

Laby, R.J. & Beer, S.V. (1990) The *hrp* gene cluster of *Erwinia amylovora* shares DNA homology with other bacteria. Phytopathology. 80: 1038–1039.

Lindgren, P.B., Panopoulos, N.J., Staskawicz, B.J. & Dahlbeck,D. (1988) Gene required for pathogenicity and hypersensitivity are conserved and interchangeable among pathovars of *Pseudomonas syringae*. Mol. Gen. Genet. 211: 4999–5006.

Malakooti, J., Komeda, Y. & Matsumara, P. (1989) DNA sequence analysis, gene product identification, and localization of flagellar motor components of *Escherichia coli*. J. Bacteriol. 171: 2727–2734.

Meadows, M.E. & Stall, R.E. (1981) Different induction periods for hypersensitivity in pepper to *Xanthomonas vesicatoria* determined with antimicrobial agents. Phytopathol. 71: 1024–1027.

Mermod, N., Ramos, J.L., Bairoch, A. & Timmis, K.N. (1987) The *xylS* gene positive regulator of TOL plasmid pWWO: identification, sequence analysis and over production leading to constitutive expression of *meta* cleavage operon. Mol. Gen. Genet. 207: 349–354.

Michiels, T., Vanooteghem, J.-C., Lambert de Rouvroit, C., China, B., Gustin, A., Boudry, P., & Cornelis, G.R. (1991) Analysis of *virC*, an operon involved in the secretion of Yop proteins by *Yersinia enterocolitica*. J. Bacteriol. 173: 4994–5009.

Palleroni N.J. & Doudoroff M. (1971) Phenotypic characterization and deoxyribonucleic acid homologies of *Pseudomonas solanacearum*. J. Bacteriol. 107: 690–696.

Plano, G.V., Barve, S.S., & Straley, S.C. (1991) LcrD, a membrane-bound regulator of the *Yersinia pestis* low-calcium response. J. Bacteriol. 173: 7293–7303.

638

Pugsley, A.P., d'Enfert, C., Reyss, I., & Kornacker, M.G. (1990) Genetics of extracellular protein secretion by Gram-negative bacteria. Annu. Rev. Genet. 24: 67–90.

Ramakrishnan, G., Zhao, J.-L., & Newton, A. (1991) The cell cycle-regulated flagella gene *flbF* of *Caulobacter crescentus* is homologous to a virulence locus (*lcrD*) of *Yersinia pestis*. J. Bacteriol. 173: 7283–7292.

Russel, M. (1991) Filamentous phage assembly. Mol. Microbiol.5: 1607–1613.

Sanders, L.A., Van Way, S., & Mullin, D.A. (1992) Characterization of the *Caulobacter crescentus flbF* promoter and identification of the inferred FlbF product as a homolog of the LcrD protein from a *Yersinia enterocolitica* virulence plasmid. J. Bacteriol. 174: 857–866.

Trigalet, A. & Démery, D. (1986) Invasiveness in tomato plant of Tn5-induced *Pseudomonas solanacearum*. Physiol. and Mol. Plant Pathol. 28: 423–430.

Venkatesan, M.M., Buysse, J.M. & Oaks, E.V. (1992) Surface presentation of *Shigella flexneri* invasion plasmid antigens required the products of the *spa* locus. J. Bacteriol. 174: 1990–2001.

Viitanen, A.-M., Toivanen P., Skurnik, M. (1990) The *lcrE* gene is part of an operon in the *lcr* region of *Yersinia enterocolitica* O:3. J. Bacteriol. 172: 3152–3162.

Vogler, A.P., Homma, M., Irikura, V.M., & Macnab, R.M. (1991) *Salmonella typhimurium* mutants defective in flagellar filament regrowth and sequence similarity of FliI to FOF1, vacuolar, and archaebacterial ATPase subunits. J. Bacteriol. 173: 3564–3572.

Wallace, R.G., Lee, N. & Fowler, A.V. (1980) The *araC* gene of *Escherichia coli*: transcriptional and translational start-points and complete nucleotide sequence. Gene 12: 179–190.

Wei, Z.M., Laby, R.J., Zumoff, C.H., Bauer, D.W., He, S. Y., Collmer, A. & Beer, S.V. (1992) Harpin, elicitor of the hypersensitive response produced by the plant pathogen *Erwinia amylovora*. Science 257: 85–88.

Willis, D.K., Rich, J.J., & Hrabak, E.M. (1991) *hrp* genes of phytopathogenic bacteria. Mol. Plant-Microbe Interact. 4: 132–138.

Wu, H.C. & Tokunaga, M. (1986) Biogenesis of lipoproteins in bacteria. Curr. Top. Microbiol. Immunol. 125: 127–157.

46. Defining the contribution of the *Agrobacterium* chromosome in crown gall tumorigenesis

TREVOR C. CHARLES and EUGENE W. NESTER

Abstract. In order to more fully understand the biology of crown gall tumorigenesis, we are searching for additional *Agrobacterium tumefaciens* chromosomal virulence genes. We reason that genes that are important in tumorigenesis will likely be either associated with the bacterial cell surface, or will be expressed under the unique environmental conditions of the plant wound. Thus, by identifying mutations in genes that encode cell surface encoding proteins and genes that are induced by the environmental conditions of the plant wound, we may discover new chromosomal virulence genes. So far, through Tn*phoA* mutagenesis we have identified two novel cell envelope protein encoding virulence genes, *chvG* and *chvH*.

Abbreviations: T-DNA, Transferred DNA

Introduction

The molecular biology of *Agrobacterium tumefaciens*, the causative agent of crown gall tumors, has been the subject of intense study over the past 15 years. In that time, this remarkable example of interkingdom genetic exchange has become a model system for interaction and communication between bacteria and eukaryotic organisms. This study has made a significant contribution to our understanding of plant development, bacterial gene regulation and signal transduction, conjugation and plasmid biology. However, many aspects of *A. tumefaciens* and crown gall biology have been neglected in recent years. In order to fully understand this complex interaction it is imperative that we have a good understanding of *A. tumefaciens* biology.

The central virulence determinant is a replicon called the Ti plasmid. This replicon is usually about 200 kb in size, depending on the isolate, and carries the T-DNA and the *vir* regulon. The interaction between *A. tumefaciens* and its plant host is initiated with deposition of viable bacterial cells into a plant wound, subsequent attachment of the bacterial cells to sites on the plant cell wall, and sensing of wound metabolites (monosaccharides, and phenolic compounds such as acetosyringone) by the bacterial cells. This sensing results in induction of expression of the *vir* regulon, whose products act to excise the T-DNA from the Ti plasmid, transfer it across the bacterial membranes and into the plant cell, and integrate it into plant nuclear DNA. The integrated T-DNA contains genes which encode the synthesis of phytohormones, which cause tumor formation,

C.I. Kado and J.H. Crosa (eds.), Molecular Mechanisms of Bacterial Virulence, 639–649.

and synthesis of opines, which the bacteria are able to use as carbon and nitrogen source (for a comprehensive review see Binns and Thomashow, 1988).

What advantages do *A. tumefaciens* cells gain from gall formation? Presumably, within the gall the bacteria are protected from adverse conditions, and are provided with a steady supply of carbon and nitrogen in the form of opines. This should enhance the survival of the bacterial cells. In addition, conjugation of the Ti plasmid is enhanced by conditions within the tumor, thus resulting in greater genetic exchange between bacterial cells (Van Larebeke *et al.*, 1975).

The 1975 review by Lippincott and Lippincott (1975) was written before the surge of molecular genetics and molecular biology and the discovery of T-DNA (plant genetic transformation) and the *vir* regulon. It provides a revealing summary of the early work on tumorigenesis. With the excitement of the discovery and analysis of interkingdom genetic transfer, the realization of the use of this system for the genetic engineering of plants, and the characterization of the signal transduction process that results in the induction of *vir* regulon expression by phenolic plant wound metabolite derivatives such as acetosyringone, we have unfortunately lost sight of some of this early work. Consequently some of the fundamental early physiological observations have yet to be addressed and confirmed conclusively by the powerful molecular techniques that are now available. For example, it was demonstrated by an elegant series of experiments in several laboratories that specific attachment of *A. tumefaciens* to the plant cell was a requirement for tumorigenesis. This attachment was independent of metabolic activity, and even cell envelope preparations retained attachment ability (reviewed in Lippincott and Lippincott, 1975). Neither the bacterial nor the plant structures for attachment have yet been identified. However, it was recently demonstrated that some component of the regenerating plant cell wall may be important for attachment (Binns, 1991), and evidence was presented that a vitronectin-like plant cell surface protein may be a specific receptor for *Agrobacterium* attachment to carrot cells (Wagner & Matthysse, 1992).

Searching for chromosomal virulence determinants by transposon mutagenesis

Although the majority of the experimental work on *A. tumefaciens* virulence functions has focused on the Ti-plasmid encoded genes (the *vir* regulon and the T-DNA), some chromosomal genes that are involved in tumorigenesis have been identified and characterized (Table 1). In some cases, researchers have searched specifically for chromosomal virulence genes. A number of approaches have been used in these searches, although all have in common mutagenesis with transposon Tn5 or derivatives. The advantages of transposon mutagenesis are many fold. Every transconjugant obtained from transposition harbors a single genetic lesion. This is especially valuable when screening for alterations that have no phenotype in the saprophytic state, and hence can only be identified by

Table 1. Agrobacterium tumefaciens chromosomal virulence and associated functions loci.

Locus	Function or phenotype	References
chvA/chvB	cyclic β-1,2 glucan synthesis and export, plant cell attachment, adaptation to conditions of low osmolarity	Cangelosi et al 1990b; Douglas et al. 1982, 1985; Puvanesarajah et al. 1985; Zorreguieta et al. 1988
exoC (pscA)	polysaccharide synthesis, including cyclic β-1,2-glucan	Cangelosi et al. 1987; Thomashow et al. 1987
att	deficient in plant cell attachment	Matthysse 1987
chvE	induction of vir regulon by monosaccharides	Cangelosi et al. 1990a; Huang et al. 1990
ivr	deficient in vir regulon induction	Metts et al. 1991
ros	repressor of virC and virD expression	Close et al. 1985, 1987; Cooley et al. 1991
miaA	tRNA::isopentenyltransferase, reduced vir regulon induction but no reduction in virulence	Gray et al. 1991

individual plant tests. Each mutation is marked by antibiotic resistance and an insertion of defined sequence, which greatly aid genetic analysis and genetic and physical manipulation of the mutation. Transposon generated mutations result from insertional inactivation of the gene, therefore reversion rates are usually much lower than for chemically-induced mutants.

Transposon mutant libraries have been screened for virulence directly on plants, or for various functions such as *vir* regulon induction in response to acetosyringone, attachment to plant cells, and ability to bind calcofluor (see below). From the mutants isolated thus far it appears that the chromosome encoded proteins play roles in the general physiology of the cell during growth in the absence of the plant as well as being involved in the tumorigenic process. However, in recent years we have learned little of the basic physiological requirements that are necessary for tumor formation.

One of the earliest studies to take advantage of transposon mutagenesis and that resulted in the identification of virulence genes was that of Garfinkel and Nester (1980). By screening prototrophic Tn5 mutants for virulence on Kalanchöe plants, several *vir* loci were identified. These included both Ti plasmid and chromosomally located insertions. Further study of the chromosomal loci resulted in the characterization of *chvA/B*, which encodes β-1,2-glucan synthesis and transport and is somehow involved in plant cell attachment (Douglas *et al.*, 1982, 1985; Puvanesarajah *et al.*, 1985; Zorreguieta *et al.*, 1988) and *chvE* (Huang *et al.*, 1990), which encodes a sugar binding protein that mediates monosaccharide induction of the *vir* regulon through interaction with VirA (Cangelosi *et al.*, 1990a). This type of approach to mutant isolation, with no enrichment for mutations in genes that may be involved in virulence prior to testing for virulence, is extremely labor intensive. Nevertheless, these studies laid the groundwork for much of the work on virulence genes that contributed to our understanding of the operation of the Ti plasmid *vir* system and gave some insight into the importance of the *A. tumefaciens* chromosome for tumorigenesis.

Following this work, several researchers attempted to identify additional chromosomal virulence genes, using various different approaches. After the discovery that expression of the *vir* genes was coordinately induced by plant phenolic wound metabolites (Stachel *et al.*, 1986), one common approach was to screen for transposon induced mutants that blocked the signal transduction pathway. This approach has been relatively fruitless, however, resulting in the characterization of only one new locus, *miaA*, encoding tRNA::isopentenyl-transferase (Gray *et al.*, 1991). Mutations in *miaA* caused a significant reduction in *vir* gene activation but did not result in a reduction in tumorigenicity. A similar approach involved screening mutants for attenuated response of *virG* to low pH and phosphate starvation (Winans *et al.*, 1988). This resulted in the identification of *chvD* mutants, which are attenuated in virulence. The role of *chvD* in virulence is not clear however, especially, in light of recent data which suggests that the level of *virG* expression has little effect on *vir* regulon induction (Chen and Winans, 1991).

Selection for mutants in which specific *vir* genes are constitutively expressed in the absence of inducing signals resulted in the identification of the *ros* chromosomal locus (Close *et al.*, 1985). In *ros* mutants, *virC* and *virD* are expressed to high levels independently of *virA* and *virG* (Close *et al.*, 1987). These mutants are also deficient in acidic exopolysaccharide production (Close *et al.*, 1987). The *ros* gene product was determined to be an autoregulated 15.5 kDa protein that probably binds at specific sites in the *virC* and *virD* promoters (Cooley *et al.*, 1991). The *ros* mutations did not affect virulence (Close *et al.*, 1985).

To characterize the genetic basis of attachment to plant cells, prototrophic Tn*5* mutants were screened for virulence and ability to attach to plant cells (Matthysse, 1987). Attachment deficient (*att*), avirulent mutants were identified, and these loci were clustered on the chromosome and distinct from *chvA/B*. Recently, using an approach very similar to that of Garfinkel *et al.* (1980), novel chromosomal virulence loci were identified (Metts *et al.*, 1991). In this study, Tn*5* mutants were first screened for virulence, and avirulent chromosomal mutants that retained normal binding to plant cells were studied further. Three such mutants (termed *ivr*) were characterized and were all apparently blocked in the signal transduction pathway.

An example of a strategy in which mutants are isolated on the basis of a property that may be important for virulence is the mutagenesis screens for lack of fluorescence under UV light on calcofluor containing media. Mutants that are deficient in exopolysaccharide (or cellulose) exhibit diminished calcofluor fluorescence. It was assumed that exopolysaccharide may be involved in virulence. Screens for calcofluor dark mutants resulted in the identification of the *A. tumefaciens exoC* (*pscA*) locus as a virulence factor. Several other calcofluor dark mutants were fully virulent, however, and it became apparent that the avirulence of the *exoC* mutants was due to a deficiency in β-1,2-glucan synthesis in these mutants (Thomashow *et al.*, 1987; Cangelosi *et al.*, 1987).

Another strategy that was utilized was based on the rationale that bacterial genes that are involved in interaction with the plant may be induced by factors that are present in plant exudate. Expression of the Ti plasmid *vir* regulon is controlled by phenolic compounds and monosaccharides that are associated with the plant wound (Stachel *et al.*, 1986; Ankenbauer and Nester, 1990). It is a reasonable assumption that additional virulence genes may be expressed in response to other plant factors. A search carried out in the laboratory of S. Gelvin for loci that were induced by carrot root extract resulted in the identification of *picA* (Rong *et al.*, 1990, 1991). Expression of this locus was induced by carrot acidic polysaccharides. Mutants of *picA* retained virulence, however, and the role of this locus in tumor formation is unclear. It is likely that there are other genes that are expressed in response to as yet unidentified plant signal molecules, and some of these may be important in virulence.

In many cases it has been difficult to define precisely what role chromosomal virulence genes play in virulence. For example, the *chvE* encoded glucose/galactose binding protein functions in monosaccharide induction of the *vir* regulon and as well is important for chemotaxis towards and uptake of monosaccharides (Cangelosi *et al.*, 1990a). The genes *chvA*, *chvB* and *exoC/pscA* encode functions required for synthesis/export of cyclic β-1,2-glucan which is required for attachment to the plant cell and adaptation to conditions of low osmolarity (Cangelosi *et al.*, 1990b). The *ivr* loci are involved somehow in the signal transduction pathway, and mutants in these loci have altered lipopolysaccharide (Metts *et al.*, 1991). The *att* locus is involved in plant cell attachment and mutants have altered outer membrane protein profiles (Matthysse, 1987). Significantly, no chromosomal mutations that specifically impair T-DNA transport functions have been isolated.

The pleiotropic nature of some of these mutations illustrates a major problem in their analysis. Is their avirulence a direct result of a single defect or is it due to a general defect in the cell's metabolism or structure which indirectly affects virulence? For example, early studies suggested that some auxotrophs are avirulent or exhibit reduced virulence (Lippincott and Lippincott, 1966), presumably because the cells must be growing in order to set up a productive infection. Other studies have suggested, however, that most auxotrophs, with the exception of those affected in tryptophan biosynthesis, are in fact virulent (Miller *et al.*, 1986; Miles *et al.*, 1987). Although we still do not have a good understanding of the general physiological requirements for tumorigenesis, analyses of chromosomal avirulent mutants have provided considerable insight into various requirements of the transformation process.

Novel approaches to the identification and study of chromosomal virulence functions

I. Identification and characterization of cell envelope virulence functions

The *A. tumefaciens* cell envelope is of central importance throughout the tumorigenic process. Presumably the cell surface structures that are involved in specific attachment to the plant cell are associated with the membrane. Sensing of the phenolic and monosaccharide inducing signals occurs by way of the cytoplasmic membrane and periplasm located proteins VirA and ChvE, and the T-DNA transport apparatus spans the cytoplasmic membrane.

Using the rationale that additional virulence functions are likely to be associated with the cell envelope, we have recently taken advantage of the extracytoplasmic domain probe transposon TnphoA (Manoil and Beckwith, 1985) to identify additional chromosomal virulence loci which encode cell envelope functions (Cangelosi et al., 1991). TnphoA is a derivative of Tn5 which contains a 5'-truncated E. coli phoA (alkaline phosphatase) gene that lacks its amino-terminal signal sequence. This 'phoA gene is inserted in IS50L of Tn5 such that it generates translational fusions when the transposon inserts in frame with an adjacent protein coding sequence. Activity of alkaline phosphatase requires traversing of the cytoplasmic membrane, thus alkaline phosphatase activity in TnphoA mutants is only observed when translational fusion occurs to extracytoplasmic domains (Manoil and Beckwith, 1985). Alkaline phosphatase activity is easily scored on media containing the chromogenic substrate X-Phos. TnphoA has been used in several Gram negative organisms, including the mammalian pathogens Vibrio cholerae and Salmonella sp., to identify genes encoding cell envelope virulence determinants (Peterson and Mekalanos, 1988; Taylor et al., 1989; Goldberg et al., 1990; Finlay et al., 1988). In the symbiotic phytobacterium Rhizobium meliloti, a closely related cousin of Agrobacterium spp., TnphoA has been used to identify cell envelope proteins involved in symbiosis (Long et al., 1988)

Table 2. Avirulent mutants obtained from TnphoA mutagenesis.

Locus	No. of Mutants
virB	3
chvB	2
chvE	1
chvG	2
chvH	1

ᵃca. 2000 PhoA⁺ mutants were screened for
virulence on Kalanchöe leaves.

In our study, over 2000 active TnphoA mutants of the Pho⁻ A. tumefaciens strain A6007 were screened for tumor forming ability on Kalanchöe leaves. Several avirulent mutants were identified, and most of these were in known chromosomal and Ti plasmid virulence genes (Table 2). Three mutants, A6340, A7678 and A6880, were not in known genes, and were studied further. All three of these mutants were found to be sensitive to detergents (Table 3), indicating that they had altered cell envelope permeability. In addition, A6340 and A7678 were ultrasensitive to several antibiotics, and were severely impaired in growth at pH 5.5 (Table 3). Strains A6340 and A7678 were complemented to wild-type phenotype by a single clone from a cosmid clone bank of A. tumefaciens

Table 3. Growth of avirulent mutants in the presence of detergent, antibiotics and acidic pH.

STRAIN	DOC[c] (1 g ℓ⁻¹)	SDS[d] (0.1 g ℓ⁻¹)	Sarkosyl[e] (2 g ℓ⁻¹)	Tetracycline (0.2 mg ℓ⁻¹)	Novobiocin (1 mg ℓ⁻¹)	Carbenicillin (2 mg ℓ⁻¹)	pH 5.5
A6007 (WT)	[f]+	+	+	+	+	+	+
A6340 (*chvG*)	[g]−	−	−	−	−	−	−
A7678 (*chvG*)	−	−	−	−	−	−	−
A6880 (*chvH*)	[h]±	−	−	+	+	ND	+

[a]Detergent or antibiotic was added to LB agar at the indicated concentration.

[b]Growth at pH 5.5 was determined on LB agar buffered with 50 mM morpholinoethanesulfonic acid.

[c]DOC = sodium deoxycholate
[d]SDS = sodium dodecyl sulfate
[e]Sarkosyl= sodium sarkosinate

[f]+ = growth after 3 days

[g]− = no growth after 3 days

[h]± = marginal growth after 3 days

chromosomal DNA, while A6880 was complemented by another clone. Subcloning and site-directed transposon mutagenesis indicated that the lesions in A6340 and A7678 were in a single transcriptional unit of approximately 2 kb. This locus was designated *chvG*, while the locus defined by A6880 was designated *chvH*. Preliminary data from DNA sequence analysis indicate that the *chvG* gene product exhibits homology to the PhoR class of regulatory proteins (Charles and Nester, unpublished). Given the pleiotropic nature of *chvG* and *chvH* mutations, caution must be taken in drawing conclusions regarding the specific roles of these loci in tumorigenesis.

II. Identification of genes that are expressed during the infection process

The results of the above experiments illustrate that it is possible to isolate mutations in additional chromosomal virulence genes by enriching for mutants in the types of genes, such as cell envelope protein encoding genes, that are more likely to play a role in virulence. Similarly, by targetting promoters that respond to the conditions that *A. tumefaciens* encounters immediately before and during infection, we may identify chromosomal genes whose expression is specifically induced under those conditions. Some of these genes are likely to play a role in tumorigenesis. Infection of a plant by *A. tumefaciens*, leading to tumor formation, requires wounding of the plant. Thus, the environment in which *vir* regulon induction initiates is that of the plant wound. Much effort has been directed towards an understanding of the signalling process that results in expression of the Ti plasmid *vir* regulon. In contrast, little is known of the changes in gene expression that occur when *A. tumefaciens* initially encounters the unique micro-environment of the plant wound.

In addition to the requirement for plant-wound-associated phenolic and monosaccharide inducing signals, *vir* regulon induction is maximal at conditions of acidic pH (Stachel *et al.*, 1985). The plant apoplast, especially after wounding, is acidic, due to leakage of acidic vacuolar contents (Grignon and Sentenac, 1991). One of the key environmental signals in the plant wound is thus likely to be acidic pH, and genes that are induced under acidic conditions may be involved in the preparation of the cell for the reception of the phenolic and monosaccharide signals. For example, the requirement for acid pH is apparently manifest on at least two levels of the signalling process during the initial stages of infection. Transcription of *virG* can be induced from a heat-shock-like promoter in a VirA-independent manner in response to environmental stress stimuli such as acidic pH (Stachel and Zambryski, 1986, Winans, Kerstetter and Nester, 1988, Mantis and Winans, 1992), but acidic pH-dependent transcription of *virG* from this promoter is not fully responsible for the acidic pH optimum for *vir* induction (Chen and Winans, 1991), although other data suggest that provision of multiple copies of *virG* can overcome the acidic pH requirement for *vir* gene induction (Liu *et al.*, 1993). It has been recently proposed that acidic pH is involved in the interaction of acetosyringone with VirA or a receptor molecule (Hess *et al.*, 1991). This is supported by earlier findings of a VirA dependent acidic pH optimum for induction (Melchers *et al.*, 1989). In addition to the very apparent role of acidic pH in *vir* regulon induction, it should be noted that the later steps of T-DNA processing and transport also occur in the plant-wound environment and may be similarly affected by pH.

One-dimensional protein electrophoresis studies indicated that the *A. tumefaciens* protein profile is significantly altered by a shift to acidic pH (Mantis and Winans, 1992), and the expression of at least one chromosomal gene is induced by acetosyringone (Engstrom *et al.*, 1987). We may draw an analogy to the response of animal pathogens to changing environmental conditions. Control of virulence by specific signal molecules has not been described in animal pathogens, as it has in *A. tumefaciens*. However, it is becoming apparent that regulation of virulence genes by environmental signals, such as pH and temperature, is a widespread phenomenon (Mekalanos, 1992). The expression of virulence genes may be coordinately modulated by conditions and signals that indicate the presence and susceptible state of a potential host (Miller *et al.*, 1989).

We feel that by targetting promoters that respond to the conditions that *A. tumefaciens* encounters immediately before and during the process of infection, we may identify chromosomal genes whose expression is specifically induced under those conditions. Some of these genes are likely to play a role in tumorigenesis, and also may be important in survival within the crown gall tumor. Several promoter probe transposons are available, (Bellofatto *et al.*, 1984; Simon *et al.*, 1989) which make possible by *in vivo* methods the identification of genes that are induced specifically under certain conditions. Such a genetic approach could prove to be fruitful in the study of bacterial interactions, pathogenic and otherwise, with eukaryotic organisms.

Conclusions

The study of crown gall tumorigenesis has resulted in the elucidation of a remarkable system of cross-kingdom genetic interaction. As our knowledge of this system increases we must learn to ask new questions and develop new methods to answer these questions. We are especially ignorant of the role of the *Agrobacterium* chromosome in tumorigenesis, and the basic physiological processes that are controlled by this replicon.

References

Ankenbauer RG & Nester EW (1990) Sugar-mediated induction of *Agrobacterium tumefaciens* virulence genes: structural specificity and activities of monosaccharides. J. Bacteriol. 172: 6443–6446.

Bellofatto V, Shapiro L & Hodgson DA (1984) Generation of a Tn5 promoter probe and its use in the study of gene expression in *Caulobacter crescentus*. Proc. Natl. Acad. Sci. USA 81: 1035–1039.

Binns AN (1991) Transformation of wall deficient cultured tobacco protoplasts by *Agrobacterium tumefaciens*. Plant Physiol. 96: 498–506.

Binns AN & Thomashow MF (1988) Cell biology of *Agrobacterium* infection and transformation of plants. Annu. Rev. Microbiol. 42: 575–606.

Cangelosi GA, Ankenbauer RG & Nester EW (1990a) Sugars induce the *Agrobacterium* virulence genes through a periplasmic binding protein and a transmembrane signal protein. Proc. Natl. Acad. Sci. USA. 87: 6708–6712.

Cangelosi GA, Best EA, Martinetti G & Nester EW (1991) Genetic analysis of *Agrobacterium*. Meth. Enzymol 204: 384–397.

Cangelosi GA, Hung L, Puvanesarajah V, Stacey G, Ozga D, Leigh J & Nester EW (1987) Common loci for *Agrobacterium tumefaciens* and *Rhizobium meliloti* exopolysaccharide synthesis and their roles in plant interactions. J. Bacteriol. 169: 2086–2091.

Cangelosi GA, Martinetti G & Nester EW (1990b) Osmosensitivity phenotypes of *Agrobacterium tumefaciens* mutants that lack periplasmic β-1,2-glucan. J. Bacteriol. 172: 2172–2174.

Chen CY & Winans SC (1991) Controlled expression of the transcriptional activator gene *virG* in *Agrobacterium tumefaciens* by using the *Escherichia coli lac* promoter. J. Bacteriol. 173: 1139–1144.

Close TJ, Rogowsky PM, Kado CI, Winans SC, Yanofsky MF & Nester EW (1987) Dual control of *Agrobacterium tumefaciens* Ti plasmid virulence genes. J. Bacteriol. 169: 5113–5118.

Close TJ, Tait RC & Kado CI (1985) Regulation of Ti plasmid virulence genes by a chromosomal locus of *Agrobacterium tumefaciens*. J. Bacteriol. 164: 774–781.

Cooley MB, D'Souza MR & Kado CI (1991) The *virC* and *virD* operons of the *Agrobacterium* Ti plasmid are regulated by the *ros* chromosomal gene: analysis of the cloned *ros* gene. J. Bacteriol. 173: 2608–2616.

Douglas CJ, Halperin W & Nester EW (1982) *Agrobacterium tumefaciens* mutants affected in attachment to plant cells. J. Bacteriol. 152: 1265–1275.

Douglas CJ, Staneloni RJ, Rubin RA & Nester EW (1985) Identification and genetic analysis of an *Agrobacterium tumefaciens* chromosomal virulence region. J. Bacteriol. 161: 850–860.

Engström P, Zambryski P, Van Montagu M & Stachel S (1987) Characterization of *Agrobacterium tumefaciens* virulence proteins induced by the plant factor acetosyringone. J. Mol. Biol 197: 635–645.

Finlay BB, Starnbach MN, Francis CL, Stocker BAD, Chatfield S, Dougan G & Falkow S (1988) Identification and characterization of TnphoA mutants of *Salmonella* that are unable to pass through a polarized MDCK epithelial cell monolayer. Mol. Microbiol. 2: 757–766.

648

Garfinkel DJ & Nester EW (1980) *Agrobacterium tumefaciens* mutants affected in crown gall tumorigenesis and octopine catabolism. J. Bacteriol. 144: 732–743.

Goldberg MB, DiRita VJ & Calderwood SB (1990) Identification of an iron-regulated virulence determinant in *Vibrio cholerae*, using Tn*phoA* mutagenesis. Infect Immun. 58: 55–60.

Gray J, Wang J & Gelvin S (1991) Mutation of the *miaA* gene of *Agrobacterium tumefaciens* results in reduced *vir* gene expression. J. Bacteriol. 174: 1086–1098.

Grignon G & Sentenac H (1991) pH and ionic conditions in the apoplast. Annu. Rev. Plant Physiol. Plant Mol. Biol. 42: 103–128.

Hess KM, Dudley MW, Lynn DG, Joerger RD & Binns AN (1991) Mechanism of phenolic activation of *Agrobacterium* virulence genes: development of a specific inhibitor of bacterial sensor/response systems. Proc. Natl. Acad. Sci. USA 88: 7854–7858.

Huang MLW, Cangelosi GA, Halperin W & Nester EW (1990) A chromosomal *Agrobacterium tumefaciens* gene required for effective plant signal transduction. J. Bacteriol. 172: 1814–1822.

Lippincott BB & Lippincott JA (1966) Characteristics of *Agrobacterium tumefaciens* auxotrophic mutant infectivity. J. Bacteriol. 92: 937–945.

Lippincott JA & Lippincott BB (1975) The genus *Agrobacterium* and plant tumorigenesis. Annu. Rev. Microbiol. 29: 377–405.

Liu C-N, Steck TR, Habeck LL, Meyer JA & Gelvin SB (1993) Multiple copies of *virG* allow induction of *Agrobacterium tumefaciens vir* genes and T-DNA processing at alkaline pH. Mol. Plant-Microbe Interact. 6: 144–156.

Long S, McCune S & Walker GC (1988) Symbiotic loci of *Rhizobium meliloti* identified by random Tn*phoA* mutagenesis. J. Bacteriol. 170: 4257–4265.

Manoil C & Beckwith J (1985) Tn*phoA*: a transposon probe for protein export signals. Proc. Natl. Acad. Sci. USA 82: 8129–8133.

Mantis NJ & Winans SC (1992) The *Agrobacterium tumefaciens vir* gene transcriptional activator *virG* is transcriptionally induced by acidic pH and other stress stimuli. J. Bacteriol. 174: 1189–1196.

Matthysse AG (1987) Characterization of nonattaching mutants of *Agrobacterium tumefaciens*. J. Bacteriol. 169: 313–323.

Mekalanos JJ (1992) Environmental signals controlling expression of virulence determinants in bacteria. J. Bacteriol. 174: 1–7.

Melchers LS, Regensburg-Tuink JGT, Bourret R, Sedee N, Schilperoort R & Hooykaas PJJ (1989) Membrane topology and functional analysis of the sensory protein VirA of *Agrobacterium tumefaciens*. EMBO J. 8: 1919–1925.

Metts J, West J, Doares SH & Matthysse AG (1991) Characterization of three *Agrobacterium tumefaciens* avirulent mutants with chromosomal mutations that affect induction of *vir* genes. J. Bacteriol. 173: 1080–1087.

Miles, CA, Mountain A & Sastry GRK (1987) Cloning of the *Agrobacterium tumefaciens* C58 *trpE* gene by complementation in *Escherichia coli*. Mol. Gen. Genet. 206: 169–173.

Miller IS, Fox D, Saeed N, Borland PA, Miles CA & Sastry GRK (1986) Enlarged map of *Agrobacterium tumefaciens* C58 and the location of chromosomal regions which affect tumorigenicity. Mol. Gen. Genet. 205: 153–159.

Miller JF, Mekalanos JJ & Falkow S (1989) Coordinate regulation and sensory transduction in the control of bacterial virulence. Science 243: 916–922.

Peterson KM & Mekalanos JJ (1988) Characterization of the *Vibrio cholerae* ToxR regulon: Identification of novel genes involved in intestinal colonization. Infect Immun. 56: 2822–2829.

Puvanesarajah V, Schell FM, Stacey G, Douglas CJ & Nester EW (1985) Role for 2-linked-β-D-glucan in the virulence of *Agrobacterium tumefaciens*. J. Bacteriol. 164: 102–106.

Rong LJ, Karcher SJ & Gelvin SB (1991) Genetic and molecular analyses of *picA*, a plant-inducible locus on the *Agrobacterium tumefaciens* chromosome. J. Bacteriol. 173: 5110–5120.

Rong LJ, Karcher SJ, O'Neal K, Hawes MC, Yerkes CD, Jayaswal RK, Hallberg CA & Gelvin SB (1990) *picA*, a novel plant-inducible locus on the *Agrobacterium tumefaciens* chromosome. J. Bacteriol. 172: 5828–5836.

Simon R, Quandt J & Klipp W (1989) New derivatives of transposon Tn5 suitable for mobilization of replicons, generation of operon fusions and induction of genes in Gram-negative bacteria. Gene 80: 161–169.

Stachel SE, Messen E, Van Montagu M & Zambryski P (1985) Identification of the signal molecules produced by wounded plant cells that activate T-DNA transfer in *Agrobacterium tumefaciens*. Nature 318: 624–629.

Stachel SE, Nester EW & Zambryski PC (1986) A plant cell factor induces *Agrobacterium tumefaciens vir* gene expression. Proc. Natl. Acad. Sci. USA 83: 379–383.

Taylor RK, Manoil C & Mekalanos JJ (1989) Broad-host-range vectors for delivery of Tn*phoA*: use in genetic analysis of secreted virulence determinants of *Vibrio cholerae*. J. Bacteriol. 171: 1870–1878.

Thomashow MF, Karlinsey JE, Marks JR & Hurlburt RE (1987) Identification of a new virulence locus in *Agrobacterium tumefaciens* that affects polysaccharide composition and plant cell attachment. J. Bacteriol 169: 3209–3216.

Van Larebeke N, Genetello C, Schell J, Schilperoort RA, Hermans AK, Hernalsteens JP & Van Montagu M (1975) Acquisition of tumor-inducing ability by non-oncogenic agrobacteria as a result of plasmid transfer. Nature, London 255: 742–743.

Wagner VT & Matthysse AG (1992) Involvement of a vitronectin-like protein in attachment of *Agrobacterium tumefaciens* to carrot suspension culture cells. J. Bacteriol. 174: 5999–6003.

Winans SC, Kerstetter RA & Nester EW (1988) Transcriptional regulation of the *virA* and *virG* genes of *Agrobacterium tumefaciens*. J. Bacteriol. 170: 4047–4054.

Zorreguita A, Geremia RA, Cavaignac S, Cangelosi GA & Nester EW (1988) Identification of the product of an *Agrobacterium tumefaciens* chromosomal virulence gene. Mol. Plant-Microbe Interact. 1: 121–127.

Index

654

Developments in Plant Pathology

1. R. Johnson and G.J. Jellis (eds.): *Breeding for Disease Resistance.* 1993
   ISBN 0-7923-1607-X
2. B. Fritig and M. Legrand (eds.): *Mechanisms of Plant Defense Responses.*
   1993   ISBN 0-7923-2154-5
3. C.I. Kado and J.H. Crosa (eds.): *Molecular Mechanisms of Bacterial
   Virulence.* 1994   ISBN 0-7923-1901-X

KLUWER ACADEMIC PUBLISHERS – DORDRECHT / BOSTON / LONDON